APPLIED LINEAR ALGEBRA
AND OPTIMIZATION
USING MATLAB®

APPLIED LINEAR ALGEBRA AND OPTIMIZATION USING MATLAB®

RIZWAN BUTT, PhD

MERCURY LEARNING AND INFORMATION
Dulles, Virginia

Publisher: David Pallai
MERCURY LEARNING AND INFORMATION
22841 Quicksilver Drive
Dulles, VA 20166
info@merclearning.com
www.merclearning.com
1-800-758-3756

This book is printed on acid-free paper.

R. Butt, PhD. Applied Linear Algebra and Optimization using MATLAB®
ISBN: 978-1-9364200-4-9

The publisher recognizes and respects all marks used by companies, manufacturers, and developers as a means to distinguish their products. All brand names and product names mentioned in this book are trademarks or service marks of their respective companies. Any omission or misuse (of any kind) of service marks or trademarks, etc. is not an attempt to infringe on the property of others.

Library of Congress Control Number: 2010941258

1112133 2 1

Our titles are available for adoption, license, or bulk purchase by institutions, corporations, etc. For additional information, please contact the Customer Service Dept. at 1-800-758-3756 (toll free).

The sole obligation of MERCURY LEARNING AND INFORMATION to the purchaser is to replace the disc, based on defective materials or faulty workmanship, but not based on the operation or functionality of the product.

Printed in India.

Dedicated to
Muhammad Sarwar Khan,
The Greatest Friend in the World

Contents

Preface

This book presents an integrated approach to numerical linear algebra and optimization theory based on a computer—in this case, using the software package MATLAB. This book has evolved over many years from lecture notes on *Numerical Linear Algebra and Optimization Theory* that accompany both graduate and post-graduate courses in mathematics at the King Saud University at Riyadh, Saudi Arabia. These courses deal with linear equations, approximations, eigenvalue problems, and linear and nonlinear optimization problems. We discuss several numerical methods for solving both linear systems of equations and optimization problems. It is generally accepted that linear algebra methods aid in finding the solution of linear and nonlinear optimization problems.

The main approach used in this book is quite different from currently available books, which are either too theoretical or too computational. The approach adopted in this book lies between the above two extremities. The book fully exploits MATLAB's symbolic, numerical, and graphical capabilities to develop a thorough understanding of linear algebra and optimization algorithms.

The book covers two distinct topics: linear algebra and optimization theory. Linear algebra plays an important role in both applied and theoretical mathematics, as well as in all of science and engineering, computer science, probability and statistics, economics, numerical analysis, and many other disciplines. Nowadays, a proper grounding in both calculus and linear algebra is an essential prerequisite for a successful career in science, engineering, and mathematics. Linear algebra can be viewed as the mathematical apparatus needed to solve potentially huge linear systems, to understand their underlying structure, and to apply what is learned in other contexts. The term *linear* is the key and, in fact, refers not just to linear algebraic equations, but also to linear differential equations, linear boundary value problems, linear iterative systems, and so on.

The other focus of this book is on optimization theory. This theory is the study of the extremal values of a function; its maxima and minima. The topics in this theory range from conditions for existence of a unique extremal value to methods—both analytic and numeric—for finding the extremal values, and for what values of the independent variables the function attains its extremes. It is a branch of mathematics that encompasses many diverse areas of optimization and minimization. The more modern term is operational research. It includes the calculus of variations, control theory, convex optimization theory, decision theory, game theory, linear and nonlinear programming, queuing systems, etc. In this book we emphasize only linear and nonlinear programming problems.

A wide range of applications appears throughout the book. They have been chosen and written to give the student a sense of the broad range of applicability of linear algebra and optimization theory. These applications range from theoretical applications such as the use of linear algebra in differential equations, difference equations, and least squares analysis.

When dealing with linear algebra or optimization theory, we often need a computer. We believe that computers can improve the conceptional understanding of mathematics, not just enable the completion of complicated calculations. We have chosen MATLAB as our standard package because it is a widely used software for working with matrices. The surge of popularity in MATLAB is related to the increasing popularity of UNIX and computer graphics. To what extent numerical computations will be programmed in MATLAB in the future is uncertain. A short introduction to

MATLAB is given in Appendix C, and the programs in the text serve as further examples.

The topics are discussed in a simplified manner with a number of examples illustrating the different concepts and applications. Most of the sections contain a fairly large number of exercises, some of which relate to real-life problems. Chapter 1 covers the basic concepts of matrices and determinants and describes the basic computational methods used to solve nonhomogeneous linear equations. Direct methods, including Cramer's rule, the Gaussian elimination method and its variants, the Gauss–Jordan method, and LU decomposition methods, are discussed. It also covers the conditioning of linear systems. Many ill-conditioned problems are discussed. The chapter closes with the many interesting applications of linear systems. In Chapter 2, we discuss iterative methods, including the Jacobi method, the Gauss–Seidel method, the SOR iterative method, the conjugate gradient method, and the residual corrector method. Chapter 3 covers the selected methods of computing matrix eigenvalues. The approach discussed here should help students understand the relationship of eigenvalues to the roots of characteristic equations. We define eigenvalues and eigenvectors and study several examples. We discuss the diagonalization of matrices and the computation of powers of diagonalizable matrices. Some interesting applications of the eigenvalues and eigenvectors of a matrix are also discussed at the end of the chapter. In Chapter 4, various numerical methods are discussed for the eigenvalues of matrices. Among them are the power iterative methods, the Jacobi method, Given's method, the Householder method, the QR iteration method, the LR method, and the singular value decomposition method. Chapter 5 describes the approximation of functions. In this chapter we also describe curve fitting of experimental data based on least squares methods. We discuss linear, nonlinear, plane, and trigonometric function least squares approximations. We use QR decomposition and singular value decomposition for the solution of the least squares problem. In Chapter 6, we describe standard linear programming formulations. The subject of linear programming, in general, involves the development algorithms and methodologies in optimization. The field, developed by George Dantzig and his associates in 1947, is now widely used in industry and has its foundation in linear algebra. In keeping with the intent of this book, this chapter presents the mathematical formulations of basic linear programming problems. In Chapter 7, we describe nonlinear programming formulations. We discuss many numerical methods for solving unconstrained and constrained problems. In the beginning of the chapter some of the basic mathematical concepts useful in developing optimization theory are presented. For unconstrained optimization problems we discuss the golden-section search method and quadratic interpolation method, which depend on the initial guesses that bracket the single optimum, and Newton's method, which is based on the idea from calculus that the minimum or maximum can be found by solving $f'(x) = 0$. For the functions of several variables, we use the steepest descent method and Newton's method. For handling nonlinear optimization problems with constraints, we discuss the generalized reduced-gradient method, Lagrange multipliers, and KT conditions. At the end of the chapter, we also discuss quadratic programming problems and the separable programming problems.

In each chapter, we discuss several examples to guide students step-by-step through the most complex topics. Since the only real way to learn mathematics is to use it, there is a list of exercises provided at the end of each chapter. These exercises range from very easy to quite difficult. This book is completely self-contained, with all the necessary mathematical background given in it. Finally, this book provides balanced convergence of the theory, application, and numerical computation of all the topics discussed.

Appendix A covers different kinds of errors that are preparatory subjects for numerical computations. To explain the sources of these errors, there is a brief discussion of Taylor's series and how numbers are computed and saved in computers. Appendix B consists of a brief introduction to vectors in space and a review of complex numbers and how to do linear algebra with them. It is also devoted to general inner product spaces and to how different notations and processes generalize. In Appendix C, we discuss the basic commands for the software package MATLAB. In Appendix D, we give answers to selected odd-numbered exercises.

Acknowledgments

I wish to express my gratitude to all those colleagues, friends, and associates of mine, without whose help this work was not possible. I am grateful, especially, to Dr. Saleem, Dr. Zafar Ellahi, Dr. Esia Al-Said, and Dr. Salah Hasan for reading earlier versions of the manuscript and for providing encouraging comments. I have written this book as the background material for an interactive first course in linear algebra and optimization. The encouragement and positive feedback that I have received during the design and development of the book have given me the energy required to complete the project.

I also want to express my heartfelt thanks to a special person who has been very helpful to me in a great many ways over the course of my career: Muhammad Sarwar Khan, of King Saud University, Riyadh, Saudi Arabia.

My sincere thanks are also due to the Deanship of the Scientific Research Center, College of Science, King Saud University, Riyadh, KSA, for financial support and for providing facilities throughout the research project No. (Math/2008/05/B).

It has taken me five years to write this book and thanks must go to my long-suffering family for my frequent unsocial behavior over these years. I am profoundly grateful to my wife Saima, and our children Fatima, Usman,

Fouzan, and Rahmah, for their patience, encouragement, and understanding throughout this project. Special thanks goes to my elder daughter, Fatima, for creating all the figures in this project.

Dr. Rizwan Butt
Department of Mathematics,
College of Science
King Saud University
August, 2010

Chapter 1

Matrices and Linear Systems

1.1 Introduction

When engineering systems are modeled, the mathematical description is frequently developed in terms of a set of algebraic simultaneous equations. Sometimes these equations are nonlinear and sometimes linear. In this chapter, we discuss systems of simultaneous linear equations and describe the numerical methods for the approximate solutions of such systems. The solution of a system of simultaneous linear algebraic equations is probably one of the most important topics in engineering computation. Problems involving simultaneous linear equations arise in the areas of elasticity, electric-circuit analysis, heat transfer, vibrations, and so on. Also, the numerical integration of some types of ordinary and partial differential equations may be reduced to the solution of such a system of equations. It has been estimated, for example, that about 75% of all scientific problems require the solution of a system of linear equations at one stage or another. It is therefore important to be able to solve linear problems efficiently and accurately.

1

Definition 1.1 (Linear Equation)

It is an equation in which the highest exponent in a variable term is no more than one. The graph of such an equation is a straight line. ●

A linear equation in two variables x_1 and x_2 is an equation that can be written in the form

$$a_1x_1 + a_2x_2 = b,$$

where a_1, a_2, and b are real numbers. Note that this is the equation of a straight line in the plane. For example, the equations

$$5x_1 + 2x_2 = 2, \qquad \frac{4}{5}x_1 + 2x_2 = 1, \qquad 2x_1 - 4x_2 = \pi$$

are all linear equations in two variables.

A linear equation in n variables $x_1, x_2, \ldots, x_n$ is an equation that can be written as

$$a_1x_1 + a_2x_2 + \cdots + a_nx_n = b,$$

where $a_1, a_2, \ldots, a_n$ are real numbers and called the *coefficients* of unknown variables $x_1, x_2, \ldots, x_n$ and the real number b, the right-hand side of the equation, is called the *constant term* of the equation.

Definition 1.2 (System of Linear Equations)

A system of linear equations (or linear system) is simply a finite set of linear equations. ●

For example,

$$\begin{aligned} 4x_1 &- 2x_2 &= 5 \\ 3x_1 &+ 2x_2 &= 4 \end{aligned}$$

is a system of two equations in two variables x_1 and x_2, and

$$\begin{aligned} 2x_1 &+ x_2 &- 5x_3 &+ 2x_4 &= 9 \\ 4x_1 &+ 3x_2 &+ 2x_3 &+ 4x_4 &= 3 \\ x_1 &+ 2x_2 &+ 3x_3 &+ 2x_4 &= 11 \end{aligned}$$

is the system of three equations in the four variables x_1, x_2, x_3, and x_4.

In order to write a general system of m linear equations in the n variables $x_1, \ldots, x_n$, we have

$$
\begin{array}{ccccccccc}
a_{11}x_1 & + & a_{12}x_2 & + & \cdots & + & a_{1n}x_n & = & b_1 \\
a_{21}x_1 & + & a_{22}x_2 & + & \cdots & + & a_{2n}x_n & = & b_2 \\
\vdots & & \vdots & & \cdots & & \vdots & & \vdots \\
a_{m1}x_1 & + & a_{m2}x_2 & + & \cdots & + & a_{mn}x_n & = & b_m
\end{array}
\tag{1.1}
$$

or, in compact form the system (1.1) can be written as

$$
\sum_{j=1}^{n} a_{ij}x_j = b_i, \qquad i = 1, 2, \ldots, m.
\tag{1.2}
$$

For such a system we seek all possible ordered sets of numbers $c_1, \ldots, c_n$ which satisfy all m equations when they are substituted for the variables $x_1, x_2, \ldots, x_n$. Any such set $\{c_1, c_2, \ldots, c_n\}$ is called a *solution* of the system of linear equations (1.1) or (1.2).

There are three possible types of linear systems that arise in engineering problems, and they are described as follows:

1. If there are more equations than unknown variables $(m > n)$, then the system is usually called *overdetermined*. Typically, an overdetermined system has no solution. For example, the following system

$$
\begin{array}{rcrcl}
4x_1 & & & = & 8 \\
3x_1 & + & 9x_2 & = & 13 \\
& & 3x_2 & = & 9
\end{array}
$$

has no solution.

2. If there are more unknown variables than the number of the equations $(n > m)$, then the system is usually called *underdetermined*. Typically, an underdetermined system has an infinite number of solutions. For example, the system

$$
\begin{array}{rcrcrcl}
x_1 & + & 5x_2 & & & = & 45 \\
& & 3x_2 & + & 4x_3 & = & 21
\end{array}
$$

has infinitely many solutions.

3. If there are the same number of equations as unknown variables ($m = n$), then the system is usually called a *simultaneous* system. It has a unique solution if the system satisfies certain conditions (which we will discuss below). For example, the system

$$
\begin{array}{rrrrrrr}
2x_1 & + & 4x_2 & + & x_3 & = & -11 \\
-x_1 & + & 3x_2 & - & 2x_3 & = & -16 \\
2x_1 & - & 3x_2 & + & 5x_3 & = & 21
\end{array}
$$

has the unique solution $x_1 = 2, x_2 = -4, x_3 = 1$.

Most engineering problems fall into this category. In this chapter, we will solve simultaneous linear systems using many numerical methods.

A simultaneous system of linear equations is said to be *linear independent* if no equation in the system can be expressed as a linear combination of the others. Under these circumstances a unique solution exists. For example, the system of linear equations

$$
\begin{array}{rrrrrrr}
2x_1 & + & x_2 & - & x_3 & = & 1 \\
x_1 & - & 2x_2 & + & 3x_3 & = & 4 \\
x_1 & + & x_2 & & & = & 1
\end{array}
$$

is linear independent and therefore has the unique solution

$$
x_1 = 1, \quad x_2 = 0, \quad x_3 = 1.
$$

However, the system

$$
\begin{array}{rrrrrrr}
5x_1 & + & x_2 & + & x_3 & = & 4 \\
3x_1 & - & x_2 & + & x_3 & = & -2 \\
x_1 & + & x_2 & & & = & 3
\end{array}
$$

does not have a unique solution since the equations are not linear independent; the first equation is equal to the second equation plus twice the third equation.

Theorem 1.1 (Solution of a Linear System)

Every system of linear equations has either no solution, exactly one solution, or infinitely many solutions. ●

For example, in the case of a system of two equations in two variables, we can have these three possibilities for the solutions of the linear system. First, the two lines (since the graph of a linear equation is a straight line) may be parallel and distinct, and in this case, there is no solution to the system because the two lines do not intersect each other at any point. For example, consider the system

$$
\begin{aligned}
x_1 + x_2 &= 1 \\
2x_1 + 2x_2 &= 3.
\end{aligned}
$$

From the graphs (Figure 1.1(a)) of the given two equations we can see that the lines are parallel, so the given system has no solution. It can be proved algebraically simply by multiplying the first equation of the system by 2 to get a system of the form

$$
\begin{aligned}
2x_1 + 2x_2 &= 2 \\
2x_1 + 2x_2 &= 3,
\end{aligned}
$$

which is not possible.

Second, the two lines may not be parallel, and they may meet at exactly one point, so in this case the system has exactly one solution. For example, consider the system

$$
\begin{aligned}
x_1 - x_2 &= -1 \\
3x_1 - x_2 &= 3.
\end{aligned}
$$

From the graphs (Figure 1.1(b)) of these two equations we can see that the lines intersect at exactly one point, namely, $(2, 3)$, and so the system has exactly one solution, $x_1 = 2$, $x_2 = 3$. To show this algebraically, if we substitute $x_2 = x_1 + 1$ in the second equation, we have $3x_1 - x_1 - 1 = 3$, or $x_1 = 2$, and using this value of x in $x_2 = x_1 + 1$ gives $x_2 = 3$.

Finally, the two lines may actually be the same line, and so in this case, every point on the lines gives a solution to the system and therefore there are infinitely many solutions. For example, consider the system

$$
\begin{aligned}
x_1 + x_2 &= 1 \\
2x_1 + 2x_2 &= 2.
\end{aligned}
$$

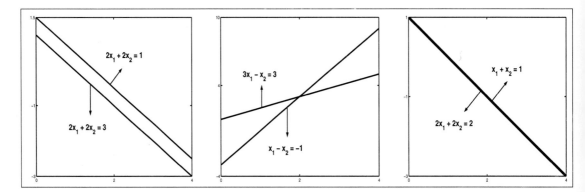

Figure 1.1: Three possible solutions of simultaneous systems.

Here, both equations have the same line for their graph (Figure 1.1(c)). So this system has infinitely many solutions because any point on this line gives a solution to this system, since any solution of the first equation is also a solution of the second equation. For example, if we set $x_2 = x_1 - 1$, and choose $x_1 = 0, x_2 = 1, x_1 = 1, x_2 = 0$, and so on.   ●

Note that a system of equations with no solution is said to be an *inconsistent system* and if it has at least one solution, it is said to be a *consistent system*.

1.1.1   Linear Systems in Matrix Notation

The general simultaneous system of n linear equations with n unknown variables $x_1, x_2, \ldots, x_n$ is

$$
\begin{aligned}
a_{11}x_1 + a_{12}x_2 + \cdots + a_{1n}x_n &= b_1 \\
a_{21}x_1 + a_{22}x_2 + \cdots + a_{2n}x_n &= b_2 \\
\vdots \qquad \vdots \qquad \vdots \qquad \vdots \qquad \vdots& \\
a_{n1}x_1 + a_{n2}x_2 + \cdots + a_{nn}x_n &= b_n.
\end{aligned}
\tag{1.3}
$$

The system of linear equations (1.3) can be written as the single matrix equation

$$
\begin{pmatrix}
a_{11} & a_{12} & \cdots & a_{1n} \\
a_{21} & a_{22} & \cdots & a_{2n} \\
\vdots & \vdots & \vdots & \vdots \\
a_{n1} & a_{n2} & \cdots & a_{nn}
\end{pmatrix}
\begin{pmatrix}
x_1 \\ x_2 \\ \vdots \\ x_n
\end{pmatrix}
=
\begin{pmatrix}
b_1 \\ b_2 \\ \vdots \\ b_n
\end{pmatrix}.
\tag{1.4}
$$

If we compute the product of the two matrices on the left-hand side of (1.9), we have

$$
\begin{pmatrix}
a_{11}x_1 + a_{12}x_2 + \cdots + a_{1n}x_n \\
a_{21}x_1 + a_{22}x_2 + \cdots + a_{2n}x_n \\
\vdots \qquad \vdots \qquad \vdots \qquad \vdots \\
a_{n1}x_1 + a_{n2}x_2 + \cdots + a_{nn}x_n
\end{pmatrix}
=
\begin{pmatrix}
b_1 \\ b_2 \\ \vdots \\ b_n
\end{pmatrix}.
\tag{1.5}
$$

But two matrices are equal if and only if their corresponding elements are equal. Hence, the single matrix equation (1.9) is equivalent to the system of the linear equations (1.3). If we define

$$
A = \begin{pmatrix}
a_{11} & a_{12} & \cdots & a_{1n} \\
a_{21} & a_{22} & \cdots & a_{2n} \\
\vdots & \vdots & \vdots & \vdots \\
a_{n1} & a_{n2} & \cdots & a_{nn}
\end{pmatrix},
\qquad
\mathbf{x} = \begin{pmatrix}
x_1 \\ x_2 \\ \vdots \\ x_n
\end{pmatrix},
\qquad
\mathbf{b} = \begin{pmatrix}
b_1 \\ b_2 \\ \vdots \\ b_n
\end{pmatrix},
$$

the coefficient matrix, the column matrix of unknowns, and the column matrix of constants, respectively, and then the system (1.3) can be written very compactly as

$$
A\mathbf{x} = \mathbf{b},
\tag{1.6}
$$

which is called the *matrix form* of the system of linear equations (1.3). The column matrices **x** and **b** are called *vectors*.

If the right-hand sides of the equal signs of (1.6) are not zero, then the linear system (1.6) is called a *nonhomogeneous system,* and we will find that all the equations must be independent to obtain a unique solution.

If the constants **b** of (1.6) are added to the coefficient matrix A as a column of elements in the position shown below

$$[A|\mathbf{b}] = \begin{pmatrix} a_{11} & a_{12} & \cdots & a_{1n} & \vdots & b_1 \\ a_{21} & a_{22} & \cdots & a_{2n} & \vdots & b_2 \\ \vdots & \vdots & \vdots & \vdots & \vdots & \vdots \\ a_{n1} & a_{n2} & \cdots & a_{nn} & \vdots & b_n \end{pmatrix}, \tag{1.7}$$

then the matrix $[A|\mathbf{b}]$ is called the *augmented matrix* of the system (1.6). In many instances, it may be convenient to operate on the augmented matrix instead of manipulating the equations. It is customary to put a bar between the last two columns of the augmented matrix to remind us where the last column came from. However, the bar is not absolutely necessary. The coefficient and augmented matrices of a linear system will play key roles in our methods of solving linear systems.

Using MATLAB commands we can define an augmented matrix as follows:

```
>> A = [1 2 3; 4 5 6; 7 8 9];
>> b = [10; 11; 12];
>> Aug = [A b]
Aug =
      1   2   3   10
      4   5   6   11
      7   8   9   12
```

Also,

```
>> Aug = [A eye(3)]
Aug =
    1  2  3  1  0  0
    4  5  6  0  1  0
    7  8  9  0  0  1
```

If all of the constant terms $b_1, b_2, \ldots, b_n$ on the right-hand sides of the equal signs of the linear system (1.6) are zero, then the system is called a *homogeneous system*, and it can be written as

$$
\begin{aligned}
a_{11}x_1 &+ a_{12}x_2 + \cdots + a_{1n}x_n = 0 \\
a_{21}x_1 &+ a_{22}x_2 + \cdots + a_{2n}x_n = 0 \\
&\vdots \\
a_{n1}x_1 &+ a_{n2}x_2 + \cdots + a_{nn}x_n = 0.
\end{aligned}
\tag{1.8}
$$

The system of linear equations (1.8) can be written as the single matrix equation

$$
\begin{pmatrix}
a_{11} & a_{12} & \cdots & a_{1n} \\
a_{21} & a_{22} & \cdots & a_{2n} \\
\vdots & \vdots & \vdots & \vdots \\
a_{n1} & a_{n2} & \cdots & a_{nn}
\end{pmatrix}
\begin{pmatrix}
x_1 \\
x_2 \\
\vdots \\
x_n
\end{pmatrix}
=
\begin{pmatrix}
0 \\
0 \\
\vdots \\
0
\end{pmatrix}.
\tag{1.9}
$$

It can also be written in more compact form as

$$
A\mathbf{x} = \mathbf{0},
\tag{1.10}
$$

where

$$
A =
\begin{pmatrix}
a_{11} & a_{12} & \cdots & a_{1n} \\
a_{21} & a_{22} & \cdots & a_{2n} \\
\vdots & \vdots & \vdots & \vdots \\
a_{n1} & a_{n2} & \cdots & a_{nn}
\end{pmatrix},
\quad
\mathbf{x} =
\begin{pmatrix}
x_1 \\
x_2 \\
\vdots \\
x_n
\end{pmatrix},
\quad
\mathbf{0} =
\begin{pmatrix}
0 \\
0 \\
\vdots \\
0
\end{pmatrix}.
$$

It can be seen by inspection of the homogeneous system (1.10) that one of its solution, is $\mathbf{x} = \mathbf{0}$; such a solution, in which all of the unknowns are zero, is called the *trivial solution* or *zero solution*. For the general nonhomogeneous linear system there are three possibilities: no solution, one solution, or infinitely many solutions. For the general homogeneous system, there are only two possibilities: either the zero solution is the only solution, or there are infinitely many solutions (called nontrivial solutions). Of course, it is usually nontrivial solutions that are of interest in physical problems. A nontrivial solution to the homogeneous system can occur with certain conditions on the coefficient matrix A, which we will discuss later.

1.2 Properties of Matrices and Determinants

To discuss the solutions of linear systems, it is necessary to introduce the basic algebraic properties of matrices that make it possible to describe linear systems in a concise way and make solving a system of n linear equations easier.

1.2.1 Introduction to Matrices

A matrix can be described as a rectangular array of elements that can be represented as follows:

$$A = \begin{pmatrix} a_{11} & a_{12} & \cdots & a_{1n} \\ a_{21} & a_{22} & \cdots & a_{2n} \\ \vdots & \vdots & \vdots & \vdots \\ a_{m1} & a_{m2} & \cdots & a_{mn} \end{pmatrix}. \tag{1.11}$$

The numbers $a_{11}, a_{12}, \ldots, a_{mn}$ that make up the array are called the elements of the matrix. The first subscript for the element denotes the row and the second denotes the column in which the element appears. The elements of a matrix may take many forms. They could be all numbers (*real* or *complex*), or variables, or functions, or integrals, or derivatives, or even matrices themselves.

The *order* or *size* of a matrix is specified by the number of rows (m) and column (n); thus, the matrix A in (1.11) is of order m by n, usually written as $m \times n$.

A vector can be considered a special case of a matrix having only one row or one column. A row vector containing n elements is a $1 \times n$ matrix, called a *row matrix*, and a column vector of n elements is an $n \times 1$ matrix, called a (*column matrix*). A matrix of order 1×1 is called a *scalar*.

Definition 1.3 (Matrix Equality)

Two matrices $A = (a_{ij})$ and $B = (b_{ij})$ are equal if they are the same size and the corresponding elements in A and B are equal, i.e.,

$$A = B, \quad \text{if and only if} \quad a_{ij} = b_{ij}$$

for $i = 1, 2, \ldots, m$ and $j = 1, 2, \ldots, n$. For example, the matrices

$$A = \begin{pmatrix} 1 & -1 & 2 \\ 1 & 3 & 2 \\ 2 & 4 & 3 \end{pmatrix} \quad \text{and} \quad B = \begin{pmatrix} 1 & -1 & z \\ 1 & 3 & 2 \\ x & y & w \end{pmatrix}$$

are equal, if and only if $x = 2, y = 4, z = 2$, and $w = 3$. ●

Definition 1.4 (Addition of Matrices)

Let $A = (a_{ij})$ and $B = (b_{ij})$ both be $m \times n$ matrices, then the sum $A + B$ of two matrices of the same size is a new matrix $C = (c_{ij})$, each of whose elements is the sum of the two corresponding elements in the original matrices, i.e.,

$$c_{ij} = a_{ij} + b_{ij}, \quad \text{for } i = 1, 2, \ldots, m, \quad \text{and} \quad j = 1, 2, \ldots, n.$$

For example, let

$$A = \begin{pmatrix} 1 & 2 \\ 3 & 4 \end{pmatrix} \quad \text{and} \quad B = \begin{pmatrix} 4 & 1 \\ 5 & 2 \end{pmatrix}.$$

Then

$$\begin{pmatrix} 1 & 2 \\ 3 & 4 \end{pmatrix} + \begin{pmatrix} 4 & 1 \\ 5 & 2 \end{pmatrix} = \begin{pmatrix} 5 & 3 \\ 8 & 6 \end{pmatrix} = C.$$

●

Using MATLAB commands and adding two matrices A and B of the same size results in the answer C, another matrix of the same size:

```
>> A = [1 2; 3 4];
>> B = [4 1; 5 2];
>> C = A + B
C =
     5   3
     8   6
```

Definition 1.5 (Difference of Matrices)

Let A and B be $m \times n$ matrices, and we write $A + (-1)B$ as $A - B$ and the difference of two matrices of the same size is a new matrix C, each of whose elements is the difference of the two corresponding elements in the original matrices. For example, let

$$A = \begin{pmatrix} 1 & 2 \\ 3 & 4 \end{pmatrix} \quad and \quad B = \begin{pmatrix} 4 & 1 \\ 5 & 2 \end{pmatrix}.$$

Then

$$\begin{pmatrix} 1 & 2 \\ 3 & 4 \end{pmatrix} - \begin{pmatrix} 4 & 1 \\ 5 & 2 \end{pmatrix} = \begin{pmatrix} -3 & 1 \\ -2 & 2 \end{pmatrix} = C.$$

Note that $(-1)B = -B$ is obtained by multiplying each entry of matrix B by (-1), the scalar multiple of matrix B by -1. The matrix $-B$ is called the negative of the matrix B. •

Definition 1.6 (Multiplication of Matrices)

The multiplication of two matrices is defined only when the number of columns in the first matrix is equal to the number of rows in the second. If an $m \times n$ matrix A is multiplied by an $n \times p$ matrix B, then the product matrix C is an $m \times p$ matrix where each term is defined by

$$c_{ij} = \sum_{k=1}^{n} a_{ik} b_{kj}$$

for each $i = 1, 2, \ldots, m$ and $j = 1, 2, \ldots, p$. For example, let

$$A = \begin{pmatrix} 1 & 2 \\ 3 & 4 \end{pmatrix} \quad and \quad B = \begin{pmatrix} 4 & 1 \\ 5 & 2 \end{pmatrix}.$$

Then

$$\begin{pmatrix} 1 & 2 \\ 3 & 4 \end{pmatrix} \begin{pmatrix} 4 & 1 \\ 5 & 2 \end{pmatrix} = \begin{pmatrix} 4+10 & 1+4 \\ 12+20 & 3+8 \end{pmatrix} = \begin{pmatrix} 14 & 5 \\ 32 & 11 \end{pmatrix} = C.$$

Note that even if AB is defined, the product BA may not be defined. Moreover, a simple multiplication of two square matrices of the same size will show that even if BA is defined, it need not be equal to AB, i.e., they do not commute. For example, if

$$A = \begin{pmatrix} 1 & 2 \\ -1 & 3 \end{pmatrix} \quad and \quad B = \begin{pmatrix} 2 & 1 \\ 0 & 1 \end{pmatrix},$$

then

$$AB = \begin{pmatrix} 2 & 3 \\ -2 & 2 \end{pmatrix} \quad while \quad BA = \begin{pmatrix} 1 & 7 \\ -1 & 3 \end{pmatrix}.$$

Thus, AB ≠ BA. •

Using MATLAB commands, matrix multiplication has the standard meaning as well. Multiplying two matrices A and B of size $m \times p$ and $p \times n$ respectively, results in the answer C, another matrix of size $m \times n$:

```
>> A = [1 2; -1 3];
>> B = [2 1; 0 1];
>> C = A * B
C =
      2   3
     -2   2
```

MATLAB also has component-wise operations for multiplication, division, and exponentiation. These three operations are a combination of a period (.) and one of the operators $*$, $/$, and $\hat{}$, which perform operations on a pair of matrices (or vectors) with equal numbers of rows and columns. For example, consider the two row vectors:

```
>> u = [1 2 3 4];
>> v = [5 3 0 2];
>> x = u. * v
x =
    5 6 0 8
```

Warning: Divide by zero.

```
>> y = u./v;
y =
    0.2000 0.6667 Inf 2.0000
```

These operations apply to matrices as well as vectors:

```
>> A = [1 2 3; 4 5 6; 7 8 9];
>> B = [9 8 7; 6 5 4; 3 2 1];
>> C = A. * B
C =
     9   16   21
    24   25   24
    21   16    9
```

Note that $A. * B$ is not the same as $A * B$.

The array exponentiation operator, $.\hat{}$, raises the individual elements of a matrix to a power:

```
>> A = [1 2 3; 4 5 6; 7 8 9];
>> D = A.^2
D =
     1    4    9
    16   25   36
    49   64   81
```

```
>> E = A.^ (1/2)
D =
      1.0000  1.4142  1.7321
      2.0000  2.2361  2.4495
      2.6458  2.8284  3.0000
```

The syntax of array operators requires the correct placement of a typographically small symbol, a period, in what might be a complex formula. Although MATLAB will catch syntax errors, it is still possible to make computational mistakes with legal operations. For example, A.^ 2 and A^ 2 are both legal, but not at all equivalent.

In linear algebra, the addition and subtraction of matrices and vectors are element-by-element operations. Thus, there are no special array operators for addition and subtraction.

1.2.2 Some Special Matrix Forms

There are many special types of matrices encountered frequently in engineering analysis. We discuss some of them in the following.

Definition 1.7 (Square Matrix)

A matrix A which has the same number of rows m and columns n, i.e., m = n, defined as

$$A = (a_{ij}), \quad for \quad i = 1, 2, \ldots, n, \quad and \quad j = 1, 2, \ldots, n$$

is called a square matrix. For example, the matrices

$$A = \begin{pmatrix} 1 & 2 \\ -1 & 3 \end{pmatrix} \quad and \quad B = \begin{pmatrix} 2 & 1 & 2 \\ 1 & 2 & 3 \\ 0 & 1 & 5 \end{pmatrix}$$

are square matrices because both have the same number of rows and columns.

•

Definition 1.8 (Null Matrix)

It is a matrix in which all elements are zero, i.e.,

$$A = (a_{ij}) = \mathbf{0}, \qquad for \quad i = 1, 2, \ldots, n \quad and \quad j = 1, 2, \ldots, n.$$

It is also called a zero matrix. It may be either rectangular or square. For example, the matrices

$$A = \begin{pmatrix} 0 & 0 & 0 \\ 0 & 0 & 0 \end{pmatrix} \quad and \quad B = \begin{pmatrix} 0 & 0 & 0 \\ 0 & 0 & 0 \\ 0 & 0 & 0 \end{pmatrix}$$

are zero matrices.   •

Definition 1.9 (Identity Matrix)

It is a square matrix in which the main diagonal elements are equal to 1. It is defined as

$$\mathbf{I} = (a_{ij}) = \begin{cases} a_{ij} & = & 0, & if & i \neq j, \\ a_{ij} & = & 1, & if & i = j. \end{cases}$$

An example of a 4×4 identity matrix may be written as

$$\mathbf{I}_4 = \begin{pmatrix} 1 & 0 & 0 & 0 \\ 0 & 1 & 0 & 0 \\ 0 & 0 & 1 & 0 \\ 0 & 0 & 0 & 1 \end{pmatrix}.$$

The identity matrix (also called a unit matrix) serves somewhat the same purpose in matrix algebra as does the number one (unity) in scalar algebra. It is called the identity matrix because multiplication of a matrix by it will result in the same matrix. For a square matrix A of order n, it can be seen that

$$\mathbf{I}_n A = A \mathbf{I}_n = A.$$

Similarly, for a rectangular matrix B of order $m \times n$, we have

$$\mathbf{I}_m B = B \mathbf{I}_n = B.$$

The multiplication of an identity matrix by itself results in the same identity matrix.   •

In MATLAB, identity matrices are created with the **eye** function, which can take either one or two input arguments:

```
>> I = eye(n)
>> I = eye(m, n)
```

Definition 1.10 (Transpose Matrix)

The transpose of a matrix A is a new matrix formed by interchanging the rows and columns of the original matrix. If the original matrix A is of order $m \times n$, then the transpose matrix, A^T, will be of the order $n \times m$, i.e.,

$$if \quad A = (a_{ij}), \quad for \quad i = 1, 2, \ldots, m \text{ and } j = 1, 2, \ldots, n,$$

then

$$A^T = (a_{ji}), \quad for \quad i = 1, 2, \ldots, n \text{ and } j = 1, 2, \ldots, m.$$

●

The transpose of a matrix A can be found by using the following MATLAB commands:

```
>> A = [1 2 3; 4 5 6; 7 8 9]
>> B = A'
B =
      1.0000   4.0000   7.0000
      2.0000   5.0000   8.0000
      3.0000   6.0000   9.0000
```

Note that

1. $(A^T)^T = A,$

2. $(A_1 + A_2)^T = A_1^T + A_2^T,$

3. $(A_1 A_2)^T = A_2^T A_1^T,$

4. $(\alpha A)^T = \alpha A^T, and \ \alpha$ is a scalar.

Definition 1.11 (Inverse Matrix)

An $n \times n$ matrix A has an inverse or is invertible if there exists an $n \times n$ matrix B such that

$$AB = BA = I_n.$$

Then the matrix B is called the inverse of A and is denoted by A^{-1}. For example, let

$$A = \begin{pmatrix} 2 & 3 \\ 2 & 2 \end{pmatrix} \quad and \quad B = \begin{pmatrix} -1 & \frac{3}{2} \\ 1 & 1 \end{pmatrix}.$$

Then we have

$$AB = BA = \mathbf{I}_2,$$

which means that B is an inverse of A. Note that the invertible matrix is also called the nonsingular matrix.   ●

To find the inverse of a square matrix A using MATLAB commands we do as follows:

```
>> A = [21 0 0; -1 2 - 1 0; 0 - 1 2 - 1; 0 0 - 1 2]
>> Ainv = INVMAT(A)
Ainv =
    0.8000   0.6000   0.4000   0.2000
    0.6000   1.2000   0.8000   0.4000
    0.4000   0.8000   1.2000   0.6000
    0.2000   0.4000   0.6000   0.8000
```

Program 1.1
MATLAB m-file for Finding the Inverse of a Matrix
function [Ainv]=INVMAT(A)
[n,n]=size(A); I=zeros(n,n);
for i=1:n; I(i,i)=1; end
m(1:n,1:n)=A; $m(1:n, n+1:2*n) = I$;
for i=1:n; $m(i, 1:2*n) = m(i, 1:2*n)/m(i,i)$;
for k=1:n; if i˜ =k
$m(k, 1:2*n) = m(k, 1:2*n) - m(k,i)*m(i,1:2*n)$;
end; end; end
$invrs = m(1:n, n+1:2*n)$;

The MATLAB built-in function **inv(A)** can be also used to calculate the inverse of a square matrix A, if A is invertible:

```
>> I = Ainv * A;
>> format short e
>> disp(I)
I =
       1.0000e + 00   −1.1102e − 16   0              0
       0               1.0000e + 00   0              0
       0               0              1.0000e + 00   2.2204e − 16
       0               0              0              1.0000e + 00
```

The values of $I(2,1)$, and $I(3,4)$ are very small, but nonzero, due to round-off errors in the computation of **Ainv** and I. It is often preferable to use rational numbers rather than decimal numbers. The function **frac**(x) returns the rational approximation to x, or we can use the other MATLAB command as follows:

```
>> format rat
```

If the matrix A is not invertible, then the matrix A is called *singular*.

There are some well-known properties of the invertible matrix which are defined as follows.

Theorem 1.2 *If the matrix A is invertible, then:*

1. *It has exactly one inverse. If B and C are the inverses of A, then $B = C$.*

2. *Its inverse matrix A^{-1} is also invertible and $(A^{-1})^{-1} = A$.*

3. *Its product with another invertible matrix is invertible, and the inverse of the product is the product of the inverses in the reverse order. If A and B are invertible matrices of the same size, then AB is invertible and $(AB)^{-1} = B^{-1}A^{-1}$.*

4. *Its transpose matrix A^T is invertible and $(A^T)^{-1} = (A^{-1})^T$.*

5. *The kA for any nonzero k is invertible, i.e., $(kA)^{-1} = \frac{1}{k}A^{-1}$.*

6. *The A^k for any k is also invertible, i.e., $(A^k)^{-1} = (A^{-1})^k$.*

7. *Its size 1×1 is invertible when it is nonzero. If $A = (a)$, then $A^{-1} = \left(\frac{1}{a}\right)$.*

8. *The formula for A^{-1} when $n = 2$ is*

$$A^{-1} = \begin{pmatrix} a_{11} & a_{12} \\ a_{21} & a_{22} \end{pmatrix}^{-1} = \frac{1}{a_{11}a_{22} - a_{12}a_{21}} \begin{pmatrix} a_{22} & -a_{12} \\ -a_{21} & a_{11} \end{pmatrix},$$

provided that $a_{11}a_{22} - a_{12}a_{21} \neq 0$.   ●

Definition 1.12 (Diagonal Matrix)

It is a square matrix having all elements equal to zero except those on the main diagonal, i.e.,

$$A = (a_{ij}) = \begin{cases} a_{ij} = 0, & if \ i \neq j \\ a_{ij} \neq 0, & if \ i = j. \end{cases}$$

Note that all diagonal matrices are invertible if all diagonal entries are nonzero.   ●

The MATLAB function **diag** is used to either create a diagonal matrix from a vector or to extract the diagonal entries of a matrix. If the input argument of the **diag** function is a vector, MATLAB uses the vector to create a diagonal matrix:

```
>> x = [2, 2, 2];
>> A = diag(x)
A =
     2   0   0
     0   2   0
     0   0   2
```

The matrix A is called the *scalar* matrix because it has all the elements on the main diagonal equal to the same scalars 2. Multiplication of a square matrix and a scalar matrix is commutative, and the product is also a diagonal matrix.

If the input argument of the **diag** function is a matrix, the result is a vector of the diagonal elements:

```
>> B = [2 − 4 1; 6 10 − 3; 0 5 8]
>> M = diag(B)
M =
     2
    10
     8
```

Definition 1.13 (Upper-Triangular Matrix)

It is a square matrix which has zero elements below and to the left of the main diagonal. The diagonal as well as the above diagonal elements can take on any value, i.e.,

$$U = (u_{ij}), \quad where \quad u_{ij} = 0, \quad if \quad i > j.$$

An example of such a matrix is

$$U = \begin{pmatrix} 1 & 2 & 3 \\ 0 & 4 & 5 \\ 0 & 0 & 6 \end{pmatrix}.$$

The upper-triangular matrix is called an upper-unit-triangular matrix if the diagonal elements are equal to one. This type of matrix is used in solving linear algebraic equations by LU decomposition with Crout's method. Also, if the main diagonal elements of the upper-triangular matrix are zero, then the matrix

$$A = \begin{pmatrix} 0 & a_{12} & a_{13} \\ 0 & 0 & a_{23} \\ 0 & 0 & 0 \end{pmatrix}$$

is called a strictly upper-triangular matrix. This type of matrix will be used in solving linear systems by iterative methods.   ●

Using the MATLAB command **triu**(A) we can create an upper-triangular matrix from a given matrix A as follows:

```
>> A = [1 2 3; 4 5 6; 7 8 9];
>> U = triu(A)
U =
     1  2  3
     0  4  5
     0  0  6
```

We can also create a strictly upper-triangular matrix, i.e., an upper-triangular matrix with zero diagonals, from a given matrix A by using the MATLAB built-in function **triu**(A,I) as follows:

```
>> A = [1 2 3; 4 5 6; 7 8 9];
>> U = triu(A, I)
U =
     0  2  3
     0  0  5
     0  0  0
```

Definition 1.14 (Lower-Triangular Matrix)

It is a square matrix which has zero elements above and to the right of the main diagonal, and the rest of the elements can take on any value, i.e.,

$$L = (l_{ij}), \quad \text{where} \quad l_{ij} = 0, \quad \text{if} \quad i < j.$$

An example of such a matrix is

$$L = \begin{pmatrix} 2 & 0 & 0 \\ 3 & 1 & 0 \\ 4 & 5 & 3 \end{pmatrix}.$$

The lower-triangular matrix is called a lower-unit-triangular matrix if the diagonal elements are equal to one. This type of matrix is used in solving linear algebraic equations by LU decomposition with Doolittle's method. Also, if the main diagonal elements of the lower-triangular matrix are zero, then the matrix

$$A = \begin{pmatrix} 0 & 0 & 0 \\ a_{21} & 0 & 0 \\ a_{31} & a_{32} & 0 \end{pmatrix}$$

is called a strictly lower-triangular matrix. We will use this type of matrix in solving the linear systems by using iterative methods. ●

In a similar way, we can create a lower-triangular matrix and a strictly lower-triangular matrix from a given matrix A by using the MATLAB built-in functions **tril**(A) and **tril**(A,I), respectively.

Note that all the triangular matrices (upper or lower) with nonzero diagonal entries are invertible.

Definition 1.15 (Symmetric Matrix)

A symmetric matrix is one in which the elements a_{ij} of a matrix A in the ith row and jth column are equal to the elements a_{ji} in the jth row and ith column, which means that

$$A^T = A, \quad \text{i.e.,} \quad a_{ij} = a_{ji}, \quad \text{for} \quad i \neq j.$$

Note that any diagonal matrix, including the identity, is symmetric. A lower- or upper-triangular matrix is symmetric if and only if it is, in fact, a diagonal matrix.

One way to generate a symmetric matrix is to multiply a matrix by its transpose, since $A^T A$ is symmetric for any A. To generate a symmetric matrix using MATLAB commands we do the following:

$$
\begin{aligned}
&>> A = [1:4; 5:8; 9:12] \\
&\%A \text{ is not symmetric} \\
&>> B = A' * A \\
&B = \\
&\quad\quad
\begin{array}{cccc}
107 & 122 & 137 & 152 \\
122 & 140 & 158 & 176 \\
137 & 158 & 179 & 200 \\
152 & 176 & 200 & 224
\end{array} \\
&>> C = A * A' \\
&C = \\
&\quad\quad
\begin{array}{ccc}
30 & 70 & 110 \\
70 & 174 & 278 \\
110 & 278 & 446
\end{array}
\end{aligned}
$$

Example 1.1 *Find all the values of a, b, and c for which the following matrix is symmetric:*

$$
A = \begin{pmatrix}
4 & a+b+c & 0 \\
-1 & 3 & b-c \\
-a+2b-2c & 1 & b-2c
\end{pmatrix}.
$$

Solution. *If the given matrix is symmetric, then $A = A^T$, i.e.,*

$$
A = \begin{pmatrix}
4 & a+b+c & 0 \\
-1 & 3 & b-c \\
-a+2b-2c & 1 & b-2c
\end{pmatrix}
$$

$$
= \begin{pmatrix}
4 & -1 & -a+2b-2c \\
a+b+c & 3 & 1 \\
0 & b-c & b-2c
\end{pmatrix} = A^T,
$$

which implies that

$$
\begin{aligned}
0 &= -a + 2b - 2c \\
-1 &= a + b + c \\
1 &= b - c.
\end{aligned}
$$

Solving the above system, we get

$$a = 2, \quad b = -1, \quad c = -2,$$

and using these values, we have the given matrix of the form

$$
A = \begin{pmatrix} 4 & -1 & 0 \\ -1 & 3 & 1 \\ 0 & 1 & 3 \end{pmatrix}.
$$

●

Theorem 1.3 *If A and B are symmetric matrices of the same size, and if k is any scalar, then:*

1. A^T *is also symmetric;*

2. $A + B$ *and* $A - B$ *are symmetric;*

3. *and* kA *is also symmetric.*

   Note that the product of symmetric matrices is not symmetric in general, but the product is symmetric if and only if the matrices commute. Also, note that if A is a square matrix, then the matrices A, AA^T, and $A^T A$ are either all nonsingular or all singular. ●

   If for a matrix A, the $a_{ij} = -a_{ji}$ for a $i \neq j$ and the main diagonal elements are not all zero, then the matrix A is called a skew matrix. If all the elements on the main diagonal of a skew matrix are zero, then the matrix is called skew symmetric, i.e.,

$$A = -A^T, \quad with \quad a_{ij} = -a_{ji} \quad for \quad i \neq j, \quad and \quad a_{ii} = 0.$$

Any square matrix may be split into the sum of a symmetric and a skew symmetric matrix. Thus,

$$A = \frac{1}{2}(A + A^T) + \frac{1}{2}(A - A^T),$$

where $\frac{1}{2}(A+A^T)$ is a symmetric matrix and $\frac{1}{2}(A-A^T)$ is a skew symmetric matrix. The matrices

$$\begin{pmatrix} 1 & 2 & 3 \\ 2 & 4 & 5 \\ 3 & 5 & 6 \end{pmatrix}, \quad \begin{pmatrix} 1 & 2 & 3 \\ -2 & 4 & -5 \\ -3 & 5 & 6 \end{pmatrix}, and \quad \begin{pmatrix} 0 & 2 & 3 \\ -2 & 0 & 5 \\ -3 & 5 & 0 \end{pmatrix}$$

are examples of symmetric, skew, and skew symmetric matrices, respectively.   ●

Definition 1.16 (Partitioned Matrix)

A matrix A is said to be partitioned if horizontal and vertical lines have been introduced, subdividing A into submatrices called blocks. Partitioning allows A to be written as a matrix A whose entries are its blocks. A simple example of a partitioned matrix may be an augmented matrix, which can be partitioned in the form

$$B = [A|\mathbf{b}].$$

It is frequently necessary to deal separately with various groups of elements, or submatrices, within a large matrix. This situation can arise when the size of a matrix becomes too large for convenient handling, and it becomes necessary to work with only a portion of the matrix at any one time. Also, there will be cases in which one part of a matrix will have a physical significance that is different from the remainder, and it is instructive to isolate that portion and identify it by a special symbol. For example, the following 4×5 matrix A has been partitioned into four blocks of elements, each of which is itself a matrix:

$$A = \left(\begin{array}{ccc:cc} a_{11} & a_{12} & a_{13} & a_{14} & a_{15} \\ a_{21} & a_{22} & a_{23} & a_{24} & a_{25} \\ a_{31} & a_{32} & a_{33} & a_{34} & a_{35} \\ \cdots & \cdots & \cdots & \cdots & \cdots \\ a_{41} & a_{42} & a_{43} & a_{44} & a_{45} \end{array} \right).$$

The partitioning lines must always extend entirely through the matrix as in the above example. If the submatrices of A are denoted by the symbols $A_{11}, A_{12}, A_{21},$ and A_{22} so that

$$A_{11} = \begin{pmatrix} a_{11} & a_{12} & a_{13} \\ a_{21} & a_{22} & a_{23} \\ a_{31} & a_{32} & a_{33} \end{pmatrix}, \qquad A_{12} = \begin{pmatrix} a_{14} & a_{15} \\ a_{24} & a_{25} \\ a_{34} & a_{35} \end{pmatrix},$$

$$A_{21} = \begin{pmatrix} a_{41} & a_{42} & a_{43} \end{pmatrix}, \qquad A_{22} = \begin{pmatrix} a_{44} & a_{45} \end{pmatrix},$$

then the original matrix can be written in the form

$$A = \begin{pmatrix} A_{11} & A_{12} \\ A_{21} & A_{22} \end{pmatrix}.$$

A partitioned matrix may be transposed by appropriate transposition and rearrangement of the submatrices. For example, it can be seen by inspection that the transpose of the matrix A is

$$A^T = \begin{pmatrix} A_{11}^T & A_{12}^T \\ A_{21}^T & A_{22}^T \end{pmatrix}.$$

Note that A^T has been formed by transposing each submatrix of A and then interchanging the submatrices on the secondary diagonal.

Partitioned matrices such as the one given above can be added, subtracted, and multiplied provided that the partitioning is performed in an appropriate manner. For the addition and subtraction of two matrices, it is necessary that both matrices be partitioned in exactly the same way. Thus, a partitioned matrix B of order 4×5 (compare with matrix A above) will be conformable for addition with A only if it is partitioned as follows:

$$B = \begin{pmatrix} b_{11} & b_{12} & b_{13} & \vdots & b_{14} & b_{15} \\ b_{21} & b_{22} & b_{23} & \vdots & b_{24} & b_{25} \\ b_{31} & b_{32} & b_{33} & \vdots & b_{34} & b_{35} \\ \cdots & \cdots & \cdots & \vdots & \cdots & \cdots \\ b_{41} & b_{42} & b_{43} & \vdots & b_{44} & b_{45} \end{pmatrix}.$$

It can be expressed in the form

$$B = \begin{pmatrix} B_{11} & B_{12} \\ B_{21} & B_{22} \end{pmatrix},$$

in which B_{11}, B_{12}, B_{21}, and B_{22} represent the corresponding submatrices. In order to add A and B and obtain a sum C, it is necessary according to the rules for addition of matrices that the following represent the sum:

$$A + B = \begin{pmatrix} A_{11} + B_{11} & A_{12} + B_{12} \\ A_{21} + B_{21} & A_{22} + B_{22} \end{pmatrix} = \begin{pmatrix} C_{11} & C_{12} \\ C_{21} & C_{22} \end{pmatrix} = C.$$

Note that like A and B, the sum matrix C will also have the same partitions.

The conformability requirement for multiplication of partitioned matrices is somewhat different from that for addition and subtraction. To show the requirement, consider again the matrix A given previously and assume that it is to be postmultiplied by a matrix D, which must have five rows but may have any number of columns. Also assume that D is partitioned into four submatrices as follows:

$$D = \begin{pmatrix} D_{11} & D_{12} \\ D_{21} & D_{22} \end{pmatrix}.$$

Then, when forming the product AD according to the usual rules for matrix multiplication, the following result is obtained:

$$\begin{aligned} M \; = \; AD &= \begin{pmatrix} A_{11} & A_{12} \\ A_{21} & A_{22} \end{pmatrix} \begin{pmatrix} D_{11} & D_{12} \\ D_{21} & D_{22} \end{pmatrix} \\[2mm] &= \begin{pmatrix} A_{11}D_{11} + A_{12}D_{21} & A_{11}D_{12} + A_{12}D_{22} \\ A_{21}D_{11} + A_{22}D_{21} & A_{21}D_{12} + A_{22}D_{22} \end{pmatrix} \\[2mm] &= \begin{pmatrix} M_{11} & M_{12} \\ M_{21} & M_{22} \end{pmatrix}. \end{aligned}$$

Thus, the multiplication of the two partitioned matrices is possible if the columns of the first partitioned matrix are partitioned in exactly the same way as the rows of the second partitioned matrix. It does not matter how the rows of the first partitioned matrix and the columns of the second partitioned matrix are partitioned. ●

Definition 1.17 (Band Matrix)

An $n \times n$ square matrix A is called a band matrix if there exists positive integers p and q, with $1 < p$ and $q < n$, such that

$$a_{ij} = 0 \quad for \quad p \leq j - i \quad or \quad q \leq i - j.$$

The number p describes the number of diagonals above, including the main diagonal on which the nonzero entries may lie. The number q describes the number of diagonals below, including the main diagonal on which the nonzero entries may lie. The number $p + q - 1$ is called the bandwidth of the matrix A, which tells us how many of the diagonals can contain nonzero entries. For example, the matrix

$$A = \begin{pmatrix} 1 & 2 & 3 & \\ 2 & 3 & 4 & 5 \\ 0 & 5 & 6 & 7 \\ 0 & 0 & 7 & 8 \end{pmatrix}$$

is banded with $p = 3$ and $q = 2$, and so the bandwidth is equal to 4. An important property of the band matrix is called the tridiagonal matrix, in this case, $p = q = 2$, i.e., all nonzero elements lie either on or directly above or below the main diagonal. For this type of matrix, Gaussian elimination is particularly simpler. In general, the nonzero elements of a tridiagonal matrix lie in three bands: the superdiagonal, diagonal, and subdiagonal. For example, the matrix

$$A = \begin{pmatrix} 1 & 2 & & & & & \\ 2 & 3 & 1 & & & & \\ & 3 & 2 & 1 & & & \\ & & 2 & 4 & 3 & & \\ & & & 1 & 2 & 3 & \\ & & & & 1 & 6 & 4 \\ & & & & & 3 & 4 \end{pmatrix}$$

is a tridiagonal matrix.

A matrix which is predominantly zero is called a sparse matrix. A band matrix or a tridiagonal matrix is a sparse matrix, but the nonzero elements of a sparse matrix are not necessarily near the diagonal. •

Definition 1.18 (Permutation Matrix)

A permutation matrix P has only 0s and 1s and there is exactly one in each row and column of P. For example, the following matrices are permutation matrices:

$$P = \begin{pmatrix} 1 & 0 & 0 \\ 0 & 0 & 1 \\ 0 & 1 & 0 \end{pmatrix}, \qquad P = \begin{pmatrix} 0 & 1 & 0 & 0 \\ 1 & 0 & 0 & 0 \\ 0 & 0 & 1 & 0 \\ 0 & 0 & 0 & 1 \end{pmatrix}.$$

The product PA has the same rows as A but in a different order (permuted), while AP is just A with the columns permuted. •

1.2.3   Solutions of Linear Systems of Equations

Here we shall discuss the familiar technique called the *method of elimination* to find the solutions of linear systems. This method starts with the augmented matrix of the given linear system and obtains a matrix of a certain form. This new matrix represents a linear system that has exactly the same solutions as the given origin system. In the following, we define two well-known forms of a matrix.

Definition 1.19 (Row Echelon Form)

An $m \times n$ matrix A is said to be in row echelon form if it satisfies the following properties:

1. *Any rows consisting entirely of zeros are at the bottom.*

2. *The first entry from the left of a nonzero row is 1. This entry is called the leading one of its row.*

3. *For each nonzero row, the leading one appears to the right and below any leading ones in preceding rows.*

Note that, in particular, in any column containing a leading one, all entries below the leading one are zero. For example, the following matrices are in row echelon form:

$$
\begin{pmatrix} 1 & 2 & 1 \\ 0 & 1 & 3 \\ 0 & 0 & 0 \end{pmatrix}, \quad
\begin{pmatrix} 1 & 0 & 2 \\ 0 & 1 & 2 \\ 0 & 0 & 1 \end{pmatrix}, \quad
\begin{pmatrix} 1 & 2 & 3 & 4 \\ 0 & 0 & 1 & 2 \\ 0 & 0 & 0 & 0 \end{pmatrix}, \quad
\begin{pmatrix} 0 & 1 & 0 & 1 \\ 0 & 0 & 1 & 0 \\ 0 & 0 & 0 & 0 \\ 0 & 0 & 0 & 0 \end{pmatrix}.
$$

Observe that a matrix in row echelon form is actually the augmented matrix of a linear system (i.e., the last column is the right-hand side of the system $A\mathbf{x} = \mathbf{b}$), and the system is quite easy to solve by backward substitution. For example, writing the first above matrix in linear system form, we have

$$
\begin{aligned}
x_1 \ + \ 2x_2 &= \ 1 \\
x_2 &= \ 3.
\end{aligned}
$$

No need to involve the last equation, which is

$$
0x_1 + 0x_2 = 0,
$$

and it satisfies for any choices of x_1 and x_2. Thus, by using backward substitution, we get

$$
x_2 = 3 \quad and \quad x_1 = (1 - 2x_2) = (1 - 2(3)) = -5.
$$

So the unique solution of the linear system is $[-5, 3]^T$.

Similarly, the linear system that corresponds to the second above matrix is

$$
\begin{aligned}
x_1 &= \ 2 \\
x_2 &= \ 2 \\
0 &= \ 1.
\end{aligned}
$$

The third equation of this system shows that

$$
0x_1 + 0x_2 = 1,
$$

which is not possible for any choices of x_1 and x_2. Hence, the system has no solution.

Finally, the linear system that corresponds to the third above matrix is

$$
\begin{aligned}
x_1 \;+\; 2x_2 \;+\; 3x_3 &= 4 \\
x_3 &= 2 \\
0x_1 \;+\; 0x_2 \;+\; 0x_3 &= 0,
\end{aligned}
$$

and by backward substitution (without using the third equation of the system), we get

$$
x_3 = 2, \quad and \quad x_1 = 4 - 2x_2 - 3x_3 = -2 - 2x_2.
$$

By choosing an arbitrary nonzero value of x_2, we will get the value of x_1, which implies that we have infinitely many solutions for such a linear system.   •

If we add one more property in the above definition of row echelon form, then we will get another well-known form of a matrix, called *reduced row echelon form*, which we define as follows.

Definition 1.20 (Reduced Row Echelon Form)

An $m \times n$ matrix A is said to be in reduced row echelon form if it satisfies the following properties:

1. *Any rows consisting entirely of zeros are at the bottom.*

2. *The first entry from the left of a nonzero row is 1. This entry is called the leading one of its row.*

3. *For each nonzero row, the leading one appears to the right and below any leading ones in preceding rows.*

4. *If a column contains a leading one, then all other entries in that column (above and below a leading one) are zeroes.*

For example, the following matrices are in reduced row echelon form:

$$\begin{pmatrix} 1 & 0 & 1 \\ 0 & 1 & 2 \\ 0 & 0 & 0 \end{pmatrix}, \quad \begin{pmatrix} 1 & 0 & 0 & 2 \\ 0 & 1 & 0 & 4 \\ 0 & 0 & 1 & 6 \end{pmatrix}, \quad \begin{pmatrix} 1 & 4 & 5 & 0 \\ 0 & 0 & 0 & 1 \\ 0 & 0 & 0 & 0 \end{pmatrix}, \quad \begin{pmatrix} 1 & 1 & 0 & 0 & 0 \\ 0 & 0 & 1 & 0 & 2 \\ 0 & 0 & 0 & 1 & 1 \end{pmatrix},$$

and the following matrices are not in reduced row echelon form:

$$\begin{pmatrix} 1 & 3 & 0 & 2 \\ 0 & 0 & 0 & 0 \\ 0 & 0 & 1 & 4 \end{pmatrix}, \quad \begin{pmatrix} 1 & 3 & 0 & 2 \\ 0 & 0 & 5 & 4 \\ 0 & 0 & 0 & 1 \end{pmatrix}, \quad \begin{pmatrix} 1 & 0 & 0 & 3 \\ 0 & 0 & 1 & 2 \\ 0 & 1 & 0 & 6 \end{pmatrix}, \quad \begin{pmatrix} 1 & 0 & 2 & 0 & 0 \\ 0 & 1 & 1 & 0 & 2 \\ 0 & 2 & 0 & 1 & 1 \end{pmatrix}.$$

Note that a useful property of matrices in reduced row echelon form is that if A is an $n \times n$ matrix in reduced row echelon form not equal to identity matrix $\mathbf{I}_n$, then A has a row consisting entirely of zeros. •

There are usually many sequences of row operations that can be used to transform a given matrix to reduced row echelon form—they all, however, lead to the same reduced row echelon form. In the following, we shall discuss how to transform a given matrix in reduced row echelon form.

Definition 1.21 (Elementary Row Operations)
It is the procedure that can be used to transform a given matrix into row echelon or reduced row echelon form. An elementary row operation on an $m \times n$ matrix A is any of the following operations:

1. Interchanging two rows of a matrix A;

2. Multiplying a row of A by a nonzero constant;

3. Adding a multiple of a row of A to another row.

Observe that when a matrix is viewed as the augmented matrix of a linear system, the elementary row operations are equivalent, respectively, to interchanging two equations, multiplying an equation by a nonzero constant, and adding a multiple of an equation to another equation. •

Example 1.2 *Consider the matrix*

$$A = \begin{pmatrix} 0 & 0 & 1 & 3 \\ 3 & 2 & 0 & 4 \\ 4 & 4 & 8 & 12 \end{pmatrix}.$$

Interchanging rows 1 and 2 gives

$$R_1 = \begin{pmatrix} 3 & 2 & 0 & 4 \\ 0 & 0 & 1 & 3 \\ 4 & 4 & 8 & 12 \end{pmatrix}.$$

Multiplying the third row of A by $\frac{1}{4}$, *we get*

$$R_2 = \begin{pmatrix} 0 & 0 & 1 & 3 \\ 3 & 2 & 0 & 4 \\ 1 & 1 & 2 & 3 \end{pmatrix}.$$

Adding (-2) *times row 2 of A to row 3 of A gives*

$$R_3 = \begin{pmatrix} 0 & 0 & 1 & 3 \\ 3 & 2 & 0 & 4 \\ -2 & 0 & 8 & 4 \end{pmatrix}.$$ $-5 -3 \ 2 \ -5$

Observe that in obtaining R_3 *from A, row 2 did not change.* •

Theorem 1.4 *Every matrix can be brought to reduced row echelon form by a series of elementary row operations.* •

Example 1.3 *Consider the matrix*

$$A = \begin{pmatrix} 1 & -3 & 0 & 0 & 1 \\ 2 & -6 & -1 & 1 & 1 \\ 3 & -9 & 2 & -1 & 5 \end{pmatrix}.$$

Using the finite sequence of elementary row operations, we get the matrix of the form

$$R_1 = \begin{pmatrix} 1 & -3 & 0 & 0 & 1 \\ 0 & 0 & 1 & -1 & 1 \\ 0 & 0 & 0 & 1 & 0 \end{pmatrix},$$

which is in row echelon form. If we continue with the matrix R_1 and make all elements above the leading one equal to zero, we obtain

$$R_2 = \begin{pmatrix} 1 & -3 & 0 & 0 & 1 \\ 0 & 0 & 1 & 0 & 1 \\ 0 & 0 & 0 & 1 & 0 \end{pmatrix},$$

which is the reduced row echelon form of the given matrix A. ●

MATLAB has a function **rref** used to arrive directly at the reduced echelon form of a matrix. For example, using the above given matrix, we do the following:

```
>> A = [1 − 3 0 0 1; 2 − 6 − 1 1 1; 3 − 9 2 − 1 5];
>> B = rref(A)
B =
    1  −3  0  0  1
    0   0  1  0  1
    0   0  0  1  0
```

Definition 1.22 (Row Equivalent Matrix)

An $m \times n$ matrix A is said to be row equivalent to an $m \times n$ matrix B if B can be obtained by applying a finite sequence of elementary row operations to the matrix A. ●

Example 1.4 *Consider the matrix*

$$A = \begin{pmatrix} 1 & 3 & 6 & 5 \\ 2 & 1 & 4 & 3 \\ 3 & -4 & 3 & 4 \end{pmatrix}.$$

If we add (-1) times row 1 of A to its third row, we get

$$R_1 = \begin{pmatrix} 1 & 3 & 6 & 5 \\ 2 & 1 & 4 & 3 \\ 2 & -7 & -3 & -1 \end{pmatrix},$$

so R_1 is row equivalent to A.

Interchanging row 2 and row 3 of the matrix R_1 gives the matrix of the form

$$R_2 = \begin{pmatrix} 1 & 3 & 6 & 5 \\ 2 & -7 & -3 & -1 \\ 2 & 1 & 4 & 3 \end{pmatrix},$$

so R_2 is row equivalent to R_1.

Multiplying row 2 of R_2 by (-2), we obtain

$$R_3 = \begin{pmatrix} 1 & 3 & 6 & 5 \\ -4 & 14 & 6 & 2 \\ 2 & 1 & 4 & 3 \end{pmatrix},$$

so R_3 is row equivalent to R_2.

It then follows that R_3 is row equivalent to the given matrix A since we obtained the matrix R_3 by applying three successive elementary row operations to A. $\qquad\bullet$

Theorem 1.5

1. *Every matrix is row equivalent to itself.*

2. *If a matrix A is row equivalent to a matrix B, then B is row equivalent to A.*

3. *If a matrix A is row equivalent to a matrix B and B is row equivalent to a matrix C, then A is row equivalent to C.* $\qquad\bullet$

Theorem 1.6 *Every $m \times n$ matrix is row equivalent to a unique matrix in reduced row echelon form.* $\qquad\bullet$

Example 1.5 *Use elementary row operations on matrices to solve the linear system*

$$\begin{array}{rcrcrcr} & - & x_2 & + & x_3 & = & 1 \\ x_1 & - & x_2 & - & x_3 & = & 1 \\ -x_1 & & & + & 3x_3 & = & -2. \end{array}$$

Solution. *The process begins with the augmented matrix form*

$$\begin{pmatrix} 0 & -1 & 1 & \vdots & 1 \\ 1 & -1 & -1 & \vdots & 1 \\ -1 & 0 & 3 & \vdots & -2 \end{pmatrix}.$$

Interchanging the first and the second rows gives

$$\begin{pmatrix} 1 & -1 & -1 & \vdots & 1 \\ 0 & -1 & 1 & \vdots & 1 \\ -1 & 0 & 3 & \vdots & -2 \end{pmatrix}.$$

Adding (1) times row 1 of the above matrix to its third row, we get

$$\begin{pmatrix} 1 & -1 & -1 & \vdots & 1 \\ 0 & -1 & 1 & \vdots & 1 \\ 0 & -1 & 2 & \vdots & -1 \end{pmatrix}.$$

Now multiplying the second row by -1 *gives*

$$\begin{pmatrix} 1 & -1 & -1 & \vdots & 1 \\ 0 & 1 & -1 & \vdots & -1 \\ 0 & -1 & 2 & \vdots & -1 \end{pmatrix}.$$

Replace row 1 with the sum of itself and (1) times row 2, and then also replace row 3 with the sum of itself and (1) times row 2, and we get the matrix of the form

$$\begin{pmatrix} 1 & 0 & -2 & \vdots & 0 \\ 0 & 1 & -1 & \vdots & -1 \\ 0 & 0 & 1 & \vdots & -2 \end{pmatrix}.$$

Replace row 1 with the sum of itself and (2) times row 3, and then replace row 2 with the sum of itself and (1) times the row 3, and we get

$$\begin{pmatrix} 1 & 0 & 0 & \vdots & -4 \\ 0 & 1 & 0 & \vdots & -3 \\ 0 & 0 & 1 & \vdots & -2 \end{pmatrix}.$$

Now by writing in equation form and using backward substitution

$$
\begin{aligned}
x_1 & & & = & -4 \\
& x_2 & & = & -3 \\
& & x_3 & = & -2,
\end{aligned}
$$

and we get the solution $[-4, -3, -2]^T$ *of the given linear system.*   ●

1.2.4   The Determinant of a Matrix

The determinant is a certain kind of a function that associates a real number with a square matrix. We will denote the determinant of a square matrix A by $\det(A)$ or $|A|$.

Definition 1.23 (Determinant of a Matrix)

Let $A = (a_{ij})$ *be an* $n \times n$ *square matrix, then a determinant of* A *is given by:*

1. $\det(A) = a_{11},$ *if* $n = 1.$

2. $\det(A) = a_{11}a_{22} - a_{12}a_{21},$ *if* $n = 2.$ ●

For example, if

$$
A = \begin{pmatrix} 4 & 2 \\ -3 & 7 \end{pmatrix} \quad \text{and} \quad B = \begin{pmatrix} 6 & 3 \\ 2 & 5 \end{pmatrix},
$$

then

$$
\det(A) = (4)(7) - (-3)(2) = 34 \quad \text{and} \quad \det(B) = (6)(5) - (3)(2) = 24.
$$

Notice that the determinant of a 2×2 matrix is given by the difference of the products of the two diagonals of a matrix. The determinant of a 3×3 matrix is defined in terms of the determinants of 2×2 matrices, and the determinant of a 4×4 matrix is defined in terms of the determinants of 3×3 matrices, and so on.

The MATLAB function **det**(A), is calculated by the determinant of the square matrix A as:

```
>> A = [2 2; 6 7];
>> B = det(A)
B =
    2.0000
```

Another way to find the determinants of only 2×2 and 3×3 matrices can be found easily and quickly using diagonals (or direct evaluation). For a 2×2 matrix, the determinant can be obtained by forming the product of the entries on the line from left to right and subtracting from this number the product of the entries on the line from right to left. For a matrix of size 3×3, the diagonals of an array consisting of the matrix with the first two columns added to the right are used. Then the determinant can be obtained by forming the sum of the products of the entries on the lines from left to right, and subtracting from this number the products of the entries on the lines from right to left, as shown in Figure (1.2).

Thus, for a 2×2 matrix

$$|A| = a_{11}a_{22} - a_{12}a_{21}$$

and for 3×3 matrix

$$|A| = \underbrace{a_{11}a_{22}a_{33} + a_{12}a_{23}a_{31} + a_{13}a_{21}a_{32}}_{\left(\text{diagonal products from left to right}\right)} - \underbrace{a_{13}a_{22}a_{31} - a_{11}a_{23}a_{32} - a_{12}a_{21}a_{33}}_{\left(\text{diagonal products from right to left}\right)}.$$

For example, the determinant of a 2×2 matrix can be computed as

$$|A| = \begin{vmatrix} 12 & 5 \\ -7 & 6 \end{vmatrix} = (12)(6) - (5)(-7) = 72 + 35 = 107,$$

and the determinant of a 3×3 matrix can be obtained as

$$
\begin{aligned}
|A| = \begin{vmatrix} 4 & 5 & 6 \\ -3 & 8 & 2 \\ 4 & 9 & 7 \end{vmatrix} &= [(4)(8)(7) + (5)(2)(4) + (6)(-3)(9)] \\
&\quad - [(6)(8)(4) + (4)(2)(9) + (5)(-3)(7)] \\
&= 102 - 159 = -57.
\end{aligned}
$$

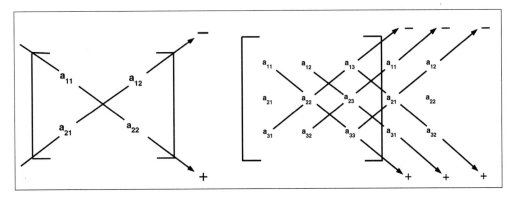

Figure 1.2: Direct evaluation of 2×2 and 3×3 determinants.

For finding the determinants of the higher-order matrices, we will define the following concepts called the *minor* and *cofactor* of matrices.

Definition 1.24 (Minor of a Matrix)

The minor M_{ij} of all elements a_{ij} of a matrix A of order $n \times n$ as the determinant of the submatrix of order $(n-1) \times (n-1)$ is obtained from A by deleting the ith row and jth column (also called the ijth minor of A). For example, let

$$A = \begin{pmatrix} 2 & 3 & -1 \\ 5 & 3 & 2 \\ 4 & -2 & 4 \end{pmatrix},$$

then the minor M_{11} will be obtained by deleting the first row and the first column of the given matrix A, i.e.,

$$M_{11} = \begin{vmatrix} 3 & 2 \\ -2 & 4 \end{vmatrix} = (3)(4) - (-2)(2) = 12 + 4 = 16.$$

Similarly, we can find the other possible minors of the given matrix as follows:

$$M_{12} = \begin{vmatrix} 5 & 2 \\ 4 & 4 \end{vmatrix} = 20 - 16 = 4$$

$$M_{13} = \begin{vmatrix} 5 & 3 \\ 4 & -2 \end{vmatrix} = -10 - 12 = -22$$

$$M_{21} = \begin{vmatrix} 3 & -1 \\ -2 & 4 \end{vmatrix} = 12 - 2 = 10$$

$$M_{22} = \begin{vmatrix} 2 & -1 \\ 4 & 4 \end{vmatrix} = 8 + 4 = 12$$

$$M_{23} = \begin{vmatrix} 2 & 3 \\ 4 & -2 \end{vmatrix} = -4 - 12 = -16$$

$$M_{31} = \begin{vmatrix} 3 & -1 \\ 3 & 2 \end{vmatrix} = 6 + 3 = 9$$

$$M_{32} = \begin{vmatrix} 2 & -1 \\ 5 & 2 \end{vmatrix} = 4 + 5 = 9$$

$$M_{33} = \begin{vmatrix} 2 & 3 \\ 5 & 3 \end{vmatrix} = 6 - 15 = -9,$$

which are the required minors of the given matrix. •

Definition 1.25 (Cofactor of a Matrix)

The cofactor A_{ij} of all elements a_{ij} of a matrix A of order $n \times n$ is given by

$$A_{ij} = (-1)^{i+j} M_{ij},$$

where M_{ij} is the minor of all elements a_{ij} of a matrix A. For example, the cofactors A_{ij} of all elements a_{ij} of the matrix

$$A = \begin{pmatrix} 2 & 3 & -1 \\ 5 & 3 & 2 \\ 4 & -2 & 4 \end{pmatrix}$$

are computed as follows:

$$
\begin{aligned}
A_{11} &= (-1)^{1+1}M_{11} = M_{11} = 16 \\
A_{12} &= (-1)^{1+2}M_{12} = -M_{12} = -4 \\
A_{13} &= (-1)^{1+3}M_{13} = M_{13} = -22 \\
A_{21} &= (-1)^{2+1}M_{21} = -M_{21} = -10 \\
A_{22} &= (-1)^{2+2}M_{22} = M_{22} = 12 \\
A_{23} &= (-1)^{2+3}M_{23} = -M_{23} = 16 \\
A_{31} &= (-1)^{3+1}M_{31} = M_{31} = 9 \\
A_{32} &= (-1)^{3+2}M_{32} = -M_{32} = -9 \\
A_{33} &= (-1)^{3+3}M_{33} = M_{33} = -9,
\end{aligned}
$$

which are the required cofactors of the given matrix.   •

To get the above results, we use the MATLAB command window as follows:

```
>> A = [2 3 − 1; 5 3 2; 4 − 2 4];
>> CofA = cofactor(A, 1, 1);
>> CofA = cofactor(A, 1, 2);
>> CofA = cofactor(A, 1, 3);
>> CofA = cofactor(A, 2, 1);
>> CofA = cofactor(A, 2, 2);
>> CofA = cofactor(A, 2, 3);
>> CofA = cofactor(A, 3, 1);
>> CofA = cofactor(A, 3, 2);
>> CofA = cofactor(A, 3, 3);
```

> # Program 1.2
> MATLAB m-file for Finding Minors and Cofactors of a Matrix
>
> ```
> function CofA = cofactor(A,i,j)
> [m,n] = size(A);
> if m ~ = n  error(Matrix must be square)  end
> A1 = A([1:i-1,i+1:n],[1:j-1,j+1:n]);
> Minor = det(A1);
> CofA = (-1)^ (i+j)*det(Minor);
> ```

Definition 1.26 (Cofactor Expansion of a Determinant of a Matrix)

Let A be a square matrix, then we define the determinant of A as the sum of the products of the elements of the first row and their cofactors. If A is a 3×3 matrix, then its determinant is defined as

$$\det(A) = |A| = a_11A_{11} + a_12A_{12} + a_13A_{13}.$$

Similarly, in general, for an $n \times n$ matrix, we define it as

$$\det(A) = |A| = \sum_{1}^{n} a_{ij}A_{ij}, \quad n > 2, \tag{1.12}$$

where the summation is on i for any fixed value of the jth column $(1 \leq j \leq n)$, or on j for any fixed value of the ith row $(1 \leq i \leq n)$, and A_{ij} is the cofactor of element a_{ij}. ●

Example 1.6 *Find the minors and cofactors of the matrix A and use them to evaluate the determinant of the matrix*

$$A = \begin{pmatrix} 3 & 1 & -4 \\ 2 & 5 & 6 \\ 1 & 4 & 8 \end{pmatrix}.$$

Solution. *The minors of A are calculated as follows:*

$$M_{11} = \begin{vmatrix} 5 & 6 \\ 4 & 8 \end{vmatrix} = 40 - 24 = 16$$

$$M_{12} = \begin{vmatrix} 2 & 6 \\ 1 & 8 \end{vmatrix} = 16 - 6 = 10$$

$$M_{13} = \begin{vmatrix} 2 & 5 \\ 1 & 4 \end{vmatrix} = 8 - 5 = 3.$$

From these values of the minors, we can calculate the cofactors of the elements of the given matrix as follows:

$$
\begin{aligned}
A_{11} &= (-1)^{1+1} M_{11} = M_{11} = 16 \\
A_{12} &= (-1)^{1+2} M_{12} = -M_{12} = -10 \\
A_{13} &= (-1)^{1+3} M_{13} = M_{13} = 3.
\end{aligned}
$$

Now by using the cofactor expansion along the first row, we can find the determinant of the matrix as follows:

$$\det(A) = a_{11}A_{11} + a_{12}A_{12} + a_{13}A_{13} = (3)(16) + (1)(-10) + (-4)(3) = 26.$$

Note that in Example 1.6, we computed the determinant of the matrix by using the cofactor expansion along the first row, but it can also be found along the first column of the matrix.

To get the results of Example 1.6, we use the MATLAB Command Window as follows:

```
>> A = [3 1 − 4; 2 5 6; 1 4 8];
>> DetA = CofFexp(A);
```

> **Program 1.3**
> MATLAB m-file for Finding the Determinant of a
> Matrix by Cofactor Expansion
> function DetA = CofFexp(A)
> [m,n] = size(A);
> if m ˜ = n error (Matrix must be square) end
> a = A(1,:);c = [ ];
> for i=1:n
> cli = cofactor(A,1,i);
> c = [c;cli]; end
> DetA = a*c;;

Theorem 1.7 (The Laplace Expansion Theorem)

The determinant of an $n \times n$ matrix $A = \{a_{ij}\}$, when $n \geq 2$, can be computed as

$$\det(A) = a_{i1}A_{i1} + a_{i2}A_{i2} + \cdots + a_{in}A_{in}$$

$$= \sum_{j=1}^{n} a_{ij}A_{ij},$$

which is called the cofactor expansion along the ith row, and also as

$$\det(A) = a_{1j}A_{1j} + a_{2j}A_{2j} + \cdots + a_{nj}A_{nj}$$

$$= \sum_{i=1}^{n} a_{ij}A_{ij}$$

and is called the cofactor expansion along the jth column. This is called the Laplace expansion theorem. ●

Note that the cofactor and minor of an element a_{ij} differs only in sign, i.e., $A_{ij} = \pm M_{ij}$. A quick way for determining whether to use the + or − is to use the fact that the sign relating A_{ij} and M_{ij} is in the *ith* row and *jth* column of the *checkerboard* array

$$\begin{pmatrix} + & - & + & - & + & \cdots \\ - & + & - & + & - & \cdots \\ + & - & + & - & + & \cdots \\ - & + & - & + & - & \cdots \\ \vdots & \vdots & \vdots & \vdots & \vdots & \ddots \end{pmatrix}.$$

For example, $A_{11} = M_{11}$, $A_{21} = -M_{21}$, $A_{12} = -M_{12}$, $A_{22} = M_{22}$, and so on.

Definition 1.27 (Cofactor Matrix)

If A is any $n \times n$ matrix and A_{ij} is the cofactor of a_{ij}, then the matrix

$$\begin{pmatrix} A_{11} & A_{12} & \cdots & A_{1n} \\ A_{21} & A_{22} & \cdots & A_{2n} \\ \vdots & \vdots & \cdots & \vdots \\ A_{n1} & A_{n2} & \cdots & A_{nn} \end{pmatrix}$$

is called the matrix of the cofactor from A. For example, the cofactor of the matrix

$$A = \begin{pmatrix} 3 & 2 & -1 \\ 1 & 6 & 3 \\ 2 & -4 & 0 \end{pmatrix}$$

can be calculated as follows:

$$\begin{array}{lll} A_{11} = 12, & A_{12} = 6, & A_{13} = -16 \\ A_{21} = 4, & A_{22} = 2, & A_{23} = 16 \\ A_{31} = 12, & A_{32} = -10, & A_{33} = 16. \end{array}$$

So that the matrix of the form

$$\begin{pmatrix} 12 & 6 & -16 \\ 4 & 2 & 16 \\ 12 & -10 & 16 \end{pmatrix}$$

is the required cofactor matrix of the given matrix. ●

$$\begin{pmatrix} + & - & + & - & + & \cdots \\ - & + & - & + & - & \cdots \\ + & - & + & - & + & \cdots \\ - & + & - & + & - & \cdots \\ \vdots & \vdots & \vdots & \vdots & \vdots & \ddots \end{pmatrix}.$$

For example, $A_{11} = M_{11}, A_{21} = -M_{21}, A_{12} = -M_{12}, A_{22} = M_{22}$, and so on.

Definition 1.27 (Cofactor Matrix)

If A is any $n \times n$ matrix and A_{ij} is the cofactor of a_{ij}, then the matrix

$$\begin{pmatrix} A_{11} & A_{12} & \cdots & A_{1n} \\ A_{21} & A_{22} & \cdots & A_{2n} \\ \vdots & \vdots & \cdots & \vdots \\ A_{n1} & A_{n2} & \cdots & A_{nn} \end{pmatrix}$$

is called the matrix of the cofactor from A. For example, the cofactor of the matrix

$$A = \begin{pmatrix} 3 & 2 & -1 \\ 1 & 6 & 3 \\ 2 & -4 & 0 \end{pmatrix}$$

can be calculated as follows:

$$\begin{array}{lll} A_{11} = 12, & A_{12} = 6, & A_{13} = -16 \\ A_{21} = 4, & A_{22} = 2, & A_{23} = 16 \\ A_{31} = 12, & A_{32} = -10, & A_{33} = 16. \end{array}$$

So that the matrix of the form

$$\begin{pmatrix} 12 & 6 & -16 \\ 4 & 2 & 16 \\ 12 & -10 & 16 \end{pmatrix}$$

is the required cofactor matrix of the given matrix. •

Program 1.2
MATLAB m-file for Finding Minors and Cofactors of a Matrix

```
function CofA = cofactor(A,i,j)
[m,n] = size(A);
if m ~ = n  error(Matrix must be square)  end
A1 = A([1:i-1,i+1:n],[1:j-1,j+1:n]);
Minor = det(A1);
CofA = (-1)^ (i+j)*det(Minor);
```

Definition 1.26 (Cofactor Expansion of a Determinant of a Matrix)

Let A be a square matrix, then we define the determinant of A as the sum of the products of the elements of the first row and their cofactors. If A is a 3×3 matrix, then its determinant is defined as

$$\det(A) = |A| = a_1 1 A_{11} + a_1 2 A_{12} + a_1 3 A_{13}.$$

Similarly, in general, for an $n \times n$ matrix, we define it as

$$\det(A) = |A| = \sum_{1}^{n} a_{ij} A_{ij}, \quad n > 2, \tag{1.12}$$

where the summation is on i for any fixed value of the jth column ($1 \leq j \leq n$), or on j for any fixed value of the ith row ($1 \leq i \leq n$), and A_{ij} is the cofactor of element a_{ij}. •

Example 1.6 *Find the minors and cofactors of the matrix A and use them to evaluate the determinant of the matrix*

$$A = \begin{pmatrix} 3 & 1 & -4 \\ 2 & 5 & 6 \\ 1 & 4 & 8 \end{pmatrix}.$$

Solution. *The minors of A are calculated as follows:*

$$M_{11} = \begin{vmatrix} 5 & 6 \\ 4 & 8 \end{vmatrix} = 40 - 24 = 16$$

$$M_{12} = \begin{vmatrix} 2 & 6 \\ 1 & 8 \end{vmatrix} = 16 - 6 = 10$$

$$M_{13} = \begin{vmatrix} 2 & 5 \\ 1 & 4 \end{vmatrix} = 8 - 5 = 3.$$

From these values of the minors, we can calculate the cofactors of the elements of the given matrix as follows:

$$A_{11} = (-1)^{1+1}M_{11} = M_{11} = 16$$
$$A_{12} = (-1)^{1+2}M_{12} = -M_{12} = -10$$
$$A_{13} = (-1)^{1+3}M_{13} = M_{13} = 3.$$

Now by using the cofactor expansion along the first row, we can find the determinant of the matrix as follows:

$$\det(A) = a_{11}A_{11} + a_{12}A_{12} + a_{13}A_{13} = (3)(16) + (1)(-10) + (-4)(3) = 26.$$

Note that in Example 1.6, we computed the determinant of the matrix by using the cofactor expansion along the first row, but it can also be found along the first column of the matrix.

To get the results of Example 1.6, we use the MATLAB Command Window as follows:

```
>> A = [3 1 −4; 2 5 6; 1 4 8];
>> DetA = CofFexp(A);
```

Program 1.3
MATLAB m-file for Finding the Determinant of a Matrix by Cofactor Expansion
```
function DetA = CofFexp(A)
[m,n] = size(A);
if m ~ = n  error (Matrix must be square)  end
a = A(1,:);c = [ ];
for i=1:n
c1i = cofactor(A,1,i);
c = [c;c1i]; end
DetA = a*c;;
```

Theorem 1.7 (The Laplace Expansion Theorem)

The determinant of an $n \times n$ matrix $A = \{a_{ij}\}$, when $n \geq$ computed as

$$\det(A) = a_{i1}A_{i1} + a_{i2}A_{i2} + \cdots + a_{in}A_{in}$$
$$= \sum_{j=1}^{n} a_{ij}A_{ij},$$

which is called the cofactor expansion along the ith row, and a

$$\det(A) = a_{1j}A_{1j} + a_{2j}A_{2j} + \cdots + a_{nj}A_{nj}$$
$$= \sum_{i=1}^{n} a_{ij}A_{ij}$$

and is called the cofactor expansion along the jth column. T the Laplace expansion theorem.

Note that the cofactor and minor of an element a_{ij} differs i.e., $A_{ij} = \pm M_{ij}$. A quick way for determining whether to us is to use the fact that the sign relating A_{ij} and M_{ij} is in the *jth* column of the *checkerboard* array

Definition 1.28 (Adjoint of a Matrix)

If A is any $n \times n$ matrix and A_{ij} is the cofactor of a_{ij} of A, then the transpose of this matrix is called the adjoint of A and is denoted by $Adj(A)$. For example, the cofactor matrix of the matrix

$$A = \begin{pmatrix} 3 & 2 & -1 \\ 1 & 6 & 3 \\ 2 & -4 & 0 \end{pmatrix}$$

is calculated as

$$\begin{pmatrix} 12 & 6 & -16 \\ 4 & 2 & 16 \\ 12 & -10 & 16 \end{pmatrix}.$$

So by taking its transpose, we get the matrix

$$\begin{pmatrix} 12 & 6 & -16 \\ 4 & 2 & 16 \\ 12 & -10 & 16 \end{pmatrix}^T = \begin{pmatrix} 12 & 4 & 12 \\ 6 & 2 & -10 \\ -16 & 16 & 16 \end{pmatrix} = Adj(A),$$

which is called the adjoint of the given matrix A. ●

Example 1.7 *Find the determinant of the following matrix using cofactor expansion and show that $\det(A) = 0$ when $x = 4$:*

$$A = \begin{pmatrix} x+2 & x & 2 \\ 1 & x-1 & 3 \\ 4 & x+1 & x \end{pmatrix}.$$

Solution. *Using the cofactor expansion along the first row, we compute the determinant of the given matrix as*

$$|A| = a_{11}C_{11} + a_{12}C_{12} + a_{13}C_{13},$$

where

$$C_{11} = (-1)^{1+1}M_{11} = M_{11} = \begin{vmatrix} x-1 & 3 \\ x+1 & x \end{vmatrix} = x^2 - 4x - 3$$

$$C_{12} = (-1)^{1+2}M_{12} = -M_{12} = -\begin{vmatrix} 1 & 3 \\ 4 & x \end{vmatrix} = -x + 12$$

$$C_{13} = (-1)^{1+3}M_{13} = M_{13} = -\begin{vmatrix} 1 & x-1 \\ 4 & x+1 \end{vmatrix} = -3x + 5.$$

Thus,

$$|A| = (x+2)[x^2 - 4x - 3] + x[-x + 12] + 2[-3x + 5] = x^3 - 3x^2 - 5x + 4.$$

Now taking $x = 4$, we get

$$|A| = (4)^3 - 3(4)^2 - 5(4) + 4 = 64 - 48 - 20 + 4 = 0,$$

which is the required determinant of the matrix at $x = 4$.   ●

The following are special properties, which will be helpful in reducing the amount of work involved in evaluating determinants.

Theorem 1.8 (Properties of the Determinant)

Let A be an $n \times n$ matrix:

1. *The determinant of a matrix A is zero if any row or column is zero or equal to a linear combination of other rows and columns.*
   For example, if
   $$A = \begin{pmatrix} 3 & 1 & 0 \\ 2 & 1 & 0 \\ 4 & 3 & 0 \end{pmatrix},$$

   then $\det(A) = 0$.

2. *A determinant of a matrix A is changed in sign if the two rows or two columns are interchanged. For example, if*
   $$A = \begin{pmatrix} 3 & 2 \\ 4 & 5 \end{pmatrix},$$

   then $\det(A) = 7$, but for the matrix
   $$B = \begin{pmatrix} 4 & 5 \\ 3 & 2 \end{pmatrix},$$

   obtained from the matrix A by interchanging its rows, we have $\det(B) = -7$.

3. *The determinant of a matrix A is equal to the determinant of its transpose. For example, if*

$$A = \begin{pmatrix} 5 & 3 \\ 4 & 4 \end{pmatrix},$$

then $\det(A) = 8$, *and for the matrix*

$$B = \begin{pmatrix} 5 & 4 \\ 3 & 4 \end{pmatrix},$$

obtained from the matrix A by taking its transpose, we have

$$\det(B) = 8 = \det(A).$$

5. *If the matrix B is obtained from the matrix A by multiplying every element in one row or in one column by k, then the determinant of the matrix B is equal to k times the determinant of A. For example, if*

$$A = \begin{pmatrix} 6 & 5 \\ 3 & 4 \end{pmatrix},$$

then $\det(A) = 9$, *but for the matrix*

$$B = \begin{pmatrix} 12 & 10 \\ 3 & 4 \end{pmatrix},$$

obtained from the matrix A by multiplying its first row by 2, we have

$$\det(B) = 18 = 2(9) = 2\det(A).$$

6. *If the matrix B is obtained from the matrix A by adding to a row (or a column) a multiple of another row (or another column) of A, then the determinant of the matrix B is equal to the determinant of A. For example, if*

$$A = \begin{pmatrix} 4 & 3 \\ 5 & 4 \end{pmatrix},$$

then $\det(A) = 1$, *and for the matrix*

$$B = \begin{pmatrix} 4 & 3 \\ 13 & 10 \end{pmatrix},$$

obtained from the matrix A *by adding to its second row* 2 *times the first row, we have*

$$\det(B) = 1 = \det(A).$$

7. *If two rows or two columns of a matrix* A *are identical, then the determinant is zero. For example, if*

$$A = \begin{pmatrix} 2 & 3 \\ 2 & 3 \end{pmatrix},$$

then $\det(A) = 0$.

8. *The determinant of a product of matrices is the product of the determinants of all matrices. For example, if*

$$A = \begin{pmatrix} 3 & 4 & 5 \\ 3 & 2 & 1 \\ 2 & 1 & 6 \end{pmatrix}, \quad B = \begin{pmatrix} 1 & 2 & 3 \\ 4 & 2 & 3 \\ 1 & 3 & 5 \end{pmatrix},$$

then $\det(A) = -36$ *and* $\det(A) = -3$. *Also,*

$$AB = \begin{pmatrix} 24 & 29 & 46 \\ 12 & 13 & 20 \\ 12 & 24 & 39 \end{pmatrix},$$

then $\det(AB) = 108$. *Thus,*

$$\det(A)\det(B) = (-36)(-3) = 108 = \det(AB).$$

9. *The determinant of a triangular matrix (upper-triangular or lower-triangular matrix) is equal to the product of all their main diagonal elements. For example, if*

$$A = \begin{pmatrix} 3 & 4 & 5 \\ 0 & 4 & 7 \\ 0 & 0 & 5 \end{pmatrix},$$

then

$$\det(A) = (3)(4)(5) = 60.$$

10. *The determinant of an $n \times n$ matrix A times the scalar multiple k is equal to k^n times the determinant of the matrix A, i.e., $\det(kA) = k^n \det(A)$. For example, if*

$$A = \begin{pmatrix} 3 & 4 & 5 \\ 2 & 3 & 6 \\ 1 & 0 & 5 \end{pmatrix},$$

then $\det(A) = 14$, and for the matrix

$$B = 2A = \begin{pmatrix} 6 & 8 & 10 \\ 4 & 6 & 12 \\ 2 & 0 & 10 \end{pmatrix},$$

obtained from the matrix A by multiplying by 2, we have

$$\det(B) = 112 = 8(14) = 2^3 \det(A).$$

11. *The determinant of the kth power of a matrix A is equal to the kth power of the determinant of the matrix A, i.e., $\det(A^k) = (\det(A))^k$. For example, if*

$$A = \begin{pmatrix} 2 & -2 & 0 \\ 2 & 3 & -1 \\ 1 & 0 & 1 \end{pmatrix},$$

then $\det(A) = 12$, and for the matrix

$$B = A^3 = \begin{pmatrix} -18 & -30 & 12 \\ 24 & -3 & -9 \\ 3 & -12 & 3 \end{pmatrix},$$

obtained by taking the cubic power of the matrix A, we have

$$\det(B) = 1728 = (12)^3 = (\det(A))^3.$$

12. *The determinant of a scalar matrix (1×1) is equal to the element itself. For example, if $A = (8)$, then $\det(A) = 8$.*

Example 1.8 *Find all the values of α for which $\det(A) = 0$, where*

$$A = \begin{pmatrix} \alpha - 3 & 1 & 0 \\ 0 & \alpha - 1 & 1 \\ 0 & 2 & \alpha \end{pmatrix}.$$

Solution. *We find the determinant of the given matrix by using the co-factor expansion along the first row, so we compute*

$$
\begin{aligned}
|A| &= a_{11}C_{11} + a_{12}C_{12} + a_{13}C_{13} \\
&= (\alpha - 3)\begin{vmatrix} \alpha - 1 & 1 \\ 2 & \alpha \end{vmatrix} - 1\begin{vmatrix} 0 & 1 \\ 0 & \alpha \end{vmatrix} + 0\begin{vmatrix} 0 & \alpha - 1 \\ 0 & 2 \end{vmatrix} \\
&= (\alpha - 3)[(\alpha - 1)(\alpha) - 2] + 1[0 - 0] + 0 \\
&= (\alpha - 3)[\alpha^2 - \alpha) - 2] \\
&= (\alpha - 3)(\alpha + 1)(\alpha - 2)
\end{aligned}
$$

given $\det(A) = 0$, *which implies that*

$$
\begin{aligned}
|A| &= 0 \\
(\alpha - 3)(\alpha + 1)(\alpha - 2) &= 0,
\end{aligned}
$$

which gives

$$\alpha = -1, \quad \alpha = 2, \quad \alpha = 3,$$

the required values of α for which $\det(A) = 0$. ●

Example 1.9 *Find all the values of α such that*

$$\det\begin{pmatrix} 4\alpha & \alpha \\ 1 & \alpha + 1 \end{pmatrix} = \det\begin{pmatrix} 3 & -1 & 0 \\ 0 & \alpha & -2 \\ -1 & 3 & \alpha + 1 \end{pmatrix}.$$

Solution. *Since*

$$\begin{vmatrix} 4\alpha & \alpha \\ 1 & \alpha + 1 \end{vmatrix} = 4\alpha(\alpha + 1) - \alpha,$$

which is equivalent to

$$\begin{vmatrix} 4\alpha & \alpha \\ 1 & \alpha + 1 \end{vmatrix} = 4\alpha^2 + 3\alpha.$$

Also,

$$\begin{vmatrix} 3 & -1 & 0 \\ 0 & \alpha & -2 \\ -1 & 3 & \alpha+1 \end{vmatrix} = 3[\alpha(\alpha+1)+6]-(-1)[(0)(\alpha+1)+2]+0[(0)(3)-(-1)(\alpha)],$$

which can be written as

$$\begin{vmatrix} 3 & -1 & 0 \\ 0 & \alpha & -2 \\ -1 & 3 & \alpha+1 \end{vmatrix} = 3\alpha^2 + 3\alpha + 16.$$

Given that

$$\begin{vmatrix} 4\alpha & \alpha \\ 1 & \alpha+1 \end{vmatrix} = \begin{vmatrix} 3 & -1 & 0 \\ 0 & \alpha & -2 \\ -1 & 3 & \alpha+1 \end{vmatrix},$$

we get

$$4\alpha^2 + 3\alpha = 3\alpha^2 + 3\alpha + 16.$$

Simplifying this quadratic polynomial, we have

$$\alpha^2 = 16 \quad or \quad \alpha^2 - 16 = 0,$$

which gives

$$\alpha = -4 \quad and \quad \alpha = 4,$$

the required values of α.   ●

Example 1.10 *Find the determinant of the matrix*

$$A = \begin{pmatrix} -5a & -5b & -5c \\ 2d-g & 2e-h & 2f-i \\ 2d & 2e & 2f \end{pmatrix},$$

if

$$\det \begin{pmatrix} a & b & c \\ d & e & f \\ g & h & i \end{pmatrix} = 4.$$

Solution. *Using the property of the determinant, we get*

$$|A| = (-5) \begin{vmatrix} a & b & c \\ 2d - g & 2e - h & 2f - i \\ 2d & 2e & 2f \end{vmatrix}.$$

Subtracting the third row from the second row gives

$$|A| = (-5) \begin{vmatrix} a & b & c \\ -g & -h & -i \\ 2d & 2e & 2f \end{vmatrix}.$$

Interchanging the last two rows, we get

$$|A| = (-5)(-1) \begin{vmatrix} a & b & c \\ 2d & 2e & 2f \\ -g & -h & -i \end{vmatrix},$$

or

$$|A| = (-5)(-1)(2)(-1) \begin{vmatrix} a & b & c \\ d & e & f \\ g & h & i \end{vmatrix}.$$

Since it is given that

$$\begin{vmatrix} a & b & c \\ d & e & f \\ g & h & i \end{vmatrix} = 4,$$

we have

$$|A| = (-5)(-1)(2)(-1)(4) = -40,$$

the required determinant of the given matrix.   ●

Elimination Method for Evaluating a Determinant

One can easily transform the given determinant into upper-triangular form by using the following row operations:

1. Add a multiple of one row to another row, and this will not affect the determinant.

2. Interchange two rows of the determinant, and this will be done by multiplying the determinant by -1.

After transforming the given determinant into upper-triangular form, then use the fact that the determinant of a triangular matrix is the product of its diagonal elements.

Example 1.11 *Find the following determinant:*

$$\begin{vmatrix} 3 & 6 & 9 \\ 6 & 2 & -7 \\ -3 & 1 & -1 \end{vmatrix}.$$

Solution. *Multiplying row 1 of the determinant by $\frac{1}{3}$ gives*

$$(3)\begin{vmatrix} 1 & 2 & 3 \\ 6 & 2 & -7 \\ -3 & 1 & -1 \end{vmatrix}.$$

Now to create the zeros below the main diagonal, column by column, we do as follows:

Replace the second row of the determinant with the sum of itself and (-6) times the first row of the determinant and then replace the third row of the determinant with the sum of itself and (3) times the first row of the determinant, which gives

$$(3)\begin{vmatrix} 1 & 2 & 3 \\ 0 & -10 & -25 \\ 0 & 7 & 8 \end{vmatrix}.$$

Multiplying row 2 of the determinant by $-\frac{1}{10}$ gives

$$(3)(-10)\begin{vmatrix} 1 & 2 & 3 \\ 0 & 1 & -5/2 \\ 0 & 7 & 8 \end{vmatrix}.$$

Replacing the third row of the determinant with the sum of itself and (-7) times the second row of the determinant, we obtain

$$(3)(-10) \begin{vmatrix} 1 & 2 & 3 \\ 0 & 1 & -5/2 \\ 0 & 0 & -19/2 \end{vmatrix} = (3)(-10)(1)(1)(-19/2) = 285,$$

which is the required value of the given determinant.   •

Theorem 1.9 *If A is an invertible matrix, then:*

1. $\det(A) \neq 0$.

2. $\det(A^{-1}) = \dfrac{1}{\det(A)}$.

3. $A^{-1} = \dfrac{Adj(A)}{\det(A)}$.

4. $(adj(A))^{-1} = \dfrac{A}{\det(A)} = adj(A^{-1})$.

5. $\det(adj(A)) = \det(A)^{n-1}$.   •

By using Theorem 1.9 we can find the inverse of a matrix by showing that the determinant of a matrix is not equal to zero and by using the adjoint and determinant of the given matrix A.

Example 1.12 *For what values of α does the following matrix have an inverse?*

$$A = \begin{pmatrix} 1 & 0 & \alpha \\ 2 & 2 & 1 \\ 0 & 2\alpha & 1 \end{pmatrix}$$

Solution. *We find the determinant of the given matrix by using cofactor expansion along the first row as follows:*

$$|A| = a_{11}C_{11} + a_{12}C_{12} + a_{13}C_{13},$$

which is equal to

$$|A| = (1)C_{11} + (0)C_{12} + (\alpha)C_{13} = C_{11} + \alpha C_{13}.$$

Now we compute the values of C_{11} and C_{13} as follows:

$$C_{11} = (-1)^{1+1}M_{11} = M_{11} = \begin{vmatrix} 2 & 1 \\ 2\alpha & 1 \end{vmatrix} = 2 - 2\alpha$$

$$C_{13} = (-1)^{1+3}M_{13} = M_{13} = - \begin{vmatrix} 2 & 2 \\ 0 & 2\alpha \end{vmatrix} = 4\alpha.$$

Thus,

$$|A| = C_{11} + \alpha C_{13} = 2 - 2\alpha + 4\alpha^2.$$

From Theorem 1.9 we know that the matrix has an inverse if $\det(A) \neq 0$, so

$$|A| = 2 - 2\alpha + 4\alpha^2 \neq 0,$$

which implies that

$$2\alpha^2 - \alpha - 1 = (2\alpha + 1)(\alpha - 1) \neq 0.$$

Hence, the given matrix has an inverse if $\alpha \neq -1/2$ and $\alpha \neq 1$. •

Example 1.13 *Use the adjoint method to compute the inverse of the following matrix:*

$$A = \begin{pmatrix} 1 & 2 & -1 \\ 2 & -1 & 1 \\ 1 & 2 & 2 \end{pmatrix}.$$

Also, find the inverse and determinant of the adjoint matrix.

Solution. *First, we compute the determinant of the given matrix as follows:*

$$|A| = a_{11}C_{11} + a_{12}C_{12} + a_{13}C_{13},$$

which gives

$$|A| = (1)(-4) - (2)(3) + (-1)(5) = -15$$

Now we compute the nine cofactors as follows:

$$C_{11} = + \begin{vmatrix} -1 & 1 \\ 2 & 2 \end{vmatrix} = -4,\ C_{12} = - \begin{vmatrix} 2 & 1 \\ 1 & 2 \end{vmatrix} = -3,\ C_{13} = + \begin{vmatrix} 2 & -1 \\ 1 & 2 \end{vmatrix} = 5,$$

$$C_{21} = - \begin{vmatrix} 2 & -1 \\ 2 & 2 \end{vmatrix} = -6,\ C_{22} = + \begin{vmatrix} 1 & -1 \\ 1 & 2 \end{vmatrix} = 3,\ C_{23} = - \begin{vmatrix} 1 & 2 \\ 1 & 2 \end{vmatrix} = 0,$$

$$C_{31} = + \begin{vmatrix} 2 & -1 \\ -1 & 1 \end{vmatrix} = 1,\ C_{32} = - \begin{vmatrix} 1 & -1 \\ 2 & 1 \end{vmatrix} = -3,\ C_{33} = + \begin{vmatrix} 1 & 2 \\ 2 & -1 \end{vmatrix} = -5.$$

Thus, the cofactor matrix has the form

$$\begin{pmatrix} -4 & -3 & 5 \\ -6 & 3 & 0 \\ 1 & -3 & -5 \end{pmatrix},$$

and the adjoint is the transpose of the cofactor matrix

$$adj(A) = \begin{pmatrix} -4 & -3 & 5 \\ -6 & 3 & 0 \\ 1 & -3 & -5 \end{pmatrix}^T = \begin{pmatrix} -4 & -6 & 1 \\ -3 & 3 & -3 \\ 5 & 0 & -5 \end{pmatrix}.$$

To get the adjoint of the matrix of Example 1.13, we use the MATLAB Command Window as follows:

```
>> A = [1 2 − 1; 2 − 1 1; 1 2 2];
>> AdjA = Adjoint(A);
```

Program 1.4
MATLAB m-file for Finding the Adjoint of a Matrix Function AdjA = Adjoint(A)
[m,n] = size(A);
if m ˜ = n error('Matrix must be square') end
A1 = [ ];
for i = 1:n
for j=1:n
A1 = [A1;cofactor(A,i,j)];end;end
AdjA = reshape(A1,n,n);

Then by using Theorem 1.9 we can have the inverse of the matrix as follows:

$$A^{-1} = \frac{Adj(A)}{\det(A)} = -\frac{1}{15}\begin{pmatrix} -4 & -6 & 1 \\ -3 & 3 & -3 \\ 5 & 0 & -5 \end{pmatrix} = \begin{pmatrix} 4/15 & 2/5 & -1/15 \\ 1/5 & -1/5 & 1/5 \\ -1/3 & 0 & 1/3 \end{pmatrix}.$$

Using Theorem 1.9 we can compute the inverse of the adjoint matrix as:

$$(adj(A))^{-1} = \frac{A}{\det(A)} = \begin{pmatrix} -1/15 & -2/15 & 1/15 \\ -2/15 & 1/15 & -1/15 \\ -1/15 & -2/15 & -2/15 \end{pmatrix},$$

and the determinant of the adjoint matrix as

$$\det(adj(A)) = (\det(A))^{3-1} = (-15)^2 = 225.$$

●

Now we consider the implementation of finding the inverse of the matrix

$$A = \begin{pmatrix} 1 & -1 & 1 & 2 \\ 1 & 0 & 1 & 3 \\ 0 & 0 & 2 & 4 \\ 1 & 1 & -1 & 1 \end{pmatrix}$$

by using the adjoint and the determinant of the matrix in the MATLAB Command Window as:

```
>> A = [1 − 1 1 2; 1 0 1 3; 0 0 2 4; 1 1 − 1 1];
```

The cofactors A_{ij} of elements of the given matrix A can also be found directly by using the MATLAB Command Window as follows:

```
>> A11 = (-1)^ (1 + 1) * det(A([2 : 4], [2 : 4]));
>> A12 = (-1)^ (1 + 2) * det(A([2 : 4], [1, 3 : 4]));
>> A13 = (-1)^ (1 + 3) * det(A([2 : 4], [1 : 2, 4]));
>> A14 = (-1)^ (1 + 4) * det(A([2 : 4], [1 : 3]));
>> A21 = (-1)^ (2 + 1) * det(A([1, 3 : 4], [2 : 4]));
>> A22 = (-1)^ (2 + 2) * det(A([1, 3 : 4], [1, 3 : 4]));
>> A23 = (-1)^ (2 + 3) * det(A([1, 3 : 4], [1 : 2, 4]));
>> A24 = (-1)^ (2 + 4) * det(A([1, 3 : 4], [1 : 3]));
>> A31 = (-1)^ (3 + 1) * det(A([1 : 2, 4], [2 : 4]));
>> A32 = (-1)^ (3 + 2) * det(A([1 : 2, 4], [1, 3 : 4]));
>> A33 = (-1)^ (3 + 3) * det(A([1 : 2, 4], [1 : 2, 4]));
>> A34 = (-1)^ (3 + 4) * det(A([1 : 2, 4], [1 : 3]));
>> A41 = (-1)^ (4 + 1) * det(A([1 : 3], [2 : 4]));
>> A42 = (-1)^ (4 + 2) * det(A([1 : 3], [1, 3 : 4]));
>> A43 = (-1)^ (4 + 3) * det(A([1 : 3], [1 : 2, 4]));
>> A44 = (-1)^ (4 + 4) * det(A([1 : 3], [1 : 3]));
```

Now form the cofactor matrix B using the A_{ij}s as follows:

```
>> B =
    [A11  A12  A13  A14;
     A21  A22  A23  A24;
     A31  A32  A33  A34;
     A41  A42  A43  A44]
```

which gives

```
B =
    -2  -4  -4   2
     6   6   8  -4
    -3  -2  -3   2
    -2  -2  -4   2
```

The adjoint matrix is the transpose of the cofactor matrix:

```
>> adj A = B'

   -2    6   -3   -2
   -4    6   -2   -2
   -4    8   -3   -4
    2   -4    2    2
```

The determinant of the matrix can be obtained as:

```
>> det(A)
ans =
     2
```

The inverse of A is the adjoint matrix divided by the determinant of A.

```
>> invA = (1/det(A)) * adj A;
invA =
   -1    3   -1.5   -1
   -2    3    -1    -1
   -2    4   -1.5   -2
    1   -2     1     1
```

Verify the results by finding A^{-1} directly using the MATLAB command:

```
>> inv(A)
```

Example 1.14 *If* $\det(A) = 3$ *and* $\det(B) = 4$, *then show that*

$$det(A^2 B^{-1} A^T B^3) = 432.$$

Solution. *By using the properties of the determinant of the matrix, we have*

$$det(A^2 B^{-1} A^T B^3) = \det(A^2)\det(B^{-1})\det(A^T)\det(B^3),$$

which can also be written as

$$det(A^2 B^{-1} A^T B^3) = (\det(A))^2 \frac{1}{\det(B)}(\det(A))(\det(B))^3.$$

Now using the given information, we get

$$det(A^2 B^{-1} A^T B^3) = (3)^2 \frac{1}{4}(3)(4)^3 = 3^3 4^2 = 432,$$

the required solution. ●

1.2.5 Homogeneous Linear Systems

We have seen that every system of linear equations has either no solution, a unique solution, or infinitely many solutions. However, there is another type of system that always has at least one solution, i.e., either a unique solution (called a zero solution or trivial solution) or infinitely many solutions (called nontrivial solutions). Such a system is called a *homogeneous linear system*.

Definition 1.29 *A system of linear equations is said to be homogeneous if all the constant terms are zero, i.e.,*

$$A\mathbf{x} = \mathbf{b} = \mathbf{0}. \tag{1.13}$$

For example,

$$\begin{aligned} x_1 + 2x_2 - x_3 &= 0 \\ 2x_1 - 3x_2 + 3x_3 &= 0 \end{aligned}$$

is a homogeneous linear system. But

$$\begin{aligned} x_1 + 2x_2 - x_3 &= 0 \\ 2x_1 - 3x_2 + 3x_3 &= 1 \end{aligned}$$

is not a homogeneous linear system.

The general homogeneous system of m linear equations with n unknown variables $x_1, x_2, \ldots, x_n$ is

$$\begin{aligned} a_{11}x_1 &+ a_{12}x_2 &+ \cdots &+ a_{1n}x_n &= 0 \\ a_{21}x_1 &+ a_{22}x_2 &+ \cdots &+ a_{2n}x_n &= 0 \\ \vdots & \vdots & \vdots & \vdots & \vdots \\ a_{m1}x_1 &+ a_{m2}x_2 &+ \cdots &+ a_{mn}x_n &= 0. \end{aligned} \tag{1.14}$$

The system of linear equations (1.14) can be written as the single matrix equation

$$
\begin{pmatrix}
a_{11} & a_{12} & \cdots & a_{1n} \\
a_{21} & a_{22} & \cdots & a_{2n} \\
\vdots & \vdots & \vdots & \vdots \\
a_{m1} & a_{m2} & \cdots & a_{mn}
\end{pmatrix}
\begin{pmatrix}
x_1 \\ x_2 \\ \vdots \\ x_n
\end{pmatrix}
=
\begin{pmatrix}
0 \\ 0 \\ \vdots \\ 0
\end{pmatrix}. \tag{1.15}
$$

If we compute the product of the two matrices on the left-hand side of (1.15), we have

$$
\begin{pmatrix}
a_{11}x_1 & + & a_{12}x_2 & + & \cdots & + & a_{1n}x_n \\
a_{21}x_1 & + & a_{22}x_2 & + & \cdots & + & a_{2n}x_n \\
\vdots & & \vdots & & \vdots & & \vdots \\
a_{m1}x_1 & + & a_{m2}x_2 & + & \cdots & + & a_{mn}x_n
\end{pmatrix}
=
\begin{pmatrix}
0 \\ 0 \\ \vdots \\ 0
\end{pmatrix}. \tag{1.16}
$$

But the two matrices are equal if and only if their corresponding elements are equal. Hence, the single matrix equation (1.15) is equivalent to the system of the linear equations (1.14). If we define

$$
A =
\begin{pmatrix}
a_{11} & a_{12} & \cdots & a_{1n} \\
a_{21} & a_{22} & \cdots & a_{2n} \\
\vdots & \vdots & \vdots & \vdots \\
a_{m1} & a_{m2} & \cdots & a_{mn}
\end{pmatrix},
\quad
\mathbf{x} =
\begin{pmatrix}
x_1 \\ x_2 \\ \vdots \\ x_n
\end{pmatrix},
\quad
\mathbf{b} =
\begin{pmatrix}
0 \\ 0 \\ \vdots \\ 0
\end{pmatrix},
$$

the coefficient matrix, the column matrix of unknowns, and the column matrix of constants, respectively, then the system (1.14) can be written very compactly as

$$ A\mathbf{x} = \mathbf{b}, \tag{1.17} $$

which is called the matrix form of the homogeneous system. ●

Note that a homogeneous linear system has an augmented matrix of the form

$$ [A|\mathbf{0}]. $$

Theorem 1.10 *Every homogeneous linear system* $A\mathbf{x} = \mathbf{0}$ *has either exactly one solution or infinitely many solutions.*   •

Example 1.15 *Solve the following homogeneous linear system:*

$$\begin{aligned}
x_1 + x_2 + 2x_3 &= 0 \\
2x_1 + 3x_2 + 4x_3 &= 0 \\
3x_1 + 4x_2 + 7x_3 &= 0.
\end{aligned}$$

Solution. *Consider the augmented matrix form of the given system as follows:*

$$[A|\mathbf{0}] = \begin{pmatrix} 1 & 1 & 2 & \vdots & 0 \\ 2 & 3 & 4 & \vdots & 0 \\ 3 & 4 & 7 & \vdots & 0 \end{pmatrix}.$$

To convert it into reduced echelon form, we first do the elementary row operations: row2 – (2)row1 and row3 – (3)row1 gives

$$\equiv \begin{pmatrix} 1 & 1 & 2 & \vdots & 0 \\ 0 & 1 & 0 & \vdots & 0 \\ 0 & 1 & 1 & \vdots & 0 \end{pmatrix}.$$

Next, using the elementary row operations: row3 – row2 and row1 – row2, we get

$$\equiv \begin{pmatrix} 1 & 0 & 2 & \vdots & 0 \\ 0 & 1 & 0 & \vdots & 0 \\ 0 & 0 & 1 & \vdots & 0 \end{pmatrix}.$$

Finally, using the elementary row operation: row1 – (2)row3, we obtain

$$\equiv \begin{pmatrix} 1 & 0 & 0 & \vdots & 0 \\ 0 & 1 & 0 & \vdots & 0 \\ 0 & 0 & 1 & \vdots & 0 \end{pmatrix}.$$

Thus,

$$x_1 = 0, \quad x_2 = 0, \quad x_3 = 0$$

is the only trivial solution of the given system. ●

Theorem 1.11 *A homogeneous linear system* $A\mathbf{x} = \mathbf{0}$ *of* m *linear equations with* n *unknowns, where* $m < n$, *has infinitely many solutions.* ●

Example 1.16 *Solve the homogeneous linear system*

$$
\begin{aligned}
x_1 + 2x_2 + x_3 &= 0 \\
2x_1 - 3x_2 + 4x_3 &= 0.
\end{aligned}
$$

Solution. *Consider the augmented matrix form of the given system as*

$$[A|\mathbf{0}] = \begin{pmatrix} 1 & 2 & 1 & \vdots & 0 \\ 2 & -3 & 4 & \vdots & 0 \end{pmatrix}.$$

To convert it into reduced echelon form, we first do the elementary row operation row2 – 2row1, and we get

$$\sim \begin{pmatrix} 1 & 2 & 1 & \vdots & 0 \\ 0 & -7 & 2 & \vdots & 0 \end{pmatrix}.$$

Doing the elementary row operation: $-\frac{1}{7}row2$ *gives*

$$\sim \begin{pmatrix} 1 & 2 & 1 & \vdots & 0 \\ 0 & 1 & -\frac{2}{7} & \vdots & 0 \end{pmatrix}.$$

Finally, using the elementary row operation row1 – 2row3, we get

$$\sim \begin{pmatrix} 1 & 2 & \frac{11}{7} & \vdots & 0 \\ 0 & 1 & -\frac{2}{7} & \vdots & 0 \end{pmatrix}.$$

Writing it in the system of equations form, we have

$$x_1 + 0x_2 + \frac{11}{7}x_3 \;=\; 0$$

$$0x_1 + x_2 - \frac{2}{7}x_3 \;=\; 0$$

and from it, we get

$$x_2 = \frac{2}{7}x_3.$$

Taking $x_3 = t$, for $t \in \mathbf{R}$ and $t \neq 0$, we get the nontrivial solution

$$[x_1, x_2, x_3]^T = [\frac{11}{7}t, \frac{2}{7}t, t]^T.$$

Thus, the given system has infinitely many solutions, and this is to be expected because the given system has three unknowns and only two equations.   ●

Example 1.17 *For what values of α does the homogeneous linear system*

$$(\alpha - 2)x_1 + x_2 \;=\; 0$$
$$x_1 + (\alpha - 2)x_2 \;=\; 0$$

have nontrivial solutions?

Solution. *The augmented matrix form of the given system is*

$$[A|\mathbf{0}] = \begin{pmatrix} (\alpha - 2) & 1 & \vdots & 0 \\ 1 & (\alpha - 2) & \vdots & 0 \end{pmatrix}.$$

By interchanging row1 by row2, we get

$$\sim \begin{pmatrix} 1 & (\alpha - 2) & \vdots & 0 \\ (\alpha - 2) & 1 & \vdots & 0 \end{pmatrix}.$$

Doing the elementary row operation: row2 – $(\alpha - 2)$ row1 gives

$$\sim \begin{pmatrix} 1 & (\alpha - 2) & \vdots & 0 \\ 0 & 1 - (\alpha - 2)^2 & \vdots & 0 \end{pmatrix}.$$

Using backward substitution, we obtain

$$
\begin{aligned}
x_1 + (\alpha - 2)x_2 &= 0 \\
0x_1 + 1 - (\alpha - 2)^2 x_2 &= 0.
\end{aligned}
$$

Notice that if $x_2 = 0$, then $x_1 = 0$, and the given system has a trivial solution, so let $x_2 \neq 0$. This implies that

$$
\begin{aligned}
1 - (\alpha - 2)^2 &= 0 \\
1 - \alpha^2 + 4\alpha - 4 &= 0 \\
\alpha^2 - 4\alpha + 3 &= 0 \\
(\alpha - 3)(\alpha - 1) &= 0,
\end{aligned}
$$

which gives

$$\alpha = 1 \quad and \quad \alpha = 3.$$

Notice that for these values of α, the given set of equations are identical, i.e.,
(for $\alpha = 1$)

$$
\begin{aligned}
-x_1 + x_2 &= 0 \\
x_1 - x_2 &= 0,
\end{aligned}
$$

and (for $\alpha = 3$)

$$
\begin{aligned}
x_1 + x_2 &= 0 \\
x_1 + x_2 &= 0.
\end{aligned}
$$

Thus, the given system has nontrivial solutions (infinitely many solutions) for $\alpha = 1$ and $\alpha = 3$. ●

The following basic theorems on the solvability of linear systems are proved in linear algebra.

Theorem 1.12 *A homogeneous system of n equations in n unknowns has a solution other than the trivial solution if and only if the determinant of the coefficients matrix A vanishes, i.e., matrix A is singular.*   •

Theorem 1.13 (Necessary and Sufficient Condition for a Unique Solution)

A nonhomogeneous system of n equations in n unknowns has a unique solution if and only if the determinant of a coefficients matrix A does not vanish, i.e., A is nonsingular.   •

1.2.6 Matrix Inversion Method

If matrix A is nonsingular, then the linear system (1.6) always has a unique solution for each $\mathbf{b}$ since the inverse matrix A^{-1} exists, so the solution of the linear system (1.6) can be formally expressed as

$$A^{-1}A\mathbf{x} = A^{-1}\mathbf{b}$$
$$\mathbf{Ix} = A^{-1}\mathbf{b}$$

or

$$\mathbf{x} = A^{-1}\mathbf{b}. \tag{1.18}$$

If A is a square invertible matrix, there exists a sequence of elementary row operations that carry A to the identity matrix $\mathbf{I}$ of the same size, i.e., $A \longrightarrow \mathbf{I}$. This same sequence of row operations carries $\mathbf{I}$ to A^{-1}, i.e., $\mathbf{I} \longrightarrow A^{-1}$. This can also be written as

$$[A|\mathbf{I}] \longrightarrow [\mathbf{I}|A^{-1}].$$

Example 1.18 *Use the matrix inversion method to find the solution of the following linear system:*

$$\begin{aligned} x_1 + 2x_2 &= 1 \\ -2x_1 + x_2 + 2x_3 &= 1 \\ -x_1 + x_2 + x_3 &= 1. \end{aligned}$$

Solution. *First, we compute the inverse of the given matrix as*

$$A = \begin{pmatrix} 1 & 2 & 0 \\ -2 & 1 & 2 \\ -1 & 1 & 1 \end{pmatrix}$$

by reducing A to the identity matrix $\mathbf{I}$ by elementary row operations and then applying the same sequence of operations to $\mathbf{I}$ to produce A^{-1}. Consider the augmented matrix

$$[A|\mathbf{I}] = \left(\begin{array}{ccc:ccc} 1 & 2 & 0 & 1 & 0 & 0 \\ -2 & 1 & 2 & 0 & 1 & 0 \\ -1 & 1 & 1 & 0 & 0 & 1 \end{array} \right).$$

Multiply the first row by -2 and -1 and then, subtracting the results from the second and third rows, respectively, we get

$$\sim \left(\begin{array}{ccc:ccc} 1 & 2 & 0 & 1 & 0 & 0 \\ 0 & 5 & 2 & 2 & 1 & 0 \\ 0 & 3 & 1 & 1 & 0 & 1 \end{array} \right).$$

Multiplying the second row by $\frac{1}{5}$, we get

$$\sim \left(\begin{array}{ccc:ccc} 1 & 2 & 0 & 1 & 0 & 0 \\ 0 & 1 & 2/5 & 2/5 & 1/5 & 0 \\ 0 & 3 & 1 & 1 & 0 & 1 \end{array} \right).$$

Multiplying the second row by 2 and 3 and then subtracting the results from the first and third rows, respectively, we get

$$\sim \left(\begin{array}{ccc:ccc} 1 & 0 & -4/5 & 1/5 & -2/5 & 0 \\ 0 & 1 & 2/5 & 2/5 & 1/5 & 0 \\ 0 & 0 & -1/5 & -1/5 & -3/5 & 1 \end{array} \right).$$

After multiplying the third row by -5, we obtain

$$\sim \begin{pmatrix} 1 & 0 & -4/5 & \vdots & 1/5 & -2/5 & 0 \\ 0 & 1 & 2/5 & \vdots & 2/5 & 1/5 & 0 \\ 0 & 0 & 1 & \vdots & 1 & 3 & -5 \end{pmatrix}.$$

Multiplying the third row by $\frac{2}{5}$ and $-\frac{4}{5}$ and then subtracting the results from the second and first rows, respectively, we get

$$\sim \begin{pmatrix} 1 & 0 & 0 & \vdots & 1 & 2 & -4 \\ 0 & 1 & 0 & \vdots & 0 & -1 & 2 \\ 0 & 0 & 1 & \vdots & 1 & 3 & -5 \end{pmatrix}.$$

Thus, the inverse of the given matrix is

$$A^{-1} = \begin{pmatrix} 1 & 2 & -4 \\ 0 & -1 & 2 \\ 1 & 3 & -5 \end{pmatrix},$$

and the unique solution of the system can be computed as

$$\mathbf{x} = A^{-1}\mathbf{b} = \begin{pmatrix} 1 & 2 & -4 \\ 0 & -1 & 2 \\ 1 & 3 & -5 \end{pmatrix} \begin{pmatrix} 1 \\ 1 \\ 1 \end{pmatrix} = \begin{pmatrix} -1 \\ 1 \\ -1 \end{pmatrix},$$

i.e.,

$$x_1 = -1, \quad x_2 = 1, \quad x_3 = -1,$$

the solution of the given system by the matrix inversion method.   ●

Thus, when the matrix inverse A^{-1} of the coefficient matrix A is computed, the solution vector $\mathbf{x}$ of the system (1.6) is simply the product of inverse matrix A^{-1} and the right-hand side vector $\mathbf{b}$.

Using MATLAB commands, the linear system of equations defined by the coefficient matrix A and the right-hand side vector $\mathbf{b}$ using the matrix inverse method is solved with:

```
>> A = [1 2 0; −2 1 2; −1 1 1];
>> b = [1; 1; 1];
>> x = A \ b
x =
      1.0000
      1.0000
     −1.0000
```

Theorem 1.14 *For an $n \times n$ matrix A, the following properties are equivalent:*

1. *The inverse of matrix A exists, i.e., A is nonsingular.*

2. *The determinant of matrix A is nonzero.*

3. *The homogeneous system $A\mathbf{x} = \mathbf{0}$ has a trivial solution $\mathbf{x} = \mathbf{0}$.*

4. *The nonhomogeneous system $A\mathbf{x} = \mathbf{b}$ has a unique solution.*   ●

Not all matrices have inverses. Singular matrices don't have inverses and thus the corresponding systems of equations do not have unique solutions. The inverse of a matrix can also be computed by using the following numerical methods for linear systems: Gauss-elimination method, Gauss–Jordan method, and LU decomposition method. But the best and simplest method for finding the inverse of a matrix is to perform the Gauss–Jordan method on the augmented matrix with an identity matrix of the same size.

1.2.7   Elementary Matrices

An $n \times n$ matrix E is called an *elementary matrix* if it can be obtained from the $n \times n$ identity matrix $\mathbf{I}_n$ by a single elementary row operation. For example, the first elementary matrix E_1 is obtained by multiplying the second row of the identity matrix by 6, i.e.,

$$\mathbf{I} = \begin{pmatrix} 1 & 0 & 0 \\ 0 & 1 & 0 \\ 0 & 0 & 1 \end{pmatrix} \longrightarrow \begin{pmatrix} 1 & 0 & 0 \\ 0 & 6 & 0 \\ 0 & 0 & 1 \end{pmatrix} = E_1.$$

The second elementary matrix E_2 is obtained by multiplying the first row of the identity matrix by -5 and adding it to the third row, i.e.,

$$\mathbf{I} = \begin{pmatrix} 1 & 0 & 0 \\ 0 & 1 & 0 \\ 0 & 0 & 1 \end{pmatrix} \longrightarrow \begin{pmatrix} 1 & 0 & 0 \\ 0 & 1 & 0 \\ -5 & 0 & 1 \end{pmatrix} = E_2.$$

Similarly, the third elementary matrix E_3 is obtained by interchanging the second and third rows of the identity matrix, i.e.,

$$\mathbf{I} = \begin{pmatrix} 1 & 0 & 0 \\ 0 & 1 & 0 \\ 0 & 0 & 1 \end{pmatrix} \longrightarrow \begin{pmatrix} 1 & 0 & 0 \\ 0 & 0 & 0 \\ 0 & 1 & 0 \end{pmatrix} = E_3.$$

Notice that elementary matrices are always square.

Theorem 1.15 *To perform an elementary row operation on the $m \times n$ matrix A, multiply A on the left by the corresponding elementary matrix.*

●

Example 1.19 *Let*

$$A = \begin{pmatrix} 1 & 2 & 3 & -5 \\ 3 & 3 & 2 & 1 \\ 4 & 1 & -2 & 4 \end{pmatrix}.$$

Find an elementary matrix E such that EA is the matrix that results by adding 5 times the first row of A to the third row.

Solution. *The matrix E must be 3×3 to conform to the product EA. So, we get E by adding 5 times the first row to the third row. This gives*

$$E = \begin{pmatrix} 1 & 0 & 0 \\ 0 & 1 & 0 \\ 5 & 0 & 1 \end{pmatrix},$$

and the product EA is given as

$$EA = \begin{pmatrix} 1 & 0 & 0 \\ 0 & 1 & 0 \\ 5 & 0 & 1 \end{pmatrix} \begin{pmatrix} 1 & 2 & 3 & -5 \\ 3 & 3 & 2 & 1 \\ 4 & 1 & -2 & 4 \end{pmatrix} = \begin{pmatrix} 1 & 2 & 3 & -5 \\ 3 & 3 & 2 & 1 \\ 9 & 11 & 13 & -21 \end{pmatrix}.$$

●

Theorem 1.16 *An elementary matrix is invertible, and the inverse is also an elementary matrix.* ●

Example 1.20 *Express the matrix*

$$A = \begin{pmatrix} 2 & 3 \\ 1 & 1 \end{pmatrix}$$

as a product of elementary matrices.

Solution. *We reduce A to identity matrix $\mathbf{I}$ and write the elementary matrix at each stage, given*

$$A = \begin{pmatrix} 2 & 3 \\ 1 & 1 \end{pmatrix}.$$

By interchanging the first and the second rows, we get

$$E_1 A = \begin{pmatrix} 1 & 1 \\ 2 & 3 \end{pmatrix}, \quad \text{where} \quad E_1 = \begin{pmatrix} 0 & 1 \\ 1 & 0 \end{pmatrix}.$$

Multiplying the second row by 2 and subtracting the result from the second row, we get

$$E_2(E_1 A) = E_2 E_1 A = \begin{pmatrix} 1 & 1 \\ 0 & 1 \end{pmatrix}, \quad \text{where} \quad E_2 = \begin{pmatrix} 1 & 0 \\ -2 & 1 \end{pmatrix}.$$

Finally, by subtracting the third row from the first row, we get

$$E_3(E_2 E_1 A) = E_3 E_2 E_1 A = \begin{pmatrix} 1 & 0 \\ 0 & 1 \end{pmatrix}, \quad \text{where} \quad E_3 = \begin{pmatrix} 1 & -1 \\ 0 & 1 \end{pmatrix}.$$

Hence,

$$E_3 E_2 E_1 A = \mathbf{I},$$

and so

$$A = (E_3 E_2 E_1)^{-1}.$$

This means that

$$A = E_1^{-1} E_2^{-1} E_3^{-1} = \begin{pmatrix} 0 & 1 \\ 1 & 0 \end{pmatrix} \begin{pmatrix} 1 & 0 \\ 2 & 1 \end{pmatrix} \begin{pmatrix} 1 & 1 \\ 0 & 1 \end{pmatrix}.$$

●

Theorem 1.17 *A square matrix A is invertible if and only if it is a product of elementary matrices.*   ●

Theorem 1.18 *An $n \times n$ matrix A is invertible if and only if:*

1. *It is row equivalent to identity matrix $\mathbf{I}_n$.*

2. *Its reduced row echelon form is identity matrix $\mathbf{I}_n$.*

3. *It is expressible as a product of elementary matrices.*

4. *It has n pivots.*   ●

In the following, we will discuss the direct methods for solving the linear systems.

1.3   Numerical Methods for Linear Systems

To solve systems of linear equations using numerical methods, there are two types of methods available. The first type of methods are called *direct methods* or *elimination methods*. The other type of numerical methods are called *iterative methods*. In this chapter we will discuss only the first type of the numerical methods, and the other type of the numerical methods will be discussed in Chapter 2. The first type of methods find the solution in a finite number of steps. These methods are guaranteed to succeed and are recommended for general use. Here, we will consider *Cramer's rule*, the *Gaussian elimination method* and its variants, the *Gauss–Jordan method*, and *LU decomposition* (by Doolittle's, Crout's, and Cholesky methods).

1.4   Direct Methods for Linear Systems

This type of method refers to a procedure for computing a solution from a form that is mathematically exact. We shall begin with a simple method called Cramer's rule with determinants. We shall then continue with the Gaussian elimination method and its variants and methods involving triangular, symmetric, and tridiagonal matrices.

1.4.1 Cramer's Rule

This is our first direct method for solving linear systems by the use of determinants. This method is one of the least efficient for solving a large number of linear equations. It is, however, very useful for explaining some problems inherent in the solution of linear equations.

Consider a system of two linear equations

$$a_{11}x_1 + a_{12}x_2 = b_1$$
$$a_{21}x_1 + a_{22}x_2 = b_2,$$

with the condition that $a_{11}a_{22} - a_{12}a_{21} \neq 0$, i.e., the determinant of the given matrix must not be equal to zero or the matrix must be nonsingular. Solving the above system using systematic elimination by multiplying the first equation of the system with a_{22} and the second equation by a_{12} and subtracting gives

$$(a_{11}a_{22} - a_{12}a_{21})x_1 = a_{22}b_1 - a_{12}b_2,$$

and now solving for x_1 gives

$$x_1 = \frac{a_{22}b_1 - a_{12}b_2}{a_{11}a_{22} - a_{12}a_{21}},$$

and putting the value of x_1 in any equation of the given system, we have x_2 as

$$x_2 = \frac{a_{22}b_2 - a_{12}b_1}{a_{11}a_{22} - a_{12}a_{21}}.$$

Then writing it in determinant form, we have

$$x_1 = \frac{|A_1|}{|A|} \quad \text{and} \quad x_2 = \frac{|A_2|}{|A|},$$

where

$$|A_1| = \begin{vmatrix} b_1 & a_{12} \\ b_2 & a_{22} \end{vmatrix}, \quad |A_2| = \begin{vmatrix} a_{11} & b_1 \\ a_{21} & b_2 \end{vmatrix}, \quad \text{and} \quad |A| = \begin{vmatrix} a_{11} & a_{12} \\ a_{21} & a_{22} \end{vmatrix}.$$

In a similar way, one can use Cramer's rule for a set of n linear equations as follows:

$$x_i = \frac{|A_i|}{|A|}, \qquad i = 1, 2, 3, \ldots, n, \tag{1.19}$$

i.e., the solution for any one of the unknown x_i in a set of simultaneous equations is equal to the ratio of two determinants; the determinant in the denominator is the determinant of the coefficient matrix A, while the determinant in the numerator is the same determinant with the *ith* column replaced by the elements from the right-hand sides of the equation.

Example 1.21 *Solve the following system using Cramer's rule:*

$$
\begin{array}{rrrrrrrr}
5x_1 & & & + & x_3 & + & 2x_4 & = & 3 \\
x_1 & + & x_2 & + & 3x_3 & + & x_4 & = & 5 \\
x_1 & + & x_2 & & & + & 2x_4 & = & 1 \\
x_1 & + & x_2 & + & x_3 & + & x_4 & = & -1.
\end{array}
$$

Solution. *Writing the given system in matrix form*

$$
\begin{pmatrix} 5 & 0 & 1 & 2 \\ 1 & 1 & 3 & 1 \\ 1 & 1 & 0 & 2 \\ 1 & 1 & 1 & 1 \end{pmatrix}
\begin{pmatrix} x_1 \\ x_2 \\ x_3 \\ x_4 \end{pmatrix}
=
\begin{pmatrix} 3 \\ 5 \\ 1 \\ -1 \end{pmatrix}
$$

gives

$$
A = \begin{pmatrix} 5 & 0 & 1 & 2 \\ 1 & 1 & 3 & 1 \\ 1 & 1 & 0 & 2 \\ 1 & 1 & 1 & 1 \end{pmatrix}
\quad and \quad
\mathbf{b} = \begin{pmatrix} 3 \\ 5 \\ 1 \\ -1 \end{pmatrix}.
$$

The determinant of the matrix A can be calculated by using cofactor expansion as follows:

$$
|A| = \begin{vmatrix} 5 & 0 & 1 & 2 \\ 1 & 1 & 3 & 1 \\ 1 & 1 & 0 & 2 \\ 1 & 1 & 1 & 1 \end{vmatrix}
$$

$$= a_{11}c_{11} + a_{12}c_{12} + a_{13}c_{13} + a_{14}c_{14} = 5(2) + 0(-2) + 1(0) + 2(0) = 10 \neq 0,$$

which shows that the given matrix A is nonsingular. Then the matrices $A_1, A_2, A_3,$ and A_4 can be computed as

$$A_1 = \begin{pmatrix} 3 & 0 & 1 & 2 \\ 5 & 1 & 3 & 1 \\ 1 & 1 & 0 & 2 \\ -1 & 1 & 1 & 1 \end{pmatrix}, \quad A_2 = \begin{pmatrix} 5 & 3 & 1 & 2 \\ 1 & 5 & 3 & 1 \\ 1 & 1 & 0 & 2 \\ 1 & -1 & 1 & 1 \end{pmatrix},$$

$$A_3 = \begin{pmatrix} 5 & 0 & 3 & 2 \\ 1 & 1 & 5 & 1 \\ 1 & 1 & 1 & 2 \\ 1 & 1 & -1 & 1 \end{pmatrix}, \quad A_4 = \begin{pmatrix} 5 & 0 & 1 & 3 \\ 1 & 1 & 3 & 5 \\ 1 & 1 & 0 & 1 \\ 1 & 1 & 1 & -1 \end{pmatrix}.$$

The determinant of the matrices $A_1, A_2, A_3,$ and A_4 can be computed as follows:

$$\begin{aligned}
|A_1| &= 3(2) + 0(18) + 1(-6) + 2(-10) = 6 + 0 - 6 - 20 = -20 \\
|A_2| &= 5(-18) + 3(-2) + 1(6) + 2(10) = -90 - 6 + 6 + 20 = -70 \\
|A_3| &= 5(6) + 0(-6) + 3(0) + 2(0) = 30 + 0 + 0 + 0 = 30 \\
|A_4| &= 5(10) + 0(-10) + 1(0) + 3(0) = 50 + 0 + 0 + 0 = 50.
\end{aligned}$$

Now applying Cramer's rule, we get

$$x_1 = \frac{|A_1|}{|A|} = -\frac{20}{10} = -2$$

$$x_2 = \frac{|A_2|}{|A|} = -\frac{70}{10} = -7$$

$$x_3 = \frac{|A_3|}{|A|} = \frac{30}{10} = 3$$

$$x_4 = \frac{|A_3|}{|A|} = \frac{50}{10} = 5,$$

which is the required solution of the given system. ●

Thus Cramer's rule is useful in hand calculations only if the determinants can be evaluated easily, i.e., for $n = 3$ or $n = 4$. The solution of a system of n linear equations by Cramer's rule will require $N = (n+1)\frac{n^3}{3}$ multiplications. Therefore, this rule is much less efficient for large values of n and is at most never used for computational purposes. When the number of equations is large ($n > 4$), other methods of solutions are more desirable.

Use MATLAB commands to find the solution of the above linear system by Cramer's rule as follows:

```
>> A = [5 0 1 2; 1 1 3 1; 1 1 0 2; 1 1 1 1];
>> b = [3; 5; 1; -1];
>> A1 = [b A(:, [2 : 4])];
>> x1 = det(A1)/det(A);
>> A2 = [A(:, 1) b A(:, [3 : 4])];
>> x2 = det(A2)/det(A);
>> A3 = [A(:, [1 : 2]) b A(:, 4)];
>> x3 = det(A3)/det(A);
>> A4 = [A(:, [1 : 3]) b];
>> x4 = det(A4)/det(A);
```

Procedure 1.1 (Cramer's Rule)

1. *Form the coefficient matrix A and column matrix $\mathbf{b}$.*

2. *Compute the determinant of A. If $\det A = 0$, then the system has no solution; otherwise, go to the next step.*

3. *Compute the determinant of the new matrix A_i by replacing the ith matrix with the column vector $\mathbf{b}$.*

4. *Repeat step 3 for $i = 1, 2, \ldots, n$.*

5. *Solve for the unknown variables $\mathbf{x}_i$ using*

$$\mathbf{x}_i = \frac{\det(A_i)}{\det(A)}, \quad for \quad i = 1, 2, \ldots, n.$$

The m-file CRule.m and the following MATLAB commands can be used to generate the solution of Example 1.21 as follows:

```
>> A = [5 0 1 2;1 1 3 1;1 1 0 2;1 1 1 1];
>> b = [3; 5; 1; −1];
>> sol = CRule(A, b);
```

Program 1.5

MATLAB m-file for Cramer's Rule for a Linear System
function sol=CRule(A,b)
[m, n] = size(A);
if m ˜ = n error('Matrix is not square.'); end
if det(A) == 0 error('Matrix is singular.');end
for i = 1:n
B = A; B(:, i) = b;
sol(i) = det(B) / det(A);end
sol = sol';

1.4.2 Gaussian Elimination Method

It is one of the most popular and widely used direct methods for solving linear systems of algebraic equations. No method of solving linear systems requires fewer operations than the Gaussian procedure. The goal of the Gaussian elimination method for solving linear systems is to convert the original system into the equivalent upper-triangular system from which each unknown is determined by backward substitution.

The Gaussian elimination procedure starts with *forward elimination*, in which the first equation in the linear system is used to eliminate the first variable from the rest of the $(n-1)$ equations. Then the new second equation is used to eliminate the second variable from the rest of the $(n-2)$ equations, and so on. If $(n-1)$ such elimination is performed, and the resulting system will be the triangular form. Once this forward elimination is complete, we can determine whether the system is overdetermined or underdetermined or has a unique solution. If it has a unique solution, then *backward substitution* is used to solve the triangular system easily and one can find the unknown variables involved in the system.

Now we shall describe the method in detail for a system of n linear equations. Consider the following system of n linear equations:

$$
\begin{aligned}
a_{11}x_1 &+ a_{12}x_2 &+ a_{13}x_3 &+ \cdots &+ a_{1n}x_n &= b_1 \\
a_{21}x_1 &+ a_{22}x_2 &+ a_{23}x_3 &+ \cdots &+ a_{2n}x_n &= b_2 \\
a_{31}x_1 &+ a_{32}x_2 &+ a_{33}x_3 &+ \cdots &+ a_{3n}x_n &= b_3 \\
\vdots \qquad & \quad \vdots \qquad & \quad \vdots \qquad & \quad \vdots \qquad & \quad \vdots \qquad & \quad \vdots \\
a_{n1}x_1 &+ a_{n2}x_2 & a_{n3}x_3 &+ \cdots &+ a_{nn}x_n &= b_n.
\end{aligned}
\tag{1.20}
$$

Forward Elimination

Consider the first equation of the given system (1.20)

$$
a_{11}x_1 + a_{12}x_2 + a_{13}x_3 + \cdots + a_{1n}x_n = b_1
\tag{1.21}
$$

as the first pivotal equation with the first pivot element a_{11}. Then the first equation times multiples $m_{i1} = (a_{i1}/a_{11})$, $i = 2, 3, \ldots, n$ is subtracted from the *i*th equation to eliminate the first variable x_1, producing an equivalent system

$$
\begin{aligned}
a_{11}x_1 &+ a_{12}x_2 &+ a_{13}x_3 &+ \cdots &+ a_{1n}x_n &= b_1 \\
&\quad a_{22}^{(1)}x_2 &+ a_{23}^{(1)}x_3 &+ \cdots &+ a_{2n}^{(1)}x_n &= b_2^{(1)} \\
&\quad a_{32}^{(1)}x_2 &+ a_{33}^{(1)}x_3 &+ \cdots &+ a_{3n}^{(1)}x_n &= b_3^{(1)} \\
&\quad \vdots & \quad \vdots & \quad \vdots & \quad \vdots & \quad \vdots \\
&\quad a_{n2}^{(1)}x_2 &+ a_{n3}^{(1)}x_3 &+ \cdots &+ a_{nn}^{(1)}x_n &= b_n^{(1)}.
\end{aligned}
\tag{1.22}
$$

Now consider a second equation of the system (1.22), which is

$$
a_{22}^{(1)}x_2 + a_{23}^{(1)}x_3 + \cdots + a_{2n}^{(1)}x_n = b_2^{(1)},
\tag{1.23}
$$

the second pivotal equation with the second pivot element $a_{22}^{(1)}$. Then the second equation times multiples $m_{i2} = (a_{i2}^{(1)}/a_{22}^{(1)})$, $i = 3, \ldots, n$ is subtracted from the *i*th equation to eliminate the second variable x_2, producing an equivalent system

$$
\begin{aligned}
a_{11}x_1 + a_{12}x_2 + a_{13}x_3 + \cdots + a_{1n}x_n &= b_1 \\
a_{22}^{(1)}x_2 + a_{23}^{(1)}x_3 + \cdots + a_{2n}^{(1)}x_n &= b_2^{(1)} \\
a_{33}^{(2)}x_3 + \cdots + a_{3n}^{(2)}x_n &= b_3^{(2)} \\
\vdots \qquad \vdots \qquad \vdots \qquad \vdots& \\
a_{n3}^{(2)}x_3 + \cdots + a_{nn}^{(2)}x_n &= b_n^{(2)}.
\end{aligned}
\tag{1.24}
$$

Now consider a third equation of the system (1.24), which is

$$
a_{33}^{(2)}x_3 + \cdots + a_{3n}^{(2)}x_n = b_3^{(2)},
\tag{1.25}
$$

the third pivotal equation with the third pivot element $a_{33}^{(2)}$. Then the third equation times multiples $m_{i3} = (a_{i3}^{(2)}/a_{33}^{(2)})$, $i = 4, \ldots, n$ is subtracted from the *i*th equation to eliminate the third variable x_3. Similarly, after *(n–1)th* steps, we have the *n*th pivotal equation which has only one unknown variable x_n, i.e.,

$$
\begin{aligned}
a_{11}x_1 + a_{12}x_2 + a_{13}x_3 + \cdots + a_{1n}x_n &= b_1 \\
+ a_{22}^{(1)}x_2 + a_{23}^{(1)}x_3 + \cdots + a_{2n}^{(1)}x_n &= b_2^{(1)} \\
+ a_{33}^{(2)}x_3 + \cdots + a_{3n}^{(2)}x_n &= b_3^{(2)} \\
\vdots \qquad \vdots& \\
a_{nn}^{(n-1)}x_n &= b_n^{(n-1)},
\end{aligned}
\tag{1.26}
$$

with the *n*th pivotal element $a_{nn}^{(n-1)}$. After getting the upper-triangular system, which is equivalent to the original system, the forward elimination is completed.

Backward Substitution

After the triangular set of equations has been obtained, the last equation of system (1.26) yields the value of x_n directly. The value is then substituted into the equation next to the last one of the system (1.26) to obtain a value of x_{n-1}, which is, in turn, used along with the value of x_n in the second to the last equation to obtain a value of x_{n-2}, and so on. A mathematical formula can be obtained for the backward substitution:

$$
\left.
\begin{aligned}
x_n &= \frac{b_n^{(n-1)}}{a_{nn}^{(n-1)}} \\[2mm]
x_{n-1} &= \frac{1}{a_{n-1n-1}^{(n-2)}} \left(b_{n-1}^{(n-2)} - a_{n-1n}^{(n-2)} x_n \right) \\
&\vdots \\
x_1 &= \frac{1}{a_{11}} \left(b_1 - \sum_{j=2}^{n} a_{1j} x_j \right)
\end{aligned}
\right\}.
\tag{1.27}
$$

The Gaussian elimination can be carried out by writing only the coefficients and the right-hand side terms in a matrix form, the augmented matrix form. Indeed, this is exactly what a computer program for Gaussian elimination does. Even for hand calculations, the augmented matrix form is more convenient than writing all sets of equations. The augmented matrix is formed as follows:

$$
\left(
\begin{array}{ccccc|c}
a_{11} & a_{12} & a_{13} & \cdots & a_{1n} & b_1 \\
a_{21} & a_{22} & a_{23} & \cdots & a_{2n} & b_2 \\
a_{31} & a_{32} & a_{33} & \cdots & a_{3n} & b_3 \\
\vdots & \vdots & \vdots & \vdots & \vdots & \\
a_{n1} & a_{n2} & a_{n3} & \cdots & a_{nn} & b_n
\end{array}
\right).
\tag{1.28}
$$

The operations used in the Gaussian elimination method can now be applied to the augmented matrix. Consequently, system (1.26) is now written directly as

$$
\left(
\begin{array}{ccccc|c}
a_{11} & a_{12} & a_{13} & \cdots & a_{1n} & b_1 \\
 & a_{22}^{(1)} & a_{23}^{(1)} & \cdots & a_{2n}^{(1)} & b_2^{(1)} \\
 & & a_{33}^{(2)} & \cdots & a_{3n}^{(2)} & b_3^{(2)} \\
 & & & \vdots & \vdots & \\
 & & & & a_{nn}^{(n-1)} & b_n^{(n-1)}
\end{array}
\right),
\tag{1.29}
$$

from which the unknowns are determined as before by using backward substitution. The number of multiplications and divisions for the Gaussian elimination method for one **b** vector is approximately

$$N = \left(\frac{n^3}{3}\right) + n^2 - \left(\frac{n}{3}\right). \tag{1.30}$$

Simple Gaussian Elimination Method

First, we will solve the linear system using the simplest variation of the Gaussian elimination method, called *simple* Gaussian elimination or Gaussian elimination *without pivoting*. The basics of this variation is that all possible diagonal elements (called *pivot elements*) should be nonzero. If at any stage an element becomes zero, then interchange that row with any row below with a nonzero element at that position. After getting the upper-triangular matrix, we use backward substitution to get the solution of the given linear system.

Example 1.22 *Solve the following linear system using the simple Gaussian elimination method:*

$$\begin{array}{rcrcrcl}
x_1 & + & 2x_2 & + & x_3 & = & 2 \\
2x_1 & + & 5x_2 & + & 2x_3 & = & 1 \\
x_1 & + & 3x_2 & + & 4x_3 & = & 5.
\end{array}$$

Solution. *The process begins with the augmented matrix form*

$$\begin{pmatrix}
1 & 2 & 1 & \vdots & 2 \\
2 & 5 & 3 & \vdots & 1 \\
1 & 3 & 4 & \vdots & 5
\end{pmatrix}.$$

Since $a_{11} = 1 \neq 0$, we wish to eliminate the elements a_{21} and a_{31} by subtracting from the second and third rows the appropriate multiples of the first row. In this case, the multiples are given as

$$m_{21} = \frac{2}{1} = 2 \quad and \quad m_{31} = \frac{1}{1} = 1.$$

Hence,

$$\begin{pmatrix} 1 & 2 & 1 & \vdots & 2 \\ 0 & 1 & 1 & \vdots & -3 \\ 0 & 1 & 3 & \vdots & 3 \end{pmatrix}.$$

Since $a_{22}^{(1)} = 1 \neq 0$, we eliminate the entry in the $a_{32}^{(1)}$ position by subtracting the multiple $m_{32} = \dfrac{1}{1} = 1$ of the second row from the third row to get

$$\begin{pmatrix} 1 & 2 & 1 & \vdots & 2 \\ 0 & 1 & 1 & \vdots & -3 \\ 0 & 0 & 2 & \vdots & 6 \end{pmatrix}.$$

Obviously, the original set of equations has been transformed to an upper-triangular form. Since all the diagonal elements of the obtaining upper-triangular matrix are nonzero, the coefficient matrix of the given system is nonsingular, and the given system has a unique solution. Now expressing the set in algebraic form yields

$$\begin{array}{rcrcrcr} x_1 & + & 2x_2 & + & x_3 & = & 2 \\ & & x_2 & + & x_3 & = & -3 \\ & & & & 2x_3 & = & 6. \end{array}$$

Now using backward substitution, we get

$$\begin{array}{llll} 2x_3 & = & 6 & \quad gives \quad & x_3 = 3 \\ x_2 & = & -x_3 - 3 = -(3) - 3 = -6 & \quad gives \quad & x_2 = -6 \\ x_1 & = & 2 - 2x_2 - x_3 = 2 - 2(-6) - 3 & \quad gives \quad & x_1 = 11, \end{array}$$

which is the required solution of the given system.   ●

The above results can be obtained using MATLAB commands as follows:

```
>> B = [1 2 1 2; 2 5 3 1; 1 3 4 5];
%B = [A|b] = Augmented matrix
>> x = WP(B);
>> disp(x)
```

Program 1.6

MATLAB m-file for Simple Gaussian Elimination Method

```
function x=WP(B)
[n,t]=size(B); U=B;
for k=1:n-1; for i=k:n-1; m=U(i+1,k)/U(k,k);
for j=1:t; U(i+1,j)=U(i+1,j)-m*U(k,j);end;end end
i=n; x(i,1)=U(i,t)/U(i,i);
for i=n-1:-1:1; s=0;
for k=n:-1:i+1; s = s + U(i,k) * x(k,1); end
x(i,1)=(U(i,t)-s)/U(i,i); end; B; U; x; end
```

In the simple description of Gaussian elimination without pivoting just given, we used the *kth* equation to eliminate the variable x_k from equations $k+1, \ldots, n$ during the *kth* step of the procedure. This is possible only if at the beginning of the *kth* step, the coefficient $a_{kk}^{(k-1)}$ of x_k in equation k is not zero. Even though these coefficients are used as denominators both in the multipliers m_{ij} and in the backward substitution equations, this does not necessarily mean that the linear system is not solvable, but that the procedure of the solution must be altered.

Example 1.23 *Solve the following linear system using the simple Gaussian elimination method:*

$$
\begin{array}{rcrcrcl}
 & & x_2 & + & x_3 & = & 1 \\
x_1 & + & 2x_2 & + & 2x_3 & = & 1 \\
2x_1 & + & x_2 & + & 2x_3 & = & 3.
\end{array}
$$

Solution. *Write the given system in augmented matrix form:*

$$
\begin{pmatrix}
0 & 1 & 1 & \vdots & 1 \\
1 & 2 & 2 & \vdots & 1 \\
2 & 1 & 2 & \vdots & 3
\end{pmatrix}.
$$

To solve this system, the simple Gaussian elimination method will fail immediately because the element in the first row on the leading diagonal, the pivot, is zero. Thus, it is impossible to divide that row by the pivot value. Clearly, this difficulty can be overcome by rearranging the order of the rows; for example, making the first row the second gives

$$\begin{pmatrix} 1 & 2 & 2 & \vdots & 1 \\ 0 & 1 & 1 & \vdots & 1 \\ 2 & 1 & 2 & \vdots & 3 \end{pmatrix}.$$

Now we use the usual elimination process. The first elimination step is to eliminate the element $a_{31} = 3$ from the third row by subtracting a multiple $m_{31} = \frac{3}{1} = 3$ of row 1 from row 3, which gives

$$\begin{pmatrix} 1 & 2 & 2 & \vdots & 1 \\ 0 & 1 & 1 & \vdots & 1 \\ 0 & -3 & -2 & \vdots & 1 \end{pmatrix}.$$

We finished with the first elimination step since the element a_{21} is already eliminated from the second row. The second elimination step is to eliminate the element $a_{32}^{(1)} = -3$ from the third row by subtracting a multiple $m_{32} = \frac{-3}{1}$ of row 2 from row 3, which gives

$$\begin{pmatrix} 1 & 2 & 2 & \vdots & 1 \\ 0 & 1 & 1 & \vdots & 1 \\ 0 & 0 & 1 & \vdots & 4 \end{pmatrix}.$$

Obviously, the original set of equations has been transformed to an upper-triangular form. Now expressing the set in algebraic form yields

$$\begin{aligned} x_1 + 2x_2 + 2x_3 &= 1 \\ x_2 + x_3 &= 1 \\ x_3 &= 4. \end{aligned}$$

Now using backward substitution, we get

$$x_3 = 4$$

$$x_2 = 1 - x_3 = 1 - 4 = -3$$

$$x_1 = 1 - 2x_2 - 2x_3 = 1 - 2(-3) - 2(4) = -1,$$

the solution of the given system.   •

Example 1.24 *Solve the following linear system using the simple Gaussian elimination method:*

$$
\begin{aligned}
x_1 &+ x_2 &+ x_3 &= 3 \\
2x_1 &+ 2x_2 &+ 3x_3 &= 7 \\
x_1 &+ 2x_2 &+ 3x_3 &= 6.
\end{aligned}
$$

Solution. *Write the given system in augmented matrix form:*

$$
\begin{pmatrix}
1 & 1 & 1 & \vdots & 3 \\
2 & 2 & 3 & \vdots & 7 \\
1 & 2 & 3 & \vdots & 6
\end{pmatrix}.
$$

The first elimination step is to eliminate the elements $a_{21} = 2$ and $a_{31} = 1$ from the second and third rows by subtracting the multiples $m_{21} = \frac{2}{1} = 2$ and $m_{31} = \frac{1}{1} = 1$ of row 1 from row 2 and row 3, respectively, which gives

$$
\begin{pmatrix}
1 & 1 & 1 & \vdots & 3 \\
0 & 0 & 1 & \vdots & 1 \\
0 & 1 & 2 & \vdots & 3
\end{pmatrix}.
$$

We finished the first elimination step. To start the second elimination step, since we know that the element $a_{22}^{(1)} = 0$, called the second pivot element, the simple Gaussian elimination cannot continue in its present form. Therefore, we interchange rows 2 and 3 to get

$$
\begin{pmatrix}
1 & 1 & 1 & \vdots & 3 \\
0 & 1 & 2 & \vdots & 3 \\
0 & 0 & 1 & \vdots & 1
\end{pmatrix}.
$$

We have finished with the second elimination step since the element $a_{32}^{(1)}$ is already eliminated from the third row. Obviously, the original set of equations has been transformed to an upper-triangular form. Now expressing the set in algebraic form yields

$$
\begin{aligned}
x_1 + x_2 + x_3 &= 3 \\
x_2 + 2x_3 &= 3 \\
x_3 &= 1.
\end{aligned}
$$

Now using backward substitution, we get

$$
x_3 = 1, \quad x_2 = 1, \quad x_1 = 1,
$$

the solution of the system.   •

Example 1.25 *Using the simple Gaussian elimination method, find all values of a and b for which the following linear system is consistent or inconsistent:*

$$
\begin{aligned}
2x_1 - x_2 + 3x_3 &= 1 \\
4x_1 + 2x_2 + 2x_3 &= 2a \\
2x_1 + x_2 + x_3 &= b.
\end{aligned}
$$

Solution. *Write the given system in augmented matrix form:*

$$
\begin{pmatrix}
2 & -1 & 3 & 1 \\
4 & 2 & 2 & 2a \\
2 & 1 & 1 & b
\end{pmatrix},
$$

in which we wish to eliminate the elements a_{21} and a_{31} by subtracting from the second and third rows the appropriate multiples of the first row. In this case, the multiples are given as

$$
m_{21} = \frac{4}{2} = 2 \quad and \quad m_{31} = \frac{2}{2} = 1.
$$

Hence,

$$
\begin{pmatrix}
2 & -1 & 3 & 1 \\
0 & 4 & -4 & 2a - 2 \\
0 & 2 & -2 & b - 1
\end{pmatrix}.
$$

We have finished the first elimination step. The second elimination step is to eliminate element $a_{32}^{(1)} = 2$ by subtracting a multiple $m_{32} = \frac{2}{4} = \frac{1}{2}$ of row 2 from row 3, which gives

$$
\begin{pmatrix}
2 & -1 & 3 & 1 \\
0 & 4 & -4 & 2a - 2 \\
0 & 0 & 0 & b - a
\end{pmatrix}.
$$

We finished the second column. So the third row of the equivalent upper-triangular system is

$$0x_1 + 0x_2 + 0x_3 = b - a. \tag{1.31}$$

First, if (1.31) has no constraint on unknowns x_1, x_2, and x_3, then the upper-triangular system represents only two nontrivial equations, namely,

$$\begin{array}{rrrrr} 2x_1 & - & x_2 & + & 3x_3 & = & 1 \\ & & 4x_2 & - & 4x_3 & = & 2a - 2 \end{array}$$

in the three unknowns. As a result, one of the unknowns can be chosen arbitrarily, say $x_3 = x_3^*$, then x_2^* and x_1^* can be obtained by using backward substitution:

$$x_2^* = 1/2a - 1/2 - x_3^*; \quad x_1^* = \frac{1}{2}(1 + 1/2a - 1/2 - 4x_3^*).$$

Hence,

$$\mathbf{x}^* = [\frac{1}{2}(1 + 1/2a - 1/2 - 4x_3^*), 1/2a - 1/2 - x_3^*, x_3^*]^T$$

is an approximation solution of the given system for any value of x_3^* for any real value of a. Hence, the given linear system is consistent (infinitely many solutions).

Second, when $b - a \neq 0$, in this case, (1.31) puts a restriction on unknowns x_1, x_2, and x_3 that is impossible to satisfy. So the given system cannot have any solutions and, therefore, is inconsistent. ●

Example 1.26 *Solve the following homogeneous linear system using the simple Gaussian elimination method:*

$$\begin{array}{rrrrrrr} x_1 & + & x_2 & - & 2x_3 & = & 0 \\ 2x_1 & + & 4x_2 & - & 3x_3 & = & 0 \\ 3x_1 & + & 7x_2 & - & 5x_3 & = & 0. \end{array}$$

Solution. *The process begins with the augmented matrix form*

$$\begin{pmatrix} 1 & 1 & -2 & \vdots & 0 \\ 2 & 4 & -3 & \vdots & 0 \\ 3 & 7 & -5 & \vdots & 0 \end{pmatrix}.$$

Using the following multiples,

$$m_{21} = \frac{2}{1} = 2 \quad and \quad m_{31} = \frac{3}{1} = 3$$

finishes the first elimination step, and we get

$$\begin{pmatrix} 1 & 1 & -2 & \vdots & 0 \\ 0 & 2 & 1 & \vdots & 0 \\ 0 & 4 & 1 & \vdots & 0 \end{pmatrix}.$$

Then using the multiple $m_{32} = \frac{4}{2} = 2$ of the second row from the third row, we get

$$\begin{pmatrix} 1 & 1 & -2 & \vdots & 0 \\ 0 & 2 & 1 & \vdots & 0 \\ 0 & 0 & -1 & \vdots & 0 \end{pmatrix}.$$

Obviously, the original set of equations has been transformed to an upper-triangular form. Thus, the system has the unique solution $[0,0,0]^T$, i.e., the system has only the trivial solution.   ●

Example 1.27 *Find the value of k for which the following homogeneous linear system has nontrivial solutions by using the simple Gaussian elimination method:*

$$\begin{array}{rrrrr} 2x_1 & - & 3x_2 & + & 5x_3 & = & 0 \\ -2x_1 & + & 6x_2 & - & x_3 & = & 0 \\ 4x_1 & - & 9x_2 & + & kx_3 & = & 0. \end{array}$$

Solution. *The process begins with the augmented matrix form*

$$\begin{pmatrix} 2 & -3 & 5 & \vdots & 0 \\ -2 & 6 & -1 & \vdots & 0 \\ 4 & -9 & k & \vdots & 0 \end{pmatrix},$$

and then using the following multiples,

$$m_{21} = \frac{-2}{2} = -1 \quad and \quad m_{31} = \frac{4}{2} = 2,$$

which gives

$$\begin{pmatrix} 2 & -3 & 5 & \vdots & 0 \\ 0 & 3 & 4 & \vdots & 0 \\ 0 & -3 & k-10 & \vdots & 0 \end{pmatrix}.$$

Also, by using the multiple $m_{32} = \frac{-3}{3} = -1$, *we get*

$$\begin{pmatrix} 2 & -3 & 5 & \vdots & 0 \\ 0 & 3 & 4 & \vdots & 0 \\ 0 & 0 & k-6 & \vdots & 0 \end{pmatrix}.$$

From the last row of the above system, we obtain

$$k - 6 = 0, \quad which \; gives \quad k = 6.$$

Also, solving the above underdetermined system

$$\begin{aligned} 2x_1 \; - \; 3x_2 \; + \; 5x_3 \; &= \; 0 \\ 3x_2 \; + \; 4x_3 \; &= \; 0 \end{aligned}$$

by taking $x_3 = 1$, *we have the nontrivial solutions*

$$x^* = \alpha[-9/2, -4/3, 1]^T, \quad for \quad \alpha \neq 0.$$

Note that if we put $x_3 = 0$, *for example, we obtain the trivial solution* $[0, 0, 0]^T$. $\qquad \bullet$

Theorem 1.19 *An upper-triangular matrix* A *is nonsingular if and only if all its diagonal elements are not zero.* $\qquad \bullet$

Example 1.28 *Use the simple Gaussian elimination method to find all the values of* α *which make the following matrix singular:*

$$A = \begin{pmatrix} 1 & -1 & \alpha \\ 2 & 2 & 1 \\ 0 & \alpha & -1.5 \end{pmatrix}.$$

Solution. *Apply the forward elimination step of the simple Gaussian elim-ination on the given matrix A and eliminate the element a_{21} by subtracting from the second row the appropriate multiple of the first row. In this case, the multiple is given as*

$$\begin{pmatrix} 1 & -1 & \alpha \\ 0 & 4 & 1-2\alpha \\ 0 & \alpha & -1.5 \end{pmatrix}.$$

We finished the first elimination step. The second elimination step is to eliminate element $a_{32}^{(1)} = \alpha$ by subtracting a multiple $m_{32} = \frac{\alpha}{4}$ of row 2 from row 3, which gives

$$\begin{pmatrix} 1 & -1 & \alpha \\ 0 & 4 & 1-2\alpha \\ 0 & 0 & -1.5 - \dfrac{\alpha(1-2\alpha)}{4} \end{pmatrix}.$$

To show that the given matrix is singular, we have to set the third diagonal element equal to zero (by Theorem 1.19), i.e.,

$$-1.5 - \frac{\alpha(1-2\alpha)}{4} = 0.$$

After simplifying, we obtain

$$2\alpha^2 - \alpha - 6 = 0.$$

Solving the above quadratic equation, we get

$$\alpha = -\frac{3}{2} \quad and \quad \alpha = 2,$$

which are the possible values of α, which make the given matrix singular.●

Example 1.29 *Use the smallest positive integer value of α to find the unique solution of the linear system $A\mathbf{x} = [1, 6, -4]^T$ by the simple Gaussian elimination method, where*

$$A = \begin{pmatrix} 1 & -1 & \alpha \\ 2 & 2 & 1 \\ 0 & \alpha & -1.5 \end{pmatrix}.$$

Solution. *Since we know from Example 1.28 that the given matrix A is singular when $\alpha = -\frac{3}{2}$ and $\alpha = 2$, to find the unique solution we take the smallest positive integer value $\alpha = 1$ and consider the augmented matrix as follows:*

$$\begin{pmatrix} 1 & -1 & 1 & \vdots & 1 \\ 2 & 2 & 1 & \vdots & 6 \\ 0 & 1 & -1.5 & \vdots & -4 \end{pmatrix}.$$

Applying the forward elimination step of the simple Gaussian elimination on the given matrix A and eliminating the element a_{21} by subtracting from the second row the appropriate multiple $m_{21} = 2$ of the first row gives

$$\begin{pmatrix} 1 & -1 & 1 & \vdots & 1 \\ 0 & 4 & -1 & \vdots & 4 \\ 0 & 1 & -1.5 & \vdots & -4 \end{pmatrix}.$$

The second elimination step is to eliminate element $a_{32}^{(1)} = 1$ by subtracting a multiple $m_{32} = \frac{1}{4}$ of row 2 from row 3, which gives

$$\begin{pmatrix} 1 & -1 & 1 & \vdots & 1 \\ 0 & 4 & -1 & \vdots & 4 \\ 0 & 0 & -5/4 & \vdots & -5 \end{pmatrix}.$$

Now expressing the set in algebraic form yields

$$\begin{aligned} x_1 - x_2 + x_3 &= 1 \\ 4x_2 - x_3 &= 4 \\ -5/4 x_3 &= -5. \end{aligned}$$

Using backward substitution, we obtain

$$x_3 = 4, \quad x_2 = 2, \quad x_1 = -1,$$

the unique solution of the given system. ●

Note that the inverse of the nonsingular matrix A can be easily determined by using the simple Gaussian elimination method. Here, we have to consider the augmented matrix as a combination of the given matrix A and the identity matrix $\mathbf{I}$ (the same size as A). To find the inverse matrix BA^{-1}, we must solve the linear system in which the *jth* column of the matrix B is the solution of the linear system with the right-hand side the *jth* column of the matrix $\mathbf{I}$.

Example 1.30 *Use the simple Gaussian elimination method to find the inverse of the following matrix:*

$$A = \begin{pmatrix} 2 & -1 & 3 \\ 4 & -1 & 6 \\ 2 & -3 & 4 \end{pmatrix}.$$

Solution. *Suppose that the inverse $A^{-1} = B$ of the given matrix exists and let*

$$AB = \begin{pmatrix} 2 & -1 & 3 \\ 4 & -1 & 6 \\ 2 & -3 & 4 \end{pmatrix} \begin{pmatrix} b_{11} & b_{12} & b_{13} \\ b_{21} & b_{22} & b_{23} \\ b_{31} & b_{32} & b_{33} \end{pmatrix} = \begin{pmatrix} 1 & 0 & 0 \\ 0 & 1 & 0 \\ 0 & 0 & 1 \end{pmatrix} = \mathbf{I}.$$

Now to find the elements of the matrix B, we apply simple Gaussian elimination on the augmented matrix:

$$[A|\mathbf{I}] = \begin{pmatrix} 2 & -1 & 3 & \vdots & 1 & 0 & 0 \\ 4 & -1 & 6 & \vdots & 0 & 1 & 0 \\ 2 & -3 & 4 & \vdots & 0 & 0 & 1 \end{pmatrix}.$$

Apply the forward elimination step of the simple Gaussian elimination on the given matrix A and eliminate the elements $a_{21} = 4$ and $a_{31} = 2$ by subtracting from the second and the third rows the appropriate multiples $m_{21} = \frac{4}{2} = 2$ and $m_{31} = \frac{2}{2} = 1$ of the first row. It gives

$$\begin{pmatrix} 2 & -1 & 3 & \vdots & 1 & 0 & 0 \\ 0 & 1 & 0 & \vdots & -2 & 1 & 0 \\ 0 & -2 & 1 & \vdots & -1 & 0 & 1 \end{pmatrix}.$$

We finished the first elimination step. The second elimination step is to eliminate element $a_{32}^{(1)} = -2$ by subtracting a multiple $m_{32} = \frac{-2}{1} = -2$ of row 2 from row 3, which gives

$$\left(\begin{array}{ccc:ccc} 2 & -1 & 3 & 1 & 0 & 0 \\ 0 & 1 & 0 & -2 & 1 & 0 \\ 0 & 0 & 1 & -5 & 2 & 1 \end{array} \right).$$

We solve the first system

$$\left(\begin{array}{ccc} 2 & -1 & 3 \\ 0 & 1 & 0 \\ 0 & 0 & 1 \end{array} \right) \left(\begin{array}{c} b_{11} \\ b_{21} \\ b_{31} \end{array} \right) = \left(\begin{array}{c} 1 \\ -2 \\ -5 \end{array} \right)$$

by using backward substitution, and we get

$$\begin{array}{rcrcrcr} 2b_{11} & - & b_{21} & + & 3b_{31} & = & 1 \\ & & b_{21} & & & = & -2 \\ & & & & b_{31} & = & -5, \end{array}$$

which gives

$$b_{11} = 7, \quad b_{21} = -2, \quad b_{31} = -5.$$

Similarly, the solution of the second linear system

$$\left(\begin{array}{ccc} 2 & -1 & 3 \\ 0 & 1 & 0 \\ 0 & 0 & 1 \end{array} \right) \left(\begin{array}{c} b_{12} \\ b_{22} \\ b_{32} \end{array} \right) = \left(\begin{array}{c} 0 \\ 1 \\ 2 \end{array} \right)$$

can be obtained as follows:

$$\begin{array}{rcrcrcr} 2b_{12} & - & b_{22} & + & 3b_{32} & = & 0 \\ & & b_{22} & & & = & 1 \\ & & & & b_{32} & = & 2, \end{array}$$

which gives

$$b_{12} = -5/2, \quad b_{22} = 1, \quad b_{32} = 2.$$

Finally, the solution of the third linear system

$$
\begin{pmatrix} 2 & -1 & 3 \\ 0 & 1 & 0 \\ 0 & 0 & 1 \end{pmatrix} \begin{pmatrix} b_{13} \\ b_{23} \\ b_{33} \end{pmatrix} = \begin{pmatrix} 0 \\ 0 \\ 1 \end{pmatrix}
$$

can be obtained as follows:

$$
\begin{aligned}
2b_{13} \;-\; b_{23} \;+\; 3b_{33} &= 0 \\
b_{23} &= 0 \\
b_{33} &= 1,
\end{aligned}
$$

and it gives

$$
b_{13} = -3/2, \quad b_{23} = 0, \quad b_{33} = 1.
$$

Hence, the elements of the inverse matrix B are

$$
B = A^{-1} = \begin{pmatrix} 7 & -\dfrac{5}{2} & -\dfrac{3}{2} \\[2mm] -2 & 1 & 0 \\[2mm] -5 & 2 & 1 \end{pmatrix},
$$

which is the required inverse of the given matrix A.   ●

Procedure 1.2 (Gaussian Elimination Method)

1. *Form the augmented matrix, $B = [A|\mathbf{b}]$.*

2. *Check the first pivot element $a_{11} \neq 0$, then move to the next step; otherwise, interchange rows so that $a_{11} \neq 0$.*

3. *Multiply row one by multiplier $m_{i1} = a_{i1}/a_{11}$ and subtract to the ith row for $i = 2, 3, \ldots, n$.*

4. *Repeat steps 2 and 3 for the remaining pivots elements unless coefficient matrix A becomes upper-triangular matrix U.*

5. *Use backward substitution to solve x_n from the nth equation $x_n = \dfrac{b_n^{n-1}}{a_{nn}}$ and solve the other $(n-1)$ unknown variables by using (1.27).*

We now introduce the most important numerical quantity associated with a matrix.

Definition 1.30 (Rank of a Matrix)

The rank of a matrix A is the number of pivots. An $m \times n$ matrix will, in general, have a rank r, where r is an integer and $r \leq min\{m, n\}$. If $r = min\{m, n\}$, then the matrix is said to be full rank. If $r < min\{m, n\}$, then the matrix is said to be rank deficient. ●

In principle, the rank of a matrix can be determined by using the Gaussian elimination process in which the coefficient matrix A is reduced to upper-triangular form U. After reducing the matrix to triangular form, we find that the rank is the number of columns with nonzero values on the diagonal of U. In practice, especially for large matrices, round-off errors during the row operation may cause a loss of accuracy in this method of rank computation.

Theorem 1.20 *For a system of n equations with n unknowns written in the form $A\mathbf{x} = \mathbf{b}$, the solution $\mathbf{x}$ of a system exists and is unique for any $\mathbf{b}$, if and only if $rank(A) = n$.* ●

Conversely, if $rank(A) < n$ for an $n \times n$ matrix A, then the system of equations $A\mathbf{x} = \mathbf{b}$ may or may not be consistent. Such a system may not have a solution, or the solution, if it exists, will not be unique.

Example 1.31 *Find the rank of the following matrix:*

$$A = \begin{pmatrix} 1 & 2 & 4 \\ 1 & 1 & 5 \\ 1 & 1 & 6 \end{pmatrix}.$$

Solution. *Apply the forward elimination step of simple Gaussian elimination on the given matrix A and eliminate the elements below the first pivot (first diagonal element) to*

$$\begin{pmatrix} 1 & 2 & 4 \\ 0 & -1 & 1 \\ 0 & -1 & 2 \end{pmatrix}.$$

We finished the first elimination step. The second pivot is in the $(2,2)$
*position, but after eliminating the element below it, we find the triangular
form to be*

$$\begin{pmatrix} 1 & 2 & 4 \\ 0 & -1 & 1 \\ 0 & 0 & 3 \end{pmatrix}.$$

Since the number of pivots are three, the rank of the given matrix is 3. *Note
that the original matrix is nonsingular since the rank of the* 3×3 *matrix
is* 3. ●

In MATLAB, the built-in **rank** function can be used to estimate the
rank of a matrix:

```
>> A = [1 2 4; 1 1 5; 1 1 6];
>> rank(A)
ans =
    3
```

Note that:
$$\begin{aligned} \text{rank}(AB) &\leq \min(\text{rank}(A), \text{rank}(B)) \\ \text{rank}(A+B) &\leq \text{rank}(A) + \text{rank}(B) \\ \text{rank}(AA^T) &= \text{rank}(A) = \text{rank}(A^T A) \end{aligned}$$

Although the rank of a matrix is very useful to categorize the behavior
of matrices and systems of equations, the rank of a matrix is usually not
computed. ●

The use of nonzero pivots is sufficient for the theoretical correctness
of simple Gaussian elimination, but more care must be taken if one is to
obtain reliable results. For example, consider the linear system

$$\begin{aligned} 0.000100x_1 &+ x_2 = 1 \\ x_1 &+ x_2 = 2, \end{aligned}$$

which has the exact solution $\mathbf{x} = [1.00010, 0.99990]^T$. Now we solve this
system by simple Gaussian elimination. The first elimination step is to
eliminate the first variable x_1 from the second equation by subtracting

multiple $m_{21} = 10000$ of the first equation from the second equation, which gives

$$
\begin{aligned}
0.000100x_1 \;+\;& x_2 \;=\; 1 \\
-\;& 10000x_2 \;=\; -10000.
\end{aligned}
$$

Using backward substitution we get the solution $\mathbf{x}^* = [0, 1]^T$. Thus, a computational disaster has occurred. But if we interchange the equations, we obtain

$$
\begin{aligned}
x_1 \;+\; x_2 \;&=\; 2 \\
0.000100x_1 \;+\; x_2 \;&=\; 1.
\end{aligned}
$$

Applying Gaussian elimination again, we get the solution $\mathbf{x}^* = [1, 1]^T$. This solution is as good as one would hope. So, we conclude from this example that it is not enough just to avoid a zero pivot, one must also avoid a relatively small one. Here we need some pivoting strategies to help us overcome the difficulties faced during the process of simple Gaussian elimination.

1.4.3 Pivoting Strategies

We know that simple Gaussian elimination is applied to a problem with no pivotal elements that are zero, but the method does not work if the first coefficient of the first equation or a diagonal coefficient becomes zero in the process of the solution, because they are used as denominators in a forward elimination.

Pivoting is used to change the sequential order of the equations for two purposes; first to prevent diagonal coefficients from becoming zero, and second, to make each diagonal coefficient larger in magnitude than any other coefficient below it, i.e., to decrease the round-off errors. The equations are not mathematically affected by changes in sequential order, but changing the order makes the coefficient become nonzero. Even when all diagonal coefficients are nonzero, the change of order increases the accuracy of the computations.

There are two standard pivoting strategies used to handle these difficulties easily. They are explained as follows.

Partial Pivoting

Here, we develop an implementation of Gaussian elimination that utilizes the pivoting strategy discussed above. In using Gaussian elimination by partial pivoting (or row pivoting), the basic approach is to use the largest (in absolute value) element on or below the diagonal in the column of current interest as the pivotal element for elimination in the rest of that column.

One immediate effect of this will be to force all the multiples used to be not greater than 1 in absolute value. This will inhibit the growth of error in the rest of the elimination phase and in subsequent backward substitution.

At stage k of forward elimination, it is necessary, therefore, to be able to identify the largest element from $|a_{kk}|, |a_{k+1,k}|, \ldots, |a_{nk}|$, where these a_{ik}s are the elements in the current partially triangularized coefficient matrix. If this maximum occurs in row p, then the *p*th and *k*th rows of the augmented matrix are interchanged and the elimination proceeds as usual. In solving n linear equations, a total of $N = n(n+1)/2$ coefficients must be examined.

Example 1.32 *Solve the following linear system using Gaussian elimination with partial pivoting:*

$$
\begin{array}{rcrcrcl}
x_1 & + & x_2 & + & x_3 & = & 1 \\
2x_1 & + & 3x_2 & + & 4x_3 & = & 3 \\
4x_1 & + & 9x_2 & + & 16x_3 & = & 11.
\end{array}
$$

Solution. *For the first elimination step, since 4 is the largest absolute coefficient of the first variable x_1, the first row and the third row are interchanged, which gives us*

$$
\begin{array}{rcrcrcl}
4x_1 & + & 9x_2 & + & 16x_3 & = & 11 \\
2x_1 & + & 3x_2 & + & 4x_3 & = & 3 \\
x_1 & + & x_2 & + & x_3 & = & 1.
\end{array}
$$

Eliminate the first variable x_1 from the second and third rows by subtracting the multiples $m_{21} = \frac{2}{4}$ and $m_{31} = \frac{1}{4}$ of row 1 from row 2 and row 3, respectively, which gives

$$
\begin{array}{rcrcrcr}
4x_1 & + & 9x_2 & + & 16x_3 & = & 11 \\
 & - & 3/2x_2 & - & 4x_3 & = & -5/2 \\
 & - & 5/4x_2 & - & x_3 & = & -7/5.
\end{array}
$$

For the second elimination step, $-\frac{3}{2}$ *is the largest absolute coefficient of the second variable* x_2, *so eliminate the second variable* x_2 *from the third row by subtracting the multiple* $m_{32} = \frac{5}{6}$ *of row 2 from row 3, which gives*

$$
\begin{array}{rcrcrcr}
4x_1 & + & 9x_2 & + & 16x_3 & = & 11 \\
 & - & 3/2x_2 & - & 4x_3 & = & -5/2 \\
 & & & & 1/3x_3 & = & 1/3.
\end{array}
$$

Obviously, the original set of equations has been transformed to an equivalent upper-triangular form. Now using backward substitution, we get

$$x_1 = 1, \quad x_2 = -1, \quad x_3 = 1,$$

which is the required solution of the given linear system. ●

The following MATLAB commands will give the same results we obtained in Example 1.32 of the Gaussian elimination method with partial pivoting:

```
>> B = [1 1 1 1; 2 3 4 3; 4 9 16 11];
>> x = PP(B);
>> disp(x)
```

Program 1.7

MATLAB m-file for Gaussian Elimination by Partial Pivoting

```
function x=PP(B)
% B = input('input matrix in form[A/b]');
[n,t] = size(B); U = B;
for M = 1:n-1
mx(M) = abs(U(M,M)); r = M;
for i = M+1:n
if mx(M) < abs(U(i,M))
mx(M)=abs(U(i,M)); r = i; end; end
rw1(1,1:t)=U(r,1:t); rw2(1,1:t)=U(M,1:t);
U(M,1:t)=rw1 ; U(r,1:t)=rw2 ;
for k=M+1:n
m=U(k,M)/U(M,M);
for j=M:t
U(k,j) = U(k,j) - m * U(M,j); end;end
i=n; x(i)=U(i,t)/U(i,i);
for i=n-1:-1:1; s=0;
for k=n:-1:i+1
s = s + U(i,k) * x(k); end
x(i)=(U(i,t)-s)/U(i,i); end; B; U; x; end
```

Procedure 1.3 (Partial Pivoting)

1. *Suppose we are about to work on the ith column of the matrix. Then we search that portion of the ith column below and including the diagonal and find the element that has the largest absolute value. Let p denote the index of the row that contains this element.*

2. *Interchange row i and p.*

3. *Proceed with elimination procedure 1.2.*

Total Pivoting

In the case of total pivoting (or complete pivoting), we search for the largest number (in absolute value) in the entire array instead of just in the first column, and this number is the pivot. This means that we shall probably need to interchange the columns as well as rows. When solving a system of equations using complete pivoting, each row interchange is equivalent to interchanging two equations, while each column interchange is equivalent to interchanging the two unknowns.

At the *kth* step, interchange both the rows and columns of the matrix so that the largest number in the remaining matrix is used as the pivot i.e., after the pivoting

$$|a_{kk}| = max|a_{ij}|, \quad \text{for} \quad i = k, k+1, \ldots, n, \quad j = k, k+1, \ldots, n.$$

There are times when the partial pivoting procedure is inadequate. When some rows have coefficients that are very large in comparison to those in other rows, partial pivoting may not give a correct solution.

Therefore, when in doubt, use total pivoting. No amount of pivoting will remove inherent ill-conditioning (we will discuss this later in the chapter) from a set of equations, but it helps to ensure that no further ill-conditioning is introduced in the course of computation.

Example 1.33 *Solve the following linear system using Gaussian elimination with total pivoting:*

$$
\begin{array}{rcrcrcl}
x_1 & + & x_2 & + & x_3 & = & 1 \\
2x_1 & + & 3x_2 & + & 4x_3 & = & 3 \\
4x_1 & + & 9x_2 & + & 16x_3 & = & 11.
\end{array}
$$

Solution. *For the first elimination step, since 16 is the largest absolute coefficient of variable x_3 in the given system, the first row and the third row are interchanged as well as the first column and third column, and we get*

$$
\begin{array}{rcrcrcl}
16x_3 & + & 9x_2 & + & 4x_1 & = & 11 \\
4x_3 & + & 3x_2 & + & 2x_1 & = & 3 \\
x_3 & + & 9x_2 & + & x_1 & = & 1.
\end{array}
$$

Then eliminate the third variable x_3 from the second and third rows by subtracting the multiples $m_{21} = \frac{4}{16}$ and $m_{31} = \frac{1}{16}$ of row 1 from rows 2 and 3, which respectively, gives

$$
\begin{aligned}
16x_3 + 9x_2 + 4x_1 &= 11 \\
\tfrac{3}{4}x_2 + x_1 &= \tfrac{1}{4} \\
\tfrac{7}{16}x_2 + 3/4x_1 &= \tfrac{5}{16}.
\end{aligned}
$$

For the second elimination step, 1 is the largest absolute coefficient of the first variable x_1 in the second row and third column, so the second and third columns are interchanged, giving us

$$
\begin{aligned}
16x_3 + 4x_1 + 9x_2 &= 11 \\
x_1 + \tfrac{3}{4}x_2 &= \tfrac{1}{4} \\
\tfrac{3}{4}x_1 + \tfrac{7}{16}x_2 &= \tfrac{5}{16}.
\end{aligned}
$$

Eliminate the first variable x_1 from the third row by subtracting the multiple $m_{32} = \frac{3}{4}$ of row 2 from row 3, which gives

$$
\begin{aligned}
16x_3 + 4x_1 + 9x_2 &= 11 \\
x_1 + \tfrac{3}{4}x_2 &= \tfrac{1}{4} \\
- \tfrac{1}{8}x_2 &= \tfrac{1}{8}.
\end{aligned}
$$

The original set of equations has been transformed to an equivalent upper-triangular form. Now using backward substitution, we get

$$
x_1 = 1, \quad x_2 = -1, \quad x_3 = 1,
$$

which is the required solution of the given linear system.   ●

Program 1.8
MATLAB m-file for the Gaussian Elimination by Total Pivoting
function x=TP(B)
% B = input('input matrix in form[A/b]');
[n,m]=size(B);U=B; w=zeros(n,n);
for i=1:n; N(i)=i; end
for M = 1:n-1; r=M; c=M;
for i = M:n; for j = M:n
if $max(M) < abs(U(i,j))$; max(M)=abs(U(i,j));
r = i; c = j; end; end; end
rw1(1,1:m)=U(r,1:m); rw2(1,1:m)=U(M,1:m);
U(M,1:m)=rw1;U(r,1:m)=rw2 ; cl1(1:n,1)= U(1:n,c);
$cl2(1 : n, 1) = U(1 : n, M); U(1 : n, M) = cl1(1 : n, 1);$
$U(1 : n, c) = cl2(1 : n, 1); p = N(M); N(M) = N(c);$
$N(c) = p; w(M, 1 : n) = N;$
for $k = M + 1 : n; e = U(k, M)/U(M, M);$
for $j = M : m; U(k, j) = U(k, j) - e * U(M, j);$ end; end
$i = n; x(i, 1) = U(i, m)/U(i, i);$
for $i = n - 1 : -1 : 1; s = 0;$
for $k = n : -1 : i + 1; s = s + U(i, k) * x(k, 1);$end
$x(i, 1) = (U(i, m) - s)/U(i, i);$ end
for i=1:n; $X(N(i), 1) = x(i, 1);$end; B;U;X; end

MATLAB can be used to get the same results we obtained in Example 1.33 of the Gaussian elimination method with total pivoting with the following command:

```
>> B = [1 1 1 1; 2 3 4 3; 4 9 16 11];
>> x = TP(B);
>> disp(x)
```

Total pivoting offers little advantage over partial pivoting and it is significantly slower, requiring $N = \frac{n(n+1)(2n+1)}{6}$ elements to be examined in total. It is rarely used in practice because interchanging columns changes the order of the xs and, consequently, add significant and usually unjustified complexity to the computer program. So for getting good results partial pivoting has shown to be a very reliable procedure.

1.4.4 Gauss–Jordan Method

This method is a modification of the Gaussian elimination method. The Gauss–Jordan method is inefficient for practical calculation, but is often useful for theoretical purposes. The basis of this method is to convert the given matrix into a diagonal form. The forward elimination of the Gauss–Jordan method is identical to the Gaussian elimination method. However, Gauss–Jordan elimination uses backward elimination rather than backward substitution. In the Gauss–Jordan method the forward elimination and backward elimination need not be separated. This is possible because a pivot element can be used to eliminate the coefficients not only below but also above at the same time. If this approach is taken, the form of the coefficients matrix becomes diagonal when elimination by the last pivot is completed. The Gauss–Jordan method simply yields a transformation of the augmented matrix of the form

$$[A|\mathbf{b}] \rightarrow [\mathbf{I}|\mathbf{c}],$$

where $\mathbf{I}$ is the identity matrix and $\mathbf{c}$ is the column matrix, which represents the possible solution of the given linear system.

Example 1.34 *Solve the following linear system using the Gauss–Jordan method:*

$$\begin{array}{rcrcrcr} x_1 & + & 2x_2 & & & = & 3 \\ -x_1 & & & - & 2x_3 & = & -5 \\ -3x_1 & - & 5x_2 & + & x_3 & = & -4. \end{array}$$

Solution. *Write the given system in the augmented matrix form*

$$\begin{pmatrix} 1 & 2 & 0 & \vdots & 3 \\ -1 & 0 & -2 & \vdots & -5 \\ -3 & -5 & 1 & \vdots & -4 \end{pmatrix}.$$

The first elimination step is to eliminate elements $a_{21} = -1$ and $a_{31} = -3$ by subtracting the multiples $m_{21} = -1$ and $m_{31} = -3$ of row 1 from rows

2 and 3, respectively, which gives

$$\begin{pmatrix} 1 & 2 & 0 & \vdots & 3 \\ 0 & 2 & -2 & \vdots & -2 \\ 0 & 1 & 1 & \vdots & 5 \end{pmatrix}.$$

The second row is now divided by 2 to give

$$\begin{pmatrix} 1 & 2 & 0 & \vdots & 3 \\ 0 & 1 & -1 & \vdots & -1 \\ 0 & 1 & 1 & \vdots & 5 \end{pmatrix}.$$

The second elimination step is to eliminate the elements in positions $a_{12}^{(1)} = 2$ and $a_{32} = 1$ by subtracting the multiples $m_{12} = 2$ and $m_{32} = 1$ of row 2 from rows 1 and 3, respectively, which gives

$$\begin{pmatrix} 1 & 0 & 2 & \vdots & 5 \\ 0 & 1 & -1 & \vdots & -1 \\ 0 & 0 & 2 & \vdots & 6 \end{pmatrix}.$$

The third row is now divided by 2 to give

$$\begin{pmatrix} 1 & 0 & 2 & \vdots & 5 \\ 0 & 1 & -1 & \vdots & -1 \\ 0 & 0 & 1 & \vdots & 3 \end{pmatrix}.$$

The third elimination step is to eliminate the elements in positions $a_{23}^{(1)} = -1$ and $a_{13} = 2$ by subtracting the multiples $m_{23} = -1$ and $m_{13} = 2$ of row 3 from rows 2 and 1, respectively, which gives

$$\begin{pmatrix} 1 & 0 & 0 & \vdots & -1 \\ 0 & 1 & 0 & \vdots & 2 \\ 0 & 0 & 1 & \vdots & 3 \end{pmatrix}.$$

Obviously, the original set of equations has been transformed to a diagonal form. Now expressing the set in algebraic form yields

$$
\begin{aligned}
x_1 &= -1 \\
x_2 &= 2 \\
x_3 &= 3,
\end{aligned}
$$

which is the required solution of the given system.   •

The above results can be obtained using MATLAB commands, as follows:

> $\gg Ab = [A|b] = [1\ 2\ 0\ 3; -1\ 0\ -2\ -5; -3\ -5\ 1\ -4];$
> $\gg GaussJ(Ab);$

Program 1.9
MATLAB m-file for the Gauss–Jordan Method
function sol=GaussJ(Ab)
[m,n]=size(Ab);
for i=1:m
$Ab(i,:) = Ab(i,:)/Ab(i,i);$
for j=1:m
if $j == i$; continue; end
$Ab(j,:) = Ab(j,:) - Ab(j,i) * Ab(i,:);$
end; end; sol=Ab;

Procedure 1.4 (Gauss–Jordan Method)

1. *Form the augmented matrix, $[A|\mathbf{b}]$.*

2. *Reduce the coefficient matrix A to unit upper-triangular form using the Gaussian procedure.*

3. *Use the nth row to reduce the nth column to an equivalent identity matrix column.*

4. *Repeat step 3 for n–1 through 2 to get the augmented matrix of the form $[\mathbf{I}|\mathbf{c}]$.*

5. *Solve for the unknown* $x_i = c_i,$ *for* $i = 1, 2, \ldots, n.$

The number of multiplications and divisions required for the Gauss–Jordan method is approximately

$$N = \left(\frac{n^3}{2}\right) - n^2 - \left(\frac{n}{2}\right),$$

which is approximately 50% larger than for the Gaussian elimination method. Consequently, the Gaussian elimination method is preferred.

The Gauss–Jordan method is particularly well suited to compute the inverse of a matrix through the transformation

$$[A|\mathbf{I}] \rightarrow [\mathbf{I}|A^{-1}].$$

Note if the inverse of the matrix can be found, then the solution of the linear system can be computed easily from the product of matrix A^{-1} and column matrix **b**, i.e.,

$$\mathbf{x} = A^{-1}\mathbf{b}. \tag{1.32}$$

Example 1.35 *Apply the Gauss–Jordan method to find the inverse of the following matrix:*

$$A = \begin{pmatrix} 10 & 1 & -5 \\ -20 & 3 & 20 \\ 5 & 3 & 5 \end{pmatrix}.$$

Then solve the system with $\mathbf{b} = [1, 2, 6]^T.$

Solution. *Consider the following augmented matrix:*

$$[A|I] = \begin{pmatrix} 10 & 1 & -5 & \vdots & 1 & 0 & 0 \\ -20 & 3 & 20 & \vdots & 0 & 1 & 0 \\ 5 & 3 & 5 & \vdots & 0 & 0 & 1 \end{pmatrix}.$$

Divide the first row by 10, which gives

$$= \begin{pmatrix} 1 & 0.1 & -0.5 & \vdots & 0.1 & 0 & 0 \\ -20 & 3 & 20 & \vdots & 0 & 1 & 0 \\ 5 & 3 & 5 & \vdots & 0 & 0 & 1 \end{pmatrix}.$$

The first elimination step is to eliminate the elements in positions $a_{21} = -20$ and $a_{31} = 5$ by subtracting the multiples $m_{21} = -20$ and $m_{31} = 5$ of row 1 from rows 2 and 3, respectively, which gives

$$= \begin{pmatrix} 1 & 0.1 & -0.5 & \vdots & 0.1 & 0 & 0 \\ 0 & 5 & 10 & \vdots & 2 & 1 & 0 \\ 0 & 2.5 & 7.5 & \vdots & -0.5 & 0 & 1 \end{pmatrix}.$$

Divide the second row by 5, which gives

$$= \begin{pmatrix} 1 & 0.1 & -0.5 & \vdots & 0.1 & 0 & 0 \\ 0 & 1 & 2 & \vdots & 0.4 & 0.2 & 0 \\ 0 & 2.5 & 7.5 & \vdots & -0.5 & 0 & 1 \end{pmatrix}.$$

The second elimination step is to eliminate the elements in positions $a_{12} = 0.1$ and $a_{32}^{(1)} = 2.5$ by subtracting the multiples $m_{12} = 0.1$ and $m_{32} = 2.5$ of row 2 from rows 1 and 3, respectively, which gives

$$= \begin{pmatrix} 1 & 0 & -0.7 & \vdots & 0.06 & -0.02 & 0 \\ 0 & 1 & 2 & \vdots & 0.4 & 0.2 & 0 \\ 0 & 0 & 2.5 & \vdots & -1.5 & -0.5 & 1 \end{pmatrix}.$$

Divide the third row by 2.5, which gives

$$= \begin{pmatrix} 1 & 0 & -0.7 & \vdots & 0.06 & -0.02 & 0 \\ 0 & 1 & 2 & \vdots & 0.4 & 0.2 & 0 \\ 0 & 0 & 1 & \vdots & -0.6 & -0.2 & 0.4 \end{pmatrix}.$$

The third elimination step is to eliminate the elements in positions $a_{23}^{(2)} = 2$ and $a_{13}^{(2)} = -0.7$ by subtracting the multiples $m_{23} = 2$ and $m_{13} = -0.7$ of row 3 from rows 2 and 1, respectively, which gives

$$= \begin{pmatrix} 1 & 0 & 0 & \vdots & -0.36 & -0.16 & 0.28 \\ 0 & 1 & 0 & \vdots & 1.6 & 0.6 & -0.8 \\ 0 & 0 & 1 & \vdots & -0.6 & -0.2 & 0.4 \end{pmatrix} = [\mathbf{I}|A^{-1}].$$

Obviously, the original augmented matrix $[A|I]$ *has been transformed to the augmented matrix of the form* $[I|A^{-1}]$. *Hence, the solution of the linear system can be obtained by the matrix multiplication (1.32) as*

$$
\begin{pmatrix} x_1 \\ x_2 \\ x_3 \end{pmatrix} = \begin{pmatrix} -0.36 & -0.16 & 0.28 \\ 1.6 & 0.6 & -0.8 \\ -0.6 & -0.2 & 0.4 \end{pmatrix} \begin{pmatrix} 1 \\ 2 \\ 6 \end{pmatrix} = \begin{pmatrix} 1 \\ -2 \\ 1.4 \end{pmatrix}.
$$

Hence, $\mathbf{x}^* = [1, -2, 1.4]^T$ *is the solution of the given system.*   •

The above results can be obtained using MATLAB, as follows:

```
>> Ab = [A|I] = [10 1 − 5 1 0 0; −20 3 20 0 1 0; 5 3 5 0 0 1];
>> [I|inv(A)] = GaussJ(Ab);
>> b = [1 2 6]';
>> x = inv(A) ∗ b;
```

1.4.5   LU Decomposition Method

This is another direct method to find the solution of a system of linear equations. *LU* decomposition (or the factorization method) is a modification of the elimination method. Here we decompose or factorize the coefficient matrix A into the product of two triangular matrices in the form

$$
A = LU, \tag{1.33}
$$

where L is a lower-triangular matrix and U is the upper-triangular matrix. Both are the same size as the coefficients matrix A. To solve a number of linear equations sets in which the coefficients matrices are all identical but the right-hand sides are different, then *LU* decomposition is more efficient than the elimination method. Specifying the diagonal elements of either L or U makes the factoring unique. The procedure based on unity elements on the diagonal of matrix L is called *Doolittle's method* (or Gauss factorization), while the procedure based on unity elements on the diagonal of matrix U is called *Crout's method*. Another method, called the *Cholesky method*, is based on the constraint that the diagonal elements of L are equal to the diagonal elements of U, i.e., $l_{ii} = u_{ii}$, for $i = 1, 2, \ldots, n$.

The general forms of L and U are written as

$$
L = \begin{pmatrix} l_{11} & 0 & \cdots & 0 \\ l_{21} & l_{22} & \cdots & 0 \\ \vdots & \vdots & \vdots & \vdots \\ l_{n1} & l_{n2} & \cdots & l_{nn} \end{pmatrix}, \quad
U = \begin{pmatrix} u_{11} & u_{12} & \cdots & u_{1n} \\ 0 & u_{22} & \cdots & u_{2n} \\ \vdots & \vdots & \vdots & \vdots \\ 0 & 0 & \cdots & u_{nn} \end{pmatrix}, \quad (1.34)
$$

such that $l_{ij} = 0$ for $i < j$ and $u_{ij} = 0$ for $i > j$.

Consider a linear system

$$
A\mathbf{x} = \mathbf{b} \tag{1.35}
$$

and let A be factored into the product of L and U, as shown by (1.34). Then the linear system (1.35) becomes

$$
LU\mathbf{x} = \mathbf{b}
$$

or can be written as

$$
L\mathbf{y} = \mathbf{b},
$$

where

$$
\mathbf{y} = U\mathbf{x}.
$$

The unknown elements of matrix L and matrix U are computed by equating corresponding elements in matrices A and LU in a systematic way. Once the matrices L and U have been constructed, the solution of system (1.35) can be computed in the following two steps:

1. Solve the system $L\mathbf{y} = \mathbf{b}$.

By using *forward elimination*, we will find the components of the unknown vector $\mathbf{y}$ by using the following steps:

$$
\left.\begin{aligned}
y_1 &= b_1, \\
y_i &= b_i - \sum_{j=1}^{i-1} l_{ij} y_j, \qquad i = 2, 3, \ldots, n
\end{aligned}\right\} . \tag{1.36}
$$

2. Solve the system $\qquad U\mathbf{x} = \mathbf{y}.$

By using *backward substitution*, we will find the components of the unknown vector $\mathbf{x}$ by using the following steps:

$$\left.\begin{array}{rcl} x_n & = & \dfrac{y_n}{u_{nn}}, \\[2ex] x_i & = & \dfrac{1}{u_{ii}}\left[y_i - \displaystyle\sum_{j=i+1}^{n} u_{ij}x_j \right], \quad i = n-1, n-2, \ldots, 1 \end{array}\right\} . \quad (1.37)$$

Thus, the relationship of the matrices L and U to the original matrix A is given by the following theorem.

Theorem 1.21 *If Gaussian elimination can be performed on the linear system $A\mathbf{x} = \mathbf{b}$ without row interchanges, then the matrix A can be factored into the product of a lower-triangular matrix L and an upper-triangular matrix U, i.e.,*

$$A = LU,$$

where the matrices L and U are the same size as A. ●

Let us consider a nonsingular system $A\mathbf{x} = \mathbf{b}$ and with the help of the simple Gauss elimination method we will convert the coefficient matrix A into the upper-triangular matrix U by using elementary row operations. If all the pivots are nonzero, then row interchanges are not necessary, and the decomposition of the matrix A is possible. Consider the following matrix:

$$A = \begin{pmatrix} 2 & 4 & 2 \\ 4 & 9 & 7 \\ -2 & -2 & 5 \end{pmatrix} .$$

To convert it into the upper-triangular matrix U, we first apply the following row operations

$$Row2 - (2)Row1 \quad \text{and} \quad Row3 + Row1,$$

which gives

$$\begin{pmatrix} 2 & 4 & 2 \\ 0 & 1 & 3 \\ 0 & 2 & 7 \end{pmatrix} .$$

Once again, applying the row operation

$$Row2 - (2)Row2,$$

we get

$$\begin{pmatrix} 2 & 4 & 2 \\ 0 & 1 & 3 \\ 0 & 0 & 1 \end{pmatrix} = U,$$

which is the required upper-triangular matrix.

Now defining the three elementary matrices (each of them can be obtained by adding a multiple of row i to row j) associated with these row operations:

$$E1 = \begin{pmatrix} 1 & 0 & 0 \\ -2 & 1 & 0 \\ 0 & 0 & 1 \end{pmatrix}, \quad E2 = \begin{pmatrix} 1 & 0 & 0 \\ 0 & 1 & 0 \\ 1 & 0 & 1 \end{pmatrix}, \quad E3 = \begin{pmatrix} 1 & 0 & 0 \\ 0 & 1 & 0 \\ 0 & -2 & 1 \end{pmatrix}.$$

Then

$$E3E2E1 = \begin{pmatrix} 1 & 0 & 0 \\ 0 & 1 & 0 \\ 0 & -2 & 1 \end{pmatrix} \begin{pmatrix} 1 & 0 & 0 \\ 0 & 1 & 0 \\ 1 & 0 & 1 \end{pmatrix} \begin{pmatrix} 1 & 0 & 0 \\ -2 & 1 & 0 \\ 0 & 0 & 1 \end{pmatrix} = \begin{pmatrix} 1 & 0 & 0 \\ -2 & 1 & 0 \\ 5 & -2 & 1 \end{pmatrix}$$

and

$$E3E2E1A = \begin{pmatrix} 1 & 0 & 0 \\ -2 & 1 & 0 \\ 5 & -2 & 1 \end{pmatrix} \begin{pmatrix} 2 & 4 & 2 \\ 4 & 9 & 7 \\ -2 & -2 & 5 \end{pmatrix} = \begin{pmatrix} 2 & 4 & 2 \\ 0 & 1 & 3 \\ 0 & 0 & 1 \end{pmatrix} = U.$$

So

$$A = E1^{-1}E2^{-1}E3^{-1} = LU,$$

where

$$E1^{-1}E2^{-1}E3^{-1} = \begin{pmatrix} 1 & 0 & 0 \\ 2 & 1 & 0 \\ 0 & 0 & 1 \end{pmatrix} \begin{pmatrix} 1 & 0 & 0 \\ 0 & 1 & 0 \\ -1 & 0 & 1 \end{pmatrix} \begin{pmatrix} 1 & 0 & 0 \\ 0 & 1 & 0 \\ 0 & 2 & 1 \end{pmatrix} = \begin{pmatrix} 1 & 0 & 0 \\ 2 & 1 & 0 \\ -1 & 2 & 1 \end{pmatrix} =$$

Thus, $A = LU$ is a product of a lower-triangular matrix L and an upper-triangular matrix U. Naturally, this is called an *LU* decomposition of A.

Theorem 1.22 *Let A be an $n \times n$ matrix that has an LU factorization, i.e.,*

$$A = LU.$$

If A has rank n (i.e., all pivots are nonzeros), then L and U are uniquely determined by A. $\bullet$

Now we will discuss all three possible variations of LU decomposition to find the solution of the nonsingular linear system in the following.

Doolittle's Method

In Doolittle's method (called Gauss factorization), the upper-triangular matrix U is obtained by forward elimination of the Gaussian elimination method and the lower-triangular matrix L containing the multiples used in the Gaussian elimination process as the elements below the diagonal with unity elements on the main diagonal.

For the matrix A in Example 1.22, we can have the decomposition of matrix A in the form

$$\begin{pmatrix} 1 & 2 & 1 \\ 2 & 5 & 3 \\ 1 & 3 & 4 \end{pmatrix} = \begin{pmatrix} 1 & 0 & 0 \\ 2 & 1 & 0 \\ 1 & 1 & 1 \end{pmatrix} \begin{pmatrix} 1 & 2 & 1 \\ 0 & 1 & 1 \\ 0 & 0 & 2 \end{pmatrix},$$

where the unknown elements of matrix L are the used multiples and the matrix U is the same as we obtained in the forward elimination process.

Example 1.36 *Construct the LU decomposition of the following matrix A by using Gauss factorization (i.e., LU decomposition by Doolittle's method). Find the value(s) of α for which the following matrix is*

$$A = \begin{pmatrix} 1 & -1 & \alpha \\ -1 & 2 & -\alpha \\ \alpha & 1 & 1 \end{pmatrix}$$

singular. Also, find the unique solution of the linear system $A\mathbf{x} = [1, 1, 2]^T$ by using the smallest positive integer value of α.

Solution. *Since we know that*

$$A = \begin{pmatrix} 1 & -1 & \alpha \\ -1 & 2 & -\alpha \\ \alpha & 1 & 1 \end{pmatrix} = \begin{pmatrix} 1 & 0 & 0 \\ m_{21} & 1 & 0 \\ m_{31} & m_{32} & 1 \end{pmatrix}, \begin{pmatrix} u_{11} & u_{12} & u_{13} \\ 0 & u_{22} & u_{23} \\ 0 & 0 & u_{33} \end{pmatrix} = LU,$$

now we will use only the forward elimination step of the simple Gaussian elimination method to convert the given matrix A into the upper-triangular matrix U. Since $a_{11} = 1 \neq 0$, we wish to eliminate the elements $a_{21} = -1$ and $a_{31} = \alpha$ by subtracting from the second and third rows the appropriate multiples of the first row. In this case, the multiples are given,

$$m_{21} = \frac{-1}{1} = -1 \quad and \quad m_{31} = \frac{\alpha}{1} = \alpha.$$

Hence,

$$\begin{pmatrix} 1 & -1 & \alpha \\ 0 & 1 & 0 \\ 0 & 1+\alpha & 1-\alpha^2 \end{pmatrix}.$$

Since $a_{22}^{(1)} = 1 \neq 0$, we eliminate the entry in the $a_{32}^{(1)} = 1 + \alpha$ position by subtracting the multiple $m_{32} = \frac{1+\alpha}{1}$ of the second row from the third row to get

$$\begin{pmatrix} 1 & -1 & \alpha \\ 0 & 1 & 0 \\ 0 & 0 & 1-\alpha^2 \end{pmatrix}.$$

Obviously, the original set of equations has been transformed to an upper-triangular form. Thus,

$$\begin{pmatrix} 1 & -1 & \alpha \\ -1 & 2 & -\alpha \\ \alpha & 1 & 1 \end{pmatrix} = \begin{pmatrix} 1 & 0 & 0 \\ -1 & 1 & 0 \\ \alpha & 1+\alpha & 1 \end{pmatrix} \begin{pmatrix} 1 & -1 & \alpha \\ 0 & 1 & 0 \\ 0 & 0 & 1-\alpha^2 \end{pmatrix},$$

which is the required decomposition of A. The matrix will be singular, if the third diagonal element $1 - \alpha^2$ of the upper-triangular U is equal to zero (Theorem 1.19), which gives $\alpha = \pm 1$.

To find the unique solution of the given system we take $\alpha = 2$, and it gives

$$\begin{pmatrix} 1 & -1 & 2 \\ -1 & 2 & -2 \\ 2 & 1 & 1 \end{pmatrix} = \begin{pmatrix} 1 & 0 & 0 \\ -1 & 1 & 0 \\ 2 & 3 & 1 \end{pmatrix} \begin{pmatrix} 1 & -1 & 2 \\ 0 & 1 & 0 \\ 0 & 0 & -3 \end{pmatrix}.$$

Now solve the first system $L\mathbf{y} = \mathbf{b}$ for unknown vector $\mathbf{y}$, i.e.,

$$\begin{pmatrix} 1 & 0 & 0 \\ -1 & 1 & 0 \\ 2 & 3 & 1 \end{pmatrix} \begin{pmatrix} y_1 \\ y_2 \\ y_3 \end{pmatrix} = \begin{pmatrix} 1 \\ 1 \\ 2 \end{pmatrix}.$$

Performing forward substitution yields

$$\begin{array}{rcllrcl} y_1 & = & 1 & gives & y_1 & = & 1, \\ -y_1 + y_2 & = & 1 & gives & y_2 & = & 2, \\ 2y_1 + 3y_2 + y_3 & = & 2 & gives & y_3 & = & -6. \end{array}$$

Then solve the second system $U\mathbf{x} = \mathbf{y}$ for unknown vector $\mathbf{x}$, i.e.,

$$\begin{pmatrix} 1 & -1 & 2 \\ 0 & 1 & 0 \\ 0 & 0 & -3 \end{pmatrix} \begin{pmatrix} x_1 \\ x_2 \\ x_3 \end{pmatrix} = \begin{pmatrix} 1 \\ 2 \\ -6 \end{pmatrix}.$$

Performing backward substitution yields

$$\begin{array}{rcllrcl} x_1 - x_2 + 2x_3 & = & 1 & gives & x_1 & = & -1 \\ x_2 & = & 2 & gives & x_2 & = & 2 \\ -3x_3 & = & -6 & gives & x_3 & = & 2, \end{array}$$

which gives

$$\begin{array}{rcl} x_1 & = & -1 \\ x_2 & = & 2 \\ x_3 & = & 2, \end{array}$$

the approximate solution of the given system. ●

We can write a MATLAB m-file to factor a nonsingular matrix A into a unit lower-triangular matrix L and an upper-triangular matrix U using the **lu − gauss** function. The following MATLAB commands can be used to reproduce the solution of the linear system of Example 1.22:

```
>> A = [1 2 0; −1 0 − 2; −3 − 5 1];
>> B = lu − gauss(A);
>> L = eye(size(B)) + tril(B, −1);
>> U = triu(A);
>> b = [3 − 5 − 4]';
>> y = L \ b;
>> x = U \ y;
```

Program 1.10
MATLAB m-file for the LU Decomposition Method
function $A = lu − gauss(A)$
% LU factorization without pivoting
[n,n] = size(A); for i=1:n-1; pivot = A(i,i);
for k=i+1:n; A(k,i)=A(k,i)/pivot;
for j=i+1:n; $A(k, j) = A(k, j) − A(k, i) * A(i, j)$;
end;end; end

There is another way to find the values of the unknown elements of the matrices L and U, which we describe in the following example.

Example 1.37 *Construct the LU decomposition of the following matrix using Doolittle's method:*

$$A = \begin{pmatrix} 1 & 2 & 4 \\ 1 & 3 & 3 \\ 2 & 2 & 2 \end{pmatrix}.$$

Solution. *Since*

$$A = LU = \begin{pmatrix} 1 & 0 & 0 \\ l_{21} & 1 & 0 \\ l_{31} & l_{32} & 1 \end{pmatrix} \begin{pmatrix} u_{11} & u_{12} & u_{13} \\ 0 & u_{22} & u_{23} \\ 0 & 0 & u_{33} \end{pmatrix},$$

performing the multiplication on the right-hand side gives

$$\begin{pmatrix} 1 & 2 & 4 \\ 1 & 3 & 3 \\ 2 & 2 & 2 \end{pmatrix} = \begin{pmatrix} u_{11} & u_{12} & u_{13} \\ l_{21}u_{11} & l_{21}u_{12} + u_{22} & l_{21}u_{13} + u_{23} \\ l_{31}u_{11} & l_{31}u_{12} + l_{32}u_{22} & l_{31}u_{13} + l_{32}u_{23} + u_{33} \end{pmatrix}.$$

Then equate elements of the first column to obtain

$$\begin{aligned}
1 &= u_{11}, & u_{11} &= 1 \\
1 &= l_{21}u_{11}, & l_{21} &= 1 \\
2 &= l_{31}u_{11}, & l_{31} &= 2.
\end{aligned}$$

Now equate elements of the second column to obtain

$$\begin{aligned}
2 &= u_{12}, & u_{12} &= 2 \\
3 &= l_{21}u_{12} + u_{22}, & u_{22} &= 3 - 2 = 1 \\
2 &= l_{31}u_{12} + l_{32}u_{22}, & l_{32} &= 2 - 4 = -2.
\end{aligned}$$

Finally, equate elements of the third column to obtain

$$\begin{aligned}
4 &= u_{13}, & u_{13} &= 4 \\
3 &= l_{21}u_{13} + u_{23}, & u_{23} &= 3 - 4 = -1 \\
2 &= l_{31}u_{13} + l_{32}u_{23} + u_{33}, & u_{33} &= 2 - 10 = -8.
\end{aligned}$$

Thus, we obtain

$$\begin{pmatrix} 1 & 2 & 4 \\ 1 & 3 & 3 \\ 2 & 2 & 2 \end{pmatrix} = \begin{pmatrix} 1 & 0 & 0 \\ 1 & 1 & 0 \\ 2 & -2 & 1 \end{pmatrix} \begin{pmatrix} 1 & 2 & 4 \\ 0 & 1 & 1 \\ 0 & 0 & -8 \end{pmatrix},$$

the factorization of the given matrix. ●

The general formula for getting elements of L and U corresponding to the coefficient matrix A for a set of n linear equations can be written as

$$
\left.
\begin{aligned}
u_{ij} &= a_{ij} - \sum_{k=1}^{i-1} l_{ik} u_{kj}, & 2 &\leq i \leq j \\[2mm]
l_{ij} &= \frac{1}{u_{ii}} \left[a_{ij} - \sum_{k=1}^{j-1} l_{ik} u_{kj} \right], & i &> j \geq 2 \\[2mm]
u_{ij} &= a_{1j}, & i &= 1 \\[2mm]
l_{ij} &= \frac{a_{i1}}{u_{11}} = \frac{a_{i1}}{a_{11}}, & j &= 1
\end{aligned}
\right\}. \tag{1.38}
$$

Example 1.38 *Solve the following linear system by LU decomposition using Doolittle's method:*

$$
A = \begin{pmatrix} 1 & 2 & 4 \\ 1 & 3 & 3 \\ 2 & 2 & 2 \end{pmatrix} \quad and \quad \mathbf{b} = \begin{pmatrix} -2 \\ 3 \\ -6 \end{pmatrix}.
$$

Solution. *The factorization of the coefficient matrix A has already been constructed in Example 1.37 as*

$$
\begin{pmatrix} 1 & 2 & 4 \\ 1 & 3 & 3 \\ 2 & 2 & 2 \end{pmatrix} = \begin{pmatrix} 1 & 0 & 0 \\ 1 & 1 & 0 \\ 2 & -2 & 1 \end{pmatrix} \begin{pmatrix} 1 & 2 & 4 \\ 0 & 1 & 1 \\ 0 & 0 & -8 \end{pmatrix}.
$$

Then solve the first system $L\mathbf{y} = \mathbf{b}$ for unknown vector $\mathbf{y}$, i.e.,

$$
\begin{pmatrix} 1 & 0 & 0 \\ 1 & 1 & 0 \\ 2 & -2 & 1 \end{pmatrix} \begin{pmatrix} y_1 \\ y_2 \\ y_3 \end{pmatrix} = \begin{pmatrix} -2 \\ 3 \\ -6 \end{pmatrix}.
$$

Performing forward substitution yields

$$
\begin{array}{rcrcll}
y_1 & & & = & -2 & \text{gives} \quad y_1 = -2, \\
y_1 &+& y_2 & = & 3 & \text{gives} \quad y_2 = 5, \\
2y_1 &-& 2y_2 + y_3 & = & -6 & \text{gives} \quad y_3 = 8.
\end{array}
$$

Then solve the second system $U\mathbf{x} = \mathbf{y}$ *for unknown vector* $\mathbf{x}$, *i.e.,*

$$
\begin{pmatrix} 1 & 2 & 4 \\ 0 & 1 & 1 \\ 0 & 0 & -8 \end{pmatrix} \begin{pmatrix} x_1 \\ x_2 \\ x_3 \end{pmatrix} = \begin{pmatrix} -2 \\ 5 \\ 8 \end{pmatrix}.
$$

Performing backward substitution yields

$$
\begin{array}{rcrclrcl}
x_1 & + & 2x_2 & + & 4x_3 & = & -2 & \textit{gives} & x_1 & = & -6 \\
 & & x_2 & + & x_3 & = & 5 & \textit{gives} & x_2 & = & 4 \\
 & & & - & 8x_3 & = & 8 & \textit{gives} & x_3 & = & -1,
\end{array}
$$

which gives

$$
\begin{array}{rcl}
x_1 & = & -6 \\
x_2 & = & 4 \\
x_3 & = & -1,
\end{array}
$$

the approximate solution of the given system. ●

We can also write the MATLAB m-file called Doolittle.m to get the solution of the linear system by LU decomposition by using Doolittle's method. In order to reproduce the above results using MATLAB commands, we do the following:

```
>> A = [1 2 4; 1 3 3; 2 2 2];
>> b = [-2 3 - 6];
>> sol = Doolittle(A, b);
```

Program 1.11
MATLAB m-file for using Doolittle's Method
function sol = Doolittle(A,b)
[n,n]=size(A); u=A;l=zeros(n,n);
for i=1:n-1; if abs(u(i,i))> 0
for i1=i+1:n; m(i1,i)=u(i1,i)/u(i,i);
for j=1:n
$u(i1, j) = u(i1, j) - m(i1, i) * u(i, j)$;end;end;end;end
for i=1:n; l(i,1)=A(i,1)/u(1,1); end
for j=2:n; for i=2:n; s=0;
for k=1:j-1; $s = s + l(i, k) * u(k, j)$; end
l(i,j)=(A(i,j)-s)/u(j,j); end; end y(1)=b(1)/l(1,1);
for k=2:n; sum=b(k);
for i=1:k-1; $sum = sum - l(k, i) * y(i)$; end
y(k)=sum/l(k,k); end
x(n)=y(n)/u(n,n);
for k=n-1:-1:1; sum=y(k);
for i=k+1:n; $sum = sum - u(k, i) * x(i)$; end
x(k)=sum/u(k,k); end; l; u; y; x

Procedure 1.5 (LU Decomposition by Doolittle's Method)

1. *Take the nonsingular matrix A.*

2. *If possible, decompose the matrix $A = LU$ using (1.38).*

3. *Solve linear system $L\mathbf{y} = \mathbf{b}$ using (1.36).*

4. *Solve linear system $U\mathbf{x} = \mathbf{y}$ using (1.37).*

The LDV Factorization

There is some asymmetry in LU decomposition because the lower-triangular matrix has 1s on its diagonal, while the upper-triangular matrix has a nonunit diagonal. This is easily remedied by factoring the diagonal entries out of the upper-triangular matrix as follows:

$$
\begin{pmatrix}
u_{11} & u_{12} & \cdots & u_{1n} \\
0 & u_{22} & \cdots & u_{2n} \\
\vdots & \vdots & \vdots & \vdots \\
0 & 0 & \cdots & u_{nn}
\end{pmatrix}
=
\begin{pmatrix}
u_{11} & & \cdots & 0 \\
0 & u_{22} & \cdots & \\
\vdots & \vdots & \vdots & \vdots \\
0 & & \cdots & u_{nn}
\end{pmatrix}
\begin{pmatrix}
1 & u_{12}/u_{11} & \cdots & u_{1n}/u_{11} \\
0 & 1 & \cdots & u_{2n}/u_{22} \\
\vdots & \vdots & \vdots & \vdots \\
0 & 0 & \cdots & 1
\end{pmatrix}.
$$

Let D denote the diagonal matrix having the same diagonal elements as the upper-triangular matrix U; in other words, D contains the pivots on its diagonal and zeros everywhere else. Let V be the redefining upper-triangular matrix obtained from the original upper-triangular matrix U by dividing each row by its pivot, so that V has all 1s on the diagonal. It is easily seen that $U = DV$, which allows any LU decomposition to be written as

$$
A = LDV,
$$

where L and V are lower- and upper-triangular matrices with 1s on both of their diagonals. This is called the LDV factorization of A.

Example 1.39 *Find the LDV factorization of the following matrix:*

$$
A = \begin{pmatrix}
1 & 2 & -1 \\
3 & 2 & 1 \\
2 & 4 & 1
\end{pmatrix}.
$$

Solution. *By using Doolittle's method, the LU decomposition of A can be obtained as*

$$
A = \begin{pmatrix}
1 & 2 & -1 \\
3 & 2 & 1 \\
2 & 4 & 1
\end{pmatrix}
=
\begin{pmatrix}
1 & 0 & 0 \\
3 & 1 & 0 \\
2 & 0 & 1
\end{pmatrix}
\begin{pmatrix}
1 & 2 & -1 \\
0 & -4 & 4 \\
0 & 0 & 3
\end{pmatrix}
= LU.
$$

Then the matrix D and the matrix V can be obtained as

$$
U = \begin{pmatrix}
1 & 2 & -1 \\
0 & -4 & 4 \\
0 & 0 & 3
\end{pmatrix}
=
\begin{pmatrix}
1 & 0 & 0 \\
0 & -4 & 0 \\
0 & 0 & 3
\end{pmatrix}
\begin{pmatrix}
1 & 2 & -1 \\
0 & 1 & 4 \\
0 & 0 & 1
\end{pmatrix}
= DV.
$$

Thus, the LDV factorization of the given matrix A is obtained as

$$A = LDV = \begin{pmatrix} 1 & 0 & 0 \\ 3 & 1 & 0 \\ 2 & 0 & 1 \end{pmatrix} \begin{pmatrix} 1 & 0 & 0 \\ 0 & -4 & 0 \\ 0 & 0 & 3 \end{pmatrix} \begin{pmatrix} 1 & 2 & -1 \\ 0 & 1 & 4 \\ 0 & 0 & 1 \end{pmatrix}.$$

●

If a given matrix A is symmetric, then there is a connection between the lower-triangular matrix L and the upper-triangular matrix U in the LU decomposition. In the first elimination step, the elements in Ls first column are obtained by dividing Us first row by the diagonal elements. Similarly, during the second elimination step, $l_{32} = \frac{u_{23}}{u_{22}}$. In general, when a symmetric matrix is decomposed without pivots, l_{ij} is related to u_{ji} through the identity

$$l_{ij} = \frac{u_{ji}}{u_{jj}}.$$

In other words, each column of a matrix L equals the corresponding row of a matrix U divided by the diagonal element. It is uniquely determined that the LDV decomposition of a symmetric matrix has the form LDL^T, since $A = LDV$. Taking the transpose of it, we get

$$(A)^T = (LDV)^T = V^T D^T L^T = V^T D L^T,$$

(the diagonal matrix D is symmetric), and the uniqueness of the LDV decomposition implies that

$$L = V^T \quad \text{and} \quad V = L^T.$$

Note that not every symmetric matrix has an LDL^T factorization. However, if $A = LDL^T$, then A must be symmetric because

$$(A)^T = (LDL^T)^T = (L^T)^T D^T L^T = LDL^T = A.$$

Example 1.40 *Find the LDL^T factorization of the following symmetric matrix:*

$$A = \begin{pmatrix} 1 & 3 & 2 \\ 3 & 4 & 1 \\ 2 & 1 & 2 \end{pmatrix}.$$

Solution. *By using Doolittle's method, the LU decomposition of A can be obtained as*

$$A = \begin{pmatrix} 1 & 3 & 2 \\ 3 & 4 & 1 \\ 2 & 1 & 2 \end{pmatrix} = \begin{pmatrix} 1 & 0 & 0 \\ 3 & 1 & 0 \\ 2 & 1 & 1 \end{pmatrix} \begin{pmatrix} 1 & 3 & 2 \\ 0 & -5 & -5 \\ 0 & 0 & 3 \end{pmatrix} = LU.$$

Then the matrix D and the matrix V can be obtained as

$$U = \begin{pmatrix} 1 & 3 & 2 \\ 0 & -5 & -5 \\ 0 & 0 & 3 \end{pmatrix} = \begin{pmatrix} 1 & 0 & 0 \\ 0 & -5 & 0 \\ 0 & 0 & 3 \end{pmatrix} \begin{pmatrix} 1 & 3 & 2 \\ 0 & 1 & 1 \\ 0 & 0 & 1 \end{pmatrix} = DV.$$

Note that

$$V = \begin{pmatrix} 1 & 3 & 2 \\ 0 & 1 & 1 \\ 0 & 0 & 1 \end{pmatrix} = L^T = \begin{pmatrix} 1 & 0 & 0 \\ 3 & 1 & 0 \\ 2 & 1 & 1 \end{pmatrix}.$$

Thus, we obtain

$$A = LDL^T = \begin{pmatrix} 1 & 0 & 0 \\ 3 & 1 & 0 \\ 2 & 1 & 1 \end{pmatrix} \begin{pmatrix} 1 & 0 & 0 \\ 0 & -5 & 0 \\ 0 & 0 & 3 \end{pmatrix} \begin{pmatrix} 1 & 3 & 2 \\ 0 & 1 & 1 \\ 0 & 0 & 1 \end{pmatrix},$$

the LDL^T factorization of the given matrix A. ●

Crout's Method

Crout's method, in which matrix U has unity on the main diagonal, is similar to Doolittle's method in all other aspects. The L and U matrices are obtained by expanding the matrix equation $A = LU$ term by term to determine the elements of the L and U matrices.

Example 1.41 *Construct the LU decomposition of the following matrix using Crout's method:*

$$A = \begin{pmatrix} 1 & 2 & 3 \\ 6 & 5 & 4 \\ 2 & 5 & 6 \end{pmatrix}.$$

Solution. *Since*

$$A = LU = \begin{pmatrix} l_{11} & 0 & 0 \\ l_{21} & l_{22} & 0 \\ l_{31} & l_{32} & l_{33} \end{pmatrix} \begin{pmatrix} 1 & u_{12} & u_{13} \\ 0 & 1 & u_{23} \\ 0 & 0 & 1 \end{pmatrix},$$

performing the multiplication on the right-hand side gives

$$\begin{pmatrix} 1 & 2 & 3 \\ 6 & 5 & 4 \\ 2 & 5 & 6 \end{pmatrix} = \begin{pmatrix} l_{11} & l_{11}u_{12} & l_{11}u_{13} \\ l_{21} & l_{21}u_{12} + l_{22} & l_{21}u_{13} + l_{22}u_{23} \\ l_{31} & l_{31}u_{12} + l_{32} & l_{31}u_{13} + l_{32}u_{23} + l_{33} \end{pmatrix}.$$

Then equate elements of the first column to obtain

$$\begin{aligned} 1 &= l_{11} \\ 6 &= l_{21} \\ 2 &= l_{31}. \end{aligned}$$

Then equate elements of the second column to obtain

$$2 = l_{11}u_{12}, \qquad u_{12} = 2$$

$$5 = l_{21}u_{12} + l_{22}, \quad l_{22} = 5 - 12 = -7$$

$$5 = l_{31}u_{12} + l_{32}, \quad l_{32} = 5 - 4 = 1.$$

Finally, equate elements of the third column to obtain

$$3 = l_{11}u_{13}, \qquad\qquad u_{13} = 3$$

$$4 = l_{21}u_{13} + l_{22}u_{23}, \qquad u_{23} = (4 - 18)/-7 = 2$$

$$6 = l_{31}u_{13} + l_{32}u_{23} + l_{33}, \quad l_{33} = (6 - 6 - 2) = -2.$$

Thus, we get

$$\begin{pmatrix} 1 & 2 & 3 \\ 6 & 5 & 4 \\ 2 & 5 & 6 \end{pmatrix} = \begin{pmatrix} 1 & 0 & 0 \\ 6 & -7 & 0 \\ 2 & 1 & -2 \end{pmatrix} \begin{pmatrix} 1 & 2 & 3 \\ 0 & 1 & 2 \\ 0 & 0 & 1 \end{pmatrix},$$

the factorization of the given matrix.　　　　　　　　　　　　　　　●

The general formula for getting elements of L and U corresponding to the coefficient matrix A for a set of n linear equations can be written as

$$\left.\begin{array}{rl} l_{ij} &= a_{ij} - \displaystyle\sum_{k=1}^{j-1} l_{ik} u_{kj}, \quad i \geq j, \quad i = 1, 2, \ldots, n \\[3ex] u_{ij} &= \dfrac{1}{l_{ii}}[a_{ij} - \displaystyle\sum_{k=1}^{i-1} l_{ik} u_{kj}], \quad i < j, \quad j = 2, 3, \ldots, n \\[3ex] l_{ij} &= a_{i1}, \quad j = 1 \\[3ex] u_{ij} &= \dfrac{a_{ij}}{a_{11}}, \quad i = 1 \end{array}\right\}. \quad (1.39)$$

Example 1.42 *Solve the following linear system by LU decomposition using Crout's method:*

$$A = \begin{pmatrix} 1 & 2 & 3 \\ 6 & 5 & 4 \\ 2 & 5 & 6 \end{pmatrix} \quad and \quad \mathbf{b} = \begin{pmatrix} 1 \\ -1 \\ 5 \end{pmatrix}.$$

Solution. *The factorization of the coefficient matrix A has already been constructed in Example (1.41) as*

$$\begin{pmatrix} 1 & 2 & 3 \\ 6 & 5 & 4 \\ 2 & 5 & 6 \end{pmatrix} = \begin{pmatrix} 1 & 0 & 0 \\ 6 & -7 & 0 \\ 2 & 1 & -2 \end{pmatrix} \begin{pmatrix} 1 & 2 & 3 \\ 0 & 1 & 2 \\ 0 & 0 & 1 \end{pmatrix}.$$

Then solve the first system $L\mathbf{y} = \mathbf{b}$ for unknown vector $\mathbf{y}$, i.e.,

$$\begin{pmatrix} 1 & 0 & 0 \\ 6 & -7 & 0 \\ 2 & 1 & -2 \end{pmatrix} \begin{pmatrix} y_1 \\ y_2 \\ y_3 \end{pmatrix} = \begin{pmatrix} 1 \\ -1 \\ 5 \end{pmatrix}.$$

Performing forward substitution yields

$$
\begin{array}{rcccccl}
y_1 & & & = & 1 & gives\ y_1 & = & 1 \\
6y_1 & - & 7y_2 & = & -1 & gives\ y_2 & = & 1 \\
2y_1 & + & y_2 - 2y_3 & = & 5 & gives\ y_3 & = & -1.
\end{array}
$$

Then solve the second system $U\mathbf{x} = \mathbf{y}$ *for unknown vector* $\mathbf{x}$, *i.e.,*

$$
\begin{pmatrix} 1 & 2 & 3 \\ 0 & 1 & 2 \\ 0 & 0 & 1 \end{pmatrix}
\begin{pmatrix} x_1 \\ x_2 \\ x_3 \end{pmatrix}
= \begin{pmatrix} 1 \\ 1 \\ -1 \end{pmatrix}.
$$

Performing backward substitution yields

$$
\begin{array}{rcccccl}
x_1 + 2x_2 + 3x_3 & = & 1 & gives & x_1 & = & -2 \\
x_2 + 2x_3 & = & 1 & gives & x_2 & = & 3 \\
x_3 & = & -1 & gives & x_3 & = & -1,
\end{array}
$$

which gives the approximate solution $\mathbf{x}^* = [-2, 3, -1]^T$.   ●

The above results can be reproduced by using MATLAB commands as follows:

```
>> A = [1 2 3; 6 5 4; 2 5 6];
>> b = [1 − 1 5];
>> sol = Crout(A, b);
```

Program 1.12

MATLAB m-file for the Crout's Method
function sol = Crout(A, b)
[n,n]=size(A); u=zeros(n,n); l=u;
for i=1:n; u(i,i)=1; end
l(1,1)=A(1,1);
for i=2:n
u(1,i)=A(1,i)/l(1,1); l(i,1)=A(i,1); end
for i=2:n; for j=2:n; s=0;
if $i <= j$; K=i-1;
else; K=j-1; end
for k=1:K; $s = s + l(i,k) * u(k,j)$; end
if $j > i$; u(i,j)=(A(i,j)-s)/l(i,i); else
l(i,j)=A(i,j)-s; end;end;end
y(1)=b(1)/l(1,1);
for k=2:n; sum=b(k);
for i=1:k-1; $sum = sum - l(k,i) * y(i)$; end
y(k)=sum/l(k,k); end
x(n)=y(n)/u(n,n);
for k=n-1:-1:1; sum=y(k);
for i=k+1:n; $sum = sum - u(k,i) * x(i)$; end
x(k)=sum/u(k,k); end; l; u; y; x;

Procedure 1.6 (LU Decomposition by Crout's Method)

1. *Take the nonsingular matrix A.*

2. *If possible, decompose the matrix $A = LU$ using (1.39).*

3. *Solve linear system $L\mathbf{y} = \mathbf{b}$ using (1.36).*

4. *Solve linear system $U\mathbf{x} = \mathbf{y}$ using (1.37).*

Note that the factorization method is also used to invert matrices. Their usefulness for this purpose is based on the fact that triangular matrices are easily inverted. Once the factorization has been affected, the inverse of a matrix A is found from the formula

$$A^{-1} = (LU)^{-1} = U^{-1}L^{-1}. \tag{1.40}$$

Then

$$U A^{-1} = L^{-1}, \qquad \text{where} \qquad LL^{-1} = I.$$

A practical way of calculating the determinant is to use the forward elimination process of Gaussian elimination or, alternatively, LU decomposition. If no pivoting is used, calculation of the determinant using LU decomposition is very easy, since by one of the properties of the determinant

$$\det(A) = \det(LU) = \det(L)\det(U).$$

So when using LU decomposition by Doolittle's method,

$$\det(A) = \det(U) = \prod_{i=1}^{n} u_{ii} = (u_{11}u_{22}\cdots u_{nn}),$$

where $\det(L) = 1$ because L is a lower-triangular matrix and all its diagonal elements are unity. For LU decomposition by Crout's method,

$$\det(A) = \det(L) = \prod_{i=1}^{n} l_{ii} = (l_{11}l_{22}\cdots l_{nn}),$$

where $\det(U) = 1$ because U is an upper-triangular matrix and all its diagonal elements are unity.

Example 1.43 *Find the determinant and inverse of the following matrix using LU decomposition by Doolittle's method:*

$$A = \begin{pmatrix} 1 & -2 & 1 \\ 1 & -1 & 1 \\ 1 & 1 & 2 \end{pmatrix}.$$

Solution. *We know that*

$$A = \begin{pmatrix} 1 & -2 & 1 \\ 1 & -1 & 1 \\ 1 & 1 & 2 \end{pmatrix} = \begin{pmatrix} 1 & 0 & 0 \\ m_{21} & 1 & 0 \\ m_{31} & m_{32} & 1 \end{pmatrix} \begin{pmatrix} u_{11} & u_{12} & u_{13} \\ 0 & u_{22} & u_{23} \\ 0 & 0 & u_{33} \end{pmatrix} = LU.$$

Now we will use only the forward elimination step of the simple Gaussian elimination method to convert the given matrix A into the upper-triangular matrix U. Since $a_{11} = 1 \neq 0$, we wish to eliminate the elements $a_{21} = 1$ and $a_{31} = 1$ by subtracting from the second and third rows the appropriate multiples of the first row. In this case, the multiples are given as

$$m_{21} = 1, \quad and \quad m_{31} = 1.$$

Hence,

$$\begin{pmatrix} 1 & -2 & 1 \\ 0 & 1 & 0 \\ 0 & 3 & 1 \end{pmatrix}.$$

Since $a_{22}^{(1)} = 1 \neq 0$, we eliminate the entry in the $a_{32}^{(1)} = 3$ position by subtracting the multiple $m_{32} = 3$ of the second row from the third row to get

$$\begin{pmatrix} 1 & -2 & 1 \\ 0 & 1 & 0 \\ 0 & 0 & 1 \end{pmatrix}.$$

Obviously, the original set of equations has been transformed to an upper-triangular form. Thus,

$$\begin{pmatrix} 1 & -2 & 1 \\ 1 & -1 & 1 \\ 1 & 1 & 2 \end{pmatrix} = \begin{pmatrix} 1 & 0 & 0 \\ 1 & 1 & 0 \\ 1 & 3 & 1 \end{pmatrix} \begin{pmatrix} 1 & -2 & 1 \\ 0 & 1 & 0 \\ 0 & 0 & 1 \end{pmatrix},$$

which is the required decomposition of A.
Now we find the determinant of matrix A as

$$\det(A) = \det(U) = u_{11}u_{22}u_{33} = (1)(1)(1) = 1.$$

To find the inverse of matrix A, first we will compute the inverse of the lower-triangular matrix L^{-1} from

$$LL^{-1} = \begin{pmatrix} 1 & 0 & 0 \\ 1 & 1 & 0 \\ 1 & 3 & 1 \end{pmatrix} \begin{pmatrix} l'_{11} & 0 & 0 \\ l'_{21} & l'_{22} & 0 \\ l'_{31} & l'_{32} & l'_{33} \end{pmatrix} = \begin{pmatrix} 1 & 0 & 0 \\ 0 & 1 & 0 \\ 0 & 0 & 1 \end{pmatrix} = I$$

by using forward substitution.
To solve the first system

$$
\begin{pmatrix} 1 & 0 & 0 \\ 1 & 1 & 0 \\ 1 & 3 & 1 \end{pmatrix} \begin{pmatrix} l'_{11} \\ l'_{21} \\ l'_{31} \end{pmatrix} = \begin{pmatrix} 1 \\ 0 \\ 0 \end{pmatrix},
$$

by using forward substitution, we get

$$
l'_{11} = 1, \quad l'_{21} = -1, \quad l'_{31} = 2.
$$

Similarly, the solution of the second linear system

$$
\begin{pmatrix} 1 & 0 & 0 \\ 1 & 1 & 0 \\ 1 & 3 & 1 \end{pmatrix} \begin{pmatrix} 0 \\ l'_{22} \\ l'_{32} \end{pmatrix} = \begin{pmatrix} 0 \\ 1 \\ 0 \end{pmatrix}
$$

can be obtained

$$
l'_{22} = 1, \quad l'_{32} = -3.
$$

Finally, the solution of the third linear system

$$
\begin{pmatrix} 1 & 0 & 0 \\ 1 & 1 & 0 \\ 1 & 3 & 1 \end{pmatrix} \begin{pmatrix} 0 \\ 0 \\ l'_{33} \end{pmatrix} = \begin{pmatrix} 0 \\ 0 \\ 1 \end{pmatrix}
$$

gives $l'_{33} = 1$.
Hence, the elements of the matrix L^{-1} are

$$
L^{-1} = \begin{pmatrix} 1 & 0 & 0 \\ -1 & 1 & 0 \\ 2 & -3 & 1 \end{pmatrix},
$$

which is the required inverse of the lower-triangular matrix L.
To find the inverse of the given matrix A, we will solve the system

$$
UA^{-1} = \begin{pmatrix} 1 & -2 & 1 \\ 0 & 1 & 0 \\ 0 & 0 & 1 \end{pmatrix} \begin{pmatrix} a'_{11} & a'_{12} & a'_{13} \\ a'_{21} & a'_{22} & a'_{23} \\ a'_{31} & a'_{32} & a'_{33} \end{pmatrix} = \begin{pmatrix} 1 & 0 & 0 \\ -1 & 1 & 0 \\ 2 & -3 & 1 \end{pmatrix} = L^{-1}
$$

by using backward substitution.
We solve the first system

$$
\begin{pmatrix} 1 & -2 & 1 \\ 0 & 1 & 0 \\ 0 & 0 & 1 \end{pmatrix} \begin{pmatrix} a'_{11} \\ a'_{21} \\ a'_{31} \end{pmatrix} = \begin{pmatrix} 1 \\ -1 \\ 2 \end{pmatrix}
$$

by using backward substitution, and we get

$$
a'_{11} = -3, \quad a'_{21} = -1, \quad a'_{31} = 2.
$$

Similarly, the solution of the second linear system

$$
\begin{pmatrix} 1 & -2 & 1 \\ 0 & 1 & 0 \\ 0 & 0 & 1 \end{pmatrix} \begin{pmatrix} a'_{12} \\ a'_{22} \\ a'_{32} \end{pmatrix} = \begin{pmatrix} 0 \\ 1 \\ -3 \end{pmatrix}
$$

can be obtained as follows:

$$
a'_{12} = 5, \quad a'_{22} = 1, \quad a'_{32} = -3.
$$

Finally, the solution of the third linear system

$$
\begin{pmatrix} 1 & -2 & 1 \\ 0 & 1 & 0 \\ 0 & 0 & 1 \end{pmatrix} \begin{pmatrix} a'_{13} \\ a'_{23} \\ a'_{33} \end{pmatrix} = \begin{pmatrix} 0 \\ 0 \\ 1 \end{pmatrix}
$$

can be obtained as follows:

$$
a'_{13} = -1, \quad a'_{23} = 0, \quad a'_{33} = 1.
$$

Hence, the elements of the inverse matrix A^{-1} are

$$
A^{-1} = \begin{pmatrix} -3 & 5 & -1 \\ -1 & 1 & 0 \\ 2 & -3 & 1 \end{pmatrix},
$$

which is the required inverse of the given matrix A. ●

For *LU* decomposition we have not used pivoting for the sake of simplicity. However, pivoting is important for the same reason as in Gaussian elimination. We know that pivoting in Gaussian elimination is equivalent to interchanging the rows of the coefficients matrix together with the terms on the right-hand side. This indicates that pivoting may be applied to *LU* decomposition as long as the interchanging is applied to the left and right terms in the same way. When performing pivoting in *LU* decomposition, the changes in the order of the rows are recorded. The same reordering is then applied to the right-hand side terms before starting the solution in accordance with the forward elimination and backward substitution steps.

Indirect LU Decomposition

It is to be noted that a nonsingular matrix A sometimes cannot be directly factored as $A = LU$. For example, the matrix in Example 1.24 is nonsingular, but it cannot be factored into the product LU. Let us assume it has a *LU* form and

$$\begin{pmatrix} 2 & 2 & -4 \\ 2 & 2 & -1 \\ 3 & 2 & -3 \end{pmatrix} = \begin{pmatrix} u_{11} & u_{12} & u_{13} \\ l_{21}u_{11} & l_{21}u_{12} + u_{22} & l_{21}u_{13} + u_{23} \\ l_{31}u_{11} & l_{31}u_{12} + l_{32}u_{22} & l_{31}u_{13} + l_{32}u_{23} + u_{33} \end{pmatrix}.$$

Then equate elements of the first column to obtain

$$\begin{aligned} 2 &= u_{11} & \text{gives} && u_{11} &= 2 \\ 2 &= l_{21}u_{11} & \text{gives} && l_{21} &= 1 \\ 3 &= l_{31}u_{11} & \text{gives} && l_{31} &= \tfrac{3}{2}. \end{aligned}$$

Then equate elements of the second column to obtain

$$\begin{aligned} 2 &= u_{12} & \text{gives} && u_{12} &= 2 \\ 2 &= l_{21}u_{12} + u_{22} & \text{gives} && u_{22} &= 0 \\ 2 &= l_{31}u_{12} + l_{32}u_{22} & \text{gives} && 0 &= -1, \end{aligned}$$

which is not possible because $0 \neq -1$, a contradiction. Hence, the matrix A cannot be directly factored into the product of L and U. The indirect factorization LU of A can be obtained by using the permutation matrix P

and replacing the matrix A by PA. For example, using the above matrix A, we have

$$PA = \begin{pmatrix} 1 & 0 & 0 \\ 0 & 0 & 1 \\ 0 & 1 & 0 \end{pmatrix} \begin{pmatrix} 2 & 2 & -4 \\ 2 & 2 & -1 \\ 3 & 2 & -3 \end{pmatrix} = \begin{pmatrix} 2 & 2 & -4 \\ 3 & 2 & -3 \\ 2 & 2 & -1 \end{pmatrix}.$$

From this multiplication we see that rows 2 and 3 of the original matrix A are interchanged, and the resulting matrix PA has a LU factorization and we have

$$\begin{pmatrix} 1 & 0 & 0 \\ 1.5 & 1 & 0 \\ 1 & 0 & 1 \end{pmatrix} \begin{pmatrix} 2 & 2 & -4 \\ 0 & -1 & 3 \\ 0 & 0 & 3 \end{pmatrix} = \begin{pmatrix} 2 & 2 & -4 \\ 3 & 2 & -3 \\ 2 & 2 & -1 \end{pmatrix}.$$

The following theorem is an extension of Theorem 1.21, which includes the case when interchanged rows are required. Thus, LU factorization can be used to find the solution to any linear system $A\mathbf{x} = \mathbf{b}$ with a nonsingular matrix A.

Theorem 1.23 *Let A be a square $n \times n$ matrix and assume that Gaussian elimination can be performed successfully to solve the linear system $A\mathbf{x} = \mathbf{b}$, but that row interchanges are required. Then there exists a permutation matrix $P = p_k, \ldots, p_2, p_1$ (where $p_1, p_2, \ldots, p_k$ are the elementary matrices corresponding to the row interchanges used) so that the PA matrix has a LU factorization, i.e.,*

$$PA = LU, \tag{1.41}$$

where PA is the matrix obtained from A by doing these interchanges to A. Note that $P = \mathbf{I_n}$ if no interchanges are used. •

When *pivoting* is used in LU decomposition, its effects should be taken into consideration. First, we recognize that LU decomposition with pivoting is equivalent to performing two separate process:

1. Transform A to A' by performing all shifting of rows.

2. Then decompose A' to LU with no pivoting.

The former step can be expressed by

$$A' = PA, \quad \text{equivalently} \quad A = P^{-1}A',$$

where P is called a permutation matrix and represents the pivoting operation. The second process is

$$A' = PA = LU$$

and so

$$A = P^{-1}LU = (P^T L)U$$

since $P^{-1} = P^T$. The determinant of A may now be written as

$$\det(A) = \det(P^{-1})\det(L)\det(U)$$

or

$$\det(A) = \beta\det(L)\det(U),$$

where $\beta = \det(P^{-1})$ equals -1 or $+1$ depending on whether the number pivoting is odd or even, respectively. •

One can use the MATLAB built-in **lu** function to obtain the permutation matrix P so that the PA matrix has a LU decomposition:

```
>> A = [0 1 2; −1 4 2; 2 2 1];
>> [L, U, P] = lu(A);
```

It will give us the permutation matrix P and the matrices L and U as follows:

$$P = \begin{pmatrix} 0 & 0 & 1 \\ 0 & 1 & 0 \\ 1 & 0 & 0 \end{pmatrix}$$

and

$$PA = \begin{pmatrix} 1 & 0 & 0 \\ -0.5 & 1 & 0 \\ 0 & 0.2 & 1 \end{pmatrix} \begin{pmatrix} 2 & 2 & 1 \\ 0 & 5 & 2.5 \\ 0 & 0 & 1.5 \end{pmatrix} = LU.$$

So

$$A = P^{-1}LU$$

or

$$A = (P^T L)U = \begin{pmatrix} 0 & 0.2 & 1 \\ -0.5 & 1 & 0 \\ 1 & 0 & 0 \end{pmatrix} \begin{pmatrix} 2 & 2 & 1 \\ 0 & 5 & 2.5 \\ 0 & 0 & 1.5 \end{pmatrix}.$$

Example 1.44 *Consider the following matrix:*

$$A = \begin{pmatrix} 0 & 3 & 2 \\ 3 & 2 & 5 \\ 6 & 2 & 4 \end{pmatrix},$$

then:

1. *Show that A does not have LU factorization;*

2. *Use Gauss elimination by partial pivoting and find the permutation matrix P as well as the LU factors such that $PA = LU$;*

3. *Use the information in $P, L,$ and U to solve the system $A\mathbf{x} = [6, 4, 3]^T$.*

Solution. *(1) I using simple Gauss elimination, since $a_{11} = 0$, from Theorem 1.21, the LU decomposition of A is not possible.*
(2) For applying Gauss elimination by partial pivoting, the interchanges of the rows between row 1 and row 3 gives

$$\begin{pmatrix} 6 & 2 & 4 \\ 3 & 2 & 5 \\ 0 & 3 & 3 \end{pmatrix},$$

and then using multiple $m_{21} = \frac{3}{6} = \frac{1}{2}$, we obtain

$$\begin{pmatrix} 6 & 2 & 4 \\ 0 & 1 & 3 \\ 0 & 3 & 3 \end{pmatrix}.$$

Now interchanging row 2 and row 3 gives

$$\begin{pmatrix} 6 & 2 & 4 \\ 0 & 3 & 3 \\ 0 & 1 & 3 \end{pmatrix}.$$

By using multiple $m_{32} = \frac{1}{3}$, we get

$$\begin{pmatrix} 6 & 2 & 4 \\ 0 & 3 & 2 \\ 0 & 0 & 2 \end{pmatrix}.$$

Note that during this elimination process two row interchanges were needed, which means we got two elementary permutation matrices of the interchanges (from Theorem 1.23), which are

$$p_1 = \begin{pmatrix} 0 & 0 & 1 \\ 0 & 1 & 0 \\ 1 & 0 & 0 \end{pmatrix} \quad and \quad p_2 = \begin{pmatrix} 1 & 0 & 0 \\ 0 & 0 & 1 \\ 0 & 1 & 0 \end{pmatrix}.$$

Thus, the permutation matrix is

$$P = p_2 p_1 = \begin{pmatrix} 1 & 0 & 0 \\ 0 & 0 & 1 \\ 0 & 1 & 0 \end{pmatrix} \begin{pmatrix} 0 & 0 & 1 \\ 0 & 1 & 0 \\ 1 & 0 & 0 \end{pmatrix} = \begin{pmatrix} 0 & 1 & 0 \\ 0 & 0 & 1 \\ 1 & 0 & 0 \end{pmatrix}.$$

If we do these interchanges to the given matrix A, the result is the matrix PA, i.e.,

$$PA = \begin{pmatrix} 0 & 1 & 0 \\ 0 & 0 & 1 \\ 1 & 0 & 0 \end{pmatrix} \begin{pmatrix} 0 & 3 & 2 \\ 3 & 2 & 5 \\ 6 & 2 & 4 \end{pmatrix} = \begin{pmatrix} 3 & 2 & 5 \\ 6 & 2 & 4 \\ 0 & 3 & 3 \end{pmatrix}.$$

Now apply LU decomposition to the matrix PA, and we will convert it to the upper-triangular matrix U by using the possible multiples

$$m_{21} = 2, \quad m_{31} = 0, \quad m_{32} = -\frac{3}{2}$$

as follows:

$$\begin{pmatrix} 3 & 2 & 5 \\ 6 & 2 & 4 \\ 0 & 3 & 3 \end{pmatrix} \rightarrow \begin{pmatrix} 3 & 2 & 5 \\ 6 & -2 & -6 \\ 0 & 3 & 3 \end{pmatrix} \rightarrow \begin{pmatrix} 3 & 2 & 5 \\ 0 & -2 & -6 \\ 0 & 0 & -6 \end{pmatrix} = U.$$

Thus, $PA = LU$, where

$$L = \begin{pmatrix} 1 & 0 & 0 \\ 2 & 1 & 0 \\ 0 & -3/2 & 1 \end{pmatrix} \quad and \quad U = \begin{pmatrix} 3 & 2 & 5 \\ 0 & -2 & -6 \\ 0 & 0 & -6 \end{pmatrix}.$$

(3) Solve the first system $L\mathbf{y} = P\mathbf{b} = [4, 3, 6]^T$ for unknown vector $\mathbf{y}$, i.e.,

$$\begin{pmatrix} 1 & 0 & 0 \\ 2 & 1 & 0 \\ 0 & -3/2 & 1 \end{pmatrix} \begin{pmatrix} y_1 \\ y_2 \\ y_3 \end{pmatrix} = \begin{pmatrix} 4 \\ 3 \\ 6 \end{pmatrix}.$$

Performing forward substitution yields

$$
\begin{array}{rcccl}
y_1 & & = & 4 & \textit{gives } y_1 = \quad 4 \\
2y_1 + & y_2 & = & 3 & \textit{gives } y_2 = \quad -5 \\
- & 3/2 y_2 + y_3 & = & 6 & \textit{gives } y_3 = -1.5.
\end{array}
$$

Then solve the second system $U\mathbf{x} = \mathbf{y}$ for the unknown vector $\mathbf{x}$, i.e.,

$$\begin{pmatrix} 3 & 2 & 5 \\ 0 & -2 & -6 \\ 0 & 0 & -6 \end{pmatrix} \begin{pmatrix} x_1 \\ x_2 \\ x_3 \end{pmatrix} = \begin{pmatrix} 4 \\ -5 \\ -1.5 \end{pmatrix}.$$

Performing backward substitution yields

$$
\begin{array}{rcccl}
3x_1 + 2x_2 + 5x_3 & = & 4 & \textit{gives} & x_1 = -0.25 \\
- 2x_2 - 6x_3 & = & -5 & \textit{gives} & x_2 = 1.75 \\
- 6x_3 & = & -1.5 & \textit{gives} & x_3 = 0.25,
\end{array}
$$

which gives the approximate solution $\mathbf{x}^* = [-0.25, 1.75, 0.25]^T$. ●

The major advantage of the *LU* decomposition methods is the efficiency when multiple unknown **b** vectors must be considered. The number of multiplications and divisions required by the complete Gaussian elimination method is $N = \left(\frac{n^3}{3}\right) + n^2 - \left(\frac{n}{3}\right)$. The forward substitution step required to solve the system $L\mathbf{y} = \mathbf{b}$ requires $N = n^2 - \left(\frac{n}{2}\right)$ operations, and the backward substitution step required to solve the system $U\mathbf{x} = \mathbf{y}$ requires $N = n^2 + \left(\frac{n}{2}\right)$ operations. Thus, the total number of multiplications and divisions required by LU decomposition, after L and U matrices have been determined, is $N = 2n^2$, which is much less work than required by the Gaussian elimination method, especially for large systems. ●

In the analysis of many physical systems, sets of linear equations arise that have coefficient matrices that are both symmetric and positive-definite. Now we factorize such a matrix A into the product of lower-triangular and upper-triangular matrices which have these two properties. Before we do the factorization, we define the following matrix.

Definition 1.31 (Positive-Definite Matrix)

The function

$$\mathbf{x}^T A \mathbf{x} = \begin{pmatrix} x_1 & x_2 & \cdots & x_n \end{pmatrix} \begin{pmatrix} a_{11} & a_{12} & \cdots & a_{1n} \\ a_{21} & a_{22} & \cdots & a_{2n} \\ a_{31} & a_{32} & \cdots & a_{3n} \\ \vdots & \vdots & \vdots & \vdots \\ a_{n1} & a_{n2} & \cdots & a_{nn} \end{pmatrix} \begin{pmatrix} x_1 \\ x_2 \\ \vdots \\ x_n \end{pmatrix}$$

or

$$\mathbf{x}^T A \mathbf{x} = \sum_{i=1}^{n} \sum_{j=1}^{n} a_{ij} x_i x_j \tag{1.42}$$

can be used to represent any quadratic polynomial in the variables $x_1, x_2, \ldots, x_n$ and is called a quadratic form. A matrix is said to be positive-definite if its quadratic form is positive for all real nonzero vectors $\mathbf{x}$, i.e.,

$$\mathbf{x}^T A \mathbf{x} > 0, \quad \text{for every n-dimensional column vector} \quad \mathbf{x} \neq 0.$$

Example 1.45 *The matrix*

$$A = \begin{pmatrix} 4 & -1 & 0 \\ -1 & 4 & -1 \\ 0 & -1 & 4 \end{pmatrix}$$

is positive-definite and suppose $\mathbf{x}$ is any nonzero three-dimensional column vector, then

$$\mathbf{x}^T A \mathbf{x} = \begin{pmatrix} x_1 & x_2 & x_3 \end{pmatrix} \begin{pmatrix} 4 & -1 & 0 \\ -1 & 4 & -1 \\ 0 & -1 & 4 \end{pmatrix} \begin{pmatrix} x_1 \\ x_2 \\ x_3 \end{pmatrix}$$

or

$$= \left(\begin{array}{ccc} x_1 & x_2 & x_3 \end{array} \right) \left(\begin{array}{ccc} 4x_1 & - & x_2 \\ -x_1 & + \ 4x_2 & - & x_3 \\ & - & x_2 & + \ 4x_3 \end{array} \right).$$

Thus,

$$\mathbf{x}^T A \mathbf{x} = 4x_1^2 - 2x_1 x_2 + 4x_2^2 - 2x_2 x_3 + 4x_3^2.$$

After rearranging the terms, we have

$$\mathbf{x}^T A \mathbf{x} = 3x_1^2 + (x_1 - x_2)^2 + 2x_2^2 + (x_2 - x_3)^2 + 3x_3^2.$$

Hence,

$$3x_1^2 + (x_1 - x_2)^2 + 2x_2^2 + (x_2 - x_3)^2 + 3x_3^2 > 0,$$

unless $x_1 = x_2 = x_3 = 0.$ ●

Symmetric positive-definite matrices occur frequently in equations derived by minimization or energy principles, and their properties can often be utilized in numerical processes.

Theorem 1.24 *If A is a positive-definite matrix, then:*

1. A is nonsingular.

2. $a_{ii} > 0,$ *for each* $i = 1, 2, \ldots, n.$

Theorem 1.25 *The symmetric matrix A is a positive-definite matrix, if and only if Gaussian elimination without row interchange can be performed on the linear system* $A\mathbf{x} = \mathbf{b}$, *with all pivot elements positive.*   ●

Theorem 1.26 *A matrix A is positive-definite if the determinant of the principal minors of A are positive.*

The principal minors of a matrix A are the square submatrices lying in the upper-left hand corner of A. An $n \times n$ matrix A has n of these principal minors. For example, for the matrix

$$A = \left(\begin{array}{ccc} 6 & 2 & 1 \\ 2 & 3 & 2 \\ 1 & 1 & 2 \end{array} \right),$$

the determinant of its principal minors are

$$\det(6) \ = \ 6 > 0,$$

$$\det \begin{pmatrix} 6 & 2 \\ 2 & 3 \end{pmatrix} \ = \ 18 - 4 = 14 > 0,$$

$$\det \begin{pmatrix} 6 & 2 & 1 \\ 2 & 3 & 2 \\ 1 & 1 & 2 \end{pmatrix} \ = \ 19 > 0.$$

Thus, the matrix A is positive-definite.   •

Theorem 1.27 *If a symmetric matrix A is diagonally dominant, then it must be positive-definite.*   •

For example, for the diagonally dominant matrix

$$A = \begin{pmatrix} 4 & -1 & 2 \\ -1 & 3 & 0 \\ 2 & 0 & 5 \end{pmatrix},$$

the determinant of its principal minors are

$$\det(4) \ = \ 4 > 0,$$

$$\det \begin{pmatrix} 4 & -1 \\ -1 & 3 \end{pmatrix} \ = \ 12 - 1 = 11 > 0,$$

$$\det \begin{pmatrix} 4 & -1 & 2 \\ -1 & 3 & 0 \\ 2 & 0 & 5 \end{pmatrix} \ = \ 43 > 0.$$

Hence, (using Theorem 1.26) matrix A is positive-definite.   •

Theorem 1.28 *If a matrix A is nonsingular, then $A^T A$ is always positive-definite.*   •

For example, for the matrix

$$A = \begin{pmatrix} 1 & 1 & 1 \\ 1 & 1 & 2 \\ 1 & 2 & 1 \end{pmatrix},$$

we can have

$$A^T A = \begin{pmatrix} 3 & 4 & 4 \\ 4 & 6 & 5 \\ 4 & 5 & 6 \end{pmatrix}.$$

Then the determinant of its principal minors are

$$\det(3) = 3 > 0,$$

$$\det \begin{pmatrix} 3 & 4 \\ 4 & 6 \end{pmatrix} = 18 - 16 = 2 > 0,$$

$$\det \begin{pmatrix} 3 & 4 & 4 \\ 4 & 6 & 5 \\ 4 & 5 & 6 \end{pmatrix} = 1 > 0.$$

Thus, matrix A is positive-definite. •

Cholesky Method

The Cholesky method (or square root method) is of the same form as Doolittle's method and Crout's method except it is limited to equations involving symmetrical coefficient matrices. In the case of a symmetric and positive-definite matrix A it is possible to construct an alternative triangular factorization with a saved number of calculations compared with previous factorizations. Here, we decompose the matrix A into the product of LL^T, i.e.,

$$A = LL^T, \tag{1.43}$$

where L is the lower-triangular matrix and L^T is its transpose. The elements of L are computed by equating successive columns in the relation

$$
\begin{pmatrix}
a_{11} & a_{12} & \cdots & a_{1n} \\
a_{21} & a_{22} & \cdots & a_{2n} \\
\vdots & \vdots & \vdots & \vdots \\
a_{n1} & a_{n2} & \cdots & a_{nn}
\end{pmatrix}
=
\begin{pmatrix}
l_{11} & 0 & \cdots & 0 \\
l_{21} & l_{22} & \cdots & 0 \\
\vdots & \vdots & \vdots & \vdots \\
l_{n1} & l_{n2} & \cdots & l_{nn}
\end{pmatrix}
\begin{pmatrix}
l_{11} & l_{21} & \cdots & l_{n1} \\
0 & l_{22} & \cdots & l_{n2} \\
\vdots & \vdots & \vdots & \vdots \\
0 & 0 & \cdots & l_{nn}
\end{pmatrix}.
$$

After constructing the matrices L and L^T, the solution of the system $A\mathbf{x} = \mathbf{b}$ can be computed in the following two steps:

1. Solve $\quad L\mathbf{y} = \mathbf{b}, \quad$ for $\quad \mathbf{y}$.
   (using forward substitution)

2. Solve $\quad L^T\mathbf{x} = \mathbf{y}, \quad$ for $\quad \mathbf{x}$.
   (using backward substitution)

In this procedure, it is necessary to take the square root of the elements on the main diagonal of the coefficient matrix. However, for a positive-definite matrix the terms on its main diagonal are positive, so no difficulty will arise when taking the square root of these terms.

Example 1.46 *Construct the LU decomposition of the following matrix using the Cholesky method:*

$$
A = \begin{pmatrix} 1 & 1 & 2 \\ 1 & 2 & 4 \\ 2 & 4 & 9 \end{pmatrix}.
$$

Solution. *Since*

$$
A = LL^T = \begin{pmatrix} l_{11} & 0 & 0 \\ l_{21} & l_{22} & 0 \\ l_{31} & l_{32} & l_{33} \end{pmatrix}\begin{pmatrix} l_{11} & l_{21} & l_{31} \\ 0 & l_{22} & l_{32} \\ 0 & 0 & l_{33} \end{pmatrix},
$$

performing the multiplication on the right-hand side gives

$$
\begin{pmatrix} 1 & 1 & 2 \\ 1 & 2 & 4 \\ 2 & 4 & 9 \end{pmatrix} = \begin{pmatrix} l_{11}^2 & l_{11}l_{21} & l_{11}l_{31} \\ l_{11}l_{21} & l_{21}^2 + l_{22}^2 & l_{21}l_{31} + l_{22}l_{32} \\ l_{11}l_{31} & l_{31}l_{21} + l_{22}l_{32} & l_{31}^2 + l_{32}^2 + l_{33}^2 \end{pmatrix}.
$$

Then equate elements of the first column to obtain

$$1 = l_{11}^2 \qquad gives \qquad l_{11} = \sqrt{1} = 1$$

$$1 = l_{11}l_{21} \qquad gives \qquad l_{21} = 1$$

$$2 = l_{11}l_{31} \qquad gives \qquad l_{31} = 2.$$

Note that l_{11} could be $-\sqrt{1}$ and so the matrix L is not (quite) unique.
Now equate elements of the second column to obtain

$$2 = l_{21}^2 + l_{22}^2 \qquad gives \qquad l_{22} = 1$$
$$4 = l_{31}l_{21} + l_{32}l_{22} \quad gives \qquad l_{32} = 2.$$

Finally, equate elements of the third column to obtain

$$9 = l_{31}^2 + l_{32}^2 + l_{33}^2 \qquad gives \quad l_{33} = 1.$$

Thus, we obtain

$$\begin{pmatrix} 1 & 1 & 2 \\ 1 & 2 & 4 \\ 2 & 4 & 9 \end{pmatrix} = \begin{pmatrix} 1 & 0 & 0 \\ 1 & 1 & 0 \\ 2 & 2 & 1 \end{pmatrix} \begin{pmatrix} 1 & 1 & 2 \\ 0 & 1 & 2 \\ 0 & 0 & 1 \end{pmatrix},$$

the factorization of the given matrix.   ●

For a general $n \times n$ matrix, the elements of the lower-triangular matrix L are constructed from

$$\left.\begin{array}{rcl} l_{11} & = & \sqrt{a_{11}} \\[2mm] l_{j1} & = & \dfrac{a_{j1}}{l_{11}}, \qquad j > 1 \\[3mm] l_{ii} & = & \sqrt{\left(a_{ii} - \displaystyle\sum_{k=1}^{i-1} l_{ik}^2 \right)}, \qquad 1 < i < n \\[4mm] l_{ji} & = & \dfrac{1}{l_{ii}} \left[a_{ji} - \displaystyle\sum_{k=1}^{i-1} l_{jk} l_{ik} \right], \qquad j > i > 1 \\[4mm] l_{nn} & = & \sqrt{\left(a_{nn} - \displaystyle\sum_{k=1}^{n-1} l_{nk}^2 \right)} \end{array}\right\}. \qquad (1.44)$$

The method fails if $l_{jj} = 0$ and the expression inside the square root is negative, in which case all of the elements in column j are purely imaginary. There is, however, a special class of matrices for which these problems don't occur.

The Cholesky method provides a convenient method for investigating the positive-definiteness of symmetric matrices. The formal definition $\mathbf{x}^T A \mathbf{x} > 0$, for all $\mathbf{x} \neq 0$, is not easy to verify in practice. However, it is relatively straightforward to attempt the construct of a Cholesky decomposition of a symmetric matrix.

Theorem 1.29 *A matrix A is positive-definite, if and only if A can be factored in the form $A = LL^T$, where L is a lower-triangular matrix with nonzero diagonal entries.* •

Example 1.47 *Show that the following matrix is positive-definite by using the Cholesky method:*

$$A = \begin{pmatrix} 9 & 3 & 6 \\ 3 & 10 & 8 \\ 6 & 8 & 9 \end{pmatrix}.$$

Solution. *Since*

$$A = LL^T = \begin{pmatrix} l_{11} & 0 & 0 \\ l_{21} & l_{22} & 0 \\ l_{31} & l_{32} & l_{33} \end{pmatrix} \begin{pmatrix} l_{11} & l_{21} & l_{31} \\ 0 & l_{22} & l_{32} \\ 0 & 0 & l_{33} \end{pmatrix},$$

performing the multiplication on the right-hand side gives

$$\begin{pmatrix} 9 & 3 & 6 \\ 3 & 10 & 8 \\ 6 & 8 & 9 \end{pmatrix} = \begin{pmatrix} l_{11}^2 & l_{11}l_{21} & l_{11}l_{31} \\ l_{11}l_{21} & l_{21}^2 + l_{22}^2 & l_{21}l_{31} + l_{22}l_{32} \\ l_{11}l_{31} & l_{31}l_{21} + l_{22}l_{32} & l_{31}^2 + l_{32}^2 + l_{33}^2 \end{pmatrix}.$$

Then equate elements of the first column to obtain

$$9 = l_{11}^2 \quad gives \quad l_{11} = \sqrt{9} = 3$$

$$3 = l_{11}l_{21} \quad gives \quad l_{21} = 1$$

$$6 = l_{11}l_{31} \quad gives \quad l_{31} = 2.$$

Now equate elements of the second column to obtain

$$10 = l_{21}^2 + l_{22}^2 \qquad gives \qquad l_{22} = 3$$
$$8 = l_{31}l_{21} + l_{32}l_{22} \quad gives \qquad l_{32} = 2.$$

Finally, equate elements of the third column to obtain

$$9 = l_{31}^2 + l_{32}^2 + l_{33}^2 \qquad gives \quad l_{33} = 1.$$

Thus, the factorization obtained as

$$A = \begin{pmatrix} 9 & 3 & 6 \\ 3 & 10 & 8 \\ 6 & 8 & 9 \end{pmatrix} = \begin{pmatrix} 3 & 0 & 0 \\ 1 & 3 & 0 \\ 2 & 2 & 1 \end{pmatrix} \begin{pmatrix} 3 & 1 & 2 \\ 0 & 3 & 2 \\ 0 & 0 & 1 \end{pmatrix} = LL^T,$$

and it shows that the given matrix is positive-definite. ●

If the symmetric coefficient matrix is not positive-definite, then the terms on the main diagonal can be zero or negative. For example, the symmetric coefficient matrix

$$A = \begin{pmatrix} 1 & 1 & 2 \\ 1 & 2 & 4 \\ 2 & 4 & 8 \end{pmatrix}$$

is not positive-definite because the Cholesky decomposition of the matrix has the form

$$
A = \begin{pmatrix} 1 & 1 & 2 \\ 1 & 2 & 4 \\ 2 & 4 & 8 \end{pmatrix} = \begin{pmatrix} 1 & 0 & 0 \\ 1 & 1 & 0 \\ 2 & 2 & 0 \end{pmatrix} \begin{pmatrix} 1 & 1 & 2 \\ 0 & 1 & 2 \\ 0 & 0 & 0 \end{pmatrix} = LL^T,
$$

which shows that one of the diagonal elements of L and L^T is zero.   ●

Example 1.48 *Solve the following linear system by LU decomposition using the Cholesky method:*

$$
A = \begin{pmatrix} 1 & 1 & 2 \\ 1 & 2 & 4 \\ 2 & 4 & 9 \end{pmatrix} \quad and \quad \mathbf{b} = \begin{pmatrix} 2 \\ 1 \\ 1 \end{pmatrix}.
$$

Solution. *The factorization of the coefficient matrix A has already been constructed in Example (1.46) as*

$$
\begin{pmatrix} 1 & 1 & 2 \\ 1 & 2 & 4 \\ 2 & 4 & 9 \end{pmatrix} = \begin{pmatrix} 1 & 0 & 0 \\ 1 & 1 & 0 \\ 2 & 2 & 1 \end{pmatrix} \begin{pmatrix} 1 & 1 & 2 \\ 0 & 1 & 2 \\ 0 & 0 & 1 \end{pmatrix}.
$$

Then solve the first system $L\mathbf{y} = \mathbf{b}$ for unknown vector $\mathbf{y}$, i.e.,

$$
\begin{pmatrix} 1 & 0 & 0 \\ 1 & 1 & 0 \\ 2 & 2 & 1 \end{pmatrix} \begin{pmatrix} y_1 \\ y_2 \\ y_3 \end{pmatrix} = \begin{pmatrix} 2 \\ 1 \\ 1 \end{pmatrix}.
$$

Performing forward substitution yields

$$
\begin{array}{rcrcrclcrcr}
y_1 & & & & & = & 2, & y_1 & = & 2 \\
y_1 & + & y_2 & & & = & 1, & y_2 & = & -1 \\
2y_1 & + & 2y_2 & + & y_3 & = & 1, & y_3 & = & -1.
\end{array}
$$

Then solve the second system $L^T\mathbf{x} = \mathbf{y}$ for unknown vector $\mathbf{x}$, i.e.,

$$
\begin{pmatrix} 1 & 1 & 2 \\ 0 & 1 & 2 \\ 0 & 0 & 1 \end{pmatrix} \begin{pmatrix} x_1 \\ x_2 \\ x_3 \end{pmatrix} = \begin{pmatrix} 2 \\ -1 \\ -1 \end{pmatrix}.
$$

Performing backward substitution yields

$$
\begin{array}{rrrrrlll}
x_1 & + & x_2 & + & 2x_3 & = & 2 & \textit{gives} & x_1 = & 3 \\
& & x_2 & + & 2x_3 & = & -1 & \textit{gives} & x_2 = & 1 \\
& & & & x_3 & = & -1 & \textit{gives} & x_3 = & -1,
\end{array}
$$

which gives the approximate solution $\mathbf{x}^* = [3, 1, -1]^T.$ ●

Now use the following MATLAB commands to obtain the above results:

```
>> A = [1 1 2; 1 2 4; 2 4 9];
>> b = [2 1 1];
>> sol = Cholesky(A, b);
```

Procedure 1.7 (LU Decomposition by the Cholesky Method)

1. *Take the positive-definite matrix A.*

2. *If possible, decompose the matrix $A = LL^T$ using (1.44).*

3. *Solve linear system $L\mathbf{y} = \mathbf{b}$ using (1.36).*

4. *Solve linear system $L^T\mathbf{x} = \mathbf{y}$ using (1.37).*

Example 1.49 *Find the bounds on α for which the Cholesky factorization of the following matrix with real elements*

$$
A = \begin{pmatrix} 1 & 2 & \alpha \\ 2 & 8 & 2\alpha \\ \alpha & 2\alpha & 9 \end{pmatrix}
$$

is possible.

Solution. *Since*

$$
A = LL^T = \begin{pmatrix} l_{11} & 0 & 0 \\ l_{21} & l_{22} & 0 \\ l_{31} & l_{32} & l_{33} \end{pmatrix} \begin{pmatrix} l_{11} & l_{21} & l_{31} \\ 0 & l_{22} & l_{32} \\ 0 & 0 & l_{33} \end{pmatrix},
$$

performing the multiplication on the right-hand side gives

$$\begin{pmatrix} 1 & 2 & \alpha \\ 2 & 8 & 2\alpha \\ \alpha & 2\alpha & 9 \end{pmatrix} = \begin{pmatrix} l_{11}^2 & l_{11}l_{21} & l_{11}l_{31} \\ l_{11}l_{21} & l_{21}^2 + l_{22}^2 & l_{21}l_{31} + l_{22}l_{32} \\ l_{11}l_{31} & l_{31}l_{21} + l_{22}l_{32} & l_{31}^2 + l_{32}^2 + l_{33}^2 \end{pmatrix}.$$

Then equate elements of the first column to obtain

$$1 = l_{11}^2 \qquad gives \qquad l_{11} = \sqrt{1} = 1$$

$$2 = l_{11}l_{21} \qquad gives \qquad l_{21} = 2$$

$$\alpha = l_{11}l_{31} \qquad gives \qquad l_{31} = \alpha.$$

Note that l_{11} could be $-\sqrt{1}$ and so matrix L is not (quite) unique. Now equate elements of the second column to obtain

$$8 = l_{21}^2 + l_{22}^2 \qquad gives \qquad l_{22} = 2$$
$$2\alpha = l_{31}l_{21} + l_{32}l_{22} \qquad gives \qquad l_{32} = 0.$$

Finally, equate elements of the third column to obtain

$$9 = l_{31}^2 + l_{32}^2 + l_{33}^2 \qquad gives \quad l_{33} = \sqrt{9 - \alpha^2},$$

which shows that the allowable values of α must satisfy $9 - \alpha^2 > 0$. Thus, α is bounded by $-3 < \alpha < 3$.   ●

Program 1.13
MATLAB m-file for the Cholesky Method
function sol = Cholesky(A, b)
[n,n]=size(A); l=zeros(n,n); u=l;
$l(1,1) = (A(1,1)) \setminus 0.5$; u(1,1)=l(1,1);
for i=2:n; u(1,i)=A(1,i)/l(1,1);
l(i,1)=A(i,1)/u(1,1); end
for i=2:n; for j=2:n; s=0;
if $i <= j$; K=i-1; else; K=j-1; end
for k=1:K; $s = s + l(i,k) * u(k,j)$; end
if $j > i$; u(i,j)=(A(i,j)-s)/l(i,i);
elseif $i == j$
$l(i,j) = (A(i,j) - s) \setminus 0.5$; u(i,j)=l(i,j);
else; l(i,j)=(A(i,j)-s)/u(j,j); end; end; end
y(1)=b(1)/l(1,1);
for k=2:n; sum=b(k);
for i=1:k-1; $sum = sum - l(k,i) * y(i)$; end
y(k)=sum/l(k,k); end
x(n)=y(n)/u(n,n);
for k=n-1:-1:1; sum=y(k);
for i=k+1:n; $sum = sum - u(k,i) * x(i)$; end
x(k)=sum/u(k,k); end; l; u; y; x;

Example 1.50 *Find the LU decomposition of the following matrix using Doolittle's, Crout's, and the Cholesky methods:*

$$A = \begin{pmatrix} 4 & -2 & 4 \\ -2 & 2 & 2 \\ 4 & 2 & 29 \end{pmatrix}.$$

Solution. *By using the simple Gauss elimination method, one can convert the given matrix into the upper-triangular matrix*

$$U = \begin{pmatrix} 4 & -2 & 4 \\ 0 & 1 & 4 \\ 0 & 0 & 9 \end{pmatrix}$$

with the help of the possible multiples

$$m_{21} = -0.5, \quad m_{31} = 1, \quad m_{32} = 4.$$

Thus, the LU decomposition of A using Doolittle's method is

$$A = LU = \begin{pmatrix} 1 & 0 & 0 \\ -0.5 & 1 & 0 \\ 1 & 4 & 1 \end{pmatrix} \begin{pmatrix} 4 & -2 & 4 \\ 0 & 1 & 4 \\ 0 & 0 & 9 \end{pmatrix}.$$

Rather than computing the next two factorizations directly, we can obtain them from Doolittle's factorization above. From Doolittle's factorization the LDV factorization of the given matrix A can be obtained as

$$A = LDV = \begin{pmatrix} 1 & 0 & 0 \\ -0.5 & 1 & 0 \\ 1 & 4 & 1 \end{pmatrix} \begin{pmatrix} 4 & 0 & 0 \\ 0 & 1 & 0 \\ 0 & 0 & 9 \end{pmatrix} \begin{pmatrix} 1 & -0.5 & 1 \\ 0 & 1 & 4 \\ 0 & 0 & 1 \end{pmatrix}.$$

By putting $\hat{L} = LD$, *i.e.*,

$$\hat{L} = LD = \begin{pmatrix} 1 & 0 & 0 \\ -0.5 & 1 & 0 \\ 1 & 4 & 1 \end{pmatrix} \begin{pmatrix} 4 & 0 & 0 \\ 0 & 1 & 0 \\ 0 & 0 & 9 \end{pmatrix} = \begin{pmatrix} 4 & 0 & 0 \\ -2 & 1 & 0 \\ 4 & 4 & 9 \end{pmatrix},$$

we can obtain Crout's factorization as follows:

$$A = LDV = \hat{L}V = \begin{pmatrix} 4 & 0 & 0 \\ -2 & 1 & 0 \\ 4 & 4 & 9 \end{pmatrix} \begin{pmatrix} 1 & -0.5 & 1 \\ 0 & 1 & 4 \\ 0 & 0 & 1 \end{pmatrix}.$$

Similarly, the Cholesky factorization is obtained by splitting diagonal matrix D into the form $D^{\frac{1}{2}}D^{\frac{1}{2}}$ *in the LDV factorization and associating one factor with L and the other with V. Thus,*

$$A = LDV = (LD^{\frac{1}{2}})(D^{\frac{1}{2}}V) = \hat{L}\hat{L}^T,$$

where

$$\hat{L} = LD^{1/2} = \begin{pmatrix} 1 & 0 & 0 \\ -0.5 & 1 & 0 \\ 1 & 4 & 1 \end{pmatrix} \begin{pmatrix} 2 & 0 & 0 \\ 0 & 1 & 0 \\ 0 & 0 & 3 \end{pmatrix} = \begin{pmatrix} 2 & 0 & 0 \\ -1 & 1 & 0 \\ 2 & 4 & 3 \end{pmatrix}$$

and

$$\hat{L}^T = D^{1/2}V = \begin{pmatrix} 2 & 0 & 0 \\ 0 & 1 & 0 \\ 0 & 0 & 3 \end{pmatrix} \begin{pmatrix} 1 & -0.5 & 1 \\ 0 & 1 & 4 \\ 0 & 0 & 1 \end{pmatrix} = \begin{pmatrix} 2 & -1 & 2 \\ 0 & 1 & 4 \\ 0 & 0 & 3 \end{pmatrix}.$$

Thus, we obtain

$$A = \hat{L}\hat{L}^T = \begin{pmatrix} 2 & 0 & 0 \\ -1 & 1 & 0 \\ 2 & 4 & 3 \end{pmatrix} \begin{pmatrix} 2 & -1 & 2 \\ 0 & 1 & 4 \\ 0 & 0 & 3 \end{pmatrix},$$

the Cholesky factorization of the given matrix A. •

The factorization of primary interest is $A = LU$, where L is a unit lower-triangular matrix and U is an upper-triangular matric. Henceforth, when we refer to a LU decomposition, we mean one in which L is a unit lower-triangular matrix.

Example 1.51 *Show that the following matrix cannot be factored as $A = LDL^T$:*

$$A = \begin{pmatrix} 1 & 2 & 1 \\ 2 & 5 & 3 \\ 1 & 3 & 2 \end{pmatrix}.$$

Solution. *By using the simple Gauss elimination method we can use the multipliers*

$$m_{21} = 2, \quad m_{31} = 1, \quad m_{32} = 1,$$

and we can convert the given matrix into an upper-triangular matrix as follows:

$$\begin{pmatrix} 1 & 2 & 1 \\ 0 & 1 & 1 \\ 0 & 1 & 1 \end{pmatrix} \rightarrow \begin{pmatrix} 1 & 2 & 1 \\ 0 & 1 & 1 \\ 0 & 0 & 0 \end{pmatrix} = U.$$

Since the element $a_{33}^{(2)} = 0$, the simple Gaussian elimination cannot continue in its present form and from Theorem 1.21, the decomposition of A is not possible. Hence, A cannot be factored as $A = LDL^T$. •

Since we know that not every matrix has a direct *LU* decomposition, we define the following matrix which gives the sufficient condition for the *LU* decomposition of the matrix. It also helps us with the convergence of the iterative methods for solving linear systems.

Definition 1.32 (Strictly Diagonally Dominant Matrix)

A square matrix is said to be Strictly Diagonally Dominant (SDD) if the absolute value of each element on the main diagonal is greater than the sum of the absolute values of all the other elements in that row. Thus, a SDD matrix is defined as

$$|a_{ii}| > \sum_{\substack{j=1 \\ j \neq i}}^{n} |a_{ij}|, \qquad for \quad i = 1, 2, \ldots, n. \tag{1.45}$$

Example 1.52 *The matrix*

$$A = \begin{pmatrix} 7 & 3 & 1 \\ 1 & 6 & 3 \\ -2 & 4 & 8 \end{pmatrix}$$

is SDD since

$$
\begin{aligned}
|7| &> |3| + |1|, & i.e., \quad 7 &> 4 \\
|6| &> |1| + |3|, & i.e., \quad 6 &> 4 \\
|8| &> |-2| + |4|, & i.e., \quad 8 &> 6,
\end{aligned}
$$

but the matrix

$$B = \begin{pmatrix} 6 & -3 & 4 \\ 3 & 7 & 3 \\ 5 & -4 & 10 \end{pmatrix}$$

is not SDD since

$$|6| > |-3| + |4|, \quad i.e., \quad 6 > 7,$$

which is not true.   ●

An SDD matrix occurs naturally in a wide variety of practical applications, and when solving an SDD system by the Gauss elimination method, partial pivoting is never required.

Theorem 1.30 *If a matrix A is strictly diagonally dominant, then:*

1. *Matrix A is nonsingular.*

2. *Gaussian elimination without row interchange can be performed on the linear system $A\mathbf{x} = \mathbf{b}$.*

3. *Matrix A has LU factorization.* •

Example 1.53 *Solve the following linear system using the simple Gaussian elimination method and also find the LU decomposition of the matrix using Doolittle's method and Crout's method:*

$$
\begin{aligned}
5x_1 + x_2 + x_3 &= 7 \\
2x_1 + 6x_2 + x_3 &= 9 \\
x_1 + 2x_2 + 9x_3 &= 12.
\end{aligned}
$$

Solution. *Start with the augmented matrix form*

$$
\begin{pmatrix}
5 & 1 & 1 & \vdots & 7 \\
2 & 6 & 1 & \vdots & 9 \\
1 & 2 & 9 & \vdots & 12
\end{pmatrix},
$$

and since $a_{11} = 5 \neq 0$, we can eliminate the elements a_{21} and a_{31} by subtracting from the second and third rows the appropriate multiples of the first row. In this case the multiples are given,

$$
m_{21} = \frac{2}{5} = 0.4 \quad and \quad m_{31} = \frac{1}{5} = 0.2.
$$

Hence,

$$
\begin{pmatrix}
5 & 1 & 1 & \vdots & 7 \\
0 & 5.6 & 0.6 & \vdots & 6.2 \\
0 & 1.8 & 8.8 & \vdots & 10.6
\end{pmatrix}.
$$

Since $a_{22}^{(1)} = 5.6 \neq 0$, we eliminate the entry in the $a_{32}^{(1)}$ position by subtracting the multiple $m_{32} = \frac{1.8}{5.6} = 0.32$ of the second row from the third row *to get*

$$\begin{pmatrix} 5 & 1 & 1 & \vdots & 7 \\ 0 & 5.6 & 0.6 & \vdots & 6.2 \\ 0 & 0 & 8.6 & \vdots & 8.6 \end{pmatrix}.$$

Obviously, the original set of equations has been transformed to an upper-triangular form. All the diagonal elements of the obtaining upper-triangular matrix are nonzero, which means that the coefficient matrix of the given system is nonsingular, therefore, the given system has a unique solution. Now expressing the set in algebraic form yields

$$\begin{aligned} 5x_1 + x_2 + x_3 &= 7 \\ 5.6x_2 + 0.6x_3 &= 6.2 \\ 8.6x_3 &= 8.6. \end{aligned}$$

Now use backward substitution to get the solution of the system as

$$\begin{aligned} 8.6x_3 &= 8.6 & \text{gives} & \quad x_3 = 1 \\ 5.6x_2 &= -0.6x_3 + 6.2 = -0.6 + 6.2 = 5.6 & \text{gives} & \quad x_2 = 1 \\ 5x_1 &= 7 - x_2 - x_3 = 7 - 1 - 1 = 5 & \text{gives} & \quad x_1 = 1. \end{aligned}$$

We know that when using LU decomposition by Doolittle's method the unknown elements of matrix L are the multiples used and the matrix U is the same as we obtained in the forward elimination process of the simple Gauss elimination. Thus, the LU decomposition of matrix A can be obtained by using Doolittle's method as follows:

$$A = \begin{pmatrix} 5 & 1 & 1 \\ 2 & 6 & 1 \\ 1 & 2 & 9 \end{pmatrix} \begin{pmatrix} 1 & 0 & 0 \\ 0.4 & 1 & 0 \\ 0.5 & 0.32 & 1 \end{pmatrix} \begin{pmatrix} 5 & 1 & 1 \\ 0 & 5.6 & 0.6 \\ 0 & 0 & 8.6 \end{pmatrix} = LU.$$

Similarly, the LU decomposition of matrix A can be obtained by using Crout's method as

$$A = \begin{pmatrix} 5 & 1 & 1 \\ 2 & 6 & 1 \\ 1 & 2 & 9 \end{pmatrix} \begin{pmatrix} 5 & 0 & 0 \\ 2 & 5.6 & 0 \\ 1 & 1.8 & 8.6 \end{pmatrix} \begin{pmatrix} 1 & 0.2 & 0.2 \\ 0 & 1 & 0.1 \\ 0 & 0 & 1 \end{pmatrix} = LU.$$

Thus, the conditions of Theorem 1.30 are satisfied.   ●

1.4.6 Tridiagonal Systems of Linear Equations

The application of numerical methods to the solution of certain engineering problems may in some cases result in a set of tridiagonal linear algebraic equations. Heat conduction and fluid flow problems are some of the many applications that generate such a system.

A tridiagonal system has a coefficients matrix T of which all elements except those on the main diagonal and the two diagonals just above and below the main diagonal (usually called superdiagonal and subdiagonal, respectively) are defined as

$$T = \begin{pmatrix} \alpha_1 & c_1 & 0 & \cdots & & 0 \\ \beta_2 & \alpha_2 & c_2 & \ddots & & 0 \\ 0 & \beta_3 & a_3 & \ddots & & 0 \\ \vdots & \ddots & \ddots & \ddots & c_{n-1} & \\ 0 & 0 & 0 & \beta_n & & \alpha_n \end{pmatrix}. \tag{1.46}$$

This type of matrix can be stored more economically, which is the case for a fully populated matrix. Obviously, one may use any one of the methods discussed in the previous sections for solving the tridiagonal system

$$T\mathbf{x} = \mathbf{b}, \tag{1.47}$$

but the linear system involving nonsingular matrices of the form T given in (1.47) are also most easily solved by the LU decomposition method just described for the general linear system. The tridiagonal matrix T can be factored into a lower-bidiagonal factor L and an upper-bidiagonal factor U having the following forms:

$$L = \begin{pmatrix} 1 & 0 & 0 & \cdots & 0 \\ l_2 & 1 & 0 & \ddots & 0 \\ 0 & l_3 & 1 & \ddots & 0 \\ \ddots & \ddots & \ddots & \ddots & 0 \\ 0 & 0 & 0 & l_n & 1 \end{pmatrix}, \quad U = \begin{pmatrix} u_1 & c_1 & 0 & \cdots & 0 \\ 0 & u_2 & c_2 & \ddots & 0 \\ 0 & 0 & u_3 & \ddots & 0 \\ \ddots & \ddots & \ddots & \ddots & c_{n-1} \\ 0 & 0 & 0 & 0 & u_n \end{pmatrix}. \tag{1.48}$$

The unknown elements l_i and u_i of matrices L and U, respectively, can be computed as a special case of Doolittle's method using the LU decomposition method,

$$\left. \begin{aligned} u_1 &= \alpha_1 \\[2mm] l_i &= \frac{\beta_i}{u_{i-1}}, & i = 2, 3, \ldots, n \\[2mm] u_i &= \alpha_i - l_i c_{i-1}, & i = 2, 3, \ldots, n. \end{aligned} \right\} . \qquad (1.49)$$

After finding the values for l_i and u_i, then they are used along with the elements c_i, to solve the tridiagonal system (1.47) by solving the first bidiagonal system

$$L\mathbf{y} = \mathbf{b}, \qquad (1.50)$$

for $\mathbf{y}$ by using forward substitution,

$$\left. \begin{aligned} y_1 &= b_1 \\ y_i &= b_i - l_i y_{i-1}, & i = 2, 3, \ldots, n \end{aligned} \right\} , \qquad (1.51)$$

followed by solving the second bidiagonal system,

$$U\mathbf{x} = \mathbf{y}, \qquad (1.52)$$

for $\mathbf{x}$ by using backward substitution,

$$\left. \begin{aligned} x_n &= y_n / u_n \\ x_i &= y_i - c_i x_{i+1}, & i = n - 1, \ldots, 1 \end{aligned} \right\} . \qquad (1.53)$$

The entire process for solving the original system (1.47) requires $3n$ additions, $3n$ multiplications, and $2n$ divisions. Thus, the total number of multiplications and divisions is approximately $5n$.

Most large tridiagonal systems are strictly diagonally dominant (defined as follows), so pivoting is not necessary. When solving systems of equations with a tridiagonal coefficients matrix T, iterative methods can sometimes be used to one's advantage. These methods are introduced in Chapter 2.

Example 1.54 *Solve the following tridiagonal system of equations using the LU decomposition method:*

$$
\begin{aligned}
x_1 + x_2 &= 1 \\
x_1 + 2x_2 + x_3 &= 0 \\
x_2 + 3x_3 + x_4 &= 1 \\
x_3 + 4x_4 &= 1.
\end{aligned}
$$

Solution. *Construct the factorization of tridiagonal matrix T as follows:*

$$
\begin{pmatrix}
1 & 1 & 0 & 0 \\
1 & 2 & 1 & 0 \\
0 & 1 & 3 & 1 \\
0 & 0 & 1 & 4
\end{pmatrix}
=
\begin{pmatrix}
1 & 0 & 0 & 0 \\
l_2 & 1 & 0 & 0 \\
0 & l_3 & 1 & 0 \\
0 & 0 & l_4 & 1
\end{pmatrix}
\begin{pmatrix}
u_1 & 1 & 0 & 0 \\
0 & u_2 & 1 & 0 \\
0 & 0 & u_3 & 1 \\
0 & 0 & 0 & u_4
\end{pmatrix}.
$$

Then the elements of the L and U matrices can be computed by using (1.48) as follows:

$$u_1 = \alpha_1 = 1$$

$$l_2 = \frac{\beta_2}{u_1} = \frac{1}{1} = 1$$

$$u_2 = \alpha_2 - l_2 c_1 = 2 - (1)1 = 1$$

$$l_3 = \frac{b_3}{u_2} = \frac{1}{1} = 1$$

$$u_3 = \alpha_3 - l_3 c_2 = 3 - (1)1 = 2$$

$$l_4 = \frac{b_4}{u_3} = \frac{1}{2}$$

$$u_4 = \alpha_4 - l_4 c_3 = 4 - (\frac{1}{2})1 = \frac{7}{2}.$$

After finding the elements of the bidiagonal matrices L and U, we solve the *first system* $L\mathbf{y} = \mathbf{b}$ *as follows:*

$$\begin{pmatrix} 1 & 0 & 0 & 0 \\ 1 & 1 & 0 & 0 \\ 0 & 1 & 1 & 0 \\ 0 & 0 & \frac{1}{2} & 1 \end{pmatrix} \begin{pmatrix} y_1 \\ y_2 \\ y_3 \\ y_4 \end{pmatrix} = \begin{pmatrix} 1 \\ 0 \\ 1 \\ 1 \end{pmatrix}.$$

Using forward substitution, we get

$$[y_1, y_2, y_3, y_4]^T = [1, -1, 2, 0]^T.$$

Now we solve the second system $U\mathbf{x} = \mathbf{y}$ *as follows:*

$$\begin{pmatrix} 1 & 1 & 0 & 0 \\ 0 & 1 & 1 & 0 \\ 0 & 0 & 2 & 1 \\ 0 & 0 & 0 & \frac{7}{2} \end{pmatrix} \begin{pmatrix} x_1 \\ x_2 \\ x_3 \\ x_4 \end{pmatrix} = \begin{pmatrix} 1 \\ -1 \\ 2 \\ 0 \end{pmatrix}.$$

Using backward substitution, we get

$$\mathbf{x}^* = [x_1, x_2, x_3, x_4]^T = [3, -2, 1, 0]^T,$$

which is the required solution of the given system. ●

The above results can be obtained using MATLAB commands. We do the following:

```
>> Tb = [T|b] = [1 1 0 0 1;1 2 1 0 0;0 1 3 1 1;0 0 1 4 1];
>> TriDLU(Tb);
```

Program 1.14
MATLAB m-file for *LU* Decomposition for a Tridiagonal System
function sol=TRiDLU(Tb)
[m,n]=size(Tb); L=eye(m); U=zeros(m);
U(1,1)=Tb(1,1);
for i=2:m
$U(i-1,i) = Tb(i-1,i);$
$L(i,i-1) = Tb(i,i-1)/U(i-1,i-1);$
$U(i,i) - L(i,i-1) * Tb(i-1,i);$ end
disp('The lower-triangular matrix') L;
disp('The upper-triangular matrix') U;
$y = inv(L) * Tb(:,n); \quad x = inv(U) * y;$

Procedure 1.8 (LU Decomposition by the Tridiagonal Method)

1. *Take the tridiagonal matrix T.*

2. *Decompose the matrix $T = LU$ using (1.49).*

3. *Solve linear system $L\mathbf{y} = \mathbf{b}$ using (1.51).*

4. *Solve linear system $U\mathbf{x} = \mathbf{y}$ using (1.53).*

1.5   Conditioning of Linear Systems

In solving the linear system numerically we have to see the problem conditioning, algorithm stability, and cost. Earlier we discussed efficient elimination schemes to solve a linear system, and these schemes are stable when pivoting is employed. But there are some ill-conditioned systems which are tough to solve by any method. These types of linear systems are identified in this chapter.

Here, we will present a parameter, the *condition number*, which quantitatively measures the conditioning of a linear system. The condition number is greater than and equal to one and as a linear system becomes more ill-conditioned, the condition number increases. After factoring a matrix, the condition number can be estimated in roughly the same time it takes to solve a few factored systems $(LU)\mathbf{x} = \mathbf{b}$. Hence, after factoring a matrix, the extra computer time needed to estimate the condition number is usually insignificant.

1.5.1   Norms of Vectors and Matrices

For solving linear systems, we discuss a method for quantitatively measuring the distance between vectors in $\mathbb{R}^n$, the set of all column vectors with real components, to determine whether the sequence of vectors that results from using a direct method converges to a solution of the system. To define a distance in $\mathbb{R}^n$, we use the notation of the *norm* of a vector.

Vector Norms

It is sometimes useful to have a scalar measure of the magnitude of a vector. Such a measure is called a *vector norm* and for a vector $\mathbf{x}$ is written as $\|\mathbf{x}\|$.

A vector norm on $\mathbb{R}^n$ is a function from $\mathbb{R}^n$ to $\mathbb{R}$ satisfying:

1. $\|\mathbf{x}\| > 0, \quad \text{for all} \quad \mathbf{x} \in \mathbb{R}^n;$

2. $\|\mathbf{x}\| = 0, \quad \text{if and only if} \quad \mathbf{x} = \mathbf{0};$

3. $\|\alpha\mathbf{x}\| = |\alpha|\|\mathbf{x}\|, \quad \text{for all} \quad \alpha \in \mathbb{R}, \quad \mathbf{x} \in \mathbb{R}^n;$

4. $\|\mathbf{x} + \mathbf{y}\| \leq \|\mathbf{x}\| + \|\mathbf{y}\|, \quad \text{for all} \quad \mathbf{x}, \mathbf{y} \in \mathbb{R}^n.$

There are three norms in $\mathbb{R}^n$ that are most commonly used in applications, called l_1-norm, l_2-norm, and l_∞-norm, and are defined for the given vectors

$\mathbf{x} = [x_1, x_2, \ldots, x_n]^T$ as

$$\|\mathbf{x}\|_1 = \sum_{i=1}^{n} |x_i|$$

$$\|\mathbf{x}\|_2 = \left(\sum_{i=1}^{n} x_i^2 \right)^{1/2}$$

$$\|\mathbf{x}\|_\infty = \max_{1 \le i \le n} |x_i|.$$

The l_1-norm is called the *absolute norm*, the l_2-norm is frequently called the *Euclidean norm* as it is just the formula for distance in ordinary three-dimensional Euclidean space extended to dimension n, and finally, the l_∞-norm is called the *maximum norm* or occasionally the *uniform norm*. All these three norms are also called the *natural norms*.

Example 1.55 *Compute the l_p-norms ($p = 1, 2, \infty$) of the vector $\mathbf{x} = [-5, 3, -2]^T$ in $\mathbb{R}^3$.*

Solution. *These l_p-norms ($p = 1, 2, \infty$) of the given vector are:*

$$\|\mathbf{x}\|_1 = |x_1| + |x_2| + |x_3| = |-5| + |3| + |-2| = 10,$$

$$\|\mathbf{x}\|_2 = (x_1^2 + x_2^2 + x_3^2)^{1/2} = \left[(-5)^2 + (3)^2 + (-2)^2 \right]^{1/2} \approx 6.16,$$

$$\|\mathbf{x}\|_\infty = \max\{|x_1|, |x_2|, |x_3|\} = \max\{|-5|, |3|, |-2|\} = 5.$$

●

In MATLAB, the built-in **norm** function computes the l_p-norms of vectors. If only one argument is passed to norm, the l_2-norm is returned and for two arguments, the second one is used to specify the value of p:

```
>> x = [-5 3  - 2];
>> v = norm(x)
v = 6.16
>> x = [-5 3  - 2];
>> v = norm(x, 2)
v = 6.16
>> x = [-5 3  - 2];
>> v = norm(x, 1)
v = 10
>> x = [-5 3  - 2];
>> v = norm(x, inf)
v = 5
```

The internal MATLAB constant *inf* is used to select the l_∞-norm.

Matrix Norms

A matrix norm is a measure of how well one matrix approximates another, or, more accurately, of how well their difference approximates the zero matrix. An iterative procedure for inverting a matrix produces a sequence of approximate inverses. Since, in practice, such a process must be terminated, it is desirable to have some measure of the error of an approximate inverse.

So a matrix norm on the set of all $n \times n$ matrices is a real-valued function, $\|.\|$, defined on this set, satisfying for all $n \times n$ matrices A and B and all real numbers α as follows:

1. $\|A\| > 0, \quad A \neq \mathbf{0}$;

2. $\|A\| = 0, \quad A = \mathbf{0}$;

3. $\|I\| = 1, \quad I \text{ is the identity matrix}$;

4. $\|\alpha A\| = |\alpha|\|A\|, \quad \text{for scalar} \quad \alpha \in \mathbb{R}$;

5. $\|A + B\| \leq \|A\| + \|B\|$;

6. $\|AB\| \leq \|A\|\|B\|$;

7. $\|A - B\| \geq \left| \|A\| - \|B\| \right|$.

Several norms for matrices have been defined, and we shall use the following three natural norms l_1, l_2, and l_∞ for a square matrix of order n:

$$\|A\|_1 = \max_j \left(\sum_{i=1}^{n} |a_{ij}| \right) = \text{maximum column sum,}$$

$$\|A\|_2 = \max_{\|x\|_2 = 1} \|A\mathbf{x}\|_2 = \text{spectral norm,}$$

$$\|A\|_\infty = \max_i \left(\sum_{j=1}^{n} |a_{ij}| \right) = \text{row sum norm.}$$

The l_1-norm and l_∞-norm are widely used because they are easy to calculate. The matrix norm $\|A\|_2$ that corresponds to the l_2-norm is related to the eigenvalues of the matrix. It sometimes has special utility because no other norm is smaller than this norm. It, therefore, provides the best measure of the size of a matrix, but is also the most difficult to compute. We will discuss this natural norm later in the chapter.

For an $m \times n$ matrix, we can paraphrase the *Frobenius* (or *Euclidean*) norm (which is not a natural norm) and define it as

$$\|A\|_F = \left(\sum_{i=1}^{m} \sum_{j=1}^{n} |a_{ij}|^2 \right)^{1/2}.$$

It can be shown that

$$\|A\|_F = \sqrt{tr(A^T A)},$$

where $tr(A^T A)$ is the *trace* of a matrix $A^T A$, i.e., the sum of the diagonal entries of $A^T A$. The Frobenius norm of a matrix is a good measure of the magnitude of a matrix. Note that $\|A\|_F \neq \|A\|_2$. For a diagonal matrix, all norms have the same values.

Example 1.56 *Compute the l_p-norms ($p = 1, \infty, F$) of the following matrix:*

$$A = \begin{pmatrix} 4 & 2 & -1 \\ 3 & 5 & -2 \\ 1 & -2 & 7 \end{pmatrix}.$$

Solution. *These norms are:*

$$\sum_{i=1}^{3} |a_{i1}| = |4| + |3| + |1| = 8,$$

$$\sum_{i=1}^{3} |a_{i2}| = |2| + |5| + |-2| = 9,$$

$$\sum_{i=1}^{3} |a_{i3}| = |-1| + |-2| + |7| = 10,$$

so

$$\|A\|_1 = max\{8, 9, 10\} = 10.$$

Also,

$$\sum_{j=1}^{3} |a_{1j}| = |4| + |2| + |-1| = 7,$$

$$\sum_{j=1}^{3} |a_{2j}| = |3| + |5| + |-2| = 10,$$

$$\sum_{j=1}^{3} |a_{3j}| = |1| + |-2| + |7| = 10,$$

so

$$\|A\|_\infty = max\{7, 10, 10\} = 10.$$

Finally, we have

$$\|A\|_F = (16 + 4 + 1 + 9 + 25 + 4 + 1 + 4 + 49)^{1/2} \approx 10.6301,$$

the Frobenius norm of the given matrix.   ●

     Like the l_p-norms of vectors, in MATLAB the built-in **norm** function can be used to compute the l_p-norms of matrices. The l_1-norm of a matrix can be computed as follows:

```
>> A = [4 2  - 1; 3 5  - 2; 1  - 2  - 7];
>> B = norm(A, 1)
B =
    10
```

The l_∞-norm of a matrix A is:

```
>> A = [4 2  - 1; 3 5  - 2; 1  - 2  - 7];
>> B = norm(A, inf)
B =
    10
```

Finally, the Frobenius norm of the matrix A is:

```
>> A = [4 2  - 1; 3 5  - 2; 1  - 2  - 7];
>> B = norm(A,' fro')
B =
    10.6301
```

1.5.2   **Errors in Solving Linear Systems**

Any computed solution of a linear system must, because of round-off and other errors, be considered an approximate solution. Here, we shall consider the most natural method for determining the accuracy of a solution of the linear system. One obvious way of estimating the accuracy of the computed solution $\mathbf{x}^*$ is to compute $A\mathbf{x}^*$ and to see how close $A\mathbf{x}^*$ comes to $\mathbf{b}$. Thus, if $\mathbf{x}^*$ is an approximate solution of the given system $A\mathbf{x} = \mathbf{b}$, we compute a vector

$$\mathbf{r} = \mathbf{b} - A\mathbf{x}^*, \tag{1.54}$$

which is called the *residual vector* and which can be easily calculated. The quantity

$$\frac{\|\mathbf{r}\|}{\|\mathbf{b}\|} = \frac{\|\mathbf{b} - A\mathbf{x}^*\|}{\|\mathbf{b}\|}$$

is called the *relative residual.* We use MATLAB as follows:

Program 1.15
MATLAB m-file for Finding the Residual Vector
function r=RES(A,b,x0)
[n,n]=size(A);
for i=1:n; R(i) = b(i);
for j=1:n
R(i)=R(i)-A(i,j)*x0(j);end
RES(i)=R(i); end
r=RES'

The smallness of the residual then provides a measure of the goodness of the approximate solution $\mathbf{x}^*$. If every component of vector $\mathbf{r}$ vanishes, then $\mathbf{x}^*$ is the exact solution. If $\mathbf{x}^*$ is a good approximation, then we would expect each component of $\mathbf{r}$ to be small, at least in a relative sense. For example, the linear system

$$
\begin{array}{rcrcl}
x_1 & + & 2x_2 & = & 3 \\
1.0001x_1 & + & 2x_2 & = & 3.0001
\end{array}
$$

has the approximate solution $\mathbf{x}^* = [3, 0]^T$. To see how good this solution is, we compute the residual, $\mathbf{r} = [0, -0.0002]^T$.

We can conclude from the residual that the approximate solution is correct to at most three decimal places. Also, the linear system

$$
\begin{array}{rcrcrcrcl}
1.0000x_1 & + & 0.9600x_2 & + & 0.8400x_3 & + & 0.6400x_4 & = & 3.4400 \\
0.9600x_1 & + & 0.9214x_2 & + & 0.4406x_3 & + & 0.2222x_4 & = & 2.5442 \\
0.8400x_1 & + & 0.4406x_2 & + & 1.0000x_3 & + & 0.3444x_4 & = & 2.6250 \\
0.6400x_1 & + & 0.2222x_2 & + & 0.3444x_3 & + & 1.0000x_4 & = & 2.2066
\end{array}
$$

has the exact solution $\mathbf{x} = [1, 1, 1, 1]^T$ and the approximate solution due to Gaussian elimination without pivoting is

$$
\mathbf{x}^* = [1.0000322, 0.99996948, 0.99998748, 1.0000113]^T,
$$

and the residual is

$$
\mathbf{r} = [0.6 \times 10^{-7}, 0.6 \times 10^{-7}, -0.53 \times 10^{-5}, -0.21 \times 10^{-4}]^T.
$$

The approximate solution due to Gaussian elimination with partial pivoting is

$$\mathbf{x}^* = [0.9999997, 0.99999997, 0.99999996, 1.0000000]^T,$$

and the residual is

$$\mathbf{r} = [0.3 \times 10^{-7}, 0.3 \times 10^{-7}, 0.6 \times 10^{-7}, 0.1 \times 10^{-8}]^T.$$

We found that all the elements of the residual for the second case (with pivoting) are less than 0.6×10^{-7}, whereas for the first case (without pivoting) they are as large as 0.2×10^{-4}. Even without knowing the exact solution, it is clear that the solution obtained in the second case is much better than the first case. The residual provides a reasonable measure of the accuracy of a solution in those cases where the error is primarily due to the accumulation of round-off errors.

Intuitively it would seem reasonable to assume that when $\|\mathbf{r}\|$ is small for a given vector norm, then the error $\|\mathbf{x} - \mathbf{x}^*\|$ would be small as well. In fact, this is true for some systems. However, there are systems of equations which do not satisfy this property. Such systems are said to be *ill-conditioned*.

These are systems in which small changes in the coefficients of the system lead to large changes in the solution. For example, consider the linear system

$$\begin{aligned} x_1 + x_2 &= 2 \\ x_1 + 1.01x_2 &= 2.01. \end{aligned}$$

The exact solution is easily verified to be $x_1 = x_2 = 1$. On the other hand, the system

$$\begin{aligned} x_1 + x_2 &= 2 \\ 1.001x_1 + x_2 &= 2.01 \end{aligned}$$

has the solution $x_1 = 10, x_2 = -8$. Thus, a change of 1% in the coefficients has changed the solution by a factor of 10. If in the above given system, we substitute $x_1 = 10, x_2 = 8$, we find that the residuals are $r_1 = 0, r_2 = 0.09$, so this solution looks reasonable, although it is grossly in error. In practical problems we can expect the coefficients in the system to be subject to small errors, either because of round-off or because of physical measurement. If the system is ill-conditioned, the resulting solution may be grossly in error. Errors of this type, unlike those caused by round-off error accumulation, cannot be avoided by careful programming.

We have seen that for ill-conditioned systems the residual is not necessarily a good measure of the accuracy of a solution. How then can we tell when a system is ill-conditioned? In the following we discuss some possible indicators of ill-conditioned systems.

Definition 1.33 (Condition Number of a Matrix)

The number $\|A\|\|A^{-1}\|$ is called the condition number of a nonsingular matrix A and is denoted by $K(A)$, i.e.,

$$cond(A) = K(A) = \|A\|\|A^{-1}\|. \tag{1.55}$$

Note that the condition number $K(A)$ for A depends on the matrix norm used and can, for some matrices, vary considerably as the matrix norm is changed. Since

$$1 = \|I\| = \|AA^{-1}\| \leq \|A\|\|A^{-1}\| = K(A),$$

the condition number is always in the range $1 \leq K(A) \leq \infty$ regardless of any natural norm. The lower limit is attained for identity matrices and $K(A) = \infty$ if A is singular. So the matrix A is *well-behaved* (or *well-conditioned*) if $K(A)$ is close to 1 and is increasingly *ill-conditioned* when $K(A)$ is significantly greater than 1, i.e., $K(A) \rightarrow \infty$. •

The condition numbers provide bounds for the sensitivity of the solution of a set of equations to changes in the coefficient matrix. Unfortunately, the evaluation of any of the condition numbers of a matrix A is not a trivial task since it is necessary first to obtain its inverse.

So if the condition number of a matrix is a very large number, then this is one of the indicators of an ill-conditioned system. Another indicator of ill-conditioning is when the pivots during the process of elimination suffer a loss of one or more significant figures. Small changes in the right-hand side terms of the system lead to large changes in the solution and give another indicator of an ill-conditioned system. Also, when the elements of the inverse of the coefficient matrix are large compared with the elements of the coefficients matrix, this also shows an ill-conditioned system.

Example 1.57 *Compute the condition number of the following matrix using the l_∞-norm:*

$$A = \begin{pmatrix} 2 & -1 & 0 \\ 2 & -4 & -1 \\ -1 & 0 & 2 \end{pmatrix}.$$

Solution. *The condition number of a matrix is defined as*

$$K(A) = \|A\|_\infty \|A^{-1}\|_\infty.$$

First, we calculate the inverse of the given matrix, which is

$$A^{-1} = \begin{pmatrix} \dfrac{8}{13} & -\dfrac{2}{13} & -\dfrac{1}{13} \\[3mm] \dfrac{3}{13} & -\dfrac{4}{13} & -\dfrac{2}{13} \\[3mm] \dfrac{4}{13} & -\dfrac{1}{13} & \dfrac{6}{13} \end{pmatrix}.$$

Now we calculate the l_∞-norm of both the matrices A and A^{-1}. Since the l_∞-norm of a matrix is the maximum of the absolute row sums, we have

$$\|A\|_\infty = max\{|2| + |-1| + |0|, |2| + |-4| + |-1|, |-1| + |0| + |2|\} = 7$$

and

$$\|A^{-1}\|_\infty = max\left\{\left|\frac{8}{13}\right| + \left|\frac{-2}{13}\right| + \left|\frac{-1}{13}\right|, \left|\frac{3}{13}\right| + \left|\frac{-4}{13}\right| + \left|\frac{-2}{13}\right|, \left|\frac{4}{13}\right| + \left|\frac{-1}{13}\right| + \left|\frac{6}{13}\right|\right\},$$

which gives

$$\|A^{-1}\|_\infty = \frac{11}{13}.$$

Therefore,

$$K(A) = \|A\|_\infty \|A^{-1}\|_\infty = (7)\left(\frac{11}{13}\right) \approx 5.9231.$$

Depending on the application, we might consider this number to be reasonably small and conclude that the given matrix A is reasonably well-conditioned. •

To get the above results using MATLAB commands, we do the following:

```
>> A = [2 − 1 0; 2 − 4 − 1; −1 0 2];
>> Ainv = inv(A)
>> K(A) = norm(A, inf) ∗ norm(Ainv, inf);
K(A) =
    5.9231
```

Some matrices are notoriously ill-conditioned. For example, consider the 4×4 Hilbert matrix

$$H = \begin{pmatrix} 1 & \frac{1}{2} & \frac{1}{3} & \frac{1}{4} \\ \frac{1}{2} & \frac{1}{3} & \frac{1}{4} & \frac{1}{5} \\ \frac{1}{3} & \frac{1}{4} & \frac{1}{5} & \frac{1}{6} \\ \frac{1}{4} & \frac{1}{5} & \frac{1}{6} & \frac{1}{7} \end{pmatrix};$$

whose entries are defined by

$$a_{ij} = \frac{1}{(i+j-1)}, \quad \text{for} \quad i, j = 1, 2, \ldots, n.$$

The inverse of the matrix H can be obtained as

$$H^{-1} = \begin{pmatrix} 16 & -120 & 240 & -140 \\ -120 & 1200 & -2700 & 1680 \\ 240 & -2700 & 6480 & -4200 \\ -140 & 1680 & -4200 & 2800 \end{pmatrix}.$$

Then the condition number of the Hilbert matrix is

$$K(H) = \|H\|_\infty \|H^{-1}\|_\infty = (2.0833)(13620) \approx 28375,$$

which is quite large. Note that the condition numbers of Hilbert matrices increase rapidly as the sizes of the matrices increase. Therefore, large Hilbert matrices are considered to be extremely ill-conditioned.

We might think that if the determinant of a matrix is close to zero, then the matrix is ill-conditioned. However, this is false. Consider the matrix

$$A = \begin{pmatrix} 10^{-7} & 0 \\ 0 & 10^{-7} \end{pmatrix},$$

for which $\det A = 10^{-14} \approx 0$. One can easily find the condition number of the given matrix as

$$K(A) = \|A\|_\infty \|A^{-1}\|_\infty = (10^{-7})(10^7) = 1.$$

The matrix A is therefore perfectly conditioned. Thus, a small determinant is necessary but not sufficient for a matrix to be ill-conditioned.

The condition number of a matrix $K(A)$ using the l_2-norm can be computed by the built-in function **cond** command in MATLAB as follows:

```
>> A = [1 − 1 2; 3 1 − 1; 2 0 1];
>> K(A) = cond(A);
K(A) =
    19.7982
```

Theorem 1.31 (Error in Linear Systems)

Suppose that $\mathbf{x}^*$ is an approximation to the solution $\mathbf{x}$ of the linear system $A\mathbf{x} = \mathbf{b}$ and A is a nonsingular matrix and $\mathbf{r}$ is the residual vector for $\mathbf{x}^*$. Then for any natural norm, the error is

$$\|\mathbf{x} - \mathbf{x}^*\| \le \|\mathbf{r}\| \|A^{-1}\|, \tag{1.56}$$

and the relative error is

$$\frac{\|\mathbf{x} - \mathbf{x}^*\|}{\|\mathbf{x}\|} \leq K(A)\frac{\|\mathbf{r}\|}{\|\mathbf{b}\|}, \quad \text{\textit{provided}} \quad \mathbf{x} \neq 0, \ \mathbf{b} \neq 0. \qquad (1.57)$$

Proof. Since $\mathbf{r} = \mathbf{b} - A\mathbf{x}^*$ and A is nonsingular, then

$$A\mathbf{x} - A\mathbf{x}^* = \mathbf{b} - (\mathbf{b} - \mathbf{r}) = \mathbf{r},$$

which implies that

$$A(\mathbf{x} - \mathbf{x}^*) = \mathbf{r} \qquad (1.58)$$

or

$$\mathbf{x} - \mathbf{x}^* = A^{-1}\mathbf{r}.$$

Taking the norm on both side gives

$$\|\mathbf{x} - \mathbf{x}^*\| = \|A^{-1}\mathbf{r}\| \leq \|A^{-1}\|\|\mathbf{r}\|.$$

Moreover, since $\mathbf{b} = A\mathbf{x}$, then

$$\|\mathbf{b}\| \leq \|A\|\|\mathbf{x}\|, \quad \text{or} \quad \|\mathbf{x}\| \geq \frac{\|\mathbf{b}\|}{\|A\|}.$$

Hence,

$$\frac{\|\mathbf{x} - \mathbf{x}^*\|}{\|\mathbf{x}\|} \leq \frac{\|A^{-1}\|\|\mathbf{r}\|}{\|\mathbf{b}\|/\|A\|} \leq K(A)\frac{\|\mathbf{r}\|}{\|\mathbf{b}\|}.$$

The inequalities (1.56) and (1.57) imply that the quantities $\|A^{-1}\|$ and $K(A)$ can be used to give an indication of the connection between the residual vector and the accuracy of the approximation. If the quantity $K(A) \approx 1$, the relative error will be fairly close to the relative residual. But if $K(A) >> 1$, then the relative error could be many times larger than the relative residual.

Example 1.58 *Consider the following linear system:*

$$\begin{array}{rrrrrrr} x_1 & + & x_2 & - & x_3 & = & 1 \\ x_1 & + & 2x_2 & - & 2x_3 & = & 0 \\ -2x_1 & + & x_2 & + & x_3 & = & -1. \end{array}$$

(a) Discuss the ill-conditioning of the given linear system.
(b) Let $\mathbf{x}^* = [2.01, 1.01, 1.98]^T$ be an approximate solution of the given system, then find the residual vector $\mathbf{r}$ and its norm $\|\mathbf{r}\|_\infty$.
(c) Estimate the relative error using (1.57).
(d) Use the simple Gaussian elimination method to find the approximate error using (1.58).

Solution. *(a) Given the matrix*

$$A = \begin{pmatrix} 1 & 1 & -1 \\ 1 & 2 & -2 \\ -2 & 1 & 1 \end{pmatrix},$$

the inverse can be computed as

$$A^{-1} = \begin{pmatrix} 2 & -1 & 0 \\ 1.5 & -0.5 & 0.5 \\ 2.5 & -1.5 & 0.5 \end{pmatrix}.$$

Then the l_∞-norms of both matrices are

$$\|A\|_\infty = 5 \quad and \quad \|A^{-1}\|_\infty = 4.5.$$

Using the values of both matrices' norms, we can find the value of the condition number of A as

$$K(A) = \|A\|_\infty \|A^{-1}\|_\infty = 22.5 >> 1,$$

which shows that the matrix is ill-conditioned. Thus, the given system is ill-conditioned.

```
>> A = [1 1 − 1; 1 2 − 2; −2 1 1];
>> K(A) = norm(A, inf) * norm(inv(A), inf);
```

(b) The residual vector can be calculated as

$$\mathbf{r} = \mathbf{b} - A\mathbf{x}^*$$

$$= \begin{pmatrix} 1 \\ 0 \\ -1 \end{pmatrix} - \begin{pmatrix} 1 & 1 & -1 \\ 1 & 2 & -2 \\ -2 & 1 & 1 \end{pmatrix} \begin{pmatrix} 2.01 \\ 1.01 \\ 1.98 \end{pmatrix}.$$

After simplifying, we get

$$\mathbf{r} = \begin{pmatrix} -0.04 \\ -0.07 \\ 0.03 \end{pmatrix},$$

and it gives

$$\|\mathbf{r}\|_\infty = 0.07.$$

```
>> A = [1 1 − 1; 1 2 − 2; −2 1 1];
>> b = [1 0 − 1]';
>> x0 = [2.01 1.01 1.98]';
>> r = RES(A, b, x0);
>> rnorm = norm(r, inf);
```

(c) *From (1.57), we have*

$$\frac{\|\mathbf{x} - \mathbf{x}^*\|}{\|\mathbf{x}\|} \le K(A)\frac{\|\mathbf{r}\|}{\|\mathbf{b}\|}.$$

By using parts (a) and (b) and the value $\|\mathbf{b}\|_\infty = 1$, *we obtain*

$$\frac{\|\mathbf{x} - \mathbf{x}^*\|}{\|\mathbf{x}\|} \le (22.5)\frac{(0.07)}{1} = 1.575.$$

```
>> RelErr = (K(A) * rnorm)/norm(b, inf);
```

(d) *Solve the linear system* $A\mathbf{e} = \mathbf{r}$, *where*

$$A = \begin{pmatrix} 1 & 1 & -1 \\ 1 & 2 & -2 \\ -2 & 1 & 1 \end{pmatrix} \quad and \quad \mathbf{r} = \begin{pmatrix} -0.04 \\ -0.07 \\ 0.03 \end{pmatrix}$$

and $\mathbf{e} = \mathbf{x} - \mathbf{x}^*$. *Write the above system in the augmented matrix form*

$$\begin{pmatrix} 1 & 1 & -1 & \vdots & -0.04 \\ 1 & 2 & -2 & \vdots & -0.07 \\ -2 & 1 & 1 & \vdots & 0.03 \end{pmatrix}.$$

After applying the forward elimination step of the simple Gauss elimination method, we obtain

$$\begin{pmatrix} 1 & 1 & -1 & \vdots & -0.04 \\ 0 & 1 & -1 & \vdots & -0.03 \\ 0 & 0 & 2 & \vdots & 0.04 \end{pmatrix}.$$

Now by using backward substitution, we obtain the solution

$$\mathbf{e}^* = [-0.01, -0.01, 0.02]^T,$$

which is the required approximation of the exact error. •

```
>> B = [1 1 − 1 − 0.04; 1 2 − 2 − 0.07; −2 1 1 0.03];
>> WP(B);
```

Conditioning

Let us consider the conditioning of the linear system

$$A\mathbf{x} = \mathbf{b}. \tag{1.59}$$

Case 1.1 *Suppose that the right-hand side term* $\mathbf{b}$ *is replaced by* $\mathbf{b} + \delta\mathbf{b}$, *where* $\delta\mathbf{b}$ *is an error in* $\mathbf{b}$. *If* $\mathbf{x} + \delta\mathbf{x}$ *is the solution corresponding to the right-hand side* $\mathbf{b} + \delta\mathbf{b}$, *then we have*

$$A(\mathbf{x} + \delta\mathbf{x}) = (\mathbf{b} + \delta\mathbf{b}), \tag{1.60}$$

which implies that

$$\begin{aligned} A\mathbf{x} + A\delta\mathbf{x} &= \mathbf{b} + \delta\mathbf{b}, \\ A\delta\mathbf{x} &= \delta\mathbf{b}. \end{aligned}$$

Multiplying by A^{-1}, *we get*

$$\delta\mathbf{x} = A^{-1}\delta\mathbf{b}.$$

Taking the norm gives

$$\|\delta\mathbf{x}\| = \|A^{-1}\delta\mathbf{b}\| \le \|A^{-1}\|\|\delta\mathbf{b}\|. \tag{1.61}$$

Thus, the change $\|\delta\mathbf{x}\|$ in the solution is bounded by $\|A^{-1}\|$ times the change $\|\delta\mathbf{b}\|$ in the right-hand side.

The conditioning of the linear system is connected with the ratio between the relative error $\dfrac{\|\delta\mathbf{x}\|}{\|\mathbf{x}\|}$ and the relative change $\dfrac{\|\delta\mathbf{b}\|}{\|\mathbf{b}\|}$ in the right-hand side, which gives

$$\frac{\|\delta\mathbf{x}\|/\|\mathbf{x}\|}{\|\delta\mathbf{b}\|/\|\mathbf{b}\|} = \frac{\|A^{-1}\delta\mathbf{b}\|/\|\mathbf{x}\|}{\|\delta\mathbf{b}\|/\|A\mathbf{x}\|} = \frac{\|A\mathbf{x}\|\|A^{-1}\delta\mathbf{b}\|}{\|\mathbf{x}\|\|\delta\mathbf{b}\|} \leq \|A\|\|A^{-1}\|,$$

which implies that

$$\frac{\|\delta\mathbf{x}\|}{\|\mathbf{x}\|} \leq K(A)\frac{\|\delta\mathbf{b}\|}{\|\mathbf{b}\|}. \tag{1.62}$$

Thus, the relative change in the solution is bounded by the condition number of the matrix times the relative change in the right-hand side. When the product in the right-hand side is small, the relative change in the solution is small.

Case 1.2 *Suppose that the matrix A is replaced by $A + \delta A$, where δA is the error in A, while the right-hand side term $\mathbf{b}$ is similar. If $\mathbf{x} + \delta\mathbf{x}$ is the solution corresponding to the matrix $A + \delta A$, then we have*

$$(A + \delta A)(\mathbf{x} + \delta\mathbf{x}) = \mathbf{b}, \tag{1.63}$$

which implies that

$$A\mathbf{x} + A\delta\mathbf{x} + \delta A(\mathbf{x} + \delta\mathbf{x}) = \mathbf{b}$$

or

$$A\delta\mathbf{x} = -\delta A(\mathbf{x} + \delta\mathbf{x}).$$

Multiplying by A^{-1}, we get

$$\delta\mathbf{x} = -A^{-1}\delta A(\mathbf{x} + \delta\mathbf{x}).$$

Taking the norm gives

$$\|\delta\mathbf{x}\| = \| -A^{-1}\delta A(\mathbf{x} + \delta\mathbf{x})\| \leq \|A^{-1}\|\|\delta A\|(\|\mathbf{x}\| + \|\delta\mathbf{x}\|)$$

or

$$\|\delta\mathbf{x}\|(1 - \|A^{-1}\|\|\delta A\|) \le \|A^{-1}\|\|\delta A\|\|\mathbf{x}\|,$$

which can be written as

$$\frac{\|\delta\mathbf{x}\|}{\|\mathbf{x}\|} \le \frac{\|A^{-1}\|\|\delta A\|}{(1 - \|A^{-1}\|\|\delta A\|)} = \frac{K(A)\|\delta A\|/\|A\|}{(1 - \|A^{-1}\|\|\delta A\|)}. \tag{1.64}$$

If the product $\|A^{-1}\|\|\delta A\|$ is much smaller than 1, the denominator in (1.64) is near 1. Consequently, when $\|A^{-1}\|\|\delta A\|$ is much smaller than 1, then (1.64) implies that the relative change in the solution is bounded by the condition number of a matrix times the relative change in the coefficient matrix.

Case 1.3 *Suppose that there is a change in the coefficient matrix A and the right-hand side term $\mathbf{b}$ together, and if $\mathbf{x} + \delta\mathbf{x}$ is the solution corresponding to the coefficient matrix $A + \delta A$ and the right-hand side $\mathbf{b} + \delta\mathbf{b}$, then we have*

$$(A + \delta A)(\mathbf{x} + \delta\mathbf{x}) = (\mathbf{b} + \delta\mathbf{b}), \tag{1.65}$$

which implies that

$$A\mathbf{x} + A\delta\mathbf{x} + \mathbf{x}\delta A + \delta A\delta\mathbf{x} = \mathbf{b} + \delta\mathbf{b}$$

or

$$A\delta\mathbf{x} + \delta\mathbf{x}\delta A = (\delta\mathbf{b} - \mathbf{x}\delta A).$$

Multiplying by A^{-1}, we get

$$\delta\mathbf{x}(I + A^{-1}\delta A) = A^{-1}(\delta\mathbf{b} - \mathbf{x}\delta A)$$

or

$$\delta\mathbf{x} = (I + A^{-1}\delta A)^{-1}A^{-1}(\delta\mathbf{b} - \mathbf{x}\delta A). \tag{1.66}$$

Since we know that if A is nonsingular and δA is the error in A, we obtain

$$\|A^{-1}\delta A\| \le \|A^{-1}\|\|\delta A\| < 1, \tag{1.67}$$

it then follows that (see Fröberg 1969) the matrix $(I + A^{-1}\delta A)$ is nonsingular and

$$\|(I + A^{-1}\delta A)^{-1}\| \le \frac{1}{1 - \|A^{-1}\delta A\|} \le \frac{1}{1 - \|A^{-1}\|\|\delta A\|}. \tag{1.68}$$

Taking the norm of (1.66) and using (1.68) gives

$$\|\delta\mathbf{x}\| \leq \frac{\|A^{-1}\|}{1 - \|A^{-1}\|\|\delta A\|}[\|\delta\mathbf{b}\| + \|\mathbf{x}\|\|\delta A\|]$$

or

$$\frac{\|\delta\mathbf{x}\|}{\|\mathbf{x}\|} \leq \frac{\|A^{-1}\|}{1 - \|A^{-1}\|\|\delta A\|}\left[\frac{\|\delta\mathbf{b}\|}{\|\mathbf{x}\|} + \|\delta A\|\right]. \tag{1.69}$$

Since we know that

$$\|\mathbf{x}\| \geq \frac{\|\mathbf{b}\|}{\|A\|}, \tag{1.70}$$

by using (1.70) in (1.69), we get

$$\frac{\|\delta\mathbf{x}\|}{\|\mathbf{x}\|} \leq \frac{K(A)}{1 - K(A)\dfrac{\|\delta A\|}{\|A\|}}\left[\frac{\|\delta A\|}{\|A\|} + \frac{\|\delta\mathbf{b}\|}{\|\mathbf{b}\|}\right]. \tag{1.71}$$

The estimate (1.71) shows that small relative changes in A and **b** *cause small relative changes in the solution* **x** *of the linear system (1.59) if the inequality*

$$\frac{K(A)}{1 - K(A)\dfrac{\|\delta A\|}{\|A\|}} \tag{1.72}$$

is not too large.   •

1.6   Applications

In this section we discuss applications of linear systems. Here, we will solve or tackle a variety of real-life problems from several areas of science.

1.6.1   Curve Fitting, Electric Networks, and Traffic Flow

Curve Fitting

The following problem occurs in many different branches of science. A set of data points

$$(x_1, y_1), (x_2, y_2), \ldots, (x_n, y_n)$$

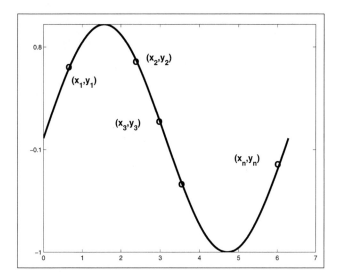

Figure 1.3: Fitting a graph to data points.

is given and it is necessary to find a polynomial whose graph passes through the points. The points are often measurements in an experiment. The x-coordinates are called *base points*. It can be shown that if the base points are all distinct, then a unique polynomial of degree $n - 1$ (or less)

$$p(x) = a_0 + a_1 x + \cdots + a_{n-2} x^{n-2} + a_{n-1} x^{n-1}$$

can be fitted to the points (Figure 1.3).

The coefficients $a_{n-1}, a_{n-2}, \ldots, a_1, a_0$ of the appropriate polynomial can be found by substituting the points into the polynomial equation and then solving a system of linear equations. It is usual to write the polynomial in terms of ascending powers of x for the purpose of finding these coefficients. The columns of the matrix of coefficients of the system of equations then often follow a pattern. More will be discussed about this in the next chapter.

We now illustrate the procedure by fitting a polynomial of degree 2, a parabola, to a set of three such data points.

Example 1.59 *Determine the equation of the polynomial of degree 2 whose graph passes through the points* $(1, 6), (2, 3),$ *and* $(3, 2)$.

Solution. *Observe that in this example we are given three points and we want to find a polynomial of degree 2 (one less than the number of data points). Let the polynomial be*

$$p(x) = a_0 + a_1 x + a_2 x^2.$$

We are given three points and shall use these three sets of information to determine the three unknowns $a_0, a_1,$ *and* a_2. *Substituting*

$$x = 1, y = 6; x = 2, y = 3; x = 3, y = 2,$$

in turn, into the polynomial leads to the following system of three linear equations in $a_0, a_1,$ *and* a_2:

$$
\begin{array}{rrrrl}
a_0 & + & a_1 & + & a_2 & = & 6 \\
a_0 & + & 2a_1 & + & 4a_2 & = & 3 \\
a_0 & + & 3a_1 & + & 9a_2 & = & 2.
\end{array}
$$

Solve this system for $a_2, a_1,$ *and* a_0 *using the Gauss elimination method:*

$$
\begin{pmatrix}
1 & 1 & 1 & \vdots & 6 \\
1 & 2 & 4 & \vdots & 3 \\
1 & 3 & 9 & \vdots & 2
\end{pmatrix}
\approx
\begin{pmatrix}
1 & 1 & 1 & \vdots & 6 \\
0 & 1 & 3 & \vdots & -3 \\
0 & 2 & 8 & \vdots & -4
\end{pmatrix}
\approx
\begin{pmatrix}
1 & 1 & 1 & \vdots & 6 \\
0 & 1 & 3 & \vdots & -3 \\
0 & 0 & 2 & \vdots & 2
\end{pmatrix}.
$$

Now use backward substitution to get the solution of the system (Figure 1.4),

$$
\begin{array}{rclccrcl}
2a_2 & = & 2 & \quad gives \quad & a_2 & = & 1 \\[2mm]
a_1 + 3a_2 & = & -3 & \quad gives \quad & a_1 & = & -6 \\[2mm]
a_0 + a_1 + a_2 & = & 6 & \quad gives \quad & a_0 & = & 11.
\end{array}
$$

Thus,

$$p(x) = 11 - 6x + x^2$$

is the required the polynomial.   ●

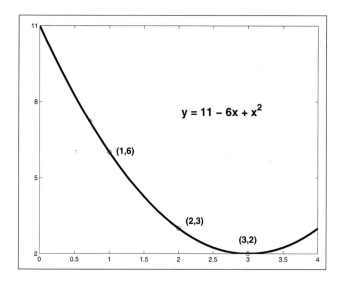

Figure 1.4: Fitting a graph to data points of Example 1.59.

Electrical Network Analysis

Systems of linear equations are used to determine the currents through various branches of electrical networks. The following two laws, which are based on experimental verification in the laboratory, lead to the equations.

Theorem 1.32 (Kirchoff's Laws)

1. Junctions: *All the current flowing into a junction must flow out of it.*
2. Paths: *The sum of the IR terms (where I denotes current and R resistance) in any direction around a closed path is equal to the total voltage in the path in that direction.* •

Example 1.60 *Consider the electric network in Figure 1.5. Let us determine the currents through each branch of this network.*

Solution. *The batteries are 8 volts and 16 volts. The resistances are 1 ohm, 4 ohms, and 2 ohms. The current entering each battery will be the*

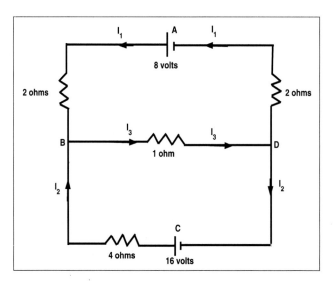

Figure 1.5: Electrical circuit.

same as that leaving it.

Let the currents in the various branches of the given circuit be $\mathbf{I_1}, \mathbf{I_2}$, and $\mathbf{I_3}$. Kirchhoff's Laws refer to junctions and closed paths. There are two junctions in these circuits, namely, the points $\mathbf{B}$ and $\mathbf{D}$. There are three closed paths, namely ABDA, CBDC, and ABCDA. Apply the laws to the junctions and paths.

Junctions

$$Junction\ B: \mathbf{I_1}\ +\ \mathbf{I_2}\ =\ \mathbf{I_3}$$
$$Junction\ D: \mathbf{I_3}\ =\ \mathbf{I_1}\ +\ \mathbf{I_2}$$

These two equations result in a single linear equation

$$\mathbf{I_1} + \mathbf{I_2} - \mathbf{I_3} = 0.$$

Paths

$$Path\ ABDA: 2\mathbf{I_1}\ +\ 1\mathbf{I_3}\ +\ 2\mathbf{I_1}\ =\ 8$$
$$Path\ CBDC: 4\mathbf{I_2}\ +\ 1\mathbf{I_3}\ \phantom{+\ 2\mathbf{I_1}}\ =\ 16$$

It is not necessary to look further at path ABCDA. We now have a system of three linear equations in three unknowns, I_1, I_2, and I_3. Path ABCDA, in fact, leads to an equation that is a combination of the last two equations; there is no new information.

The problem thus reduces to solving the following system of three linear equations in three variables I_1, I_2, and I_3:

$$\begin{array}{rcrcrcl} I_1 & + & I_2 & - & I_3 & = & 0 \\ 4I_1 & & & + & I_3 & = & 8 \\ & & 4I_2 & + & I_3 & = & 16. \end{array}$$

Solve this system for I_1, I_2, and I_3 using the Gauss elimination method:

$$\begin{pmatrix} 1 & 1 & -1 & \vdots & 0 \\ 4 & 0 & 1 & \vdots & 8 \\ 0 & 4 & 1 & \vdots & 16 \end{pmatrix} \approx \begin{pmatrix} 1 & 1 & -1 & \vdots & 0 \\ 0 & -4 & 5 & \vdots & 8 \\ 0 & 4 & 1 & \vdots & -4 \end{pmatrix} \approx \begin{pmatrix} 1 & 1 & -1 & \vdots & 0 \\ 0 & -4 & 5 & \vdots & 8 \\ 0 & 0 & 6 & \vdots & 24 \end{pmatrix}.$$

Now use backward substitution to get the solution of the system:

$$6I_3 = 24 \qquad gives \qquad I_3 = 4$$

$$-4I_2 + 5I_3 = 8 \qquad gives \qquad I_2 = 3$$

$$I_1 + I_2 - I_3 = 0 \qquad gives \qquad I_1 = 1.$$

Thus, the currents are $I_1 = 1, I_2 = 3$, and $I_3 = 4$. The units are amps. The solution is unique, as is to be expected in this physical situation. ●

Traffic Flow

Network analysis, as we saw in the previous discussion, plays an important role in electrical engineering. In recent years, the concepts and tools of network analysis have been found to be useful in many other fields, such as information theory and the study of transportation systems. The following analysis of traffic flow through a road network during peak periods illustrates how systems of linear equations with many solutions can arise in practice.

Consider the typical road network in Figure 1.6. It represents an area of downtown Jacksonville, Florida. The streets are all one-way with the arrows indicating the direction of traffic flow. The flow of traffic in and out of the network is measured in vehicles per hour (vph). The figures given here are based on midweek peak traffic hours, 7 A.M. to 9 A.M. and 4 P.M. to 6 P.M. An increase of 2% in the overall flow should be allowed for during Friday evening traffic flow. Let us construct a mathematical model that can be used to analyze this network. Let the traffic flows along the

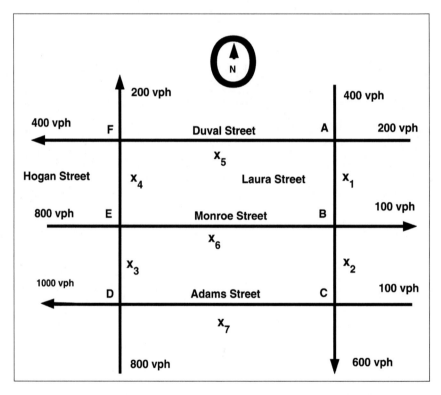

Figure 1.6: Downtown Jacksonville, Florida, USA.

various branches be $x_1, \ldots, x_7$ as shown in Figure 1.6.

Theorem 1.33 (Traffic Law)

All traffic entering a junction must leave that junction. •

This conservation of flow constraint (compare it to the first of Kirchhoff's Laws for electrical networks) leads to a system of linear equations:

$$
\begin{aligned}
\textit{Junction } A : \quad &\textit{Traffic entering } = & 400 + 200 \\
&\textit{Traffic leaving } = & x_1 + x_5 \\
&\textit{Thus} & x_1 + x_5 = 600 \\
\textit{Junction } B : \quad &\textit{Traffic entering } = & x_1 + x_6 \\
&\textit{Traffic leaving } = & x_2 + 100 \\
&\textit{Thus,} & x_1 + x_6 = x_2 + 100.
\end{aligned}
$$

Continuing thus for each junction and writing the resulting equations in convenient form with variables on the left and constraints on the right, we get the following system of linear equations:

$$
\begin{array}{llrrrrrrl}
\textit{Junction } A : & x_1 & & & & +x_5 & & & = 600 \\
\textit{Junction } B : & x_1 & -x_2 & & & & +x_6 & & = 100 \\
\textit{Junction } C : & & x_2 & & & & & -x_7 & = 500 \\
\textit{Junction } D : & & & -x_3 & & & & +x_7 & = 200 \\
\textit{Junction } E : & & & -x_3 & +x_4 & & +x_6 & & = 800 \\
\textit{Junction } F : & & & & x_4 & +x_5 & & & = 600
\end{array}
$$

The Gauss–Jordan elimination method is used to solve this system of equations. Observe that the augmented matrix contains many zeros. These zeros greatly reduce the amount of computation involved. In practice, networks are much larger than the one we have illustrated here, and the systems of linear equations that describe them are thus much larger. The systems are solved on a computer, however, the augmented matrices of all such systems contain many zeros.

Solve this system for $x_1, x_2, \ldots, x_7$ using the Gauss–Jordan elimination method:

$$
\left(
\begin{array}{ccccccc:c}
1 & 0 & 0 & 0 & 1 & 0 & 0 & 600 \\
1 & -1 & 0 & 0 & 0 & 1 & 0 & 100 \\
0 & 1 & 0 & 0 & 0 & 0 & -1 & 500 \\
0 & 0 & -1 & 0 & 0 & 0 & 1 & 200 \\
0 & 0 & -1 & 0 & 0 & 0 & 1 & 800 \\
0 & 0 & 0 & 1 & 1 & 0 & 0 & 600
\end{array}
\right) \approx \cdots \approx
$$

$$\approx \begin{pmatrix} 1 & 0 & 0 & 0 & 0 & 1 & -1 & \vdots & 600 \\ 0 & 1 & 0 & 0 & 0 & 0 & -1 & \vdots & 500 \\ 0 & 0 & 1 & 0 & 0 & 0 & -1 & \vdots & -200 \\ 0 & 0 & 0 & 1 & 0 & 1 & -1 & \vdots & 600 \\ 0 & 0 & 0 & 0 & 1 & -1 & 1 & \vdots & 000 \\ 0 & 0 & 0 & 0 & 0 & 0 & 0 & \vdots & 000 \end{pmatrix}.$$

The system of equations that corresponds to this form is:

$$
\begin{aligned}
x_1 && +x_6 & -x_7 &= 600 \\
x_2 && & -x_7 &= 500 \\
x_3 && & -x_7 &= -200 \\
x_4 && +x_6 & -x_7 &= 600 \\
x_5 && -x_6 & +x_7 &= 000.
\end{aligned}
$$

Expressing each leading variable in terms of the remaining variables, we get

$$
\begin{aligned}
x_1 &= -x_6 + x_7 + 600 \\
x_2 &= x_7 + 500 \\
x_3 &= x_7 - 200 \\
x_4 &= -x_6 + x_7 + 600 \\
x_5 &= x_6 - x_7
\end{aligned}.
$$

As was perhaps to be expected, the system of equations has many solutions—there are many traffic flows possible. One does have a certain amount of choice at intersections.

Let us now use this mathematical model to arrive at information. Suppose it becomes necessary to perform road work on the stretch of Adams Street between Laura and Hogan. It is desirable to have as small a flow of traffic as possible along this stretch of road. The flows can be controlled along various branches by means of traffic lights at junctions. What is the minimum flow possible along Adams that would not lead to traffic congestion? What are the flows along the other branches when this is attained? Our model will enable us to answer these questions.

Minimizing the flow along Adams corresponds to minimizing x_7. Since all traffic flows must be greater than or equal to zero, the third equation implies that the minimum value of x_7 is 200, otherwise, x_3 could become negative. (A negative flow would be interpreted as traffic moving in the opposite direction to the one permitted on a one-way street.) Thus, the road work must allow for a flow of at least 200 cars per hour on the branch CD in the peak period.

Let us now examine what the flows in the other branches will be when this minimum flow along Adams is attained, when x_7 gives

$$
\begin{aligned}
x_1 &= -x_6 + 800 \\
x_2 &= + 700 \\
x_3 &= \ 000 \\
x_4 &= -x_6 + 800 \\
x_5 &= x_6 - 200.
\end{aligned}
$$

Since $x_7 = 200$ implies that $x_3 = 0$ and vice-versa, we see that the minimum flow in branch x_7 can be attained by making $x_3 = 0$; i.e., by closing branch DE to traffic. ●

1.6.2   Heat Conduction

Another typical application of linear systems is in heat-transfer problems in physics and engineering.

Suppose we have a thin rectangular metal plate whose edges are kept at fixed temperatures. As an example, let the left edge be $0°C$, the right edge $2°C$, and the top and bottom edges $1°C$ (Figure 1.7). We want to know the temperature inside the plate. There are several ways of approaching this kind of problem. The simplest approach of interest to us will be the following type of approximation: we shall overlay our plate with finer and finer grids, or meshes. The intersections of the mesh lines are called mesh points. Mesh points are divided into boundary and interior points, depending on whether they lie on the boundary or the interior of the plate. We may consider these points as heat elements, such that each influences its neighboring points. We need the temperature of the interior points,

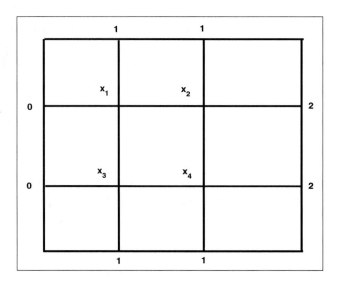

Figure 1.7: Heat-transfer problem.

given the temperature of the boundary points. It is obvious that the finer the grid, the better the approximation of the temperature distribution of the plate. To compute the temperature of the interior points, we use the following principle.

Theorem 1.34 (Mean Value Property for Heat Conduction)

The temperature at any interior point is the average of the temperatures of its neighboring points.   ●

Suppose, for simplicity, we have only four interior points with unknown temperatures x_1, x_2, x_3, x_4, and 12 boundary points (not named) with the temperatures indicated in Figure 1.7.

Example 1.61 *Compute the unknown temperatures x_1, x_2, x_3, x_4 using Figure 1.7.*

Solution. *According to the mean value property, we have*

$$x_1 = \frac{1}{4}(x_2 + x_3 + 1)$$

$$x_2 = \frac{1}{4}(x_1 + x_4 + 3)$$

$$x_3 = \frac{1}{4}(x_1 + x_4 + 1)$$

$$x_4 = \frac{1}{4}(x_2 + x_3 + 3).$$

The problem thus reduces to solving the following system of four linear equations in four variables x_1, x_2, x_3, and x_4:

$$
\begin{array}{rrrrcl}
4x_1 & - & x_2 & - & x_3 & & & = & 1 \\
-x_1 & + & 4x_2 & & & - & x_4 & = & 3 \\
-x_1 & & & + & 4x_3 & - & x_4 & = & 1 \\
& - & x_2 & - & x_3 & + & 4x_4 & = & 3.
\end{array}
$$

Solve this system for x_1, x_2, x_3, and x_4 using the Gauss elimination method:

$$
\begin{pmatrix}
4 & -1 & -1 & 0 & \vdots & 1 \\
-1 & 4 & 0 & -1 & \vdots & 3 \\
-1 & 0 & 4 & -1 & \vdots & 1 \\
0 & -1 & -1 & 4 & \vdots & 3
\end{pmatrix}
\approx \cdots \approx
\begin{pmatrix}
4 & -1 & -1 & 0 & \vdots & 1 \\
0 & \dfrac{15}{4} & -\dfrac{1}{4} & -1 & \vdots & \dfrac{13}{4} \\
0 & 0 & \dfrac{56}{15} & -\dfrac{16}{15} & \vdots & \dfrac{22}{15} \\
0 & 0 & 0 & \dfrac{24}{7} & \vdots & \dfrac{30}{7}
\end{pmatrix}.
$$

Now use backward substitution to get the solution of the system:

$$\frac{24}{7}x_4 = \frac{30}{7} \qquad gives \qquad x_4 = \frac{5}{4}$$

$$\frac{56}{15}x_3 - \frac{16}{15}x_4 = \frac{22}{15} \qquad gives \qquad x_3 = \frac{3}{4}$$

$$\frac{15}{4}x_2 - \frac{1}{4}x_3 - x_4 = \frac{13}{4} \qquad gives \qquad x_2 = \frac{5}{4}$$

$$4x_1 - x_2 - x_3 = 1 \qquad gives \qquad x_1 = \frac{3}{4}.$$

Thus, the temperatures are $x_1 = \frac{3}{4}, x_2 = \frac{5}{4}, x_3 = \frac{3}{4}$, and $x_4 = \frac{5}{4}$.   ●

1.6.3 Chemical Solutions and Balancing Chemical Equations

Example 1.62 (Chemical Solutions) *It takes three different ingredients, A, B, and C, to produce a certain chemical substance. A, B, and C have to be dissolved in water separately before they interact to form the chemical. The solution containing A at 2.5g per cubic centimeter (g/cm^3) combined with the solution containing B at $4.2g/cm^3$, combined with the solution containing C at $5.6g/cm^3$, makes 26.50g of the chemical. If the proportions for A, B, C in these solutions are changed to $3.4, 4.7$, and $2.8g/cm^3$, respectively (while the volumes remain the same), then 22.86g of the chemical is produced. Finally, if the proportions are changed to $3.7, 6.1$, and $3.7g/cm^3$, respectively, then 29.12g of the chemical is produced. What are the volumes in cubic centimeters of the solutions containing A, B, and C?*

Solution. *Let x, y, z be the cubic centimeters of the corresponding volumes of the solutions containing A, B, and C. Then $2.5x$ is the mass of A in the first case, $4.2y$ is the mass of B, and $5.6z$ is the mass of C. Added together, the three masses should be 26.50. So $2.5x + 4.2y + 5.6z = 26.50$. The same reasoning applies to the other two cases, and we get the system*

$$
\begin{aligned}
2.5x &+& 4.2y &+& 5.6z &=& 26.50 \\
3.4x &+& 4.7y &+& 2.8z &=& 22.86 \\
3.6x &+& 6.1y &+& 3.7z &=& 29.12.
\end{aligned}
$$

Solve this system for x, y, and z using the Gauss elimination method:

$$
\begin{pmatrix}
2.5 & 4.2 & 5.6 & \vdots & 26.50 \\
3.4 & 4.7 & 2.8 & \vdots & 22.86 \\
3.6 & 6.1 & 3.7 & \vdots & 29.12
\end{pmatrix}
$$

$$
\approx
\begin{pmatrix}
2.5 & 4.2 & 5.6 & \vdots & 26.50 \\
0 & -1.012 & -4.816 & \vdots & -13.18 \\
0 & 0.052 & -4.364 & \vdots & -9.04
\end{pmatrix}
$$

$$\approx \begin{pmatrix} 2.5 & 4.2 & 5.6 & \vdots & 26.50 \\ 0 & -1.012 & -4.816 & \vdots & -13.18 \\ 0 & 0 & -4.612 & \vdots & -9.717 \end{pmatrix}.$$

Now use backward substitution to get the solution of the system:

$$\begin{array}{lll} -4.612z = -9.717 & gives & z = 2.107 \\ -1.012y - 4.816z = -13.18 & gives & y = 2.996 \\ 2.5x + 4.2y + 5.6z = 26.50 & gives & x = 0.847. \end{array}$$

Hence, the volumes of the solutions containing $A, B,$ and C are, respectively, $0.847cm^3, 2.996cm^3,$ and $2.107cm^3$. •

Balancing Chemical Equations

When a chemical reaction occurs, certain molecules (the reactants) combine to form new molecules (the products). A balanced chemical equation is an algebraic equation that gives the relative numbers of reactants and products in the reaction and has the same number of atoms of each type on the left- and right-hand sides. The equation is usually written with the reactants on the left, the products on the right, and an arrow in between to show the direction of the reaction.

For example, for the reaction in which hydrogen gas (H_2) and oxygen (O_2) combine to form water (H_2O), a balanced chemical equation is

$$2H_2 + O_2 \longrightarrow 2H_2O,$$

indicating that two molecules of hydrogen combine with one molecule of oxygen to form two molecules of water. Observe that the equation is balanced, since there are four hydrogen atoms and two oxygen atoms on each side. Note that there will never be a unique balanced equation for a reaction, since any positive integer multiple of a balanced equation will also be balanced. For example, $6H_2 + 3O_2 \longrightarrow 6H_2O$ is also balanced. Therefore, we usually look for the simplest balanced equation for a given reaction. Note that the process of balancing chemical equations really involves solving a homogeneous system of linear equations.

Example 1.63 (Balancing Chemical Equations) *The combustion of ammonia (NH_3) in oxygen produces nitrogen (N_2) and water. Find a balanced chemical equation for this reaction.*

Solution. *Let w, x, y, and z denote the numbers of molecules of ammonia, oxygen, nitrogen, and water, respectively, then we are seeking an equation of the form*

$$wNH_3 + xO_2 \longrightarrow yN_2 + zH_2O.$$

Comparing the number of nitrogen, hydrogen, and oxygen atoms in the reactants and products, we obtain three linear equations:

$$
\begin{array}{rrcl}
Nitrogen: & w & = & 2y \\
Hydrogen: & 3w & = & 2z \\
Oxygen: & 2x & = & z.
\end{array}
$$

Rewriting these equations in standard form gives us a homogeneous system of three equations in four variables:

$$
\begin{array}{rcrcrcl}
w & & & - & 2y & & & = & 0 \\
3w & & & & & - & 2z & = & 0 \\
& & 2x & & & - & z & = & 0.
\end{array}
$$

The augmented matrix form of the system is

$$
\left(
\begin{array}{cccc:c}
1 & 0 & -2 & 0 & 0 \\
3 & 0 & 0 & -2 & 0 \\
0 & 2 & 0 & -1 & 0
\end{array}
\right).
$$

Solve this system for w, x, y, and z using the Gauss elimination method with partial pivoting:

$$
\left(
\begin{array}{cccc:c}
3 & 0 & 0 & -2 & 0 \\
0 & 0 & -2 & \frac{2}{3} & 0 \\
0 & 2 & 0 & -1 & 0
\end{array}
\right)
\approx
\left(
\begin{array}{cccc:c}
3 & 0 & 0 & -2 & 0 \\
0 & 2 & 0 & -1 & 0 \\
0 & 0 & -2 & \frac{2}{3} & 0
\end{array}
\right).
$$

Now use backward substitution to get the solution of the homogeneous system:

$$-2y \quad + \quad \tfrac{2}{3}z = 0 \qquad gives \quad y = \tfrac{1}{3}z$$

$$2x \quad - \quad z = 0 \qquad gives \quad x = \tfrac{1}{2}z$$

$$3w \quad - \quad 2z = 0 \qquad gives \quad w = \tfrac{2}{3}z.$$

The smallest positive value of z that will produce integer values for all four variables is the least common denominator of the fractions $\tfrac{2}{3}, \tfrac{1}{2}$, and $\tfrac{1}{3}$—namely, 6—which gives

$$w = 4, \quad x = 3, \quad y = 2, \quad z = 6.$$

Therefore,

$$4NH_3 + 3O_2 \longrightarrow 2N_2 + 6H_2O$$

is the balanced chemical equation.   ●

1.6.4   Manufacturing, Social, and Financial Issues

Example 1.64 (Manufacturing) *Sun Microsystems manufactures three types of personal computers: The Cyclone, the Cyclops, and the Cycloid. It takes 15 hours to assemble the Cyclone, 4 hours to test its hardware, and 5 hours to install its software. The hours required for the Cyclops are 12 hours to assemble, 4.5 hours to test, and 2.5 hours to install. The Cycloid, being the lower end of the line, requires 10 hours to assemble, 3 hours to test, and 2.5 hours to install. If the company's factory can afford 1250 labor hours per month for assembling, 400 hours for testing, and 320 hours for installation, how many PCs of each kind can be produced in a month?*

Solution. *Let x, y, z be the number of Cyclones, Cyclops, and Cycloids produced each month. Then it takes $15x + 12y + 10z$ hours to assemble the computers. Hence, $15x + 12y + 10z = 1250$. Similarly, we get equations for testing and installing. The resulting system is*

$$
\begin{array}{rcrcrcl}
15x & + & 12y & + & 10z & = & 1250 \\
4x & + & 4.5y & + & 3z & = & 400 \\
5x & + & 2.5y & + & 2.5z & = & 320.
\end{array}
$$

Solve this system for $x, y,$ and z using the Gauss elimination method:

$$
\begin{pmatrix}
15 & 12 & 10 & \vdots & 1250 \\
4 & 4.5 & 3 & \vdots & 400 \\
5 & 2.5 & 2.5 & \vdots & 320
\end{pmatrix}
\approx
\begin{pmatrix}
15 & 12 & 10 & \vdots & 1250 \\
0 & \dfrac{13}{10} & \dfrac{1}{3} & \vdots & \dfrac{200}{3} \\
0 & -\dfrac{3}{2} & -\dfrac{5}{6} & \vdots & -\dfrac{290}{3}
\end{pmatrix}
$$

$$
\approx
\begin{pmatrix}
15 & 12 & 10 & \vdots & 1250 \\
0 & \dfrac{13}{10} & \dfrac{1}{3} & \vdots & \dfrac{200}{3} \\
0 & 0 & -\dfrac{35}{78} & \vdots & -\dfrac{770}{39}
\end{pmatrix}.
$$

Now use backward substitution to get the solution of the system:

$$-\frac{35}{78}z = -\frac{770}{39} \qquad gives \quad z = 44$$

$$\frac{13}{10}y + \frac{1}{3}z = \frac{200}{3} \qquad gives \quad y = 40$$

$$15x + 12y + 10z = 1250 \qquad gives \quad x = 20.$$

Hence, 20 Cyclones, 40 Cyclops, and 44 Cycloids can be manufactured monthly. ●

Example 1.65 (Weather) *The average of the temperature for the cities of Jeddah, Makkah, and Riyadh was 50°C during a given summer day. The temperature in Makkah was 5°C higher than the average of the temperatures of the other two cities. The temperature in Riyadh was 5°C lower than the average temperature of the other two cities. What was the temperature in each of the cities?*

Solution. *Let x, y, z be the temperatures in Jeddah, Makkah, and Riyadh, respectively. The average temperature of all three cities is $\frac{(x+y+z)}{3}$, which is*

$50°C$. *On the other hand, the temperature in Makkah exceeds the average temperature of Jeddah and Riyadh,* $\frac{(x+z)}{2}$, *by* $5°C$. *So,* $y = \frac{(x+z)}{2+5}$. *Likewise, we have* $z = \frac{(x+y)}{2-5}$. *So, the system becomes*

$$\frac{(x+y+z)}{3} = 50$$

$$y = \frac{(x+z)}{2} + 5$$

$$z = \frac{(x+y)}{2} - 5.$$

Rewriting the above system in standard form, we get

$$\begin{array}{rcrcrcr}
x & + & y & + & z & = & 150 \\
-x & + & 2y & - & z & = & 10 \\
-x & - & y & + & 2z & = & -10.
\end{array}$$

Solve this system for x, y, and z using the Gauss elimination method:

$$\begin{pmatrix} 1 & 1 & 1 & \vdots & 150 \\ -1 & 2 & -1 & \vdots & 10 \\ -1 & -1 & 2 & \vdots & -10 \end{pmatrix} \approx \begin{pmatrix} 1 & 1 & 1 & \vdots & 150 \\ 0 & 3 & 0 & \vdots & 160 \\ 0 & 0 & 3 & \vdots & 140 \end{pmatrix}.$$

Now use backward substitution to get the solution of the system:

$$\begin{array}{lcl}
3z = 140 & gives & z = 46.667 \\
3y = 160 & gives & y = 53.333 \\
x + y + z = 150 & gives & x = 50.
\end{array}$$

Thus, the temperature in Jeddah was $50°C$ and the temperatures in Makkah and Riyadh were approximately, $53°C$ and $47°C$, respectively. ●

Example 1.66 (Foreign Currency Exchange) *An international business person needs, on the average, fixed amounts of Pakistani rupees, English pounds, and Saudi riyals during each of his business trips. He traveled three times this year. The first time he exchanged a total of $26000 at the* *following rates: the dollar was 60 rupees, 0.6 pounds, and 3.75 riyals. The second time he exchanged a total of $25500 at these rates: the dollar was 65 rupees, 0.56 pounds, and 3.76 riyals. The third time he exchanged again a total of $25500 at these rates: the dollar was 65 rupees, 0.6 pounds, and 3.75 riyals. How many rupees, pounds, and riyals did he buy each time?*

Solution. *Let x, y, z be the fixed amounts of rupees, pounds, and riyals he purchases each time. Then the first time he spent $(\frac{1}{60})x$ dollars to buy rupees, $(\frac{1}{0.6})y$ dollars to buy pounds, and $(\frac{1}{3.75})z$ dollars to buy riyals. Hence,*

$$\left(\frac{1}{60}\right)x + \left(\frac{1}{0.8}\right)y + \left(\frac{1}{3.75}\right)z = 26000.$$

The same reasoning applies to the other two purchases, and we get the system

$$\frac{1}{60}x + \frac{5}{3}y + \frac{4}{15}z = 26000$$

$$\frac{1}{65}x + \frac{25}{14}y + \frac{25}{94}z = 25500$$

$$\frac{1}{65}x + \frac{5}{3}y + \frac{4}{15}z = 25500.$$

Solve this system for $x, y,$ and z using the Gauss elimination method:

$$\begin{pmatrix} \frac{1}{60} & \frac{5}{3} & \frac{4}{15} & \vdots & 26000 \\ \frac{1}{65} & \frac{25}{14} & \frac{25}{94} & \vdots & 25500 \\ \frac{1}{65} & \frac{5}{3} & \frac{4}{15} & \vdots & 25500 \end{pmatrix} \approx \begin{pmatrix} \frac{1}{60} & \frac{5}{3} & \frac{4}{15} & \vdots & 26000 \\ 0 & \frac{45}{182} & \frac{121}{6110} & \vdots & 1500 \\ 0 & \frac{5}{39} & \frac{4}{195} & \vdots & 1500 \end{pmatrix}$$

$$\approx \begin{pmatrix} \dfrac{1}{60} & \dfrac{5}{3} & \dfrac{4}{15} & \vdots & 26000 \\[3mm] 0 & \dfrac{45}{182} & \dfrac{121}{6110} & \vdots & 1500 \\[3mm] 0 & 0 & \dfrac{13}{1269} & \vdots & \dfrac{6500}{9} \end{pmatrix}.$$

Now use backward substitution to get the solution of the system:

$$\frac{13}{1269}z = \frac{6500}{9} \qquad \textit{gives} \qquad z = 70500$$

$$\frac{45}{182}y + \frac{121}{6110}z = 1500 \qquad \textit{gives} \qquad y = 420$$

$$\frac{1}{60}x + \frac{5}{3}y + \frac{4}{15}z = 26000 \qquad \textit{gives} \qquad x = 390000.$$

Therefore, each time he bought 390000 rupees, 420 pounds, and 70500 riyals for his trips. •

Example 1.67 (Inheritance) *A father plans to distribute his estate, worth SR234,000, between his four daughters as follows:* $\frac{2}{3}$ *of the estate is to be split equally among the daughters. For the rest, each daughter is to receive SR3,000 for each year that remains until her 21st birthday. Given that the daughters are all 3 years apart, how much would each receive from her father's estate? How old are the daughters now?*

Solution. *Let x, y, z, and w be the amounts of money that each daughter will receive from the splitting of $\frac{1}{3}$ of the estate, according to age, starting with the oldest one. Then $x + y + z + w = \frac{1}{3}(234,000) = 78,000$. On the other hand, $w - z = 3(3000)$, $z - y = 3(3000)$, and $y - x = 3(3000)$. The problem thus reduces to solving the following system of four linear equations in four variables $x, y, z,$ and w:*

$$\begin{aligned} x + y + z + w &= 78,000 \\ -z + w &= 9,000 \\ -y + z &= 9,000 \\ -x + y &= 9,000. \end{aligned}$$

Solve this system for x_1, x_2, x_3, and x_4 using the Gauss elimination method with partial pivoting:

$$
\left(\begin{array}{cccc:c}
1 & 1 & 1 & 1 & 78,000 \\
0 & 0 & -1 & 1 & 9,000 \\
0 & -1 & 1 & 0 & 9,000 \\
-1 & 1 & 0 & 0 & 9,000
\end{array}\right)
\approx
\left(\begin{array}{cccc:c}
1 & 1 & 1 & 1 & 78,000 \\
0 & 0 & -1 & 1 & 9,000 \\
0 & -1 & 1 & 0 & 9,000 \\
0 & 2 & 1 & 1 & 87,000
\end{array}\right)
$$

and

$$
\left(\begin{array}{cccc:c}
1 & 1 & 1 & 1 & 78,000 \\
0 & 2 & 1 & 1 & 87,000 \\
0 & 0 & \frac{3}{2} & \frac{1}{2} & 52,500 \\
0 & 0 & -1 & 1 & 9,000
\end{array}\right)
\approx
\left(\begin{array}{cccc:c}
1 & 1 & 1 & 1 & 78,000 \\
0 & 2 & 1 & 1 & 87,000 \\
0 & 0 & \frac{3}{2} & \frac{1}{2} & 52,500 \\
0 & 0 & 0 & \frac{4}{3} & 44,000
\end{array}\right).
$$

Now use backward substitution to get the solution of the system:

$$
\frac{4}{3}w = 44,000 \qquad \text{gives} \quad w \;=\; 33,000
$$

$$
\frac{3}{2}z + \frac{1}{2}w = 52,500 \qquad \text{gives} \quad z \;=\; 24,000
$$

$$
2y + z + w = 87,000 \qquad \text{gives} \quad y \;=\; 15,000
$$

$$
x + y + z + w = 78,000 \qquad \text{gives} \quad x \;=\; 6,000.
$$

One-quarter of two-thirds of the estate is worth $\frac{1}{4}(\frac{2}{3}(234,000)) = SR39,000$. So, the youngest daughter will receive $(33,000 + 39,000) = SR72,000$, the next one $(24,000 + 39,000) = SR63,000$, the next one $(15,000 + 39,000) = SR54,000$, and the first one $(6,000 + 39,000) = SR45,000$. The oldest daughter will receive $6,000 = 2(3,000)$, so she is currently $21 - 2 = 19$. The second one is 16, the third one is 13, and the last one is 10 years old.

●

1.6.5   Allocation of Resources

A great many applications of systems of linear equations involve allocating limited resources subject to a set of constraints.

Example 1.68 *A dietitian is to arrange a special diet composed of four foods $A, B, C,$ and D. The diet is to include 72 units of calcium, 45 units of iron, 42 units of vitamin A, and 60 units of vitamin B. The following table shows the amount of calcium, iron, vitamin A, and vitamin B (in*

Food	Calcium	Iron	Vitamin A	Vitamin B
A	18	6	6	6
B	9	6	12	9
C	9	9	6	9
D	12	12	9	18

units) per ounce in foods $A, B, C,$ and D. Find, if possible, the amount of foods $A, B, C,$ and D that can be included in the special diet to conform to the dietitian's recommendations.

Solution. *Let $x, y, z,$ and w be the ounces of foods $A, B, C,$ and D, respectively. Then we have the system of equations*

$$
\begin{array}{rcrcrcrcr}
18x & + & 9y & + & 9z & + & 12w & = & 72 \\
6x & + & 6y & + & 9z & + & 12w & = & 45 \\
6x & + & 12y & + & 6z & + & 9w & = & 42 \\
6x & + & 9y & + & 9z & + & 18w & = & 60.
\end{array}
$$

Solve this system for $x, y, z,$ and w using the Gauss elimination method:

$$
\left(
\begin{array}{cccc:c}
18 & 9 & 9 & 12 & 72 \\
6 & 6 & 9 & 12 & 45 \\
6 & 12 & 6 & 9 & 42 \\
6 & 9 & 9 & 18 & 60
\end{array}
\right)
\approx
\left(
\begin{array}{cccc:c}
18 & 9 & 9 & 12 & 72 \\
0 & 3 & 6 & 8 & 21 \\
0 & 9 & 3 & 5 & 18 \\
0 & 6 & 6 & 14 & 36
\end{array}
\right)
$$

and

$$
\left(\begin{array}{cccc:c}
18 & 9 & 9 & 12 & 72 \\
0 & 3 & 6 & 8 & 21 \\
0 & 0 & -15 & -19 & -45 \\
0 & 0 & -6 & -2 & 36
\end{array} \right)
\approx
\left(\begin{array}{cccc:c}
18 & 9 & 9 & 12 & 72 \\
0 & 3 & 6 & 8 & 21 \\
0 & 0 & -15 & -19 & -45 \\
0 & 0 & 0 & \dfrac{28}{5} & 12
\end{array} \right).
$$

Now use backward substitution to get the solution of the system:

$$\frac{28}{5}w = 12 \qquad gives \qquad w = \frac{15}{7}$$

$$-15z - 19w = -45 \qquad gives \qquad z = \frac{2}{7}$$

$$3y + 6z + 8w = 21 \qquad gives \qquad y = \frac{5}{7}$$

$$18x + 9y + 9z + 12w = 72 \qquad gives \qquad x = \frac{29}{14}.$$

Thus, the amount in ounces of foods $A, B, C,$ and D are $x = \frac{29}{14}, y = \frac{5}{7}, z = \frac{2}{7},$ and $w = \frac{15}{7}$, respectively.   ●

1.7 Summary

The basic methods for solving systems of linear algebraic equations were discussed in this chapter. Since these methods use matrices and determinants, the basic properties of matrices and determinants were presented.

Several direct solution methods were also discussed. Among them were Cramer's rule, Gaussian elimination and its variants, the Gauss–Jordan method, and the LU decomposition method. Cramer's rule is impractical for solving systems with more than three or four equations. Gaussian elimination is the best choice for solving linear systems. For systems of equations having a constant coefficients matrix but many right-hand side vectors, LU decomposition is the method of choice. The LU decomposition method has been used for the solution of tridiagonal systems. Direct methods are generally used when the number of equations is small, or most of the coefficients of the equations are nonzero, or the system of equations is not diagonally dominant, or the system of equations is ill-conditioned. But these methods are generally impractical when a large number of equations must be solved simultaneously. In this chapter we also discussed conditioning of linear systems by using a parameter called the condition number. Many ill-conditioned systems were discussed. The coefficient matrix A of an ill-conditioned system $A\mathbf{x} = \mathbf{b}$ has a large condition number. The numerical solution to a linear system is less reliable when A has a large condition number than when A has a small condition number. The numerical solution $\mathbf{x}^*$ of $A\mathbf{x} = \mathbf{b}$ is different from the exact solution $\mathbf{x}$ because of round-off errors in all stages of the solution process. The round-off errors occur in the elimination or factorization of A and during backward substitution to compute $\mathbf{x}^*$. The degree to which perturbation in A and $\mathbf{b}$ affect the numerical solution is determined by the value of the condition number $K(A)$. A large value of $K(A)$ indicates that A is close to being singular. When $K(A)$ is large, matrix A is said to be ill-conditioned and small perturbations in A and $\mathbf{b}$ cause relatively large differences between $\mathbf{x}$ and $\mathbf{x}^*$. If $K(A)$ is small, any stable algorithm will return a solution with small residual $\mathbf{r}$, while if $K(A)$ is large, then the return solution may have large errors even though the residuals are small. The best way to deal with ill-conditioning is to avoid it by reformulating the problem.

At the end of the chapter we discussed many applications of linear systems. Fitting a polynomial of degree $(n-1)$ to n data points leads to a system of linear equations that has a unique solution. The analysis of electric networks and traffic flow give rise to systems that have unique solutions and many solutions. The model for traffic flow is similar to that of electric networks, but it has fewer restrictions, leading to more freedom and thus many solutions in place of a unique solution. Applications to heat conduction, chemical reactions, balancing equations, manufacturing, social and financial issues, and allocation of resources were also covered.

1.8　Problems

1. Determine the matrix C given by the following expression

$$C = 2A - 3B,$$

if the matrices A and B are

$$A = \begin{pmatrix} 2 & -1 & 1 \\ -1 & 2 & 3 \\ 2 & 1 & 2 \end{pmatrix}, \qquad B = \begin{pmatrix} 1 & 1 & 1 \\ 0 & 1 & 3 \\ 2 & 1 & 4 \end{pmatrix}.$$

2. Find the product AB and BA for the matrices of Problem 1.

3. Show that the product AB of the following rectangular matrices is a singular matrix:

$$A = \begin{pmatrix} 6 & -3 \\ 1 & 4 \\ -2 & 1 \end{pmatrix}, \qquad B = \begin{pmatrix} 2 & -1 & -2 \\ 3 & -4 & -1 \end{pmatrix}.$$

4. Let

$$A = \begin{pmatrix} 1 & 2 & 3 \\ 0 & -1 & 2 \\ 2 & 0 & 2 \end{pmatrix}, \quad B = \begin{pmatrix} 1 & 1 & 2 \\ -1 & 1 & -1 \\ 1 & 0 & 2 \end{pmatrix}, \quad C = \begin{pmatrix} 1 & 0 & 1 \\ 0 & 1 & 2 \\ 2 & 0 & 1 \end{pmatrix}.$$

   (a) Compute AB and BA and show that $AB \neq BA$.
   (b) Find $(A + B) + C$ and $A + (B + C)$.
   (c) Show that $(AB)^T = B^T A^T$.

5. Find a value of x and y such that $AB^T = C^T$, where

$$A = \begin{pmatrix} 1 & 2 & 3 \\ 4 & 2 & 0 \\ 2 & 1 & 3 \end{pmatrix}, \quad B = [1 \; x \; 1], \quad C = [-2 \; -2 \; y].$$

6. Find the values of a and b such that each of the following matrices is symmetric:

(a) $A = \begin{pmatrix} 1 & 3 & 5 \\ a+2 & 5 & 6 \\ b+1 & 6 & 7 \end{pmatrix}$, (b) $B = \begin{pmatrix} -2 & a+b & 2 \\ 3 & 4 & 2a+b \\ 2 & 5 & -3 \end{pmatrix}$,

(c) $C = \begin{pmatrix} 1 & 4 & a-b \\ 4 & 2 & a+3b \\ 7 & 3 & 4 \end{pmatrix}$, (d) $D = \begin{pmatrix} 1 & a-4b & 2 \\ 2 & 8 & 6 \\ 7 & a-7b & 8 \end{pmatrix}$.

7. Which of the following matrices are skew symmetric?

(a)
$$A = \begin{pmatrix} 1 & -5 \\ 5 & 0 \end{pmatrix}, \qquad B = \begin{pmatrix} 0 & -4 \\ 4 & 0 \end{pmatrix},$$

(b)
$$C = \begin{pmatrix} 1 & 9 \\ -9 & 7 \end{pmatrix}, \qquad D = \begin{pmatrix} 1 & 6 \\ -6 & 2 \end{pmatrix},$$

(c)
$$E = \begin{pmatrix} 0 & 2 & -2 \\ -2 & 0 & 4 \\ 2 & -4 & 0 \end{pmatrix}, \qquad F = \begin{pmatrix} 3 & -3 & -3 \\ 3 & 3 & -3 \\ 3 & 3 & 3 \end{pmatrix},$$

(d)
$$G = \begin{pmatrix} 1 & -5 & 1 \\ 5 & 1 & 4 \\ -1 & -4 & 1 \end{pmatrix}, \qquad H = \begin{pmatrix} 2 & 8 & 6 \\ -8 & 4 & 2 \\ -6 & -2 & 5 \end{pmatrix}.$$

8. Determine whether each of the following matrices is in row echelon form, reduced row echelon form, or neither:

(a)
$$A = \begin{pmatrix} 1 & 0 & 0 \\ 0 & 1 & 0 \\ 0 & 0 & 3 \end{pmatrix}, \qquad B = \begin{pmatrix} 1 & 0 & 8 \\ 0 & 1 & 2 \\ 0 & 0 & 0 \end{pmatrix},$$

(b)

$$C = \begin{pmatrix} 1 & 2 & 3 & 0 \\ 0 & 0 & 0 & 1 \\ 0 & 0 & 0 & 1 \end{pmatrix}, \quad D = \begin{pmatrix} 1 & 2 & 0 & 0 & 1 \\ 0 & 0 & 1 & 0 & 1 \\ 0 & 0 & 0 & 1 & 0 \end{pmatrix},$$

(c)

$$E = \begin{pmatrix} 1 & 4 & 5 & 6 \\ 0 & 1 & 7 & 8 \\ 0 & 0 & 1 & 9 \\ 0 & 0 & 0 & 0 \end{pmatrix}, \quad F = \begin{pmatrix} 1 & 0 & 0 & 0 & 3 \\ 0 & 0 & 1 & 0 & 4 \\ 0 & 0 & 0 & 1 & 5 \end{pmatrix},$$

(d)

$$G = \begin{pmatrix} 1 & 0 & 0 & 3 \\ 0 & 1 & 0 & 4 \\ 0 & 0 & 0 & 5 \\ 0 & 0 & 0 & 6 \end{pmatrix}, \quad H = \begin{pmatrix} 0 & 0 & 0 & 0 & 0 \\ 0 & 0 & 1 & 2 & 4 \\ 0 & 0 & 0 & 1 & 0 \\ 0 & 0 & 0 & 0 & 0 \end{pmatrix}.$$

9. Find the row echelon form of each of the following matrices using elementary row operations, and then solve the linear system:

(a)

$$A = \begin{pmatrix} 0 & 1 & 2 \\ 2 & 3 & 4 \\ 1 & 3 & 2 \end{pmatrix}, \quad \mathbf{b} = \begin{pmatrix} 1 \\ -1 \\ 2 \end{pmatrix}.$$

(b)

$$A = \begin{pmatrix} 1 & 2 & 3 \\ 0 & 3 & 1 \\ -1 & 4 & 5 \end{pmatrix}, \quad \mathbf{b} = \begin{pmatrix} 1 \\ 0 \\ -3 \end{pmatrix}.$$

(c)

$$A = \begin{pmatrix} 0 & -1 & 0 \\ 3 & 0 & 1 \\ 0 & 1 & 1 \end{pmatrix}, \quad \mathbf{b} = \begin{pmatrix} 1 \\ 3 \\ 2 \end{pmatrix}.$$

(d)

$$A = \begin{pmatrix} 0 & -1 & 2 & 4 \\ 2 & 3 & 5 & 6 \\ 1 & 3 & -2 & 4 \\ 1 & 2 & -1 & 3 \end{pmatrix}, \quad \mathbf{b} = \begin{pmatrix} 2 \\ 1 \\ -1 \\ 2 \end{pmatrix}.$$

10. Find the row echelon form of each of the following matrices using elementary row operations, and then solve the linear system:

(a)

$$A = \begin{pmatrix} 1 & 4 & -2 \\ 2 & 3 & 2 \\ 6 & 4 & 1 \end{pmatrix}, \qquad b = \begin{pmatrix} 5 \\ -3 \\ 4 \end{pmatrix}.$$

(b)

$$A = \begin{pmatrix} 2 & 2 & 7 \\ 0 & 3 & 2 \\ 3 & 2 & 1 \end{pmatrix}, \qquad b = \begin{pmatrix} 3 \\ 2 \\ 5 \end{pmatrix}.$$

(c)

$$A = \begin{pmatrix} 0 & -1 & 0 \\ 5 & 0 & 2 \\ -1 & 1 & 4 \end{pmatrix}, \qquad b = \begin{pmatrix} 1 \\ 1 \\ 1 \end{pmatrix}.$$

(d)

$$A = \begin{pmatrix} 1 & 1 & 2 & 4 \\ 1 & 3 & 4 & 5 \\ 1 & 4 & 2 & 4 \\ 2 & 2 & -1 & 3 \end{pmatrix}, \qquad b = \begin{pmatrix} 11 \\ 7 \\ 6 \\ 4 \end{pmatrix}.$$

11. Find the reduced row echelon form of each of the following matrices using elementary row operations, and then solve the linear system:

(a)

$$A = \begin{pmatrix} 1 & 2 & 3 \\ -1 & 2 & 1 \\ 0 & 1 & 2 \end{pmatrix}, \qquad b = \begin{pmatrix} 4 \\ 3 \\ 1 \end{pmatrix}.$$

(b)

$$A = \begin{pmatrix} 0 & 1 & 4 \\ 2 & 1 & -1 \\ 1 & 3 & 4 \end{pmatrix}, \qquad b = \begin{pmatrix} 1 \\ 1 \\ -1 \end{pmatrix}.$$

(c)

$$A = \begin{pmatrix} 0 & -1 & 3 & 2 \\ 3 & 2 & 5 & 4 \\ -1 & 3 & 1 & 2 \\ 2 & 3 & 4 & 1 \end{pmatrix}, \quad b = \begin{pmatrix} 6 \\ 4 \\ 4 \\ 4 \end{pmatrix}.$$

(d)

$$A = \begin{pmatrix} 1 & 2 & -4 & 1 \\ -2 & 0 & 2 & 3 \\ 0 & 1 & -1 & 2 \\ 2 & 3 & 0 & -1 \end{pmatrix}, \quad b = \begin{pmatrix} 1 \\ -1 \\ 2 \\ 4 \end{pmatrix}.$$

12. Compute the determinant of each of the following matrices using cofactor expansion along any row or column:

$$A = \begin{pmatrix} \cos x & \sin x & 1 \\ 0 & 3\cos x & -3\sin x \\ 0 & 2\sin x & 2\cos x \end{pmatrix}, \quad B = \begin{pmatrix} x & y & z \\ 0 & x^2 & y \\ 0 & y^2 & x \end{pmatrix}, \quad C = \begin{pmatrix} 2x & 0 & z \\ 0 & 2y & -z \\ z & -z & 2z \end{pmatrix}$$

13. Compute the determinant of each of the following matrices using cofactor expansion along any row or column:

$$A = \begin{pmatrix} 3 & 7 & 6 \\ 0 & 3 & 5 \\ 7 & 4 & 3 \end{pmatrix}, \quad B = \begin{pmatrix} 11 & -6 & 4 \\ -16 & 8 & 6 \\ 5 & 7 & 12 \end{pmatrix}, \quad C = \begin{pmatrix} 4 & -8 & 11 \\ 10 & 1 & 4 \\ 7 & 10 & 8 \end{pmatrix}.$$

14. Let

$$A = \begin{pmatrix} 1 & 1 \\ 0 & 1 \end{pmatrix}, \quad B = \begin{pmatrix} 1 & 0 \\ 1 & 1 \end{pmatrix},$$

then show that $(AB)^{-1} = B^{-1}A^{-1}$.

15. Evaluate the determinant of each of the following matrices using the Gauss elimination method:

$$A = \begin{pmatrix} 3 & 1 & -1 \\ 2 & 0 & 4 \\ 1 & -5 & 1 \end{pmatrix}, \quad B = \begin{pmatrix} 4 & 1 & 6 \\ -3 & 6 & 4 \\ 5 & 0 & 9 \end{pmatrix}, \quad C = \begin{pmatrix} 17 & 46 & 7 \\ 20 & 49 & 8 \\ 23 & 52 & 19 \end{pmatrix}.$$

16. Evaluate the determinant of each of the following matrices using the Gauss elimination method:

$$
A = \begin{pmatrix} 4 & 2 & 5 & -1 \\ 2 & 5 & 4 & 6 \\ 4 & 5 & 1 & 3 \\ 11 & 7 & 1 & 1 \end{pmatrix}, \quad
B = \begin{pmatrix} 4 & -2 & 5 & -3 \\ 1 & 8 & 12 & 7 \\ 1 & 4 & 3 & 6 \\ 5 & 3 & -3 & 6 \end{pmatrix},
$$

$$
C = \begin{pmatrix} 13 & 22 & -12 & 8 \\ 15 & 10 & 33 & 4 \\ 9 & -12 & 5 & 7 \\ 15 & 33 & -19 & 26 \end{pmatrix}, \quad
D = \begin{pmatrix} 9 & 11 & 2 & 8 \\ 15 & 1 & 3 & 12 \\ 9 & -12 & 5 & 17 \\ 13 & 17 & 21 & 15 \end{pmatrix}.
$$

17. Find all zeros (values of x such that $f(x) = 0$) of polynomial $f(x) = \det(A)$, where

$$
A = \begin{pmatrix} x-1 & 3 & 2 \\ 3 & x & 1 \\ 2 & 1 & x-2 \end{pmatrix}.
$$

18. Find all zeros (values of x such that $f(x) = 0$) of polynomial $f(x) = \det(A)$, where

$$
A = \begin{pmatrix} x & 0 & 1 \\ 2 & 1 & 3 \\ 0 & x & 2 \end{pmatrix}.
$$

19. Find all zeros (values of x such that $f(x) = 0$) of polynomial $f(x) = \det(A)$, where

$$
A = \begin{pmatrix} x & -8 & 5 & 2 \\ -3 & x & 2 & 1 \\ 3 & 4 & x & 1 \\ 3 & 6 & -5 & 17 \end{pmatrix}.
$$

20. (a) The matrix

$$
A = \begin{pmatrix} -x & 1 & 0 \\ 0 & -x & 1 \\ -c_0 & -c_1 & -c_2 \end{pmatrix}
$$

is called the *companion matrix* of the polynomial $(-1)(c_2x^2+c_1x+c_0)$. Show that

$$|A| = \begin{vmatrix} -x & 1 & 0 \\ 0 & -x & 1 \\ -c_0 & -c_1 & -c_2 \end{vmatrix} = (-1)(c_2x^2 + c_1x + c_0).$$

(b) The matrix

$$A = \begin{pmatrix} 1 & 1 & 1 \\ x_1 & x_2 & x_3 \\ x_1^2 & x_2^2 & x_3^2 \end{pmatrix}$$

is called the *vandermonde matrix*. It is a square matrix and it is famously ill-conditioned. Show that

$$|A| = \begin{vmatrix} 1 & 1 & 1 \\ x_1 & x_2 & x_3 \\ x_1^2 & x_2^2 & x_3^2 \end{vmatrix} = (x_1 - x_2)(x_2 - x_3)(x_3 - x_1).$$

(c) A square matrix A is said to be a *nilpotent matrix*, if $A^k = 0$ for some positive integer k. Prove that if A is nilpotent, then the determinant of A is zero.

(d) A square matrix A is said to be an *idempotent matrix*, if $A^2 = A$. Prove that if A is idempotent, then either $det(A) = 1$ or $det(A) = 0$.

(e) A square matrix A is said to be an *involution matrix*, if $A^2 = \mathbf{I}$. Give an example of a 3×3 matrix that is an involution matrix.

21. Compute the adjoint of each matrix A, and find the inverse of it, if it exists:

$$\textbf{(a)} \quad A = \begin{pmatrix} 1 & 2 \\ -3 & 4 \end{pmatrix}, \quad \textbf{(b)} \quad A = \begin{pmatrix} 1 & 2 & -1 \\ 2 & 1 & 4 \\ 1 & 5 & -8 \end{pmatrix},$$

$$(c) \quad A = \begin{pmatrix} 1 & 1 & 0 \\ 1 & 0 & 1 \\ 0 & 1 & 1 \end{pmatrix}.$$

22. Show that $A(Adj\ A) = (Adj\ A)A = \det(A)\mathbf{I}_3$, if

$$A = \begin{pmatrix} 2 & 1 & 3 \\ -1 & 2 & 0 \\ 3 & -2 & 1 \end{pmatrix}.$$

23. Find the inverse and determinant of the adjoint matrix of each of the following matrices:

$$A = \begin{pmatrix} 4 & 1 & 5 \\ 5 & 6 & 3 \\ 5 & 4 & 4 \end{pmatrix}, \quad B = \begin{pmatrix} 3 & 4 & -2 \\ 2 & 5 & 4 \\ 7 & -3 & 4 \end{pmatrix}, \quad C = \begin{pmatrix} 1 & 2 & 4 \\ 1 & 4 & 0 \\ 3 & 1 & 1 \end{pmatrix}.$$

24. Find the inverse and determinant of the adjoint matrix of each of the following matrices:

$$A = \begin{pmatrix} 3 & 2 & 5 \\ 2 & 5 & 4 \\ 5 & 4 & 6 \end{pmatrix}, \quad B = \begin{pmatrix} 5 & 3 & -2 \\ 3 & 5 & 6 \\ -2 & 6 & 5 \end{pmatrix}, \quad C = \begin{pmatrix} 1 & 2 & 3 \\ 4 & 5 & 6 \\ 7 & 8 & 8 \end{pmatrix}.$$

25. Find the inverse of each of the following matrices using the determinant:

$$A = \begin{pmatrix} 0 & 1 & 5 \\ 3 & 1 & 2 \\ 2 & 3 & 4 \end{pmatrix}, \quad B = \begin{pmatrix} 2 & 4 & -2 \\ -4 & 7 & 5 \\ 5 & -4 & 4 \end{pmatrix}, \quad C = \begin{pmatrix} 0 & 4 & 2 & -4 \\ 6 & 1 & 4 & -3 \\ 4 & 3 & 1 & 3 \\ 8 & 4 & -3 & 2 \end{pmatrix}.$$

26. Solve each of the following homogeneous linear systems:

(a)

$$\begin{aligned} x_1 &- 2x_2 &+ x_3 &= 0 \\ x_1 &+ x_2 &+ 3x_3 &= 0 \\ 2x_1 &+ 3x_2 &- 5x_3 &= 0. \end{aligned}$$

(b)

$$
\begin{aligned}
x_1 &- 5x_2 + 3x_3 &= 0 \\
2x_1 &+ 3x_2 + 2x_3 &= 0 \\
x_1 &- 2x_2 - 4x_3 &= 0.
\end{aligned}
$$

(c)

$$
\begin{aligned}
3x_1 &+ 4x_2 - 2x_3 &= 0 \\
2x_1 &- 5x_2 - 4x_3 &= 0 \\
3x_1 &- 2x_2 + 3x_3 &= 0.
\end{aligned}
$$

(d)

$$
\begin{aligned}
x_1 &+ x_2 + 3x_3 - 2x_4 &= 0 \\
x_1 &+ 2x_2 + 5x_3 + x_4 &= 0 \\
x_1 &- 3x_2 + x_3 + 2x_4 &= 0.
\end{aligned}
$$

27. Find value(s) of α such that each of the following homogeneous linear systems has a nontrivial solution:

(a)

$$
\begin{aligned}
2x_1 &- (1 - 3\alpha)x_2 &= 0 \\
x_1 &+ \alpha x_2 &= .
\end{aligned}
$$

(b)

$$
\begin{aligned}
2x_1 &+ 2\alpha x_2 - x_3 &= 0 \\
x_1 &- 2x_2 + x_3 &= 0 \\
\alpha x_1 &+ 2x_2 - 3x_3 &= 0.
\end{aligned}
$$

(c)

$$
\begin{aligned}
x_1 &+ 2x_2 + 4x_3 &= 0 \\
3x_1 &+ 7x_2 + \alpha x_3 &= 0 \\
3x_1 &+ 3x_2 + 15x_3 &= 0.
\end{aligned}
$$

(d)

$$
\begin{aligned}
x_1 &+ x_2 + 2x_3 - 3x_4 &= 0 \\
x_1 &+ 2x_2 + x_3 - 2x_4 &= 0 \\
3x_1 &+ x_2 + \alpha x_3 + 3x_4 &= 0 \\
3x_1 &+ x_2 + \alpha x_3 + 3x_4 &= 0 \\
2x_1 &+ 3x_2 + x_3 + \alpha x_4 &= 0.
\end{aligned}
$$

28. Using the matrices in Problem 15, solve the following systems using the matrix inversion method:

(a) $A\mathbf{x} = [1, 1, -3]^T$,   (b) $B\mathbf{x} = [2, 1, 3]^T$,   (c) $C\mathbf{x} = [1, 0, 1]^T$.

29. Solve the following systems using the matrix inversion method:

(a)
$$
\begin{aligned}
x_1 + 3x_2 - x_3 &= 4 \\
5x_1 - 2x_2 - x_3 &= -2 \\
2x_1 + 2x_2 + x_3 &= 9.
\end{aligned}
$$

(b)
$$
\begin{aligned}
x_1 + x_2 + 3x_3 &= 2 \\
5x_1 + 3x_2 + x_3 &= 3 \\
2x_1 + 3x_2 + x_3 &= -1.
\end{aligned}
$$

(c)
$$
\begin{aligned}
4x_1 + x_2 - 3x_3 &= -1 \\
3x_1 + 2x_2 - 6x_3 &= -2 \\
x_1 - 5x_2 + 3x_3 &= -3.
\end{aligned}
$$

(d)
$$
\begin{aligned}
7x_1 + 11x_2 - 15x_3 &= 21 \\
3x_1 + 22x_2 - 18x_3 &= 12 \\
2x_1 - 13x_2 + 9x_3 &= 16.
\end{aligned}
$$

30. Solve the following systems using the matrix inversion method:

(a)
$$
\begin{aligned}
3x_1 - 2x_2 - 4x_3 &= 7 \\
5x_1 - 2x_2 - 3x_3 &= 8 \\
7x_1 + 4x_2 + 2x_3 &= 9.
\end{aligned}
$$

(b)
$$
\begin{aligned}
-3x_1 + 4x_2 + 3x_3 &= 11 \\
5x_1 + 3x_2 + x_3 &= 12 \\
x_1 + x_2 + 5x_3 &= 10.
\end{aligned}
$$

(c)

$$\begin{aligned}
x_1 + 4_2 - 8x_3 &= 7 \\
2x_1 + 7x_2 - 5x_3 &= -5 \\
3x_1 - 6x_2 + 6x_3 &= 4.
\end{aligned}$$

(d)

$$\begin{aligned}
17x_1 + 18x_2 - 19x_3 &= 10 \\
43x_1 + 22x_2 - 14x_3 &= 11 \\
25x_1 - 33x_2 + 21x_3 &= 12.
\end{aligned}$$

31. Solve the following systems using the matrix inversion method:

(a)

$$\begin{aligned}
2x_1 + 3x_2 - 4x_3 + 4x_4 &= 11 \\
x_1 + 3x_2 - 4x_3 + 2x_4 &= 12 \\
4x_1 + 3x_2 + 2x_3 + 3x_4 &= 14 \\
3x_1 - 4x_2 + 5x_3 + 6x_4 &= 15.
\end{aligned}$$

(b)

$$\begin{aligned}
7x_1 + 13x_2 + 12x_3 + 9x_4 &= 21 \\
3x_1 + 23x_2 - 5x_3 + 2x_4 &= 10 \\
4x_1 - 7x_2 + 22x_3 + 3x_4 &= 11 \\
3x_1 - 4x_2 + 25x_3 + 16x_4 &= 10.
\end{aligned}$$

(c)

$$\begin{aligned}
12x_1 + 6x_2 + 5x_3 - 2x_4 &= 21 \\
11x_1 + 13x_2 + 7x_3 + 2x_4 &= 22 \\
14x_1 + 9x_2 + 2x_3 - 6x_4 &= 23 \\
7x_1 - 24x_2 - 7x_3 + 8x_4 &= 24.
\end{aligned}$$

(d)

$$\begin{aligned}
15x_1 - 26x_2 + 15x_3 - 11x_4 &= 17 \\
14x_1 + 15x_2 + 7x_3 + 7x_4 &= 18 \\
17x_1 + 14x_2 - 22x_3 - 16x_4 &= 19 \\
21x_1 - 12x_2 - 7x_3 + 8x_4 &= 20.
\end{aligned}$$

32. In each case, factor the matrix as a product of elementary matrices:

(a) $\begin{pmatrix} 1 & 1 \\ 3 & 1 \end{pmatrix}$, (b) $\begin{pmatrix} 3 & 2 \\ 1 & 2 \end{pmatrix}$, (c) $\begin{pmatrix} 1 & 1 \\ -2 & 4 \end{pmatrix}$,

$$(\text{d}) \begin{pmatrix} 1 & 0 & 1 \\ 0 & 2 & 1 \\ 2 & 2 & 3 \end{pmatrix}, (\text{e}) \begin{pmatrix} 1 & 1 & 2 \\ 0 & 1 & 2 \\ 1 & 2 & 3 \end{pmatrix}, (\text{f}) \begin{pmatrix} 1 & -3 & 5 \\ -2 & 2 & -4 \\ 4 & 7 & 9 \end{pmatrix}.$$

33. Solve Problem 30 using Cramer's rule.

34. Solve the following systems using Cramer's rule:

(a)

$$\begin{aligned} 3x_1 + 4x_2 + 5x_3 &= 1 \\ 3x_1 + 2x_2 + x_3 &= 2 \\ 4x_1 + 3x_2 + 5x_3 &= 3. \end{aligned}$$

(b)

$$\begin{aligned} x_1 - 4x_2 + 2x_3 &= 4 \\ -4x_1 + 5x_2 + 6x_3 &= 0 \\ 7x_1 - 3x_2 + 5x_3 &= 4. \end{aligned}$$

(c)

$$\begin{aligned} 6x_1 + 7x_2 + 8x_3 &= 1 \\ -5x_1 + 3x_2 + 2x_3 &= 1 \\ x_1 + 2x_2 + 3x_3 &= 1. \end{aligned}$$

(d)

$$\begin{aligned} x_1 + 3x_2 - 4x_3 + 5x_4 &= 2 \\ 6x_1 - x_2 + 6x_3 + 3x_4 &= -3 \\ 2x_1 + x_2 + 3x_3 + 2x_4 &= 4 \\ x_1 + 5x_2 + 6x_3 + 7x_4 &= 2. \end{aligned}$$

35. Solve the following systems using Cramer's rule:

(a)

$$\begin{aligned} 2x_1 - 2x_2 + 8x_3 &= 1 \\ 5x_1 + 6x_2 + 5x_3 &= 2 \\ 7x_1 + 7x_2 + 9x_3 &= 3. \end{aligned}$$

(b)

$$\begin{aligned} 3x_1 - 3x_2 + 12x_3 &= 14 \\ -4x_1 + 5x_2 + 16x_3 &= 18 \\ x_1 - 15x_2 + 24x_3 &= 19. \end{aligned}$$

(c)

$$9x_1 - 11x_2 + 12x_3 = 3$$
$$-5x_1 + 3x_2 + 2x_3 = 4$$
$$7x_1 - 12x_2 + 13x_3 = 5.$$

(d)

$$11x_1 + 3x_2 - 13x_3 + 15x_4 = 22$$
$$26x_1 - 5x_2 + 6x_3 + 13x_4 = 23$$
$$22x_1 + 6x_2 + 13x_3 + 12x_4 = 24$$
$$17x_1 - 25x_2 + 16x_3 + 27x_4 = 25.$$

36. Use the simple Gaussian elimination method to show that the following system does not have a solution:

$$3x_1 + x_2 = 1.5$$
$$2x_1 - x_2 - x_3 = 2$$
$$4x_1 + 3x_2 + x_3 = 0.$$

37. Solve Problem 34 using the simple Gaussian elimination method.

38. Solve the following systems using the simple Gaussian elimination method:

(a)

$$x_1 - x_2 = -2$$
$$-x_1 + 2x_2 - x_3 = 5$$
$$4x_1 - x_2 + 4x_3 = 1.$$

(b)

$$3x_1 + x_2 - x_3 = 5$$
$$5x_1 - 3x_2 + 2x_3 = 7$$
$$2x_1 - x_2 + x_3 = 3.$$

(c)

$$3x_1 + x_2 + x_3 = 2$$
$$2x_1 + 2x_2 + 4x_3 = 3$$
$$4x_1 + 9x_2 + 16x_3 = 1.$$

(d)

$$
\begin{aligned}
2x_1 + x_2 + x_3 - x_4 &= 9 \\
x_1 + 9x_2 + 8x_3 + 4x_4 &= 11 \\
-x_1 + 3x_2 + 5x_3 + 2x_4 &= 10 \\
5x_1 + x_2 \quad\quad + x_4 &= 12.
\end{aligned}
$$

39. Solve the following systems using the simple Gaussian elimination method:

(a)

$$
\begin{aligned}
2x_1 + 5x_2 - 4x_3 &= 3 \\
2x_1 + 2x_2 - x_3 &= 1 \\
3x_1 + 2x_2 - 3x_3 &= -5.
\end{aligned}
$$

(b)

$$
\begin{aligned}
2x_2 - x_3 &= 1 \\
3x_1 - x_2 + 2x_3 &= 4 \\
x_1 + 3x_2 - 5x_3 &= 1.
\end{aligned}
$$

(c)

$$
\begin{aligned}
x_1 + 2x_2 \quad\quad &= 3 \\
-x_1 \quad\quad - 2x_3 &= -5 \\
-3x_1 - 5x_2 + x_3 &= -4.
\end{aligned}
$$

(d)

$$
\begin{aligned}
3x_1 + 2x_2 + 4x_3 - x_4 &= 2 \\
x_1 + 4x_2 + 5x_3 + x_4 &= 1 \\
4x_1 + 5x_2 + 4x_3 + 3x_4 &= 5 \\
2x_1 + 3x_2 + 2x_3 + 4x_4 &= 6.
\end{aligned}
$$

40. For what values of a and b does the following linear system have no solution or infinitely many solutions:

(a)

$$
\begin{aligned}
2x_1 + x_2 + x_3 &= 2 \\
-2x_1 + x_2 + 3x_3 &= a \\
2x_1 \quad\quad - x_3 &= b.
\end{aligned}
$$

(b)

$$\begin{aligned}
2x_1 + 3x_2 - x_3 &= 1 \\
x_1 - x_2 + 3x_3 &= a \\
3x_1 + 7x_2 - 5x_3 &= b.
\end{aligned}$$

(c)

$$\begin{aligned}
2x_1 - x_2 + 3x_3 &= 3 \\
3x_1 + x_2 - 5x_3 &= a \\
-5x_1 - 5x_2 + 21x_3 &= b.
\end{aligned}$$

(d)

$$\begin{aligned}
2x_1 - x_2 + 3x_3 &= 5 \\
4x_1 + 2x_2 + bx_3 &= 6 \\
-2x_1 + ax_2 + 3x_3 &= 4.
\end{aligned}$$

41. Find the value(s) of α so that each of the following linear systems has a nontrivial solution:

(a)

$$\begin{aligned}
2x_1 + 2x_2 + 3x_3 &= 1 \\
3x_1 + \alpha x_2 + 5x_3 &= 3 \\
x_1 + 7x_2 + 3x_3 &= 2.
\end{aligned}$$

(b)

$$\begin{aligned}
x_1 + 2x_2 + x_3 &= 2 \\
x_1 + 3x_2 + 6x_3 &= 5 \\
2x_1 + 3x_2 + \alpha x_3 &= 6.
\end{aligned}$$

(c)

$$\begin{aligned}
\alpha x_1 + x_2 + x_3 &= 7 \\
x_1 + x_2 - x_3 &= 2 \\
x_1 + x_2 + \alpha x_3 &= 1.
\end{aligned}$$

(d)

$$\begin{aligned}
2x_1 + \alpha x_2 + 3x_3 &= 9 \\
3x_1 - 4x_2 - 5x_3 &= 11 \\
4x_1 + 5x_2 + \alpha x_3 &= 12.
\end{aligned}$$

42. Find the inverse of each of the following matrices by using the simple Gauss elimination method:

$$A = \begin{pmatrix} 3 & 3 & 3 \\ 0 & 2 & 2 \\ 2 & 4 & 5 \end{pmatrix}, \quad B = \begin{pmatrix} 5 & 3 & 2 \\ 3 & 2 & 2 \\ 2 & 6 & 5 \end{pmatrix}, \quad C = \begin{pmatrix} 1 & 2 & 3 \\ 2 & 5 & 2 \\ 3 & 4 & 3 \end{pmatrix}.$$

43. Find the inverse of each of the following matrices by using the simple Gauss elimination method:

$$A = \begin{pmatrix} 3 & 2 & 3 \\ 4 & 2 & 2 \\ 2 & 4 & 3 \end{pmatrix}, \quad B = \begin{pmatrix} 1 & -3 & 2 \\ 3 & 2 & 6 \\ 2 & -6 & 5 \end{pmatrix}, \quad C = \begin{pmatrix} 5 & 2 & 3 \\ 2 & 5 & 5 \\ 3 & 2 & 4 \end{pmatrix}.$$

44. Determine the rank of each of the following matrices:

$$A = \begin{pmatrix} 3 & 1 & -1 \\ 2 & 0 & 4 \\ 1 & -5 & 1 \end{pmatrix}, \quad B = \begin{pmatrix} 4 & 1 & 6 \\ -3 & 6 & 4 \\ 5 & 0 & 9 \end{pmatrix}, \quad C = \begin{pmatrix} 17 & 46 & 7 \\ 20 & 49 & 8 \\ 23 & 52 & 9 \end{pmatrix}.$$

45. Determine the rank of each matrix:

$$A = \begin{pmatrix} 2 & -1 & 0 \\ 2 & -1 & 1 \\ 1 & 1 & -1 \end{pmatrix}, \quad B = \begin{pmatrix} 0.1 & 0.2 & 0.3 \\ 0.4 & 0.5 & 0.6 \\ 0.7 & 0.8 & 0.91 \end{pmatrix}, \quad C = \begin{pmatrix} 1 & 2 & 3 & 4 \\ 2 & 4 & 6 & 8 \\ 3 & 5 & 7 & 9 \\ 4 & 6 & 8 & 10 \end{pmatrix}.$$

46. Let A be an $m \times n$ matrix and B be an $n \times p$ matrix. Show that the rank of AB is less than or equal to the rank of A.

47. Solve Problem 38 using Gaussian elimination with partial pivoting.

48. Solve the following linear systems using Gaussian elimination with partial and without pivoting:

(a)

$$\begin{aligned} 1.001x_1 + 1.5x_2 &= 0 \\ 2x_1 + 3x_2 &= 1. \end{aligned}$$

(b)

$$\begin{aligned}
x_1 + 1.001x_2 &= 2.001 \\
x_1 + x_2 &= 2.
\end{aligned}$$

(c)

$$\begin{aligned}
6.122x_1 + 1500.5x_2 &= 1506.622 \\
2000x_1 + 3x_2 &= 2003.
\end{aligned}$$

49. The elements of matrix A, the Hilbert matrix, are defined by

$$a_{ij} = 1/(i+j-1), \quad \text{for} \quad i,j = 1,2,\ldots,n.$$

Find the solution of the system $A\mathbf{x} = \mathbf{b}$ for $n = 4$ and $\mathbf{b} = [1,2,3,4]^T$ using Gaussian elimination by partial pivoting.

50. Solve the following systems using the Gauss–Jordan method:

(a)

$$\begin{aligned}
x_1 + 4x_2 + x_3 &= 1 \\
2x_1 + 4x_2 + x_3 &= 9 \\
3x_1 + 5x_2 - 2x_3 &= 11.
\end{aligned}$$

(b)

$$\begin{aligned}
x_1 + x_2 + x_3 &= 1 \\
2x_1 - x_2 + 3x_3 &= 4 \\
3x_1 + 2x_2 - 2x_3 &= -2.
\end{aligned}$$

(c)

$$\begin{aligned}
2x_1 + 3x_2 + 6x_3 + x_4 &= 2 \\
x_1 + x_2 - 2x_3 + 4x_4 &= 1 \\
3x_1 + 5x_2 - 2x_3 + 2x_4 &= 11 \\
2x_1 + 2x_2 + 2x_3 - 3x_4 &= 2.
\end{aligned}$$

51. The following sets of linear equations have a common coefficients matrix but different right-side terms:

(a)

$$\begin{aligned}
2x_1 + 3x_2 + 5x_3 &= 0 \\
3x_1 + x_2 - 2x_3 &= -2 \\
x_1 + 3x_2 + 4x_3 &= -3.
\end{aligned}$$

(b)

$$\begin{aligned} 2x_1 &+ 3x_2 &+ 5x_3 &= 1 \\ 3x_1 &+ x_2 &- 2x_3 &= 2 \\ x_1 &+ 3x_2 &+ 4x_3 &= 4. \end{aligned}$$

(c)

$$\begin{aligned} 2x_1 &+ 3x_2 &+ 5x_3 &= -5 \\ 3x_1 &+ x_2 &- 2x_3 &= 6 \\ x_1 &+ 3x_2 &+ 4x_3 &= -1. \end{aligned}$$

The coefficients and the three sets of right-side terms may be combined into an augmented matrix of the form

$$\begin{pmatrix} 2 & 3 & 5 & : & 0 & 1 & -5 \\ 3 & 1 & -2 & : & -2 & 2 & 6 \\ 1 & 3 & 4 & : & -3 & 4 & -1 \end{pmatrix}.$$

If we apply the Gauss–Jordan method to this augmented matrix form and reduce the first three columns to the unity matrix form, the solution for the three problems are automatically obtained in the fourth, fifth, and sixth columns when elimination is completed. Calculate the solution in this way.

52. Calculate the inverse of each matrix using the Gauss–Jordan method:

(a) $\begin{pmatrix} 3 & -9 & 5 \\ 0 & 5 & 1 \\ -1 & 6 & 3 \end{pmatrix}$, **(b)** $\begin{pmatrix} 1 & 4 & 5 \\ 2 & 1 & 2 \\ 8 & 1 & 1 \end{pmatrix}$, **(c)** $\begin{pmatrix} 5 & -2 & 0 & 0 \\ -2 & 5 & -2 & 0 \\ 0 & -2 & 5 & -2 \\ 0 & 0 & -2 & 5 \end{pmatrix}.$

53. Find the inverse of the Hilbert matrix of size 4×4 using the Gauss–Jordan method. Then solve the linear system $A\mathbf{x} = [1, 2, 3, 4]^T$.

54. Find the LU decomposition of each matrix A using Doolittle's method and then solve the systems:

(a)

$$A = \begin{pmatrix} 2 & -1 & 1 \\ -3 & 4 & -1 \\ 1 & -1 & 1 \end{pmatrix}, \qquad \mathbf{b} = \begin{pmatrix} 4 \\ 5 \\ 6 \end{pmatrix}.$$

(b)

$$A = \begin{pmatrix} 7 & 6 & 5 \\ 5 & 4 & 3 \\ 3 & 7 & 6 \end{pmatrix}, \qquad \mathbf{b} = \begin{pmatrix} 2 \\ 1 \\ 2 \end{pmatrix}.$$

(c)

$$A = \begin{pmatrix} 2 & 2 & 2 \\ 1 & 2 & 1 \\ 3 & 3 & 4 \end{pmatrix}, \qquad \mathbf{b} = \begin{pmatrix} 0 \\ -4 \\ 1 \end{pmatrix}.$$

(d)

$$A = \begin{pmatrix} 2 & 4 & -6 \\ 1 & 5 & 3 \\ 1 & 3 & 2 \end{pmatrix}, \qquad \mathbf{b} = \begin{pmatrix} -4 \\ 10 \\ 5 \end{pmatrix}.$$

(e)

$$A = \begin{pmatrix} 1 & -1 & 0 \\ 2 & -1 & 1 \\ 2 & -2 & -1 \end{pmatrix}, \qquad \mathbf{b} = \begin{pmatrix} 2 \\ 4 \\ 3 \end{pmatrix}.$$

(f)

$$A = \begin{pmatrix} 1 & 5 & 3 \\ 2 & 4 & 6 \\ 1 & 3 & 2 \end{pmatrix}, \qquad \mathbf{b} = \begin{pmatrix} 4 \\ 11 \\ 5 \end{pmatrix}.$$

55. Find the LU decomposition of each matrix A using Doolittle's method, and then solve the systems:

(a)

$$A = \begin{pmatrix} 3 & -2 & 1 & 1 \\ -3 & 7 & 4 & -3 \\ 2 & -5 & 3 & 4 \\ 7 & -3 & 2 & 4 \end{pmatrix}, \qquad \mathbf{b} = \begin{pmatrix} 3 \\ 2 \\ 1 \\ 2 \end{pmatrix}.$$

(b)

$$A = \begin{pmatrix} 2 & -4 & 5 & 3 \\ 3 & 5 & -4 & 3 \\ 1 & 6 & 2 & 6 \\ 7 & 2 & 5 & 1 \end{pmatrix}, \qquad \mathbf{b} = \begin{pmatrix} 6 \\ 5 \\ 2 \\ 4 \end{pmatrix}.$$

(c)

$$A = \begin{pmatrix} 2 & 2 & 3 & -2 \\ 10 & 2 & 13 & 11 \\ 2 & 5 & 4 & 6 \\ 1 & -4 & -2 & 7 \end{pmatrix}, \qquad b = \begin{pmatrix} 10 \\ 14 \\ 11 \\ 9 \end{pmatrix}.$$

(d)

$$A = \begin{pmatrix} 5 & 12 & 4 & -11 \\ 21 & 15 & 13 & 23 \\ 31 & 33 & 12 & 22 \\ -17 & 15 & 14 & 11 \end{pmatrix}, \qquad b = \begin{pmatrix} 44 \\ 33 \\ 55 \\ 22 \end{pmatrix}.$$

(e)

$$A = \begin{pmatrix} 1 & -1 & 10 & 8 \\ 12 & -17 & 11 & 22 \\ 22 & 31 & 13 & -1 \\ 8 & 24 & 13 & 9 \end{pmatrix}, \qquad b = \begin{pmatrix} -2 \\ 7 \\ 6 \\ 5 \end{pmatrix}.$$

(f)

$$A = \begin{pmatrix} 41 & 25 & 23 & -18 \\ 2 & 13 & -16 & 12 \\ 11 & 13 & 9 & 7 \end{pmatrix}, \qquad b = \begin{pmatrix} 41 \\ 1 \\ 15 \\ 13 \end{pmatrix}.$$

56. For the value(s) of α of each of the following matrices, if A is singular, using Doolittle's method:

(a) $A = \begin{pmatrix} 1 & -1 & 2 \\ -1 & 3 & -1 \\ \alpha & -2 & 3 \end{pmatrix}$.

(b) $A = \begin{pmatrix} 1 & 5 & 7 \\ 4 & 4 & \alpha \\ -2 & \alpha & 9 \end{pmatrix}$.

(c) $A = \begin{pmatrix} 2 & -4 & \alpha \\ 2 & 4 & 3 \\ 4 & -2 & 5 \end{pmatrix}$.

(d) $A = \begin{pmatrix} 2 & \alpha & 1-\alpha \\ 2 & 5 & -2 \\ 2 & 5 & 4 \end{pmatrix}$.

(e) $A = \begin{pmatrix} 1 & -1 & 3 \\ 3 & 2 & 3 \\ 4 & \alpha-2 & 7 \end{pmatrix}$.

(f) $A = \begin{pmatrix} 1 & 5 & \alpha \\ 1 & 4 & \alpha-2 \\ 1 & -2 & 8 \end{pmatrix}$.

57. Find the determinant of each of the following matrices using LU decomposition by Doolittle's method:

(a) $A = \begin{pmatrix} 2 & 3 & -1 \\ 1 & 2 & 1 \\ 2 & 1 & -6 \end{pmatrix}$,   (b) $A = \begin{pmatrix} 1 & -2 & 2 \\ 2 & 1 & 1 \\ 1 & 0 & 1 \end{pmatrix}$,

(c) $A = \begin{pmatrix} 2 & 4 & 1 \\ 3 & 3 & 2 \\ 4 & 1 & 4 \end{pmatrix}$,   (d) $A = \begin{pmatrix} 2 & 4 & -6 \\ 1 & 5 & 3 \\ 1 & 3 & 2 \end{pmatrix}$,

(e) $A = \begin{pmatrix} 1 & -1 & 0 \\ 2 & -1 & 1 \\ -2 & 2 & 1 \end{pmatrix}$,   (f) $A = \begin{pmatrix} 1 & 5 & 3 \\ 1 & 2 & 3 \\ 1 & 3 & 2 \end{pmatrix}$.

58. Use the smallest positive integer to find the unique solution of each of the linear systems of Problem 56 using LU decomposition by Doolittle's method:

(a) $A\mathbf{x} = [2, 3, 2]^T$.

(b) $A\mathbf{x} = [5, -6, 2]^T$.

(c) $A\mathbf{x} = [11, 13, 10]^T$.

(d) $A\mathbf{x} = [-8, 11, 8]^T$.

(e) $A\mathbf{x} = [32, 23, 12]^T$.

(f) $A\mathbf{x} = [-11, 43, 22]^T$.

59. Find the LDV factorization of each of the following matrices:

(a)
$$A = \begin{pmatrix} 3 & 4 & 3 \\ 2 & 3 & 3 \\ 1 & 3 & 5 \end{pmatrix}, \qquad B = \begin{pmatrix} 4 & -2 & 3 \\ 5 & 2 & -3 \\ 4 & 3 & 6 \end{pmatrix}.$$

(b)
$$A = \begin{pmatrix} 2 & 5 & 4 \\ 2 & 1 & 6 \\ 3 & 2 & 7 \end{pmatrix}, \qquad B = \begin{pmatrix} 3 & 2 & -6 \\ 2 & 2 & -5 \\ 3 & 4 & 7 \end{pmatrix}.$$

(c)
$$A = \begin{pmatrix} 1 & -5 & 4 \\ 2 & 3 & -4 \\ 3 & 2 & 6 \end{pmatrix}, \qquad B = \begin{pmatrix} 4 & 7 & -6 \\ 5 & 5 & -5 \\ 6 & -4 & 9 \end{pmatrix}.$$

(d)
$$A = \begin{pmatrix} 3 & -1 & 4 \\ 2 & 2 & -1 \\ 3 & 2 & 2 \end{pmatrix}, \qquad B = \begin{pmatrix} 2 & 3 & 4 & 5 \\ 3 & 1 & 2 & 4 \\ 3 & 1 & 1 & 1 \\ 4 & 3 & 1 & 2 \end{pmatrix}.$$

60. Find the LDL^T factorization of each of the following matrices:

   (a)
   $$A = \begin{pmatrix} 2 & 3 & 4 \\ 3 & 5 & 2 \\ 4 & 2 & 6 \end{pmatrix}.$$

   (b)
   $$A = \begin{pmatrix} 3 & -2 & 4 \\ -2 & 2 & 1 \\ 4 & 1 & 3 \end{pmatrix}.$$

   (c)
   $$A = \begin{pmatrix} 2 & 1 & -1 \\ 1 & 3 & 2 \\ -1 & 2 & 2 \end{pmatrix}.$$

   (d)
   $$A = \begin{pmatrix} 1 & -2 & 3 & 4 \\ -2 & 3 & 4 & 5 \\ 3 & 4 & 5 & -6 \\ 4 & 5 & -6 & 7 \end{pmatrix}.$$

61. Solve Problem 54 by LU decomposition using Crout's method.

62. Find the determinant of each of the following matrices using LU decomposition by Crout's method:

   (a) $A = \begin{pmatrix} 2 & 2 & -1 \\ 1 & 2 & 1 \\ 2 & 1 & -4 \end{pmatrix}$, (b) $A = \begin{pmatrix} 2 & -1 & 1 \\ 1 & 2 & 2 \\ 2 & 0 & 2 \end{pmatrix}$,

   (c) $A = \begin{pmatrix} 4 & 4 & 1 \\ 5 & 4 & 2 \\ 1 & 4 & 4 \end{pmatrix}$, (d) $A = \begin{pmatrix} 2 & 4 & 5 \\ 3 & 5 & 3 \\ 4 & 3 & 2 \end{pmatrix}$,

   (e) $A = \begin{pmatrix} 1 & -1 & 2 \\ 2 & -1 & 1 \\ -2 & 2 & 4 \end{pmatrix}$, (f) $A = \begin{pmatrix} 1 & 5 & 3 \\ 1 & 2 & 3 \\ 1 & 3 & 4 \end{pmatrix}.$

63. Solve the following systems by *LU* decomposition using the Cholesky method:

(a)

$$A = \begin{pmatrix} 1 & -1 & 1 \\ -1 & 5 & -1 \\ 1 & -1 & 10 \end{pmatrix}, \qquad b = \begin{pmatrix} 2 \\ 2 \\ 2 \end{pmatrix}.$$

(b)

$$A = \begin{pmatrix} 10 & 2 & 1 \\ 2 & 10 & 3 \\ 1 & 3 & 10 \end{pmatrix}, \qquad b = \begin{pmatrix} 7 \\ -4 \\ 3 \end{pmatrix}.$$

(c)

$$A = \begin{pmatrix} 4 & 2 & 3 \\ 2 & 17 & 1 \\ 3 & 1 & 5 \end{pmatrix}, \qquad b = \begin{pmatrix} 1 \\ 2 \\ 5 \end{pmatrix}.$$

(d)

$$A = \begin{pmatrix} 3 & 4 & -6 & 0 \\ 4 & 5 & 3 & 1 \\ -6 & 3 & 3 & 1 \\ 0 & 1 & 1 & 3 \end{pmatrix}, \qquad b = \begin{pmatrix} 4 \\ 5 \\ 2 \\ 3 \end{pmatrix}.$$

64. Solve the following systems by *LU* decomposition using the Cholesky method:

(a)

$$A = \begin{pmatrix} 5 & -1 & 1 \\ -3 & 5 & -2 \\ 2 & -1 & 7 \end{pmatrix}, \qquad b = \begin{pmatrix} 5 \\ 7 \\ 9 \end{pmatrix}.$$

(b)

$$A = \begin{pmatrix} 6 & 2 & -3 \\ 3 & 12 & -4 \\ 6 & 3 & 13 \end{pmatrix}, \qquad b = \begin{pmatrix} 5 \\ 2 \\ 4 \end{pmatrix}.$$

(c)

$$A = \begin{pmatrix} 5 & 2 & -5 \\ 2 & 4 & 4 \\ -3 & -2 & 7 \end{pmatrix}, \qquad b = \begin{pmatrix} 3 \\ 11 \\ 14 \end{pmatrix}.$$

(d)

$$A = \begin{pmatrix} 1 & 4 & -6 & 0 \\ 2 & 2 & 3 & 3 \\ -3 & 6 & 7 & 1 \\ 0 & 2 & -3 & 5 \end{pmatrix}, \qquad b = \begin{pmatrix} 12 \\ 13 \\ 14 \\ 15 \end{pmatrix}.$$

65. Solve the following tridiagonal systems using LU decomposition:

(a)

$$A = \begin{pmatrix} 3 & -1 & 0 \\ -1 & 3 & -1 \\ 0 & -1 & 3 \end{pmatrix}, \qquad b = \begin{pmatrix} 1 \\ 1 \\ 1 \end{pmatrix}.$$

(b)

$$A = \begin{pmatrix} 2 & 3 & 0 & 0 \\ 3 & 2 & 3 & 0 \\ 0 & 3 & 2 & 3 \\ 0 & 0 & 3 & 2 \end{pmatrix}, \qquad b = \begin{pmatrix} 6 \\ 7 \\ 5 \\ 3 \end{pmatrix}.$$

(c)

$$A = \begin{pmatrix} 4 & -1 & 0 & 0 \\ -1 & 4 & -1 & 0 \\ 0 & -1 & 4 & -1 \\ 0 & 0 & -1 & 4 \end{pmatrix}, \qquad b = \begin{pmatrix} 1 \\ 1 \\ 1 \\ 1 \end{pmatrix}.$$

(d)

$$A = \begin{pmatrix} 2 & 3 & 0 & 0 \\ 3 & 5 & 4 & 0 \\ 0 & 4 & 6 & 3 \\ 0 & 0 & 3 & 4 \end{pmatrix}, \qquad b = \begin{pmatrix} 1 \\ 2 \\ 3 \\ 4 \end{pmatrix}.$$

66. Solve the following tridiagonal systems using LU decomposition:

(a)

$$A = \begin{pmatrix} 4 & -2 & 0 \\ -2 & 5 & -2 \\ 0 & -2 & 6 \end{pmatrix}, \qquad b = \begin{pmatrix} 5 \\ 6 \\ 7 \end{pmatrix}.$$

(b)

$$A = \begin{pmatrix} 8 & 1 & 0 & 0 \\ 1 & 8 & 1 & 0 \\ 0 & 1 & 8 & 1 \\ 0 & 0 & 1 & 8 \end{pmatrix}, \qquad b = \begin{pmatrix} 2 \\ 2 \\ 2 \\ 2 \end{pmatrix}.$$

(c)

$$A = \begin{pmatrix} 5 & -3 & 0 & 0 \\ -3 & 6 & -2 & 0 \\ 0 & -2 & 7 & -5 \\ 0 & 0 & -5 & 8 \end{pmatrix}, \qquad b = \begin{pmatrix} 7 \\ -5 \\ 4 \\ 2 \end{pmatrix}.$$

(d)

$$A = \begin{pmatrix} 2 & -4 & 0 & 0 \\ -4 & 5 & 7 & 0 \\ 0 & 7 & 6 & 2 \\ 0 & 0 & 2 & 8 \end{pmatrix}, \qquad b = \begin{pmatrix} 11 \\ 12 \\ 13 \\ 14 \end{pmatrix}.$$

67. Find $\|x\|_1$, $\|x\|_2$, and $\|x\|_\infty$ for the following vectors:

(a)

$$[2, -1, -6, 3]^T.$$

(b)

$$[\sin k, \cos k, 3^k]^T, \quad \text{for a fixed integer } k.$$

68. Find $\|.\|_1$, $\|.\|_\infty$ and $\|.\|_e$ for the following matrices:

$$A = \begin{pmatrix} 3 & 1 & -1 \\ 2 & 0 & 4 \\ 1 & -5 & 1 \end{pmatrix}, \qquad B = \begin{pmatrix} 4 & 1 & 6 \\ -3 & 6 & 4 \\ 5 & 0 & 9 \end{pmatrix},$$

$$C = \begin{pmatrix} 17 & 46 & 7 \\ 20 & 49 & 8 \\ 23 & 52 & 9 \end{pmatrix}, \qquad D = \begin{pmatrix} 3 & 11 & -5 & 2 \\ 6 & 8 & -11 & 6 \\ -4 & -8 & 10 & 14 \\ 13 & 14 & -12 & 9 \end{pmatrix}.$$

69. Consider the following matrices:

$$A = \begin{pmatrix} -11 & 7 & -8 \\ 5 & 9 & 6 \\ 6 & 3 & 7 \end{pmatrix}, \quad B = \begin{pmatrix} 6 & 2 & 7 \\ -12 & 10 & 8 \\ 3 & -15 & 14 \end{pmatrix},$$

$$C = \begin{pmatrix} 5 & -6 & 4 \\ -7 & 8 & 5 \\ 3 & -9 & 12 \end{pmatrix}, \quad D = \begin{pmatrix} 2 & 1 & -1 & 1 \\ 1 & 3 & 5 & 2 \\ -2 & -3 & 4 & 5 \\ 3 & 4 & -2 & 4 \end{pmatrix}.$$

Find $\|.\|_1$ and $\|.\|_\infty$ for (a) A^3, (b) $A^2 + B^2 + C^2 + D^2$, (c) BC, (d) $C^2 + D^2$.

70. The $n \times n$ Hilbert matrix $H^{(n)}$ is defined by

$$H_{ij}^{(n)} = \frac{1}{i+j-1}, \quad 1 \le i, \ j \le n.$$

Find the l_∞-norm of the 10×10 Hilbert matrix.

71. Compute the condition numbers of the following matrices relative to $\|.\|_\infty$:

(a) $\begin{pmatrix} \dfrac{1}{3} & \dfrac{1}{2} & \dfrac{1}{5} \\ \dfrac{1}{2} & \dfrac{1}{5} & \dfrac{1}{3} \\ \dfrac{1}{5} & \dfrac{1}{3} & \dfrac{1}{2} \end{pmatrix}$, (b) $\begin{pmatrix} 0.03 & 0.01 & -0.02 \\ 0.15 & 0.51 & -0.11 \\ 1.11 & 2.22 & 3.33 \end{pmatrix}$, (c) $\begin{pmatrix} 1.11 & 1.98 & 2.01 \\ 1.01 & 1.05 & 2.05 \\ 0.85 & 0.45 & 1.25 \end{pmatrix}$

72. The following linear systems have $\mathbf{x}$ as the exact solution and $\mathbf{x}^*$ is an approximate solution. Compute $\|\mathbf{x} - \mathbf{x}^*\|_\infty$ and $K(A)\dfrac{\|\mathbf{r}\|_\infty}{\|A\|_\infty}$, where $\mathbf{r} = \mathbf{b} - A\mathbf{x}^*$ is the residual vector:

(a)

$$\begin{aligned} 0.89x_1 &+ 0.53x_2 &= 0.36 \\ 0.47x_1 &+ 0.28x_2 &= 0.19 \end{aligned}$$

$$\mathbf{x} = [1, -1]^T$$
$$\mathbf{x}^* = [0.702, -0.500]^T$$

(b)

$$0.986x_1 + 0.579x_2 = 0.235$$
$$0.409x_1 + 0.237x_2 = 0.107$$

$$\mathbf{x} = [2, -3]^T$$
$$\mathbf{x}^* = [2.110, -3.170]^T$$

(c)

$$1.003x_1 + 58.090x_2 = 68.12$$
$$5.550x_1 + 321.8x_2 = 377.3$$

$$\mathbf{x} = [10, 1]^T$$
$$\mathbf{x}^* = [-10, 1]^T$$

73. Discuss the ill-conditioning (stability) of the linear system

$$1.01x_1 + 0.99x_2 = 2$$
$$0.99x_1 + 1.01x_2 = 2.$$

If $\mathbf{x}^* = [2, 0]^t$ is an approximate solution of the system, then find the residual vector $\mathbf{r}$ and estimate the relative error.

74. Show that if B is singular, then

$$\frac{1}{K(A)} \leq \frac{\|A - B\|}{\|A\|}.$$

75. Consider the following matrices:

$$A = \begin{pmatrix} 0.06 & 0.01 & 0.02 \\ 0.13 & 0.05 & 0.11 \\ 1.01 & 2.02 & 3.03 \end{pmatrix}, \qquad B = \begin{pmatrix} 0.1 & 0.2 & 0.12 \\ 0.1 & 0.4 & 0.2 \\ 0.2 & 0.05 & 0.1 \end{pmatrix}.$$

Using Problem 74, compute the approximation of the condition number of the matrix A relative to $\|.\|_\infty$.

76. Let A and B be nonsingular $n \times n$ matrices. Show that

(a)
$$K(A) \geq 1 \quad \text{and} \quad K(B) \geq 1.$$

(b)
$$K(AB) \leq K(A)K(B).$$

77. The exact solution of the linear system

$$
\begin{aligned}
x_1 + x_2 &= 1 \\
x_1 + 1.01x_2 &= 2
\end{aligned}
$$

is $\mathbf{x} = [-99, 100]^T$. Change the coefficient matrix slightly to

$$\delta A = \begin{pmatrix} 1 & 1 \\ 1 & 0.99 \end{pmatrix},$$

and consider the linear system

$$
\begin{aligned}
x_1 + x_2 &= 1 \\
x_1 + 0.99x_2 &= 2.
\end{aligned}
$$

Compute the changed solution $\delta \mathbf{x}$ of the system. Is the matrix A ill-conditioned?

78. Using Problem 77, compute the relative error and the relative residual.

79. The exact solution of the linear system

$$
\begin{aligned}
x_1 + 3x_2 &= 4 \\
1.0001x_1 + 3x_2 &= 4.0001
\end{aligned}
$$

is $\mathbf{x} = [1, 1]^T$. Change the right-hand vector $\mathbf{b}$ slightly to $\delta \mathbf{b} = [4.0001, 4.0003]^T$, and consider the linear system

$$
\begin{aligned}
x_1 + 3x_2 &= 4.0001 \\
1.0001x_1 + 3x_2 &= 4.0003.
\end{aligned}
$$

Compute the changed solution $\delta \mathbf{x}$ of the system. Is the matrix A ill-conditioned?

80. If $\|A\| < 1$, then show that the matrix $(\mathbf{I} - A)$ is nonsingular and

$$\|(\mathbf{I} - A)^{-1}\| \leq \frac{1}{1 - \|A\|}.$$

81. The exact solution of the linear system

$$\begin{array}{rcl} x_1 \; + \; x_2 & = & 3 \\ x_1 \; + \; 1.0005x_2 & = & 3.0010 \end{array}$$

is $\mathbf{x} = [1, 2]^T$. Change the coefficient matrix and the right-hand vector $\mathbf{b}$ slightly to

$$\delta A = \begin{pmatrix} 1 & 1 \\ 1 & 1.001 \end{pmatrix} \quad \text{and} \quad \delta \mathbf{b} = \begin{pmatrix} 2.99 \\ 3.01 \end{pmatrix},$$

and consider the linear system

$$\begin{array}{rcl} x_1 \; + \; x_2 & = & 2.99 \\ x_1 \; + \; 1.001x_2 & = & 3.01 \end{array}$$

Compute the changed solution $\delta \mathbf{x}$ of the system. Is the matrix A ill-conditioned?

82. Find the condition number of the following matrix:

$$A_n = \begin{pmatrix} 1 & 1 \\ 1 & 1 - \dfrac{1}{n} \end{pmatrix}.$$

Solve the linear system $A_4\mathbf{x} = [2, 2]^T$ and compute the relative residual.

83. Determine equations of the polynomials of degree two whose graphs pass through the given points.

(a) $(1, 2), (2, 2), (3, 4)$.
(b) $(1, 14), (2, 22), (3, 32)$.
(c) $(1, 5), (2, 7), (3, 9)$.
(d) $(-1, -1), (0, 1), (1, -3)$.
(e) $(1, 8), (3, 26), (5, 60)$.

84. Find an equation of the polynomial of degree three whose graph passes through the points $(1, -3), (2, -1), (3, 9), (4, 33)$.

85. Determine the currents through the various branches of the electrical network in Figure 1.8:

    (a) When battery C is 9 volts.
    (b) When battery C is 23 volts.

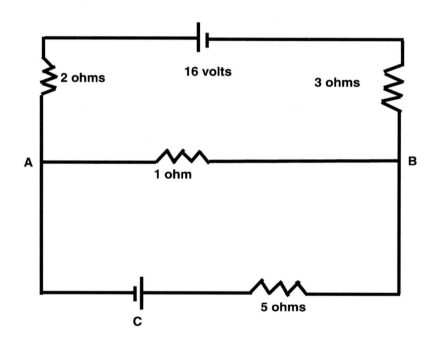

Figure 1.8: Electrical circuit.

Note how the current through the branch AB is reversed in (b). What would the voltage of C have to be for no current to pass through AB?

86. Construct a mathematical model that describes the traffic flow in the road network of Figure 1.9. All streets are one-way streets in the directions indicated. The units are in vehicles per hour. Give two distinct possible flows of traffic. What is the minimum possible flow that can be expected along branch AB?

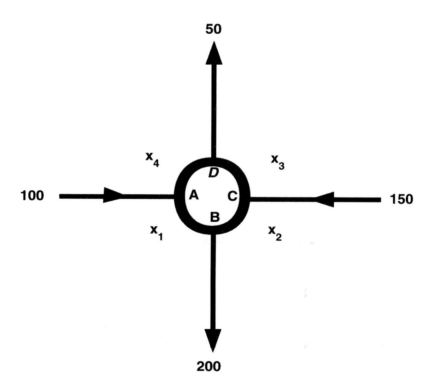

Figure 1.9: Traffic flow.

87. Figure 1.10 represents the traffic entering and leaving a "roundabout" road junction. Such junctions are very common in Europe. Construct a mathematical model that describes the flow of traffic along the various branches. What is the minimum flow theoretically possible along the branch BC? Is this flow ever likely to be realized in practice?

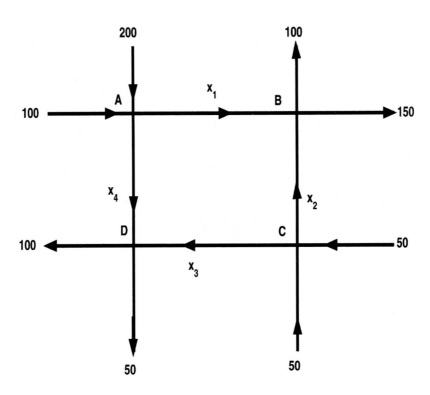

Figure 1.10: Traffic flow.

88. Find the temperatures at x_1, x_2, x_3, and x_4 of the triangular metal plate shown in Figure 1.11, given that the temperature of each interior point is the average of its four neighboring points.

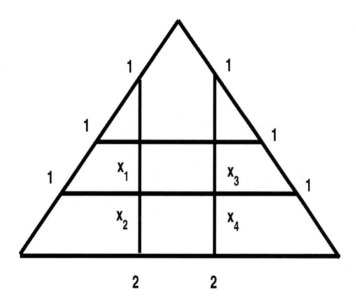

Figure 1.11: Heat Conduction.

89. Find the temperatures at x_1, x_2, and x_3 of the triangular metal plate shown in Figure 1.12, given that the temperature of each interior point is the average of its four neighboring points.

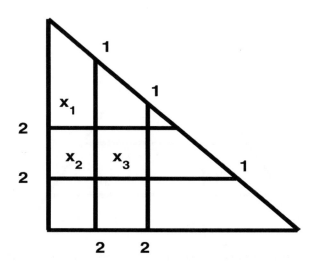

Figure 1.12: Heat conduction.

90. It takes three different ingredients, A, B, and C, to produce a certain chemical substance. A, B, and C have to be dissolved in water separately before they interact to form the chemical. The solution containing A at $2.2g/cm^3$ combined with the solution containing B at $2.5g/cm^3$, combined with the solution containing C at $4.6g/cm^3$, makes $18.25g$ of the chemical. If the proportions for A, B, C in these solutions are changed to $2.4, 3.5$, and $5.8g/cm^3$, respectively (while the volumes remain the same), then $21.26g$ of the chemical is produced. Finally, if the proportions are changed to $1.7, 2.1$, and $3.9g/cm^3$, respectively, then $15.32g$ of the chemical is produced. What are the volumes in cubic centimeters of the solutions containing A, B, and C?

91. Find a balanced chemical equation for each reaction:

(a) $FeS_2 + O_2 \longrightarrow Fe_2O_3 + SO_2$.

(b) $CO_2 + H_2O \longrightarrow C_6H_{12}O_6 + O_2$ (This reaction takes place when a green plant converts carbon dioxide and water to glucose and oxygen during photosynthesis.)

(c) $C_4H_{10} + O_2 \longrightarrow CO_2 + H_2O$ (This reaction occurs when butane, C_4H_{10}, burns in the presence of oxygen to form carbon dioxide and water.)

(d) $C_5H_{11}OH + O_2 \longrightarrow H_2O + CO_2$ (This reaction represents the combustion of amyl alcohol.)

92. Find a balanced chemical equation for each reaction:

(a) $C_7H_6O_2 + O_2 \longrightarrow H_2O + CO_2$.

(b) $HClO_4 + P_4O_{10} \longrightarrow H_3PO_4 + Cl_2O_7$.

(c) $Na_2CO_3 + C + N_2 \longrightarrow NaCN + CO$.

(d) $C_2H_2Cl_4 + Ca(OH)_2 \longrightarrow C_2HCl_3 + CaCl_2 + H_2O$.

93. A manufacturing company produces three products, **I, II**, and **III**. It uses three machines, A, B, and C, for $350, 150$, and 100 hours, respectively. Making one thousand atoms of type **I** requires $30, 10$, and 5 hours on machines A, B, and C, respectively. Making one thousand atoms of type **II** requires $20, 10$, and 10 hours on machines A, B, and C, respectively. Making one thousand atoms of type **III** requires $30, 30$, and 5 hours on machines A, B, and C, respectively. Find the number of items of each type of product that can be produced if the machines are used at full capacity.

94. The average of the temperature for the cities of Jeddah, Makkah, and Riyadh was $15°C$ during a given winter day. The temperature in Makkah was $6°C$ higher than the average of the temperatures of the other two cities. The temperature in Riyadh was $6°C$ lower than the average temperature of the other two cities. What was the temperature in each one of the cities?

95. An international business person needs, on the average, fixed amounts of Japanese yen, French francs, and German marks during each of his business trips. He traveled three times this year. The first time he exchanged a total of $2,400 at the following rates: the dollar was 100 yen, 1.5 francs, and 1.2 marks. The second time he exchanged a total of $2,350 at these rates: the dollar was 100 yen, 1.2 francs, and 1.5 marks. The third time he exchanged a total of $2,390 at these rates: the dollar was 125 yen, 1.2 francs, and 1.2 marks. How many yen, francs, and marks did he buy each time?

96. A father plans to distribute his estate, worth SR1000,000, between his four sons as follows: $\frac{3}{4}$ of the estate is to be split equally among the sons. For the rest, each son is to receive SR5,000 for each year that remains until his 25th birthday. Given that the sons are all 4 years apart, how much would each receive from his father's estate?

97. A biologist has placed three strains of bacteria (denoted by **I, II**, and **III**) in a test tube, where they will feed on three different food sources $(A, B,$ and $C)$. Each day 2300 units of A, 800 units of B, and 1500 units of C are placed in the test tube, and each bacterium consumes a certain number of units of each food per day, as shown in the given table. How many bacteria of each strain can coexist in the test tube and consume all the food?

Food	Bacteria Strain **I**	Bacteria Strain **II**	Bacteria Strain **III**
Food A	2	2	4
Food B	1	2	0
Food C	1	3	1

98. Al-karim hires three types of laborers, **I**, **II**, and **III**, and pays them SR20, SR15, and SR10 per hour, respectively. If the total amount paid is SR20,000 for a total of 300 hours of work, find the possible number of hours put in by the three categories of workers if the category **III** workers must put in the maximum amount of hours.

Chapter 2

Iterative Methods for Linear Systems

2.1 Introduction

The methods discussed in Chapter 1 for the solution of the system of linear equations have been direct, which required a finite number of arithmetic operations. The elimination methods for solving such systems usually yield sufficiently accurate solutions for approximately 20 to 25 simultaneous equations, where most of the unknowns are present in all of the equations. When the coefficients matrix is sparse (has many zeros), a considerably large number of equations can be handled by the elimination methods. But these methods are generally impractical when many hundreds or thousands of equations must be solved simultaneously.

There are, however, several methods that can be used to solve large numbers of simultaneous equations. These methods, called *iterative methods*, are methods by which an approximation to the solution of a system of linear equations may be obtained. The iterative methods are used most often for large, sparse systems of linear equations and they are efficient in terms of computer storage and time requirements. Systems of this type arise frequently in the numerical solutions of boundary value problems and partial differential equations. Unlike the direct methods, the iterative methods may not always yield a solution, even if the determinant of the coefficients matrix is not zero.

The iterative methods to solve the system of linear equations

$$A\mathbf{x} = \mathbf{b} \qquad (2.1)$$

start with an initial approximation $\mathbf{x}^{(0)}$ to the solution $\mathbf{x}$ of the linear system (2.1) and generate a sequence of vectors $\{\mathbf{x}^{(k)}\}_{k=0}^{\infty}$ that converges to $\mathbf{x}$. Most of these iterative methods involve a process that converts the system (2.1) into an equivalent system of the form

$$\mathbf{x} = T\mathbf{x} + \mathbf{c} \qquad (2.2)$$

for some square matrix T and vector $\mathbf{c}$. After the initial vector $\mathbf{x}^{(0)}$ is selected, the sequence of approximate solution vectors is generated by computing

$$\mathbf{x}^{(k+1)} = T\mathbf{x}^{(k)} + \mathbf{c}, \qquad \text{for} \quad k = 0, 1, 2, \ldots. \qquad (2.3)$$

The sequence is terminated when the error is sufficiently small, i.e.,

$$\|\mathbf{x}^{(k+1)} - \mathbf{x}^{(k)}\| < \epsilon, \quad \text{for small positive} \quad \epsilon. \qquad (2.4)$$

Among them, the most useful methods are the *Jacobi* method, the *Gauss–Seidel* method, the *Successive Over-Relaxation* (SOR) method, and the *conjugate gradient* method.

Before discussing these methods, it is convenient to introduce notations for some matrices. The matrix A is written as

$$A = L + D + U, \qquad (2.5)$$

where L is strictly lower-triangular, U is strictly upper-triangular, and D is the diagonal parts of the coefficients matrix A, i.e.,

$$
L = \begin{pmatrix} 0 & 0 & 0 & \cdots & 0 \\ a_{21} & 0 & 0 & \cdots & 0 \\ a_{31} & a_{32} & 0 & \cdots & 0 \\ \vdots & \vdots & \vdots & \vdots & \vdots \\ a_{n1} & a_{n2} & a_{n3} & \cdots & 0 \end{pmatrix}, \quad U = \begin{pmatrix} a_{11} & a_{12} & a_{13} & \cdots & a_{1n} \\ 0 & 0 & a_{23} & \cdots & a_{2n} \\ 0 & 0 & 0 & \cdots & a_{3n} \\ \vdots & \vdots & \vdots & \vdots & \vdots \\ 0 & 0 & 0 & \cdots & 0 \end{pmatrix},
$$

and

$$
D = \begin{pmatrix} a_{11} & 0 & 0 & \cdots & 0 \\ 0 & a_{22} & 0 & \cdots & 0 \\ 0 & 0 & a_{33} & \cdots & 0 \\ \vdots & \vdots & \vdots & \vdots & \vdots \\ 0 & 0 & 0 & \cdots & a_{nn} \end{pmatrix}.
$$

Then (2.1) can be written as

$$(L + D + U)\mathbf{x} = \mathbf{b}. \tag{2.6}$$

Now we discuss our first iterative method to solve the linear system (2.6).

2.2 Jacobi Iterative Method

This is one of the easiest iterative methods to find the approximate solution of the system of linear equations (2.1). To explain its procedure, consider a system of three linear equations as follows:

$$
\begin{aligned}
a_{11}x_1 + a_{12}x_2 + a_{13}x_3 &= b_1 \\
a_{21}x_1 + a_{22}x_2 + a_{23}x_3 &= b_2 \\
a_{31}x_1 + a_{32}x_2 + a_{33}x_3 &= b_3.
\end{aligned}
$$

The solution process starts by solving for the first variable x_1 from the first equation, the second variable x_2 from the second equation and the third variable x_3 from the third equation, which gives

$$
\begin{aligned}
a_{11}x_1 &= b_1 - a_{12}x_2 - a_{13}x_3 \\
a_{22}x_2 &= b_2 - a_{21}x_1 - a_{23}x_3 \\
a_{33}x_3 &= b_3 - a_{31}x_1 - a_{32}x_2
\end{aligned}
$$

or in matrix form

$$D\mathbf{x} = \mathbf{b} - (L + U)\mathbf{x}.$$

Divide both sides of the above three equations by their diagonal elements, $a_{11}, a_{22},$ and a_{33}, respectively, to get

$$x_1 = \frac{1}{a_{11}}\left[b_1 - a_{12}x_2 - a_{13}x_3\right]$$

$$x_2 = \frac{1}{a_{22}}\left[b_2 - a_{21}x_1 - a_{23}x_3\right]$$

$$x_3 = \frac{1}{a_{33}}\left[b_3 - a_{31}x_1 - a_{32}x_2\right],$$

which can be written in the matrix form

$$\mathbf{x} = D^{-1}[\mathbf{b} - (L + U)\mathbf{x}].$$

Let $\mathbf{x}^{(0)} = \left[x_1^{(0)}, x_2^{(0)}, x_3^{(0)}\right]^T$ be an initial solution of the exact solution $\mathbf{x}$ of the linear system (2.1). Then define an iterative sequence

$$x_1^{(k+1)} = \frac{1}{a_{11}}\left[b_1 - a_{12}x_2^{(k)} - a_{13}x_3^{(k)}\right]$$

$$x_2^{(k+1)} = \frac{1}{a_{22}}\left[b_2 - a_{21}x_1^{(k)} - a_{23}x_3^{(k)}\right] \qquad (2.7)$$

$$x_3^{(k+1)} = \frac{1}{a_{33}}\left[b_3 - a_{31}x_1^{(k)} - a_{32}x_2^{(k)}\right]$$

or in matrix form

$$\mathbf{x}^{(k+1)} = D^{-1}[\mathbf{b} - (L + U)\mathbf{x}^{(k)}], \qquad k = 0, 1, 2, \ldots, \qquad (2.8)$$

where k is the number of iterative steps. Then the form (2.7) is called the Jacobi formula for the system of three equations and (2.8) is called its matrix form. For a general system of n linear equations, the Jacobi method is defined by

$$x_i^{(k+1)} = \frac{1}{a_{ii}} \left[b_i - \sum_{j=1}^{i-1} a_{ij} x_j^{(k)} - \sum_{j=i+1}^{n} a_{ij} x_j^{(k)} \right] \qquad (2.9)$$

$$i = 1, 2, \ldots, n, \quad k = 0, 1, 2, \ldots$$

provided that the diagonal elements $a_{ii} \neq 0$, for each $i = 1, 2, \ldots, n$. If the diagonal elements equal zero, then reordering of the equations can be performed so that no element in the diagonal position equals zero. The matrix form of the Jacobi iterative method (2.9) can be written as

$$\mathbf{x}^{(k+1)} = \mathbf{c} + T_J \mathbf{x}^{(k)}, \qquad k = 0, 1, 2, \ldots \qquad (2.10)$$

or

$$\begin{pmatrix} x_1 \\ x_2 \\ \vdots \\ x_n \end{pmatrix}_{(k+1)} = \begin{pmatrix} c_1 \\ c_2 \\ \vdots \\ c_n \end{pmatrix} + \begin{pmatrix} 0 & -t_{12} & \cdots & -t_{1n} \\ -t_{21} & 0 & \cdots & -t_{2n} \\ \vdots & \vdots & \vdots & \vdots \\ -t_{n1} & -t_{n2} & \cdots & 0 \end{pmatrix} \begin{pmatrix} x_1 \\ x_2 \\ \vdots \\ x_n \end{pmatrix}_{(k)},$$

$$(2.11)$$

where the *Jacobi iteration matrix* T_J and vector $\mathbf{c}$ are defined as follows:

$$T_J = -D^{-1}(L + U) \qquad \text{and} \qquad \mathbf{c} = D^{-1}\mathbf{b}, \qquad (2.12)$$

and their elements are defined by

$$t_{ij} = \frac{a_{ij}}{a_{ii}}, \qquad i, j = 1, 2, \ldots, n, \quad i \neq j$$

$$t_{ij} = 0, \qquad i = j$$

$$c_i = \frac{b_i}{a_{ii}}, \qquad i = 1, 2, \ldots, n.$$

The Jacobi iterative method is sometimes called the method of *simultaneous iterations*, because all values of x_i are iterated simultaneously. That is, all values of $x_i^{(k+1)}$ depend only on the values of $x_i^{(k)}$.

Note that the diagonal elements of the Jacobi iteration matrix T_J are always zero. As usual with iterative methods, an initial approximation $x_i^{(0)}$ must be supplied. If we don't have knowledge of the exact solution, it is conventional to start with $x_i^{(0)} = \mathbf{0}$, for all i. The iterations defined by (2.9) are stopped when

$$\|\mathbf{x}^{(k+1)} - \mathbf{x}^{(k)}\| < \epsilon, \tag{2.13}$$

or by using other possible stopping criteria

$$\frac{\|\mathbf{x}^{(k+1)} - \mathbf{x}^{(k)}\|}{\|\mathbf{x}^{(k+1)}\|} < \epsilon \tag{2.14}$$

where ϵ is a preassigned small positive number. For this purpose, any convenient norm can be used, the most common being the l_∞-norm.

Example 2.1 *Solve the following system of equations using the Jacobi iterative method using $\epsilon = 10^{-5}$ in the l_∞-norm:*

$$
\begin{array}{rcrcrcrcr}
15x_1 & - & x_2 & - & 2x_3 & - & 3x_4 & = & 11 \\
-x_1 & + & 15x_2 & - & 2x_3 & - & 3x_4 & = & 22 \\
-x_1 & - & 2x_2 & + & 15x_3 & - & 3x_4 & = & 33 \\
-x_1 & - & 2x_2 & - & 3x_3 & + & 15x_4 & = & 44.
\end{array}
$$

Start with the initial solution $\mathbf{x}^{(0)} = [0, 0, 0, 0]^T$.

Solution. *The Jacobi method for the given system is*

$$x_1^{(k+1)} = \frac{1}{15}\left[15 + x_2^{(k)} + 2x_3^{(k)} + 3x_4^{(k)}\right]$$

$$x_2^{(k+1)} = \frac{1}{15}\left[22 + x_1^{(k)} + 2x_3^{(k)} + 3x_4^{(k)}\right]$$

$$x_3^{(k+1)} = \frac{1}{15}\left[33 + x_1^{(k)} + 2x_2^{(k)} + 3x_4^{(k)}\right]$$

$$x_4^{(k+1)} = \frac{1}{15}\left[44 + x_1^{(k)} + 2x_2^{(k)} + 3x_3^{(k)}\right],$$

Table 2.1: Solution of Example 2.1.

k	$x_1^{(k)}$	$x_2^{(k)}$	$x_3^{(k)}$	$x_4^{(k)}$
0	0.00000	0.00000	0.00000	0.00000
1	0.73333	1.46667	2.20000	2.93333
2	1.71111	2.39556	3.03111	3.61778
3	2.02074	2.70844	3.35704	3.97304
4	2.15611	2.84366	3.49045	4.10058
5	2.20842	2.89592	3.54300	4.15431
6	2.22966	2.91716	3.56421	4.17528
7	2.23810	2.92560	3.57266	4.18377
8	2.24148	2.92898	3.57604	4.18715
9	2.24283	2.93033	3.57739	4.18851
10	2.24338	2.93088	3.57793	4.18905
11	2.24359	2.93109	3.57815	4.18926
12	2.24368	2.93118	3.57824	4.18935
13	2.24371	2.93121	3.57827	4.18938
14	2.24373	2.93123	3.57829	4.18940
15	2.24373	2.93123	3.57829	4.18940

and starting with the initial approximation $x_1^{(0)} = 0, x_2^{(0)} = 0, x_3^{(0)} = 0, x_4^{(0)} = 0$, then for $k = 0$, we obtain

$$x_1^{(1)} = \frac{1}{15}\left[15 + x_2^{(0)} + 2x_3^{(0)} + 3x_4^{(0)}\right] = 0.73333$$

$$x_2^{(1)} = \frac{1}{15}\left[22 + x_1^{(0)} + 2x_3^{(0)} + 3x_4^{(0)}\right] = 1.46667$$

$$x_3^{(1)} = \frac{1}{15}\left[33 + x_1^{(0)} + 2x_2^{(0)} + 3x_4^{(0)}\right] = 2.20000$$

$$x_4^{(1)} = \frac{1}{15}\left[44 + x_1^{(0)} + 2x_2^{(0)} + 3x_3^{(0)}\right] = 2.93333.$$

The first and subsequent iterations are listed in Table 2.1.

Note that the Jacobi method converged and after 15 iterations we obtained the good approximation $[2.24373, 2.93123, 3.57829, 4.18940]^T$ of the given system having the exact solution $[2.24374, 2.93124, 3.57830, 4.18941]^T$. Ideally, the iterations should stop automatically when we obtain the required accuracy using one of the stopping criteria mentioned in (2.13) or (2.14). •

The above results can be obtained using MATLAB commands, as follows:

```
>> Ab = [15 − 1 − 2 − 3 11; −1 15 − 2 − 3 22; −1 − 2 15 − 3 33; ...
−1 − 2 − 3 44];
>> x = [0 0 0 0];
>> acc = 1e − 05;
>> JacobiM(Ab, x, acc);
```

Example 2.2 *Solve the following system of equations using the Jacobi iterative method:*

$$
\begin{array}{rcrcrcrcr}
-x_1 & + & 15x_2 & - & 2x_3 & - & 3x_4 & = & 22 \\
15x_1 & - & x_2 & - & 2x_3 & - & 3x_4 & = & 11 \\
-x_1 & - & 2x_2 & + & 15x_3 & - & 3x_4 & = & 33 \\
-x_1 & - & 2x_2 & - & 3x_3 & + & 15x_4 & = & 44.
\end{array}
$$

Start with the initial solution $\mathbf{x}^{(0)} = [0, 0, 0, 0]^T$.

Solution. *Results for this linear system are listed in Table 2.2. Note that in this case the Jacobi method diverges rapidly. Although the given linear system is the same as the linear system of Example 2.1, the first and second equations are interchanged. From this example we conclude that the Jacobi iterative method is not always convergent.*

Table 2.2: Solution of Example 2.2.

k	$x_1^{(k)}$	$x_2^{(k)}$	$x_3^{(k)}$	$x_4^{(k)}$
0	0.000000	0.000000	0.000000	0.00000
1	−2.2000e+001	−1.1000e+001	2.2000e+000	2.9333e+000
2	−2.0020e+002	−3.5420e+002	−1.4667e-001	4.4000e-001
3	−5.3360e+003	−3.0150e+003	−5.8285e+001	−5.7669e+001
4	−4.4958e+004	−7.9762e+004	−7.6707e+002	−7.6646e+002
5	−1.1926e+006	−6.7054e+005	−1.3783e+004	−1.3783e+004
6	−9.9893e+006	−1.7820e+007	−1.7167e+005	−1.7167e+005
7	−2.6645e+008	−1.4898e+008	−3.0763e+006	−3.0763e+006
8	−2.2193e+009	−3.9813e+009	−3.8242e+007	−3.8242e+007
9	−5.9529e+010	−3.3099e+010	−6.8645e+008	−6.8645e+008
10	−4.9305e+011	−8.8950e+011	−8.5190e+009	−8.5190e+009

Program 2.1

MATLAB m-file for the Jacobi Iterative Method

```
function x=JacobiM(Ab,x,acc)   % Ab = [A  b]
[n,t]=size(Ab); b=Ab(1:n,t); R=1; k=1;
d(1,1:n+1)=[0 x]; while R > acc
for i=1:n
sum=0;
for j=1:n; if j ~ =i
sum = sum + Ab(i,j) * d(k,j + 1); end;
x(1,i) = (1/Ab(i,i)) * (b(i,1) − sum); end;end
k=k+1; d(k,1:n+1)=[k-1 x];
R=max(abs((d(k,2:n+1)-d(k-1,2:n+1))));
if k > 10 & R > 100
('Jacobi Method diverges')
break; end; end; x=d;
```

Procedure 2.1 (Jacobi Method)

1. *Check that the coefficient matrix A is strictly diagonally dominant (for guaranteed convergence).*

2. *Initialize the first approximation* $\mathbf{x}^{(0)}$ *and preassigned accuracy* ϵ.

3. *Compute the constant* $\mathbf{c} = D^{-1}\mathbf{b} = \dfrac{b_i}{a_{ii}}, \quad for \quad i = 1, 2, \ldots, n.$

4. *Compute the Jacobi iteration matrix* $T_J = -D^{-1}(L + U).$

5. *Solve for the approximate solutions* $\mathbf{x}_i^{(k+1)} = T_J\mathbf{x}_i^{(k)} + \mathbf{c}, \quad i = 1, 2, \ldots, n$
    $k = 0, 1, \ldots.$

6. *Repeat step* **5** *until* $\|\mathbf{x}_i^{(k+1)} - \mathbf{x}_i^{(k)}\| < \epsilon.$

2.3   Gauss–Seidel Iterative Method

This is one of the most popular and widely used iterative methods for finding the approximate solution of the system of linear equations. This iterative method is a modification of the Jacobi iterative method and gives us good accuracy by using the most recently calculated values.

From the Jacobi iterative formula (2.9), it is seen that the new estimates for solution $\mathbf{x}$ are computed from the old estimates and only when all the new estimates have been determined are they then used in the right-hand side of the equation to perform the next iteration. But the Gauss–Seidel method is used to make use of the new estimates in the right-hand side of the equation as soon as they become available. For example, the Gauss–Seidel formula for the system of three equations can be defined as an iterative sequence:

$$x_1^{(k+1)} = \frac{1}{a_{11}}\left[b_1 - a_{12}x_2^{(k)} - a_{13}x_3^{(k)}\right]$$

$$x_2^{(k+1)} = \frac{1}{a_{22}}\left[b_2 - a_{21}x_1^{(k+1)} - a_{23}x_3^{(k)}\right] \qquad (2.15)$$

$$x_3^{(k+1)} = \frac{1}{a_{33}}\left[b_3 - a_{31}x_1^{(k+1)} - a_{32}x_2^{(k+1)}\right].$$

For a general system of n linear equations, the Gauss–Seidel iterative method is defined as

$$x_i^{(k+1)} = \frac{1}{a_{ii}}\left[b_i - \sum_{j=1}^{i-1}a_{ij}x_j^{(k+1)} - \sum_{j=i+1}^{n}a_{ij}x_j^{(k)}\right], \qquad (2.16)$$

$$i = 1, 2, \ldots, n, \quad k = 0, 1, 2, \ldots$$

and in matrix form, can be represented by

$$\mathbf{x}^{(k+1)} = (D+L)^{-1}[\mathbf{b} - U\mathbf{x}^{(k)}], \qquad \text{for each} \quad k = 0, 1, 2, \ldots. \qquad (2.17)$$

For the lower-triangular matrix $(D+L)$ to be nonsingular, it is necessary and sufficient that the diagonal elements $a_{ii} \neq 0$, for each $i = 1, 2, \ldots, n$. By comparing (2.3) and (2.17), we obtain

$$T_G = -(D+L)^{-1}U \qquad \text{and} \qquad \mathbf{c} = (D+L)^{-1}\mathbf{b}, \qquad (2.18)$$

which are called the *Gauss–Seidel iteration matrix* and the vector, respectively.

The Gauss–Seidel iterative method is sometimes called the method of *successive iteration*, because the most recent values of all $\mathbf{x}_i$ are used in the calculation.

Example 2.3 *Solve the following system of equations using the Gauss–Seidel iterative method, with $\epsilon = 10^{-5}$ in the l_∞-norm:*

$$\begin{array}{rcrcrcrcr}
15x_1 & - & x_2 & - & 2x_3 & - & 3x_4 & = & 11 \\
-x_1 & + & 15x_2 & - & 2x_3 & - & 3x_4 & = & 22 \\
-x_1 & - & 2x_2 & + & 15x_3 & - & 3x_4 & = & 33 \\
-x_1 & - & 2x_2 & - & 3x_3 & + & 15x_4 & = & 44.
\end{array}$$

Start with the initial solution $\mathbf{x}^{(0)} = [0, 0, 0, 0]^T$.

Solution. *The Gauss–Seidel method for the given system is*

$$x_1^{(k+1)} = \frac{1}{15}\left[15 + x_2^{(k)} + 2x_3^{(k)} + 3x_4^{(k)}\right]$$

$$x_2^{(k+1)} = \frac{1}{15}\left[22 + x_1^{(k+1)} + 2x_3^{(k)} + 3x_4^{(k)}\right]$$

$$x_3^{(k+1)} = \frac{1}{15}\left[33 + x_1^{(k+1)} + 2x_2^{(k+1)} + 3x_4^{(k)}\right]$$

$$x_4^{(k+1)} = \frac{1}{15}\left[44 + x_1^{(k+1)} + 2x_2^{(k+1)} + 3x_3^{(k+1)}\right],$$

and starting with the initial approximation $x_1^{(0)} = 0, x_2^{(0)} = 0, x_3^{(0)} = 0, x_4^{(0)} = 0$, then for $k = 0$, we obtain

$$x_1^{(1)} = \frac{1}{15}\left[15 + x_2^{(0)} + 2x_3^{(0)} + 3x_4^{(0)}\right] = 0.73333$$

$$x_2^{(1)} = \frac{1}{15}\left[22 + x_1^{(1)} + 2x_3^{(0)} + 3x_4^{(0)}\right] = 1.51556$$

$$x_3^{(1)} = \frac{1}{15}\left[33 + x_1^{(1)} + 2x_2^{(1)} + 3x_4^{(0)}\right] = 2.45096$$

$$x_4^{(1)} = \frac{1}{15}\left[44 + x_1^{(1)} + 2x_2^{(1)} + 3x_3^{(1)}\right] = 3.67449.$$

The first and subsequent iterations are listed in Table 2.3.

The above results can be obtained using MATLAB commands as follows:

```
>> Ab = [15 − 1 − 2 − 3 11; −1 15 − 2 − 3 22; −1 − 2 15 − 3 33; ...
−1 − 2 − 3 44];
>> x = [0 0 0 0];
>> acc = 1e − 5;
>> GaussSM(Ab, x, acc);
```

Table 2.3: Solution of Example 2.3.

k	$x_1^{(k)}$	$x_2^{(k)}$	$x_3^{(k)}$	$x_4^{(k)}$
0	0.00000	0.00000	0.00000	0.00000
1	0.73333	1.51556	2.45096	3.67449
2	1.89606	2.65476	3.41527	4.09676
3	2.18504	2.88706	3.54996	4.17394
4	2.23392	2.92371	3.57354	4.18681
5	2.24208	2.92997	3.57749	4.18897
6	2.24346	2.93102	3.57816	4.18933
7	2.24369	2.93120	3.57827	4.18939
8	2.24373	2.93123	3.57829	4.18940
9	2.24374	2.93123	3.57830	4.18941

Note that the Gauss–Seidel method converged for the given system and required nine iterations to obtain the approximate solution $[2.24374, 2.93123,$ $3.57830, 4.18941]^T$, which is equal to the exact solution $[2.24374, 2.93124,$ $3.57830, 4.18941]^T$ up to six significant digits, which is six iterations less than required by the Jacobi method for the same linear system. •

Example 2.4 *Solve the following system of equations using the Gauss–Seidel iterative method, with $\epsilon = 10^{-5}$ in the l_∞-norm:*

$$
\begin{array}{rrrrr}
-x_1 & + \ 15x_2 & - \ 2x_3 & - \ 3x_4 & = \ 22 \\
15x_1 & - \ x_2 & - \ 2x_3 & - \ 3x_4 & = \ 11 \\
-x_1 & - \ x_2 & + \ 15x_3 & - \ 3x_4 & = \ 33 \\
-x_1 & - \ 2x_2 & - \ 3x_3 & + \ 15x_4 & = \ 44.
\end{array}
$$

Start with the initial solution $\mathbf{x}^{(0)} = [0, 0, 0, 0]^T$.

Solution. *Results for this linear system are listed in Table 2.4. Note that in this case the Gauss–Seidel method diverges rapidly. Although the given linear system is the same as the linear system of the previous Example 2.3, the first and second equations are interchanged. From this example we conclude that the Gauss–Seidel iterative method is not always convergent.*•

Table 2.4: Solution of Example 2.4.

k	$x_1^{(k)}$	$x_2^{(k)}$	$x_3^{(k)}$	$x_4^{(k)}$
0	0.00000	0.00000	0.00000	0.0000
1	$-2.2000e+001$	$-3.4100e+002$	$-4.4733e+001$	$-5.2947e+001$
2	$-4.8887e+003$	$-7.3093e+004$	$-1.0080e+004$	$-1.2085e+004$
3	$-1.0400e+006$	$-1.5544e+007$	$-2.1442e+006$	$-2.5707e+006$
4	$-2.2115e+008$	$-3.3053e+009$	$-4.5597e+008$	$-5.4665e+008$
5	$-4.7028e+010$	$-7.0287e+011$	$-9.6960e+010$	$-1.1624e+011$
6	$-1.0000e+013$	$-1.4946e+014$	$-2.0618e+013$	$-2.4719e+013$
7	$-2.1265e+015$	$-3.1783e+016$	$-4.3844e+015$	$-5.2564e+015$
8	$-4.5220e+017$	$-6.7585e+018$	$-9.3233e+017$	$-1.1177e+018$
9	$-9.6160e+019$	$-1.4372e+021$	$-1.9826e+020$	$-2.3769e+020$

Program 2.2

MATLAB m-file for the Gauss–Seidel Iterative Method

```
function x=GaussSM(Ab,x,acc)     % Ab = [A  b]
[n,t]=size(Ab); b=Ab(1:n,t);R=1; k=1;
d(1,1:n+1)=[0  x]; k=k+1; while R > acc
for i=1:n; sum=0; for j=1:n
if j <= i - 1; sum = sum + Ab(i,j) * d(k,j + 1);
elseif j >= i + 1
sum = sum + Ab(i,j) * d(k - 1,j + 1); end; end
x(1,i) = (1/Ab(i,i)) * (b(i,1) - sum);
d(k,1)=k-1; d(k,i+1)=x(1,i); end
R=max(abs((d(k,2:n+1)-d(k-1,2:n+1))));
k=k+1; if R > 100 & k > 10 ('Gauss–Seidel method Diverges')
break ;end;end;x=d;
```

Procedure 2.2 (Gauss–Seidel Method)

1. *Check that the coefficient matrix A is strictly diagonally dominant (for guaranteed convergence).*

2. *Initialize the first approximation* $\mathbf{x}^{(0)} \in \mathbb{R}$ *and preassigned accuracy* ϵ.

3. *Compute the constant* $\mathbf{c} = (D+L)^{-1}\mathbf{b}$.

4. *Compute the Gauss–Seidel iteration matrix* $T_G = -(D+L)^{-1}U$.

5. *Solve for the approximate solutions* $x_i^{(k+1)} = T_G x_i^{(k)} + \mathbf{c}$, $i = 1, 2, \ldots, n, k = 0, 1, \ldots$.

6. *Repeat step* **5** *until* $\|\mathbf{x}_i^{(k+1)} - \mathbf{x}_i^{(k)}\| < \epsilon$.

From Example 2.1 and Example 2.3, we note that the solution by the Gauss–Seidel method converges more quickly than the Jacobi method. In general, we may state that **if both the Jacobi method and the Gauss–Seidel method converge, then the Gauss–Seidel method will converge more quickly**. This is generally the case but is not always true. In fact, there are some linear systems for which the Jacobi method converges but the Gauss–Seidel method does not, and others for which the Gauss–Seidel method converges but the Jacobi method does not.

Example 2.5 *Solve the following system of equations using the Jacobi and Gauss–Seidel iterative methods, using $\epsilon = 10^{-5}$ in the l_∞-norm and taking the initial solution $\mathbf{x}^{(0)} = [0, 0, 0, 0]^T$:*

$$
\begin{array}{rcrcrcrcl}
7x_1 & + & x_2 & & & + & x_4 & = & 2 \\
2x_1 & + & 5x_2 & + & x_3 & + & 3x_4 & = & 2 \\
& & 4x_2 & + & 5x_3 & + & 2x_4 & = & 3 \\
x_1 & + & 3x_2 & + & 2x_3 & + & 6x_4 & = & 4.
\end{array}
$$

Solution. *First, we solve by the Jacobi method and for the given system, the Jacobi formula is*

$$
x_1^{(k+1)} = \frac{1}{7}\left[2 - 11x_2^{(k)} - x_4^{(k)}\right]
$$

$$
x_2^{(k+1)} = \frac{1}{5}\left[2 - 2x_1^{(k)} - x_3^{(k)} - 3x_4^{(k)}\right]
$$

$$
x_3^{(k+1)} = \frac{1}{2}\left[3 - 4x_2^{(k)} - 2x_4^{(k)}\right]
$$

$$
x_4^{(k+1)} = \frac{1}{6}\left[4 - x_1^{(k)} - 3x_2^{(k)} - 2x_3^{(k)}\right],
$$

Table 2.5: Solution by the Jacobi method.

k	$x_1^{(k)}$	$x_2^{(k)}$	$x_3^{(k)}$	$x_4^{(k)}$
0	0.00000	0.00000	0.00000	0.00000
1	0.28571	0.40000	0.60000	0.66667
2	−0.43810	−0.23429	0.01333	0.21905
3	0.62259	0.44114	0.69981	0.85238
4	−0.52928	−0.50043	−0.09387	0.10906
5	1.05652	0.56505	0.95672	1.03638
6	−0.75027	−0.83578	−0.26659	−0.11085
7	1.61492	0.81994	1.31296	1.29847
8	−1.18825	−1.28764	−0.57535	−0.45011
9	2.37345	1.26043	1.81015	1.70031
10	−1.93787	−1.93160	−1.08847	−0.96251

and starting with the initial approximation $x_1^{(0)} = 0, x_2^{(0)} = 0, x_3^{(0)} = 0, x_4^{(0)} = 0$, then for $k = 0$, we obtain

$$x_1^{(1)} = \frac{1}{7}\left[2 - 11x_2^{(0)} - x_4^{(0)}\right] = 0.28571$$

$$x_2^{(1)} = \frac{1}{5}\left[2 - 2x_1^{(0)} - x_3^{(0)} - 3x_4^{(0)}\right] = 0.40000$$

$$x_3^{(1)} = \frac{1}{2}\left[3 - 4x_2^{(0)} - 2x_4^{(0)}\right] = 0.60000$$

$$x_4^{(1)} = \frac{1}{6}\left[4 - x_1^{(0)} - 3x_2^{(0)} - 2x_3^{(0)}\right] = 0.66667.$$

The first and subsequent iterations are listed in Table 2.5. Now we solve the same system by the Gauss–Seidel method and for the given system, the

Gauss–Seidel formula is

$$x_1^{(k+1)} = \frac{1}{7}\left[2 - 11x_2^{(k)} - x_4^{(k)}\right]$$

$$x_2^{(k+1)} = \frac{1}{5}\left[2 - 2x_1^{(k+1)} - x_3^{(k)} - 3x_4^{(k)}\right]$$

$$x_3^{(k+1)} = \frac{1}{2}\left[3 - 4x_2^{(k+1)} - 2x_4^{(k)}\right]$$

$$x_4^{(k+1)} = \frac{1}{6}\left[4 - x_1^{(k+1)} - 3x_2^{(k+1)} - 2x_3^{(k+1)}\right],$$

and starting with the initial approximation $x_1^{(0)} = 0, x_2^{(0)} = 0, x_3^{(0)} = 0, x_4^{(0)} = 0$, *then for* $k = 0$, *we obtain*

$$x_1^{(1)} = \frac{1}{7}\left[2 - 11x_2^{(0)} - x_4^{(0)}\right] = 0.28571$$

$$x_2^{(1)} = \frac{1}{5}\left[2 - 2x_1^{(1)} - x_3^{(0)} - 3x_4^{(0)}\right] = 0.28571$$

$$x_3^{(1)} = \frac{1}{2}\left[3 - 4x_2^{(1)} - 2x_4^{(0)}\right] = 0.37143$$

$$x_4^{(1)} = \frac{1}{6}\left[4 - x_1^{(1)} - 3x_2^{(1)} - 2x_3^{(1)}\right] = 0.35238.$$

Table 2.6: Solution by the Gauss–Seidel method.

k	$x_1^{(k)}$	$x_2^{(k)}$	$x_3^{(k)}$	$x_4^{(k)}$
0	0.00000	0.00000	0.00000	0.00000
1	0.28571	0.28571	0.37143	0.35238
2	−0.21361	0.19973	0.29927	0.50265
3	−0.09995	0.07854	0.33611	0.53202
4	0.08630	−0.02095	0.40395	0.52811
5	0.24319	−0.09493	0.46470	0.51870
6	0.36080	−0.14848	0.51130	0.51034
7	0.44613	−0.18692	0.54540	0.50397
8	0.50745	−0.21444	0.56996	0.49933
9	0.55136	−0.23413	0.58758	0.49598
10	0.58278	−0.24822	0.60018	0.49358
⋮	⋮	⋮	⋮	⋮
25	0.66118	−0.28335	0.63163	0.48760
26	0.66132	−0.28342	0.63169	0.48759
27	0.66143	−0.28346	0.63174	0.48758
28	0.66150	−0.28350	0.63177	0.48758

The first and subsequent iterations are listed in Table 2.6. Note that the Jacobi method diverged and the Gauss–Seidel method converged after 28 iterations with the approximate solution $[0.66150, -0.28350, 0.63177, 0.48758]^T$ of the given system, which has the exact solution $[0.66169, -0.28358, 0.63184, 0.48756]^T$. •

Example 2.6 *Solve the following system of equations using the Jacobi and Gauss–Seidel iterative methods, using $\epsilon = 10^{-5}$ in the l_∞-norm and taking the initial solution $\mathbf{x}^{(0)} = [0,0,0,0]^T$:*

$$\begin{aligned} x_1 + 2x_2 - 2x_3 &= 1 \\ x_1 + x_2 + x_3 &= 2 \\ 2x_1 + 2x_2 + x_3 &= 3 \\ x_1 + x_2 + x_3 + x_4 &= 4. \end{aligned}$$

Start with the initial solution $\mathbf{x}^{(0)} = [0,0,0,0]^T$.

Solution. *First, we solve by the Jacobi method and for the given system, the Jacobi formula is*

$$x_1^{(k+1)} = \frac{1}{1}\left[1 - 2x_2^{(k)} + 2x_3^{(k)}\right]$$

$$x_2^{(k+1)} = \frac{1}{1}\left[2 - x_1^{(k)} - x_3^{(k)}\right]$$

$$x_3^{(k+1)} = \frac{1}{1}\left[3 - 2x_1^{(k)} - 2x_2^{(k)}\right]$$

$$x_4^{(k+1)} = \frac{1}{1}\left[4 - x_1^{(k)} - x_2^{(k)} - x_3^{(k)}\right],$$

and starting with initial approximation $x_1^{(0)} = 0, x_2^{(0)} = 0, x_3^{(0)} = 0, x_4^{(0)} = 0,$
then for $k = 0$, *we obtain*

$$x_1^{(1)} = \frac{1}{1}\left[1 - 2x_2^{(0)} + 2x_3^{(0)}\right] = 1.0000$$

$$x_2^{(1)} = \frac{1}{1}\left[2 - x_1^{(0)} - x_3^{(0)}\right] = 2.0000$$

$$x_3^{(1)} = \frac{1}{1}\left[3 - 2x_1^{(0)} - 2x_2^{(0)}\right] = 3.0000$$

$$x_4^{(1)} = \frac{1}{1}\left[4 - x_1^{(0)} - x_2^{(0)} - x_3^{(0)}\right] = 4.0000.$$

The first and subsequent iterations are listed in Table 2.7. Now we solve the same system by the Gauss–Seidel method and for the given system, the

Table 2.7: Solution by the Jacobi method.

k	$x_1^{(k)}$	$x_2^{(k)}$	$x_3^{(k)}$	$x_4^{(k)}$
0	0.0000	0.0000	0.0000	0.0000
1	1.0000	2.0000	3.0000	4.0000
2	3.0000	−2.0000	−3.0000	−2.0000
3	−1.0000	2.0000	1.0000	6.0000
4	−1.0000	2.0000	1.0000	2.0000
5	−1.0000	2.0000	1.0000	2.0000

Gauss–Seidel formula is

$$x_1^{(k+1)} = \frac{1}{1}\left[1 - 2x_2^{(k)} + 2x_3^{(k)} \right]$$

$$x_2^{(k+1)} = \frac{1}{1}\left[2 - x_1^{(k+1)} - x_3^{(k)} \right]$$

$$x_3^{(k+1)} = \frac{1}{1}\left[3 - 2x_1^{(k+1)} - 2x_2^{(k+1)} \right]$$

$$x_4^{(k+1)} = \frac{1}{1}\left[4 - x_1^{(k+1)} - x_2^{(k+1)} - x_3^{(k+1)} \right],$$

and starting with the initial approximation $x_1^{(0)} = 0, x_2^{(0)} = 0, x_3^{(0)} = 0, x_4^{(0)} = 0,$ *then for* $k = 0,$ *we obtain*

$$x_1^{(1)} = \frac{1}{1}\left[1 - 2x_2^{(0)} + 2x_3^{(0)} \right] = 1.0000$$

$$x_2^{(1)} = \frac{1}{1}\left[2 - x_1^{(1)} - x_3^{(0)} \right] = 1.0000$$

$$x_3^{(1)} = \frac{1}{1}\left[3 - 2x_1^{(1)} - 2x_2^{(1)} \right] = -1.0000$$

$$x_4^{(1)} = \frac{1}{1}\left[4 - x_1^{(1)} - x_2^{(1)} - x_3^{(1)} \right] = 3.0000.$$

The first and subsequent iterations are listed in Table 2.8. Note that the

Table 2.8: Solution by the Gauss–Seidel method.

k	$x_1^{(k)}$	$x_2^{(k)}$	$x_3^{(k)}$	$x_4^{(k)}$
0	0.0000	0.0000	0.0000	0.0000
1	1.0000	1.0000	-1.0000	3.0000
2	-3.0000	6.0000	-3.0000	4.0000
3	-17.0000	22.0000	-7.0000	6.0000
4	-57.0000	66.0000	-15.0000	10.0000
5	-417.0000	450.0000	-63.0000	34.0000
7	-1025.0000	1090.0000	-127.0000	66.0000
8	-2433.0000	2562.0000	-255.0000	130.0000
9	-5633.0000	5890.0000	-511.0000	258.0000

Jacobi method converged quickly (only five iterations) but the Gauss–Seidel method diverged for the given system. •

Example 2.7 *Consider the system:*

$$
\begin{array}{rcrcrcr}
6x_1 & + & 2x_2 & & & = & 1 \\
x_1 & + & 7x_2 & - & 2x_3 & = & 2 \\
3x_1 & - & 2x_2 & + & 9x_3 & = & -1.
\end{array}
$$

(a) Find the matrix form of the iterative (Jacobi and Gauss–Seidel) methods.

(b) If $\mathbf{x}^{(k)} = [x_1^{(k)}, x_2^{(k)}, x_3^{(k)}]^T$, then write the iterative forms of part (a) in component forms and find the exact solution of the given system.

(c) Find the formulas for the error $\mathbf{e}^{(k+1)}$ in the $(n+1)$th step.

(d) Find the second approximation of the error $e^{(2)}$ using part (c) if $\mathbf{x}^{(0)} = [0, 00]^T$.

Solution. *Since the given matrix A is*

$$
A = \begin{pmatrix} 6 & 2 & 0 \\ 1 & 7 & -2 \\ 3 & -2 & 9 \end{pmatrix},
$$

$$A = L + U + D = \begin{pmatrix} 0 & 0 & 0 \\ 1 & 0 & 0 \\ 3 & -2 & 0 \end{pmatrix} + \begin{pmatrix} 0 & 2 & 0 \\ 0 & 0 & -2 \\ 0 & 0 & 0 \end{pmatrix} + \begin{pmatrix} 6 & 0 & 0 \\ 0 & 7 & 0 \\ 0 & 0 & 9 \end{pmatrix}.$$

Jacobi Iterative Method

(a) Since the matrix form of the Jacobi iterative method can be written as

$$\mathbf{x}^{(k+1)} = T_J \mathbf{x}^{(k)} + \mathbf{c}, \qquad k = 0, 1, 2, \ldots$$

where

$$T_J = -D^{-1}(L + U) \qquad and \qquad \mathbf{c} = D^{-1}\mathbf{b}$$

one can easily compute the Jacobi iteration matrix T_J and the vector $\mathbf{c}$ as follows:

$$T_J = \begin{pmatrix} 0 & -\dfrac{2}{6} & 0 \\ -\dfrac{1}{7} & 0 & \dfrac{2}{7} \\ -\dfrac{3}{9} & \dfrac{2}{9} & 0 \end{pmatrix} \qquad and \qquad \mathbf{c} = \begin{pmatrix} \dfrac{1}{6} \\ \dfrac{2}{7} \\ -\dfrac{1}{9} \end{pmatrix}.$$

Thus, the matrix form of the Jacobi iterative method is

$$\mathbf{x}^{(k+1)} = \begin{pmatrix} 0 & -\dfrac{2}{6} & 0 \\ -\dfrac{1}{7} & 0 & \dfrac{2}{7} \\ -\dfrac{3}{9} & \dfrac{2}{9} & 0 \end{pmatrix} \mathbf{x}^{(k)} + \begin{pmatrix} \dfrac{1}{6} \\ \dfrac{2}{7} \\ -\dfrac{1}{9} \end{pmatrix}, \qquad k = 0, 1, 2.$$

(b) Now by writing the above iterative matrix form in component form, we

have

$$\begin{pmatrix} x_1 \\ x_2 \\ x_3 \end{pmatrix} = \begin{pmatrix} 0 & -\dfrac{1}{3} & 0 \\ -\dfrac{1}{7} & 0 & \dfrac{2}{7} \\ -\dfrac{1}{3} & \dfrac{2}{9} & 0 \end{pmatrix} \begin{pmatrix} x_1 \\ x_2 \\ x_3 \end{pmatrix} + \begin{pmatrix} \dfrac{1}{6} \\ \dfrac{2}{7} \\ -\dfrac{1}{9} \end{pmatrix},$$

and it is equivalent to

$$
\begin{array}{rcl}
x_1 & = & -\dfrac{1}{3}x_2 + \dfrac{1}{6} \\
x_2 & = & -\dfrac{1}{7}x_1 + \dfrac{2}{7}x_3 + \dfrac{2}{7} \\
x_3 & = & -\dfrac{1}{3}x_1 + \dfrac{2}{9}x_2 - \dfrac{1}{9}.
\end{array}
$$

Now solving for x_1, x_2, and x_3, we get

$$\begin{pmatrix} x_1 \\ x_2 \\ x_3 \end{pmatrix} = \begin{pmatrix} \dfrac{1}{12} \\ \dfrac{1}{4} \\ -\dfrac{1}{12} \end{pmatrix},$$

which is the exact solution of the given system.

(c) Since the error in the $(n+1)$th step is defined as

$$\mathbf{e}^{(k+1)} = \mathbf{x} - \mathbf{x}^{(k+1)},$$

we have

$$\mathbf{e}^{(k+1)} = \begin{pmatrix} \dfrac{1}{12} \\ \dfrac{1}{4} \\ -\dfrac{1}{12} \end{pmatrix} - \begin{pmatrix} 0 & -\dfrac{2}{6} & 0 \\ -\dfrac{1}{7} & 0 & \dfrac{2}{7} \\ -\dfrac{3}{9} & \dfrac{2}{9} & 0 \end{pmatrix} \mathbf{x}^{(k)} + \begin{pmatrix} \dfrac{1}{6} \\ \dfrac{2}{7} \\ -\dfrac{1}{9} \end{pmatrix}.$$

This can also be written as

$$
\mathbf{e}^{(k+1)} = \begin{pmatrix} \dfrac{1}{12} \\[2mm] \dfrac{1}{4} \\[2mm] -\dfrac{1}{12} \end{pmatrix} - \begin{pmatrix} 0 & -\dfrac{2}{6} & 0 \\[2mm] -\dfrac{1}{7} & 0 & \dfrac{2}{7} \\[2mm] -\dfrac{3}{9} & \dfrac{2}{9} & 0 \end{pmatrix} \begin{pmatrix} \dfrac{1}{12} \\[2mm] \dfrac{1}{4} \\[2mm] -\dfrac{1}{12} \end{pmatrix}
$$

$$
+ \begin{pmatrix} 0 & -\dfrac{2}{6} & 0 \\[2mm] -\dfrac{1}{7} & 0 & \dfrac{2}{7} \\[2mm] -\dfrac{3}{9} & \dfrac{2}{9} & 0 \end{pmatrix} \mathbf{e}^{(k)}
$$

$$
+ \begin{pmatrix} \dfrac{1}{6} \\[2mm] \dfrac{2}{7} \\[2mm] -\dfrac{1}{9} \end{pmatrix}
$$

or

$$
\mathbf{e}^{(k+1)} = \begin{pmatrix} 0 & -\dfrac{2}{6} & 0 \\[2mm] -\dfrac{1}{7} & 0 & \dfrac{2}{7} \\[2mm] -\dfrac{3}{9} & \dfrac{2}{9} & 0 \end{pmatrix} \mathbf{e}^{(k)},
$$

*(because $\mathbf{x}^{(k)} = \mathbf{e}^{(k)} - \mathbf{x}$) which is the required error in the $(n+1)$th step.
(d) Now finding the first approximation of the error, we have to compute*

the following:

$$\mathbf{e}^{(1)} = \begin{pmatrix} 0 & -\dfrac{2}{6} & 0 \\[2ex] -\dfrac{1}{7} & 0 & \dfrac{2}{7} \\[2ex] -\dfrac{3}{9} & \dfrac{2}{9} & 0 \end{pmatrix} \mathbf{e}^{(0)},$$

where

$$\mathbf{e}^{(0)} = \mathbf{x} - \mathbf{x}^{(0)}.$$

Using $\mathbf{x}^{(0)} = [0, 0, 0]^T$, *we have*

$$\mathbf{e}^{(0)} = \begin{pmatrix} \dfrac{1}{12} \\[2ex] \dfrac{1}{4} \\[2ex] -\dfrac{1}{12} \end{pmatrix} - \begin{pmatrix} 0 \\[1ex] 0 \\[1ex] 0 \end{pmatrix} = \begin{pmatrix} \dfrac{1}{12} \\[2ex] \dfrac{1}{4} \\[2ex] -\dfrac{1}{12} \end{pmatrix}.$$

Thus,

$$\mathbf{e}^{(1)} = \begin{pmatrix} 0 & -\dfrac{2}{6} & 0 \\[2ex] -\dfrac{1}{7} & 0 & \dfrac{2}{7} \\[2ex] -\dfrac{3}{9} & \dfrac{2}{9} & 0 \end{pmatrix} \begin{pmatrix} \dfrac{1}{12} \\[2ex] \dfrac{1}{4} \\[2ex] -\dfrac{1}{12} \end{pmatrix} = \begin{pmatrix} -\dfrac{1}{12} \\[2ex] -\dfrac{1}{28} \\[2ex] \dfrac{1}{36} \end{pmatrix}.$$

Similarly, for the second approximation of the error, we have to compute the following:

$$\mathbf{e}^{(2)} = \begin{pmatrix} 0 & -\dfrac{2}{6} & 0 \\[2ex] -\dfrac{1}{7} & 0 & \dfrac{2}{7} \\[2ex] -\dfrac{3}{9} & \dfrac{2}{9} & 0 \end{pmatrix} \mathbf{e}^{(1)}$$

or

$$
\mathbf{e}^{(2)} = \begin{pmatrix} 0 & -\dfrac{2}{6} & 0 \\[2mm] -\dfrac{1}{7} & 0 & \dfrac{2}{7} \\[2mm] -\dfrac{3}{9} & \dfrac{2}{9} & 0 \end{pmatrix} \begin{pmatrix} -\dfrac{1}{12} \\[2mm] -\dfrac{1}{28} \\[2mm] \dfrac{1}{36} \end{pmatrix} = \begin{pmatrix} \dfrac{1}{84} \\[2mm] \dfrac{5}{252} \\[2mm] \dfrac{5}{252} \end{pmatrix},
$$

which is the required second approximation of the error.

Gauss–Seidel Iterative Method

(a) Now by using the Gauss–Seidel method, first we compute the Gauss–Seidel iteration matrix T_G and the vector $\mathbf{c}$ as follows:

$$
T_G = \begin{pmatrix} 0 & -\dfrac{1}{3} & 0 \\[2mm] 0 & \dfrac{1}{21} & \dfrac{2}{7} \\[2mm] 0 & \dfrac{23}{189} & \dfrac{4}{63} \end{pmatrix} \quad and \quad \mathbf{c} = \begin{pmatrix} \dfrac{1}{6} \\[2mm] \dfrac{11}{42} \\[2mm] -\dfrac{41}{378} \end{pmatrix}.
$$

Thus, the matrix form of the Gauss–Seidel iterative method is

$$
\mathbf{x}^{(k+1)} = \begin{pmatrix} 0 & -\dfrac{1}{3} & 0 \\[2mm] 0 & \dfrac{1}{21} & \dfrac{2}{7} \\[2mm] 0 & \dfrac{23}{189} & \dfrac{4}{63} \end{pmatrix} \mathbf{x}^{(k)} + \begin{pmatrix} \dfrac{1}{6} \\[2mm] \dfrac{11}{42} \\[2mm] -\dfrac{41}{378} \end{pmatrix}, \quad k = 0, 1, 2.
$$

(b) *Writing the above iterative form in component form, we get*

$$
\begin{pmatrix} x_1 \\ x_2 \\ x_3 \end{pmatrix} = \begin{pmatrix} 0 & -\dfrac{1}{3} & 0 \\ 0 & \dfrac{1}{21} & \dfrac{2}{7} \\ 0 & \dfrac{23}{189} & \dfrac{4}{63} \end{pmatrix} \begin{pmatrix} x_1 \\ x_2 \\ x_3 \end{pmatrix} + \begin{pmatrix} \dfrac{1}{6} \\ \dfrac{11}{42} \\ -\dfrac{41}{378} \end{pmatrix},
$$

and it is equivalent to

$$
\begin{aligned}
x_1 &= & & -\frac{1}{3}x_2 &+ \frac{1}{6} \\
x_2 &= & \frac{1}{21}x_2 &+ \frac{2}{7}x_3 &+ \frac{11}{42} \\
x_3 &= & \frac{23}{189}x_2 &+ \frac{4}{63}x_2 &- \frac{41}{378}.
\end{aligned}
$$

Now solving for $x_1, x_2,$ and x_3, we get

$$
\begin{pmatrix} x_1 \\ x_2 \\ x_3 \end{pmatrix} = \begin{pmatrix} \dfrac{1}{12} \\ \dfrac{1}{4} \\ -\dfrac{1}{12} \end{pmatrix},
$$

which is the exact solution of the given system.
(c) The error in the $(n+1)$th step can be easily computed as

$$
\mathbf{e}^{(k+1)} = \begin{pmatrix} 0 & -\dfrac{1}{3} & 0 \\ 0 & \dfrac{1}{21} & \dfrac{2}{7} \\ 0 & \dfrac{23}{189} & \dfrac{4}{63} \end{pmatrix} \mathbf{e}^{(k)}.
$$

(d) The first and second approximations of the error can be calculated as follows:

$$
\mathbf{e}^{(1)} = \begin{pmatrix} 0 & -\dfrac{1}{3} & 0 \\[2mm] 0 & \dfrac{1}{21} & \dfrac{2}{7} \\[2mm] 0 & \dfrac{23}{189} & \dfrac{4}{63} \end{pmatrix} \quad \mathbf{e}^{(0)} = [-\dfrac{1}{12}, -\dfrac{1}{84}, \dfrac{19}{756}]^{T}
$$

and

$$
\mathbf{e}^{(2)} = \begin{pmatrix} 0 & -\dfrac{1}{3} & 0 \\[2mm] 0 & \dfrac{1}{21} & \dfrac{2}{7} \\[2mm] 0 & \dfrac{23}{189} & \dfrac{4}{63} \end{pmatrix} \quad \mathbf{e}^{(1)} = [\dfrac{1}{252}, \dfrac{5}{756}, \dfrac{1}{6804}]^{T},
$$

which is the required second approximation of the error.   ●

2.4   Convergence Criteria

Since we noted that the Jacobi method and the Gauss–Seidel method do not always converge to the solution of the given system of linear equations, here, we need some conditions which make both methods converge. The sufficient condition for the convergence of both methods is defined in the following theorem.

Theorem 2.1 (Sufficient Condition for Convergence)

If a matrix A is strictly diagonally dominant, then for any choice of initial approximation $\mathbf{x}^{(0)} \in \mathbb{R}$, both the Jacobi method and the Gauss–Seidel method give the sequence $\{x^{(k)}\}_{k=0}^{\infty}$ of approximations that converge to the solution of a linear system.   ●

There is another sufficient condition for the convergence of both iterative methods, which is defined in the following theorem.

Theorem 2.2 (Sufficient Condition for Convergence)

For any initial approximation $\mathbf{x}^{(0)} \in \mathbb{R}$, *the sequence* $\{x^{(k)}\}_{k=0}^{\infty}$ *of approximations defined by*

$$\mathbf{x}^{(k+1)} = T\mathbf{x}^{(k)} + \mathbf{c}, \quad \text{for each} \quad k \geq 0, \quad \text{and} \quad \mathbf{c} \neq 0 \qquad (2.19)$$

converges to the unique solution of $\mathbf{x} = T\mathbf{x} + \mathbf{c}$ *if* $\|T\| < 1$, *for any natural matrix norm, and the following error bounds hold:*

$$\|\mathbf{x} - \mathbf{x}^{(k)}\| \leq \|T\|^k \|\mathbf{x}^{(0)} - \mathbf{x}\|$$

$$\|\mathbf{x} - \mathbf{x}^{(k)}\| \leq \frac{\|T\|^k}{1 - \|T\|} \|\mathbf{x}^{(1)} - \mathbf{x}^{(0)}\|. \qquad (2.20)$$

●

Note that the smaller the value of $\|T\|$, the faster the convergence of the iterative methods.

Example 2.8 *Show that for the nonhomogeneous linear system* $A\mathbf{x} = \mathbf{b}$, *with the matrix* A

$$A = \begin{pmatrix} 5 & 0 & -1 \\ -1 & 3 & 0 \\ 0 & -1 & 4 \end{pmatrix},$$

the Gauss–Seidel iterative method converges faster than the Jacobi iterative method.

Solution. *Here we will show that the* l_∞-*norm of the Gauss–Seidel iteration matrix* T_G *is less than the* l_∞-*norm of the Jacobi iteration matrix* T_J, *i.e.,*

$$\|T_G\| < \|T_J\|.$$

The Jacobi iteration matrix T_J *can be obtained from the given matrix* A *as follows:*

$$T_J = -D^{-1}(L+U) = - \begin{pmatrix} 5 & 0 & 0 \\ 0 & 3 & 0 \\ 0 & 0 & 4 \end{pmatrix}^{-1} \begin{pmatrix} 0 & 0 & -1 \\ -1 & 0 & 0 \\ 0 & -1 & 0 \end{pmatrix} = \begin{pmatrix} 0 & 0 & \frac{1}{5} \\ \frac{1}{3} & 0 & 0 \\ 0 & -\frac{1}{4} & 0 \end{pmatrix}.$$

Then the l_∞-norm of the matrix T_J is

$$\|T_J\|_\infty = \max\left\{\frac{1}{5}, \frac{1}{3}, \frac{1}{4}\right\} = \frac{1}{3} = 0.3333 < 1.$$

The Gauss–Seidel iteration matrix T_G is defined as

$$T_G = -(D+L)^{-1}U = - \begin{pmatrix} 5 & 0 & 0 \\ -1 & 3 & 0 \\ 0 & -1 & 4 \end{pmatrix} \begin{pmatrix} 0 & 0 & -1 \\ 0 & 0 & 0 \\ 0 & 0 & 0 \end{pmatrix},$$

and it gives

$$T_G = \begin{pmatrix} 0 & 0 & \frac{1}{5} \\ 0 & 0 & -\frac{1}{15} \\ 0 & 0 & \frac{1}{60} \end{pmatrix}.$$

Then the l_∞-norm of the matrix T_G is

$$\|T_G\|_\infty = \max\left\{\frac{1}{5}, \frac{1}{15}, \frac{1}{60}\right\} = \frac{1}{5} = 0.2000 < 1,$$

which shows that the Gauss–Seidel method will converge faster than the Jacobi method for the given linear system.   ●

Note that the condition $\|T\| < 1$ is equivalent to the condition that a matrix A is to be strictly diagonally dominant.

For the Jacobi method for a general matrix A, the norm of the Jacobi iteration matrix is defined as

$$\|T_J\| = \max_{1 \le i \le n} \sum_{\substack{j=1 \\ j \ne i}}^{n} \left| \frac{a_{ij}}{a_{ii}} \right|.$$

Thus, $\|T_J\| < 1$ is equivalent to requiring

$$\sum_{\substack{j=1 \\ j \ne i}}^{n} |a_{ij}| < |a_{ii}|,$$

i.e., the matrix A is strictly diagonally dominant.

Example 2.9 *Consider the following linear system of equations:*

$$
\begin{array}{rrrrcr}
10x_1 & + & 2x_2 & + & x_3 & + & x_4 & = & 5 \\
x_1 & + & 12x_2 & + & x_3 & + & 2x_4 & = & 9 \\
2x_1 & + & x_2 & + & 13x_3 & + & 3x_4 & = & 1 \\
x_1 & + & 2x_2 & + & x_3 & + & 15x_4 & = & 13.
\end{array}
$$

(a) *Show that both iterative methods (Jacobi and Gauss–Seidel) will converge by using $\|T\|_\infty < 1$.*
(b) *Find the second approximation $\mathbf{x}^{(2)}$ when the initial solution is $\mathbf{x}^{(0)} = [0, 0, 0, 0]^T$.*
(c) *Compute the error bounds for your approximations.*
(d) *How many iterations are needed to get an accuracy within 10^{-4}?*

Solution. *Since the given matrix A is*

$$A = \begin{pmatrix} 10 & 2 & 1 & 1 \\ 1 & 12 & 1 & 2 \\ 2 & 1 & 13 & 3 \\ 1 & 2 & 1 & 15 \end{pmatrix}$$

from (2.5), we have

$$A = L + U + D = \begin{pmatrix} 0 & 0 & 0 & 0 \\ 1 & 0 & 0 & 0 \\ 2 & 1 & 0 & 0 \\ 1 & 2 & 1 & 0 \end{pmatrix} + \begin{pmatrix} 10 & 0 & 0 & 0 \\ 0 & 12 & 0 & 0 \\ 0 & 0 & 13 & 0 \\ 0 & 0 & 0 & 15 \end{pmatrix} + \begin{pmatrix} 0 & 2 & 1 & 1 \\ 0 & 0 & 1 & 2 \\ 0 & 0 & 0 & 3 \\ 0 & 0 & 0 & 0 \end{pmatrix}.$$

Jacobi Iterative Method

(a) Since the Jacobi iteration matrix is defined as

$$T_J = -D^{-1}(L+U),$$

and computing the right-hand side, we get

$$T_J = - \begin{pmatrix} \frac{1}{10} & 0 & 0 & 0 \\ 0 & \frac{1}{12} & 0 & 0 \\ 0 & 0 & \frac{1}{13} & 0 \\ 0 & 0 & 0 & \frac{1}{15} \end{pmatrix} \begin{pmatrix} 0 & 2 & 1 & 1 \\ 1 & 0 & 1 & 2 \\ 2 & 1 & 0 & 3 \\ 1 & 2 & 1 & 0 \end{pmatrix} = \begin{pmatrix} 0 & -\frac{2}{10} & -\frac{1}{10} & -\frac{1}{10} \\ -\frac{1}{12} & 0 & -\frac{1}{12} & -\frac{2}{12} \\ -\frac{2}{13} & -\frac{1}{13} & 0 & -\frac{3}{13} \\ -\frac{1}{15} & -\frac{2}{15} & -\frac{1}{15} & 0 \end{pmatrix},$$

then the l_∞ norm of the matrix T_J is

$$\|T_J\|_\infty = \max\left\{ \frac{4}{10}, \frac{4}{12}, \frac{6}{13}, \frac{4}{15} \right\} = \frac{6}{13} = 0.46154 < 1.$$

Thus, the Jacobi method will converge for the given linear system.

(b) The Jacobi method for the given system is

$$x_1^{(k+1)} = \frac{1}{10}\left[5 \quad - \quad 2x_2^{(k)} \quad - \quad x_3^{(k)} \quad - \quad x_4^{(k)} \right]$$

$$x_2^{(k+1)} = \frac{1}{12}\left[9 \quad - \quad x_1^{(k)} \quad - \quad x_3^{(k)} \quad - \quad 2x_4^{(k)} \right]$$

$$x_3^{(k+1)} = \frac{1}{13}\left[1 \quad - \quad 2x_1^{(k)} \quad - \quad x_2^{(k)} \quad - \quad 3x_4^{(k)} \right]$$

$$x_4^{(k+1)} = \frac{1}{15}\left[13 \quad - \quad x_1^{(k)} \quad - \quad 2x_2^{(k)} \quad - \quad x_3^{(k)} \right].$$

Starting with an initial approximation $x_1^{(0)} = 0, x_2^{(0)} = 0, x_3^{(0)} = 0, x_4^{(0)} = 0,$
and for $k = 0, 1,$ *we obtain the first and the second approximations as follows:*

$$\mathbf{x}^{(1)} = [0.5, 0.75, 0.07692, 0.86667]^T, \quad \mathbf{x}^{(2)} = [0.25564, 0.62970, -0.12436, 0.72821]^T.$$

(c) Using the error bound formula (2.20), we obtain

$$\|\mathbf{x} - \mathbf{x}^{(2)}\| \leq \frac{(\frac{6}{13})^2}{1 - \frac{6}{13}} \left\| \begin{pmatrix} 0.5 \\ 0.75 \\ 0.07692 \\ 0.86667 \end{pmatrix} - \begin{pmatrix} 0 \\ 0 \\ 0 \\ 0 \end{pmatrix} \right\|$$

or

$$\|\mathbf{x} - \mathbf{x}^{(2)}\| \leq \frac{0.21302}{0.53846}(0.86667) = 0.34286.$$

(d) To find the number of iterations, we use formula (2.20) as

$$\|\mathbf{x} - \mathbf{x}^{(k)}\| \leq \frac{\|T_J\|^k}{1 - \|T_J\|}\|\mathbf{x}^{(1)} - \mathbf{x}^{(0)}\| \leq 10^{-4}.$$

It gives

$$\frac{(\frac{6}{13})^k}{\frac{7}{13}}(0.86667) \leq 10^{-4}$$

or

$$(\frac{6}{13})^k \leq \frac{(\frac{7}{13}) \times 10^{-4}}{0.86667},$$

which gives

$$(0.46154)^k \leq (6.21) \times 10^{-5}.$$

Taking ln *on both sides, we obtain*

$$k \ln(\frac{2}{3}) \leq \ln\left((6.21) \times 10^{-5}\right)$$

or

$$k(-0.77327) \leq (-9.68621),$$

and it gives

$$k \geq 12.5263 \quad or \quad k = 13,$$

which is the required number of iterations.

Gauss–Seidel Iterative Method

(a) Since the Gauss–Seidel iteration matrix is defined as

$$T_G = -(D + L)^{-1}U,$$

and computing the right-hand side, we have

$$T_G = - \begin{pmatrix} \frac{1}{10} & 0 & 0 & 0 \\ -\frac{1}{20} & \frac{1}{12} & 0 & 0 \\ -\frac{23}{1560} & -\frac{1}{156} & \frac{1}{13} & 0 \\ -\frac{107}{23400} & -\frac{5}{468} & -\frac{1}{195} & \frac{1}{15} \end{pmatrix} \begin{pmatrix} 0 & 2 & 1 & 1 \\ 0 & 0 & 1 & 2 \\ 0 & 0 & 0 & 3 \\ 0 & 0 & 0 & 0 \end{pmatrix},$$

and it gives

$$T_G = \begin{pmatrix} 0 & -\frac{1}{5} & -\frac{1}{10} & -\frac{1}{10} \\ 0 & \frac{1}{60} & -\frac{3}{40} & -\frac{19}{120} \\ 0 & \frac{23}{780} & \frac{11}{520} & -\frac{317}{1560} \\ 0 & \frac{107}{11700} & \frac{119}{7800} & \frac{136}{3291} \end{pmatrix},$$

then the l_∞ norm of the matrix T_G is

$$\|T_G\|_\infty = \max\{0.4, 0.25058, 0.2539, 0.0657\} = 0.4 < 1.$$

Thus, the Gauss–Seidel method will converge for the given linear system.

(b) The Gauss–Seidel method for the given system is

$$x_1^{(k+1)} = \frac{1}{10}\left[5 \quad - \quad 2x_2^{(k)} \quad - \quad x_3^{(k)} \quad - \quad x_4^{(k)}\right]$$

$$x_2^{(k+1)} = \frac{1}{12}\left[9 \quad - \quad x_1^{(k+1)} \quad - \quad x_3^{(k)} \quad - \quad 2x_4^{(k)}\right]$$

$$x_3^{(k+1)} = \frac{1}{13}\left[1 \quad - \quad 2x_1^{(k+1)} \quad - \quad x_2^{(k+1)} \quad - \quad 3x_4^{(k)}\right]$$

$$x_4^{(k+1)} = \frac{1}{15}\left[13 \quad - \quad x_1^{(k+1)} \quad - \quad 2x_2^{(k+1)} \quad - \quad x_3^{(k+1)}\right].$$

Starting with an initial approximation $x_1^{(0)} = 0, x_2^{(0)} = 0, x_3^{(0)} = 0, x_4^{(0)} = 0$, and for $k = 0, 1$, we obtain the first and the second approximations as follows:

$$\mathbf{x}^{(1)} = [0.5, 0.70833, -0.05449, 0.74252]^T$$

$$\mathbf{x}^{(2)} = [0.28953, 0.66854, -0.07616, 0.76330]^T.$$

(c) Using error bound formula (2.20), we obtain

$$\|\mathbf{x} - \mathbf{x}^{(2)}\| \leq \frac{(0.4)^2}{1 - 0.4}\left\|\begin{pmatrix} 0.5 \\ 0.70833 \\ -0.05449 \\ 0.74252 \end{pmatrix} - \begin{pmatrix} 0 \\ 0 \\ 0 \\ 0 \end{pmatrix}\right\|$$

or

$$\|\mathbf{x} - \mathbf{x}^{(2)}\| \leq \frac{0.16}{0.6}(0.74252) = 0.19801.$$

(d) To find the number of iterations, we use formula (2.20) as

$$\|\mathbf{x} - \mathbf{x}^{(k)}\| \leq \frac{\|T_J\|^k}{1 - \|T_J\|}\|\mathbf{x}^{(1)} - \mathbf{x}^{(0)}\| \leq 10^{-4}.$$

It gives

$$\frac{(0.4)^k}{0.6}(0.74252) \leq 10^{-4}$$

or

$$(0.4)^k \leq (8.08 \times 10^{-5}).$$

Taking ln *on both sides, we obtain*

$$k \ln(0.4) \leq \ln(8.08 \times 10^{-5})$$

or

$$k(-0.91629) \leq (-9.4235),$$

and it gives

$$k \geq 10.28441 \quad or \quad k = 11,$$

which is the required number of iterations.   ●

Theorem 2.3 *If A is a symmetric positive-definite matrix with positive diagonal entries, then the Gauss–Seidel method converges to a unique solution of the linear system $A\mathbf{x} = \mathbf{b}$.*   ●

Example 2.10 *Solve the following system of linear equations using Gauss–Seidel iterative methods, using $\epsilon = 10^{-5}$ in the l_∞-norm and taking the initial solution $\mathbf{x}^{(0)} = [0,0,0,0]^T$:*

$$
\begin{array}{rrrrcr}
5x_1 & & - & x_3 & & = & 1 \\
& 14x_2 & - & x_3 & - & x_4 & = & 1 \\
-x_1 & - & x_2 & + & 13x_3 & & = & 4 \\
& - & x_2 & & + & 9x_4 & = & 3.
\end{array}
$$

Solution. *The matrix*

$$
A = \begin{pmatrix}
5 & 0 & -1 & 0 \\
0 & 14 & -1 & -1 \\
-1 & -1 & 13 & 0 \\
0 & -1 & 0 & 9
\end{pmatrix}
$$

of the given system is symmetric positive-definite with positive diagonal entries, and the Gauss–Seidel formula for the system is

$$x_1^{(k+1)} = \frac{1}{5}\left[1 \qquad\qquad + \quad x_3^{(k)} \qquad\qquad\right]$$

$$x_2^{(k+1)} = \frac{1}{14}\left[1 \qquad\qquad + \quad x_3^{(k)} + x_4^{(k)}\right]$$

$$x_3^{(k+1)} = \frac{1}{13}\left[4 + x_1^{(k+1)} + x_2^{(k+1)} \qquad\qquad\right]$$

$$x_4^{(k+1)} = \frac{1}{9}\left[3 \qquad\qquad + \quad x_2^{(k+1)} \qquad\qquad\right].$$

So starting with an initial approximation $x_1^{(0)} = 0, x_2^{(0)} = 0, x_3^{(0)} = 0, x_4^{(0)} = 0,$ *and for* $k = 0$, *we get*

$$x_1^{(1)} = \frac{1}{5}\left[1 \qquad + \quad x_3^{(0)} \qquad\qquad\right] = 0.200000$$

$$x_2^{(1)} = \frac{1}{14}\left[1 \qquad + \quad x_3^{(0)} + x_4^{(0)}\right] = 0.071429$$

$$x_3^{(1)} = \frac{1}{13}\left[4 + x_1^{(1)} + x_2^{(1)} \qquad\qquad\right] = 0.328571$$

$$x_4^{(1)} = \frac{1}{9}\left[3 \qquad + \quad x_2^{(1)} \qquad\qquad\right] = 0.341270.$$

The first and subsequent iterations are listed in Table 2.9. ●

Note that the Gauss–Seidel method converged very fast (only five iterations) and the approximate solution of the given system $[0.267505, 0.120302, 0.337524, 0.346700]^T$ is equal to the exact solution $[0.267505, 0.120302, 0.337524, 0.346700]^T$ up to six decimal places.

Table 2.9: Solution by the Gauss–Seidel method.

k	$x_1^{(k)}$	$x_2^{(k)}$	$x_3^{(k)}$	$x_4^{(k)}$
0	0.000000	0.000000	0.000000	0.000000
1	0.200000	0.071429	0.328571	0.341270
2	0.265714	0.119274	0.337307	0.346586
3	0.267461	0.120278	0.337518	0.346698
4	0.267504	0.120301	0.337524	0.346700
5	0.267505	0.120302	0.337524	0.346700

2.5 Eigenvalues and Eigenvectors

Here, we will briefly discuss the eigenvalues and eigenvectors of an $n \times n$ matrix. We also show how they can be used to describe the solutions of linear systems.

Definition 2.1 *An $n \times n$ matrix A is said to have an eigenvalue λ of A if there exists a nonzero vector, called an eigenvector $\mathbf{x}$, such that*

$$A\mathbf{x} = \lambda\mathbf{x}. \tag{2.21}$$

Then the relation (2.21) represents the eigenvalue problem, and we refer to $(\lambda, \mathbf{x})$ as an eigenpair. •

The equivalent form of (2.21) is

$$(A - \lambda\mathbf{I})\mathbf{x} = \mathbf{0}, \tag{2.22}$$

where $\mathbf{I}$ is an $n \times n$ identity matrix. The system of equations (2.22) has the nontrivial solution $\mathbf{x}$ if, and only if, $A - \lambda\mathbf{I}$ is singular or, equivalently,

$$\det(A - \lambda\mathbf{I}) = |A - \lambda\mathbf{I}| = 0. \tag{2.23}$$

The above relation (2.23) represents a polynomial equation in λ of degree n which in principle could be used to obtain the eigenvalues of the matrix A. This equation is called the *characteristic equation* of A. There are n roots of (2.23), which we will denote by $\lambda_1, \lambda_2, \ldots, \lambda_n$. For a given eigenvalue λ_i, the corresponding eigenvector $\mathbf{x}_i$ is not uniquely determined. If $\mathbf{x}$ is an eigenvector, then so is $\alpha\mathbf{x}$, where α is any nonzero scalar.

Example 2.11 *Find the eigenvalues and eigenvectors of the following matrix:*

$$A = \begin{pmatrix} -6 & 0 & 0 \\ 11 & -3 & 0 \\ -3 & 6 & 7 \end{pmatrix}.$$

Solution. *To find the eigenvalues of the given matrix A by using (2.23), we have*

$$\begin{vmatrix} -6-\lambda & 0 & 0 \\ 11 & -3-\lambda & 0 \\ -3 & 6 & 7-\lambda \end{vmatrix} = 0,$$

which gives a characteristic equation of the form

$$\lambda^3 + 2\lambda^2 - 45\lambda - 126 = 0.$$

It factorizes to

$$(-6-\lambda)(-3-\lambda)(7-\lambda) = 0$$

and gives us the eigenvalues $\lambda = -6, \lambda = -3$, and $\lambda = 7$ of the given matrix A. Note that the sum of these eigenvalues is -2, and this agrees with the trace of A. After finding the eigenvalues of the matrix we turn to the problem of finding eigenvectors. The eigenvectors of A corresponding to the eigenvalues λ are the nonzero vectors $\mathbf{x}$ that satisfy (2.22). Equivalently, the eigenvectors corresponding to λ are the nonzero vectors in the solution space of (2.22). We call this solution space the eigenspace of A corresponding to λ.

To find the eigenvectors of the above given matrix A corresponding to each of these eigenvalues, we substitute each of these three eigenvalues in (2.22). When $\lambda = -6$, we have

$$\begin{pmatrix} 0 & 0 & 0 \\ 11 & 3 & 0 \\ -3 & 6 & 13 \end{pmatrix} \begin{pmatrix} x_1 \\ x_2 \\ x_3 \end{pmatrix} = \begin{pmatrix} 0 \\ 0 \\ 0 \end{pmatrix},$$

which implies that

$$\begin{array}{ccccccc} 0x_1 & + & 0x_2 & + & 0x_3 & = & 0 \\ 11x_1 & + & 3x_2 & + & 0x_3 & = & 0 \\ -3x_1 & + & 6x_2 & + & 13x_3 & = & 0. \end{array}$$

Solving this system, we get $x_1 = 3, x_2 = -11$, and $x_3 = 75$. Hence, the eigenvector $\mathbf{x}^{(1)}$ corresponding to the first eigenvalue, $\lambda_1 = -6$, is

$$\mathbf{x}^{(1)} = \alpha[3, -11, 75]^T, \quad where \quad \alpha \in \mathbb{R}, \quad \alpha \neq 0.$$

When $\lambda = -3$, we have

$$\begin{pmatrix} -3 & 0 & 0 \\ 11 & 0 & 0 \\ -3 & 6 & 10 \end{pmatrix} \begin{pmatrix} x_1 \\ x_2 \\ x_3 \end{pmatrix} = \begin{pmatrix} 0 \\ 0 \\ 0 \end{pmatrix},$$

which implies that

$$\begin{array}{rcrcrcl} 3x_1 & + & 0x_2 & + & 0x_3 & = & 0 \\ 11x_1 & + & 0x_2 & + & 0x_3 & = & 0 \\ -3x_1 & + & 6x_2 & + & 10x_3 & = & 0, \end{array}$$

which gives the solution, $x_1 = 0, x_2 = 5$, and $x_3 = -3$. Hence, the eigenvector $\mathbf{x}^{(2)}$ corresponding to the second eigenvalue, $\lambda_2 = -3$, is

$$\mathbf{x}^{(2)} = \alpha[0, 5, -3]^T, \quad where \quad \alpha \in \mathbb{R}, \quad \alpha \neq 0.$$

Finally, when $\lambda = 7$, we have

$$\begin{pmatrix} -13 & 0 & 0 \\ 11 & -10 & 0 \\ -3 & 6 & 0 \end{pmatrix} \begin{pmatrix} x_1 \\ x_2 \\ x_3 \end{pmatrix} = \begin{pmatrix} 0 \\ 0 \\ 0 \end{pmatrix},$$

which implies that

$$\begin{array}{rcrcrcl} -13x_1 & + & 0x_2 & + & 0x_3 & = & 0 \\ 11x_1 & - & 10x_2 & + & 0x_3 & = & 0 \\ -3x_1 & + & 6x_2 & + & 0x_3 & = & 0, \end{array}$$

which gives $x_1 = x_2 = 0$, and $x_3 = 1$. Hence,

$$\mathbf{x}^{(3)} = \alpha[0, 0, 1]^T, \quad where \quad \alpha \in \mathbb{R}, \quad \alpha \neq 0$$

is the eigenvector $\mathbf{x}^{(3)}$ corresponding to the third eigenvalue, $\lambda_3 = 7$. ●

The MATLAB command **eig** is the basic eigenvalue and eigenvector routine. The command

$$>> D = eig(A);$$

returns a vector containing all the eigenvalues of the matrix A. If the eigenvectors are also wanted, the syntax

$$>> [X, D] = eig(A);$$

will return a matrix X whose columns are eigenvectors of A corresponding to the eigenvalues in the diagonal matrix D. To get the results of Example 2.11, we use the MATLAB Command Window as follows:

```
>> A = [-6 0 0; 11 - 3 0; -3 6 7];
>> P = poly(A);
>> PP = poly2sym(P);
>> [X, D] = eig(A);
>> eigenvalues = diag(D);
```

Definition 2.2 (Spectral Radius of a Matrix)

Let A be an $n \times n$ matrix. Then the spectral radius $\rho(A)$ of a matrix A is defined as

$$\rho(A) = \max_{1 \leq i \leq n} |\lambda_i|,$$

where λ_i are the eigenvalues of a matrix A. •

For example, the matrix

$$A = \begin{pmatrix} 4 & 1 & -3 \\ 0 & 0 & 2 \\ 0 & 0 & -3 \end{pmatrix}$$

has the characteristic equation of the form

$$\det(A - \lambda \mathbf{I}) = -\lambda^3 + \lambda^2 + 12\lambda = 0,$$

which gives the eigenvalues $\lambda = 4, 0, -3$ of A. Hence, the spectral radius of A is

$$\rho(A) = \max\{|4|, |0|, |-3|\} = 4.$$

The spectral radius of a matrix A may be found using MATLAB commands as follows:

```
>> A = [4 1 − 3; 0 0 2; 0 0 − 3];
>> B = max(eig(A))
B =
    4
```

Example 2.12 *For the matrix*

$$A = \begin{pmatrix} a & b \\ c & d \end{pmatrix},$$

if the eigenvalues of the Jacobi iteration matrix and the Gauss–Seidel iteration matrix are λ_i and μ_i, respectively, then show that $\mu_{max} = \lambda_{max}^2$.

Solution. *Decompose the given matrix into the following form:*

$$A = L + D + U = \begin{pmatrix} 0 & 0 \\ c & 0 \end{pmatrix} + \begin{pmatrix} a & 0 \\ 0 & d \end{pmatrix} + \begin{pmatrix} 0 & b \\ 0 & 0 \end{pmatrix}.$$

First, we define the Jacobi iteration matrix as

$$T_J = -D^{-1}(L + U),$$

and computing the right-hand side, we get

$$T_J = - \begin{pmatrix} \dfrac{1}{a} & 0 \\ 0 & \dfrac{1}{d} \end{pmatrix} \begin{pmatrix} 0 & b \\ c & 0 \end{pmatrix} = \begin{pmatrix} 0 & -\dfrac{b}{a} \\ -\dfrac{c}{d} & 0 \end{pmatrix}.$$

To find the eigenvalues of the matrix T_J, we do as follows:

$$\det(T_J - \lambda\mathbf{I}) = \begin{vmatrix} -\lambda & -\dfrac{b}{a} \\ -\dfrac{c}{d} & -\lambda \end{vmatrix} = 0$$

gives

$$\lambda_1 = -\sqrt{\frac{cb}{ad}}, \qquad \lambda_2 = \sqrt{\frac{cb}{ad}}$$

and

$$\lambda_{max} = \sqrt{\frac{cb}{ad}} .$$

Similarly, we can find the Gauss–Seidel iteration matrix as

$$T_G = -(L+D)^{-1}U,$$

and computing the right-hand side, we get

$$T_G = \begin{pmatrix} 0 & -\dfrac{b}{a} \\[2mm] 0 & \dfrac{cb}{ad} \end{pmatrix} .$$

To find the eigenvalues of the matrix T_G, *we do as follows:*

$$\det(T_G - \lambda\mathbf{I}) = \begin{vmatrix} -\mu & -\dfrac{b}{a} \\[3mm] 0 & \dfrac{cb}{ad} - \mu \end{vmatrix} = 0,$$

which gives

$$\mu_1 = 0, \qquad \mu_2 = \frac{cb}{ad}$$

and

$$\mu_{max} = \frac{cb}{ad} .$$

Thus,

$$\mu_{max} = \left(\sqrt{\frac{cb}{ad}}\right)^2 = \lambda_{max}^2 ,$$

which is the required result. ●

The necessary and sufficient condition for the convergence of the Jacobi iterative method and the Gauss–Seidel iterative method is defined in the following theorem.

Theorem 2.4 (Necessary and Sufficient Condition for Convergence)

For any initial approximation $\mathbf{x}^{(0)} \in \mathbb{R}$, *the sequence* $\{x^{(k)}\}_{k=0}^{\infty}$ *of approximations defined by*

$$\mathbf{x}^{(k+1)} = T\mathbf{x}^{(k)} + \mathbf{c}, \quad \text{for each} \quad k \geq 0, \quad \text{and} \quad \mathbf{c} \neq 0 \qquad (2.24)$$

converges to the unique solution of $\mathbf{x} = T\mathbf{x} + \mathbf{c}$, *if and only if* $\rho(T) < 1$.

Note that the condition $\rho(T) < 1$ *is satisfied when* $\|T\| < 1$ *because* $\rho(T) \leq \|T\|$ *for any natural norm.* •

No general results exist to help us choose between the Jacobi method or the Gauss–Seidel method to solve an arbitrary linear system. However, the following theorem is suitable for the special case.

Theorem 2.5 *If* $a_{ii} \leq 0$, *for each* $i \neq j$, *and* $a_{ii} > 0$, *for each* $i = 1, 2, \ldots, n$, *then one and only one of the following statements holds:*

1. $0 \leq \rho(T_G) < \rho(T_J) < 1$.

2. $1 < \rho(T_J) < \rho(T_G)$.

3. $\rho(T_J) = \rho(T_G) = 0$.

4. $\rho(T_J) = \rho(T_G) = 1$. •

Example 2.13 *Find the spectral radius of the Jacobi and the Gauss–Seidel iteration matrices using each of the following matrices:*

$$\textbf{(a)} \quad A = \begin{pmatrix} 2 & 0 & -1 \\ -1 & 3 & 0 \\ 0 & -1 & 4 \end{pmatrix}, \quad \textbf{(b)} \quad A = \begin{pmatrix} 1 & -1 & 1 \\ -2 & 2 & -1 \\ 0 & 1 & 5 \end{pmatrix},$$

$$\textbf{(c)} \quad A = \begin{pmatrix} 1 & 0 & 0 \\ -1 & 2 & 0 \\ 0 & -1 & 3 \end{pmatrix}, \quad \textbf{(d)} \quad A = \begin{pmatrix} 1 & 0 & -1 \\ 1 & 1 & 0 \\ 0 & 1 & 1 \end{pmatrix}.$$

Solution. (a) *The Jacobi iteration matrix T_J for the given matrix A can be obtained as*

$$T_J = \begin{pmatrix} 0 & 0 & \dfrac{1}{2} \\[2mm] \dfrac{1}{3} & 0 & 0 \\[2mm] 0 & \dfrac{1}{4} & 0 \end{pmatrix},$$

and the characteristic equation of the matrix T_J is

$$\det(T_J - \lambda\mathbf{I}) = -\lambda^3 + \frac{1}{24} = 0.$$

Solving this cubic polynomial, the maximum eigenvalue (in absolute) of T_J is $\dfrac{329}{949}$, i.e.,

$$\rho(T_J) = \frac{329}{949} = 0.3467.$$

Also, the Gauss–Seidel iteration matrix T_G for the given matrix A is

$$T_G = \begin{pmatrix} 0 & 0 & \dfrac{1}{2} \\[2mm] 0 & 0 & \dfrac{1}{6} \\[2mm] 0 & 0 & \dfrac{1}{24} \end{pmatrix}$$

and has the characteristic equation of the form

$$\det(T_G - \lambda\mathbf{I}) = -\lambda^3 + \frac{1}{24}\lambda^2 = 0.$$

Solving this cubic polynomial, we obtain the maximum eigenvalue of T_G, $\dfrac{1}{24}$, i.e.,

$$\rho(T_G) = \frac{1}{24} = 0.0417.$$

283

(b) *The Jacobi iteration matrix T_J for the given matrix A is*

$$T_J = \begin{pmatrix} 0 & 1 & -1 \\ 1 & 0 & \frac{1}{2} \\ 0 & -\frac{1}{5} & 0 \end{pmatrix},$$

with the characteristic equation of the form

$$\det(T_J - \lambda \mathbf{I}) = -\lambda^3 + \frac{9}{20}\lambda^2 + \frac{1}{15} = 0,$$

and it gives

$$\rho(T_J) = \frac{1098}{1051} = 1.0447.$$

The Gauss–Seidel iteration matrix T_G is

$$T_G = \begin{pmatrix} 0 & 1 & -1 \\ 0 & 1 & -\frac{1}{2} \\ 0 & -\frac{1}{5} & \frac{1}{10} \end{pmatrix},$$

with the characteristic equation of the form

$$\det(T_G - \lambda \mathbf{I}) = -\lambda^3 + \frac{11}{10}\lambda^2 = 0,$$

and it gives

$$\rho(T_G) = \frac{11}{10} = 1.1000.$$

*Similarly, for the matrices for **(c)** and **(d)**, we have*

$$T_J = \begin{pmatrix} 0 & 0 & 0 \\ \frac{1}{2} & 0 & 0 \\ 0 & \frac{1}{3} & 0 \end{pmatrix}, \quad T_G = \begin{pmatrix} 0 & 0 & 0 \\ 0 & 0 & 0 \\ 0 & 0 & 0 \end{pmatrix},$$

with

$$\rho(T_J) = 0.0000 \quad and \quad \rho(T_G) = 0.0000$$

and

$$T_J = \begin{pmatrix} 0 & 0 & 1 \\ -1 & 0 & 0 \\ 0 & -1 & 0 \end{pmatrix}, \quad T_G = \begin{pmatrix} 0 & 0 & 1 \\ 0 & 0 & -1 \\ 0 & 0 & 1 \end{pmatrix},$$

with

$$\rho(T_J) = 1.0000 \quad and \quad \rho(T_G) = 1.0000,$$

respectively. ●

Definition 2.3 (Convergent Matrix)

An $n \times n$ matrix is called a convergent matrix if

$$\lim_{k \to \infty} (A^k)_{ij} = 0, \quad for \ each \quad i, j = 1, 2, \ldots, n.$$

●

Example 2.14 *Show that the matrix*

$$A = \begin{pmatrix} \dfrac{1}{3} & 0 \\[2mm] \dfrac{1}{9} & \dfrac{1}{3} \end{pmatrix}$$

is the convergent matrix.

Solution. *By computing the powers of the given matrix, we obtain*

$$A^2 = \begin{pmatrix} \dfrac{1}{9} & 0 \\[2mm] \dfrac{2}{27} & \dfrac{1}{9} \end{pmatrix}, \quad A^3 = \begin{pmatrix} \dfrac{1}{27} & 0 \\[2mm] \dfrac{3}{81} & \dfrac{1}{27} \end{pmatrix}, \quad A^4 = \begin{pmatrix} \dfrac{1}{81} & 0 \\[2mm] \dfrac{4}{243} & \dfrac{1}{81} \end{pmatrix}.$$

Then in general, we have

$$A^k = \begin{pmatrix} \left(\dfrac{1}{3}\right)^k & 0 \\[3mm] \dfrac{k}{3^{k+1}} & \left(\dfrac{1}{3}\right)^k \end{pmatrix},$$

and it gives

$$\lim_{k\to\infty}\left(\frac{1}{3}\right)^k = 0 \quad and \quad \lim_{k\to\infty}\left(\frac{k}{3^{k+1}}\right) = 0.$$

Hence, the given matrix A is convergent.   ●

Since the above matrix has the eigenvalue $\frac{1}{3}$ of order two, its spectral radius is $\frac{1}{3}$. This shows the important relation existing between the spectral radius of a matrix and the convergent of a matrix.

Theorem 2.6 *The following statements are equivalent:*

1. *A is a convergent matrix.*

2. *$\lim_{n\to\infty} \|A^n\| = 0, \quad$ for all natural norms.*

3. *$\rho(A) < 1$.*

4. *$\lim_{n\to\infty} A^n\mathbf{x} = \mathbf{0}, \quad$ for every $\mathbf{x}$.*   ●

Example 2.15 *Show that the matrix*

$$A = \begin{pmatrix} 1 & 1 & 0 & 1 \\ 1 & 1 & 1 & 0 \\ 0 & 1 & 1 & 1 \\ 1 & 0 & 1 & 1 \end{pmatrix}$$

is not the convergent matrix.

Solution. *First, we shall find the eigenvalues of the given matrix A by computing the characteristic equation of the matrix as follows:*

$$\det(A - \lambda\mathbf{I}) = \lambda^4 - 4\lambda^3 + 2\lambda^2 + 4\lambda - 3 = 0,$$

which factorizes to

$$(\lambda + 1)(\lambda - 3)(\lambda - 1)^2 = 0$$

and gives the eigenvalues 3, 1, 1, and –1 of the given matrix A. Hence, the spectral radius of A is

$$\rho(A) = \max\{|3|, |1|, |1|, |-1|\} = 3,$$

which shows that the given matrix is not convergent.   ●

We will discuss some very important results concerning the eigenvalue problems. The proofs of all the results are beyond the scope of this text and will be omitted. However, they are very easily understood and can be used.

Theorem 2.7 *If A is an $n \times n$ matrix, then*

1. $[\rho(A^T A)]^{1/2} = \|A\|_2$, *and*

2. $\rho(A) \leq \|A\|$, *for any natural norm* $\|.\|$.

Example 2.16 *Consider the matrix*

$$A = \begin{pmatrix} -2 & 1 & 2 \\ 1 & 0 & 0 \\ 0 & 1 & 0 \end{pmatrix},$$

which gives a characteristic equation of the form

$$det(A - \lambda\mathbf{I}) = -\lambda^3 - 2\lambda^2 + \lambda + 2 = 0.$$

Solving this cubic equation, the eigenvalues of A are –2, –1, and 1. Thus the spectral radius of A is

$$\rho(A) = max\{|-2|, |-1|, |1|\} = 2.$$

Also,

$$A^T A = \begin{pmatrix} -2 & 1 & 0 \\ 1 & 0 & 1 \\ 2 & 0 & 0 \end{pmatrix} \begin{pmatrix} -2 & 1 & 2 \\ 1 & 0 & 0 \\ 0 & 1 & 0 \end{pmatrix} = \begin{pmatrix} 5 & -2 & -4 \\ -2 & 2 & 2 \\ -4 & 2 & 4 \end{pmatrix},$$

and a characteristic equation of $A^T A$ is

$$-\lambda^3 + 11\lambda^2 - 14\lambda + 4 = 0,$$

which gives the eigenvalues 0.4174, 1, and 9.5826. Therefore, the spectral radius of $A^T A$ is 9.5826. Hence,

$$\|A\|_2 = \sqrt{\rho(A^T A)} = \sqrt{9.5826} \approx 3.0956.$$

From this we conclude that

$$\rho(A) = 2 < 3.0956 \approx \|A\|_2.$$

One can also show that

$$\begin{aligned} \rho(A) &= 2 < 5 = \|A\|_\infty \\ \rho(A) &= 2 < 3 = \|A\|_1, \end{aligned}$$

which satisfies Theorem 2.7.   ●

The spectral norm of a matrix A may be found using MATLAB commands as follows:

```
>> A = [−2 1 2; 1 0 0; 0 1 0];
>> B = sqrt(max(eig(A' * A)))
B =
    3.0956
```

Theorem 2.8 *If A is a symmetric matrix then*

$$\|A\|_2 = \sqrt{\rho(A^T A)} = \rho(A).$$

●

Example 2.17 *Consider a symmetric matrix*

$$A = \begin{pmatrix} 3 & 0 & 1 \\ 0 & -3 & 0 \\ 1 & 0 & 3 \end{pmatrix},$$

which has a characteristic equation of the form

$$-\lambda^3 + 4\lambda^2 + 9\lambda - 36 = 0.$$

Solving this cubic equation, we have the eigenvalues 4, –3, and 3 of the given matrix A. Therefore, the spectral radius of A is 4. Since A is symmetric,

$$A^T A = A^2 = \begin{pmatrix} 10 & 0 & 6 \\ 0 & 9 & 0 \\ 6 & 0 & 10 \end{pmatrix}.$$

Since we know that the eigenvalues of A^2 are the eigenvalues of A raised to the power of 2, the eigenvalues of $A^T A$ are 16, 9, and 9, and its spectral radius is $\rho(A^T A) = \rho(A^2) = [\rho(A)]^2 = 16$. Hence,

$$\|A\|_2 = \sqrt{\rho(A^T A)} = \sqrt{16} = 4 = \rho(A),$$

which satisfies Theorem 2.8. ●

Theorem 2.9 *If A is a nonsingular matrix, then for any eigenvalue of A*

$$\frac{1}{\|A^{-1}\|_2} \le |\lambda| \le \|A\|_2.$$

Note that this result is also true for any natural norm. ●

Example 2.18 *Consider the matrix*

$$A = \begin{pmatrix} 2 & 1 \\ 3 & 2 \end{pmatrix},$$

and its inverse matrix is

$$A^{-1} = \begin{pmatrix} 2 & -1 \\ -3 & 2 \end{pmatrix}.$$

First, we find the eigenvalues of the matrix

$$A^T A = \begin{pmatrix} 13 & 8 \\ 8 & 5 \end{pmatrix},$$

which can be obtained by solving the characteristic equation

$$\det(A^T A - \lambda I) = \begin{vmatrix} 13 - \lambda & 8 \\ 8 & 5 - \lambda \end{vmatrix} = \lambda^2 - 18\lambda + 1 = 0,$$

which gives the eigenvalues 17.96 and 0.04. The spectral radius of $A^T A$ is 17.96. Hence,

$$\|A\|_2 = \sqrt{\rho(A^T A)} = \sqrt{17.96} \approx 4.24.$$

Since a characteristic equation of $(A^{-1})^T (A^{-1})$ is

$$\det[(A^{-1})^T (A^{-1}) - \lambda I] = \begin{vmatrix} 13 - \lambda & 4 \\ 4 & 5 - \lambda \end{vmatrix} = \lambda^2 - 18\lambda + 49 = 0,$$

which gives the eigenvalues 14.64 and 3.36 of $(A^{-1})^T(A^{-1})$, its spectral radius 14.64. Hence,

$$\|A^{-1}\|_2 = \sqrt{\rho((A^{-1})^T(A^{-1}))} = \sqrt{14.64} \approx 3.83.$$

Note that the eigenvalues of A are 3.73 and 0.27, therefore, its spectral radius is 3.73. Hence,

$$\frac{1}{3.83} < |3.73| < 4.24,$$

which satisfies Theorem 2.9.   •

2.6   Successive Over-Relaxation Method

We have seen that the Gauss–Seidel method uses updated information immediately and converges more quickly than the Jacobi method, but in some large systems of equations the Gauss–Seidel method converges at a very slow rate. Many techniques have been developed in order to improve the convergence of the Gauss–Seidel method. Perhaps one of the simplest and most widely used methods is *Successive Over-Relaxation* (SOR). A useful modification to the Gauss–Seidel method is defined by the iterative scheme:

$$x_i^{(k+1)} = (1-\omega)x_i^{(k)} + \frac{\omega}{a_{ii}}\left[b_i - \sum_{j=1}^{i-1} a_{ij}x_j^{(k+1)} - \sum_{j=i+1}^{n} a_{ij}x_j^{(k)}\right], \quad (2.25)$$

$$i = 1, 2, \ldots, n, \quad k = 1, 2, \ldots$$

which can be written as

$$x_i^{(k+1)} = x_i^{(k)} + \frac{\omega}{a_{ii}}\left[b_i - \sum_{j=1}^{i-1} a_{ij}x_j^{(k+1)} - \sum_{j=i}^{n} a_{ij}x_j^{(k)}\right]. \quad (2.26)$$

$$i = 1, 2, \ldots, n, \quad k = 1, 2, \ldots$$

The matrix form of the SOR method can be represented by

$$\mathbf{x}^{(k+1)} = (D+\omega L)^{-1}[(1-\omega)D+\omega U]\mathbf{x}^{(k)} + \omega(D-\omega L)^{-1}\mathbf{b}, \quad (2.27)$$

which is equivalent to

$$\mathbf{x}^{(k+1)} = T_\omega \mathbf{x}^{(k)} + \mathbf{c}, \qquad (2.28)$$

where

$$T_\omega = (D + \omega L)^{-1}[(1-\omega)D - \omega U] \quad \text{and} \quad \mathbf{c} = \omega(D - \omega L)^{-1}\mathbf{b} \qquad (2.29)$$

are called the *SOR iteration matrix* and the vector, respectively.

The quantity ω is called the relaxation factor. It can be formally proved that convergence can be obtained for values of ω in the range $0 < \omega < 2$. For $\omega = 1$, the SOR method (2.25) is simply the Gauss–Seidel method. The methods involving (2.25) are called *relaxation methods*. For the choices of $0 < \omega < 1$, the procedures are called *under-relaxation methods* and can be used to obtain convergence of some systems that are not convergent by the Gauss–Seidel method. For choices $1 < \omega < 2$, the procedures are called *over-relaxation methods*, which can be used to accelerate the convergence for systems that are convergent by the Gauss–Seidel method. The SOR methods are particularly useful for solving linear systems that occur in the numerical solutions of certain partial differential equations.

Example 2.19 *Find the l_∞-norm of the SOR iteration matrix T_ω, when $\omega = 1.005$, by using the following matrix:*

$$A = \begin{pmatrix} 5 & -1 \\ -1 & 10 \end{pmatrix}.$$

Solution. *Since the SOR iteration matrix is*

$$T_\omega = (D + \omega L)^{-1}[(1-\omega)D - \omega U],$$

where

$$L = \begin{pmatrix} 0 & 0 \\ -1 & 0 \end{pmatrix}, \quad U = \begin{pmatrix} 0 & -1 \\ 0 & 0 \end{pmatrix}, \quad D = \begin{pmatrix} 5 & 0 \\ 0 & 10 \end{pmatrix},$$

then

$$T_\omega = \left[\begin{pmatrix} 5 & 0 \\ 0 & 10 \end{pmatrix} + 1.005 \begin{pmatrix} 0 & 0 \\ -1 & 0 \end{pmatrix} \right]^{-1}$$

$$\left[(1-1.005)\begin{pmatrix} 5 & 0 \\ 0 & 10 \end{pmatrix} - 1.005 \begin{pmatrix} 0 & -1 \\ 0 & 0 \end{pmatrix} \right],$$

which is equal to

$$T_\omega = \begin{pmatrix} 5 & 0 \\ -1.005 & 10 \end{pmatrix}^{-1} \begin{pmatrix} -0.025 & 1.005 \\ 0 & -0.05 \end{pmatrix}.$$

Thus,

$$T_\omega = \begin{pmatrix} 0.2 & 0 \\ 0.0201 & 0.1 \end{pmatrix} \begin{pmatrix} -0.025 & 1.005 \\ 0 & -0.05 \end{pmatrix}$$

or

$$T_\omega = \begin{pmatrix} -0.005 & 0.201 \\ -0.0005 & 0.0152 \end{pmatrix}.$$

The l_∞-norm of the matrix T_ω is

$$\|T_\omega\|_\infty = \max\{0.206, 0.0157\} = 0.206.$$

●

Example 2.20 *Solve the following system of linear equations, taking an initial approximation $\mathbf{x}^{(0)} = [0,0,0,0]^T$ and with $\epsilon = 10^{-4}$ in the l_∞-norm:*

$$\begin{array}{rcrcrcrcr}
2x_1 & + & 8x_2 & & & & & = & 1 \\
5x_1 & - & x_2 & + & x_3 & & & = & 2 \\
-x_1 & + & x_2 & + & 4x_3 & + & x_4 & = & 12 \\
& & x_2 & + & x_3 & + & 5x_4 & = & 12.
\end{array}$$

(a) Using the Gauss–Seidel method.
(b) Using the SOR method with $\omega = 0.33$.

Solution. *(a) The Gauss–Seidel method for the given system is*

$$x_1^{(k+1)} = \frac{1}{2}\left[1 - 8x_2^{(k)}\right]$$

$$x_2^{(k+1)} = \frac{1}{-1}\left[2 - 5x_1^{(k+1)} - x_3^{(k)}\right]$$

$$x_3^{(k+1)} = \frac{1}{4}\left[12 + x_1^{(k+1)} - x_2^{(k+1)} - x_4^{(k)}\right]$$

$$x_4^{(k+1)} = \frac{1}{5}\left[12 - x_2^{(k+1)} - x_3^{(k+1)}\right].$$

Table 2.10: Solution of Example 2.20 by the Gauss–Seidel method.

k	$x_1^{(k)}$	$x_2^{(k)}$	$x_3^{(k)}$	$x_4^{(k)}$
0	0.00000	0.00000	0.00000	0.00000
1	5.0000e–001	5.0000e–001	3.0000e+000	1.7000e+000
2	–1.5000e+000	–6.5000e+000	3.8250e+000	2.9350e+000
3	2.6500e+001	1.3432e+002	–2.4690e+001	–1.9527e+001
4	–5.3680e+002	–2.7107e+003	5.5135e+002	4.3427e+002
5	1.0843e+004	5.4766e+004	–1.1086e+004	–8.7335e+003
6	–2.1906e+005	–1.1064e+006	2.2402e+005	1.7648e+005
7	4.4256e+006	2.2352e+007	–4.5257e+006	–3.5653e+006
8	–8.9408e+007	–4.5157e+008	9.1431e+007	7.2027e+007
9	1.8063e+009	9.1227e+009	–1.8471e+009	–1.4551e+009

Starting with an initial approximation $\mathbf{x}^{(0)} = [0,0,0,0]^T$, *and for* $k = 0$, *we obtain*

$$x_1^{(1)} = \frac{1}{2}\left[1 \quad - \quad 8x_2^{(0)}\right] = 0.5$$

$$x_2^{(1)} = \frac{1}{-1}\left[2 \quad - \quad 5x_1^{(1)} \quad - \quad x_3^{(0)}\right] = 0.5$$

$$x_3^{(1)} = \frac{1}{4}\left[12 \quad + \quad x_1^{(1)} \quad - \quad x_2^{(1)} \quad - \quad x_4^{(0)}\right] = 3.0$$

$$x_4^{(1)} = \frac{1}{5}\left[12 \quad\quad - \quad x_2^{(1)} \quad - \quad x_3^{(1)}\right] = 1.7.$$

The first and subsequent iterations are listed in Table 2.10.

(b) Now the SOR method for the given system is

$$x_1^{(k+1)} = (1-\omega)x_1^{(k)} + \frac{\omega}{2}\left[1 - 8x_2^{(k)}\right]$$

$$x_2^{(k+1)} = (1-\omega)x_2^{(k)} + \frac{\omega}{-1}\left[2 - 5x_1^{(k+1)} - x_3^{(k)}\right]$$

$$x_3^{(k+1)} = (1-\omega)x_3^{(k)} + \frac{\omega}{4}\left[12 + x_1^{(k+1)} - x_2^{(k+1)} - x_4^{(k)}\right]$$

$$x_4^{(k+1)} = (1-\omega)x_4^{(k)} + \frac{\omega}{5}\left[12 - x_2^{(k+1)} - x_3^{(k+1)}\right].$$

Starting with an initial approximation $\mathbf{x}^{(0)} = [0,0,0,0]^T$, $\omega = 0.33$, *and for* $k = 0$, *we obtain*

$$x_1^{(1)} = (1-\omega)x_1^{(0)} + \frac{\omega}{2}\left[1 - 8x_2^{(0)}\right] = 0.16500$$

$$x_2^{(1)} = (1-\omega)x_2^{(0)} + \frac{\omega}{-1}\left[2 - 5x_1^{(1)} - x_3^{(0)}\right] = -0.387750$$

$$x_3^{(1)} = (1-\omega)x_3^{(0)} + \frac{\omega}{4}\left[12 + x_1^{(1)} - x_2^{(1)} - x_4^{(0)}\right] = 1.03560$$

$$x_4^{(1)} = (1-\omega)x_4^{(0)} + \frac{\omega}{5}\left[12 - x_2^{(1)} - x_3^{(1)}\right] = 0.74924.$$

The first and subsequent iterations are listed in Table 2.11. Note that the Gauss–Seidel method diverged for the given system, but the SOR method converged very slowly for the given system. ●

Example 2.21 *Solve the following system of linear equations using the SOR method, with* $\epsilon = 0.5 \times 10^{-6}$ *in the* l_∞-*norm:*

$$\begin{array}{rcrcrcrcl}
2x_1 & + & x_2 & & & & & = & 4 \\
x_1 & + & 2x_2 & + & x_3 & & & = & 8 \\
& & x_2 & + & 2x_3 & + & x_4 & = & 12 \\
& & & & x_3 & + & 2x_4 & = & 11.
\end{array}$$

Start with an initial approximation $\mathbf{x}^{(0)} = [0,0,0,0]^T$ *and take* $\omega = 1.27$.

Table 2.11: Solution of Example 2.20 by the SOR Method.

k	$x_1^{(k)}$	$x_2^{(k)}$	$x_3^{(k)}$	$x_4^{(k)}$
0	0.00000	0.00000	0.00000	0.00000
1	0.16500	−0.38775	1.03560	0.74924
2	0.66376	0.51715	1.63414	1.15201
3	−0.26301	−0.20820	1.98531	1.44656
4	0.02493	−0.10321	2.21139	1.62205
5	0.05030	0.08360	2.33506	1.71914
6	−0.19531	−0.15568	2.40939	1.79508
7	−0.05655	−0.06251	2.45669	1.83669
8	−0.09343	−0.04533	2.48049	1.86186
9	−0.14497	−0.11101	2.49552	1.88207
10	−0.09613	−0.06948	2.50453	1.89227
$\vdots$	$\vdots$	$\vdots$	$\vdots$	$\vdots$
21	−0.11932	−0.08436	2.51291	1.91401
22	−0.11939	−0.08427	2.51284	1.91410

Solution. *For the given system, the SOR method with $\omega = 1.27$ is*

$$x_1^{(k+1)} = (1 - \omega)x_1^{(k)} + \frac{\omega}{2}\left[4 - x_2^{(k)}\right]$$

$$x_2^{(k+1)} = (1 - \omega)x_2^{(k)} + \frac{\omega}{2}\left[8 - x_1^{(k+1)} - x_3^{(k)}\right]$$

$$x_3^{(k+1)} = (1 - \omega)x_3^{(k)} + \frac{\omega}{2}\left[12 - x_2^{(k+1)} - x_4^{(k)}\right]$$

$$x_4^{(k+1)} = (1 - \omega)x_4^{(k)} + \frac{\omega}{2}\left[11 - x_3^{(k+1)}\right].$$

Starting with an initial approximation $\mathbf{x}^{(0)} = [0, 0, 0, 0]^T$, *and for $k = 0$,*

Table 2.12: Solution of Example 2.21 by the SOR method.

k	$x_1^{(k)}$	$x_2^{(k)}$	$x_3^{(k)}$	$x_4^{(k)}$
0	0.000000	0.000000	0.000000	0.000000
1	2.540000	3.467100	5.418392	3.544321
2	−0.34741	0.923809	3.319772	3.919978
3	2.047182	1.422556	3.331152	3.811324
4	1.083938	1.892328	3.098770	3.988224
5	1.045709	1.937328	3.020607	3.990094
6	1.027456	1.986402	3.009361	3.996730
⋮	⋮	⋮	⋮	⋮
15	0.999999	2.000000	3.000000	4.000000
16	1.000000	2.000000	3.000000	4.000000

we obtain

$$x_1^{(1)} = (1-1.27)x_1^{(0)} + \frac{1.27}{2}[4 - x_2^{(0)}] = 2.54$$

$$x_2^{(1)} = (1-1.27)x_2^{(0)} + \frac{1.27}{2}[8 - x_1^{(1)} - x_3^{(0)}] = 3.4671$$

$$x_3^{(1)} = (1-1.27)x_3^{(0)} + \frac{1.27}{2}[12 - x_2^{(1)} - x_4^{(0)}] = 5.418392$$

$$x_4^{(1)} = (1-1.27)x_4^{(0)} + \frac{1.27}{2}[11 - x_3^{(1)}] = 3.544321.$$

The first and subsequent iterations are listed in Table 2.12.

To get these results using MATLAB commands, we do the following:

```
>> Ab = [2 1 0 0 4; 1 2 1 0 8; 0 1 2 1 12; 0 0 1 2 11];
>> x = [0 0 0 0];
>> w = 1.27; acc = 0.5e − 6;
>> SORM(Ab, x, w, acc);
```

Table 2.13: Solution of Example 2.21 by the Gauss–Seidel method.

k	$x_1^{(k)}$	$x_2^{(k)}$	$x_3^{(k)}$	$x_4^{(k)}$
0	0.000000	0.000000	0.000000	0.000000
1	2.000000	3.000000	4.500000	3.250000
2	0.500000	1.500000	3.625000	3.687500
3	1.250000	1.562500	3.375000	3.812500
4	1.218750	1.703125	3.242188	3.878906
5	1.148438	1.804688	3.158203	3.920898
6	1.097656	1.872070	3.103516	3.948242
$\vdots$	$\vdots$	$\vdots$	$\vdots$	$\vdots$
35	1.000000	1.999999	3.000000	4.000000
36	1.000000	2.000000	3.000000	4.000000

We note that the SOR method converges and required 16 iterations to obtain what is obviously the correct solution for the given system. If we solve Example 2.21 using the Gauss–Seidel method, we find that this method also converges, but very slowly because it needed 36 iterations to obtain the correct solution, shown by Table 2.13, which is 20 iterations more than required by the SOR method. Also, if we solve the same example using the Jacobi method, we will find that it needs 73 iterations to get the correct solution. Comparing the SOR method with the Gauss–Seidel method, a large reduction in the number of iterations can be achieved, given an efficient choice of ω.

In practice, ω should be chosen in the range $1 < \omega < 2$, but the precise choice of ω is a major problem. Finding the optimum value for ω depends on the particular problem (size of the system of equations and the nature of the equations) and often requires careful work. A detailed study for the optimization of ω can be found in Isaacson and Keller (1966). The following theorems can be used in certain situations for the convergence of the SOR method.

Theorem 2.10 *If all the diagonal elements of a matrix A are nonzero, i.e., $a_{ii} \neq 0$, for each $i = 1, 2, \ldots, n$, then*

$$\rho(T_\omega) = |\omega - 1|.$$

This implies that the SOR method converges only if $0 < \omega < 2$. ●

Theorem 2.11 *If A is a positive-definite matrix and $0 < \omega < 2$, then the SOR method converges for any choice of initial approximation vector $\mathbf{x}^{(0)} \in \mathbb{R}$.* ●

Theorem 2.12 *If A is a positive-definite and tridiagonal matrix, then*

$$\rho(T_G) = [\rho(T_J)]^2 < 1,$$

and the optimal choices of relaxation factor ω for the SOR method is

$$\omega = \frac{2}{1 + \sqrt{1 - [\rho(T_J)]^2}}, \tag{2.30}$$

where T_G and T_J are the Gauss–Seidel iteration and the Jacobi iteration matrices, respectively. With this choice of relaxation factor ω, we can have the spectral radius of the SOR iteration matrix T_ω as

$$\rho(T_\omega) = \omega - 1.$$

Example 2.22 *Find the optimal choice for the relaxation factor ω for using it in the SOR method for solving the linear system $A\mathbf{x} = \mathbf{b}$, where the coefficient matrix A is given as follows:*

$$A = \begin{pmatrix} 2 & -1 & 0 \\ -1 & 2 & -1 \\ 0 & -1 & 2 \end{pmatrix}.$$

Solution. *Since the given matrix A is positive-definite and tridiagonal, we can use Theorem 2.12 to find the optimal choice for ω. Using matrix A, we can find the Jacobi iteration matrix T_J as follows:*

$$T_J = \begin{pmatrix} 0 & \frac{1}{2} & 0 \\ \frac{1}{2} & 0 & \frac{1}{2} \\ 0 & \frac{1}{2} & 0 \end{pmatrix}.$$

Now to find the spectral radius of the Jacobi iteration matrix T_J, we use the characteristic equation

$$\det(T_J - \lambda \mathbf{I}) = |T_J - \lambda \mathbf{i}| = -\lambda^3 + \frac{\lambda}{2},$$

which gives the eigenvalues of matrix T_J, as $\lambda = 0, \pm \frac{1}{\sqrt{2}}$. Thus,

$$\rho(T_J) = \frac{1}{\sqrt{2}} = 0.707107,$$

and the optimal value of ω is

$$\omega = \frac{2}{1 + \sqrt{1 - (0.707107)^2}} = 1.171573.$$

Also, note that the Gauss–Seidel iteration matrix T_G has a the form

$$T_G = \begin{pmatrix} 0 & \dfrac{1}{2} & 0 \\[2mm] 0 & \dfrac{1}{4} & \dfrac{1}{2} \\[2mm] 0 & \dfrac{1}{8} & \dfrac{1}{4} \end{pmatrix},$$

and its characteristic equation is

$$\det(T_G - \lambda \mathbf{I}) = |T_G - \lambda \mathbf{I}| = -\lambda^3 + \frac{\lambda^2}{2}.$$

Thus,

$$\rho(T_G) = \frac{1}{2} = 0.50000 = (\rho(T_J))^2,$$

which agrees with Theorem 2.12. ●

Note that the optimal value of ω can also be found by using (2.30) if the eigenvalues of the Jacobi iteration matrix T_J are real and $0 < \rho(T_J) < 1$. ●

Example 2.23 *Find the optimal choice for the relaxation factor* ω *by using the matrix*

$$A = \begin{pmatrix} 5 & -1 & -1 & -1 \\ 2 & 5 & -1 & 0 \\ -1 & -1 & 5 & -1 \\ -1 & -1 & -1 & 5 \end{pmatrix}.$$

Solution. *Using the given matrix* A, *we can find the Jacobi iteration matrix* T_J *as*

$$T_J = \begin{pmatrix} 0 & \dfrac{1}{5} & \dfrac{1}{5} & \dfrac{1}{5} \\[2mm] \dfrac{2}{5} & 0 & \dfrac{1}{5} & 0 \\[2mm] \dfrac{1}{5} & \dfrac{1}{5} & 0 & \dfrac{1}{5} \\[2mm] \dfrac{1}{5} & \dfrac{1}{5} & \dfrac{1}{5} & 0 \end{pmatrix}.$$

Now to find the spectral radius of the Jacobi iteration matrix T_J, *we use the characteristic equation*

$$\det(T_J - \lambda \mathbf{I}) = 0,$$

and get the following polynomial equation:

$$-\lambda^4 - \frac{6}{25}\lambda^2 - \frac{8}{125}\lambda - \frac{8}{125} = (5\lambda - 3) * (5\lambda + 1)^3 = 0.$$

Solving the above polynomial equation, we obtain

$$\lambda = \frac{3}{5}, -\frac{1}{5}, -\frac{1}{5}, -\frac{1}{5},$$

which are the eigenvalues of the matrix T_J. *From this we get*

$$\rho(T_J) = \frac{3}{5} = 0.6,$$

the spectral radius of the matrix T_J.
Since the value of $\rho(T_J)$ is less than 1, we can use formula (2.30) and get

$$\omega = \frac{2}{1 + \sqrt{1 - (0.6)^2}} = 1.1111,$$

the optimal value of ω. ●

Since the rate of convergence of an iterative method depends on the spectral radius of the matrix associated with the method, one way to choose a method to accelerate convergence is to choose a method whose associated matrix T has a minimal spectral radius.

Example 2.24 *Compare the convergence of the Jacobi, Gauss–Seidel, and SOR iterative methods for the system of linear equations $A\mathbf{x} = \mathbf{b}$, where the coefficient matrix A is given as*

$$A = \begin{pmatrix} 4 & -1 & 0 & 0 \\ -1 & 4 & -1 & 0 \\ 0 & -1 & 4 & -1 \\ 0 & 0 & -1 & 4 \end{pmatrix}.$$

Solution. *First, we compute the Jacobi iteration matrix by using*

$$T_J = -D^{-1}(L + U).$$

Since

$$D = \begin{pmatrix} 4 & 0 & 0 & 0 \\ 0 & 4 & 0 & 0 \\ 0 & 0 & 4 & 0 \\ 0 & 0 & 0 & 4 \end{pmatrix} \quad and \quad L + U = \begin{pmatrix} 0 & -1 & 0 & 0 \\ -1 & 0 & -1 & 0 \\ 0 & -1 & 0 & -1 \\ 0 & 0 & -1 & 0 \end{pmatrix},$$

$$T_J = - \begin{pmatrix} \frac{1}{4} & 0 & 0 & 0 \\ 0 & \frac{1}{4} & 0 & 0 \\ 0 & 0 & \frac{1}{4} & 0 \\ 0 & 0 & 0 & \frac{1}{4} \end{pmatrix} \begin{pmatrix} 0 & -1 & 0 & 0 \\ -1 & 0 & -1 & 0 \\ 0 & -1 & 0 & -1 \\ 0 & 0 & -1 & 0 \end{pmatrix}$$

$$= \begin{pmatrix} 0 & \frac{1}{4} & 0 & 0 \\ \frac{1}{4} & 0 & \frac{1}{4} & 0 \\ 0 & \frac{1}{4} & 0 & \frac{1}{4} \\ 0 & 0 & \frac{1}{4} & 0 \end{pmatrix}.$$

To find the eigenvalues of the Jacobi iteration matrix T_J, we evaluate the determinant as

$$\begin{vmatrix} -\lambda & \frac{1}{4} & 0 & 0 \\ \frac{1}{4} & -\lambda & \frac{1}{4} & 0 \\ 0 & \frac{1}{4} & -\lambda & \frac{1}{4} \\ 0 & 0 & \frac{1}{4} & -\lambda \end{vmatrix} = 0,$$

which gives the characteristic equation of the form

$$\lambda^4 - 0.1875\lambda^2 + 1/256 = 0.$$

Solving this fourth-degree polynomial equation, we get the eigenvalues

$$\lambda = -0.4045, \quad \lambda = -0.1545, \quad \lambda = 0.1545, \quad \lambda = 0.4045$$

of the matrix T_J. The spectral radius of the matrix T_J is

$$\rho(T_J) = 0.4045 < 1,$$

which shows that the Jacobi method will converge for the given linear system.

Since the given matrix is positive-definite and tridiagonal, by using Theorem 2.12 we can compute the spectral radius of the Gauss–Seidel iteration matrix with the help of the spectral radius of the Jacobi iteration matrix, i.e.,

$$\rho(T_G) = [\rho(T_J)]^2 = (0.4045)^2 = 0.1636 < 1,$$

which shows that the Gauss–Seidel method will also converge, and faster, than the Jacobi method. Also, from Theorem 2.12, we have

$$\rho(T_\omega) = \omega - 1.$$

Now to find the spectral radius of the SOR iteration matrix T_ω, we have to calculate first the optimal value of ω by using

$$\omega = \frac{2}{1 + \sqrt{1 - [\rho(T_J)]^2}}.$$

So using $\rho(T_J) = 0.4045$, we get

$$\omega = \frac{2}{1 + \sqrt{1 - [0.4045]^2}} = 1.045.$$

Using this optimal value of ω, we can compute the spectral radius of the SOR iteration matrix T_ω as follows:

$$\rho(T_\omega) = \omega - 1 = 1.045 - 1 = 0.045 < 1.$$

Thus the SOR method will also converge for the given system, and faster than the other two methods, because

$$\rho(T_\omega) < \rho(T_G) < \rho(T_J).$$

●

Program 2.3

MATLAB m-file for the SOR Iterative Method

```
function sol=SORM(Ab,x,w,acc)     % Ab = [A  b]
[n,t]=size(Ab); b=Ab(1:n,t); R=1; k=1;
d(1,1:n+1)=[0  x];
k=k+1; while R > acc
for i=1:n
sum=0;
for j=1:n
if j <= i − 1; sum = sum + Ab(i, j) * d(k, j + 1);
elseif j >= i + 1; sum = sum + Ab(i, j) * d(k − 1, j + 1);
end;end
x(1, i) = (1 − w) * d(k − 1, i + 1) + (w/Ab(i, i)) * (b(i, 1) −
sum);
d(k, 1) = k − 1; d(k, i + 1) = x(1, i); end
R = max(abs((d(k, 2 : n + 1) − d(k − 1, 2 : n + 1))));
if R > 100 & k > 10; break; end
k=k+1; end; x=d;
```

Procedure 2.3 (SOR Method)

1. *Find or take ω in the interval $(0, 2)$ (for guaranteed convergence).*

2. *Initialize the first approximation $\mathbf{x}^{(0)}$ and preassigned accuracy ϵ.*

3. *Compute the constant $\mathbf{c} = \omega(D - \omega L)^{-1}\mathbf{b}$.*

4. *Compute the SOR iteration matrix $T_\omega = (D+\omega L)^{-1}[(1-\omega)D-\omega U]$.*

5. *Solve for the approximate solutions $x_i^{(k+1)} = T_\omega \mathbf{x}_i^{(k)} + \mathbf{c}, \quad i = 1, 2, \ldots, n,$ and $k = 0, 1, \ldots$.*

6. *Repeat step 5 until $\|\mathbf{x}_i^{(k+1)} - \mathbf{x}_i^{(k)}\| < \epsilon$.*

2.7   Conjugate Gradient Method

So far, we have discussed two broad classes of methods for solving linear systems. The first, known as *direct methods* (Chapter 1), are based on some version of Gaussian elimination or *LU* decomposition. Direct methods eventually obtain the exact solution but must be carried through to completion before any useful information is obtained. The second class contains the *iterative methods* discussed in the present chapter that lead to closer and closer approximations to the solution, but almost never reach the exact value.

Now we discuss a method, called the *conjugate gradient* method, which was developed as long ago as 1952. It was originally developed as a direct method designed to solve an $n \times n$ positive-definite linear system. As a direct method it is generally inferior to Gaussian elimination with pivoting since both methods require n major steps to determine a solution, and the steps of the conjugate gradient method are more computationally expansive than those in Gaussian elimination. However, the conjugate gradient method is very useful when employed as an iterative approximation method for solving large sparse systems.

Actually, this method is rarely used as a primary method for solving linear systems, rather, its more common applications arise in solving differential equations and when other iterative methods converge very slowly. We assume the coefficient matrix A of the linear system $A\mathbf{x} = \mathbf{b}$ is positive-definite and orthogonality with respect to the inner product notation

$$< \mathbf{x}, \mathbf{y} >= \mathbf{x}^T A \mathbf{y},$$

where $\mathbf{x}$ and $\mathbf{y}$ are n-dimensional vectors. Also, we have for each $\mathbf{x}$ and $\mathbf{y}$,

$$< \mathbf{x}, A\mathbf{y} >=< A\mathbf{x}, \mathbf{y} > .$$

The conjugate gradient method is a variational approach in which we seek the vector $\mathbf{x}^*$ as a solution to the linear system $A\mathbf{x} = \mathbf{b}$, if and only if $\mathbf{x}^*$ minimizes

$$E(\mathbf{x}) =< \mathbf{x}, A\mathbf{x} > -2 < \mathbf{x}, \mathbf{b} > . \tag{2.31}$$

In addition, for any $\mathbf{x}$ and $\mathbf{v} \neq \mathbf{0}$, the function $E(\mathbf{x} + t\mathbf{v})$ has its minimum when

$$t = \frac{< \mathbf{v}, \mathbf{b} - A\mathbf{x} >}{< \mathbf{v}, A\mathbf{v} >}.$$

The process is started by specifying an initial estimate $\mathbf{x}^{(0)}$ at iteration zero, and by computing the initial residual vector from

$$\mathbf{r}^{(0)} = \mathbf{b} - A\mathbf{x}^{(0)}.$$

We then obtain improved estimates $\mathbf{x}^{(k)}$ from the iterative process

$$\mathbf{x}^{(k)} = \mathbf{x}^{(k-1)} + t_k \mathbf{v}^{(k)}, \tag{2.32}$$

where $\mathbf{v}^{(k)}$ is a search direction expressed as a vector and the value of

$$t_k = \frac{< \mathbf{v}^{(k)}, \mathbf{b} - A\mathbf{x}^{(k-1)} >}{< \mathbf{v}^{(k)}, A\mathbf{v}^{(k)} >}$$

is chosen to minimize the value of $E(\mathbf{x}^{(k)})$.

In a related method, called the method of *steepest descent*, $\mathbf{v}^{(k)}$ is chosen as the residual vector

$$\mathbf{v}^{(k)} = \mathbf{r}^{(k-1)} = \mathbf{b} - A\mathbf{x}^{(k-1)}.$$

This method has merit for nonlinear systems and optimization problems, but it is not used for linear systems because of slow convergence. An alternative approach uses a set of nonzero direction vectors $\{\mathbf{v}^{(1)}, \ldots, \mathbf{v}^{(n)}\}$ that satisfy

$$< \mathbf{v}^{(i)}, A\mathbf{v}^{(j)} >= 0, \qquad \text{if} \quad i \neq j.$$

This is called an A-*orthogonality condition*, and the set of vectors $\{\mathbf{v}^{(1)}, \ldots, \mathbf{v}^{(n)}\}$ is said to be A-*orthogonal*.

In the conjugate gradient method, we use $\mathbf{v}^{(1)}$ equal to $\mathbf{r}^{(0)}$ *only* at the beginning of the process. For all later iterations, we choose

$$\mathbf{v}^{(k+1)} = \mathbf{r}^{(k)} + \frac{\|\mathbf{r}^{(k)}\|^2}{\|\mathbf{r}^{(k-1)}\|^2} \mathbf{v}^{(k)}$$

to be conjugate to all previous direction vectors.

Note that the initial approximation $\mathbf{x}^{(0)}$ can be chosen by the user, with $\mathbf{x}^{(0)} = \mathbf{0}$ as the default. The number of iterations, $m \leq n$, can be chosen by the user in advance; alternatively, one can impose a stopping criterion based on the size of the residual vector, $\|\mathbf{r}^{(k)}\|$, or, alternatively, the distance between successive iterates, $\|\mathbf{x}^{(k+1)} - \mathbf{x}^{(k)}\|$. If the process is carried on to the bitter end, i.e., $m = n$, then, in the absence of round-off errors, the results will be the exact solution to the linear system. More iterations than n may be required in practical applications because of the introduction of round-off errors.

Example 2.25 *The linear system*

$$
\begin{array}{rcrcrcl}
2x_1 & - & x_2 & & & = & 1 \\
-x_1 & + & 2x_2 & - & x_3 & = & 0 \\
& - & x_2 & + & x_3 & = & 1
\end{array}
$$

has the exact solution $\mathbf{x} = [2, 3, 4]^T$. *Solve the system by the conjugate gradient method.*

Solution. *Start with an initial approximation* $\mathbf{x}^{(0)} = [0, 0, 0]^T$ *and find the residual vector as*

$$\mathbf{r}^{(0)} = \mathbf{b} - A\mathbf{x}^{(0)} = \mathbf{b} = [1, 0, 1]^T.$$

The first conjugate direction is $\mathbf{v}^{(1)} = \mathbf{r}^{(0)} = [1, 0, 1]^T$. Since $\|\mathbf{r}^{(0)}\| = \sqrt{2}$ and $< \mathbf{v}^{(1)}, \mathbf{v}^{(1)} >= [\mathbf{v}^{(1)}]^T A \mathbf{v}^{(1)} = 3$, we use (2.32) to obtain the updated approximation to the solution

$$\mathbf{x}^{(1)} = \mathbf{x}^{(0)} + \frac{\|\mathbf{r}^{(0)}\|^2}{< \mathbf{v}^{(1)}, \mathbf{v}^{(1)} >}\mathbf{v}^{(1)} = \frac{2}{3}\begin{pmatrix} 1 \\ 0 \\ 1 \end{pmatrix} = \begin{pmatrix} \frac{2}{3} \\ 0 \\ \frac{2}{3} \end{pmatrix}.$$

Now we compute the next residual vector as

$$\mathbf{r}^{(1)} = \mathbf{b} - A\mathbf{x}^{(1)} = \mathbf{b} = [-\frac{1}{3}, \frac{4}{3}, \frac{4}{3}]^T,$$

and the conjugate direction as

$$\mathbf{v}^{(2)} = \mathbf{r}^{(1)} + \frac{\|\mathbf{r}^{(1)}\|^2}{\|\mathbf{r}^{(0)}\|^2}\mathbf{v}^{(1)} = \begin{pmatrix} -\frac{1}{3} \\ \frac{4}{3} \\ \frac{1}{3} \end{pmatrix} + \frac{2}{2}\begin{pmatrix} 1 \\ 0 \\ 1 \end{pmatrix} = \begin{pmatrix} \frac{2}{3} \\ \frac{4}{3} \\ \frac{4}{3} \end{pmatrix},$$

which satisfies the conjugacy condition $< \mathbf{v}^{(1)}, \mathbf{v}^{(2)} >= [\mathbf{v}^{(1)}]^T A \mathbf{v}^{(2)} = 0$. Now we get the new approximation as

$$\mathbf{x}^{(2)} = \mathbf{x}^{(1)} + \frac{\|\mathbf{r}^{(1)}\|^2}{< \mathbf{v}^{(2)}, \mathbf{v}^{(2)} >}\mathbf{v}^{(2)} = \begin{pmatrix} \frac{2}{3} \\ 0 \\ \frac{2}{3} \end{pmatrix} + \frac{2}{\frac{8}{9}}\begin{pmatrix} \frac{2}{3} \\ \frac{4}{3} \\ \frac{4}{3} \end{pmatrix} = \begin{pmatrix} \frac{13}{6} \\ 3 \\ \frac{11}{3} \end{pmatrix}.$$

Since we are dealing with a 3×3 system, we will recover the exact solution by one more iteration of the method. The new residual vector is

$$\mathbf{r}^{(2)} = \mathbf{b} - A\mathbf{x}^{(2)} = \mathbf{b} = \left[-\frac{1}{3}, -\frac{1}{6}, \frac{1}{3}\right]^T,$$

and the final conjugate direction is

$$
\mathbf{v}^{(3)} = \mathbf{r}^{(2)} + \frac{\|\mathbf{r}^{(2)}\|^2}{\|\mathbf{r}^{(1)}\|^2} \mathbf{v}^{(2)} = \begin{pmatrix} -\dfrac{1}{3} \\[2mm] -\dfrac{1}{6} \\[2mm] \dfrac{1}{3} \end{pmatrix} + \dfrac{\tfrac{1}{4}}{2} \begin{pmatrix} \dfrac{2}{3} \\[2mm] \dfrac{4}{3} \\[2mm] \dfrac{4}{3} \end{pmatrix} = \begin{pmatrix} -\dfrac{1}{4} \\[2mm] 0 \\[2mm] \dfrac{1}{2} \end{pmatrix},
$$

which, as one can check, is conjugate to both $\mathbf{v}^{(1)}$ *and* $\mathbf{v}^{(2)}$. *Thus, the solution is obtained from*

$$
\mathbf{x}^{(3)} = \mathbf{x}^{(2)} + \frac{\|\mathbf{r}^{(2)}\|^2}{<\mathbf{v}^{(3)}, \mathbf{v}^{(3)}>} \mathbf{v}^{(3)} = \begin{pmatrix} \dfrac{13}{6} \\[2mm] 3 \\[2mm] \dfrac{11}{3} \end{pmatrix} + \dfrac{\tfrac{1}{4}}{\tfrac{3}{8}} \begin{pmatrix} -\dfrac{1}{4} \\[2mm] 0 \\[2mm] \dfrac{1}{2} \end{pmatrix} = \begin{pmatrix} 2 \\[2mm] 3 \\[2mm] 4 \end{pmatrix}.
$$

Since we applied the method $n = 3$ *times, this is the actual solution.* ●

Note that in larger examples, one would not carry through the method to the bitter end, since an approximation to the solution is typically obtained with only a few iterations. The result can be a substantial saving in computational time and effort required to produce an approximation to the solution.

To get the above results using MATLAB commands, we do the following:

```
>> A = [2 − 1 0; −1 2 − 1; 0 − 1 1];
>> b = [1 0 1]';
>> x0 = [0 0 0]';
>> acc = 0.5e − 6;
>> maxI = 3;
>> CONJG(A, b, x0, acc, maxI);
```

Program 2.4

MATLAB m-file for the Conjugate Gradient Method
  function x=CONJG(A,b,x0,acc,maxI)
$x = x0; r = b - A*x0; v = r;$
alpha=$r' * r$; iter=0;flag=0;
normb=norm(b); if normb < eps; normb=1; end
while (norm(r)/normb > acc)
$u = A*v; t = alpha/(u'*v); x = x + t*v;$
$r = r - t*u; beta = r'*r;$
$v = r + beta/alpha*v; alpha = beta;$
iter = iter+1; if (*iter == maxI*); flag= 1;
break; end; end

Procedure 2.4 (Conjugate Gradient Method)

1. *Initialize the first approximation* $\mathbf{x}^{(0)} = \mathbf{0}$.

2. *Compute* $\mathbf{r}^{(0)}$ *and set* $\mathbf{v}^{(1)}$ *equal to* $\mathbf{r}^{(0)}$.

3. *For iterations* k *equal to* $(1, 2, \ldots)$ *until convergence:*

   (a) *Compute* $\mathbf{v}^{(k+1)} = \mathbf{r}^{(k)} + \dfrac{\|\mathbf{r}^{(k)}\|^2}{\|\mathbf{r}^{(k-1)}\|^2}\mathbf{v}^{(k)}$.

   (b) *Compute* $\mathbf{x}^{(k+1)} = \mathbf{x}^{(k)} + \dfrac{\|\mathbf{r}^{(k)}\|^2}{[\mathbf{v}^{(k+1)}]^T A \mathbf{v}^{(k+1)}}\mathbf{v}^{(k+1)}$.

2.8   Iterative Refinement

In those cases when the left-hand side coefficients a_{ij} of the system are exact, but the system is ill-conditioned, an approximate solution can be improved by an iterative technique called the method of **residual correction**. The procedure of the method is defined below.

Let $\mathbf{x}^{(1)}$ be an approximate solution to the system

$$A\mathbf{x} = \mathbf{b} \tag{2.33}$$

and let $\mathbf{y}$ be a correction to $\mathbf{x}^{(1)}$ so that the exact solution $\mathbf{x}$ satisfies

$$\mathbf{x} = \mathbf{x}^{(1)} + \mathbf{y}.$$

Then by substituting into (2.33), we find that $\mathbf{y}$ must satisfy

$$A\mathbf{y} = \mathbf{r} = \mathbf{b} - A\mathbf{x}^{(1)}, \tag{2.34}$$

where $\mathbf{r}$ is the residual. The system (2.34) can now be solved to give correction $\mathbf{y}$ to the approximation $\mathbf{x}^{(1)}$. Thus, the new approximation

$$\mathbf{x}^{(2)} = \mathbf{x}^{(1)} + \mathbf{y}$$

will be closer to the solution than $\mathbf{x}^{(1)}$. If necessary, we compute the new residual

$$\mathbf{r} = \mathbf{b} - A\mathbf{x}^{(2)}$$

and solve system (2.34) again to get new corrections. Normally, two or three iterations are enough to get an exact solution. This iterative method can be used to obtain an improved solution whenever an approximate solution has been obtained by any means.

Example 2.26 *The linear system*

$$
\begin{array}{rcrcl}
1.01x_1 & + & 0.99x_2 & = & 2 \\
0.99x_1 & + & 1.01x_2 & = & 2
\end{array}
$$

has the exact solution $\mathbf{x} = [1, 1]^T$. The approximate solution using the Gaussian elimination method is $\mathbf{x}^{(1)} = [1.01, 1.01]^T$ and residual $\mathbf{r}^{(1)} = [-0.02, -0.02]^T$. Then the solution to the system

$$A\mathbf{y} = \mathbf{r}^{(1)},$$

using the simple Gaussian elimination method, is $\mathbf{y}^{(1)} = [-0.01, -0.01]^T$. So the new approximation is

$$\mathbf{x}^{(2)} = \mathbf{x}^{(1)} + \mathbf{y}^{(1)} = [1, 1]^T,$$

which is equal to the exact solution after just one iteration. •

For MATLAB commands for the above iterative method, the two m-files RES.m and WP.m have been used, then the first iteration is easily performed by the following sequence of MATLAB commands:

```
>> A = [1.01 0.99; 0.99 1.01];
>> b = [2 2]';
>> x0 = [1.01 1.01]';
>> r = RES(A, b, x0);
>> B = [A r];
>> y = WP(B);
>> x0 = x0 + y;
```

If needed, the last four commands can be repeated to generate the subsequent iterates.

Procedure 2.5 (Iterative Refinement Method)

1. *Find or use the given initial approximation* $\mathbf{x}^{(1)} \in \mathbb{R}$.

2. *Compute the residual vector* $\mathbf{r}^{(1)} = \mathbf{b} - A\mathbf{x}^{(1)}$.

3. *Solve the linear system* $A\mathbf{y} = \mathbf{r}^{(1)}$ *for the unknown* $\mathbf{y}$.

4. *Set* $\mathbf{x}^{(k+1)} = \mathbf{x}^{(k)} + \mathbf{y}$, *for* $k = 1, \ldots$.

5. *Repeat steps* **2** *to* **4**, *unless the best approximation is achieved.*

2.9 Summary

Several iterative methods were discussed. Among them were the Jacobi method, the Gauss–Seidel method, and the SOR method. All methods converge if the coefficient matrix is strictly diagonally dominant. The SOR is the best method of choice. Although the determination of the optimum value of the relaxation factor ω is difficult, it is generally worthwhile if the system of equations is to be solved many times for right-hand side vectors. The need for estimating parameters is removed in the conjugate gradient method, which, although more complicated to code, can rival the SOR method in efficiency when dealing with large, sparse systems. Iterative methods are generally used when the number of equations is large, and the coefficient matrix is strictly diagonally dominant. At the end of this chapter we discussed the residual corrector method, which improved the approximate solution.

2.10　Problems

1. Find the Jacobi iteration matrix and its l_∞-norm for each of the following matrices:

(a) $\begin{pmatrix} 11 & -3 & 2 \\ 4 & 10 & 3 \\ -2 & 5 & 9 \end{pmatrix}$,　(b) $\begin{pmatrix} 7 & 1 & 1 \\ 3 & 13 & 2 \\ -4 & 3 & 14 \end{pmatrix}$,

(c) $\begin{pmatrix} 11 & -3 & 2 \\ 4 & 10 & 3 \\ -2 & 5 & 9 \end{pmatrix}$,　(d) $\begin{pmatrix} 7 & 1 & 1 \\ 3 & 13 & 2 \\ -4 & 3 & 14 \end{pmatrix}$,

(e) $\begin{pmatrix} 8 & 1 & -1 & 0 \\ 2 & 13 & -2 & 1 \\ -1 & 3 & 15 & 2 \\ 1 & 4 & 5 & 20 \end{pmatrix}$,　(f) $\begin{pmatrix} 7 & 1 & -3 & 1 \\ 1 & 10 & 2 & -3 \\ 1 & -5 & 25 & 4 \\ 1 & 2 & 3 & 17 \end{pmatrix}$.

2. Find the Jacobi iteration matrix and its l_2-norm for each of the following matrices:

(a) $\begin{pmatrix} 5 & 2 & 3 \\ -3 & 6 & 4 \\ 2 & 5 & 8 \end{pmatrix}$,　(b) $\begin{pmatrix} 6 & 3 & 0 \\ 4 & 7 & 5 \\ -3 & 2 & 11 \end{pmatrix}$,

(c) $\begin{pmatrix} 21 & -13 & 6 \\ 5 & 15 & 2 \\ 11 & 2 & 19 \end{pmatrix}$,　(d) $\begin{pmatrix} 9 & 0 & 11 \\ 1 & 11 & 3 \\ -1 & 0 & 4 \end{pmatrix}$,

(e) $\begin{pmatrix} 18 & 2 & -3 & 2 \\ 6 & 17 & -2 & 1 \\ 1 & 13 & 25 & 2 \\ -6 & 8 & 7 & 21 \end{pmatrix}$,　(f) $\begin{pmatrix} 5 & 4 & 3 & 0 \\ 2 & 10 & 3 & 3 \\ 3 & 4 & 12 & -3 \\ 2 & 0 & -1 & 7 \end{pmatrix}$.

3. Solve the following linear systems using the Jacobi method. Start with initial approximation $\mathbf{x}^{(0)} = 0$ and iterate until $\|\mathbf{x}^{(k+1)} - x^{(k)}\|_\infty \leq 10^{-5}$ for each system:

(a)

$$
\begin{aligned}
4x_1 - x_2 + x_3 &= 7 \\
4x_1 - 8x_2 + x_3 &= -21 \\
-2x_1 + x_2 + 5x_3 &= 15
\end{aligned}
$$

(b)

$$
\begin{aligned}
3x_1 + x_2 + x_3 &= 5 \\
2x_1 + 6x_2 + x_3 &= 9 \\
x_1 + x_2 + 4x_3 &= 6
\end{aligned}
$$

(c)

$$
\begin{aligned}
4x_1 + 2x_2 + x_3 &= 1 \\
x_1 + 7x_2 + x_3 &= 4 \\
x_1 + x_2 + 20x_3 &= 7
\end{aligned}
$$

(d)

$$
\begin{aligned}
5x_1 + 2x_2 - x_3 &= 6 \\
x_1 + 6x_2 - 3x_3 &= 4 \\
2x_1 + x_2 + 4x_3 &= 7
\end{aligned}
$$

(e)

$$
\begin{aligned}
6x_1 - x_2 + 3x_3 &= -2 \\
3x_2 + x_3 &= 1 \\
-2x_1 + x_2 + 5x_3 &= 5
\end{aligned}
$$

(f)

$$
\begin{aligned}
4x_1 + x_2 &= -1 \\
2x_1 + 5x_2 + x_3 &= 0 \\
-x_1 + 2x_2 + 4x_3 &= 3
\end{aligned}
$$

(g)

$$
\begin{aligned}
5x_1 - x_2 + x_3 &= 1 \\
3x_2 - x_3 &= -1 \\
x_1 + 2x_2 + 4x_3 &= 2
\end{aligned}
$$

(h)

$$
\begin{aligned}
9x_1 &+& x_2 &+& x_3 &=& 10 \\
2x_1 &+& 10x_2 &+& 3x_3 &=& 19 \\
3x_1 &+& 4x_2 &+& 11x_3 &=& 0
\end{aligned}
$$

4. Consider the following system of equations:

$$
\begin{aligned}
4x_1 &+& 2x_2 &+& x_3 &=& 1 \\
x_1 &+& 7x_2 &+& x_3 &=& 4 \\
x_1 &+& x_2 &+& 20x_3 &=& 7.
\end{aligned}
$$

   (a) Show that the Jacobi method converges by using $\|T_J\|_\infty < 1$.
   (b) Compute the second approximation $\mathbf{x}^{(2)}$, starting with $\mathbf{x}^{(0)} = [0,0,0]^T$.
   (c) Compute an error estimate $\|\mathbf{x} - \mathbf{x}^{(2)}\|_\infty$ for your approximation.

5. If

$$
A = \begin{pmatrix} 4 & 1 & 0 \\ 1 & 3 & -1 \\ 0 & -1 & 4 \end{pmatrix}, \qquad
\mathbf{b} = \begin{pmatrix} 3 \\ 4 \\ 5 \end{pmatrix}.
$$

   Find the Jacobi iteration matrix T_J. If the first approximate solution of the given linear system by the Jacobi method is $[\frac{3}{4}, \frac{4}{3}, \frac{5}{4}]^T$, using $\mathbf{x}^{(0)} = [0,0,0]^T$, then estimate the number of iterations necessary to obtain approximations accurate to within 10^{-6}.

6. Consider the linear system $A\mathbf{x} = \mathbf{b}$, where

$$
A = \begin{pmatrix} -5 & 1 & 0 \\ 1 & 5 & -1 \\ 0 & 1 & 2 \end{pmatrix}, \qquad
\mathbf{b} = \begin{pmatrix} 4 \\ -2 \\ -5 \end{pmatrix}.
$$

   Find the Jacobi iteration matrix T_J and show that $\|T_J\| < 1$. Use the Jacobi method to find the first approximate solution $\mathbf{x}^{(1)}$ of the linear system by using $\mathbf{x}^{(0)} = [0,0,0]^T$. Also, compute the error bound $\|\mathbf{x} - \mathbf{x}^{(10)}\|$. Compute the number of steps needed to get accuracy within 10^{-5}.

7. Consider a linear system $A\mathbf{x} = \mathbf{b}$, where

$$A = \begin{pmatrix} 4 & 2 & 1 \\ 1 & 7 & 1 \\ 1 & 1 & 20 \end{pmatrix}, \qquad \mathbf{b} = \begin{pmatrix} 1 \\ 4 \\ 7 \end{pmatrix}.$$

Show that the Jacobi method converges for the given linear system. If the first approximate solution of the system by the Jacobi method is $\mathbf{x}^{(1)} = [0.25, 0.57, 0.35]^T$, by using an initial approximation $\mathbf{x}^{(0)} = [0, 0, 0]^T$, compute an error bound $\|\mathbf{x} - \mathbf{x}^{(20)}\|$.

8. Consider a linear system $A\mathbf{x} = \mathbf{b}$, where

$$A = \begin{pmatrix} 3 & 2 \\ 2 & 50 \end{pmatrix}, \qquad \mathbf{b} = \begin{pmatrix} 1 \\ 1 \end{pmatrix}.$$

Using the Jacobi method and the Gauss–Seidel method, which one will converge faster and why? If an approximate solution of the system is $[0.1, 0.4]^T$, then find the upper bound for the relative error in solving the given linear system.

9. Rearrange the following system such that convergence of the Gauss–Seidel method is guaranteed. Then use $x^{(0)} = [0, 0, 0]^T$ to find the first approximation by the Gauss–Seidel method. Also, compute an error bound $\|\mathbf{x} - \mathbf{x}^{(10)}\|$.

$$A = \begin{pmatrix} 1 & 2 & 4 \\ 5 & -1 & 1 \\ 0 & 3 & -1 \end{pmatrix}, \qquad \mathbf{b} = \begin{pmatrix} 2 \\ 1 \\ -1 \end{pmatrix}.$$

10. Solve Problem 1 using the Gauss–Seidel method.

11. Consider the following system of equations:

$$\begin{aligned} 4x_1 &+ 2x_2 &+ x_3 &= 11 \\ -x_1 &+ 2x_2 & &= 3 \\ 2x_1 &+ x_2 &+ 4x_3 &= 16. \end{aligned}$$

(a) Show that the Gauss–Seidel method converges by using $\|T_G\|_\infty < 1$.

(b) Compute the second approximation $\mathbf{x}^{(2)}$, starting with $\mathbf{x}^{(0)} = [1, 1, 1]^T$.

(c) Compute an error estimate $\|\mathbf{x} - \mathbf{x}^{(2)}\|_\infty$ for your approximation. Consider the linear system $A\mathbf{x} = \mathbf{b}$, where

$$A = \begin{pmatrix} -5 & 2 & 1 \\ 1 & -10 & 1 \\ 1 & 1 & -4 \end{pmatrix}, \qquad \mathbf{b} = \begin{pmatrix} -3 \\ 27 \\ 4 \end{pmatrix}.$$

Find the Gauss–Seidel iteration matrix T_G and show that $\|T_G\| < 1$. Use the Gauss–Seidel method to find the second approximate solution $\mathbf{x}^{(1)}$ of the linear system by using $\mathbf{x}^{(0)} = [-0.5, -2.5, -1.5]^T$. Also, compute the error bound.

12. Consider the following linear system $A\mathbf{x} = \mathbf{b}$, where

$$A = \begin{pmatrix} -5 & 2 & 1 \\ 1 & 10 & 1 \\ 1 & 1 & -4 \end{pmatrix}, \qquad \mathbf{b} = \begin{pmatrix} -3 \\ 27 \\ 4 \end{pmatrix}.$$

Show that the Gauss–Seidel method converges for the given linear system. If the first approximate solution of the given linear system by the Gauss–Seidel method is $\mathbf{x}^{(1)} = [0.6, -2.7, -1]^T$, by using the initial approximation $\mathbf{x}^{(0)} = [0, 0, 0]^T$, then compute the number of steps needed to get accuracy within 10^{-4}. Also, compute an upper bound for the relative error in solving the given linear system.

13. Consider the following system:

$$\begin{array}{rcrcrcl} 4x_1 & + & x_2 & - & 2x_3 & = & 4 \\ 2x_1 & + & 9x_2 & - & 3x_3 & = & 3 \\ x_1 & - & 2x_2 & + & 8x_3 & = & 2. \end{array}$$

(a) Find the matrix form of both the iterative (Jacobi and Gauss–Seidel) methods.

(b) If $\mathbf{x}^{(k)} = [x_1^{(k)}, x_2^{(k)}, x_3^{(k)}]^T$, then write the iterative forms of part (a) in component form and find the exact solution of the given system.

(c) Find the formulas for the error $\mathbf{e}^{(k+1)}$ in the $(n+1)th$ step.

(d) Find the second approximation of the error $e^{(2)}$ using part (c), if $\mathbf{x}^{(0)} = [0,00]^T$.

14. Consider the following system:

$$
\begin{array}{rcrcrcl}
16x_1 & - & 3x_2 & + & 2x_3 & = & 11 \\
x_1 & + & 15x_2 & - & 3x_3 & = & 12 \\
5x_1 & - & 3x_2 & + & 14x_3 & = & 13.
\end{array}
$$

(a) Find the matrix form of both the iterative (Jacobi and Gauss–Seidel) methods.

(b) If $\mathbf{x}^{(k)} = [x_1^{(k)}, x_2^{(k)}, x_3^{(k)}]^T$, then write the iterative forms of part (a) in component form and find the exact solution of the given system.

(c) Find the formulas for the error $\mathbf{e}^{(k+1)}$ in the $(n+1)th$ step.

(d) Find the second approximation of the error $e^{(2)}$ using part (c), if $\mathbf{x}^{(0)} = [0,00]^T$.

15. Which of the following matrices is convergent?

(a) $\begin{pmatrix} 2 & 2 & 3 \\ 1 & 2 & 1 \\ 2 & -2 & 1 \end{pmatrix}$, (b) $\begin{pmatrix} 1 & 0 & 0 \\ -1 & 3 & 0 \\ 3 & 2 & -2 \end{pmatrix}$, (c) $\begin{pmatrix} 1 & 0 & -1 & 1 \\ 2 & 2 & -1 & 1 \\ 0 & 1 & 3 & -2 \\ 1 & 0 & 1 & 4 \end{pmatrix}$.

16. Find the eigenvalues and their associated eigenvectors of the matrix

$$
A = \begin{pmatrix} 2 & -2 & 3 \\ 0 & 3 & -2 \\ 0 & -1 & 2 \end{pmatrix}.
$$

Also, show that $\|A\|_2 > \rho(A)$.

17. Find the l_2-norm of each of the following matrices:

(a) $\begin{pmatrix} 2 & 1 & 3 \\ 1 & 2 & 4 \\ 2 & -2 & 1 \end{pmatrix}$, (b) $\begin{pmatrix} 1 & 2 & 0 \\ 1 & 2 & 0 \\ 1 & 2 & -2 \end{pmatrix}$, (c) $\begin{pmatrix} 1 & 3 & 2 & 1 \\ 2 & 2 & -1 & 1 \\ 0 & 1 & 2 & -2 \\ 1 & 3 & 1 & 5 \end{pmatrix}$.

18. Find the l_2-norm of each of the following matrices:

(a) $\begin{pmatrix} 2 & 1 & 3 \\ 1 & 2 & -2 \\ 3 & -2 & 1 \end{pmatrix}$, (b) $\begin{pmatrix} 1 & 2 & 0 \\ 2 & 2 & -1 \\ 0 & -1 & -2 \end{pmatrix}$, (c) $\begin{pmatrix} 1 & 3 & 2 & 1 \\ 3 & 2 & 1 & 1 \\ 2 & 1 & 3 & -2 \\ 1 & 1 & -2 & 4 \end{pmatrix}$.

19. Find the eigenvalues μ of the matrix $B = \mathbf{I} - \frac{1}{4}A$ and show that $\mu = \mathbf{I} - \frac{1}{4}\lambda$, where λ are the eigenvalues of the following matrix:

$$A = \begin{pmatrix} 4 & -1 & -1 & -1 \\ -1 & 4 & -1 & -1 \\ -1 & -1 & 4 & -1 \\ -1 & -1 & -1 & 4 \end{pmatrix}.$$

20. Solve Problem 1 using the SOR method by taking $\omega = 1.02$ for each system.

21. Write the Jacobi, Gauss–Seidel, and SOR iteration matrices for the following matrix:

$$A = \begin{pmatrix} 4 & -1 & -1 & -1 \\ -1 & 4 & -1 & -1 \\ -1 & -1 & 4 & -1 \\ -1 & -1 & -1 & 4 \end{pmatrix}.$$

22. Use the given parameter ω to solve each of the following linear systems by using the SOR method within accuracy 10^{-6} in the l_∞-norm, starting with $\mathbf{x}^{(0)} = \mathbf{0}$:

(a) $\omega = 1.323$

$$\begin{aligned} 6x_1 &- 2x_2 + x_3 &= 9 \\ 2x_1 &+ 7x_2 + 3x_3 &= 11 \\ -4x_1 &+ 5x_2 + 15x_3 &= 10 \end{aligned}$$

(b) $\omega = 1.201$

$$\begin{aligned} 2x_1 &- 3x_2 + x_3 &= 6 \\ 2x_1 &+ 16x_2 + 5x_3 &= -5 \\ 2x_1 &+ 3x_2 + 8x_3 &= 3 \end{aligned}$$

(c) $\omega = 1.110$

$$
\begin{aligned}
4x_1 + 3x_2 &= 11 \\
x_1 + 7x_2 - 3x_3 &= 13 \\
2x_1 + 7x_2 + 20x_3 &= 20
\end{aligned}
$$

(d) $\omega = 1.543$

$$
\begin{aligned}
x_1 - 2x_2 &= -3 \\
-2x_1 + 5x_2 - x_3 &= 5 \\
- x_2 + 2x_3 - 0.5x_4 &= 2 \\
- 0.5x_3 + 1.25x_4 &= 3.5
\end{aligned}
$$

23. Consider the following linear system $A\mathbf{x} = \mathbf{b}$, where

$$
A = \begin{pmatrix} 9 & 4 & -1 \\ 4 & 5 & 1 \\ -1 & 1 & 5 \end{pmatrix}, \qquad \mathbf{b} = \begin{pmatrix} 11 \\ 10 \\ 3 \end{pmatrix}.
$$

Use the Gauss–Seidel and SOR methods (using the optimal value of ω) to get the solution accurate to four significant digits, starting with $\mathbf{x}^{(0)} = [0, 0, 0]^T$.

24. Consider the following linear system with $A\mathbf{x} = \mathbf{b}$, where

$$
A = \begin{pmatrix} 4 & -2 & 0 & 0 \\ -2 & 3 & 2 & 0 \\ 0 & 2 & 3 & -1 \\ 0 & 0 & -1 & 4 \end{pmatrix}, \qquad \mathbf{b} = \begin{pmatrix} 1 \\ 2 \\ -1 \\ -1 \end{pmatrix}.
$$

Use the Jacobi, Gauss–Seidel, and SOR methods (taking $\omega = 1.007$) to get the solution accurate to four significant digits, with $\mathbf{x}^{(0)} = [2.5, 5.5, -4.5, -0.5]^T$.

25. Find the optimal choice for ω and use it to solve the linear system by the SOR method within accuracy 10^{-4} in the l_∞-norm, starting with $\mathbf{x}^{(0)} = \mathbf{0}$. Also, find how many iterations are needed by using the Gauss–Seidel and the Jacobi methods.

$$
\begin{aligned}
4x_1 + 2x_2 &= 5 \\
2x_1 + 4x_2 - 2x_3 &= 3 \\
- x_2 + 4x_3 &= 6.
\end{aligned}
$$

26. Consider the following system:

$$
\begin{array}{rcrcrcrcl}
4x_1 & - & 2x_2 & - & x_3 & & & = & 1 \\
-x_1 & + & 4x_2 & & & - & x_4 & = & 2 \\
-x_1 & & & + & 4x_3 & - & x_4 & = & 0 \\
& - & x_2 & - & x_3 & + & 4x_4 & = & 1.
\end{array}
$$

Using $\mathbf{x}^{(0)} = \mathbf{0}$, how many iterations are required to approximate the solution to within five decimal places using the (a) Jacobi method, (b) Gauss–Seidel method, and (c) SOR method (take $\omega = 1.1$)?

27. Find the spectral radius of the Jacobi, the Gauss–Seidel, and the SOR iteration matrices for each of the following matrices:

(a) $\omega = 1.25962$

$$
A = \begin{pmatrix}
2 & 1 & 0 & 0 \\
1 & 2 & 1 & 0 \\
0 & 1 & 2 & 1 \\
0 & 0 & 1 & 2
\end{pmatrix}.
$$

(b) $\omega = 1.15810$

$$
A = \begin{pmatrix}
4 & -1 & 0 & 0 \\
-1 & 4 & -1 & 0 \\
0 & -1 & 4 & -1 \\
0 & 0 & -1 & 4
\end{pmatrix}.
$$

28. Perform only two steps of the conjugate gradient method for the following linear systems, starting with $\mathbf{x}^{(0)} = \mathbf{0}$:

(a)

$$
\begin{array}{rcrcrcl}
3x_1 & - & x_2 & + & x_3 & = & 7 \\
-x_1 & + & 3x_2 & + & 2x_3 & = & 1 \\
x_1 & + & 2x_2 & + & 5x_3 & = & 5
\end{array}
$$

(b)

$$
\begin{array}{rcrcrcl}
3x_1 & - & 2x_2 & + & x_3 & = & 5 \\
-2x_1 & + & 6x_2 & - & x_3 & = & 9 \\
x_1 & - & x_2 & + & 4x_3 & = & 6
\end{array}
$$

(c)

$$
\begin{aligned}
4x_1 &- 2x_2 + x_3 = 1 \\
-2x_1 &+ 7x_2 + x_3 = 4 \\
x_1 &+ x_2 + 20x_3 = 1
\end{aligned}
$$

(d)

$$
\begin{aligned}
5x_1 &- 3x_2 - x_3 = 6 \\
-3x_1 &+ 6x_2 - 3x_3 = 4 \\
-x_1 &- 3x_2 + 4x_3 = 7
\end{aligned}
$$

(e)

$$
\begin{aligned}
9x_1 &- 3x_2 - x_3 + 2x_4 = 11 \\
3x_1 &- 10x_2 - 2x_3 + x_4 = 9 \\
2x_1 &+ 3x_2 - 11x_3 + 3x_4 = 15 \\
x_1 &- 3x_2 - 2x_3 + 12x_4 = 8
\end{aligned}
$$

29. Perform only two steps of the conjugate gradient method for the following linear systems, starting with $\mathbf{x}^{(0)} = \mathbf{0}$:

(a)

$$
\begin{aligned}
6x_1 &+ 2x_2 + x_3 = 1 \\
2x_1 &+ 3x_2 - x_3 = 0 \\
x_1 &- x_2 + 2x_3 = -2
\end{aligned}
$$

(b)

$$
\begin{aligned}
5x_1 &- 2x_2 + x_3 = 3 \\
-2x_1 &+ 4x_2 - x_3 = 2 \\
x_1 &- x_2 + 3x_3 = 1
\end{aligned}
$$

(c)

$$
\begin{aligned}
6x_1 &- x_2 - x_3 + 5x_4 = 1 \\
-x_1 &+ 7x_2 + x_3 - x_4 = 2 \\
-x_1 &+ x_2 + 3x_3 - 3x_4 = 0 \\
5x_1 &- x_2 - 3x_3 + 6x_4 = -1
\end{aligned}
$$

(d)

$$
\begin{aligned}
3x_1 &- 2x_2 - x_3 + 3x_4 = 1 \\
-2x_1 &+ 7x_2 + x_3 - x_4 = 0 \\
-x_1 &+ x_2 + 3x_3 - 3x_4 = 0 \\
3x_1 &- x_2 - 3x_3 + 6x_4 = 0
\end{aligned}
$$

30. Find the approximate solution of the linear system

$$
\begin{aligned}
x_1 &+& 2x_2 &=& 1 \\
2x_1 &+& 4.0001x_2 &=& 1.9999
\end{aligned}
$$

using simple Gaussian elimination, and then use the residual correction method (two iterations only) to improve the approximate solution.

31. The following linear system has the exact solution $\mathbf{x} = [10, 1]^T$. Find the approximate solution of the system

$$
\begin{aligned}
0.03x_1 &+& 58.9x_2 &=& 59.2 \\
5.31x_1 &-& 6.10x_2 &=& 47.0
\end{aligned}
$$

by using simple Gaussian elimination, and then use the residual correction method (one iteration only) to improve the approximate solution.

32. The following linear system has the exact solution $\mathbf{x} = [1, 1]^T$. Find the approximate solution of the system

$$
\begin{aligned}
x_1 &+& 2x_2 &=& 3 \\
x_1 &+& 2.01x_2 &=& 3.01
\end{aligned}
$$

by using simple Gaussian elimination, and then use the residual correction method (one iteration only) to improve the approximate solution.

Chapter 3

The Eigenvalue Problems

3.1 Introduction

In this chapter we describe numerical methods for solving eigenvalue problems that arise in many branches of science and engineering and seem to be a very fundamental part of the structure of the universe. Eigenvalue problems are important in a less direct manner in numerical applications. For example, discovering the condition factor in the solution of a set of linear algebraic equations involves finding the ratio of the largest to the smallest eigenvalue values of the underlying matrix. Also, the eigenvalue problem is involved when establishing the stiffness of ordinary differential equations problems. In solving eigenvalue problems, we are mainly concerned with the task of finding the values of the parameter λ and vector $\mathbf{x}$, which satisfy a set of equations of the form

$$A\mathbf{x} = \lambda\mathbf{x}. \tag{3.1}$$

The linear equation (3.1) represents the eigenvalue problem, where A is an $n \times n$ coefficient matrix, also called the system matrix, $\mathbf{x}$ is an unknown column vector, and λ is an unknown scalar. If the set of equations has a zero on the right-hand side, then a very important special case arises. For such a case, one solution of (3.1) for a real square matrix A is the trivial solution, $\mathbf{x} = \mathbf{0}$. However, there is a set of values for the parameter λ for which nontrivial solutions for the vector $\mathbf{x}$ exist. These nontrivial solutions are called eigenvectors, characteristic vectors, or latent vectors of a matrix A, and the corresponding values of the parameter λ are called eigenvalues, characteristic values, or latent roots of A. The set of all eigenvalues of A is called the *spectrum* of A. Eigenvalues may be real or complex, distinct or multiple. From (3.1), we deduce

$$A\mathbf{x} = \lambda \mathbf{I}\mathbf{x},$$

which gives

$$(A - \lambda \mathbf{I})\mathbf{x} = \mathbf{0}, \tag{3.2}$$

where $\mathbf{I}$ is an $n \times n$ identity matrix. The matrix $(A - \lambda \mathbf{I})$ appears as

$$\begin{pmatrix} (a_{11} - \lambda) & a_{12} & \cdots & a_{1n} \\ a_{21} & (a_{22} - \lambda) & \cdots & a_{2n} \\ \vdots & \vdots & \ddots & \vdots \\ a_{n1} & a_{n2} & \cdots & (a_{nn} - \lambda) \end{pmatrix},$$

and the result of the multiplication of (3.2) is a set of homogeneous equations of the form

$$\begin{aligned} (a_{11} - \lambda)x_1 + a_{12}x_2 + \cdots + a_{1n}x_n &= 0 \\ a_{21}x_1 + (a_{22} - \lambda)x_2 + \cdots + a_{2n}x_n &= 0 \\ \vdots \qquad\qquad \vdots \qquad\qquad \ddots \qquad \vdots \qquad &\ \ \vdots \\ a_{n1}x_1 + a_{n2}x_2 + \cdots + (a_{nn} - \lambda)x_n &= 0. \end{aligned} \tag{3.3}$$

Then by using Cramer's rule, we see that the determinant of the denominator, namely, the determinant of the matrix of the system (3.3) must vanish if there is to be a nontrivial solution, i.e., a solution other than $\mathbf{x} = \mathbf{0}$. Geometrically, $A\mathbf{x} = \lambda\mathbf{x}$ says that under transformation by A, eigenvectors experience only changes in magnitude or sign—the orientation of $A\mathbf{x}$ in $\mathbb{R}^n$ is the same as that of $\mathbf{x}$. The eigenvalue λ is simply the amount of "stretch" or "shrink" to which the eigenvector $\mathbf{x}$ is subjected when transformed by A (Figure 3.1).

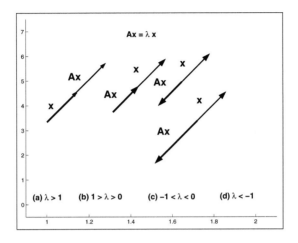

Figure 3.1: The situation in $\mathbb{R}^2$.

Definition 3.1 (Trace of a Matrix)
For an $n \times n$ matrix $A = (a_{ij})$, we define the trace of A to be the sum of the diagonal elements of A, i.e.,

$$trace(A) = a_{11} + a_{22} + \cdots + a_{nn}.$$

For example, the trace of the matrix

$$A = \begin{pmatrix} 7 & 3 & 1 \\ 2 & 6 & 2 \\ 1 & 4 & 3 \end{pmatrix}$$

is defined as

$$trace(A) = 7 + 6 + 3 = 16. \qquad \bullet$$

Theorem 3.1 *If A and B are square matrices of the same size, then:*

1. *trace* (A^T) = *trace* (A).

2. *trace* (kA) = k *trace* (A).

3. *trace* $(A + B)$ = *trace* (A) + *trace* (B).

4. *trace* $(A - B)$ = *trace* (A) – *trace* (B).

5. *trace* (AB) = *trace* (BA).   ●

For example, consider the following matrices:

$$A = \begin{pmatrix} 5 & -4 & 2 \\ 1 & 4 & 3 \\ 3 & 2 & 7 \end{pmatrix} \quad \text{and} \quad B = \begin{pmatrix} 9 & 3 & -4 \\ 1 & 5 & 2 \\ 3 & -2 & 8 \end{pmatrix}.$$

Then

$$A^T = \begin{pmatrix} 5 & 1 & 3 \\ -4 & 4 & 2 \\ 2 & 3 & 7 \end{pmatrix}$$

and

$$trace(A^T) = 16 = trace(A).$$

Also,

$$4A = \begin{pmatrix} 20 & -16 & 8 \\ 4 & 16 & 12 \\ 12 & 8 & 28 \end{pmatrix}$$

and

$$trace(4A) = 64 = 4(16) = 4trace(A).$$

The sum of the above two matrices is defined as

$$A + B = \begin{pmatrix} 14 & -1 & -2 \\ 2 & 9 & 5 \\ 6 & 0 & 15 \end{pmatrix}$$

and

$$trace(A + B) = 38 = 16 + 22 = trace(A) + trace(B).$$

Similarly, the difference of the above two matrices is defined as

$$A - B = \begin{pmatrix} -4 & -7 & 6 \\ 0 & -1 & 1 \\ 0 & 4 & -1 \end{pmatrix},$$

and

$$trace(A - B) = -6 = 16 - 22 = trace(A) - trace(B).$$

Finally, the product of the above two matrices is defined as

$$AB = \begin{pmatrix} 47 & -9 & -12 \\ 22 & 17 & 28 \\ 50 & 5 & 48 \end{pmatrix}$$

and

$$BA = \begin{pmatrix} 36 & -32 & -1 \\ 16 & 20 & 31 \\ 37 & -4 & 56 \end{pmatrix}.$$

Then

$$trace(AB) = 112 = trace(BA).$$

To get these results, we use the MATLAB Command Window as follows:

```
>> A = [5 − 4 2; 1 4 3; 3 2 7];
>> B = [9 3 − 4; 1 5 2; 3 − 2 8];
>> C = A + B;
>> D = A * B;
>> trac(C);
>> trac(D);
```

Program 3.1
MATLAB m-file for Finding Trace of The Matrix
function [trc]=trac(A)
n=max(size(A)); trc=0;
for i=1:n; for k=1:n
if i==k tracc=A(i,k);   trc=trc+tracc;   else
trc=trc;
end; end; end

There should be no confusion between the diagonal entries and eigenvalues. For a triangular matrix, they are the same but that is exceptional. Normally, the pivots and diagonal entries and eigenvalues are completely different.

The *classical method* of finding the eigenvalues of a matrix A is to estimate the roots of a characteristic equation of the form

$$p(\lambda) = \det(A - \lambda\mathbf{I}) = |A - \lambda\mathbf{I}| = \mathbf{0}. \tag{3.4}$$

Then the eigenvectors are determined by setting one of the nonzero elements of $\mathbf{x}$ to unity and calculating the remaining elements by equating coefficients in the relation (3.2).

Eigenvalues of 2 × 2 Matrices

Let λ_1 and λ_2 be the eigenvalues of a 2×2 matrix A, then a quadratic polynomial $p(\lambda)$ is defined as

$$\begin{aligned} p(\lambda) &= (\lambda - \lambda_1)(\lambda - \lambda_2) \\ &= \lambda^2 - (\lambda_1 + \lambda_2)\lambda + \lambda_1\lambda_2. \end{aligned}$$

Note that

$$\begin{aligned} trace(A) &= \lambda_1 + \lambda_2 \\ \det(A) &= \lambda_1\lambda_2. \end{aligned}$$

So

$$p(\lambda) = \lambda^2 - trace(A)\lambda + \det(A).$$

For example, if the given matrix is

$$A = \begin{pmatrix} 5 & 4 \\ 3 & 4 \end{pmatrix},$$

then

$$A - \lambda\mathbf{I} = \begin{pmatrix} 5 - \lambda & 4 \\ 3 & 4 - \lambda \end{pmatrix}$$

and

$$p(\lambda) = (5 - \lambda)(4 - \lambda) - 12 = \lambda^2 - 9\lambda + 8.$$

By solving the above quadratic polynomial, we get

$$\lambda_1 = 8, \qquad \lambda_2 = 1,$$

the possible eigenvalues of the given matrix.

Note that

$$
\begin{aligned}
trace(A) &= \lambda_1 + \lambda_2 = 9, \\
\det(A) &= \lambda_1\lambda_2 = 8,
\end{aligned}
$$

which satisfies the above result. $\bullet$

The *discriminant* of a 2×2 matrix is defined as

$$D = [trace(A)]^2 - 4\det(A).$$

For example, the discriminant of the matrix

$$A = \begin{pmatrix} 8 & 7 \\ 4 & 6 \end{pmatrix}$$

can be calculated as

$$D = [14]^2 - 4(20) = 116,$$

where the trace of A is 14 and the determinant of A is 20. $\bullet$

Theorem 3.2 *If D is discriminant of a 2×2 matrix, then the following statements hold:*

1. *The eigenvalues of A are real and distinct when $D > 0$.*

2. *The eigenvalues of A are a complex conjugate pair when $D < 0$.*

3. *The eigenvalues of A are real and equal when $D = 0$.* ●

For example, the eigenvalues of the matrix

$$A = \begin{pmatrix} 3 & 2 \\ 2 & 4 \end{pmatrix}$$

are real and distinct because

$$D = [7]^2 - 4(8) = 17 > 0.$$

Also, the eigenvalues of the matrix

$$A = \begin{pmatrix} 3 & -5 \\ 1 & 2 \end{pmatrix}$$

are a complex conjugate pair since

$$D = [5]^2 - 4(11) = -19 < 0.$$

Finally, the matrix

$$A = \begin{pmatrix} 1 & 2 \\ 0 & 1 \end{pmatrix}$$

has real and equal eigenvalues because

$$D = [2]^2 - 4(1) = 0.$$ ●

Note that the eigenvectors of a 2×2 matrix A corresponding to each of the eigenvalues of a matrix A can be found easily by substituting each eigenvalue in (3.2).

Example 3.1 *Find the eigenvalues and eigenvectors of the following matrix:*

$$A = \begin{pmatrix} 6 & -3 \\ -4 & 5 \end{pmatrix}.$$

Solution. *The eigenvalues of the given matrix are real and distinct because*

$$D = [11]^2 - 4(18) = 49 > 0.$$

Then

$$A - \lambda I = \begin{pmatrix} 6 - \lambda & -3 \\ -4 & 5 - \lambda \end{pmatrix}$$

and

$$p(\lambda) = (6 - \lambda)(5 - \lambda) - 12 = \lambda^2 - 11\lambda + 18.$$

By solving the above quadratic polynomial, we get

$$\lambda_1 = 9, \qquad \lambda_2 = 2,$$

the possible eigenvalues of the given matrix.

Note that

$$\begin{aligned} trace(A) &= \lambda_1 + \lambda_2 = 11 \\ \det(A) &= \lambda_1\lambda_2 = 18. \end{aligned}$$

Now to find the eigenvectors of the given matrix A corresponding to each of these eigenvalues, we substitute each of these two eigenvalues in (3.2). When $\lambda_1 = 9$, we have

$$\begin{pmatrix} -3 & -3 \\ -4 & -4 \end{pmatrix}\begin{pmatrix} x_1 \\ x_2 \end{pmatrix} = \begin{pmatrix} 0 \\ 0 \end{pmatrix},$$

which implies that

$$-3x_1 - 3x_2 = 0 \quad gives \quad x_1 = -x_2,$$

hence, the eigenvector $\mathbf{x}_1$ corresponding to the first eigenvalue, 9, by choosing $x_2 = 1$, is

$$\mathbf{x}_1 = \alpha[-1, 1]^T, \quad where \quad \alpha \in \mathbb{R}, \quad \alpha \neq 0.$$

When $\lambda_2 = 2$, we have

$$\begin{pmatrix} 4 & -3 \\ -4 & 3 \end{pmatrix} \begin{pmatrix} x_1 \\ x_2 \end{pmatrix} = \begin{pmatrix} 0 \\ 0 \end{pmatrix}.$$

From it, we obtain

$$4x_1 - 3x_2 = 0 \quad gives \quad x_1 = \frac{3}{4}x_2,$$

and

$$-4x_1 + 3x_2 = 0 \quad gives \quad x_1 = \frac{3}{4}x_2.$$

Thus, choosing $x_2 = 4$, we obtain

$$\mathbf{x}_2 = \alpha[3, 4]^T, \quad where \quad \alpha \in \mathbb{R}, \quad \alpha \neq 0,$$

which is the second eigenvector $\mathbf{x}_2$ corresponding to the second eigenvalue, 2. ●

To get the results of Example 3.1, we use the MATLAB Command Window as follows:

```
>> A = [6 − 3; −4 5];
>> EigTwo(A);
```

Program 3.2
MATLAB m-file for Finding Eigenvalues of a 2×2 Matrix
function [Lambda,x]= EigTwo(A)
$det = A(1,1) * A(2,2) - A(1,2) * A(2,1);$
trace = A(1,1) + A(2,2);
$L1 = (trace + sqrt(trace\char`^ 2 - 4 * det))/2;$
$L2 = (trace - sqrt(trace\char`^ 2 - 4 * det))/2;$
if A(1,2) $\cong$ 0
x1 = [A(1,2); L1-A(1,1)];
x2 = [A(1,2); L2-A(1,1)];
elseif A(2,1) $\cong$ 0
x1 = [L1-A(2,2); A(2,1)];
x2 = [L2-A(2,2); A(2,1)];
else x1 = [1; 0]; x2 = [0; 1];end
disp(['$L\char`^$ $2-'num2str(trace)'$ *   L   +'
$num2str(det)' = 0'$])
L1 x1 L2 x2

For larger size matrices, there is no doubt that the eigenvalue problem is computationally more difficult than the linear system $A\mathbf{x} = \mathbf{b}$. With a linear system, a finite number of elimination steps produces the exact answer in a finite time. In the case of an eigenvalue, no such steps and no such formula can exist. The characteristic polynomial of a 5×5 matrix is a quintic, and it is proved there can be no algebraic form for the roots of a fifth degree polynomial, although there are a few simple checks on the eigenvalues, after they have been computed, and we mention here two of them:

1. The *sum* of the n eigenvalues of a matrix A equals the sum of n diagonal entries, i.e.,

$$\sum_{i=1}^{n} \lambda_i = \sum_{i=1}^{n} a_{ii} = (a_{11} + a_{22} + \cdots + a_{nn}),$$

which is the trace of A.

2. The *product* of n eigenvalues of a matrix A equals the determinant of A, i.e.,

$$\prod_{i=1}^{n} \lambda_i = \det(A) = |A| = (\lambda_1 \lambda_2 \cdots \lambda_n).$$

It should be noted that the system matrix A of (3.1) may be real and symmetric, or real and nonsymmetric, or complex with symmetric real and skew symmetric imaginary parts. These different types of a matrix A are explained as follows:

1. If the given matrix A is a *real symmetric matrix*, then the eigenvalues of A are real but not necessarily positive, and the corresponding eigenvectors are also real. Also, if λ_i, $\mathbf{x}_i$ and λ_j, $\mathbf{x}_j$ satisfy the eigenvalue problem (3.1) and λ_i and λ_j are distinct, then

$$\mathbf{x}_i^T \mathbf{x}_j = 0, \qquad i \neq j \tag{3.5}$$

and

$$\mathbf{x}_i^T A \mathbf{x}_j = 0, \qquad i \neq j. \tag{3.6}$$

Equations (3.5) and (3.6) represent the *orthogonality* relationships. Note that if $i = j$, then in general, $\mathbf{x}_i^T \mathbf{x}_i$ and $\mathbf{x}_i^T A \mathbf{x}_i$ are not zero. Recalling that $\mathbf{x}_i$ includes an arbitrary scaling factor, then the product $\mathbf{x}_i^T \mathbf{x}_i$ must also be arbitrary. However, if the arbitrary scaling factor is adjusted so that

$$\mathbf{x}_i^T \mathbf{x}_j = 1, \tag{3.7}$$

then

$$\mathbf{x}_i^T A \mathbf{x}_j = \lambda_i, \tag{3.8}$$

and the eigenvectors are known to be *normalized*.

Sometimes the eigenvalues are not distinct and the eigenvectors associated with these equal or repeated eigenvalues are not, of necessity, orthogonal. If $\lambda_i = \lambda_j$ and the other eigenvalues, λ_k, are distinct, then

$$\mathbf{x}_i^T \mathbf{x}_k = 0, \quad k = 1, 2, \cdots, n \quad k \neq i, \quad k \neq j \tag{3.9}$$

and

$$x_j^T x_k = 0, \quad k = 1, 2, \cdots, n \quad k \neq i, \quad k. \neq j. \tag{3.10}$$

When $\lambda_i = \lambda_j$, the eigenvectors $\mathbf{x}_i$ and $\mathbf{x}_j$ are not unique and a linear combination of them, i.e., $a\mathbf{x}_i + b\mathbf{x}_j$, where a and b are arbitrary constants, also satisfies the eigenvalue problems. One important result is that *a symmetric matrix of order n always has n distinct eigenvectors even if some of the eigenvalues are repeated.*

2. If a given A is a *real nonsymmetric matrix*, then a pair of related eigenvalue problems can arise as follows:

$$A\mathbf{x} = \lambda\mathbf{x} \tag{3.11}$$

and

$$A^T\mathbf{y} = \beta\mathbf{y}. \tag{3.12}$$

By taking the transpose of (3.12), we have

$$\mathbf{y}^T A = \beta\mathbf{y}^T. \tag{3.13}$$

The vectors $\mathbf{x}$ and $\mathbf{y}$ are called the right-hand and left-hand vectors of A, respectively. The eigenvalues of A and A^T are identical, i.e., $\lambda_i = \beta_i$, but the eigenvectors $\mathbf{x}$ and $\mathbf{y}$ will, in general, differ from each other. The eigenvalues and eigenvectors of a nonsymmetric real matrix are either real or pairs of complex conjugates. If $\lambda_i, \mathbf{x}_i, \mathbf{y}_i$, and $\lambda_j, \mathbf{x}_j, \mathbf{y}_j$ are solutions that satisfy the eigenvalue problems of (3.11) and (3.12) and λ_i and λ_j are distinct, then

$$\mathbf{y}_j^T \mathbf{x}_i = 0, \quad i \neq j \tag{3.14}$$

and

$$\mathbf{y}_j^T A\mathbf{x}_i = 0, \quad i \neq j. \tag{3.15}$$

Equations (3.14) and (3.15) are called *bi-orthogonal* relationships. Note that if, in these equations, $i = j$, then, in general, $\mathbf{y}_i^T \mathbf{x}_i$ and $\mathbf{y}_i^T A\mathbf{x}_i$ are not zero. The eigenvectors $\mathbf{x}_i$ and $\mathbf{y}_i$ include arbitrary scaling factors, and so the product of these vectors will also be arbitrary. However, if the vectors are adjusted so that

$$\mathbf{y}_i^T \mathbf{x}_i = 1, \tag{3.16}$$

then

$$\mathbf{y}_i^T A \mathbf{x}_i = \lambda_i. \tag{3.17}$$

We can, in these circumstances, describe neither $\mathbf{x}_i$ nor $\mathbf{y}_i$ as normalized; the vectors still include an arbitrary scaling factor, only their product is uniquely chosen. If for a nonsymmetric matrix $\lambda_i = \lambda_j$ and the remaining eigenvalues, λ_k, are distinct, then

$$\mathbf{y}_j^T \mathbf{x}_k = 0, \quad \mathbf{y}_i^T \mathbf{x}_k = 0 \qquad k = 1, 2, \cdots, n \quad k \neq i, \quad k \neq j \tag{3.18}$$

and

$$\mathbf{x}_j^T \mathbf{x}_k = 0, \quad \mathbf{y}_k^T \mathbf{x}_j = 0 \qquad k = 1, 2, \cdots, n \quad k \neq i, \quad k \neq j. \tag{3.19}$$

For certain matrices with repeated eigenvalues, the eigenvectors may also be repeated; consequently, for an *nth*-order matrix of this type we may have less than n distinct eigenvectors. This type of matrix is called *deficient*.

3. Let us consider the case when the given A is a *complex matrix*. The properties of one particular complex matrix is an Hermitian matrix, which is defined as

$$H = A + \mathbf{i}B, \tag{3.20}$$

where A and B are real matrices such that $A = A^T$ and $B = -B^T$. Hence, A is symmetric and B is skew symmetric with zero terms on the leading diagonal. Thus, by definition of an Hermitian matrix, H has a symmetric real part and a skew symmetric imaginary part, making H equal to the transpose of its complex conjugate, denoted by H^*. Consider now the eigenvalue problem

$$H\mathbf{x} = \lambda\mathbf{x}. \tag{3.21}$$

If $\lambda_i, \mathbf{x}_i$ are solutions of (3.21), then $\mathbf{x}_i$ is complex but λ_i is real. Also, if $\lambda_i, \mathbf{x}_i$, and $\lambda_j, \mathbf{x}_j$ satisfy the eigenvalue problem (3.21) and λ_i and λ_j are distinct, then

$$\mathbf{x}_i^* \mathbf{x}_j = 0, \qquad i \neq j \tag{3.22}$$

and

$$\mathbf{x}_i^* H \mathbf{x}_j = 0, \qquad i \neq j, \tag{3.23}$$

where $\mathbf{x}_i^*$ is the transpose of the complex conjugate of $\mathbf{x}_i$. As before, $\mathbf{x}_i$ includes an arbitrary scaling factor and the product $\mathbf{x}_i^* \mathbf{x}_i$ must also be arbitrary. However, if the arbitrary scaling factor is adjusted so that

$$\mathbf{x}_i^* \mathbf{x}_i = 1, \tag{3.24}$$

then

$$\mathbf{x}_i^* H \mathbf{x}_i = \lambda_i, \tag{3.25}$$

and the eigenvectors are then said to be normalized.

A large number of numerical techniques have been developed to solve the eigenvalue problems. Before discussing all these numerical techniques, we shall start with a hand calculation, mainly to reinforce the definition and solve the following examples.

Example 3.2 *Find the eigenvalues and eigenvectors of the following matrix:*

$$A = \begin{pmatrix} 3 & 0 & 1 \\ 0 & -3 & 0 \\ 1 & 0 & 3 \end{pmatrix}.$$

Solution. *First, we shall find the eigenvalues of the given matrix A. From (3.2), we have*

$$\begin{pmatrix} 3 & 0 & 1 \\ 0 & -3 & 0 \\ 1 & 0 & 3 \end{pmatrix} \begin{pmatrix} x_1 \\ x_2 \\ x_3 \end{pmatrix} = \begin{pmatrix} 0 \\ 0 \\ 0 \end{pmatrix}.$$

For nontrivial solutions, using (3.4), we get

$$\begin{vmatrix} 3 - \lambda & 0 & 1 \\ 0 & -3 - \lambda & 0 \\ 1 & 0 & 3 - \lambda \end{vmatrix} = 0,$$

which gives a characteristic equation of the form

$$\lambda^3 - 3\lambda^2 - 10\lambda + 24 = 0,$$

which factorizes to

$$(\lambda + 3)(\lambda - 2)(\lambda - 4) = 0,$$

which gives the eigenvalues 4, –3, and 2 of the given matrix A. One can note that the sum of these eigenvalues is 3, and this agrees with the trace of A. •

The characteristic equation of the given matrix can be obtained by using the following MATLAB commands:

```
>> syms lambd
>> A = [3 0 1; 0 − 3 0; 1 0 3];
>> determ(A − lambd * eye(3));
>> ans =
      −24 + 10 * lambd + 3 * lambd^ 2−lambd^ 3;
>> factor(ans);
      −(lambd − 2) * (lambd − 4) * (3 + lambd)
```

After finding the eigenvalues of the matrix A, we turn to the problem of finding the corresponding eigenvectors. The eigenvectors of A corresponding to the eigenvalues λ are the nonzero vectors $\mathbf{x}$ that satisfy (3.2). Equivalently, the eigenvectors corresponding to λ are the nonzero vectors in the solution space of (3.2). We call this solution space the *eigenspace* of A corresponding to λ.

Now to find the eigenvectors of the given matrix A corresponding to each of these eigenvalues, we substitute each of these three eigenvalues in (3.2). When $\lambda_1 = 4$, we have

$$\begin{pmatrix} -1 & 0 & 1 \\ 0 & -7 & 0 \\ 1 & 0 & -1 \end{pmatrix} \begin{pmatrix} x_1 \\ x_2 \\ x_3 \end{pmatrix} = \begin{pmatrix} 0 \\ 0 \\ 0 \end{pmatrix},$$

which implies that

$$\begin{array}{rcrcrclclcl} -x_1 & + & 0x_2 & + & x_3 & = 0 & \Rightarrow & x_1 & = & x_3 \\ 0x_1 & - & 7x_2 & + & 0x_3 & = 0 & \Rightarrow & x_2 & = & 0 \\ x_1 & + & 0x_2 & - & x_3 & = 0 & \Rightarrow & x_1 & = & x_3. \end{array}$$

Solving this system, we get $x_1 = x_3 = \infty$, and $x_2 = 0$. Hence, the eigenvector $\mathbf{x_1}$ corresponding to the first eigenvalue, 4, by choosing $x_3 = 1$, is

$$\mathbf{x_1} = \alpha[1, 0, 1]^T, \quad \text{where} \quad \alpha \in \mathbb{R}, \quad \alpha \neq 0.$$

When $\lambda_2 = -3$, we have

$$\begin{pmatrix} 6 & 0 & 1 \\ 0 & 0 & 0 \\ 1 & 0 & 6 \end{pmatrix} \begin{pmatrix} x_1 \\ x_2 \\ x_3 \end{pmatrix} = \begin{pmatrix} 0 \\ 0 \\ 0 \end{pmatrix},$$

which implies that

$$
\begin{array}{llll}
6x_1 & + \ 0x_2 & + \quad x_3 & = 0 \\
0x_1 & + \ 0x_2 & + \ 0x_3 & = 0 \\
x_1 & + \ 0x_2 & + \ 6x_3 & = 0
\end{array}
\qquad
\begin{array}{ll}
\Rightarrow & x_1 = -\frac{1}{6}x_3 \\
\Rightarrow & x_2 = \quad \infty \\
\Rightarrow & x_1 = \ -6x_3,
\end{array}
$$

which gives the solution, $x_1 = -6x_3 = 0$ and $x_2 = \infty$. Hence, the eigenvector $\mathbf{x_2}$ corresponding to the second eigenvalue, –3, by choosing $x_2 = 1$, is

$$\mathbf{x_2} = \alpha[0, 1, 0]^T, \quad \text{where} \quad \alpha \in \mathbb{R}, \quad \alpha \neq 0.$$

Finally, when $\lambda_3 = 2$, we have

$$\begin{pmatrix} 1 & 0 & 1 \\ 0 & -5 & 0 \\ 1 & 0 & 1 \end{pmatrix} \begin{pmatrix} x_1 \\ x_2 \\ x_3 \end{pmatrix} = \begin{pmatrix} 0 \\ 0 \\ 0 \end{pmatrix},$$

which implies that

$$
\begin{array}{llll}
x_1 & + \ 0x_2 & + \ x_3 & = 0 \\
0x_1 & - \ 5x_2 & + \ 0x_3 & = 0 \\
x_1 & + \ 0x_2 & + \ x_3 & = 0
\end{array}
\qquad
\begin{array}{ll}
\Rightarrow & x_1 = \ -x_3 \\
\Rightarrow & x_2 = \ 0 \\
\Rightarrow & x_1 = \ -x_3,
\end{array}
$$

which gives $x_1 = -x_3 = \infty$ and $x_2 = 0$. Hence, by choosing $x_1 = 1$, we obtain

$$\mathbf{x_3} = \alpha[1, 0, -1]^T, \quad \text{where} \quad \alpha \in \mathbb{R}, \quad \alpha \neq 0,$$

the third eigenvector $\mathbf{x_3}$ corresponding to the third eigenvalue, 2, of the matrix. $\bullet$

MATLAB can handle eigenvalues, eigenvectors, and the characteristic polynomial. The built-**poly** function in MATLAB computes the characteristic polynomial of the matrix:

> ```
> >> A = [3 0 1; 0 - 3 0; 1 0 3];
> >> P = poly(A);
> >> PP = poly2sym(P);
> >> PP =
>      x^3 - 3x^2 - 10x + 24
> ```

The elements of vector P are arranged in decreasing power of x. To solve the characteristic equation (in order to obtain the eigenvalues of A), ask for the roots of P:

> ```
> >> roots(P);
> ```

If all we require are the eigenvalues of A, we can use the MATLAB command **eig**, which is the basic eigenvalue and eigenvector routine. The command

> ```
> >> d = eig(A);
> ```

returns a vector containing all the eigenvalues of a matrix A. If the eigenvectors are also wanted, the syntax

> ```
> >> [X, D] = eig(A);
> ```

will return a matrix X whose columns are the eigenvectors of A corresponding to the eigenvalues in the diagonal matrix D.

To get the results of Example 3.2, we use the MATLAB Command Window as follows:

```
>> A = [3 0 1; 0 − 3 0; 1 0 3];
>> P = poly(A);
>> PP = poly2sym(P);
>> [X, D] = eig(A);
>> λ = diag(D);
>> x1 = X(:, 1); x2 = X(:, 2); x3 = X(:, 3);
```

Example 3.3 *Find the eigenvalues and eigenvectors of the following matrix:*

$$A = \begin{pmatrix} 1 & 2 & 2 \\ 0 & 3 & 3 \\ -1 & 1 & 1 \end{pmatrix}.$$

Solution. *From (3.2), we have*

$$\begin{pmatrix} 1 & 2 & 2 \\ 0 & 3 & 3 \\ -1 & 1 & 1 \end{pmatrix} \begin{pmatrix} x_1 \\ x_2 \\ x_3 \end{pmatrix} = \begin{pmatrix} 0 \\ 0 \\ 0 \end{pmatrix}.$$

For nontrivial solutions, using (3.4), we get

$$\begin{vmatrix} 1 - \lambda & 2 & 2 \\ 0 & 3 - \lambda & 3 \\ -1 & 1 & 1 - \lambda \end{vmatrix} = 0,$$

which gives a characteristic equation of the form

$$-\lambda^3 + 5\lambda^2 - 6\lambda = 0.$$

It factorizes to

$$\lambda(\lambda - 2)(\lambda - 3) = 0,$$

which gives the eigenvalues 0, 2, and 3 of the given matrix A. One can note that the sum of these three eigenvalues is 5, and this agrees with the trace of A.

To find the eigenvectors corresponding to each of these eigenvalues, we substitute each of the three eigenvalues of A in (3.2). When $\lambda = 0$, we have

$$\begin{pmatrix} 1 & 2 & 2 \\ 0 & 3 & 3 \\ -1 & 1 & 1 \end{pmatrix} \begin{pmatrix} x_1 \\ x_2 \\ x_3 \end{pmatrix} = \begin{pmatrix} 0 \\ 0 \\ 0 \end{pmatrix}.$$

The augmented matrix form of the system is

$$\begin{pmatrix} 1 & 2 & 2 & 0 \\ 0 & 3 & 3 & 0 \\ -1 & 1 & 1 & 0 \end{pmatrix},$$

which can be reduced to

$$\begin{pmatrix} 1 & 2 & 2 & 0 \\ 0 & 1 & 1 & 0 \\ 0 & 0 & 0 & 0 \end{pmatrix}.$$

Thus, the components of an eigenvector must satisfy the relation

$$\begin{array}{rcrcrcl} x_1 & + & 2x_2 & + & 2x_3 & = & 0 \\ 0x_1 & + & x_2 & + & x_3 & = & 0 \\ 0x_1 & + & 0x_2 & + & 0x_3 & = & 0. \end{array}$$

This system has an infinite set of solutions. Arbitrarily, we choose $x_2 = 1$, then x_3 can be equal to -1, whence $x_1 = 0$. This gives solutions of the first eigenvector of the form $\mathbf{x}_1 = \alpha[0, 1, -1]^T$, with $\alpha \in \mathbb{R}$ and $\alpha \neq 0$. Thus, $\mathbf{x}_1 = \alpha[0, 1, -1]^T$ is the most general eigenvector corresponding to eigenvalue 0.

A similar procedure can be applied to the other two eigenvalues. The result is that we have two other eigenvectors $\mathbf{x}_2 = \alpha[4, 3, -1]^T$ and $\mathbf{x}_3 = \alpha[1, 1, 0]^T$ corresponding to the eigenvalues 2 and 3, respectively. ●

Example 3.4 *Find the eigenvalues and eigenvectors of the following matrix:*

$$A = \begin{pmatrix} 3 & 2 & -1 \\ 2 & 6 & -2 \\ -1 & -2 & 3 \end{pmatrix}.$$

Solution. *From (3.2), we have*

$$\begin{pmatrix} 3 & 2 & -1 \\ 2 & 6 & -2 \\ -1 & -2 & 3 \end{pmatrix} \begin{pmatrix} x_1 \\ x_2 \\ x_3 \end{pmatrix} = \begin{pmatrix} 0 \\ 0 \\ 0 \end{pmatrix}.$$

For nontrivial solutions, using (3.4), we get

$$\begin{vmatrix} 3-\lambda & 2 & -1 \\ 2 & 6-\lambda & -2 \\ -1 & -2 & 3-\lambda \end{vmatrix} = 0,$$

which gives a characteristic equation

$$\lambda^3 - 12\lambda^2 + 36\lambda - 32 = 0.$$

It factorizes to

$$(\lambda - 2)^2(\lambda - 8) = 0$$

and gives the eigenvalue 2 of multiplicity 2 and the eigenvalue 8 of multiplicity 1, and the sum of these three eigenvalues is 12, which agrees with the trace of A. When $\lambda = 2$, we have

$$(A - 2I) = \begin{pmatrix} 1 & 2 & -1 \\ 2 & 4 & -2 \\ -1 & -2 & 1 \end{pmatrix}.$$

and so from (3.2), we have

$$
\begin{array}{rrrrrl}
x_1 & + & 2x_2 & - & x_3 & = & 0 \\
2x_1 & + & 4x_2 & - & 2x_3 & = & 0 \\
-x_1 & - & 2x_2 & + & x_3 & = & 0.
\end{array}
$$

Let $x_2 = s$ and $x_3 = t$, then the solution to this system is

$$\begin{pmatrix} x_1 \\ x_2 \\ x_3 \end{pmatrix} = \begin{pmatrix} -2s+t \\ s \\ t \end{pmatrix} = s\begin{pmatrix} -2 \\ 1 \\ 0 \end{pmatrix} + t\begin{pmatrix} 1 \\ 0 \\ 1 \end{pmatrix}.$$

So the two eigenvectors of A are $\mathbf{x}_1 = \alpha[-2, 1, 0]^T$ and $\mathbf{x}_2 = \alpha[1, 0, 1]^T$, corresponding to the eigenvalue 2, with $s, t \in \mathbb{R}$ and $s, t \neq 0$.

Similarly, we can find the third eigenvector, $\mathbf{x}_3 = \alpha[0.5, 1, -0.5]^T$, of A corresponding to the other eigenvalue, 8. ●

Note that in all the above three examples and any other example, there is always an infinite number of choices for each eigenvector. We arbitrarily choose a simple one by setting one or more of the elements x_is equal to a convenient number. Here we have set one of the elements x_is equal to 1. ●

3.2 Linear Algebra and Eigenvalues Problems

The solutions of many physical problems require the calculation of the eigenvalues and the corresponding eigenvectors of a matrix associated with a linear system of equations. Since a matrix A of order n has n, not necessarily distinct, eigenvalues, which are the roots of a characteristic equation (3.4), theoretically, the eigenvalues of A can be obtained by finding the n roots of a characteristic polynomial $p(\lambda)$, and then the associated linear system can be solved to determine the corresponding eigenvectors. The polynomial $p(\lambda)$ is difficult to obtain except for small values of n. So for a large value of n, it is necessary to construct approximation techniques for finding the eigenvalues of A.

Before discussing such approximation techniques for finding eigenvalues and eigenvectors of a given matrix A, we need some definitions and results from linear algebra.

Definition 3.2 (Real Vector Spaces)

A vector space consists of a nonempty set V of objects (called vectors) that can be added, that can be multiplied by a real number (called a scalar), and for which certain axioms hold. If **u** *and* **v** *are two vectors in V, their sum is expressed as* **u** + **v**, *and the scalar product of* **u** *by a real number α is denoted by α**u**. These operations are called vector addition and scalar multiplication, respectively, and the following axioms are assumed to hold:*

Axioms for Vector Addition

1. *If* **u** *and* **v** *are in V, then* **u** + **v** *is in V.*

2. **u** + **v** = **v** + **u**, *for all* **u** *and* **v** *in V.*

3. **u** + (**v** + **w**) = (**u** + **v**) + **w**, *for all* **u**, **v**, *and* **w** *in V.*

4. *There exists an element* **0** *in V, called a zero vector, such that* **u** + **0** = **u**.

5. *For each* **u** *in* V, *there is an element* $-\mathbf{u}$ *(called the negative of* **u***) in* V *such that* $\mathbf{u} + (-\mathbf{u}) = \mathbf{0}$.

Axioms for Scalar Multiplication

1. *If* **v** *is in* V, *then* $\alpha\mathbf{u}$ *is in* V, *for all* $\alpha \in \mathbb{R}$.

2. $\alpha(\mathbf{u} + \mathbf{v}) = \alpha\mathbf{u} + \alpha\mathbf{v}$, *for all* **u**, **v** $\in V$ *and* $\alpha \in \mathbb{R}$.

3. $(\alpha + \beta)\mathbf{u} = \alpha\mathbf{u} + \beta\mathbf{u}$, *for all* **u** $\in V$ *and* $\alpha, \beta \in \mathbb{R}$.

4. $\alpha(\beta\mathbf{u}) = (\alpha + \beta)\mathbf{u}$, *for all* **u** $\in V$ *and* $\alpha, \beta \in \mathbb{R}$.

5. $1\mathbf{u} = \mathbf{u}$.   ●

For example, a real vector space is the space $\mathbb{R}^n$ *consisting of column vectors or n-tuples of real numbers* $\mathbf{u} = (u_1, u_2, \ldots, u_n)^T$. *Vector addition and scalar multiplication are done in the usual manner:*

$$\mathbf{u} + \mathbf{v} = \begin{pmatrix} u_1 + v_1 \\ u_2 + v_2 \\ \vdots \\ u_n + v_n \end{pmatrix}, \quad \alpha\mathbf{u} = \begin{pmatrix} \alpha u_1 \\ \alpha u_2 \\ \vdots \\ \alpha u_n \end{pmatrix},$$

whenever

$$\mathbf{u} = \begin{pmatrix} u_1 \\ u_2 \\ \vdots \\ u_n \end{pmatrix}, \quad \mathbf{v} = \begin{pmatrix} v_1 \\ v_2 \\ \vdots \\ v_n \end{pmatrix}.$$

The zero vector is $\mathbf{0} = (0, 0, \ldots, 0)^T$. *The fact that vectors in* $\mathbb{R}^n$ *satisfy all of the vector space axioms is an immediate consequence of the laws of vector addition and scalar multiplication.*   ●

The following theorem presents several useful properties common to all vector spaces.

Theorem 3.3 *If* V *is a vector space, then:*

1. $0\mathbf{u} = \mathbf{0}$, *for all* $\mathbf{u} \in V$.

2. $\alpha \mathbf{0} = \mathbf{0}$, *for all* $\alpha \in \mathbb{R}$.

3. *If* $\alpha \mathbf{u} = \mathbf{0}$, *then* $\alpha = 0$ *or* $\mathbf{u} = \mathbf{0}$.

4. $(-1)\mathbf{u} = \mathbf{u}$, *for all* $\mathbf{u} \in V$.   ●

Definition 3.3 (Subspaces)

Let V be a vector space and W be a nonempty subset of V. If W is a vector space with respect to the operations in V, then W is called a subspace of V.

For example, every vector space has at least two subspaces, itself and the subspace$\{\mathbf{0}\}$ (called the zero subspace) consisting of only the zero vector. ●

Theorem 3.4 *A nonempty subset $W \subset V$ of a vector space is a subspace, if and only if:*

1. *For every $\mathbf{u}, \mathbf{v} \in W$, then the sum $\mathbf{u} + \mathbf{v} \in W$.*

2. *For every $\mathbf{u} \in W$ and every $\alpha \in \mathbb{R}$, then the scalar product $\alpha \mathbf{u} \in W$.*
   ●

Definition 3.4 (Basis of Vector Space)

*Let V be a vector space. A finite set $S = \{\mathbf{v}_1, \mathbf{v}_2, \ldots, \mathbf{v}_n\}$ of vectors in V is a **basis** for V, if and only if any vector $\mathbf{v}$ in V can be written, in a unique way, as a linear combination of the vectors in S, i.e., if and only if any vector $\mathbf{v}$ has the form*

$$\mathbf{v} = k_1\mathbf{v}_1 + k_2\mathbf{v}_2 + \cdots + k_n\mathbf{v}_n, \tag{3.26}$$

for one and only one set of real numbers $k_1, k_2, \ldots, k_n$.   ●

Definition 3.5 (Linearly Independent Vectors)

The vectors $\mathbf{v}_1, \mathbf{v}_2, \ldots, \mathbf{v}_n$ are said to be linearly independent, if whenever

$$k_1\mathbf{v}_1 + k_2\mathbf{v}_2 + \cdots + k_n\mathbf{v}_n = \mathbf{0}, \tag{3.27}$$

then all of the coefficients $k_1, k_2, \ldots, k_n$ must be equal to zero, i.e.,

$$k_1 = k_2 = \cdots = k_n = 0. \qquad (3.28)$$

If the vectors $\mathbf{v}_1, \mathbf{v}_2, \ldots, \mathbf{v}_n$ are not linearly independent, then we say that they are linearly dependent. In other words, the vectors $\mathbf{v}_1, \mathbf{v}_2, \ldots, \mathbf{v}_n$ are linearly dependent, if and only if there exist numbers $k_1, k_2, \ldots, k_n$, not all zero, for which

$$k_1 \mathbf{v}_1 + k_2 \mathbf{v}_2 + \cdots + k_n \mathbf{v}_n = \mathbf{0}.$$

Sometimes we say that the set $\{\mathbf{v}_1, \mathbf{v}_2, \ldots, \mathbf{v}_n\}$ is linearly independent (or linearly dependent) instead of saying that the vectors $\mathbf{v}_1, \mathbf{v}_2, \ldots, \mathbf{v}_n$ are linearly independent (or linearly dependent). •

Example 3.5 *Let us consider the vectors $\mathbf{v}_1 = (1, 2)$ and $\mathbf{v}_2 = (-1, 1)$ in $\mathbb{R}^2$. To show that the vectors are linearly independent, we write*

$$
\begin{aligned}
k_1 v_1 + k_2 v_2 &= \mathbf{0} \\
k_1(1, 2) + k_2(-1, 1) &= (0, 0) \\
(k_1 - k_2, 2k_1 + k_2) &= (0, 0),
\end{aligned}
$$

showing that

$$
\begin{aligned}
k_1 - k_2 &= 0 \\
2k_1 + k_2 &= 0,
\end{aligned}
$$

and the only solution to the system is a trivial solution, i.e., $k_1 = k_2 = 0$. Thus, the vectors are linearly independent. •

The above results can be obtained using the MATLAB Command Window as follows:

```
>> v1 = [1 2]';
>> v2 = [-1 1]';
>> A = [v1 v2];
>> null(A);
```

Note that using the MATLAB command, we obtained

$$ans =$$
$$\text{Empty matrix: } 2 - by - 0,$$

which means that the only solution to the homogeneous system $A\mathbf{k} = \mathbf{0}$ is the zero solution $\mathbf{k} = \mathbf{0}$.

Example 3.6 *Consider the following functions:*

$$
\begin{aligned}
p_1(t) &= t^2 + t + 2 \\
p_2(t) &= 2t^2 + t + 3 \\
p_3(t) &= 3t^2 + 2t + 2.
\end{aligned}
$$

Show that the set $\{p_1(t), p_2(t), p_3(t)\}$ is linearly independent.

Solution. *Suppose a linear combination of these given polynomials vanishes, i.e.,*

$$k_1(t^2 + t + 2) + k_2(2t^2 + t + 3) + k_3(3t^2 + 2t + 2) = \mathbf{0}.$$

By equating the coefficients of $2, 1,$ and 0 degrees, we get the following linear system:

$$
\begin{aligned}
k_1 + 2k_2 + 3k_3 &= 0 \\
k_1 + k_2 + 2k_3 &= 0 \\
2k_1 + 3k_2 + 2k_3 &= 0.
\end{aligned}
$$

Solving this homogenous linear system

$$
\begin{pmatrix} 1 & 2 & 3 \\ 1 & 1 & 2 \\ 2 & 3 & 2 \end{pmatrix} \begin{pmatrix} k_1 \\ k_2 \\ k_3 \end{pmatrix} = \begin{pmatrix} 0 \\ 0 \\ 0 \end{pmatrix},
$$

we get

$$
\begin{pmatrix} 1 & 0 & 0 \\ 0 & 1 & 0 \\ 0 & 0 & 1 \end{pmatrix} \begin{pmatrix} k_1 \\ k_2 \\ k_3 \end{pmatrix} = \begin{pmatrix} 0 \\ 0 \\ 0 \end{pmatrix}.
$$

Since the only solution to the above system is a trivial solution

$$k_1 = k_2 = k_3 = 0,$$

the given functions are linearly independent.   ●

Example 3.7 *Suppose that the set* $\{\mathbf{v}_1, \mathbf{v}_2, \mathbf{v}_3\}$ *is linearly independent in a vector space* V. *Show that the set* $\{\mathbf{v}_1+\mathbf{v}_2+\mathbf{v}_3, \mathbf{v}_1-\mathbf{v}_2-\mathbf{v}_3, \mathbf{v}_1+\mathbf{v}_2-\mathbf{v}_3\}$ *is also linearly independent.*

Solution. *Suppose a linear combination of these given vectors* $\mathbf{v}_1 + \mathbf{v}_2 + \mathbf{v}_3, \mathbf{v}_1 - \mathbf{v}_2 - \mathbf{v}_3$, *and* $\mathbf{v}_1 + \mathbf{v}_2 - \mathbf{v}_3$ *vanishes, i.e.,*

$$k_1(\mathbf{v}_1 + \mathbf{v}_2 + \mathbf{v}_3) + k_2(\mathbf{v}_1 - \mathbf{v}_2 - \mathbf{v}_3) + k_3(\mathbf{v}_1 + \mathbf{v}_2 - \mathbf{v}_3) = \mathbf{0}.$$

We must deduce that $k_1 = k_2 = k_3 = 0$. *By equating the coefficients of* $\mathbf{v}_1, \mathbf{v}_2$, *and* $\mathbf{v}_3$, *we obtain*

$$(k_1 + k_2 + k_3)\mathbf{v}_1 + (k_1 - k_2 + k_3)\mathbf{v}_2 + (k_1 - k_2 - k_3)\mathbf{v}_3 = \mathbf{0}.$$

Since $\{\mathbf{v}_1, \mathbf{v}_2, \mathbf{v}_3\}$ *is linearly independent, we have*

$$
\begin{aligned}
k_1 &+ k_2 &+ k_3 &= 0 \\
k_1 &- k_2 &+ k_3 &= 0 \\
k_1 &- k_2 &- k_3 &= 0.
\end{aligned}
$$

Solving this linear system, we get the unique solution

$$k_1 = 0, \quad k_2 = 0, \quad k_3 = 0,$$

which means that the set $\{\mathbf{v}_1 + \mathbf{v}_2 + \mathbf{v}_3, \mathbf{v}_1 - \mathbf{v}_2 - \mathbf{v}_3, \mathbf{v}_1 + \mathbf{v}_2 - \mathbf{v}_3\}$ *is also linearly independent.* •

Theorem 3.5 *If* $\{\mathbf{v}_1, \mathbf{v}_2, \ldots, \mathbf{v}_n\}$ *is a set of* n *linearly independent vectors in* $\mathbb{R}^n$, *then any vector* $\mathbf{x} \in \mathbb{R}^n$ *can be written uniquely as*

$$\mathbf{x} = k_1\mathbf{v}_1 + k_2\mathbf{v}_2 + \cdots + k_n\mathbf{v}_n, \tag{3.29}$$

for some collection of constants $k_1, k_2, \ldots, k_n$. •

Example 3.8 *Consider the vectors* $\mathbf{v}_1 = (1, 2, 1)$, $\mathbf{v}_2 = (1, 3, -2)$, *and* $\mathbf{v}_3 = (0, 1, -3)$ *in* $\mathbb{R}^3$. *If* k_1, k_2, *and* k_3 *are numbers with*

$$k_1\mathbf{v}_1 + k_2\mathbf{v}_2 + k_3\mathbf{v}_3 = \mathbf{0},$$

this is equivalent to

$$k_1(1, 2, 1) + k_2(1, 3, -2) + k_3(0, 1, -3) = (0, 0, 0).$$

Thus, we have the system

$$
\begin{array}{rcrcrcl}
k_1 & + & k_2 & + & 0k_3 & = & 0 \\
2k_1 & + & 3k_2 & + & k_3 & = & 0 \\
k_1 & - & 2k_2 & - & 3k_3 & = & 0.
\end{array}
$$

This system has infinitely many solutions, one of which is $k_1 = 1$, $k_2 = -1$, and $k_3 = 1$. So,

$$\mathbf{v}_1 - \mathbf{v}_2 + \mathbf{v}_3 = \mathbf{0}.$$

Thus, the vectors $\mathbf{v}_1, \mathbf{v}_2$, and $\mathbf{v}_3$ are linearly dependent.   ●

The above results can be obtained using the MATLAB Command Window as follows:

```
>> v1 = [1 2 1]';
>> v2 = [1 3 − 2]';
>> v3 = [0 1 − 3]';
>> A = [v1 v2 v3];
>> null(A);
```

By using this MATLAB command, the answer we obtained means that there is a nonzero solution to the homogeneous system $A\mathbf{k} = \mathbf{0}$.

Example 3.9 *Find the value of α for which the set $\{(1, -2), (4, -\alpha)\}$ is linearly dependent.*

Solution. *Suppose a linear combination of these given vectors $(1, -2)$ and $(4, -\alpha)$ vanishes, i.e.,*

$$k_1(1, -2) + k_2(4, -\alpha) = \mathbf{0}.$$

It can be written in the linear system form as

$$
\begin{array}{rcrcl}
k_1 & + & 4k_2 & = & 0 \\
-2k_1 & - & \alpha k_2 & = & 0
\end{array}
$$

or

$$\begin{pmatrix} 1 & 4 \\ -2 & \alpha \end{pmatrix} \begin{pmatrix} k_1 \\ k_2 \end{pmatrix} = \begin{pmatrix} 0 \\ 0 \end{pmatrix}.$$

By solving this system, we obtain

$$\begin{pmatrix} 1 & 4 \\ 0 & 8-\alpha \end{pmatrix} \begin{pmatrix} k_1 \\ k_2 \end{pmatrix} = \begin{pmatrix} 0 \\ 0 \end{pmatrix},$$

and it shows that the system has infinitely many solutions for $\alpha = 8$. Thus, the given set $\{(1,-2),(4,-\alpha)\}$ is linearly dependent for $\alpha = 8$. ●

Theorem 3.6 *Let the set $\{\mathbf{v}_1, \mathbf{v}_2, \ldots, \mathbf{v}_n\}$ be linearly dependent in a vector space V. Any set of vectors in V that contains these vectors will also be linearly dependent.* ●

Note that any collection of n linearly independent vectors in $\mathbb{R}^n$ is a basis for $\mathbb{R}^n$.

Theorem 3.7 *If A is an $n \times n$ matrix and $\lambda_1, \ldots, \lambda_n$ are distinct eigenvalues of A, with associated eigenvectors $\mathbf{v}_1, \ldots, \mathbf{v}_n$, then the set $\{\mathbf{v}_1, \ldots, \mathbf{v}_n\}$ is linearly independent.* ●

Definition 3.6 (Orthogonal Vectors)

A set of vectors $\{\mathbf{v}_1, \mathbf{v}_2, \ldots, \mathbf{v}_n\}$ is called orthogonal, if

$$\mathbf{v}_i^T \mathbf{v}_j = 0, \qquad \text{for all} \qquad i \neq j. \tag{3.30}$$

If, in addition

$$\mathbf{v}_i^T \mathbf{v}_i = 1, \qquad \text{for all} \quad i = 1, 2, \ldots, n, \tag{3.31}$$

then the set is called orthonormal. ●

Theorem 3.8 *An orthogonal set of vectors that does not contain the zero vectors is linearly independent.* ●

The proof of this theorem is beyond the scope of this text and will be omitted. However, the result is extremely important and can be easily understood and used. We illustrate this result by considering the matrix

$$A = \begin{pmatrix} 6 & -2 & 2 \\ -2 & 5 & 0 \\ 2 & 0 & 7 \end{pmatrix},$$

which has the eigenvalues 3, 6, and 9. The corresponding eigenvectors of A are $[2, 2, -1]^T$, $[-1, 2, 2]^T$, and $[2, -1, 2]^T$, and they form an orthogonal set. To show that the vectors are linearly independent, we write

$$k_1\mathbf{v}_1 + k_2\mathbf{v}_2 + k_3\mathbf{v}_3 = \mathbf{0},$$

then the equation

$$k_1(2, 2, -1) + k_2(-1, 2, 2) + k_3(2, -1, 2) = (0, 0, 0)$$

leads to the homogeneous system of three equations in three unknowns, k_1, k_2, and k_3:

$$\begin{array}{rcrcrcl} 2k_1 & - & k_2 & + & 2k_3 & = & 0 \\ 2k_1 & + & 2k_2 & - & k_3 & = & 0 \\ -k_1 & + & 2k_2 & + & 2k_3 & = & 0. \end{array}$$

Thus, the vectors will be linearly independent, if and only if the above system has a trivial solution. By writing the above system as an augmented matrix form and then row-reducing, we get:

$$\left(\begin{array}{ccc|c} 2 & -1 & 2 & 0 \\ 2 & 2 & -1 & 0 \\ -1 & 2 & 2 & 0 \end{array} \right) \longrightarrow \left(\begin{array}{ccc|c} 1 & 0 & 0 & 0 \\ 0 & 1 & 0 & 0 \\ 0 & 0 & 1 & 0 \end{array} \right),$$

which gives $k_1 = 0$, $k_2 = 0$, $k_3 = 0$. Hence, the vectors are linearly independent. •

Theorem 3.9 *The determinant of a matrix is zero, if and only if the rows (or columns) of the matrix form a linearly dependent set.* •

3.3 **Diagonalization of Matrices**

Of special importance for the study of eigenvalues are diagonal matrices. These will be denoted by

$$
D = \begin{pmatrix}
\lambda_1 & 0 & 0 & \cdots & 0 \\
0 & \lambda_2 & 0 & \cdots & 0 \\
\vdots & \vdots & \vdots & \ddots & \vdots \\
0 & 0 & 0 & \cdots & \lambda_n
\end{pmatrix}
$$

and are called *spectral matrices*, i.e., all the diagonal elements of D are the eigenvalues of A. This simple but useful result makes it desirable to find ways to transform a general $n \times n$ matrix A into a diagonal matrix having the same eigenvalues. Unfortunately, the elementary operations that can be used to reduce $A \to D$ are not suitable, because the scale and subtract operations alter eigenvalues. Here, what we needed are *similarity transformations*. Similarity transformations occur frequently in the context of relating coordinate systems.

Definition 3.7 (Similar Matrix)

Let A and B be square matrices of the same size. A matrix B is said to be similar to A (i.e., $A \equiv B$), if there exists a nonsingular matrix Q such that $B = Q^{-1}AQ$. The transformation of a matrix A into the matrix B in this manner is called a similarity transformation. •

Example 3.10 *Consider the following matrices A and Q, and Q is nonsingular. Use the similarity transformation $Q^{-1}AQ$ to transform A into a matrix B.*

$$
A = \begin{pmatrix} 0 & 0 & -2 \\ 1 & 2 & 1 \\ 1 & 0 & 3 \end{pmatrix}, \qquad
Q = \begin{pmatrix} -1 & 0 & -2 \\ 0 & 1 & 1 \\ 1 & 0 & 1 \end{pmatrix}.
$$

Solution. *Let*

$$
B = Q^{-1}AQ = \begin{pmatrix} -1 & 0 & -2 \\ 0 & 1 & 1 \\ 1 & 0 & 1 \end{pmatrix}^{-1} \begin{pmatrix} 0 & 0 & -2 \\ 1 & 2 & 1 \\ 1 & 0 & 3 \end{pmatrix} \begin{pmatrix} -1 & 0 & -2 \\ 0 & 1 & 1 \\ 1 & 0 & 1 \end{pmatrix}
$$

$$
= \begin{pmatrix} 1 & 0 & 2 \\ 1 & 1 & 1 \\ -1 & 0 & -1 \end{pmatrix} \begin{pmatrix} 0 & 0 & -2 \\ 1 & 2 & 1 \\ 1 & 0 & 3 \end{pmatrix} \begin{pmatrix} -1 & 0 & -2 \\ 0 & 1 & 1 \\ 1 & 0 & 1 \end{pmatrix}
$$

$$
= \begin{pmatrix} 2 & 0 & 0 \\ 0 & 2 & 0 \\ 0 & 0 & 1 \end{pmatrix}.
$$

In Example 3.10, the matrix A is transformed into a diagonal matrix B. Not every square matrix can be "diagonalized" in this manner. Here, we will discuss conditions under which a matrix can be diagonalized and when it can, ways of constructing an approximate transforming matrix Q. We will find that eigenvalues and eigenvectors play a key role in this discussion.
•

Theorem 3.10 *Let A, B, and C be n × n matrices:*

1. *A ≡ A.*

2. *If A ≡ B, then B ≡ A.*

3. *If A ≡ B and B ≡ C, then A ≡ C.*   •

Theorem 3.11 *Let A and B be n × n matrices with A ≡ B, then:*

1. *det(A) = det(B).*

2. *A is invertible, if and only if B is invertible.*

3. *A and B have the same rank.*

4. *A and B have the same characteristic polynomial.*

5. *A and B have the same eigenvalues.*   •

Note that Theorem 3.11 gives the sufficient conditions for the similar matrices. For example, for the matrices

$$A = \begin{pmatrix} 1 & 0 \\ 0 & 1 \end{pmatrix} \quad \text{and} \quad B = \begin{pmatrix} 1 & 1 \\ 0 & 1 \end{pmatrix},$$

then

$$\begin{aligned} \det(A) &= \det(B) \\ \text{rank}A &= 1 = \text{rank}B \\ (\lambda_1 = 1, \lambda_2 = 1) &= (\lambda_1 = 1, \lambda_2 = 1). \end{aligned}$$

But A is not similar to B, since

$$Q^{-1}AQ = Q^{-1}IQ = Q^{-1}Q = I \neq B$$

for any invertible matrix Q.

Theorem 3.12 *Similar matrices have the same eigenvalues.*

Proof. *Let A and B be similar matrices. Hence, there exists a matrix Q such that $B = Q^{-1}AQ$. The characteristic polynomial of B is $|B - \lambda I|$. Substituting for B and using the multiplicative properties of determinants, we get*

$$\begin{aligned} |B - \lambda I| &= |Q^{-1}AQ - \lambda I| = |Q^{-1}(A - \lambda I)Q| \\ &= |Q^{-1}||A - \lambda I||Q| = |(A - \lambda I)|Q^{-1}||Q| \\ &= |A - \lambda I||Q^{-1}Q| = |A - \lambda I||I| \\ &= |A - \lambda I|. \end{aligned}$$

The characteristic polynomials of A and B are identical. This means that their eigenvalues are the same.   ●

Definition 3.8 (Diagonalizable Matrix)

A square matrix A is called diagonalizable if there exists an invertible matrix Q such that

$$D = Q^{-1}AQ \qquad\qquad (3.32)$$

is a diagonal matrix. Note that all the diagonal elements of it are the eigenvalues of A, and a invertible matrix Q can be written as

$$Q = \left(\mathbf{x}_1 | \mathbf{x}_2 | \cdots | \mathbf{x}_n \right)$$

and is called its model matrix because its columns contain, $\mathbf{x}_1, \mathbf{x}_2, \ldots, \mathbf{x}_n$, which are the eigenvectors of A corresponding to the eigenvalues $\lambda_1, \ldots, \lambda_n$.
•

Theorem 3.13 *Any matrix having linearly independent eigenvectors corresponding to distinct and real eigenvalues is diagonalizable, i.e.,*

$$Q^{-1}AQ = D,$$

where D is a diagonal matrix and Q is an invertible matrix.

Proof. *Let $\lambda_1, \ldots, \lambda_n$ be the eigenvalues of a matrix A, with corresponding linearly independent eigenvectors $\mathbf{x}_1, \ldots, \mathbf{x}_n$. Let Q be the matrix having $\mathbf{x}_1, \ldots, \mathbf{x}_n$ as column vectors, i.e.,*

$$Q = (\mathbf{x}_1 \cdots \mathbf{x}_n).$$

Since $A\mathbf{x}_1 = \lambda\mathbf{x}_1, \ldots, A\mathbf{x}_n = \lambda_n\mathbf{x}_n$, matrix multiplication in terms of *columns gives*

$$AQ = (A\mathbf{x}_1 \cdots A\mathbf{x}_n)$$

$$= (\lambda_1\mathbf{x}_1 \cdots \lambda_n\mathbf{x}_n)$$

$$= (\mathbf{x}_1 \cdots \mathbf{x}_n) \begin{pmatrix} \lambda_1 & & 0 \\ & \ddots & \\ 0 & & \lambda_n \end{pmatrix}$$

$$= Q \begin{pmatrix} \lambda_1 & & 0 \\ & \ddots & \\ 0 & & \lambda_n \end{pmatrix}.$$

Since the columns of Q are linearly independent, Q is invertible. Thus,

$$Q^{-1}AQ = \begin{pmatrix} \lambda_1 & & 0 \\ & \ddots & \\ 0 & & \lambda_n \end{pmatrix} = D.$$

Therefore, if a square matrix A has n linearly independent eigenvectors, these eigenvectors can be used as the columns of a matrix Q that diagonalizes A. The diagonal matrix has the eigenvalues of A as diagonal elements.

Note that the converse of the above theorem also exists, i.e., if A is diagonalizable, then it has n linearly independent eigenvectors. •

Example 3.11 *Consider the matrix*

$$A = \begin{pmatrix} 0 & 0 & 1 \\ 3 & 7 & -9 \\ 0 & 2 & -1 \end{pmatrix},$$

which has a characteristic

$$\lambda^3 - 6\lambda^2 + 11\lambda - 6 = 0,$$

and this cubic factorizes to give

$$(\lambda - 1)(\lambda - 2)(\lambda - 3) = 0.$$

The eigenvalues of A, therefore, are 1, 2, and 3, with a sum 6, which agrees with the trace of A. Corresponding to these eigenvalues, the eigenvectors of A are $\mathbf{x}_1 = [1,1,1]^T$, $\mathbf{x}_2 = [1,3,2]^T$, *and* $\mathbf{x}_3 = [1,6,3]^T$. *Thus, the nonsingular matrix Q is given by*

$$Q = \begin{pmatrix} 1 & 1 & 1 \\ 1 & 3 & 6 \\ 1 & 2 & 3 \end{pmatrix},$$

and the inverse of this matrix is given by

$$Q^{-1} = \begin{pmatrix} 3 & 1 & -3 \\ -3 & -2 & 5 \\ 1 & 1 & -2 \end{pmatrix}.$$

Thus,

$$Q^{-1}AQ = \begin{pmatrix} 3 & 1 & -3 \\ -3 & -2 & 5 \\ 1 & 1 & -2 \end{pmatrix} \begin{pmatrix} 0 & 0 & 1 \\ 3 & 7 & -9 \\ 0 & 2 & -1 \end{pmatrix} \begin{pmatrix} 1 & 1 & 1 \\ 1 & 3 & 6 \\ 1 & 2 & 3 \end{pmatrix},$$

which implies that

$$Q^{-1}AQ = \begin{pmatrix} 1 & 0 & 0 \\ 0 & 2 & 0 \\ 0 & 0 & 3 \end{pmatrix} = D.$$

●

The above results can be obtained using the MATLAB Command Window as follows:

```
>> A = [0 0 1; 3 7 - 9; 0 2 - 1];
>> P = poly(A);
>> PP = poly2sym(P);
>> [X, D] = eig(A);
>> eigenvalues = diag(D);
>> x1 = X(:, 1); x2 = X(:, 2); x3 = X(:, 3);
>> Q = [x1 x2 x3];
>> D = inv(Q) * A * Q;
```

It is possible that independent eigenvectors may exist even though the eigenvalues are not distinct, though no theorem exists to show under what conditions they may do so. The following example shows the situation that can arise.

Example 3.12 *Consider the matrix*

$$A = \begin{pmatrix} 2 & 1 & 1 \\ 2 & 3 & 2 \\ 3 & 3 & 4 \end{pmatrix},$$

which has a characteristic equation

$$\lambda^3 - 9\lambda^2 + 15\lambda - 7 = 0,$$

and it can be easily factorized to give

$$(\lambda - 7)(\lambda - 1)^2 = 0.$$

The eigenvalues of A are 7 of multiplicity one and 1 of multiplicity two. The eigenvectors corresponding to these eigenvalues are $\mathbf{x}_1 = [1, 2, 3]^T$, $\mathbf{x}_2 = [1, 0, -1]^T$, *and* $\mathbf{x}_3 = [0, 1, -1]^T$. *Thus, the nonsingular matrix Q is given by*

$$Q = \begin{pmatrix} 1 & 1 & 0 \\ 2 & 0 & 1 \\ 3 & -1 & -1 \end{pmatrix},$$

and the inverse of this matrix is

$$Q^{-1} = \frac{1}{6} \begin{pmatrix} 1 & -1 & -1 \\ -5 & 1 & 1 \\ 2 & -4 & 2 \end{pmatrix}.$$

Thus,

$$Q^{-1}AQ = \begin{pmatrix} 7 & 0 & 0 \\ 0 & 1 & 0 \\ 0 & 0 & 1 \end{pmatrix} = D.$$

●

Computing Powers of a Matrix

There are numerous problems in applied mathematics that require the computation of higher powers of a square matrix. Now we shall show how diagonalization can be used to simplify such computations for diagonalizable matrices.

If A is a square matrix and Q is an invertible matrix, then

$$(Q^{-1}AQ)^2 = Q^{-1}AQQ^{-1}AQ = Q^{-1}AIAQ = Q^{-1}A^2Q.$$

More generally, for any positive integer k, we have

$$(Q^{-1}AQ)^k = Q^{-1}A^kQ.$$

It follows from this equation that if A is diagonalizable and $Q^{-1}AQ = D$ is a diagonal matrix, then

$$Q^{-1}A^kQ = (Q^{-1}AQ)^k = D^k.$$

Solving this equation for A^k yields

$$A^k = QD^kQ^{-1}. \tag{3.33}$$

Therefore, in order to compute the *kth* power of A, all we need to do is compute the *kth* power of a diagonal matrix D and then form the matrices Q and Q^{-1} as indicated in (3.33). But taking the *kth* power of a diagonal matrix is easy, for it simply amounts to taking the *kth* power of each of the entries on the main diagonal.

Example 3.13 *Consider the matrix*

$$A = \begin{pmatrix} 1 & 1 & -4 \\ 2 & 0 & -4 \\ -1 & 1 & -2 \end{pmatrix},$$

which has a characteristic equation

$$\lambda^3 + \lambda^2 - 4\lambda - 4 = 0$$

and factorizes to

$$(\lambda + 1)(\lambda + 2)(\lambda - 2) = 0.$$

It gives eigenvalues 2, –2, and –1 of the given matrix A with the corresponding eigenvectors $[1, 2, 1]^T$, $[1, 1, 1]^T$, and $[1, 1, 0]^T$. Then the factorization

$$A = QDQ^{-1}$$

becomes

$$\begin{pmatrix} 1 & 1 & -4 \\ 2 & 0 & -4 \\ -1 & 1 & -2 \end{pmatrix} = \begin{pmatrix} 1 & 1 & 1 \\ 2 & 1 & 1 \\ 1 & 1 & 0 \end{pmatrix} \begin{pmatrix} -1 & 0 & 0 \\ 0 & -2 & 0 \\ 0 & 0 & 2 \end{pmatrix} \begin{pmatrix} -1 & 1 & 0 \\ 1 & -1 & 1 \\ 1 & 0 & -1 \end{pmatrix},$$

and from (3.33), we have

$$A^k = \begin{pmatrix} 1 & 1 & 1 \\ 2 & 1 & 1 \\ 1 & 1 & 0 \end{pmatrix} \begin{pmatrix} (-1)^k & 0 & 0 \\ 0 & (-2)^k & 0 \\ 0 & 0 & (2)^k \end{pmatrix} \begin{pmatrix} -1 & 1 & 0 \\ 1 & -1 & 1 \\ 1 & 0 & -1 \end{pmatrix},$$

which implies that

$$A^k = \begin{pmatrix} -(-1)^k + (-2)^k + 2^k & (-1)^k - (-2)^k & (-2)^k - 2^k \\ -2(-1)^k + (-2)^k + 2^k & 2(-1)^k - (-2)^k & (-2)^k - 2^k \\ -(-1)^k + (-2)^k & (-1)^k - (-2)^k & (-2)^k \end{pmatrix}.$$

For this formula, we can easily compute any power of a given matrix A. For example, if $k = 10$, then

$$A^{10} = \begin{pmatrix} 2047 & -1023 & 0 \\ 2046 & -1022 & 0 \\ 1023 & -1023 & 1024 \end{pmatrix},$$

the required 10th power of the matrix.

●

The above results can be obtained using the MATLAB Command Window as follows:

```
>> A = [1 1 − 4; 2 0 − 4; −1 1 − 2];
>> P = poly(A);
>> [X, D] = eig(A);
>> eigenvalues = diag(D);
>> x1 = X(:, 1); x2 = X(:, 2); x3 = X(:, 3);
>> Q = [x1 x2 x3];
>> A10 = Q ∗ D^ 10∗inv(Q);
```

Example 3.14 *Show that the following matrix A is not diagonalizable:*

$$A = \begin{pmatrix} 5 & -3 \\ 3 & -1 \end{pmatrix}.$$

Solution. *To compute the eigenvalues and corresponding eigenvectors of the given matrix A, we have the characteristic equation of the form*

$$|A - \lambda \mathbf{I}| = 0$$
$$\lambda^2 - 4\lambda + 4 = 0,$$

which factorizes to

$$(\lambda - 2)(\lambda - 2) = 0$$

and gives repeated eigenvalues 2 and 2 of the given matrix A. To find the corresponding eigenvectors, we solve (3.2) for $\lambda = 2$, and we get

$$\begin{pmatrix} 3 & -3 \\ 3 & -3 \end{pmatrix} \begin{pmatrix} x_1 \\ x_2 \end{pmatrix} = \mathbf{0}.$$

Solving the above homogeneous system gives $3x_1 - 3x_2 = 0$, and we have $x_1 = x_2 = \alpha$. Thus, the eigenvectors are nonzero vectors of the form

$$\alpha \begin{pmatrix} 1 \\ 1 \end{pmatrix}.$$

The eigenspace is a one-dimensional space. A is a 2×2 matrix, but it does not have two linearly independent eigenvectors. Thus, A is not diagonalizable. ●

Definition 3.9 (Orthogonal Matrix)

It is a square matrix whose inverse can be determined by transposing it, i.e.,

$$A^{-1} = A^T.$$

Such matrices do occur in some engineering problems. The matrix used to obtain rotation of coordinates about the origin of a Cartesian system is one example of an orthogonal matrix. For example, consider the square matrix

$$A = \begin{pmatrix} 0.6 & -0.8 \\ 0.8 & 0.6 \end{pmatrix}.$$

One can easily verify that the given matrix A is orthogonal because

$$A^{-1} = \begin{pmatrix} 0.6 & 0.8 \\ -0.8 & 0.6 \end{pmatrix} = A^T.$$

•

Orthogonal Diagonalization

Let Q be an orthogonal matrix, i.e., $Q^{-1} = Q^T$. Thus, if such a matrix is used in a similarity transformation, the transformation becomes $D = Q^T A Q$. This type of similarity transformation is much easier to calculate because its inverse is simply its transpose. There is therefore considerable advantage in searching for situations where a reduction to a diagonal matrix, using an orthogonal matrix, is possible.

Definition 3.10 (Orthogonally Diagonalizable Matrix)

A square matrix A is said to be orthogonally diagonalizable if there exists an orthogonal matrix Q such that

$$D = Q^{-1} A Q = Q^T A Q$$

is a diagonal matrix. $\hspace{6cm}$ •

The following theorem tells us that the set of orthogonally diagonalizable matrices is in fact the set of symmetric matrices.

Theorem 3.14 *A square matrix A is orthogonally diagonalizable if it is a symmetric matrix.*

Proof. *Suppose that a matrix A is orthogonally diagonalizable, then there exists an orthogonal matrix Q such that*

$$D = Q^T A Q.$$

Therefore,

$$A = Q D Q^T.$$

Taking its transpose gives

$$A^T = (Q D Q^T)^T = (Q^T)^T (Q D)^T = Q D Q^T = A.$$

Thus, A is symmetric.

The converse of this theorem is also true, but it is beyond the scope of this text and will be omitted.   •

Symmetric Matrices

Now our next goal is to devise a procedure for orthogonally diagonalizing a symmetric matrix, but before we can do so, we need an important theorem about eigenvalues and eigenvectors of symmetric matrices.

Theorem 3.15 *If A is a symmetric matrix, then:*

(a) The eigenvalues of A are all real numbers.

(b) Eigenvectors from distinct eigenvalues are orthogonal.   •

Theorem 3.16 *The following conditions are equivalent for an $n \times n$ matrix Q:*

(a) Q is invertible and $Q^{-1} = Q^T$.

(b) The rows of Q are orthonormal.

(c) The columns of Q are orthonormal.   •

Diagonalization of Symmetric Matrices

As a consequence of the preceding theorem we obtain the following procedure for orthogonally diagonalizing a symmetric matrix:

1. Find a basis for each eigenspace of A.

2. Find an orthonormal basis for each eigenspace.

3. Form the matrix Q whose columns are these orthonormal vectors.

4. The matrix $D = Q^T A Q$ will be a diagonal matrix.

Example 3.15 *Consider the matrix*

$$A = \begin{pmatrix} 3 & -1 & 0 \\ -1 & 2 & -1 \\ 0 & -1 & 3 \end{pmatrix},$$

which has a characteristic equation

$$\lambda^3 - 8\lambda^2 + 19\lambda - 12 = 0,$$

and it gives the eigenvalues 1, 3, and 4 for the given matrix A. Corresponding to these eigenvalues, the eigenvectors of A are $\mathbf{x}_1 = [1,2,1]^T$, $\mathbf{x}_2 = [1,0,-1]^T$, and $\mathbf{x}_3 = [1,-1,1]^T$, and they form an orthogonal set. Note that the following vectors

$$\mathbf{u}_1 = \frac{\mathbf{x}1}{\|\mathbf{x}_1\|_2} = \tfrac{1}{\sqrt{6}}[1,2,1]^T$$

$$\mathbf{u}_2 = \frac{\mathbf{x}_2}{\|\mathbf{x}_2\|_2} = \tfrac{1}{\sqrt{2}}[1,0,-1]^T$$

$$\mathbf{u}_3 = \frac{\mathbf{x}_3}{\|\mathbf{x}_3\|_2} = \tfrac{1}{\sqrt{3}}[1,-1,1]^T$$

form an orthonormal set, since they inherit orthogonality from $\mathbf{x}_1, \mathbf{x}_2$, and $\mathbf{x}_3$, and in addition

$$\|\mathbf{u}_1\|_2 = \|\mathbf{u}_2\|_2 = \|\mathbf{u}_3\|_2 = 1.$$

Then an orthogonal matrix Q forms from an orthonormal set of vectors as

$$Q = \begin{pmatrix} \dfrac{1}{\sqrt{6}} & \dfrac{1}{\sqrt{2}} & \dfrac{1}{\sqrt{3}} \\[3mm] \dfrac{2}{\sqrt{6}} & 0 & -\dfrac{1}{\sqrt{3}} \\[3mm] \dfrac{1}{\sqrt{6}} & -\dfrac{1}{\sqrt{2}} & \dfrac{1}{\sqrt{3}} \end{pmatrix}$$

and

$$Q^T A Q = \begin{pmatrix} \dfrac{1}{\sqrt{6}} & \dfrac{2}{\sqrt{6}} & \dfrac{1}{\sqrt{6}} \\[3mm] \dfrac{1}{\sqrt{2}} & 0 & -\dfrac{1}{\sqrt{2}} \\[3mm] \dfrac{1}{\sqrt{3}} & -\dfrac{1}{\sqrt{3}} & \dfrac{1}{\sqrt{3}} \end{pmatrix} \begin{pmatrix} 3 & -1 & 0 \\ -1 & 2 & -1 \\ 0 & -1 & 3 \end{pmatrix} \begin{pmatrix} \dfrac{1}{\sqrt{6}} & \dfrac{1}{\sqrt{2}} & \dfrac{1}{\sqrt{3}} \\[3mm] \dfrac{2}{\sqrt{6}} & 0 & -\dfrac{1}{\sqrt{3}} \\[3mm] \dfrac{1}{\sqrt{6}} & -\dfrac{1}{\sqrt{2}} & \dfrac{1}{\sqrt{3}} \end{pmatrix},$$

which implies that

$$Q^T A Q = \begin{pmatrix} 1 & 0 & 0 \\ 0 & 3 & 0 \\ 0 & 0 & 4 \end{pmatrix} = D.$$

Note that the eigenvalues $1, 3$, and 4 of the matrix A are real and its eigenvectors form an orthonormal set, since they inherit orthogonally from $\mathbf{x}_1, \mathbf{x}_2$, and $\mathbf{x}_3$, which satisfy the preceding theorem.   ●

The results of Example 3.15 can be obtained using the MATLAB Command Window as follows:

```
>> A = [3 − 1 0; −1 2 − 1; 0 − 1 3];
>> P = poly(A);
>> PP = poly2sym(P);
>> [X, D] = eig(A);
>> λ = diag(D);
>> x1 = X(:, 1); x2 = X(:, 2); x3 = X(:, 3);
>> u1 = x1/norm(x1); u2 = x2/norm(x2); u3 = x3/norm(x3);
>> Q = [u1 u2 u3];
>> D = Q' ∗ A ∗ Q;
```

Theorem 3.17 (Principal Axis Theorem)

The following conditions are equivalent for an $n \times n$ matrix A:

(a) A has an orthonormal set of n eigenvectors.

(b) A is orthogonally diagonalizable.

(c) A is symmetric.   ●

In the following section we shall discuss some extremely important properties of the eigenvalue problem. Before this, we will discuss some special matrices, as follows.

Definition 3.11 (Conjugate of a Matrix)

If the entries of an $n \times n$ matrix A are complex numbers, we can write

$$A = (a_{ij}) = (b_{ij} + \mathbf{i}c_{ij}),$$

where b_{ij} and c_{ij} are real numbers. The conjugate of a matrix A is a matrix

$$\bar{A} = (\bar{a}_{ij}) = (b_{ij} - \mathbf{i}c_{ij}).$$

For example, the conjugate of

$$A = \begin{pmatrix} 2 & \pi & \mathbf{i} \\ 3 & 7 & 0 \\ 4 & 1 - \mathbf{i} & 4 \end{pmatrix} \quad is \quad \bar{A} = \begin{pmatrix} 2 & \pi & -\mathbf{i} \\ 3 & 7 & 0 \\ 4 & 1 + \mathbf{i} & 4 \end{pmatrix}.$$

●

Definition 3.12 (Hermitian Matrix)

It is a square matrix $A = (a_{ij})$ that is equal to its conjugate transpose

$$A = A^* = \bar{A}^T,$$

i.e., whenever $a_{ij} = \bar{a}_{ji}$. This is the complex analog of symmetry. For example, the following matrix A is Hermitian if it has the form

$$A = \begin{pmatrix} a & b + ic \\ b - ic & d \end{pmatrix},$$

where a, b, c, d are real. An Hermitian matrix may be or may not be symmetric. For example, the matrices

$$A = \begin{pmatrix} 1 & 2 + 4i & 1 - 3i \\ 2 - 4i & 3 & 8 + 6i \\ 1 + 3i & 8 - 6i & 5 \end{pmatrix} \quad and \quad B = \begin{pmatrix} 1 & 2 + 4i & 1 - 3i \\ 2 + 4i & 3 & 8 + 6i \\ 1 - 3i & 8 + 6i & 5 \end{pmatrix}$$

are Hermitian where the matrix A is symmetric but the matrix B is not. Note that:

1. Every diagonal matrix is Hermitian, if and only if it is real.
2. The square matrix A is said to be a skew Hermitian when

$$A = -A^* = -\bar{A}^T,$$

i.e., whenever $a_{ij} = -\bar{a}_{ji}$. This is the complex analog of skew symmetry. For example, the following matrix A is skew Hermitian if it has the form

$$A = \begin{pmatrix} 0 & 1 + i \\ -1 + i & i \end{pmatrix}.$$

●

Definition 3.13 (Unitary Matrix)

Let $A = (a_{ij})$ be the square matrix, then if

$$AA^* = A^*A = \mathbf{I}_n,$$

where $\mathbf{I}_n$ *is an* $n \times n$ *identity matrix, then* A *is called the unitary matrix. For example, for any real number* θ, *the following matrix*

$$A = \begin{pmatrix} \cos\theta & -\sin\theta \\ \sin\theta & \cos\theta \end{pmatrix}$$

is unitary. ●

Note that:
1. The identity matrix is unitary.
2. The inverse of a unitary matrix is unitary.
3. A product of a unitary matrix is unitary.
4. A real matrix A is unitary, if and only if $A^T = A^{-1}$.
5. A square matrix A is *unitarily similar* to the square matrix B, if and only if there is an unitary matrix Q the same size as A and B such that

$$A = QBQ^{-1}.$$

A square matrix A is *unitarily diagonalizable*, if and only if it is unitarily similar to a diagonal matrix.

Theorem 3.18 *A square matrix A is unitarily diagonalizable, if and only if there is a unitary matrix Q of the same size whose columns are eigenvectors of A.* ●

Definition 3.14 (Normal Matrix)

A square matrix A is normal, if and only if it commutes with its conjugate transpose, i.e.,

$$AA^* = A^*A.$$

For example, if a and b are real numbers, then the following matrix

$$A = \begin{pmatrix} a & b \\ -b & a \end{pmatrix}$$

is normal because

$$AA^* = \begin{pmatrix} a^2 + b^2 & 0 \\ 0 & a^2 + b^2 \end{pmatrix} = A^*A.$$

However, its eigenvalues are $a + \mathbf{i}b$. Note that all the Hermitian, skew Hermitian and unitary matrices are normal matrices. ●

3.4   Basic Properties of Eigenvalue Problems

1. A square matrix A is *singular*, if and only if at least one of its eigenvalues is zero. It can be easily proved, since for $\lambda = 0$, we have (3.4) of the form

$$|A - \lambda I| = |A| = 0.$$

Example 3.16 *Consider the following matrix:*

$$A = \begin{pmatrix} 3 & 1 & 0 \\ 2 & -1 & -1 \\ 4 & 3 & 1 \end{pmatrix}.$$

Then the characteristic equation of A takes the form

$$-\lambda^3 + 3\lambda^2 = 0.$$

By solving this cubic equation, the eigenvalues of A are 0, 0, and 3. Hence, the given matrix is singular because two of its eigenvalues are zero.   ●

2. The eigenvalues of a matrix A and its transpose A^T are identical.

   It is well known that the determinant of a matrix A and of its A^T are the same. Therefore, they must have the same characteristic equation and the same eigenvalues.

Example 3.17 *Consider a matrix A and its transpose matrix A^T as*

$$A = \begin{pmatrix} 1 & 1 & 0 \\ 3 & 0 & 3 \\ 2 & -1 & 3 \end{pmatrix} \quad and \quad A^T = \begin{pmatrix} 1 & 3 & 2 \\ 1 & 0 & -1 \\ 0 & 3 & 3 \end{pmatrix}.$$

The characteristic equations of A and A^T are the same, which are

$$-\lambda^3 + 4\lambda^2 - 3\lambda = 0.$$

Solving this cubic polynomial equation, we have the eigenvalues $0, 1, 3$ of both the matrices, A and its transpose A^T.   ●

3. The eigenvalues of an inverse matrix A^{-1}, provided that A^{-1} exists, are the inverses of the eigenvalues of A.

To prove this, let us consider λ is an eigenvalue of A and using (3.4), gives

$$|A - \lambda I| = |A - \lambda A A^{-1}| = |A(I - \lambda A^{-1})| = |\lambda||A||\frac{1}{\lambda}I - A^{-1}| = 0.$$

Since the matrix A is nonsingular, $|A| \neq 0$, and also, $\lambda \neq 0$. Hence,

$$\left|\frac{1}{\lambda}I - A^{-1}\right| = 0,$$

which shows that $\frac{1}{\lambda}$ is an eigenvalue of a matrix A^{-1}.

Example 3.18 *Consider a matrix A and its inverse matrix A^{-1} as*

$$A = \begin{pmatrix} 3 & 0 & 1 \\ 0 & -3 & 0 \\ 1 & 0 & 3 \end{pmatrix} \quad and \quad A^{-1} = \begin{pmatrix} \frac{3}{8} & 0 & -\frac{1}{8} \\ 0 & -\frac{1}{3} & 0 \\ -\frac{1}{8} & 0 & \frac{3}{8} \end{pmatrix}.$$

Then a characteristic equation of A has the form

$$\lambda^3 - 3\lambda^2 - 10\lambda + 24 = 0,$$

which gives the eigenvalues 4, –3, and 2 of A. Also, the characteristic equation of A^{-1} is

$$\lambda^3 - \frac{5}{12}\lambda^2 - \frac{1}{8}\lambda + \frac{1}{24} = 0,$$

and it gives the eigenvalues

$$\frac{1}{4}, -\frac{1}{3}, \quad and \quad \frac{1}{2},$$

which are reciprocals to the eigenvalues $4, -3, 2$ of the matrix A. ●

4. The eigenvalues of A^k (k is an integer) are eigenvalues of A raised to the *kth* power.

   To prove this, consider the characteristic equation of a matrix A

   $$|A - \lambda I| = 0,$$

   which can also be written as

   $$0 = |A - \lambda I| = |A - \lambda I||A + \lambda I| = |(A - \lambda I)(A + \lambda I)| = |A^2 - \lambda^2 I|.$$

   Example 3.19 *Consider the matrix*

   $$A = \begin{pmatrix} 1 & 1 & 0 \\ 3 & 0 & 3 \\ 2 & -1 & 3 \end{pmatrix},$$

   which has the eigenvalues, 0, 1, and 3. Now

   $$AA = A^2 = \begin{pmatrix} 4 & 1 & 3 \\ 9 & 0 & 9 \\ 5 & -1 & 6 \end{pmatrix}$$

   has the characteristic equation of the form

   $$\lambda^3 - 10\lambda^2 + 9\lambda = 0.$$

   Solving this cubic equation, the eigenvalues of A^2 are, 0, 1, and 9, which are double the eigenvalues $0, 1, 3$ of A.   ●

5. The eigenvalues of a diagonal matrix or a triangular (upper or lower) matrix are their diagonal elements.

   Example 3.20 *Consider the following matrices:*

   $$A = \begin{pmatrix} 1 & 0 & 0 \\ 0 & 2 & 0 \\ 0 & 0 & 3 \end{pmatrix}, \qquad B = \begin{pmatrix} 2 & 2 & 3 \\ 0 & 3 & 3 \\ 0 & 0 & 4 \end{pmatrix}.$$

The characteristic equation of A is

$$\lambda^3 - 6\lambda^2 + 11\lambda - 6 = (1 - \lambda)(2 - \lambda)(3 - \lambda) = 0,$$

and it gives eigenvalues $1, 2, 3$, which are the diagonal elements of the given matrix A. Similarly, the characteristic equation of B is

$$\lambda^3 - 9\lambda^2 + 26\lambda - 24 = (2 - \lambda)(3 - \lambda)(4 - \lambda) = 0,$$

and it gives eigenvalues $2, 3, 4$, which are the diagonal elements of the given matrix B. Hence, the eigenvalues of a diagonal matrix A and the upper-triangular matrix B are their diagonal elements. ●

6. Every square matrix satisfies its own characteristic equation.

This is a well-known theorem called the *Cayley–Hamilton Theorem*. If a characteristic equation of A is

$$\lambda^n + \alpha_{n-1}\lambda^{n-1} + \alpha_{n-2}\lambda^{n-2} + \cdots + \alpha_1\lambda + \alpha_0 = 0,$$

the matrix itself satisfies the same equation, namely,

$$A^n + \alpha_{n-1}A^{n-1} + \alpha_{n-2}A^{n-2} + \cdots + \alpha_1 A + \alpha_0 = 0. \qquad (3.34)$$

Multiplying each term in (3.34) by A^{-1}, when A^{-1} exists and thus $\alpha_0 \neq 0$, gives an important relationship for the inverse of a matrix:

$$A^{n-1} + \alpha_{n-1}A^{n-2} + \alpha_{n-2}A^{n-3} + \cdots + \alpha_1 I + \alpha_0 A^{-1} = 0$$

or

$$A^{-1} = -\frac{1}{\alpha_0}[A^{n-1} + \alpha_{n-1}A^{n-2} + \alpha_{n-2}A^{n-3} + \cdots + \alpha_1 I].$$

Program 3.3

MATLAB m-file for Using the Cayley–Hamilton Theorem
function [c,Ainv]= chim(A)
n=max(size(A));
for i=1:n; for j=1:n
I(i,j)=0; I(i,i)=1; end; end
AA=A; AAA=A; c=[1];
for k=1:n; traceA=0;
for g=1:n % Loop to find the trace of matrix A.
traceA=traceA+A(g,g); end
$cc = -1/k * traceA$; % To find coefficients of the polynomial.
$c = [c, cc]; if k < n$;
for i=1:n; for j=1:n; $b(i,j) = A(i,j) + cc * I(i,j)$; end; end
for i=1:n; for j=1:n; s=0;
for m=1:n; $ss = AA(i,m) * b(m,j)$; s=s+ss; end;
A(i,j)=s; end; end; end
for i=1:n; for j=1:n
$su1(i,j) = c(n) * I(i,j) + c(n-1) * AA(i,j)$; su2(i,j)=0; end; end
if $n > 2$
for z=2:n-1; for i=1:n; for j=1:n; s=0;
for m=1:n; $ss = AAA(i,m) * AA(m,j); s = s + ss$; end
am(i,j)=s; end; end; AAA=am;
for i=1:n; for j=1:n
$su2(i,j) = su2(i,j) + c(n-z) * AAA(i,j)$; end; end;end;end
for i=1:n; for j=1:n
su(i,j)=su1(i,j)+su2(i,j); $Ainv(i,j) = -1/c(n+1) * su(i,j)$;
end;end

Example 3.21 *Consider the square matrix*

$$A = \begin{pmatrix} 2 & 1 & 2 \\ 0 & 2 & 3 \\ 0 & 0 & 5 \end{pmatrix},$$

which has a characteristic equation of the form

$$p(\lambda) = \lambda^3 - 9\lambda^2 + 24\lambda - 20 = 0,$$

and one can write

$$p(A) = A^3 - 9A^2 + 24A - 20I = 0.$$

Then the inverse of A can be obtained as

$$A^2 - 9A + 24I - 20A^{-1} = 0,$$

which gives

$$A^{-1} = \frac{1}{20}[A^2 - 9A + 24I].$$

Computing the right-hand side, we have

$$A^{-1} = \frac{1}{20} \begin{pmatrix} 10 & -5 & -1 \\ 0 & 10 & -6 \\ 0 & 0 & 4 \end{pmatrix}.$$

Similarly, one can also find the higher power of the given matrix A. For example, one can compute the value of the matrix A^5 by solving the expression

$$A^5 = 9A^4 - 24A^3 + 20A^2,$$

and it gives

$$A^5 = \begin{pmatrix} 32 & 80 & 3013 \\ 0 & 32 & 3093 \\ 0 & 0 & 3125 \end{pmatrix}.$$

•

To find the coefficients of a characteristic equation and the inverse of a matrix A by the Cayley–Hamilton theorem using MATLAB commands we do as follows:

```
>> A = [2 1 2; 0 2 3; 0 0 5];
c = chim(A);
```

7. The eigenvectors of A^{-1} are the same as the eigenvectors of A.

   Let $\mathbf{x}$ be an eigenvector of A that satisfies the equation

   $$A\mathbf{x} = \lambda\mathbf{x},$$

   then

   $$\frac{1}{\lambda}A^{-1}A\mathbf{x} = A^{-1}\mathbf{x}.$$

   Hence,

   $$A^{-1}\mathbf{x} = \frac{1}{\lambda}\mathbf{x},$$

   which shows that $\mathbf{x}$ is also an eigenvector of A^{-1}.

8. The eigenvectors of the matrix (kA) are identical to the eigenvectors of A, for any scalar k.

   Since the eigenvalues of (kA) are k times the eigenvalues of A, if $A\mathbf{x} = \lambda\mathbf{x}$, then

   $$(kA)\mathbf{x} = (k\lambda)\mathbf{x}.$$

9. A symmetric matrix A is positive-definite, if and only if all the eigenvalues of A are positive.

Example 3.22 *Consider the matrix*

$$A = \begin{pmatrix} 2 & 0 & 1 \\ 0 & 2 & 0 \\ 1 & 0 & 2 \end{pmatrix},$$

which has the characteristic equation of the form

$$\lambda^3 - 6\lambda^2 + 11\lambda - 6 = 0,$$

and it gives the eigenvalues 3, 2, and 1 of A. Since all the eigenvalues of the matrix A are positive, A is positive-definite.   ●

10. For any $n \times n$ matrix A, we have

$$
\left.
\begin{aligned}
B_0 &= I \\
A_k &= AB_{k-1} \\
c_k &= -\frac{1}{k}\mathrm{tr}(A_k) \\
B_k &= A_k + c_k I
\end{aligned}
\right\} \quad k = 1, 2, \ldots, n,
$$

where $\mathrm{tr}(A)$ is a trace of a matrix A. Then a characteristic polynomial of A is

$$
p(\lambda) = |A - \lambda I| = \lambda^n + c_1 \lambda^{n-1} + \cdots + c_{n-1}\lambda + c_n.
$$

If $c_n \neq 0$, then the inverse of A can be obtained as

$$
A^{-1} = -\frac{1}{c_n} B_{n-1}.
$$

This is called the *Sourian–Frame Theorem*.

Example 3.23 *Find a characteristic polynomial of the matrix*

$$
A = \begin{pmatrix} 5 & -2 & -4 \\ -2 & 2 & 2 \\ -4 & 2 & 5 \end{pmatrix},
$$

and then find A^{-1} by using the Sourian–Frame theorem.

Solution. *Since the given matrix is of size 3×3, the possible form of a characteristic polynomial will be*

$$
p(\lambda) = \lambda^3 + c_1 \lambda^2 + c_2 \lambda + c_3.
$$

The values of the coefficients c_1, c_2, and c_3 of the characteristic polynomial can be computed as

$$
A_1 = AB_0 = AI = A,
$$

so

$$
c_1 = -\frac{1}{1}\mathrm{tr}(A_1) = -12.
$$

Now

$$B_1 = A_1 + c_1 I = A_1 - 12I = \begin{pmatrix} -7 & -2 & -4 \\ -2 & -10 & 2 \\ -4 & 2 & -7 \end{pmatrix}$$

and

$$A_2 = AB_1 = \begin{pmatrix} -15 & 2 & -4 \\ 2 & -12 & -2 \\ 4 & -2 & -15 \end{pmatrix},$$

so

$$c_2 = -\frac{1}{2} tr(A_2) = 21.$$

Now

$$B_2 = A_2 + c_2 I = A_2 + 21I = \begin{pmatrix} 6 & 2 & 4 \\ 2 & 9 & -2 \\ 4 & -2 & 6 \end{pmatrix}$$

and

$$A_3 = AB_2 = \begin{pmatrix} 10 & 0 & 0 \\ 0 & 10 & 0 \\ 0 & 0 & 10 \end{pmatrix},$$

so

$$c_3 = -\frac{1}{3} tr(A_3) = -10.$$

Thus,

$$p(\lambda) = \lambda^3 - 12\lambda^2 + 21\lambda - 10$$

and

$$A^{-1} = -\frac{1}{c_3} B_2 = \frac{1}{10} \begin{pmatrix} 6 & 2 & 4 \\ 2 & 9 & -2 \\ 4 & -2 & 6 \end{pmatrix}$$

is the inverse of the given matrix A.   ●

These results can be obtained by using the MATLAB Command Window as follows:

```
>> A = [5 − 2 − 4; −2 2 2; −4 2 5];
>> [c, Ainv] = sourian(A);
```

Program 3.4
MATLAB m-file for Using the Sourian–Frame Theorem
function [c,Ainv]=sourian(A)
[n,n]=size(A);
for i=1:n; for j=1:n; b0(i,j)=0;b0(i,i)=1;end;end
AA=A; c[1]; for k = 1:n;
traceA=0; for i=1:n; traceA=traceA+A(i,i);end;
$cc = -1/k * tracA; c = [c, cc]; if\ k < n;$
for i=1:n; for j=1:n
$b(i, j) = A(i, j) + cc * b0(i, j);$end;end
for i=1:n; for j=1:n; s = 0; for m =1:n;
$ss = AA(i, m) * b(m, j); s = s + ss; end;$
$A(i, j) = s; end; end; end; end;$
for i=1:n; for j=1:n;
ai(i,j)=-1/c(n+1)*b(i,j); end; end
disp('Coefficients of the polynomial') c
disp('Inverse of the matrix A=ai') ai

---

11. If a characteristic equation of an $n \times n$ matrix A is

$$\lambda^n + \alpha_{n-1}\lambda^{n-1} + \alpha_{n-2}\lambda^{n-2} + \alpha_{n-3}\lambda^{n-3} + \cdots + \alpha_1\lambda + \alpha_0 = 0,$$

the values of the coefficients of a characteristic polynomial are then found from the following sequence of computations:

$$\alpha_{n-1} = -\text{tr}(A)$$
$$\alpha_{n-2} = -\frac{1}{2}[\alpha_{n-1}\text{tr}(A) + \text{tr}(A^2)]$$

$$\alpha_{n-3} = -\frac{1}{3}[\alpha_{n-2}\text{tr}(A) + \alpha_{n-1}\text{tr}(A^2) + \text{tr}(A^3)]$$
$$\vdots \qquad \vdots$$
$$\alpha_0 = -\frac{1}{n}[\alpha_1\text{tr}(A) + \alpha_2\text{tr}(A^2) + \cdots + \text{tr}(A^n)].$$

This formula is called *Bocher's formula*, which can be used to find the coefficients of a characteristic equation of a square matrix.

Example 3.24 *Find a characteristic equation of the following matrix by using Bocher's formula:*

$$A = \begin{pmatrix} 1 & 1 & 0 \\ 3 & 0 & 3 \\ 2 & -1 & 3 \end{pmatrix}.$$

Solution. *Since the size of the given matrix is 3×3, we have to find the coefficients $\alpha_2, \alpha_1, \alpha_0$ of the characteristic equation*

$$\lambda^3 + \alpha_2 \lambda^2 + \alpha_1 \lambda + \alpha_0 = 0,$$

where

$$\alpha_2 = -tr(A)$$

$$\alpha_1 = -\frac{1}{2}[\alpha_2 tr(A) + tr(A^2)]$$

$$\alpha_0 = -\frac{1}{3}[\alpha_1 tr(A) + \alpha_2 tr(A^2) + tr(A^3)].$$

In order to find the values of the above coefficients, we must compute the powers of matrix A as follows:

$$A^2 = \begin{pmatrix} 4 & 1 & 3 \\ 9 & 0 & 9 \\ 5 & -1 & 6 \end{pmatrix} \quad and \quad A^3 = \begin{pmatrix} 13 & 1 & 12 \\ 27 & 0 & 27 \\ 14 & -1 & 15 \end{pmatrix}.$$

By using these matrices, we can find the coefficients of the characteristic equation as

$$\alpha_2 = -(4) = -4$$

$$\alpha_1 = -\frac{1}{2}[-4(4) + 10] = 3$$

$$\alpha_0 = -\frac{1}{3}[3(4) + (-4)(10) + 28] = 0.$$

Hence, the characteristic equation of A is

$$\lambda^3 - 4\lambda^2 + 3\lambda = 0.$$

●

To find the coefficients of the characteristic equation by Bocher's theorem using MATLAB commands we do as follows:

```
>> A = [1 1 0; 3 0 1; 2 -1 3];
>> c = BOCH(A);
```

Program 3.5
MATLAB m-file for Using Bocher's Theorem
```
function c=BOCH(A)
[n,n]=size(A);
for i=1:n; for j=1:n; I(i,i)=1; end; end
A(n-1)=-trace(A);
for i=2:n; s=0; p=1;
for k=i-1:-1:1
s = s + A(n-k)*trace(A^ p); p = p + 1; end
if i≅n; A(n-i) = (-1/i)*(s+trace(A^ i)); else
Ao = (-1/i)*(s+trace(A^ i)); end; end
coeff=[Ao A]; if ao≅0; s=A ^ (n-1);
for i=1:n-2
s = s + a(n-i)*A^ (n-(i+1)); end
s = s + A(1)*I; Ainv = -s/Ao; else; end
```

12. For an $n \times n$ matrix A with a characteristic equation

$$|A - \lambda I| = \lambda^n + \alpha_{n-1}\lambda^{n-1} + \alpha_{n-2}\lambda^{n-2} + \cdots + \alpha_1\lambda + \alpha_0 = 0,$$

the unknown coefficients can be computed as

$$\alpha_{n-1} = -\text{tr}(AD_1)$$

$$\alpha_{n-2} = -\frac{1}{2}(AD_2)$$

$$\alpha_{n-3} = -\frac{1}{3}(AD_3)$$

$$\vdots \qquad \qquad \vdots$$

$$\alpha_0 = -\frac{1}{n}(AD_n),$$

where

$$
\begin{aligned}
D_1 &= I \\
D_2 &= AD_1 + \alpha_{n-1}I = A + \alpha_{n-1}I \\
D_3 &= AD_2 + \alpha_{n-2}I = A^2 + \alpha_{n-1}A + \alpha_{n-2}I \\
&\vdots \\
D_n &= AD_{n-1} + \alpha_1 I = A^{n-1} + \alpha_{n-1}A^{n-2} + \cdots + \alpha_2 A + \alpha_1 I
\end{aligned}
$$

and also

$$D_{n+1} = AD_n + \alpha_0 I = 0.$$

Then the determinant of A is

$$\det(A) = |A| = (-1)^n \alpha_0,$$

the adjoint of A is

$$adj(A) = (-1)^{n+1} D_n,$$

and the inverse of A is

$$A^{-1} = -\frac{1}{\alpha_0} D_n.$$

Note that a singular matrix is indicated by $\alpha_0 = 0$.

This result is known as the Faddeev–Leverrier method. This method, which is recursive, yields a characteristic equation for a square matrix, the adjoint of a matrix, and its inverse (if it exists). The determinant of a matrix, being the negative of the last coefficient in the characteristic equation, is also computed.

Example 3.25 *Find the characteristic equation, determinant, adjoint, and inverse of the following matrix using the Faddeev–Leverrier method:*

$$A = \begin{pmatrix} 2 & 2 & -1 \\ -1 & 0 & 4 \\ 3 & -1 & -3 \end{pmatrix}.$$

Solution. *Since the given matrix is of order 3×3, the possible characteristic equation will be of the form*

$$|A - \lambda I| = \lambda^3 + \alpha_2 \lambda^2 + \alpha_1 \lambda + \alpha_0 = 0.$$

The values of the unknown coefficients $\alpha_2, \alpha_1,$ and α_0 can be computed as

$$\alpha_2 = -tr(AD_1) = -tr(A) = 1$$

and

$$D_2 = AD_1 + \alpha_2 I = A + I = \begin{pmatrix} 3 & 2 & -1 \\ -1 & 1 & 4 \\ 3 & -1 & -2 \end{pmatrix}.$$

Also

$$AD_2 = \begin{pmatrix} 1 & 7 & 8 \\ 9 & -6 & -7 \\ 1 & 8 & -1 \end{pmatrix},$$

so

$$\alpha_1 = -\frac{1}{2} tr(AD_2) = -\frac{1}{2}(-6) = 3.$$

Similarly, we have

$$D_3 = AD_2 + \alpha_1 I = AD_2 + 3I = \begin{pmatrix} 4 & 7 & 8 \\ 9 & -3 & -7 \\ 1 & 8 & 2 \end{pmatrix}$$

and

$$AD_3 = \begin{pmatrix} 25 & 0 & 0 \\ 0 & 25 & 0 \\ 0 & 0 & 25 \end{pmatrix},$$

which gives

$$\alpha_0 = -\frac{1}{3} tr(AD_3) = -\frac{1}{3}(25) = -25,$$

which shows that the given matrix is nonsingular. Hence, the characteristic equation is

$$|A - \lambda I| = \lambda^3 + \lambda^2 + 3\lambda - 25 = 0.$$

Thus, the determinant of A is

$$|A| = (-1)^3(-25) = 25,$$

and the adjoint of A is

$$adj(A) = (-1)^4(D_3) = D_3 = \begin{pmatrix} 4 & 7 & 8 \\ 9 & -3 & -7 \\ 1 & 8 & 2 \end{pmatrix}.$$

Finally, the inverse of A is

$$A^{-1} = -\frac{1}{\alpha_0} D_3 = \frac{1}{25} \begin{pmatrix} 4 & 7 & 8 \\ 9 & -3 & -7 \\ 1 & 8 & 2 \end{pmatrix}.$$

●

13. All the eigenvalues of a Hermitian matrix are real.

   To prove this, consider (3.1), which is

$$A\mathbf{x} = \lambda\mathbf{x},$$

   and it implies that

$$\mathbf{x}^* A \mathbf{x} = \lambda \mathbf{x}^* \mathbf{x}.$$

   Since A is Hermitian, i.e., $A = A^*$ and $\mathbf{x}^* A \mathbf{x}$ is a scalar,

$$\mathbf{x}^* A \mathbf{x} = \mathbf{x}^* A^* \mathbf{x} = (\mathbf{x}^* A \mathbf{x})^* = \overline{\mathbf{x} A \mathbf{x}}.$$

   Thus, the scalar is equal to its own conjugate and hence, real. Therefore, λ is real.

Example 3.26 *Consider the Hermitian matrix*

$$A = \begin{pmatrix} 2 & -i & 0 \\ i & 2 & 0 \\ 0 & 0 & 3 \end{pmatrix},$$

which has a characteristic equation

$$\lambda^3 - 7\lambda^2 + 15\lambda - 9 = 0,$$

and it gives the real eigenvalues 1, 3, and 3 for the given matrix A.●

14. A matrix that is unitarily similar to a Hermitian matrix is itself Hermitian.

    To prove this, assume that $B^* = B$ and $A = QBQ^{-1}$, where $Q^{-1} = Q^*$. Then

    $$\begin{aligned} A^* &= (QBQ^{-1})^* \\ &= (QBQ^*)^* \\ &= Q^{**}B^*Q^* \\ &= QBQ^{-1} = A. \end{aligned}$$

    This shows that matrix A is Hermitian.

15. If A is a Hermitian matrix with distinct eigenvalues, then it is unitary similarly to a diagonal matrix, i.e., there exists a unitary matrix μ such that

    $$\mu^T A \mu = D,$$

    a diagonal matrix.

Example 3.27 *Consider the matrix*

$$A = \begin{pmatrix} 11 & 2 & 8 \\ 2 & 2 & -10 \\ 8 & -10 & 5 \end{pmatrix},$$

which has the characteristic equation

$$\lambda^3 - 18\lambda^2 - 81\lambda + 1458 = 0.$$

It can be easily factorized to give

$$(\lambda + 9)(\lambda - 9)(\lambda - 18) = 0,$$

and the eigenvalues of A are, -9, 9, and 18. The eigenvectors corresponding to these eigenvalues are

$$\mathbf{x}_1 = [-1, 2, 2]^T, \quad \mathbf{x}_2 = [2, 2, -1]^T, \quad \mathbf{x}_3 = [2, -1, 2]^T,$$

and they form an orthogonal set. Note that the vectors

$$\mathbf{u}_1 = \frac{\mathbf{x}_1}{\|\mathbf{x}_1\|_2} = \tfrac{1}{3}[-1, 2, 2]^T$$

$$\mathbf{u}_2 = \frac{\mathbf{x}_2}{\|\mathbf{x}_2\|_2} = \tfrac{1}{3}[2, 2, -1]^T$$

$$\mathbf{u}_3 = \frac{\mathbf{x}_3}{\|\mathbf{x}_3\|_2} = \tfrac{1}{3}[2, -1, 2]^T$$

form an orthonormal set, since they inherit orthogonality from $\mathbf{x}_1, \mathbf{x}_2$, and $\mathbf{x}_3$, and in addition

$$\|\mathbf{u}_1\|_2 = \|\mathbf{u}_2\|_2 = \|\mathbf{u}_3\|_2 = 1.$$

Thus, the unitary matrix μ is given by

$$\mu = \frac{1}{3} \begin{pmatrix} -1 & 2 & 2 \\ 2 & 2 & -1 \\ 2 & -1 & 2 \end{pmatrix}$$

and

$$\mu^T A \mu = \frac{1}{9} \begin{pmatrix} -1 & 2 & 2 \\ 2 & 2 & -1 \\ 2 & -1 & 2 \end{pmatrix} \begin{pmatrix} 11 & 2 & 8 \\ 2 & 2 & -10 \\ 8 & -10 & 5 \end{pmatrix} \begin{pmatrix} -1 & 2 & 2 \\ 2 & 2 & -1 \\ 2 & -1 & 2 \end{pmatrix},$$

which implies that

$$\mu^T A \mu = \begin{pmatrix} -9 & 0 & 0 \\ 0 & 9 & 0 \\ 0 & 0 & 18 \end{pmatrix} = D.$$

●

16. A matrix Q is *unitary*, if and only if its conjugate transpose is its inverse:

$$Q^* = Q^{-1}.$$

Note that a real matrix Q is unitary, if and only if $Q^T = Q^{-1}$.
For any square matrix A, there is a unitary matrix μ such that

$$\mu^{-1}A\mu = T,$$

an upper-triangular matrix whose diagonal entries consist of the eigenvalues of A. This is a well-known lemma called *Shur's lemma*.

Example 3.28 *Consider the matrix*

$$A = \begin{pmatrix} 2 & -1 \\ 1 & 0 \end{pmatrix},$$

which has an eigenvalue 1 of multiplicity 2. The eigenvector corresponding to this eigenvalue is $[1,1]^T$. Thus, the first column of a unitary matrix μ is $[\frac{1}{\sqrt{2}}, \frac{1}{\sqrt{2}}]^T$, and the other column is orthogonal to it, i.e., $[\frac{1}{\sqrt{2}}, -\frac{1}{\sqrt{2}}]^T$. So

$$\mu^{-1}A\mu = \begin{pmatrix} \dfrac{1}{\sqrt{2}} & \dfrac{1}{\sqrt{2}} \\ \dfrac{1}{\sqrt{2}} & -\dfrac{1}{\sqrt{2}} \end{pmatrix} \begin{pmatrix} 2 & -1 \\ 1 & 0 \end{pmatrix} \begin{pmatrix} \dfrac{1}{\sqrt{2}} & \dfrac{1}{\sqrt{2}} \\ \dfrac{1}{\sqrt{2}} & -\dfrac{1}{\sqrt{2}} \end{pmatrix},$$

which gives

$$\mu^{-1}A\mu = \begin{pmatrix} 1 & 2 \\ 0 & 1 \end{pmatrix} = T.$$

17. A matrix that is unitarily similar to a normal matrix is itself normal. Assume that $BB^* = B^*B$ and $A = QBQ^{-1}$, where $Q^{-1} = Q^*$. Then

$$\begin{aligned} A^* &= (QBQ^{-1})^* \\ &= (QBQ^*)^* \\ &= Q^{**}B^*Q^* \\ &= QB^*Q^{-1}. \end{aligned}$$

So

$$
\begin{aligned}
AA^* &= (QBQ^{-1})(QB^*Q^{-1}) \\
&= (QBB^*Q^{-1}) \\
&= (QB^*BQ^{-1}) \\
&= (QB^*Q^{-1}QBQ^{-1}) \\
&= (QB^*Q^{-1})(QBQ^{-1}) = A^*A.
\end{aligned}
$$

This shows that the matrix A is normal.

18. The value of the exponential of a matrix A can be calculated from

$$
e^A = Q(exp\Lambda)Q^{-1},
$$

where $exp\Lambda$ is a diagonal matrix whose elements are the exponential of successive eigenvalues, and Q is a matrix of the eigenvectors of A.

Example 3.29 *Consider the matrix*

$$
A = \begin{pmatrix} 0.1 & 0.1 \\ 0.0 & 0.2 \end{pmatrix}.
$$

In order to find the value of the exponential of the given matrix A, we have to find the eigenvalues of A and also the eigenvectors of A. The eigenvalues of A are 0.1 and 0.2, and the corresponding eigenvectors are $[1,0]^T$ and $[1,1]^T$. Then the matrix Q is

$$
Q = \begin{pmatrix} 1 & 1 \\ 0 & 1 \end{pmatrix}.
$$

Its inverse can be found as

$$
Q^{-1} = \begin{pmatrix} 1 & -1 \\ 0 & 1 \end{pmatrix}.
$$

Thus,

$$
e^A = Q(exp\Lambda)Q^{-1} = \begin{pmatrix} 1 & 1 \\ 0 & 1 \end{pmatrix} \begin{pmatrix} e^{0.1} & 0 \\ 0 & e^{0.2} \end{pmatrix} \begin{pmatrix} 1 & -1 \\ 0 & 1 \end{pmatrix},
$$

which gives

$$
e^A = \begin{pmatrix} 1.105 & 0.116 \\ 0.000 & 1.221 \end{pmatrix}.
$$

In the following section we discuss some very important results concerning eigenvalue problems. The proofs of all the results are beyond the scope of this text and will be omitted. However, they are very easily understood and can be used. We shall discuss these results by using the different matrices.

3.5   Some Results of Eigenvalues Problems

1. If A is a Hermitian matrix, then

$$\|A\|_2 = \sqrt{\rho(A^H A)} = \rho(A).$$

Example 3.30 *Consider the Hermitian matrix*

$$A = \begin{pmatrix} 0 & \bar{\alpha} \\ \alpha & 0 \end{pmatrix}.$$

Then the characteristic equation of A is

$$\lambda^2 - \alpha\bar{\alpha} = \lambda^2 - |\alpha|^2 = 0,$$

which implies that

$$\lambda = |\alpha| = \rho(A).$$

Also,

$$A^H A = \begin{pmatrix} 0 & \bar{\alpha} \\ \alpha & 0 \end{pmatrix} \begin{pmatrix} 0 & \bar{\alpha} \\ \alpha & 0 \end{pmatrix} = \begin{pmatrix} \alpha\bar{\alpha} & 0 \\ 0 & \alpha\bar{\alpha} \end{pmatrix},$$

and the characteristic equation of $A^H A$ is

$$(\alpha\bar{\alpha} - \lambda)^2 = 0,$$

which implies that

$$\|A\|_2 = [\rho(A^H A)]^{1/2} = |\alpha| = \rho(A).$$

2. For an arbitrary nonsingular matrix A

$$\frac{1}{\rho(A^{-1})} \geq \min_i \left(|a_{ii}| - \sum_{j=1}^{n} |a_{ij}| \right), \quad j \neq i. \qquad (3.35)$$

Example 3.31 *Consider the matrix*

$$A = \begin{pmatrix} 0 & \sqrt{3} \\ \sqrt{3} & 2 \end{pmatrix}.$$

To satisfy the above relation (3.35), first, we compute the inverse of the given matrix A, which is

$$A^{-1} = \begin{pmatrix} \dfrac{2}{3} & \dfrac{\sqrt{3}}{3} \\ \dfrac{\sqrt{3}}{3} & 0 \end{pmatrix}.$$

Now to find the eigenvalues of the above inverse matrix, we solve the characteristic equation of the form as follows:

$$\det(A^{-1} - \lambda I) = \begin{vmatrix} \dfrac{2}{3} - \lambda & \dfrac{\sqrt{3}}{3} \\ \dfrac{\sqrt{3}}{3} & -\lambda \end{vmatrix} = 3\lambda^2 - 2\lambda - 1 = 0,$$

which gives the eigenvalues $-1, \frac{1}{3}$ of A^{-1}. Hence, the spectral radius of the matrix A^{-1} is

$$\rho(A^{-1}) = \max\{|-1|, |\tfrac{1}{3}|\} = 1.$$

Thus,
$$1 > \min\{-1.7321, -0.2680\},$$

which satisfies the relation (3.35). ●

3. Let A be a symmetric matrix with $\|.\| = \|.\|_2$, then

$$\text{cond}A = \frac{\text{largest } |\lambda_i|}{\text{smallest } |\lambda_i|}.$$

It is a well-known theorem, called the *spectral radius theorem*, and it shows that for a symmetric matrix A, ill-conditioning corresponds to A having eigenvalues of both large and small magnitude. It is most commonly used to define the condition number of a matrix. As we discussed in Chapter 2, a matrix is ill-conditioned if its condition number is large. Strictly speaking, a matrix has many condition numbers, and the word "large" is not itself well defined in this context. To have an idea of what "large" means is to deal with the Hilbert matrix. For example, in dealing with a 3×3 Hilbert matrix, we have

```
>> A = hilb(3);
A =
      1.0000   0.5000   0.3333
      0.5000   0.3333   0.2500
      0.3333   0.2500   0.2000
```

and one can find the condition number of this Hilbert matrix as

```
>> cond(A)
ans =
   524.0568
```

By adapting this result, we can easily confirm that the condition numbers of Hilbert matrices increase rapidly as the sizes of the matrices increase. Large Hilbert matrices are therefore considered to be extremely ill-conditioned.

Example 3.32 *Find the conditioning of the matrix*

$$A = \begin{pmatrix} 0 & \sqrt{3} \\ \sqrt{3} & 2 \end{pmatrix}.$$

Solution. *The following*

$$\det(A - \lambda I) = \begin{vmatrix} -\lambda & \sqrt{3} \\ \sqrt{3} & 2 - \lambda \end{vmatrix} = 0$$

gives a characteristic equation

$$\lambda^2 - 2\lambda - 3 = 0.$$

Solving the above equation gives the solutions 3 and −1, which are called the eigenvalues of matrix A. Thus, the largest eigenvalue of A is 3, and the smallest one is 1. Hence, the condition number of matrix A is

$$cond\ A = \frac{3}{1} = 3.$$

Since 3 is of the order of magnitude of 1, A is well-conditioned. •

4. Let A be a nonsymmetric matrix A, with $\|.\| = \|.\|_2$, then

$$cond\ A = \left(\frac{\text{largest eigenvalue } |\lambda_i| \text{ of } A^T A}{\text{smallest eigenvalue} |\lambda_i| \text{ of } A^T A} \right)^{1/2}.$$

Example 3.33 *Find the conditioning of the matrix*

$$A = \begin{pmatrix} 4 & 5 \\ 3 & 2 \end{pmatrix}.$$

Solution. *Since*

$$\det(A^T A - \lambda I) = \begin{vmatrix} 25 - \lambda & 26 \\ 26 & 29 - \lambda \end{vmatrix} = 0,$$

solving the above equation gives

$$\lambda^2 - 54\lambda + 49 = 0.$$

The solutions 53.08 and 0.92 of the above equation are called the eigenvalues of the matrix $A^T A$. Thus, the conditioning of the given matrix can be obtained as

$$cond A = \left(\frac{53.08}{0.92} \right)^{1/2} \approx 7.6,$$

which shows that the given matrix A is not ill-conditioned. •

3.6 Applications of Eigenvalue Problems

Here, we will deal with two important applications of eigenvalues and eigenvectors. They are systems of differential equations and difference equations. The techniques used in these applications are important in science and engineering. One should master them and be able to use them whenever the need arises. We first introduce the idea of a system of differential equations.

3.6.1 System of Differential Equations

Solving a variety of problems, particularly in science and engineering, comes down to solving a differential equation or a system of differential equations. Linear algebra is helpful in the formulation and solution of differential equations. Here, we provide only a brief survey of the approach.

The general problem is to find differentiable functions $f_1(t), f_2(t), \ldots, f_n(t)$ that satisfy a system of equations of the form

$$
\begin{aligned}
f_1'(t) &= a_{11}f_1(t) + a_{12}f_2(t) + \cdots + a_{1n}f_n(t) \\
f_2'(t) &= a_{21}f_1(t) + a_{22}f_2(t) + \cdots + a_{2n}f_n(t) \\
&\ \ \vdots \qquad\quad \vdots \qquad\quad \vdots \qquad\quad \vdots \qquad\quad \vdots \\
f_n'(t) &= a_{n1}f_1(t) + a_{n2}f_2(t) + \cdots + a_{nn}f_n(t),
\end{aligned}
\tag{3.36}
$$

where the a_{ij} are known constants. This is called a *linear system of differential equations*. To write (3.36) in matrix form, we have

$$
\mathbf{f}(t) = \begin{pmatrix} f_1(t) \\ f_2(t) \\ \vdots \\ f_n(t) \end{pmatrix}, \quad
\mathbf{f}'(t) = \begin{pmatrix} f_1'(t) \\ f_2'(t) \\ \vdots \\ f_n'(t) \end{pmatrix}, \quad
A = \begin{pmatrix} a_{11} & a_{12} & \cdots & a_{1n} \\ a_{21} & a_{22} & \cdots & a_{2n} \\ \vdots & \vdots & \vdots & \vdots \\ a_{n1} & a_{n2} & \cdots & a_{nn} \end{pmatrix}.
$$

Then the system (3.36) can be written as

$$
\mathbf{f}'(t) = A\mathbf{f}(t).
\tag{3.37}
$$

With this notation, an $n-$ vector function

$$\mathbf{f}(t) = \begin{pmatrix} f_1(t) \\ f_2(t) \\ \vdots \\ f_n(t) \end{pmatrix}$$

satisfying (3.37) is called a *solution* to the given system. It can be shown that the set of all solutions to the linear system of differential equations (3.36) is a subspace of the vector space of differentiable real-valued $n-$ vector functions. One can also easily verify that if $\mathbf{f}^{(1)}(t), \mathbf{f}^{(2)}(t), \ldots, \mathbf{f}^{(n)}(t)$ are all solutions to (3.37), then

$$\mathbf{f}(t) = c_1\mathbf{f}^{(1)}(t) + c_2\mathbf{f}^{(2)}(t) + \cdots + c_n\mathbf{f}^{(n)}(t) \qquad (3.38)$$

is also a solution to (3.37).

A set of vector functions $\{\mathbf{f}^{(1)}(t), \mathbf{f}^{(2)}(t), \ldots, \mathbf{f}^{(n)}(t)\}$ is said to be a *fundamental system* for (3.36) if every solution to (3.36) can be written in the form (3.38). In this case, the right side of (3.38), where $c_1, c_2, \ldots, c_n$ are arbitrary constants, is said to be the general solution to (3.37).

If the general solution to (3.38) is known, then the *initial-value problem* can be solved by setting $t = 0$ in (3.38) and determining the constants $c_1, c_2, \ldots, c_n$ so that

$$\mathbf{f}_0 = \mathbf{f}(0) = c_1\mathbf{f}^{(1)}(0) + c_2\mathbf{f}^{(2)}(0) + \cdots + c_n\mathbf{f}^{(n)}(0), \qquad (3.39)$$

where $\mathbf{f}_0$ is a given vector, called an *initial condition*. It is easily seen that this is actually an $n \times n$ linear system with unknowns $c_1, c_2, \ldots, c_n$. This linear system can also be written as

$$B\mathbf{c} = \mathbf{f}_0, \qquad (3.40)$$

where

$$\mathbf{c} = \begin{pmatrix} c_1 \\ c_2 \\ \vdots \\ c_n \end{pmatrix},$$

and B is the $n \times n$ matrix whose columns are $\mathbf{f}^{(1)}(0), \mathbf{f}^{(2)}(0), \ldots, \mathbf{f}^{(n)}(0)$, respectively.

Note that, if $\mathbf{f}^{(1)}(t), \mathbf{f}^{(2)}(t), \ldots, \mathbf{f}^{(n)}(t)$ form a fundamental system for (3.36), then B is nonsingular, so (3.40) always has a unique solution.

Example 3.34 *The simplest system of the form (3.36) is the single equation*

$$\frac{df}{dt} = \alpha f, \tag{3.41}$$

where α is a constant. The general solution to (3.41) is

$$f(t) = ce^{\alpha t}. \tag{3.42}$$

To get the particular solution to (3.41), we have to solve the initial-value problem

$$\frac{df}{dt} = \alpha f, \quad f(0) = f_0$$

and set $t = 0$ in (3.42) and get $c = f_0$. Thus, the solution to the initial-value problem is

$$f(t) = f_0 e^{\alpha t}.$$

•

The system (3.37) is said to be *diagonal* if the matrix A is diagonal. The system (3.36) can be rewritten as

$$
\begin{aligned}
f_1'(t) &= a_{11} f_1(t) \\
f_2'(t) &= a_{22} f_2(t) \\
&\vdots \\
f_n'(t) &= a_{nn} f_n(t).
\end{aligned}
\tag{3.43}
$$

The solution of the system (3.43) can be found easily as

$$
\begin{aligned}
f_1(t) &= c_1 e^{a_{11} t} \\
f_2(t) &= c_2 e^{a_{22} t} \\
&\vdots \qquad \vdots \\
f_n(t) &= c_n e^{a_{nn} t},
\end{aligned}
\tag{3.44}
$$

where $c_1, c_2, \ldots, c_n$ are arbitrary constants. Writing (3.44) in vector form yields

$$\mathbf{f}(t) = \begin{pmatrix} c_1 e^{a_{11}t} \\ c_2 e^{a_{22}t} \\ \vdots \\ c_n e^{a_{nn}t} \end{pmatrix} = c_1 \begin{pmatrix} 1 \\ 0 \\ 0 \\ \vdots \\ 0 \end{pmatrix} e^{a_{11}t} + c_2 \begin{pmatrix} 0 \\ 1 \\ 0 \\ \vdots \\ 0 \end{pmatrix} e^{a_{22}t} + \cdots + c_n \begin{pmatrix} 0 \\ 0 \\ 0 \\ \vdots \\ 1 \end{pmatrix} e^{a_{nn}t}.$$

This implies that the vector functions

$$\mathbf{f}^{(1)}(t) = \begin{pmatrix} 1 \\ 0 \\ 0 \\ \vdots \\ 0 \end{pmatrix} e^{a_{11}t}, \quad \mathbf{f}^{(2)}(t) = \begin{pmatrix} 0 \\ 1 \\ 0 \\ \vdots \\ 0 \end{pmatrix} e^{a_{22}t}, \ldots, \mathbf{f}^{(n)}(t) = \begin{pmatrix} 0 \\ 0 \\ 0 \\ \vdots \\ 1 \end{pmatrix} e^{a_{nn}t}$$

form a fundamental system for the diagonal system (3.43).

Example 3.35 *Find the general solution to the diagonal system*

$$\begin{pmatrix} f_1'(t) \\ f_2'(t) \\ f_3'(t) \end{pmatrix} = \begin{pmatrix} 2 & 0 & 0 \\ 0 & 5 & 0 \\ 0 & 0 & 1 \end{pmatrix} \begin{pmatrix} f_1 \\ f_2 \\ f_3 \end{pmatrix}.$$

Solution. *The given system can be written as three equations of the form*

$$\begin{array}{rcl} f_1'(t) &=& 2f_1(t) \\ f_2'(t) &=& 5f_2(t) \\ f_3'(t) &=& f_3(t). \end{array}$$

Solving these equations, we get

$$f_1(t) = c_1 e^{2t}, \quad f_2(t) = c_2 e^{5t}, \quad f_3(t) = c_3 e^{t},$$

where $c_1, c_2, \ldots, c_n$ are arbitrary constants. Thus,

$$\mathbf{f}(t) = \begin{pmatrix} c_1 e^{2t} \\ c_2 e^{5t} \\ c_3 e^{t} \end{pmatrix} = c_1 \begin{pmatrix} 1 \\ 0 \\ 0 \end{pmatrix} e^{2t} + c_2 \begin{pmatrix} 0 \\ 1 \\ 0 \end{pmatrix} e^{5t} + c_3 \begin{pmatrix} 0 \\ 0 \\ 1 \end{pmatrix} e^{t}$$

is the general solution to the given system of differential equations, and the functions

$$\mathbf{f}^{(1)}(t) = \begin{pmatrix} 1 \\ 0 \\ 0 \end{pmatrix} e^{2t}, \quad \mathbf{f}^{(2)}(t) = \begin{pmatrix} 0 \\ 1 \\ 0 \end{pmatrix} e^{5t}, \quad \mathbf{f}^{(3)}(t) = \begin{pmatrix} 0 \\ 0 \\ 1 \end{pmatrix} e^{t}$$

form a fundamental system for the given diagonal system. ●

If the system (3.37) is not diagonal, then there is an extension of the method discussed in the preceding example that yields the general solution in the case where A is *diagonalizable*. Suppose that A is diagonalizable and Q is a nonsingular matrix such that

$$D = Q^{-1}AQ, \tag{3.45}$$

where D is diagonal. Then by multiplying Q^{-1} to the system (3.37), we get

$$Q^{-1}\mathbf{f}'(t) = Q^{-1}A\mathbf{f}(t)$$

or

$$Q^{-1}\mathbf{f}'(t) = Q^{-1}AQQ^{-1}\mathbf{f}(t) = (Q^{-1}AQ)(Q^{-1}\mathbf{f}(t)), \tag{3.46}$$

(since $QQ^{-1} = \mathbf{I}_n$). Let

$$\mathbf{g}(t) = Q^{-1}\mathbf{f}(t), \tag{3.47}$$

and by taking its derivative, we have

$$\mathbf{g}'(t) = Q^{-1}\mathbf{f}'(t). \tag{3.48}$$

Since Q^{-1} is a constant matrix, using (3.48) and (3.45), we can write (3.46) as

$$\mathbf{g}'(t) = D\mathbf{g}(t). \tag{3.49}$$

Then the system (3.49) is a diagonal system and can be solved by the method just discussed. Since the matrix D is a diagonal matrix and all its diagonal elements are also called the eigenvalues $\lambda_1, \lambda_2, \ldots, \lambda_n$ of A, it can be written as

$$D = \begin{pmatrix} \lambda_1 & 0 & \cdots & 0 \\ 0 & \lambda_2 & \cdots & 0 \\ \vdots & \vdots & \vdots & \vdots \\ 0 & 0 & \cdots & \lambda_n \end{pmatrix}.$$

The columns of Q are linearly independent eigenvectors of A associated, respectively, with $\lambda_1, \lambda_2, \ldots, \lambda_n$. Thus, the general solution to (3.37) is

$$\mathbf{g}(t) = \begin{pmatrix} c_1 e^{\lambda_1 t} \\ c_2 e^{\lambda_2 t} \\ \vdots \\ c_n e^{\lambda_n t} \end{pmatrix} = c_1 \mathbf{g}^{(1)}(t) + c_2 \mathbf{g}^{(2)}(t) + c_3 \mathbf{g}^{(3)}(t),$$

where

$$\mathbf{g}^{(1)}(t) = \begin{pmatrix} 1 \\ 0 \\ \vdots \\ 0 \end{pmatrix} e^{\lambda_1 t}, \quad \mathbf{g}^{(2)}(t) = \begin{pmatrix} 0 \\ 1 \\ \vdots \\ 0 \end{pmatrix} e^{\lambda_2 t}, \quad \ldots, \mathbf{g}^{(n)}(t) = \begin{pmatrix} 0 \\ 0 \\ \vdots \\ 1 \end{pmatrix} e^{\lambda_n t},$$

$$(3.50)$$

and $c_1, c_2, \ldots, c_n$ are arbitrary constants. The system (3.47) can also be written as

$$\mathbf{f}(t) = Q\mathbf{g}(t). \tag{3.51}$$

So the general solution to the given system (3.37) is

$$\mathbf{f}(t) = Q\mathbf{g}(t) = c_1 Q\mathbf{g}^{(1)}(t) + c_2 Q\mathbf{g}^{(2)}(t) + \cdots + c_n Q\mathbf{g}^{(n)}(t). \tag{3.52}$$

However, since the constant vectors in (3.50) are the columns of the identity matrix and $Q\mathbf{I}_n = Q$, (3.52) can be rewritten as

$$\mathbf{f}(t) = c_1 \mathbf{q}_1 e^{\lambda_1 t} + c_2 \mathbf{q}_2 e^{\lambda_2 t} + \cdots + c_n \mathbf{q}_n e^{\lambda_n t}, \tag{3.53}$$

where $\mathbf{q}_1, \mathbf{q}_2, \ldots, \mathbf{q}_n$ are the columns of Q, and are therefore, the eigenvectors of A associated with the eigenvalues $\lambda_1, \lambda_2, \ldots, \lambda_n$, respectively.

Theorem 3.19 *Consider a linear system*

$$\mathbf{f}'(t) = A\mathbf{f}(t)$$

of differential equations, where A is an $n \times n$ diagonalizable matrix. Let $Q^{-1}AQ$ be diagonal, where Q is given in terms of its columns

$$Q = [\mathbf{q}_1, \mathbf{q}_2, \ldots, \mathbf{q}_n],$$

and $\mathbf{q}_1, \mathbf{q}_2, \ldots, \mathbf{q}_n$ are independent eigenvectors of A. If $\mathbf{q}_i$ corresponds to the eigenvalues λ_i for each i, then

$$\{\mathbf{q}_1 e^{\lambda_1 t}, \mathbf{q}_2 e^{\lambda_2 t}, \ldots, \mathbf{q}_n e^{\lambda_n t}\}$$

is a basis for the space of solutions of $\mathbf{f}'(t) = A\mathbf{f}(t)$. ●

Example 3.36 *Find the general solution to the system*

$$\begin{array}{rcrcrcr} f_1' &=& 3f_1 &-& f_2 &-& f_3 \\ f_2' &=& -12f_1 & & &+& 5f_3 \\ f_3' &=& 4f_1 &-& 2f_2 &-& f_3. \end{array}$$

Then find the solution to the initial-value problem determined by the given initial conditions

$$f_1(0) = 1, \quad f_2(0) = 4, \quad f_3(0) = 3.$$

Solution. *Writing the given system in (3.37) form, we obtain*

$$\mathbf{f}' = \begin{pmatrix} f_1' \\ f_2' \\ f_3' \end{pmatrix} = \begin{pmatrix} 3 & -1 & -1 \\ -12 & 0 & 5 \\ 4 & -2 & -1 \end{pmatrix} \begin{pmatrix} f_1 \\ f_2 \\ f_3 \end{pmatrix} = A\mathbf{f}.$$

The characteristic polynomial of A is

$$0 = f(\lambda) = det(A - \lambda \mathbf{I}) = -\lambda^3 + 2\lambda^2 + \lambda - 2$$

or

$$0 = f(\lambda) = -(\lambda + 1)(\lambda - 1)(\lambda - 2).$$

So the eigenvalues of A are $\lambda_1 = -1, \lambda_2 = 1$, and $\lambda_3 = 2$, and the associated eigenvectors are

$$\begin{pmatrix} \frac{1}{2} \\ 1 \\ 1 \end{pmatrix}, \quad \begin{pmatrix} 3 \\ -1 \\ 7 \end{pmatrix}, \quad \begin{pmatrix} 1 \\ -1 \\ 2 \end{pmatrix},$$

respectively. The general solution is then given by

$$\mathbf{f} = c_1 \begin{pmatrix} \frac{1}{2} \\ 1 \\ 1 \end{pmatrix} e^{-t} + c_2 \begin{pmatrix} 3 \\ -1 \\ 7 \end{pmatrix} e^t + c_3 \begin{pmatrix} 1 \\ -1 \\ 2 \end{pmatrix} e^{2t},$$

where $c_1, c_2, \ldots, c_n$ are arbitrary constants.

Now we write our general solution in the form (3.51) as

$$\mathbf{f}(t) = \begin{pmatrix} f_1(t) \\ f_2(t) \\ f_3(t) \end{pmatrix} = \begin{pmatrix} \frac{1}{2} & 3 & 1 \\ 1 & -1 & -1 \\ 1 & 7 & 2 \end{pmatrix} \begin{pmatrix} c_1 e^{-t} \\ c_2 e^t \\ c_3 e^{2t} \end{pmatrix}.$$

Now taking $t = 0$, we obtain

$$\mathbf{f}(0) = \begin{pmatrix} 1 \\ 4 \\ 2 \end{pmatrix} = \begin{pmatrix} \frac{1}{2} & 3 & 1 \\ 1 & -1 & -1 \\ 1 & 7 & 2 \end{pmatrix} \begin{pmatrix} c_1 e^0 \\ c_2 e^0 \\ c_3 e^0 \end{pmatrix}$$

or

$$\begin{pmatrix} \frac{1}{2} & 3 & 1 \\ 1 & -1 & -1 \\ 1 & 7 & 2 \end{pmatrix} \begin{pmatrix} c_1 \\ c_2 \\ c_3 \end{pmatrix} = \begin{pmatrix} 1 \\ 4 \\ 3 \end{pmatrix}.$$

Solving this system for c_1, c_2, and c_3 using the Gauss elimination method, we obtain

$$c_1 = 2, \quad c_2 = 1, \quad c_3 = -3.$$

Therefore, the solution to the initial-value problem is

$$\mathbf{f}(t) = 2 \begin{pmatrix} \frac{1}{2} \\ 1 \\ 1 \end{pmatrix} e^{-t} + \begin{pmatrix} 3 \\ -1 \\ 7 \end{pmatrix} e^t - 3 \begin{pmatrix} 1 \\ -1 \\ 2 \end{pmatrix} e^{2t}.$$

3.6.2 Difference Equations

It often happens that a problem can be solved by finding a sequence of numbers $a_0, a_1, a_2, \ldots$, where the first few are known, and subsequent numbers are given in terms of earlier values.

Let $a_0, a_1, a_2, \ldots$ be a sequence of real numbers. Such a sequence may be defined by giving its *nth* term. For example, suppose

$$a_n = n^3 + 2. \tag{3.54}$$

Letting $n = 0, 1, 2, \ldots$, we get the terms of the sequence (3.54) as

$$a_0 = 2, a_1 = 3, a_2 = 10, \ldots.$$

Furthermore, any specific term of the sequence (3.54) can be found. For example, if we want a_{10}, then we let $n = 10$ in (3.54) and get

$$a_{10} = (10)^3 + 2 = 1000 + 2 = 1002.$$

When sequences arise in applications, they are often initially defined by a relationship between consecutive terms, with some initial terms known, rather than defined, by the *nth* term. For example, a sequence might be defined, by the relationship

$$a_n = 2a_{n-1} + 8a_{n-2}, \quad n = 2, 3, 4, \ldots, \tag{3.55}$$

where $a_0 = 1$ and $a_1 = 2$. Such an equation is called a *difference equation* (or *recurrence relation*), and the given terms of the sequence are called *initial conditions*. Further terms of the sequence can be found from the difference equation and initial conditions. For example, letting $n = 2, 3$, and 4, we obtain

$$
\begin{aligned}
a_2 &= 2a_1 + 8a_0 = 2(2) + 8(1) = 12 \\
a_3 &= 2a_2 + 8a_1 = 2(12) + 8(2) = 40 \\
a_4 &= 2a_3 + 8a_2 = 2(40) + 8(12) = 176.
\end{aligned}
$$

Thus, the sequence is $1, 2, 12, 40, 176, \ldots$.

However, if we want to find a specific term such as the *20th* term of the sequence (3.55), this method of using the difference equation to first find all the preceding terms is impractical. Here we need an expression for the *nth* term of the sequence. The expression for the *nth* term is called the *solution* to the difference equation. Now we discuss how the tools of linear algebra can be used to solve this problem.

Consider the difference equation of the form

$$a_n = p a_{n-1} + q a_{n-2}, \quad n = 2, 3, 4, \ldots, \tag{3.56}$$

where p and q are fixed numbers and a_0 and a_1 are the given initial conditions. This equation is called a *linear difference equation* (because each a_i appears to the first power) and is of *order* 2 (because a_n is expressed in terms of two preceding terms a_{n-1} and a_{n-2}). To solve (3.56), let us introduce a relation $b_n = a_{n-1}$. So we get the system

$$\begin{aligned} a_n &= p a_{n-1} + q b_{n-1} \\ b_n &= a_{n-1}, \quad n = 2, 3, 4, \ldots. \end{aligned} \tag{3.57}$$

To write (3.57) in matrix form, we obtain

$$\begin{pmatrix} a_n \\ b_n \end{pmatrix} = \begin{pmatrix} p & q \\ 1 & 0 \end{pmatrix} \begin{pmatrix} a_{n-1} \\ b_{n-1} \end{pmatrix}, \quad n = 2, 3, 4, \ldots. \tag{3.58}$$

Let

$$V_n = \begin{pmatrix} a_n \\ b_n \end{pmatrix} \quad \text{and} \quad A = \begin{pmatrix} p & q \\ 1 & 0 \end{pmatrix},$$

then

$$V_n = A V_{n-1}, \quad n = 2, 3, 4, \ldots. \tag{3.59}$$

Thus,

$$V_n = A V_{n-1} = A^2 V_{n-2} = A^3 V_{n-3} = \cdots = A^{n-1} V_1, \tag{3.60}$$

where

$$V_1 = \begin{pmatrix} a_1 \\ b_1 \end{pmatrix} = \begin{pmatrix} a_1 \\ a_0 \end{pmatrix}.$$

In most applications, a matrix A has distinct eigenvalues λ_1 and λ_2, and the corresponding linearly independent eigenvectors can be diagonalized using a similarity transformation. Let Q be a matrix whose columns are linearly independent eigenvectors of A and let

$$D = Q^{-1}AQ = \begin{pmatrix} \lambda_1 & 0 \\ 0 & \lambda_2 \end{pmatrix}. \tag{3.61}$$

Then

$$A^{n-1} = \left(QDQ^{-1}\right)^{n-1} = \left(QDQ^{-1}\right)\left(QDQ^{-1}\right)\cdots\left(QDQ^{-1}\right) = QD^{n-1}Q^{-1}$$

or

$$A^{n-1} = Q\begin{pmatrix} \lambda_1 & 0 \\ 0 & \lambda_2 \end{pmatrix}^{n-1}Q^{-1} = Q\begin{pmatrix} (\lambda_1)^{n-1} & 0 \\ 0 & (\lambda_2)^{n-1} \end{pmatrix}Q^{-1}.$$

This gives

$$V_n = Q\begin{pmatrix} (\lambda_1)^{n-1} & 0 \\ 0 & (\lambda_2)^{n-1} \end{pmatrix}Q^{-1}V_1. \tag{3.62}$$

Example 3.37 *Solve the difference equation*

$$a_n = 2a_{n-1} + 8a_{n-2}, \quad n = 2, 3, 4, \ldots$$

using initial conditions $a_0 = 1$ and $a_1 = 2$. Use the solution to find a_{20}.

Solution. *Construct the system*

$$\begin{aligned} a_n &= 2a_{n-1} + 8b_{n-1} \\ b_n &= a_{n-1}, \quad n = 2, 3, 4, \ldots. \end{aligned}$$

The matrix form of the system is

$$\begin{pmatrix} a_n \\ b_n \end{pmatrix} = \begin{pmatrix} 2 & 8 \\ 1 & 0 \end{pmatrix}\begin{pmatrix} a_{n-1} \\ b_{n-1} \end{pmatrix}, \quad n = 2, 3, 4, \ldots.$$

Since the matrix A is

$$A = \begin{pmatrix} 2 & 8 \\ 1 & 0 \end{pmatrix},$$

and its eigenvalues can be obtained by solving the characteristic polynomial

$$p(\lambda) = \lambda^2 - 2\lambda - 8 = (\lambda - 4)(\lambda + 2) = 0,$$

which gives the eigenvalues of A

$$\lambda_1 = 4 \quad and \quad \lambda_2 = -2,$$

one can easily find the eigenvectors of A corresponding to the eigenvalues λ_1 and λ_2 as

$$\mathbf{v}_1 = \begin{pmatrix} 2 \\ \dfrac{1}{2} \end{pmatrix} \quad and \quad \mathbf{v}_2 = \begin{pmatrix} -1 \\ \dfrac{1}{2} \end{pmatrix},$$

respectively. Consider

$$Q = \begin{pmatrix} 2 & -1 \\ \dfrac{1}{2} & \dfrac{1}{2} \end{pmatrix},$$

and its inverse can be obtained as

$$Q^{-1} = \begin{pmatrix} \dfrac{1}{3} & \dfrac{2}{3} \\ -\dfrac{1}{3} & \dfrac{4}{3} \end{pmatrix}.$$

Then

$$\begin{pmatrix} a_n \\ b_n \end{pmatrix} = Q \begin{pmatrix} (\lambda_1)^{n-1} & 0 \\ 0 & (\lambda_2)^{n-1} \end{pmatrix} Q^{-1} \begin{pmatrix} a_1 \\ b_1 \end{pmatrix},$$

which is

$$\begin{pmatrix} a_n \\ b_n \end{pmatrix} = \begin{pmatrix} 2 & -1 \\ \dfrac{1}{2} & \dfrac{1}{2} \end{pmatrix} \begin{pmatrix} (4)^{n-1} & 0 \\ 0 & (-2)^{n-1} \end{pmatrix} \begin{pmatrix} \dfrac{1}{3} & \dfrac{2}{3} \\ -\dfrac{1}{3} & \dfrac{4}{3} \end{pmatrix} \begin{pmatrix} 2 \\ 1 \end{pmatrix}$$

$(b_1 = a_0 = 1)$. *After simplifying, we obtain*

$$\begin{pmatrix} a_n \\ b_n \end{pmatrix} = \begin{pmatrix} \dfrac{2}{3}(4)^n + \dfrac{1}{3}(-2)^n \\[3mm] \dfrac{2}{3}(4)^{n-1} + \dfrac{1}{3}(-2)^{n-1} \end{pmatrix}.$$

Thus, the solution is

$$a_n = \frac{2}{3}(4)^n + \frac{1}{3}(-2)^n, \qquad n = 2, 3, 4, \ldots,$$

which gives $a_0 = 1$ and $a_1 = 2$ by taking $n = 0, 1$ and it agrees with the given initial conditions. Now taking $n = 20$, we get

$$a_{20} = \frac{2}{3}(4)^{20} + \frac{1}{3}(-2)^{20} = \frac{2}{3}(1099511627776) + \frac{1}{3}(1048576),$$

so

$$a_{20} = 733008101376. \qquad \bullet$$

3.7   Summary

In this chapter we discussed the approximation of eigenvalues and eigenvectors. We discussed similar, unitary, and diagonalizable matrices. The set of diagonalizable matrices includes matrices with n distinct eigenvalues and symmetric matrices. Matrices that are not diagonalizable are sometimes referred to as defective matrices.

We discussed the Cayley–Hamilton theorem for finding the power and inverse of a matrix. We also discussed the Sourian–Frame theorem, Bocher's theorem, and the Faddeev–Laverrier theorem for computing the coefficients of the characteristic polynomial $p(\lambda)$ of a matrix A. There are no restrictions on A. In theory, the eigenvalues of A can be obtained by factoring $p(\lambda)$ using polynomial-root-finding techniques. However, this approach is practical only for relatively small values of n. The chapter closed with two important applications.

3.8   Problems

1. Find the characteristic polynomial, eigenvalues, and eigenvectors of each matrix:

(a) $\begin{pmatrix} 3 & 2 & 1 \\ 2 & 1 & 1 \\ 1 & 1 & 1 \end{pmatrix}$,   (b) $\begin{pmatrix} -2 & 1 & 1 \\ -6 & 1 & 3 \\ -12 & -2 & 8 \end{pmatrix}$,   (c) $\begin{pmatrix} 2 & -1 & -1 \\ -1 & 2 & -1 \\ -1 & -1 & 2 \end{pmatrix}$,

(d) $\begin{pmatrix} 1 & 1 & 1 \\ 1 & 1 & 0 \\ 1 & 0 & 1 \end{pmatrix}$,   (e) $\begin{pmatrix} 3 & 2 & -2 \\ -3 & -1 & 3 \\ 1 & 2 & 0 \end{pmatrix}$,   (f) $\begin{pmatrix} 4 & 3 & 2 & 1 \\ 3 & 3 & 2 & 1 \\ 2 & 2 & 2 & 1 \\ 1 & 1 & 1 & 1 \end{pmatrix}$.

2. Determine whether each of the given sets of vectors is linearly dependent or independent:

(a) $(-3, 4, 2)$, $(7, -1, 3)$,   and $(1, 1, 8)$.
(b) $(1, 0, 2)$, $(2, 6, 4)$,   and $(1, 12, 2)$.
(c) $(1, -2, 1, 1), (3, 0, 2, -2), (0, 4, -1, 1)$,   and $(5, 0, 3, -1)$.
(d) $(3, -2, 4, 5), (0, 2, 3, -4), (0, 0, 2, 7)$,   and $(0, 0, 0, 4)$.

3. Determine whether each of the following sets $\{p_1, p_2, p_3\}$ of functions is linearly dependent or independent:

(a) $p_1 = 3x^2 - 1$, $p_2 = x^2 + 2x - 1$, $p_3 = x^2 - 4x + 3$.
(b) $p_1 = x^2 + 5x + 12$, $p_2 = 3x^2 + 5x - 3$, $p_3 = 4x^2 - 3x + 7$.
(c) $p_1 = 2x^2 - 8x + 9$, $p_2 = 6x^2 + 13x - 22$, $p_3 = 4x^2 - 11x + 2$.
(d) $p_1 = -2x^2 + 3x$, $p_2 = 7x^2 - 5x - 10$, $p_3 = -3x^2 + 9x - 13$.

4. For what values of k are the following vectors in $\mathbb{R}^3$ linearly independent?
(a) $(-1, 0, -1)$, $(2, 1, 2)$,   and $(1, 1, k)$.
(b) $(1, 2, 3)$, $(2, -1, 4)$,   and $(3, k, 4)$.
(c) $(2, k, 1)$, $(1, 0, 1)$,   and $(0, 1, 3)$.
(d) $(k, \frac{1}{2}, \frac{1}{2})$, $(\frac{1}{2}, k, \frac{1}{2})$,   and $(\frac{1}{2}, \frac{1}{2}, k)$.

5. Show that the vectors $(1, a, a^2), (1, b, b^2)$, and $(1, c, c^2)$ are linearly independent, if $a \neq b, a \neq c, b \neq c$.

6. Determine whether each of the given matrices is diagonalizable:

(a) $\begin{pmatrix} 1 & 1 & -4 \\ 2 & 0 & -4 \\ -1 & 1 & -2 \end{pmatrix}$, (b) $\begin{pmatrix} 1 & 1 & 0 \\ 3 & 0 & 3 \\ 2 & -1 & 3 \end{pmatrix}$, (c) $\begin{pmatrix} -46 & 6 & 18 \\ 0 & 2 & 0 \\ -120 & 15 & 47 \end{pmatrix}$,

(d) $\begin{pmatrix} -5 & -25 & 6 \\ 10 & 12 & 9 \\ -3 & -9 & 4 \end{pmatrix}$, (e) $\begin{pmatrix} 10 & 11 & 3 \\ -3 & -4 & -3 \\ -8 & -8 & 11 \end{pmatrix}$, (f) $\begin{pmatrix} 1 & 0 & 0 & 2 \\ 0 & 1 & 3 & 0 \\ 0 & 0 & 2 & 0 \\ 0 & 0 & 0 & 2 \end{pmatrix}$.

7. Find 3×3 nondiagonal matrix whose eigenvalues are $-2, -2$ and 3, and associated eigenvectors are

$$\begin{pmatrix} 1 \\ 0 \\ 1 \end{pmatrix}, \quad \begin{pmatrix} 0 \\ 1 \\ 1 \end{pmatrix}, \quad \begin{pmatrix} 1 \\ 1 \\ 1 \end{pmatrix}.$$

8. Find a nonsingular matrix Q such that $Q^{-1}AQ$ is a diagonal matrix, using Problem 1.

9. Find the formula for the *kth* power of each matrix considered in Problem 5, and then compute A^5.

10. Show that the following matrices are similar:

$$A = \begin{pmatrix} 1 & 0 & 0 \\ 0 & 1 & 0 \\ 1 & 0 & 1 \end{pmatrix} \quad \text{and} \quad B = \begin{pmatrix} 1 & 0 & 0 \\ 1 & 1 & 0 \\ 0 & 0 & 1 \end{pmatrix}.$$

11. Consider the diagonalizable matrix

$$A = \begin{pmatrix} 1 & 0 & 0 & 2 \\ 0 & 1 & 3 & 0 \\ 0 & 0 & 2 & 0 \\ 0 & 0 & 0 & 2 \end{pmatrix}.$$

Find a formula for the kth power of the matrix and compute A^{10}.

12. Prove that

    (a) The matrix A is similar to itself.

    (b) If A is similar to B, then B is also similar to A.

    (c) If A is similar to B, and B is similar to C, then A is similar to C.

    (d) If A is similar to B, then $det(A) = det(B)$.

    (e) If A is similar to B, then A^2 is similar to B^2.

    (f) If A is noninvertible and B is similar to A, then B is also noninvertible.

13. Find a diagonal matrix that is similar to the given matrix:

$$
\textbf{(a)} \begin{pmatrix} 3 & -1 & -1 \\ -12 & 0 & 5 \\ 4 & -2 & -1 \end{pmatrix}, \quad
\textbf{(b)} \begin{pmatrix} -5 & 0 & 0 \\ -4 & 0 & -4 \\ -7 & -7 & 0 \end{pmatrix}, \quad
\textbf{(c)} \begin{pmatrix} -3 & 2 & -3 \\ 1 & -2 & -3 \\ 1 & -1 & 3 \end{pmatrix}
$$

14. Show that each of the given matrices is not diagonalizable:

$$
\textbf{(a)} \begin{pmatrix} 10 & 11 & 3 \\ -3 & -4 & -3 \\ -8 & -8 & -1 \end{pmatrix}, \quad
\textbf{(b)} \begin{pmatrix} 3 & 3 & 3 \\ 2 & 2 & 2 \\ 1 & 1 & 1 \end{pmatrix}, \quad
\textbf{(c)} \begin{pmatrix} 2 & 0 & 0 \\ 3 & 2 & 0 \\ 0 & 0 & 5 \end{pmatrix}.
$$

15. Find the orthogonal transformations matrix Q to reduce the given matrices to diagonal matrices:

$$
\textbf{(a)} \begin{pmatrix} 1 & 0 & 0 \\ 0 & 0 & 1 \\ 0 & 1 & 0 \end{pmatrix}, \quad
\textbf{(b)} \begin{pmatrix} -1 & 2 & 2 \\ 2 & -1 & 2 \\ 2 & 2 & -1 \end{pmatrix}, \quad
\textbf{(c)} \begin{pmatrix} 3 & 2 & 2 \\ 2 & 2 & 0 \\ 2 & 0 & 4 \end{pmatrix},
$$

$$
\textbf{(d)} \begin{pmatrix} 8 & -1 & 1 \\ -1 & 8 & 1 \\ 1 & 1 & 8 \end{pmatrix}, \quad
\textbf{(e)} \begin{pmatrix} 5 & -2 & -4 \\ -2 & 8 & -2 \\ -4 & -2 & 5 \end{pmatrix}, \quad
\textbf{(f)} \begin{pmatrix} -2 & 3 & 0 \\ 3 & 4 & 0 \\ 0 & 0 & 2 \end{pmatrix}.
$$

16. Find the characteristic polynomial and inverse of each of the matrices considered in Problem 5 by using the Cayley–Hamilton theorem.

17. Use the Cayley–Hamilton theorem to compute the characteristic polynomial, powers A^3, A^4, and inverse matrices A^{-1}, A^{-2} for the each of the given matrices:

(a) $\begin{pmatrix} 2 & 3 \\ -4 & 5 \end{pmatrix}$, (b) $\begin{pmatrix} 2 & -2 & 1 \\ -2 & 4 & 2 \\ 1 & 2 & 3 \end{pmatrix}$, (c) $\begin{pmatrix} 1 & 1 & 0 \\ -1 & 0 & 1 \\ -2 & 1 & 0 \end{pmatrix}$.

18. Find the characteristic polynomial and inverse for each of the following matrices by using the Sourian–Frame theorem:

(a) $\begin{pmatrix} 5 & 0 & 0 \\ 2 & 1 & 2 \\ 0 & 1 & 1 \end{pmatrix}$, (b) $\begin{pmatrix} 2 & 2 & 1 \\ -2 & 1 & 2 \\ 1 & -2 & 2 \end{pmatrix}$, (c) $\begin{pmatrix} 1 & 1 & -1 \\ 3 & 1 & 0 \\ 1 & -2 & 1 \end{pmatrix}$,

(d) $\begin{pmatrix} 1 & 2 & 1 \\ 1 & 2 & 0 \\ 1 & 4 & 2 \end{pmatrix}$, (e) $\begin{pmatrix} 4 & 0 & 0 & 2 \\ 0 & 3 & 0 & -1 \\ 1 & 0 & 2 & 2 \\ 0 & 0 & 0 & 2 \end{pmatrix}$, (f) $\begin{pmatrix} 1 & 0 & 2 & -1 & 4 \\ 5 & 3 & -1 & 0 & 1 \\ 8 & 5 & -3 & -1 & 4 \\ 6 & 2 & 0 & 0 & 1 \\ 0 & 1 & 4 & 2 & 0 \end{pmatrix}$.

19. Find the characteristic polynomial and inverse of the following matrix by using the Sourian–Frame theorem:

$$A = \begin{pmatrix} 1 & 0 & 0 & 0 \\ -2 & 1 & 0 & 0 \\ 5 & -4 & 1 & 0 \\ -5 & 3 & 0 & 1 \end{pmatrix}.$$

20. Find the characteristic polynomial and inverse of each of the given matrices considered in Problem 11 by using the Sourian–Frame theorem.

21. Use Bocher's formula to find the coefficients of the characteristic equation of each of the matrices considered in Problem 1.

22. Find the characteristic equation, determinant, adjoint, and inverse of each of the given matrices using the Faddeev–Leverrier method:

$$\text{(a)} \begin{pmatrix} 3 & 5 & 0 \\ 4 & -2 & 1 \\ 6 & -3 & 4 \end{pmatrix}, \quad \text{(b)} \begin{pmatrix} 5 & 5 & 5 \\ 2 & 10 & 1 \\ 6 & 3 & -9 \end{pmatrix}, \quad \text{(c)} \begin{pmatrix} 1 & -3 & 0 & -2 \\ 3 & -12 & -2 & -6 \\ -2 & 10 & 2 & 5 \\ -1 & 6 & 1 & 3 \end{pmatrix}.$$

23. Find the exponential of each of the matrices considered in Problem 1.

24. Find the general solution to the following system. Then find the solution to the initial-value problem determined by the given initial condition.

$$\begin{aligned} f_1' &= -f_1 + 5f_2 \\ f_2' &= f_1 + 3f_2 \end{aligned}$$

$$f_1(0) = 1, \quad f_2(0) = -1.$$

25. Find the general solution to the following system. Then find the solution to the initial-value problem determined by the given initial condition.

$$\begin{aligned} f_1' &= 9f_1 + 2f_2 \\ f_2' &= 3f_1 + 5f_2 \end{aligned}$$

$$f_1(0) = 2, \quad f_2(0) = 4.$$

26. Find the general solution to each of the following systems. Then find the solution to the initial-value problem determined by the given initial condition.

(a)

$$\begin{aligned} f_1' &= 2f_1 + f_2 + 2f_3 \\ f_2' &= 2f_1 + 2f_2 - 2f_3 \\ f_3' &= 3f_1 + f_2 + f_3 \end{aligned}$$

$$f_1(0) = 1, \quad f_2(0) = 1, \quad f_3(0) = 1.$$

(b)

$$\begin{aligned}
f_1' &= 3f_1 + 2f_2 + 3f_3 \\
f_2' &= -f_1 + f_2 + 4f_3 \\
f_3' &= f_1 + f_2 + 2f_3
\end{aligned}$$

$$f_1(0) = 4, \quad f_2(0) = 2, \quad f_3(0) = -1.$$

(c)

$$\begin{aligned}
f_1' &= 5f_1 + 8f_2 + 16f_3 \\
f_2' &= 4f_1 + f_2 + 8f_3 \\
f_3' &= -4f_1 - 4f_2 - 11f_3
\end{aligned}$$

$$f_1(0) = 1, \quad f_2(0) = 1, \quad f_3(0) = 1.$$

(d)

$$\begin{aligned}
f_1' &= 2f_1 - 3f_2 - 2f_3 \\
f_2' &= -2f_1 \qquad\quad + 3f_3 \\
f_3' &= f_1 - 6f_2 + 9f_3
\end{aligned}$$

$$f_1(0) = 3, \quad f_2(0) = 2, \quad f_3(0) = 1.$$

27. Find the general solution to each of the following systems. Then find the solution to the initial-value problem determined by the given initial condition.

(a)

$$\begin{aligned}
f_1' &= 5f_1 \\
f_2' &= \qquad\quad + 7f_2 \\
f_3' &= \qquad\qquad\qquad + 6f_3
\end{aligned}$$

$$f_1(0) = 1, \quad f_2(0) = 2, \quad f_3(0) = 3.$$

(b)

$$\begin{aligned}
f_1' &= 3f_1 \qquad\qquad + 2f_3 \\
f_2' &= 4f_1 - 3f_2 + f_3 \\
f_3' &= 2f_1 + 2f_2 + 4f_3
\end{aligned}$$

$$f_1(0) = 3, \quad f_2(0) = 3, \quad f_3(0) = 3.$$

(c)

$$\begin{aligned}
f_1' &= 2f_1 + f_2 + 2f_3 \\
f_2' &= 1f_1 + 5f_2 + 5f_3 \\
f_3' &= 2f_1 + f_2 + f_3
\end{aligned}$$

$$f_1(0) = -1, \quad f_2(0) = 2, \quad f_3(0) = 5.$$

(d)

$$\begin{aligned}
f_1' &= f_1 \quad\quad\quad + f_3 \\
f_2' &= 2f_1 + f_2 + 3f_3 \\
f_3' &= f_1 + 3f_2 + 5f_3
\end{aligned}$$

$$f_1(0) = 2, \quad f_2(0) = 1, \quad f_3(0) = 6.$$

28. Solve each of the following difference equations and use the solution to find the given term.

     (a) $a_n = a_{n-1} + 2a_{n-2}$, $a_0 = 1, a_1 = 2$.   *Find* a_{20}.
     (b) $a_n = 4a_{n-1} + 5a_{n-2}$, $a_0 = 0, a_1 = 1$.   *Find* a_{15}.
     (c) $a_n = 3a_{n-1} + 4a_{n-2}$, $a_0 = 1, a_1 = -1$.   *Find* a_{15}.
     (d) $a_n = a_{n-1} + 12a_{n-2}$, $a_0 = 2, a_1 = 3$.   *Find* a_{12}.

29. Solve each of the following difference equations and use the solution to find the given term.

     (a) $a_n = 3a_{n-1} + 4a_{n-2}$, $a_0 = -1, a_1 = 1$.   *Find* a_{15}.
     (b) $a_n = a_{n-1} + 2a_{n-2}$, $a_0 = 2, a_1 = 3$.   *Find* a_{20}.
     (c) $a_n = 6a_{n-1} + 7a_{n-2}$, $a_0 = 2, a_1 = -2$.   *Find* a_{12}.
     (d) $a_n = 4a_{n-1} + 5a_{n-2}$, $a_0 = 3, a_1 = 2$.   *Find* a_{18}.

Chapter 4

Numerical Computation of Eigenvalues

4.1 Introduction

The importance of the eigenvalues of a square matrix in a broad range of applications has been amply demonstrated in the previous chapters. However, finding the eigenvalues and associated eigenvectors is not such an easy task. At this point, the only method we have for computing the eigenvalues of a matrix is to solve the characteristic equation. However, there are several problems with this method that render it impractical in all but small examples. The first problem is that it depends on the computation of a determinant, which is a very time-consuming process for large matrices. The second problem is that the characteristic equation is a polynomial equation, and there are no formulas for solving polynomial equations of degrees higher than 4 (polynomials of degrees 2, 3, and 4 can be solved using the quadratic formula and its analogues). Thus, we are forced to approximate eigenvalues in most practical problems. We are in need of a completely new idea if we have any hope of designing efficient numerical techniques. Unfortunately, techniques for approximating the roots of a polynomial are quite sensitive to round-off error and are therefore unreliable. Here, we will discuss a few of the most basic numerical techniques for computing eigenvalues and eigenvectors.

One class of techniques, called iterative methods, can be used to find some or all of the eigenvalues and eigenvectors of a given matrix. They start with an arbitrary approximation to one of the eigenvectors and successively improve this until the required accuracy is obtained. Among them is the power method of inverse iterations, which is used to find all of the eigenvectors of a matrix from known approximations to its eigenvalues.

The other class of techniques, which can only be applied to symmetric matrices, include the Jacobi, Given's, and Householder's methods, which reduce a given symmetric matrix to a special form whose eigenvalues are readily computed. For general matrices (symmetric or nonsymmetric matrices), the QR method and the LR method are the most widely used techniques for solving eigenvalue problems. Most of these procedures make use of a series of similarity transformations.

4.2   Vector Iterative Methods for Eigenvalues

So far we have discussed classical methods for evaluating the eigenvalues and eigenvectors for different matrices. It is evident that these methods become impractical as the matrices involved become large. Consequently, iterative methods are used for that purpose, such as the power methods. These methods are an easy means to compute eigenvalues and eigenvectors of a given matrix.

The power methods include three versions. First, is the *regular power method* or *simple iterations* based on the power of a matrix. Second, is the *inverse power method* which is based on the inverse power of a matrix. Third, is the *shifted inverse power method* in which the given matrix A is replaced by $(A - \mu \mathbf{I})$ for any given scalar μ. Following, we discuss all of these methods in some detail.

4.2.1 Power Method

The basic power method can be used to compute the eigenvalue of the largest modules and the corresponding eigenvector of a general matrix. The eigenvalue of the largest magnitude is often called the *dominant eigenvalue*. The implication of the power method is that if we assume a vector $\mathbf{x}_k$, then a new vector $\mathbf{x}_{k+1}$ can be calculated. The new vector is normalized by factoring its largest coefficient. This coefficient is then taken as a first approximation to the largest eigenvalue, and the resulting vector represents the first approximation to the corresponding eigenvector. This process is continued by substituting the new eigenvector and determining a second approximation until the desired accuracy is achieved.

Consider an $n \times n$ matrix A, then the eigenvalues and eigenvectors satisfy

$$A\mathbf{v}_i = \lambda_i \mathbf{v}_i, \tag{4.1}$$

where λ_i is the *ith* eigenvalue and $\mathbf{v}_i$ is the corresponding *ith* eigenvector of A. The power method can be used on both symmetric and nonsymmetric matrices. If A is a symmetric matrix, then all the eigenvalues are real. If A is a nonsymmetric, then there is a possibility that there is not a single real dominant eigenvalue but a complex conjugate pair. Under these conditions the power method does not converge. We assume that the largest eigenvalue is real and not repeated and that eigenvalues are numbered in increasing order, i.e.,

$$|\lambda_1| > |\lambda_2| \geq |\lambda_3| \cdots \geq |\lambda_{n-1}| \geq |\lambda_n|. \tag{4.2}$$

The power method starts with an initial guess for the eigenvector $\mathbf{x}_0$, which can be any nonzero vector. The power method is defined by the iteration

$$\mathbf{x}_{k+1} = A\mathbf{x}_k, \qquad \text{for} \quad k = 0, 1, 2, \ldots, \tag{4.3}$$

and it gives

$$\begin{aligned}
\mathbf{x}_1 &= A\mathbf{x}_0 \\
\mathbf{x}_2 &= A\mathbf{x}_1 = A^2\mathbf{x}_0 \\
\mathbf{x}_3 &= A\mathbf{x}_2 = A^3\mathbf{x}_0.
\end{aligned}$$

$$\vdots$$

Thus,

$$\mathbf{x}_k = A^k\mathbf{x}_0, \qquad \text{for} \quad k = 1, 2, \dots.$$

The vector $\mathbf{x}_0$ is an unknown linear combination of all the eigenvectors of the system, provided they are linearly independent. Thus,

$$\mathbf{x}_0 = \alpha_1\mathbf{v}_1 + \alpha_2\mathbf{v}_2 + \cdots + \alpha_n\mathbf{v}_n.$$

Let

$$\begin{aligned}
\mathbf{x}_1 = A\mathbf{x}_0 &= A(\alpha_1\mathbf{v}_1 + \alpha_2\mathbf{v}_2 + \cdots + \alpha_n\mathbf{v}_n) \\
&= \alpha_1 A\mathbf{v}_1 + \alpha_2 A\mathbf{v}_2 + \cdots + \alpha_n A\mathbf{v}_n \\
&= \alpha_1\lambda_1\mathbf{v}_1 + \alpha_2\lambda_2\mathbf{v}_2 + \cdots + \alpha_n\lambda_n\mathbf{v}_n ,
\end{aligned}$$

since from the definition of an eigenvector, $A\mathbf{v}_i = \lambda_i\mathbf{v}_i$. Similarly,

$$\begin{aligned}
\mathbf{x}_2 = A\mathbf{x}_1 &= A(\alpha_1\lambda_1\mathbf{v}_1 + \alpha_2\lambda_2\mathbf{v}_2 + \cdots + \alpha_n\lambda_n\mathbf{v}_n) \\
&= \alpha_1\lambda_1 A\mathbf{v}_1 + \alpha_2\lambda_2 A\mathbf{v}_2 + \cdots + \alpha_n\lambda_n A\mathbf{v}_n \\
&= \alpha_1\lambda_1^2\mathbf{v}_1 + \alpha_2\lambda_2^2\mathbf{v}_2 + \cdots + \alpha_n\lambda_n^2\mathbf{v}_n.
\end{aligned}$$

Continuing in this way, we get

$$\mathbf{x}_k = \alpha_1\lambda_1^k\mathbf{v}_1 + \alpha_2\lambda_2^k\mathbf{v}_2 + \cdots + \alpha_n\lambda_n^k\mathbf{v}_n, \tag{4.4}$$

which may be written as

$$\mathbf{x}_k = A^k\mathbf{x}_0 = \lambda_1^k \left[\alpha_1\mathbf{v}_1 + \alpha_2 \left(\frac{\lambda_2}{\lambda_1} \right)^k \mathbf{v}_2 + \alpha_3 \left(\frac{\lambda_3}{\lambda_1} \right)^k \mathbf{v}_3 + \cdots + \alpha_n \left(\frac{\lambda_n}{\lambda_1} \right)^k \mathbf{v}_n \right].$$

All of the terms except the first in the above relation (4.4) converge to the zero vector as $k \to \infty$, since $|\lambda_1| > |\lambda_i|$ for $i \neq 1$. Hence,

$$\mathbf{x}_k \approx (\lambda_1^k \alpha_1)\mathbf{v}_1, \quad \text{for large } k, \quad \text{provided that} \quad \alpha_1 \neq 0.$$

Since $\lambda_1^k \alpha_1 \mathbf{v}_1$ is a scalar multiple of $\mathbf{v}_1$, $\mathbf{x}_k = A^k \mathbf{x}_0$ will approach an eigenvector for the dominant eigenvalue λ_1, i.e.,

$$A\mathbf{x}_k \approx \lambda_1 \mathbf{x}_k,$$

so if $\mathbf{x}_k$ is scaled and its dominant component is 1, then

$$(\text{dominant component of } A\mathbf{x}_k) \approx \lambda_1 \times (\text{dominant component of } \mathbf{x}_k) = \lambda_1 \times 1 = \lambda_1.$$

The rate of convergence of the power method is primarily dependent on the distribution of the eigenvalues; the smaller the ratio $|\frac{\lambda_i}{\lambda_1}|$, (for $i = 2, 3, \ldots, n$), the faster the convergence; in particular, this rate depends upon the ratio $|\frac{\lambda_2}{\lambda_1}|$. The number of iterations required to get a desired degree of convergence depends upon both the rate of convergence and how large λ_1 is compared with the other λ_i, the latter depending, in turn, on the choice of initial approximation $\mathbf{x}_0$.

Example 4.1 *Find the first five iterations obtained by the power method applied to the following matrix using the initial approximation* $\mathbf{x}_0 = [1, 1, 1]^T$:

$$A = \begin{pmatrix} 1 & 1 & 2 \\ 1 & 2 & 1 \\ 1 & 1 & 0 \end{pmatrix}.$$

Solution. *Starting with an initial vector* $\mathbf{x}_0 = [1, 1, 1]^T$, *we have*

$$A\mathbf{x}_0 = \begin{pmatrix} 1 & 1 & 2 \\ 1 & 2 & 1 \\ 1 & 1 & 0 \end{pmatrix} \begin{pmatrix} 1 \\ 1 \\ 1 \end{pmatrix} = \begin{pmatrix} 4.0000 \\ 4.0000 \\ 2.0000 \end{pmatrix},$$

which gives

$$A\mathbf{x}_0 = \lambda_1 \mathbf{x}_1 = 4.0000 \begin{pmatrix} 1.0000 \\ 1.0000 \\ 0.5000 \end{pmatrix}.$$

Similarly, the other possible iterations are as follows:

$$A\mathbf{x}_1 = \begin{pmatrix} 1 & 1 & 2 \\ 1 & 2 & 1 \\ 1 & 1 & 0 \end{pmatrix} \begin{pmatrix} 1.0000 \\ 1.0000 \\ 0.5000 \end{pmatrix} = \begin{pmatrix} 3.0000 \\ 3.5000 \\ 2.0000 \end{pmatrix}$$

$$= 3.5000 \begin{pmatrix} 0.8571 \\ 1.0000 \\ 0.5714 \end{pmatrix} = \lambda_2 \mathbf{x}_2.$$

$$A\mathbf{x}_2 = \begin{pmatrix} 1 & 1 & 2 \\ 1 & 2 & 1 \\ 1 & 1 & 0 \end{pmatrix} \begin{pmatrix} 0.8571 \\ 1.0000 \\ 0.5714 \end{pmatrix} = \begin{pmatrix} 3.0000 \\ 3.4286 \\ 1.8571 \end{pmatrix}$$

$$= 3.4286 \begin{pmatrix} 0.8750 \\ 1.0000 \\ 0.5417 \end{pmatrix} = \lambda_3 \mathbf{x}_3.$$

$$A\mathbf{x}_3 = \begin{pmatrix} 1 & 1 & 2 \\ 1 & 2 & 1 \\ 1 & 1 & 0 \end{pmatrix} \begin{pmatrix} 0.8750 \\ 1.0000 \\ 0.5417 \end{pmatrix} = \begin{pmatrix} 2.9583 \\ 3.4167 \\ 1.8750 \end{pmatrix}$$

$$= 3.4167 \begin{pmatrix} 0.8659 \\ 1.0000 \\ 0.5488 \end{pmatrix} = \lambda_4 \mathbf{x}_4.$$

$$A\mathbf{x}_4 = \begin{pmatrix} 1 & 1 & 2 \\ 1 & 2 & 1 \\ 1 & 1 & 0 \end{pmatrix} \begin{pmatrix} 0.8659 \\ 1.0000 \\ 0.5488 \end{pmatrix} = \begin{pmatrix} 2.9634 \\ 3.4146 \\ 1.8659 \end{pmatrix}$$

$$= 3.4146 \begin{pmatrix} 0.8679 \\ 1.0000 \\ 0.5464 \end{pmatrix} = \lambda_5 \mathbf{x}_5.$$

Since the eigenvalues of the given matrix A are $3.4142, 0.5858$ and -1.0000, the approximation of the dominant eigenvalue after the five iterations is $\lambda_5 = 3.4146$, and the corresponding eigenvector is $[0.8679, 1.0000, 0.5464]^T$.

•

To get the above results using the MATLAB Command Window, we do the following:

```
>> A = [1 1 2; 1 2 1; 1 1 0];
>> X = [1 1 1]';
>> maxI = 5;
>> sol = PM(A, X, maxI);
```

Program 4.1
MATLAB m-file for the Power method
function sol=PM $(A, X, maxI)$
$[n, n] = size(A);$
for k=1:maxI; for i=1:n; s=0;
for j=1:n; $ss = A(i, j) * X(j, 1); s = s + ss$;end
$XX(i, 1) = s$;end; $X = XX; y = max(X);$
for i=1:n; $X(i, 1) = 1/y * X(i, 1)$;end; yy=abs(y-y1);
if $(yy <= 1e - 6)$; break; end; y; end

The power method has the disadvantage that it is unknown at the outset whether or not a matrix has a single dominant eigenvalue. Nor is it known how an initial vector $\mathbf{x}_0$ should be chosen to ensure that its representation in terms of the eigenvectors of a matrix will contain a nonzero contribution from the eigenvector associated with the dominant eigenvalue, should it exist.

Note that the dominant eigenvalue λ of a matrix can also be obtained from two successive iterations, by dividing the corresponding elements of vectors $\mathbf{x}_n$ and $\mathbf{x}_{n-1}$.

Example 4.2 *Find the dominant eigenvalue of the matrix*

$$A = \begin{pmatrix} 4 & 1 & 3 \\ 9 & 0 & 9 \\ 5 & -1 & 6 \end{pmatrix}.$$

Solution. *Let us consider an arbitrary vector* $\mathbf{x}_0 = [1, 0, 0]^T$, *then*

$$\mathbf{x}_1 = A\mathbf{x}_0 = \begin{pmatrix} 4 & 1 & 3 \\ 9 & 0 & 9 \\ 5 & -1 & 6 \end{pmatrix} \begin{pmatrix} 1 \\ 0 \\ 0 \end{pmatrix} = \begin{pmatrix} 4 \\ 9 \\ 5 \end{pmatrix}$$

$$\mathbf{x}_2 = A\mathbf{x}_1 = \begin{pmatrix} 4 & 1 & 3 \\ 9 & 0 & 9 \\ 5 & -1 & 6 \end{pmatrix} \begin{pmatrix} 4 \\ 9 \\ 5 \end{pmatrix} = \begin{pmatrix} 40 \\ 81 \\ 41 \end{pmatrix}$$

$$\mathbf{x}_3 = A\mathbf{x}_2 = \begin{pmatrix} 4 & 1 & 3 \\ 9 & 0 & 9 \\ 5 & -1 & 6 \end{pmatrix} \begin{pmatrix} 40 \\ 81 \\ 41 \end{pmatrix} = \begin{pmatrix} 364 \\ 729 \\ 365 \end{pmatrix}.$$

Then the dominant eigenvalue can be obtained as

$$\lambda_1 \approx \frac{\mathbf{x}_3}{\mathbf{x}_2} \approx \frac{364}{40} \approx \frac{729}{81} \approx \frac{365}{41} \approx 9.$$

●

Power Method and Symmetric Matrices

The power method will converge if the given $n \times n$ matrix A has linearly independent eigenvectors and a symmetric matrix satisfies this property. Now we will discuss the power method for finding the dominant eigenvalue of a symmetric matrix only.

Theorem 4.1 (Power Method with Euclidean Scaling)

Let A be a symmetric $n \times n$ matrix with a positive dominant eigenvalue λ. If $\mathbf{x}_0$ is a unit vector in $\mathbf{R}^n$ that is not orthogonal to the eigenspace corresponding to λ, then the normalized power sequence

$$\mathbf{x}_0, \quad \mathbf{x}_1 = \frac{A\mathbf{x}_0}{\|A\mathbf{x}_0\|}, \quad \mathbf{x}_2 = \frac{A\mathbf{x}_1}{\|A\mathbf{x}_1\|}, \dots, \quad \mathbf{x}_k = \frac{A\mathbf{x}_{k-1}}{\|A\mathbf{x}_{k-1}\|}, \quad \dots \qquad (4.5)$$

converges to a unit dominant eigenvector, and the eigenvalues

$$A\mathbf{x}_1.\mathbf{x}_1, \quad A\mathbf{x}_2.\mathbf{x}_2, \quad \dots, \quad A\mathbf{x}_k.\mathbf{x}_k, \quad \dots. \qquad (4.6)$$

●

The basic steps for the power method with Euclidean scaling are:

1. Choose an arbitrary nonzero vector and normalize it to obtain a unit vector $\mathbf{x}_0$.

2. Compute $A\mathbf{x}_0$ and normalize it to obtain the first approximation $\mathbf{x}_1$ to a dominant unit eigenvector. Compute $A\mathbf{x}_1.\mathbf{x}_1$ to obtain the first approximation to the dominant eigenvalue.

3. Compute $A\mathbf{x}_1$ and normalize it to obtain the second approximation $\mathbf{x}_2$ to a dominant unit eigenvector. Compute $A\mathbf{x}_2.\mathbf{x}_2$ to obtain the second approximation to the dominant eigenvalue.

4. Continuing in this way we will create a sequence of increasingly closer approximations to the dominant eigenvalue and a corresponding eigenvector.

Example 4.3 *Apply the power method with Euclidean scaling to the matrix*

$$A = \begin{pmatrix} 2 & 2 \\ 2 & 5 \end{pmatrix},$$

with $\mathbf{x}_0 = [0, 1]^T$ to get the first four approximations to the dominant unit eigenvector and the dominant eigenvalue.

Solution. *Starting with the unit vector $\mathbf{x}_0 = [0, 1]^T$, we get the first approximation of the dominant unit eigenvector as follows:*

$$A\mathbf{x}_0 = \begin{pmatrix} 2 & 2 \\ 2 & 5 \end{pmatrix} \begin{pmatrix} 0 \\ 1 \end{pmatrix} = \begin{pmatrix} 2 \\ 5 \end{pmatrix}, \quad \mathbf{x}_1 = \frac{A\mathbf{x}_0}{\max(A\mathbf{x}_0)} = \frac{1}{\sqrt{29}} \begin{pmatrix} 2 \\ 5 \end{pmatrix} \approx \begin{pmatrix} 0.3714 \\ 0.9285 \end{pmatrix}.$$

Similarly, for the second, third, and fourth approximations of the dominant unit eigenvector, we find

$$A\mathbf{x}_1 = \begin{pmatrix} 2 & 2 \\ 2 & 5 \end{pmatrix} \begin{pmatrix} 0.3714 \\ 0.9285 \end{pmatrix} = \begin{pmatrix} 2.5997 \\ 5.3852 \end{pmatrix},$$

$$\mathbf{x}_2 = \frac{A\mathbf{x}_1}{\max(A\mathbf{x}_1)} \approx \frac{1}{5.9799} \begin{pmatrix} 2.5997 \\ 5.3852 \end{pmatrix} \approx \begin{pmatrix} 0.4347 \\ 0.9006 \end{pmatrix}.$$

$$Ax_2 = \begin{pmatrix} 2 & 2 \\ 2 & 5 \end{pmatrix} \begin{pmatrix} 0.4347 \\ 0.9006 \end{pmatrix} = \begin{pmatrix} 2.6706 \\ 5.3723 \end{pmatrix},$$

$$x_3 = \frac{Ax_2}{\max(Ax_2)} \approx \frac{1}{5.9994} \begin{pmatrix} 2.6706 \\ 5.3723 \end{pmatrix} \approx \begin{pmatrix} 0.4451 \\ 0.8955 \end{pmatrix}.$$

$$Ax_3 = \begin{pmatrix} 2 & 2 \\ 2 & 5 \end{pmatrix} \begin{pmatrix} 0.4451 \\ 0.8955 \end{pmatrix} = \begin{pmatrix} 2.6812 \\ 5.3676 \end{pmatrix},$$

$$x_4 = \frac{Ax_3}{\max(Ax_3)} \approx \frac{1}{6} \begin{pmatrix} 2.6812 \\ 5.3676 \end{pmatrix} \approx \begin{pmatrix} 0.4469 \\ 0.8946 \end{pmatrix}.$$

Now we find the approximations of the dominant eigenvalue of the given matrix as follows:

$$\lambda_1 = \frac{Ax_1.x_1}{x_1.x_1} = \frac{(Ax_1)^T x_1}{x_1^T x_1} = (2.5997 \;\; 5.3852) \begin{pmatrix} 0.3714 \\ 0.9285 \end{pmatrix} \approx 5.9655,$$

$$\lambda_2 = \frac{Ax_2.x_2}{x_2.x_2} = \frac{(Ax_2)^T x_2}{x_2^T x_2} = (2.6706 \;\; 5.3723) \begin{pmatrix} 0.4347 \\ 0.9006 \end{pmatrix} \approx 5.9992,$$

$$\lambda_3 = \frac{Ax_3.x_3}{x_3.x_3} = \frac{(Ax_3)^T x_3}{x_3^T x_3} = (2.6812 \;\; 5.3676) \begin{pmatrix} 0.4451 \\ 0.8955 \end{pmatrix} \approx 6.0001,$$

$$\lambda_4 = \frac{Ax_4.x_4}{x_4.x_4} = \frac{(Ax_4)^T x_4}{x_4^T x_4} = (2.6830 \;\; 5.3668) \begin{pmatrix} 0.4469 \\ 0.8946 \end{pmatrix} \approx 6.0002.$$

These are the required approximations of the dominant eigenvalue of A. Notice that the exact dominant eigenvalue of the given matrix is $\lambda = 6$, with the corresponding dominant unit eigenvector $x = [0.4472, 0.8945]^T$. •

Now we will consider the power method using a symmetric matrix in such a way that each iterate is scaled to make its largest entry a 1, rather than being normalized.

Theorem 4.2 (Power Method with Maximum Entry Scaling)

Let A be a symmetric $n \times n$ matrix with a positive dominant eigenvalue λ. If $\mathbf{x}_0$ is a nonzero vector in $\mathbf{R}^n$ that is not orthogonal to the eigenspace corresponding to λ, then the normalized power sequence

$$\mathbf{x}_0, \quad \mathbf{x}_1 = \frac{A\mathbf{x}_0}{\max(A\mathbf{x}_0)}, \quad \mathbf{x}_2 = \frac{A\mathbf{x}_1}{\max(A\mathbf{x}_1)}, \dots, \quad \mathbf{x}_k = \frac{A\mathbf{x}_{k-1}}{\max(A\mathbf{x}_{k-1})}, \quad \dots \tag{4.7}$$

converges to an eigenvector corresponding to eigenvalue λ, and the sequence

$$\frac{A\mathbf{x}_1.\mathbf{x}_1}{\mathbf{x}_1.\mathbf{x}_1}, \quad \frac{A\mathbf{x}_2.\mathbf{x}_2}{\mathbf{x}_2.\mathbf{x}_2}, \quad \dots, \quad \frac{A\mathbf{x}_k.\mathbf{x}_k}{\mathbf{x}_k.\mathbf{x}_k}, \quad \dots \tag{4.8}$$

converges to λ.   •

In using the power method with maximum entry scaling, we have to do the following steps:

1. Choose an arbitrary nonzero vector $\mathbf{x}_0$.

2. Compute $A\mathbf{x}_0$ and divide it by the factor $\max \mathbf{x}_0$ to obtain the first approximation $\mathbf{x}_1$ to a dominant eigenvector. Compute $\frac{A\mathbf{x}_1.\mathbf{x}_1}{\mathbf{x}_1.\mathbf{x}_1}$ to obtain the first approximation to the dominant eigenvalue.

3. Compute $A\mathbf{x}_1$ and divide it by the factor $\max \mathbf{x}_1$ to obtain the second approximation $\mathbf{x}_2$ to a dominant eigenvector. Compute $\frac{A\mathbf{x}_2.\mathbf{x}_2}{\mathbf{x}_2.\mathbf{x}_2}$ to obtain the second approximation to the dominant eigenvalue.

4. Continuing in this way we will create a sequence of increasingly closer approximations to the dominant eigenvalue and a corresponding eigenvector.

Example 4.4 *Apply the power method with maximum entry scaling to the matrix*

$$A = \begin{pmatrix} 2 & 2 \\ 2 & 5 \end{pmatrix},$$

with $\mathbf{x}_0 = [0, 1]^T$, to get the first four approximations to the dominant eigenvector and the dominant eigenvalue.

Solution. *Starting with* $\mathbf{x}_0 = [0, 1]^T$, *we get the first approximation of the dominant eigenvector as follows:*

$$A\mathbf{x}_0 = \begin{pmatrix} 2 & 2 \\ 2 & 5 \end{pmatrix} \begin{pmatrix} 0 \\ 1 \end{pmatrix} = \begin{pmatrix} 2 \\ 5 \end{pmatrix}, \quad \mathbf{x}_1 = \frac{A\mathbf{x}_0}{\max(A\mathbf{x}_0)} = \frac{1}{5} \begin{pmatrix} 2 \\ 5 \end{pmatrix} = \begin{pmatrix} 0.4000 \\ 1.0000 \end{pmatrix}.$$

Similarly, for the second, third, and fourth approximations of the dominant eigenvector, we find

$$A\mathbf{x}_1 = \begin{pmatrix} 2 & 2 \\ 2 & 5 \end{pmatrix} \begin{pmatrix} 0.4000 \\ 1.0000 \end{pmatrix} = \begin{pmatrix} 2.8000 \\ 5.8000 \end{pmatrix},$$

$$\mathbf{x}_2 = \frac{A\mathbf{x}_1}{\max(A\mathbf{x}_1)} = \frac{1}{5.8000} \begin{pmatrix} 2.8000 \\ 5.8000 \end{pmatrix} \approx \begin{pmatrix} 0.4828 \\ 1.0000 \end{pmatrix}.$$

$$A\mathbf{x}_2 = \begin{pmatrix} 2 & 2 \\ 2 & 5 \end{pmatrix} \begin{pmatrix} 0.4828 \\ 1.0000 \end{pmatrix} \approx \begin{pmatrix} 2.9655 \\ 5.9655 \end{pmatrix},$$

$$\mathbf{x}_3 = \frac{A\mathbf{x}_2}{\max(A\mathbf{x}_2)} = \frac{1}{5.9655} \begin{pmatrix} 2.9655 \\ 5.9655 \end{pmatrix} \approx \begin{pmatrix} 0.4971 \\ 1.0000 \end{pmatrix}.$$

$$A\mathbf{x}_3 = \begin{pmatrix} 2 & 2 \\ 2 & 5 \end{pmatrix} \begin{pmatrix} 0.4971 \\ 1.0000 \end{pmatrix} \approx \begin{pmatrix} 2.9942 \\ 5.9942 \end{pmatrix},$$

$$\mathbf{x}_4 = \frac{A\mathbf{x}_3}{\max(A\mathbf{x}_3)} = \frac{1}{5.9942} \begin{pmatrix} 2.9942 \\ 5.9942 \end{pmatrix} \approx \begin{pmatrix} 0.4995 \\ 1.0000 \end{pmatrix},$$

which are the required first four approximations of the dominant eigenvector.

Now we find the approximations of the dominant eigenvalue of the given matrix as follows:

$$\lambda_1 = \frac{A\mathbf{x}_1 . \mathbf{x}_1}{\mathbf{x}_1 . \mathbf{x}_1} = \frac{(A\mathbf{x}_1)^T \mathbf{x}_1}{\mathbf{x}_1^T \mathbf{x}_1} \approx \frac{6.9200}{1.1600} \approx 5.9655,$$

$$\lambda_2 = \frac{A\mathbf{x}_2.\mathbf{x}_2}{\mathbf{x}_2.\mathbf{x}_2} = \frac{(A\mathbf{x}_2)^T\mathbf{x}_2}{\mathbf{x}_2^T\mathbf{x}_2} \approx \frac{7.3972}{1.2331} \approx 5.9989,$$

$$\lambda_3 = \frac{A\mathbf{x}_3.\mathbf{x}_3}{\mathbf{x}_3.\mathbf{x}_3} = \frac{(A\mathbf{x}_3)^T\mathbf{x}_3}{\mathbf{x}_3^T\mathbf{x}_3} \approx \frac{7.4826}{1.2471} \approx 6.0000,$$

$$\lambda_4 = \frac{A\mathbf{x}_4.\mathbf{x}_4}{\mathbf{x}_4.\mathbf{x}_4} = \frac{(A\mathbf{x}_4)^T\mathbf{x}_4}{\mathbf{x}_4^T\mathbf{x}_4} \approx \frac{7.4970}{1.2495} \approx 6.0000.$$

These are the required approximations of the dominant eigenvalue of A. Notice that the exact dominant eigenvalue of the given matrix is $\lambda = 6$, with the corresponding dominant eigenvector $\mathbf{x} = [0.5, 1]^T$. •

Notice that the main difference between the power method with Euclidean scaling and the power method with maximum entry scaling is that the Euclidean scaling gives a sequence that approaches a unit dominant eigenvector, whereas maximum entry scaling gives a sequence that approaches a dominant eigenvector whose largest component is 1.

4.2.2 Inverse Power Method

The power method can be modified by replacing the given matrix A with its inverse matrix A^{-1}, and this is called the inverse power method. Since we know that the eigenvalues of A^{-1} are reciprocals of A, the power method applied to A^{-1} will find the smallest eigenvalue of A. Thus, the smallest (or least) value of the eigenvalue for A will become the maximum value for A^{-1}. Of course, we must assume that the smallest eigenvalue of A is real and not repeated, otherwise, the method does not work.

In this method the solution procedure is a little more involved than finding the largest eigenvalue of the given matrix. Fortunately, it is just as straight forward. Consider

$$A\mathbf{x} = \lambda\mathbf{x}, \tag{4.9}$$

and multiplying by A^{-1}, we have

$$A^{-1}A\mathbf{x} = \lambda A^{-1}\mathbf{x}$$

or

$$A^{-1}\mathbf{x} = \frac{1}{\lambda}\mathbf{x}. \tag{4.10}$$

The solution procedure is initiated by starting with an initial guess for the vector $\mathbf{x}_i$ and improving the solution by getting a new vector $\mathbf{x}_{i+1}$, and so on until the vector $\mathbf{x}_i$ is approximately equal to $\mathbf{x}_{i+1}$.

Example 4.5 *Use the inverse power method to find the first seven approximations of the least dominant eigenvalue and the corresponding eigenvector of the following matrix using an initial approximation* $\mathbf{x}_0 = [0, 1, 2]^T$:

$$A = \begin{pmatrix} 3 & 0 & 1 \\ 0 & -3 & 0 \\ 1 & 0 & 3 \end{pmatrix}.$$

Solution. *The inverse of the given matrix A is*

$$A^{-1} = \begin{pmatrix} \dfrac{3}{8} & 0 & -\dfrac{1}{8} \\ 0 & -\dfrac{1}{3} & 0 \\ -\dfrac{1}{8} & 0 & \dfrac{3}{8} \end{pmatrix}.$$

Starting with the given initial vector $\mathbf{x}_0 = [0, 1, 2]^T$, we have

$$A^{-1}\mathbf{x}_0 = \begin{pmatrix} \dfrac{3}{8} & 0 & -\dfrac{1}{8} \\ 0 & -\dfrac{1}{3} & 0 \\ -\dfrac{1}{8} & 0 & \dfrac{3}{8} \end{pmatrix} \begin{pmatrix} 0 \\ 1 \\ 2 \end{pmatrix} = \begin{pmatrix} -0.2500 \\ -0.3333 \\ 0.7500 \end{pmatrix}$$

$$= 0.75 \begin{pmatrix} -0.3333 \\ -0.4444 \\ 1.0000 \end{pmatrix} = \lambda_1 \mathbf{x}_1.$$

Similarly, the other possible iterations are as follows:

$$A^{-1}\mathbf{x}_1 = \begin{pmatrix} \dfrac{3}{8} & 0 & -\dfrac{1}{8} \\ 0 & -\dfrac{1}{3} & 0 \\ -\dfrac{1}{8} & 0 & \dfrac{3}{8} \end{pmatrix} \begin{pmatrix} -0.3333 \\ -0.4444 \\ 1.0000 \end{pmatrix} = \begin{pmatrix} -0.2500 \\ 0.1481 \\ 0.4167 \end{pmatrix}$$

$$= 0.4167 \begin{pmatrix} -0.6000 \\ 0.3558 \\ 1.0000 \end{pmatrix} = \lambda_2 \mathbf{x}_2.$$

$$A^{-1}\mathbf{x}_2 = \begin{pmatrix} \dfrac{3}{8} & 0 & -\dfrac{1}{8} \\ 0 & -\dfrac{1}{3} & 0 \\ -\dfrac{1}{8} & 0 & \dfrac{3}{8} \end{pmatrix} \begin{pmatrix} -0.6000 \\ 0.3558 \\ 1.0000 \end{pmatrix} = \begin{pmatrix} -0.3500 \\ -0.1185 \\ 0.4500 \end{pmatrix}$$

$$= 0.4500 \begin{pmatrix} -0.7778 \\ -0.2634 \\ 1.0000 \end{pmatrix} = \lambda_3 \mathbf{x}_3.$$

$$A^{-1}\mathbf{x}_3 = \begin{pmatrix} \dfrac{3}{8} & 0 & -\dfrac{1}{8} \\ 0 & -\dfrac{1}{3} & 0 \\ -\dfrac{1}{8} & 0 & \dfrac{3}{8} \end{pmatrix} \begin{pmatrix} -0.7778 \\ -0.2634 \\ 1.0000 \end{pmatrix} = \begin{pmatrix} -0.4167 \\ 0.0878 \\ 0.4722 \end{pmatrix}$$

$$= 0.4722 \begin{pmatrix} -0.8824 \\ 0.1859 \\ 1.0000 \end{pmatrix} = \lambda_4 \mathbf{x}_4.$$

$$A^{-1}\mathbf{x}_4 = \begin{pmatrix} \dfrac{3}{8} & 0 & -\dfrac{1}{8} \\ 0 & -\dfrac{1}{3} & 0 \\ -\dfrac{1}{8} & 0 & \dfrac{3}{8} \end{pmatrix} \begin{pmatrix} -0.8824 \\ 0.1859 \\ 1.0000 \end{pmatrix} = \begin{pmatrix} -0.4559 \\ -0.0620 \\ 0.4853 \end{pmatrix}$$

$$= 0.4853 \begin{pmatrix} -0.9394 \\ -0.1277 \\ 1.0000 \end{pmatrix} = \lambda_5 \mathbf{x}_5.$$

$$A^{-1}\mathbf{x}_5 = \begin{pmatrix} \dfrac{3}{8} & 0 & -\dfrac{1}{8} \\ 0 & -\dfrac{1}{3} & 0 \\ -\dfrac{1}{8} & 0 & \dfrac{3}{8} \end{pmatrix} \begin{pmatrix} -0.9394 \\ -0.1277 \\ 1.0000 \end{pmatrix} = \begin{pmatrix} -0.4773 \\ -0.0426 \\ 0.4924 \end{pmatrix}$$

$$= 0.4924 \begin{pmatrix} -0.9692 \\ -0.0864 \\ 1.0000 \end{pmatrix} = \lambda_6 \mathbf{x}_6.$$

$$A^{-1}\mathbf{x}_6 = \begin{pmatrix} \dfrac{3}{8} & 0 & -\dfrac{1}{8} \\ 0 & -\dfrac{1}{3} & 0 \\ -\dfrac{1}{8} & 0 & \dfrac{3}{8} \end{pmatrix} \begin{pmatrix} -0.9692 \\ 0.0864 \\ 1.0000 \end{pmatrix} = \begin{pmatrix} -0.4885 \\ -0.0288 \\ 0.4962 \end{pmatrix}$$

$$= 0.4962 \begin{pmatrix} -0.9845 \\ -0.0581 \\ 1.0000 \end{pmatrix} = \lambda_7 \mathbf{x}_7.$$

Since the eigenvalues of the given matrix A are $-3.0000, 2.0000$, and 4.0000, the dominant eigenvalue of A^{-1} after the seven iterations is $\lambda_7 = 0.4962$ and converges to $\frac{1}{2}$ and so the smallest dominant eigenvalue of the given matrix A is the reciprocal of the dominant eigenvalue $\frac{1}{2}$ of the matrix A^{-1}, i.e., 2 and the corresponding eigenvector is $[-0.9845, -0.0581, 1.0000]^T$.

To get the above results using the MATLAB Command Window, we do:

```
>> A = [3 0 1; 0 − 3 0; 1 0 3];
>> X = [0 1 2]';
>> maxI = 7;
>> sol = IPM(A, X, maxI);
```

Program 4.2

MATLAB m-file for using the Inverse Power method
function sol=IPM $(A, X, maxI)$
$[n, n] = size(A); B = inv(A);$
for k=1:maxI; for i=1:n; s=0;
for j=1:n; $ss = B(i, j) * X(j, 1); s = s + ss$;end
$XX(i, 1) = s$;end; $X = XX; y = max(X)$;
for i=1:n; $X(i, 1) = 1/y * X(i, 1)$;end; yy=abs(y-y1);
if $(yy <= 1e − 6)$; break; end; y; end

4.2.3   Shifted Inverse Power Method

Another modification of the power method consists of replacing the given matrix A with $(A - \mu I)$, for any scalar μ, i.e.,

$$(A - \mu I)\mathbf{x} = (\lambda - \mu)\mathbf{x}, \qquad (4.11)$$

and it follows that the eigenvalues of $(A - \mu I)$ are the same as those of A except that they have all been shifted by an amount μ. The eigenvectors remain unaffected by the shift.

The shifted inverse power method is to apply the power method to the system

$$(A - \mu I)^{-1}\mathbf{x} = \frac{1}{(\lambda - \mu)}\mathbf{x}. \qquad (4.12)$$

Thus the iteration of $(A - \mu I)^{-1}$ leads to the largest value of $\frac{1}{(\lambda - \mu)}$, i.e., the smallest value of $(\lambda - \mu)$. The smallest value of $(\lambda - \mu)$ implies that the value of λ will be the value closest to μ. Thus, by a suitable choice of μ we have a procedure for finding the subdominant eigensolutions. So, $(A - \mu I)^{-1}$ has the same eigenvectors as A but with eigenvalues $\frac{1}{(\lambda - \mu)}$.

In practice, the inverse of $(A - \mu I)$ is never actually computed, especially if the given matrix A is a large sparse matrix. It is computationally more efficient if $(A - \mu I)$ is decomposed into the product of a lower-triangular matrix L and an upper-triangular matrix U. If $\mathbf{u}_s$ is an initial vector for the solution of (4.12), then

$$(A - \mu I)^{-1}\mathbf{u}_s = \mathbf{v}_s \tag{4.13}$$

and

$$\mathbf{u}_{s+1} = \frac{\mathbf{v}_s}{max(\mathbf{v}_s)}. \tag{4.14}$$

By rearranging (4.13), we obtain

$$\begin{aligned}\mathbf{u}_s &= (A - \mu I)\mathbf{v}_s \\ &= LU\mathbf{v}_s.\end{aligned}$$

Let

$$U\mathbf{v}_s = \mathbf{z}, \tag{4.15}$$

then

$$L\mathbf{z} = \mathbf{u}_s. \tag{4.16}$$

By using an initial value, we can find $\mathbf{z}$ from (4.16) by applying forward substitution, and knowing $\mathbf{z}$ we can find $\mathbf{v}_s$ from (4.15) by applying backward substitution. The new estimate for the vector $\mathbf{u}_{s+1}$ can then be found from (4.14). The iteration is terminated when $\mathbf{u}_{s+1}$ is sufficiently close to $\mathbf{u}_s$, and it can be easily shown when convergence is completed.

Let λ_μ be an eigenvalue of A nearest to μ, then

$$\lambda_\mu = \frac{1}{\text{dominant eigenvalue of } (A - \mu I)^{-1}} + \mu. \tag{4.17}$$

The shifted inverse power method uses the power method as a basis but gives faster convergence. Convergence is to the eigenvalue λ that is closest to μ, and if this eigenvalue is extremely close to μ, the rate of convergence will be very rapid. Inverse iteration therefore provides a means of determining an eigenvector of a matrix for which the corresponding eigenvalue has already been determined to moderate accuracy by an alternative method, such as the QR method or the Strum sequence iteration, which we will discuss later in this chapter.

When inverse iteration is used to determine eigenvectors corresponding to known eigenvalues, the matrix to be inverted, even if it is symmetric, will not normally be positive-definite, and if it is nonsymmetric, will not normally be diagonally dominant. The computation of an eigenvector corresponding to a complex conjugate eigenvalue by inverse iteration is more difficult than for a real eigenvalue.

Example 4.6 *Use the shifted inverse power method to find the first five approximations of the eigenvalue nearest $\mu = 6$ of the following matrix using the initial approximation $\mathbf{x}_0 = [1, 1]^T$:*

$$A = \begin{pmatrix} 4 & 2 \\ 3 & 5 \end{pmatrix}.$$

Solution. *Consider*

$$B = (A - 6I) = \begin{pmatrix} -2 & 2 \\ 3 & -1 \end{pmatrix}.$$

The inverse of B is

$$B^{-1} = (A - 3I)^{-1} = \begin{pmatrix} \dfrac{1}{4} & \dfrac{1}{2} \\ \dfrac{3}{4} & \dfrac{1}{2} \end{pmatrix}.$$

Now applying the power method, we obtain the following iterations:

$$B^{-1}\mathbf{x}_0 = \begin{pmatrix} \dfrac{1}{4} & \dfrac{1}{2} \\[2mm] \dfrac{3}{4} & \dfrac{1}{2} \end{pmatrix} \begin{pmatrix} 1 \\ 1 \end{pmatrix} = \begin{pmatrix} 0.7500 \\ 1.2500 \end{pmatrix}$$

$$= 1.2500 \begin{pmatrix} 0.6000 \\ 1.0000 \end{pmatrix} = \lambda_1 \mathbf{x}_1.$$

Similarly, the other approximations can be computed as

$$B^{-1}\mathbf{x}_1 = \begin{pmatrix} \dfrac{1}{4} & \dfrac{1}{2} \\[2mm] \dfrac{3}{4} & \dfrac{1}{2} \end{pmatrix} \begin{pmatrix} 0.6000 \\ 1.0000 \end{pmatrix} = \begin{pmatrix} 0.6500 \\ 0.9500 \end{pmatrix}$$

$$= 0.9500 \begin{pmatrix} 0.6842 \\ 1.0000 \end{pmatrix} = \lambda_2 \mathbf{x}_2.$$

$$B^{-1}\mathbf{x}_2 = \begin{pmatrix} \dfrac{1}{4} & \dfrac{1}{2} \\[2mm] \dfrac{3}{4} & \dfrac{1}{2} \end{pmatrix} \begin{pmatrix} 0.6842 \\ 1.0000 \end{pmatrix} = \begin{pmatrix} 0.6711 \\ 1.0132 \end{pmatrix}$$

$$= 1.0132 \begin{pmatrix} 0.6623 \\ 1.0000 \end{pmatrix} = \lambda_3 \mathbf{x}_3.$$

$$B^{-1}\mathbf{x}_3 = \begin{pmatrix} \dfrac{1}{4} & \dfrac{1}{2} \\[2mm] \dfrac{3}{4} & \dfrac{1}{2} \end{pmatrix} \begin{pmatrix} 0.6623 \\ 1.0000 \end{pmatrix} = \begin{pmatrix} 0.6656 \\ 0.9968 \end{pmatrix}$$

$$= 0.9968 \begin{pmatrix} 0.6678 \\ 1.0000 \end{pmatrix} = \lambda_4 \mathbf{x}_4.$$

$$B^{-1}\mathbf{x}_4 = \begin{pmatrix} \dfrac{1}{4} & \dfrac{1}{2} \\[2mm] \dfrac{3}{4} & \dfrac{1}{2} \end{pmatrix} \begin{pmatrix} 0.6678 \\ 1.0000 \end{pmatrix} = \begin{pmatrix} 0.6669 \\ 1.0008 \end{pmatrix}$$

$$= 1.0008 \begin{pmatrix} 0.6664 \\ 1.0000 \end{pmatrix} = \lambda_5 \mathbf{x}_5.$$

Thus, the fifth approximation of the dominant eigenvalue of $B^{-1} = (A - 3I)^{-1}$ is $\lambda_5 = 1.0008$, and it is converges to 1 with the eigenvector $[1.0000, 0.7000]^T$. Hence, the eigenvalue λ_μ of A nearest to $\mu = 6$ is

$$\lambda_\mu = \frac{1}{1} + 6 = 7.$$

To get the above results using the MATLAB Command Window, we do the following:

```
>> A = [4 2; 3 5];
>> mu = 6;
>> X = [1 1]';
>> maxI = 5;
>> sol = SIPM(A, X, mu, maxI);
```

Program 4.3
MATLAB m-file for Using Shifted Inverse Power method
function sol=SIPM $(A, X, mu, maxI)$
$[n, n] = size(A); B = A - mu * eye(n); C = inv(B);$
for k=1:maxI; for i=1:n; s=0;
for j=1:n; $ss = C(i, j) * X(j, 1); s = s + ss$;end
$XX(i, 1) = s$;end; $X = XX; y = max(X)$;
for i=1:n; $X(i, 1) = 1/y * X(i, 1)$;end; yy=abs(y-y1);
if $(yy <= 1e - 6)$; break; end; $lmu = (1/y) + mu$; end

4.3   Location of the Eigenvalues

Here, we discuss two well-known theorems that are some of the more important among the many theorems that deal with the location of eigenvalues
of both symmetric and nonsymmetric matrices, i.e., the location of zeros of
the characteristic polynomial. The eigenvalues of a nonsymmetric matrix
could, of course, be complex, in which case the theorems give us a means
of locating these numbers in the complex plane. The theorems can also be
used to estimate the magnitude of the largest and smallest eigenvalues and
thus to estimate the *spectral radius* $\rho(A)$ of A and the *condition number*
of A. Such estimates can be used to generate initial approximations to be
used in iterative methods for determining eigenvalues.

4.3.1   Gerschgorin Circles Theorem

Let A be an $n \times n$ matrix, and R_i denote the circles in the complex plane
C, with center a_{ii} and radius $\sum\limits_{\substack{j=1 \\ j \neq i}}^{n} |a_{i,j}|$, i.e.,

$$R_i = \{z \in C: \quad |z - a_{ii}| \leq \sum_{\substack{j=1 \\ j \neq i}}^{n} |a_{ij}|\}, \quad i = 1, 2, \cdots, n, \qquad (4.18)$$

where the variable z is complex valued.

    The eigenvalues of A are contained within $R = \cup_{i=1}^{n} R_i$ and the union
of any k of these circles that do not intersect the remaining $(n - k)$ must
contain precisely k (counting multiplication) of the eigenvalues.   ●

Example 4.7 *Consider the matrix*

$$A = \begin{pmatrix} 10 & 1 & 1 & 2 \\ 1 & 5 & 1 & 0 \\ 1 & 1 & -5 & 0 \\ 2 & 0 & 0 & -10 \end{pmatrix},$$

which is symmetric and has only real eigenvalues. The Gerschgorin circles associated with A are given by

$$\begin{aligned} R_1 &= \{z \in C| \; |z - 10| \le 4\} \\ R_2 &= \{z \in C| \; |z - 5| \le 2\} \\ R_3 &= \{z \in C| \; |z + 5| \le 2\} \\ R_4 &= \{z \in C| \; |z + 10| \le 2\}. \end{aligned}$$

These circles are illustrated in Figure 4.1, and Gerschgorin's theorem indicates that the eigenvalues of A lie inside the circles. The circles are about

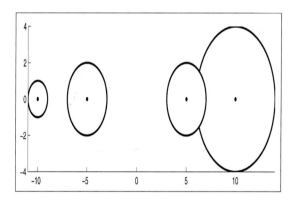

Figure 4.1: Circles for Example 4.7.

-10 and -5 each and must contain an eigenvalue. The other eigenvalues must lie in the interval $[3, 14]$. By using the shifted inverse power method, with $\epsilon = 0.000005$, with initial approximations of $10, 5, -5$, and -10, leads to approximations of

$$\begin{aligned} \lambda_1 &= 10.4698, & \lambda_2 &= 4.8803 \\ \lambda_3 &= -5.1497, & \lambda_4 &= -10.2004, \end{aligned}$$

respectively. The number of iterations required ranges from 9 to 13. ●

Example 4.8 *Consider the matrix*

$$A = \begin{pmatrix} 1 & 2 & -1 \\ 2 & 7 & 0 \\ -1 & 0 & -5 \end{pmatrix},$$

which is symmetric and so has only real eigenvalues. The Gerschgorin circles are

$$\begin{array}{lll} C_1: & |z - 1| & \leq & 3 \\ C_2: & |z - 7| & \leq & 2 \\ C_3: & |z + 5| & \leq & 1. \end{array}$$

These circles are illustrated in Figure 4.2, and Gerschgorin's theorem indicates that the eigenvalues of A lie inside the circles.

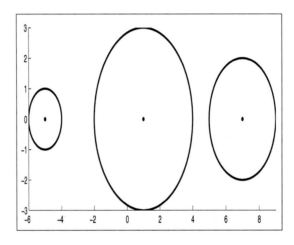

Figure 4.2: Circles for Example 4.8.

Then by Gerschgorin's theorem, any eigenvalues of A must lie in one of the three intervals $[-2, 4]$, $[5, 9]$, and $[-6, -4]$. Since the eigenvalues of A are 0, 5, and 9, $\lambda_1 = 0$ lies in circle C_1 and $\lambda_2 = 5$ and $\lambda_3 = 9$ lie in circle C_2.•

4.3.2 Rayleigh Quotient

The shifted inverse power method requires the input of an initial approximation μ for the eigenvalue λ of a matrix A. It can be obtained by the

Rayleigh quotient as

$$\mu = \frac{\mathbf{x}^T A \mathbf{x}}{\mathbf{x}^T \mathbf{x}}. \tag{4.19}$$

The maximum eigenvalue λ_1 can be obtained when $\mathbf{x}$ is the corresponding vector, as in

$$\lambda_1 = \max_{\mathbf{x} \neq 0} \frac{\mathbf{x}^T A \mathbf{x}}{\mathbf{x}^T \mathbf{x}}. \tag{4.20}$$

In the case where λ_1 is the dominant eigenvalue of a matrix A, and $\mathbf{x}$ is the corresponding eigenvector, then the Rayleigh quotient is

$$\frac{\mathbf{x}^T A \mathbf{x}}{\mathbf{x}^T \mathbf{x}} = \frac{\mathbf{x}^T \lambda_1 \mathbf{x}}{\mathbf{x}^T \mathbf{x}} = \frac{\lambda_1 (\mathbf{x}^T \mathbf{x})}{\mathbf{x}^T \mathbf{x}} = \lambda_1.$$

Thus, if $\mathbf{x}_k$ converges to a dominant eigenvector $\mathbf{x}$, then it seems reasonable that

$$\frac{\mathbf{x}_k{}^T A \mathbf{x}_k}{\mathbf{x}_k{}^T \mathbf{x}_k}$$

converges to

$$\frac{\mathbf{x}^T A \mathbf{x}}{\mathbf{x}^T \mathbf{x}} = \lambda_1,$$

which is the dominant eigenvalue.

Theorem 4.3 (Rayleigh Quotient Theorem)

If the eigenvalues of a real symmetric matrix A are

$$\lambda_1 \geq \lambda_2 \geq \lambda_3 \cdots \geq \lambda_n, \tag{4.21}$$

and if $\mathbf{x}$ is any nonzero vector, then

$$\lambda_n \leq \frac{\mathbf{x}^T A \mathbf{x}}{\mathbf{x}^T \mathbf{x}} \leq \lambda_1. \tag{4.22}$$

●

Example 4.9 *Consider the symmetric matrix*

$$A = \begin{pmatrix} 2 & -1 & 3 \\ -1 & 1 & -2 \\ 3 & -2 & 1 \end{pmatrix},$$

and the vector $\mathbf{x}$ *as*

$$\mathbf{x} = \begin{pmatrix} 1 \\ 1 \\ 1 \end{pmatrix}.$$

Then

$$\mathbf{x}^T\mathbf{x} = \begin{pmatrix} 1 & 1 & 1 \end{pmatrix} \begin{pmatrix} 1 \\ 1 \\ 1 \end{pmatrix} = 3$$

and

$$\mathbf{x}^T A\mathbf{x} = \begin{pmatrix} 1 & 1 & 1 \end{pmatrix} \begin{pmatrix} 2 & -1 & 3 \\ -1 & 1 & -2 \\ 3 & -2 & 1 \end{pmatrix} \begin{pmatrix} 1 \\ 1 \\ 1 \end{pmatrix}$$

$$= \begin{pmatrix} 1 & 1 & 1 \end{pmatrix} \begin{pmatrix} 4 \\ -2 \\ 2 \end{pmatrix} = 4.$$

Thus,

$$\lambda_3 \leq \frac{4}{3} \leq \lambda_1.$$

If μ is close to an eigenvalue λ_1, then convergence will be quite rapid.

●

4.4 Intermediate Eigenvalues

Once the largest eigenvalue is determined, then there is a method to obtain the approximations to the other possible eigenvalues of a matrix. This method is called *matrix deflation* and it is applicable to both symmetrical and nonsymmetrical coefficients matrices. The deflation method involves forming a new matrix B whose eigenvalues are the same as those of A, except that the dominant eigenvalue of A is replaced by the eigenvalue zero in B.

It is evident that this process can be continued until all of the eigenvalues have been extracted. Although this method shows promise, it does have a significant drawback, i.e., at each iteration performed in deflating the original matrix, any errors in the computed eigenvalues and eigenvectors will be passed on to the next eigenvectors. This could result in serious inaccuracy, especially when dealing with large eigenvalue problems. This is precisely why this method is generally used for small eigenvalue problems.

The following preliminary results are essential in using this technique.

Theorem 4.4 *If a matrix A has eigenvalues λ_i corresponding to eigenvectors $\mathbf{x}_i$, then $Q^{-1}AQ$ has the same eigenvalues as A but with eigenvectors $Q^{-1}\mathbf{x}_i$ for any nonsingular matrix Q.* ●

Theorem 4.5 *Let*

$$B = \begin{pmatrix} \lambda_1 & a_{12} & a_{13} & \cdots & a_{1n} \\ 0 & c_{22} & c_{23} & \cdots & c_{2n} \\ 0 & c_{32} & c_{33} & \cdots & c_{3n} \\ \vdots & \vdots & \vdots & \cdots & \vdots \\ 0 & c_{n2} & c_{n3} & \cdots & c_{nn} \end{pmatrix}, \tag{4.23}$$

and let C be an $(n-1)\times(n-1)$ matrix obtained by deleting the first row and first column of a matrix B. The matrix B has eigenvalues λ_1 together with the $(n-1)$ eigenvalues of C. Moreover, if $(\beta_2, \beta_3, \ldots, \beta_n)^T$ is an eigenvector of C with eigenvalue $\mu \neq \lambda_1$, then the corresponding eigenvector of B is $(\beta_1, \beta_2, \ldots, \beta_n)^T$, with

$$\beta_1 = \frac{\sum\limits_{j=2}^{n} a_{1j}\beta_j}{\mu - \lambda_1}. \tag{4.24}$$

Note that eigenvectors $\mathbf{x}_i$ of A can be recovered by premultiplication by Q. ●

Example 4.10 *Consider the matrix*

$$A = \begin{pmatrix} 10 & -6 & -4 \\ -6 & 11 & 2 \\ -4 & 2 & 6 \end{pmatrix},$$

which has the dominant eigenvalue $\lambda_1 = 18$, with the corresponding eigen-vector $\mathbf{x}_1 = [1, -1, -\frac{1}{2}]^T$. Use the deflation method to find the other eigenvalues and eigenvectors of A.

Solution. *The transformation matrix is given as*

$$Q = \begin{pmatrix} 1 & 0 & 0 \\ -1 & 1 & 0 \\ -\dfrac{1}{2} & 0 & 1 \end{pmatrix}.$$

Then

$$B = Q^{-1}AQ = \begin{pmatrix} 1 & 0 & 0 \\ 1 & 1 & 0 \\ \dfrac{1}{2} & 0 & 1 \end{pmatrix} \begin{pmatrix} 10 & -6 & -4 \\ -6 & 11 & 2 \\ -4 & 2 & 6 \end{pmatrix} \begin{pmatrix} 1 & 0 & 0 \\ -1 & 1 & 0 \\ -\dfrac{1}{2} & 0 & 1 \end{pmatrix}.$$

After simplifying, we get

$$B = \begin{pmatrix} 18 & -6 & -4 \\ 0 & 5 & -2 \\ 0 & -1 & 4 \end{pmatrix}.$$

So the deflated matrix is

$$C = \begin{pmatrix} 5 & -2 \\ -1 & 4 \end{pmatrix}.$$

Now we can easily find the eigenvalues of C, which are 6 and 3, with the corresponding eigenvectors $[1, -\frac{1}{2}]^T$ and $[1, 1]^T$ respectively. Thus, the other two eigenvalues of A are 6 and 3. Now we calculate the eigenvectors of A corresponding to these two eigenvalues. First, we calculate the eigenvectors of B corresponding to $\lambda = 6$ from the system

$$\begin{pmatrix} 18 & -6 & -4 \\ 0 & 5 & -2 \\ 0 & -1 & 4 \end{pmatrix} \begin{pmatrix} \beta_1 \\ 1 \\ -\dfrac{1}{2} \end{pmatrix} = 6 \begin{pmatrix} \beta_1 \\ 1 \\ -\dfrac{1}{2} \end{pmatrix}.$$

Then by solving the above system, we have

$$18\beta_1 - 4 = 6\beta_1,$$

which gives $\beta_1 = \frac{1}{3}$. Similarly, we can find the value of β_1 corresponding to $\lambda = 3$ by using the system as

$$\begin{pmatrix} 18 & -6 & -4 \\ 0 & 5 & -2 \\ 0 & -1 & 4 \end{pmatrix} \begin{pmatrix} \beta_1 \\ 1 \\ 1 \end{pmatrix} = 3 \begin{pmatrix} \beta_1 \\ 1 \\ 1 \end{pmatrix},$$

which gives $\beta_1 = \frac{2}{3}$. Thus, the eigenvectors of B are $\mathbf{v}_1 = [\frac{1}{3}, 1, -\frac{1}{2}]^T$ and $\mathbf{v}_2 = [\frac{2}{3}, 1, 1]^T$.

Now we find the eigenvectors of the original matrix A, which can be obtained by premultiplying the vectors of B by nonsingular matrix Q. First, the second eigenvector of A can be found as

$$\mathbf{x}_2 = Q\mathbf{v}_1 = \begin{pmatrix} 1 & 0 & 0 \\ -1 & 1 & 0 \\ -\frac{1}{2} & 0 & 1 \end{pmatrix} \begin{pmatrix} \frac{1}{3} \\ 1 \\ -\frac{1}{2} \end{pmatrix} = \begin{pmatrix} \frac{1}{3} \\ \frac{2}{3} \\ -\frac{2}{3} \end{pmatrix},$$

or, equivalently, $\mathbf{x}_2 = [\frac{1}{2}, 1, -1]^T$. Similarly, the third eigenvector of the given matrix A can be computed as

$$\mathbf{x}_3 = Q\mathbf{v}_2 = \begin{pmatrix} 1 & 0 & 0 \\ -1 & 1 & 0 \\ -\frac{1}{2} & 0 & 1 \end{pmatrix} \begin{pmatrix} \frac{2}{3} \\ 1 \\ 1 \end{pmatrix} = \begin{pmatrix} \frac{2}{3} \\ \frac{1}{3} \\ \frac{2}{3} \end{pmatrix},$$

or, equivalently, $\mathbf{x}_3 = [1, \frac{1}{2}, 1]^T$. ●

Note that in this example the deflated matrix C is nonsymmetric even though the original matrix A is symmetric. We deduce that the property of symmetry is not preserved in the deflation process. Also, note that the method of deflation fails whenever the first element of given vector $\mathbf{x}_1$ is zero, since $\mathbf{x}_1$ cannot then be scaled so that this number is one.

The above results can be reproduced using MATLAB commands as follows:

```
>> A = [10 − 6 − 4; −6 11 2; −4 2 6];
>> Lamda = 18;
>> XA = [1 − 1 − 0.5]';
>> [Lamda, X] = DEFLATED(A, Lamda, XA);
```

Program 4.4
MATLAB m-file for Using the Deflation method
function [Lamda,X]=DEFLATED(A,Lamda,XA)
[n,n]=size(**a**); Q=eye(n);
$Q(:, 1) = XA(:, 1); B = inv(Q) * A * Q;$ c=B(2:n,2:n);
$[xv, ev] = eig(c,' nobalance');$
for i=1:n-1
$b = -(B(1, 2 : n) * xv(:, i))/(Lamda - ev(i, i));$
$Xb(:, i) = [bxv(:, i)']'; XA(:, i + 1) = Q * Xb(:, i);$ end
Lamda=[Lamda;diag(ev)]; Lamda; $XA;$ end

4.5 Eigenvalues of Symmetric Matrices

In the previous section we discussed the power methods for finding individual eigenvalues. The regular power method can be used to find the distinct eigenvalue with the largest magnitude, i.e., the dominant eigenvalue, and the inverse power method can find the eigenvalue called the smallest eigenvalue, and the shifted inverse power method can find the subdominant eigenvalues. In this section we develop some methods to find all eigenvalues of a given matrix. The basic approach is to find a sequence of similarity transformations that transform the original matrix into a simple form. Clearly, the best form for the transformed matrix would be a diagonal one, but this is not always possible, since some transformed matrices would be tridiagonal. Furthermore, these techniques are generally limited to symmetrical matrices with real coefficients.

Before we discuss these methods, we define some special matrices, which are very useful in discussing these methods.

Definition 4.1 (Orthogonally Similar Matrix)

A matrix A is said to be orthogonally similar to a matrix B, if there is an orthogonal matrix Q for which

$$A = QBQ^T. \tag{4.25}$$

If A is a symmetric and $B = Q^{-1}AQ$, then

$$B^T = (Q^T A Q)^T = Q^T A Q = B.$$

Thus, similarity transformations on symmetric matrices that use orthogonal matrices produce matrices which are again symmetric. ●

Definition 4.2 (Rotation Matrix)

A rotation matrix Q is an orthogonal matrix that differs from the identity matrix in, at most, four elements. These four elements at the vertices of the rectangle have been replaced by $\cos\theta$, $-\sin\theta$, $\sin\theta$, and $\cos\theta$ in the positions pp, pq, qp, qq, respectively. For example, the matrix

$$B = \begin{pmatrix} 1 & 0 & 0 & 0 & 0 \\ 0 & \cos\theta & 0 & -\sin\theta & 0 \\ 0 & 0 & 1 & 0 & 0 \\ 0 & \sin\theta & 0 & \cos\theta & 0 \\ 0 & 0 & 0 & 0 & 1 \end{pmatrix} \tag{4.26}$$

is the rotation matrix, where $p = 2$ and $q = 4$. Note that a rotation matrix is also an orthogonal matrix, i.e., $B^T B = I$. ●

4.5.1 Jacobi Method

This method can be used to find all the eigenvalues and eigenvectors of a symmetric matrix by performing a series of similarity transformations. The Jacobi method permits the transformation of a symmetric matrix into a diagonal one having the same eigenvalues as the original matrix. This can be done by eliminating off-diagonal elements in a systematic way. The method requires an infinite number of iterations to produce the diagonal form. This is because the reduction of a given element to zero in a matrix will most likely introduce a nonzero element into a previous zero coefficient. Hence, the method can be viewed as an iterative procedure that can approach a diagonal form using a finite number of steps. The implication is that the off-diagonal coefficients will be close to zero rather than exactly equal to zero.

Consider the eigenvalue problem

$$A\mathbf{v} = \lambda\mathbf{v}, \tag{4.27}$$

where A is a symmetric matrix of order $n \times n$, and let the solution of (4.27) give the eigenvalues $\lambda_1, \ldots, \lambda_n$ and the corresponding eigenvectors $\mathbf{v}_1, \ldots, \mathbf{v}_n$ of A. Since the eigenvectors of a symmetric matrix are orthogonal, i.e.,

$$\mathbf{v}^T = \mathbf{v}^{-1}, \tag{4.28}$$

by using (4.28), we can write (4.27) as

$$\mathbf{v}^T A\mathbf{v} = \lambda. \tag{4.29}$$

The basic procedure for the Jacobi method is as follows.

Assume that

$$A_1 \ = \ Q_1^T A Q_1$$

$$A_2 \ = \ Q_2^T A_1 Q_2 = Q_2^T Q_1^T A Q_1 Q_2$$

$$A_3 \ = \ Q_3^T A_2 Q_3 = Q_3^T Q_2^T Q_1^T A Q_1 Q_2 Q_3$$

$$\vdots \qquad \vdots$$

$$A_k = Q_k^T \cdots Q_1^T A Q_1 \cdots Q_k.$$

We see that as $k \to \infty$, then

$$A_k \to \lambda \quad \text{and} \quad Q_1 Q_2 \cdots Q_k \to \mathbf{v}. \tag{4.30}$$

The matrix $Q_i(i = 1, 2, \ldots, k)$ is a rotation matrix that is constructed in such a way that off-diagonal coefficients in matrix A_k are reduced to zero. In other words, in a rotation matrix

$$Q_k = \begin{pmatrix} 1 & & & & & \\ & \ddots & & & & \\ & & cos\theta & -sin\theta & & \\ & & & & & \\ & & sin\theta & cos\theta & & \\ & & & & \ddots & \\ & & & & & 1 \end{pmatrix},$$

the value of θ is selected in such a way that the a_{pq} coefficient in A_k is reduced to zero, i.e.,

$$\tan 2\theta = \frac{2a_{pq}}{a_{pp} - a_{qq}}. \tag{4.31}$$

Theoretically, there are an infinite number of θ values corresponding to the infinite matrices A_k. However, as θ approaches zero, a rotation matrix Q_k becomes an identity matrix and no further transformations are required.

There are three strategies for annihilating off-diagonals. The first is called the *serial method*, which selects the elements in row order, i.e., in the positions $(1, 2), \ldots, (1, n)$; $(2, 3), \ldots, (2, n)$; $\ldots$; $(n-1, n)$ in turn, which is then repeated. The second method is called the *natural method*, which searches through all of the off-diagonals and annihilates the elements of the largest modules at each stage. Although this method converges faster than the serial method, it is not recommended for large values of n, since the actual search procedure itself can be extremely time consuming. The third method is known as the *threshold serial method,* in which the off-diagonals are cycled in row order as in the serial method, omitting transformations on any element whose magnitude is below some threshold value. This value is usually decreased after each cycle. The advantage of this approach is that zeros are only created in positions where it is worthwhile to do so, without the need for a lengthy search. Here, we shall use only the natural method for annihilating the off-diagonal elements.

Theorem 4.6 *Consider a matrix A and a rotation matrix Q as*

$$A = \begin{pmatrix} a_{11} & a_{12} \\ a_{12} & a_{22} \end{pmatrix} \quad and \quad Q = \begin{pmatrix} p_{11} & p_{12} \\ p_{21} & p_{22} \end{pmatrix}.$$

Then there exists θ such that:

1. $Q^T Q = \mathbf{I},$
2. $Q^T A Q = D,$

where $\mathbf{I}$ is an identity matrix and D is a diagonal matrix, and its diagonal elements, λ_1 and λ_2, are the eigenvalues of A.

Proof. To convert the given matrix A into a diagonal matrix D, we have to make off-diagonal element a_{12} of A zero, i.e., $p = 1$ and $q = 2$. Consider $p_{11} = \cos\theta = p_{22}$ and $p_{12} = -p_{21} = \sin\theta$, then the matrix Q has the form

$$Q = \begin{pmatrix} \cos\theta & -\sin\theta \\ \sin\theta & \cos\theta \end{pmatrix}.$$

The corresponding matrix A_1 can be constructed as

$$A_1 = Q_1^T A Q_1$$

or

$$\begin{pmatrix} a_{11}^* & a_{12}^* \\ a_{12}^* & a_{22}^* \end{pmatrix} = \begin{pmatrix} \cos\theta & \sin\theta \\ -\sin\theta & \cos\theta \end{pmatrix} \begin{pmatrix} a_{11} & a_{12} \\ a_{21} & a_{22} \end{pmatrix} \begin{pmatrix} \cos\theta & -\sin\theta \\ \sin\theta & \cos\theta \end{pmatrix}.$$

Since our task is to reduce a_{12}^* to zero, carrying out the multiplication on the right-hand side and using matrix equality gives

$$a_{12}^* = 0 = -(\sin\theta\cos\theta)a_{11} + (\cos^2\theta)a_{12} - (\sin^2\theta)a_{12} + (\cos\theta\sin\theta)a_{22}.$$

Simplifying and rearranging gives

$$\frac{\sin\theta\cos\theta}{\cos^2\theta - \sin^2\theta} = \frac{a_{12}}{a_{11} - a_{22}}$$

$$\frac{\sin 2\theta}{2\cos 2\theta} = \frac{a_{12}}{a_{11} - a_{22}}$$

$$\frac{\sin 2\theta}{\cos 2\theta} = \frac{2a_{12}}{a_{11} - a_{22}},$$

or more simply

$$\tan 2\theta = \frac{2a_{12}}{a_{11} - a_{22}}, \qquad a_{11} \neq a_{22}.$$

Note that if $a_{11} = a_{22}$, this implies that $\theta = \frac{\pi}{4}$. We found that for a 2×2 matrix, it required only one iteration to convert the given matrix A to a diagonal matrix D.

Similarly, for a higher order matrix, a diagonal matrix D can be obtained by a number of such multiplications, i.e.,

$$Q_k^T Q_{k-1}^T \cdots Q_1^T A Q_1 \cdots Q_{k-1} Q_k = D.$$

The diagonal elements of D are all the eigenvalues λ of A and the corresponding eigenvectors $\mathbf{v}$ of A can be obtained as

$$Q_1 Q_2 \cdots Q_k = \mathbf{v}.$$

•

Example 4.11 *Use the Jacobi method to find the eigenvalues and the eigenvectors of the matrix*

$$A = \begin{pmatrix} 3.0 & 0.01 & 0.02 \\ 0.01 & 2.0 & 0.1 \\ 0.02 & 0.1 & 1.0 \end{pmatrix}.$$

Solution. *The largest off-diagonal entry of the given matrix A is $a_{23} = 0.1$, so we begin by reducing element a_{23} to zero. Since $p = 2$ and $q = 3$, the first orthogonal transformation matrix has the form*

$$Q_1 = \begin{pmatrix} 1 & 0 & 0 \\ 0 & c & -s \\ 0 & s & c \end{pmatrix}.$$

The values of $c = \cos\theta$ and $s = \sin\theta$ can be obtained as follows:

$$\theta = \frac{1}{2}\arctan\left(\frac{2a_{23}}{a_{22} - a_{33}}\right) = \frac{1}{2}\arctan\left(\frac{2(0.1)}{2 - 1}\right) \approx 6.2833$$

$$\cos\theta \approx 0.9951 \quad and \quad \sin\theta \approx 0.0985.$$

Then

$$Q_1 = \begin{pmatrix} 1 & 0 & 0 \\ 0 & 0.9951 & -0.0985 \\ 0 & 0.0985 & 0.9951 \end{pmatrix} \quad and \quad Q_1^T = \begin{pmatrix} 1 & 0 & 0 \\ 0 & 0.9951 & 0.0985 \\ 0 & -0.0985 & 0.9951 \end{pmatrix}$$

and

$$A_1 = Q_1^T A Q_1 = \begin{pmatrix} 3.0 & 0.0119 & 0.0189 \\ 0.0119 & 2.0099 & 0 \\ 0.0189 & 0 & 0.9901 \end{pmatrix}.$$

Note that the rotation makes a_{32} and a_{23} zero, increasing slightly a_{21} and a_{12}, and decreasing the second dominant off-diagonal entries a_{13} and a_{31}.

Now the largest off-diagonal element of the matrix A_1 is $a_{13} = 0.0189$, so to make this position zero, we consider the second orthogonal matrix of the form

$$Q_2 = \begin{pmatrix} c & 0 & -s \\ 0 & 1 & 0 \\ s & 0 & c \end{pmatrix},$$

and the values of c and s can be obtained as follows:

$$\theta = \frac{1}{2} \arctan\left(\frac{2a_{13}}{a_{11} - a_{33}}\right) = \frac{1}{2} \arctan\left(\frac{2(0.0189)}{3 - 0.9901}\right) \approx 0.5984$$

$$\cos\theta \approx 0.9999 \quad and \quad \sin\theta \approx 0.0094.$$

Then

$$Q_2 = \begin{pmatrix} 0.9999 & 0 & -0.0094 \\ 0 & 1 & 0 \\ 0.0094 & 0 & 0.9999 \end{pmatrix} \quad and \quad Q_2^T = \begin{pmatrix} 0.9999 & 0 & 0.0094 \\ 0 & 1 & 0 \\ -0.0094 & 0 & 0.9999 \end{pmatrix}.$$

Hence,

$$A_2 = Q_2^T A_1 Q_2 = Q_2^T Q_1^T A Q_1 Q_2 = \begin{pmatrix} 3.0002 & 0.0119 & 0 \\ 0.0119 & 2.0099 & -0.0001 \\ 0 & -0.0001 & 0.9899 \end{pmatrix}.$$

Similarly, to make off-diagonal element $a_{12} = 0.0119$ of the matrix A_2 zero, we consider the third orthogonal matrix of the form

$$Q_3 = \begin{pmatrix} c & -s & 0 \\ s & c & 0 \\ 0 & 0 & 1 \end{pmatrix}$$

and

$$\theta = \frac{1}{2} \arctan\left(\frac{2a_{12}}{a_{11} - a_{22}}\right) = \frac{1}{2} \arctan\left(\frac{2(0.0119)}{3.0002 - 2.0099}\right) \approx 0.7638$$

$$\cos\theta \approx 0.9999 \quad and \quad \sin\theta \approx 0.0120.$$

Then

$$Q_3 = \begin{pmatrix} 0.9999 & -0.0120 & 0 \\ 0.0120 & 0.9999 & 0 \\ 0 & 0 & 1 \end{pmatrix} \quad and$$

$$Q_3^T = \begin{pmatrix} 0.9999 & 0.0120 & 0 \\ -0.0120 & 0.9999 & 0 \\ 0 & 0 & 1 \end{pmatrix}.$$

Hence,

$$A_3 = Q_3^T Q_2^T Q_1^T A Q_1 Q_2 Q_3 = \begin{pmatrix} 3.0003 & 0 & -1.35E-6 \\ 0 & 2.00 & -1.122E-4 \\ -1.35E-6 & -1.122E-4 & 0.9899 \end{pmatrix},$$

which gives the diagonal matrix D, and its diagonal elements converge to 3, 2, and 1, which are the eigenvalues of the original matrix A. The corresponding eigenvectors can be computed as follows:

$$\mathbf{v} = Q_1 Q_2 Q_3 = \begin{pmatrix} 0.9998 & -0.0121 & -0.0094 \\ 0.0111 & 0.9951 & -0.0985 \\ 0.0106 & 0.0984 & 0.9951 \end{pmatrix}.$$

To reproduce the above results by using the Jacobi method and the MATLAB Command Window, we do the following:

```
>> A = [3 0.01 0.02; 0.01 2 0.1; 0.02 0.1 1];
>> sol = JOBM(A);
```

Program 4.5

MATLAB m-file for the Jacobi method

```
function sol=JOBM(A)
[n,n]=size(A); QQ=[ ];
for u = 1 : .5 * n * (n - 1); for i=1:n; for j=1:n
if (j > i); aa(i,j)=A(i,j); else; aa(i,j)=0;
end; end; end
aa=abs(aa); mm=max(aa); m=max(mm);
[i,j]=find(aa==m); i=i(1); j=j(1);
t = .5 * atan(2 * A(i, j)/(A(i, i) - A(j, j))); c=cos(t); s=sin(t);
for ii=1:n; for jj=1:n; Q(ii,jj)=0.0;
if (ii==jj); Q(ii,jj)=1.0; end; end; end
Q(i,i)=c; Q(i,j)=-s; Q(j,i)=s; Q(j,j)=c;
for i1=1:n; for j1=1:n;
QT(i1,j1)=Q(j1,i1); end; end
for i2=1:n; for j2=1:n; s=0;
for k = 1 : n; ss = QT(i2, k) * A(k, j2);
s=s+ss; end; QTA(i2,j2)=s; end; end
for i3=1:n; for j3=1:n; s=0;
for k=1:n; ss = QTA(i3, k) * Q(k, j3); s=s+ss; end
A(i3,j3)=s; end; end; QQ=[QQ,Q]; end; D=A
y=[]; for k=1:n; yy=A(k,k); y=[y;yy]; end; eigvals=y
x=Q; if (n > 2) % Compute eigenvectors
x(1:n,1:n)=QQ(1:n,1:n);
for c = n + 1 : n : n * n; xx(1:n,1:n)=QQ(1:n,c:n+c-1);
for i=1:n; for j=1:n; s=0;
for k=1:n; ss = x(i, k) * xx(k, j); s=s+ss; end
x1(i,j)=s; end; end; x=x1; end; end
```

4.5.2 Sturm Sequence Iteration

When a symmetric matrix is tridiagonal, then the eigenvalues of a tridiagonal matrix can be computed to any specified precision using a simple method called the Sturm sequence iteration. In the following sections we will discuss two methods that will convert the given symmetric matrices into symmetrical tridiagonal forms by using similarity transformations. The Sturm sequence iteration below can, therefore, be used in the calculation of eigenvalues of any symmetric tridiagonal matrix. Consider a symmetric tridiagonal matrix of order 4×4 as

$$A = \begin{pmatrix} a_1 & b_2 & 0 & 0 \\ b_2 & a_2 & b_3 & 0 \\ 0 & b_3 & a_3 & b_4 \\ 0 & 0 & b_4 & a_4 \end{pmatrix},$$

and assume that $b_i \neq 0$, for each $i = 2, 3, 4$. Then one can define the characteristic polynomial of a given matrix A as

$$f_4(\lambda) = det(A - \lambda I), \tag{4.32}$$

which is equivalent to

$$f_4(\lambda) = \begin{vmatrix} a_1 - \lambda & b_2 & 0 & 0 \\ b_2 & a_2 - \lambda & b_3 & 0 \\ 0 & b_3 & a_3 - \lambda & b_4 \\ 0 & 0 & b_4 & a_4 - \lambda \end{vmatrix}.$$

We expand by minors in the last row as

$$f_4(\lambda) = (a_4 - \lambda) \begin{vmatrix} a_1 - \lambda & b_2 & 0 \\ b_2 & a_2 - \lambda & b_3 \\ 0 & b_3 & a_3 - \lambda \end{vmatrix} - b_4 \begin{vmatrix} a_1 - \lambda & b_2 & 0 \\ b_2 & a_2 - \lambda & 0 \\ 0 & b_3 & b_4 \end{vmatrix}$$

or

$$f_4(\lambda) = (a_4 - \lambda) f_3(\lambda) - b_4^2 f_2(\lambda). \tag{4.33}$$

The recurrence relation (4.33) is true for a matrix of any order $r \times r$, i.e.,

$$f_r(\lambda) = (a_r - \lambda) f_{r-1}(\lambda) - b_r^2 f_{r-2}(\lambda), \tag{4.34}$$

provided that we define $f_0(\lambda) = 1$ and evaluate $f_1(\lambda) = a_1 - \lambda$.

The sequence $\{f_0, f_1, \ldots, f_r, \ldots\}$ is known as the Sturm sequence. So starting with $f_0(\lambda) = 1$, we can eventually find a characteristic polynomial of A by using

$$f_n(\lambda) = 0. \tag{4.35}$$

Example 4.12 *Use the Sturm sequence iteration to find the eigenvalues of the symmetric tridiagonal matrix*

$$A = \begin{pmatrix} 1 & 2 & 0 & 0 \\ 2 & 4 & 1 & 0 \\ 0 & 1 & 5 & -1 \\ 0 & 0 & -1 & 3 \end{pmatrix}.$$

Solution. *We compute the Sturm sequences as follows:*

$$\begin{aligned} f_0(\lambda) &= 1 \\ f_1(\lambda) &= (a_1 - \lambda) \\ &= 1 - \lambda. \end{aligned}$$

The second sequence is

$$\begin{aligned} f_2(\lambda) &= (a_2 - \lambda)f_1(\lambda) - b_2^2 f_0(\lambda) \\ &= (4 - \lambda)(1 - \lambda) - 4(1) \\ &= \lambda^2 - 5\lambda. \end{aligned}$$

and the third sequence is

$$\begin{aligned} f_3(\lambda) &= (a_3 - \lambda)f_2(\lambda) - b_3^2 f_1(\lambda) \\ &= (5 - \lambda)(\lambda^2 - 5\lambda) - (1)^2(1 - \lambda) \\ &= -\lambda^3 + 10\lambda^2 - 24\lambda - 1. \end{aligned}$$

Finally, the fourth sequence is

$$\begin{aligned} f_4(\lambda) &= (a_4 - \lambda)f_3(\lambda) - b_4^2 f_2(\lambda) \\ &= (3 - \lambda)(-\lambda^3 + 10\lambda^2 - 24\lambda - 1) - (-1)^2(\lambda^2 - 5\lambda) \\ &= \lambda^4 - 13\lambda^3 - 53\lambda^2 - 66\lambda - 3. \end{aligned}$$

Thus,

$$f_4(\lambda) = \lambda^4 - 13\lambda^3 - 53\lambda^2 - 66\lambda - 3 = 0.$$

Solving the above equation, we have the eigenvalues $6.11, 4.41, 2.54,$ *and* -0.04 *of the given symmetric tridiagonal matrix.*   ●

To get the above results using the MATLAB Command Window, we do the following:

```
>> A = [1 2 0 0; 2 4 1 0; 0 1 5 − 1; 0 0 − 1 3];
>> sol = SturmS(A);
```

Program 4.6
MATLAB m-file for the Sturm Sequence method
function sol=SturmS(A)
% This evaluates the eigenvalues of a tridiagonal symmetric matrix
[n,n] = size(A);
ff(1,:)=[1 0 0 0 0]; ff(2,:)=[A(1,1) -1 0 0 0];
for i=3:n+1; h=[A(i-1,i-1) -1];
$ff(i, 1) = h(1) * ff(i − 1, 1) − A(i − 1, i − 2)^\wedge 2 * ff(i − 2, 1);$
for z=2:n+1
$ff(i, z) = h(1)*ff(i−1, z)+h(2)*ff(i−1, z−1)−A(i−1, i−2)^\wedge$
$2 * ff(i − 2, z);$ end; end
for i=1:n+1; y(i)=ff(n+1,n+2-i);end; eigval=roots(y)

Theorem 4.7 *For any real number λ^*, the number of agreements in signs of successive terms of the Sturm sequence $\{f_0(\lambda^*), f_1(\lambda^*), \ldots, f_n(\lambda^*)\}$ is equal to the number of eigenvalues of the tridiagonal matrix A greater than λ^*. The sign of a zero is taken to be opposite to that of the previous term.*
●

Example 4.13 *Find the number of eigenvalues of the matrix*

$$A = \begin{pmatrix} 3 & -1 & 0 \\ -1 & 2 & -1 \\ 0 & -1 & 3 \end{pmatrix}$$

lying in the interval $(0, 4)$.

Solution. *Since the given matrix is of size 3×3, we have to compute the Sturm sequences $f_3(0)$ and $f_3(4)$. First, for $\lambda^* = 0$, we have*

$$f_0(0) = 1 \quad and \quad f_1(0) = (a_1 − 0) = (3 − 0) = 3.$$

Also,

$$f_2(0) = (a_2 - 0)f_1(0) - b_2^2 f_0(0)$$

$$= (2)(3) - (-1)^2(1) = 5.$$

Finally, we have

$$f_3(0) = (a_3 - 0)f_2(0) - b_3^2 f_1(0)$$

$$= (3)(5) - (-1)^2(3) = 12,$$

which have signs $+ + ++$, with three agreements. So all three eigenvalues are greater than $\lambda^* = 0$.

Similarly, we can calculate for $\lambda^* = 4$. The Sturm sequences are

$$f_0(4) = 1 \quad and \quad f_1(4) = (a_1 - 4) = (3 - 4) = -1.$$

Also,

$$f_2(4) = (a_2 - 4)f_1(4) - b_2^2 f_0(4)$$

$$= (2 - 4)(-1) - (-1)^2(1) = 1.$$

In the last, we have

$$f_3(4) = (a_3 - 4)f_2(4) - b_3^2 f_1(4)$$

$$= (3 - 4)(1) - (-1)^2(-1) = 0,$$

which have signs $+ - +-$, with no agreements. So no eigenvalues are greater than $\lambda^* = 4$. Hence, there are exactly three eigenvalues in $[0, 4]$. Furthermore, since $f_3(0) \neq 0$ and also, $f_3(4) = 0$, we deduce that no eigenvalue is exactly equal to 0 but one eigenvalue is exactly equal to 4, because $f_3(\lambda^*) = \det(A - \lambda^* \mathbf{I})$, the characteristic polynomial of A. Therefore, there are three eigenvalues in the half-open interval $(0, 4]$ and two eigenvalues are in the open interval $(0, 4)$. Since the given matrix A is positive-definite, therefore, by a well-known result, all of the eigenvalues of A must be strictly positive. Note that the eigenvalues of the given matrix A are $1, 3$, and 4.●

Note that if the sign pattern is $+ + + - -$, for a 4×4 matrix for $\lambda = c$, then there are three eigenvalues greater than $\lambda = c$.

If the sign pattern is $+ - + - -$, for a 4×4 matrix for $\lambda = c$, then there is one eigenvalue greater than $\lambda = c$.

If the sign pattern is $+ - 0 + +$, for a 4×4 matrix for $\lambda = c$, then there are two eigenvalues greater than $\lambda = c$.

4.5.3   Given's Method

This method is also based on similarity transformations of the same type as those used for the Jacobi method. The zeros created are retained, and the symmetric matrix is reduced to a symmetric tridiagonal matrix C rather than a diagonal form using a finite number of orthogonal similarity transformations. The eigenvalues of the original matrix A are the same as those of the symmetric tridiagonal matrix C. Given's method is generally preferable to the Jacobi method in that it requires a finite number of iterations.

For Given's method, the angle θ is chosen to create zeros, not in the (p, q) and (q, p) positions as in the Jacobi method, but in the $(p - 1, q)$ and $(q, p - 1)$ positions. This is because zeros can be created in row order without destroying those previously obtained.

In the first stage of Given's method we annihilate elements along the first row (and by symmetry, down the first column) in the positions $(1, 3)$, $\ldots, (1, n)$ using the rotation matrices $Q_{23}, \ldots, Q_{2n}$ in turn. Once a zero has been created in positions $(1, j)$, subsequent transformations use matrices Q_{pq} with $p, q \neq 1, j$ and so zeros are not destroyed. In the second stage we annihilate elements in the positions $(2, 4), \ldots, (2, n)$ using $Q_{34}, \ldots, Q_{3n}$. Again, any zeros produced by these transformations are not destroyed as subsequent zeros are created along the second row. Furthermore, zeros previously obtained in the first row are also preserved. The process continues until a zero is created in the position $(n - 2, n)$ using Q_{n-1n}. The original matrix can, therefore, be converted into a symmetric tridiagonal matrix C

in exactly

$$(n-2) + (n-3) + \cdots + 1 \equiv \frac{1}{2}(n-1)(n-2)$$

steps. This method also uses rotation matrices as the Jacobi method does, but in the following form:

$$\cos\theta = (p-1, p-1), \quad \sin\theta = (p-1, q), \quad -\sin\theta = (q, p-1), \quad \cos\theta = (q, q)$$

and

$$\theta = -\arctan\left(\frac{a_{p-1q}}{a_{p-1p}}\right).$$

We can also find the values of $\cos\theta$ and $\sin\theta$ by using

$$\cos\theta = \frac{|a_{p-1p}|}{R} \quad \text{and} \quad \sin\theta = \frac{|a_{p-1q}|}{R},$$

where

$$R = \sqrt{(a_{p-1p})^2 + (a_{p-1q})^2}.$$

Example 4.14 *Use Given's method to reduce the matrix*

$$A = \begin{pmatrix} 2 & -1 & 1 & 4 \\ -1 & 3 & 1 & 2 \\ 1 & 1 & 5 & -3 \\ 4 & 2 & -3 & 6 \end{pmatrix}$$

to a symmetric tridiagonal form and then find the eigenvalues of A.

Solution.
Step I. *Create a zero in the $(1,3)$ position by using the first orthogonal transformation matrix as*

$$Q_{23} = \begin{pmatrix} 1 & 0 & 0 & 0 \\ 0 & c & s & 0 \\ 0 & -s & c & 0 \\ 0 & 0 & 0 & 1 \end{pmatrix}.$$

To find the value of the $\cos\theta$ *and* $\sin\theta$, *we have*

$$\theta = -\arctan\left(\frac{a_{13}}{a_{12}}\right) = -\arctan\left(\frac{1}{-1}\right) \approx 50.0000$$

$$\cos\theta = 0.7071, \quad and \quad \sin\theta = 0.7071.$$

Then

$$Q_{23} = \begin{pmatrix} 1.0 & 0 & 0 & 0 \\ 0 & 0.7071 & 0.7071 & 0 \\ 0 & -0.7071 & 0.7071 & 0 \\ 0 & 0 & 0 & 1.0 \end{pmatrix} \quad and$$

$$Q_{23}^T = \begin{pmatrix} 1 & 0 & 0 & 0 \\ 0 & 0.7071 & -0.7071 & 0 \\ 0 & 0.7071 & 0.7071 & 0 \\ 0 & 0 & 0 & 1 \end{pmatrix},$$

which gives

$$A_1 = Q_{23}^T A Q_{23} = \begin{pmatrix} 2.0 & -1.4142 & 0 & 4.0 \\ -1.4142 & 3.0 & -1.0 & 3.535 \\ 0 & -1.0 & 5.0 & -0.7071 \\ 4.0 & 3.535 & -0.7071 & 6.0 \end{pmatrix}.$$

Note that because of the symmetry matrix, the lower part of A_1 *is the same as the upper part.*

Step II. *Create a zero in the* $(1,4)$ *position by using the second orthogonal transformation matrix as*

$$Q_{24} = \begin{pmatrix} 1 & 0 & 0 & 0 \\ 0 & c & 0 & s \\ 0 & 0 & 1 & 0 \\ 0 & -s & 0 & c \end{pmatrix}$$

and

$$\theta = -\arctan\left(\frac{a_{14}}{a_{12}}\right) = -\arctan\left(\frac{4}{-1.4142}\right) \approx 78.3658$$

$$\cos\theta = 0.3333, \quad and \quad \sin\theta = 0.9428.$$

Then

$$Q_{24} = \begin{pmatrix} 1 & 0 & 0 & 0 \\ 0 & 0.3333 & 0 & 0.9428 \\ 0 & 0 & 1 & 0 \\ 0 & -0.9428 & 0 & 0.3333 \end{pmatrix} \quad \text{and} \quad Q_{24}^T = \begin{pmatrix} 1 & 0 & 0 & 0 \\ 0 & 0.3333 & 0 & -0.9428 \\ 0 & 0 & 1 & 0 \\ 0 & 0.9428 & 0 & 0.3333 \end{pmatrix},$$

which gives

$$A_2 = Q_{24}^T A_1 Q_{24} = \begin{pmatrix} 2.0 & -4.2426 & 0 & 0 \\ -4.2426 & 3.4444 & 0.3333 & -3.6927 \\ 0 & 0.3333 & 5.0 & -1.1785 \\ 0 & -3.6927 & -1.7185 & 5.5556 \end{pmatrix}.$$

Step III. *Create a zero in the* $(2,4)$ *position by using the third orthogonal transformation matrix as*

$$Q_{34} = \begin{pmatrix} 1 & 0 & 0 & 0 \\ 0 & 1 & 0 & 0 \\ 0 & 0 & c & s \\ 0 & 0 & -s & c \end{pmatrix}$$

and

$$\theta = -\arctan\left(\frac{a_{24}}{a_{23}}\right) = -\arctan\left(\frac{-3.6927}{0.3333}\right) \approx 94.2695$$

$$\cos\theta = 0.0899, \quad \text{and} \quad \sin\theta = 0.9960.$$

Then

$$Q_{34} = \begin{pmatrix} 1 & 0 & 0 & 0 \\ 0 & 1 & 0 & 0 \\ 0 & 0 & 0.0899 & 0.9960 \\ 0 & 0 & -0.9960 & 0.0899 \end{pmatrix} \quad \text{and} \quad Q_{34}^T = \begin{pmatrix} 1 & 0 & 0 & 0 \\ 0 & 1 & 0 & 0 \\ 0 & 0 & 0.0899 & -0.9960 \\ 0 & 0 & 0.9960 & 0.0899 \end{pmatrix},$$

which gives

$$A_3 = Q_{34}^T A_2 Q_{34} = \begin{pmatrix} 2.0 & -4.2426 & 0 & 0 \\ -4.2426 & 3.4444 & 3.7077 & 0 \\ 0 & 3.7077 & 5.7621 & 1.1097 \\ 0 & 0 & 1.1097 & 4.7934 \end{pmatrix} = C.$$

By using the Sturm sequence iteration, the eigenvalues of the symmetric tridiagonal matrix C are $9.621, 5.204, 3.560$, and -2.385, which are also the eigenvalues of A.   ●

To get the above results using the MATLAB Command Window, we do the following:

>> $A = [2 \ -1 \ 1 \ 4; -1 \ 3 \ 1 \ 2; 1 \ 1 \ 5 \ 3; 4 \ 2 \ -3 \ 6];$
>> $sol = Given(A);$

Program 4.7
MATLAB m-file for Given's method
function sol=Given(A)
$[n, n] = size(A); t = 0;$ for i=1:n; for j=1:n
if i==j; Q(i,j)=1; else; Q(i,j)=0; end; end;end
for i=1:n-2; for j=i+2:n; t=t+1;
for f=1:n; for g=1:n
$Q(f, t * n + g) = Q(f, g);$ end; end;
theta=atan(A(i,j)/A(i,i+1));
$Q(i+1, t*n+i+1) = cos(theta); Q(i+1, t*n+j) = -sin(theta);$
$Q(j, t * n + i + 1) = sin(theta); Q(j, t * n + j) = cos(theta);$
for f=1:n; for g=1:n; sum=0; for l=1:n
$sum = sum + a(f, l) * Q(l, t*n+g);$;end; aa(f,g)=sum; end; end
for f=1:n; for g=1:n; sum=0; for l=1:n
$sum = sum + Q(l, t * n + f) * aa(l, g);$ end
A(f,g)=sum; end; end; end; end T=A
% Solve the tridiagonal matrix using Sturm sequence method
ff(1,:)=[1 0 0 0 0]; ff(2,:)=[A(1,1) -1 0 0 0];
for i=3:n+1; h=[A(i-1,i-1) -1];
$ff(i, 1) = h(1) * ff(i - 1, 1) - A(i - 1, i - 2)\hat{} \ 2*ff(i - 2, 1);$
for z=2:n+1
$ff(i, z) = h(1)*ff(i-1, z)+h(2)*ff(i-1, z-1)-A(i-1, i-2)$
$\hat{} \ 2*ff(i - 2, z);$ end;end
for i=1:n+1; $y(i) = ff(n + 1, n + 2 - i);$ end; $eigval = roots(y)$

4.5.4 Householder's Method

This method is a variation of Given's method and enables us to reduce a symmetric matrix A to a symmetric tridiagonal matrix form C having the same eigenvalues. It reduces a given matrix into a symmetric tridiagonal form with about half as much computation as Given's method requires. This method is used to reduce a whole row and column (except for the tridiagonal elements) to zero at a time. Note that the symmetric tridiagonal matrix form by Given's method and Householder's method may be different, but the eigenvalues will be same.

Definition 4.3 (Householder Matrix)

A Householder matrix H_w is a matrix of the form

$$H_w = I - 2\mathbf{w}\mathbf{w}^T = I - \left(\frac{2}{\mathbf{w}^T\mathbf{w}}\right)\mathbf{w}\mathbf{w}^T,$$

where I is an $n \times n$ identity matrix and $\mathbf{w}$ is some $n \times 1$ vector satisfying

$$\mathbf{w}^T\mathbf{w} = \sum_{k=1}^{n} w_k^2 = 1,$$

i.e., the vector $\mathbf{w}$ has unit length.   •

It is easy to verify that a Householder matrix H_w is symmetric, i.e.,

$$H_w = \begin{pmatrix} 1 - 2w_1^2 & -2w_1w_2 & \cdots & -2w_1w_n \\ -2w_1w_2 & 1 - 2w_2^2 & \cdots & -2w_2w_n \\ \vdots & \vdots & \vdots & \vdots \\ -2w_1w_n & -2w_2w_n & \cdots & 1 - 2w_n^2 \end{pmatrix} = H_w^T$$

and is orthogonal, i.e.,

$$\begin{aligned} H_w^2 &= (I - 2\mathbf{w}\mathbf{w}^T)(I - 2\mathbf{w}\mathbf{w}^T) \\ &= I - 4\mathbf{w}\mathbf{w}^T + 4\mathbf{w}\mathbf{w}^T\mathbf{w}\mathbf{w}^T \\ &= I - 4\mathbf{w}\mathbf{w}^T + 4\mathbf{w}\mathbf{w}^T \\ &= I, \qquad\qquad\qquad\qquad \text{(since } \mathbf{w}^T\mathbf{w} = 1) \end{aligned}$$

Thus,

$$H_w = H_w^{-1} = H_w^T,$$

which shows that H_w is symmetric. Note that the determinant of a Householder matrix H_w is always equal to -1.

Example 4.15 *Consider a vector* $\mathbf{w} = [1, 2]^T$, *then*

$$H_w = I - \frac{2}{5}\mathbf{w}\mathbf{w}^T,$$

so

$$H_w = \begin{pmatrix} 1 & 0 \\ 0 & 1 \end{pmatrix} - \frac{2}{5}\begin{pmatrix} 1 & 2 \\ 2 & 4 \end{pmatrix} = \begin{pmatrix} \dfrac{3}{5} & -\dfrac{4}{5} \\ -\dfrac{4}{5} & -\dfrac{3}{5} \end{pmatrix},$$

which shows that the given Householder matrix H_w is symmetric and orthogonal and the determinant of H_w is -1. ●

A Householder matrix H_w corresponding to a given $\mathbf{w}$ may be generated using the MATLAB Command Window as follows:

```
>> w = [1 2]';
>> w = w/norm(w);
>> Hw = eye(2) - 2 * w * w';
```

The basic steps of Householder's method that require us to convert the symmetric matrix into a symmetric tridiagonal matrix are as follows:

$$A_1 = A$$

$$A_2 = Q_1^T A_1 Q_1$$

$$A_3 = Q_2^T A_2 Q_2$$

$$\vdots \quad \vdots$$

$$A_{k+1} = Q_k^T A_k Q_k,$$

where Q_k matrices are the Householder transformation matrices and can be constructed as

$$Q_k = I - s_k \mathbf{w}_k \mathbf{w}_k^T$$

and

$$s_k = \frac{2}{\mathbf{w}_k^T \mathbf{w}_k}.$$

The coefficients of a vector $\mathbf{w}_k$ are defined in terms of a matrix A as

$$w_{ik} = \begin{cases} 0, & \text{for} \quad i = 1, 2, \ldots, k \\ a_{ik}, & \text{for} \quad i = k+2, k+3, \ldots, n \end{cases}$$

and

$$w_{k+1k} = a_{k+1k} \pm \sqrt{\sum_{i=k+1}^{n} a_{ik}^2}.$$

The positive sign or negative sign of w_{k+1k} can be taken depending on the sign of a coefficient a_{k+1k} of a given matrix A.

Householder's method transforms a given $n \times n$ symmetric matrix to a symmetric tridiagonal matrix in exactly $(\text{n} -2)$ steps. Each step of the method creates a zero in a complete row and column. The first step annihilates elements in the positions $(1, 3), (1, 4), \ldots, (1, n)$ simultaneously. Similarly, step r annihilates elements in the positions $(r, r + 2), (r, r + 3), \ldots, (r, n)$ simultaneously. Once a symmetric tridiagonal form has been achieved, then the eigenvalues of a given matrix can be calculated by using the Sturm sequence iteration. After calculating the eigenvalues, the shifted inverse power method can be used to find the eigenvectors of a symmetric tridiagonal matrix and then the eigenvectors of the original matrix A can be found by premultiplying these eigenvectors (of a symmetric tridiagonal matrix) by the product of successive transformation matrices.

Example 4.16 *Reduce the matrix*

$$A = \begin{pmatrix} 30 & 6 & 5 \\ 6 & 30 & 9 \\ 5 & 9 & 30 \end{pmatrix}$$

to a symmetric tridiagonal form using Householder's method.

Solution. *Since the given matrix is of size 3×3, only one iteration is required in order to reduce the given symmetric matrix into symmetric tridiagonal form. Thus, for $k = 1$, we construct the elements of the vector $\mathbf{w}_1$ as follows:*

$$
\begin{aligned}
w_{11} &= 0 \\
w_{31} &= a_{31} = 5 \\
w_{21} &= a_{21} \pm \sqrt{a_{21}^2 + a_{31}^2} = 6 \pm \sqrt{6^2 + 5^2} = 6 \pm 7.81.
\end{aligned}
$$

Since the given coefficient a_{21} is positive, the positive sign must be used for w_{21}, i.e.,

$$w_{21} = 13.81.$$

Therefore, the vector $\mathbf{w}_1$ is now determined to be

$$\mathbf{w}_1 = [0, 13.81, 5]^T$$

and

$$s_1 = \frac{2}{(0)^2 + (13.81)^2 + (5)^2} = 0.0093.$$

Thus, the first transformation matrix Q_1 for the first iteration is

$$
Q_1 = \begin{pmatrix} 1 & 0 & 0 \\ 0 & 1 & 0 \\ 0 & 0 & 1 \end{pmatrix} - 0.009 \begin{pmatrix} 0 \\ 13.81 \\ 5 \end{pmatrix} \begin{pmatrix} 0 & 13.81 & 5 \end{pmatrix},
$$

and it gives

$$
Q_1 = \begin{pmatrix} 1 & 0 & 0 \\ 0 & -0.7682 & -0.6402 \\ 0 & -0.6402 & 0.7682 \end{pmatrix}.
$$

Therefore,

$$
A_2 = Q_1^T A_1 Q_1 = \begin{pmatrix} 30.0 & -7.810 & 0 \\ -7.810 & 38.85 & -1.622 \\ 0 & -1.622 & 21.15 \end{pmatrix},
$$

which is the symmetric tridiagonal form.   ●

To get the above results using the MATLAB Command Window, we do the following:

```
>> A = [30 6 5; 6 30 9; 5 9 30];
>> sol = HHHM(A);
```

Program 4.8
MATLAB m-file for Householder's method
function sol=HHHM(A)
$[n, n] = size(A); Q = eye(n);$ for k=1:n-2
$alfa = sign(A(k+1, k))*sqrt(A((k+1) : n, k)'*A((k+1) : n, k));$
$w = zeros(n, 1);$
$w(k + 1, 1) = A(k + 1, k) + alfa; w((k + 2) : n, 1) = A((k + 2) : n, k);$
$P = eye(n) - 2 * w * w'/(w' * w); Q = Q * P; A = P * A * P; end$
T=A % this is the tridiagonal matrix
% using Sturm sequence method
ff(1,:)=[1 0 0 0 0]; ff(2,:)=[A(1,1) -1 0 0 0];
for i=3:n+1
$h = [A(i-1, i-1)-1]; ff(i, 1) = h(1)*ff(i-1, 1)-A(i-1, i-2)$
$\hat{} \ 2*ff(i - 2, 1);$
for z=2:n+1
$ff(i, z) = h(1)*ff(i-1, z)+h(2)*ff(i-1, z-1)-A(i-1, i-2)\hat{}$
$2*ff(i - 2, z);$ end; end
for i=1:n+1; y(i)=ff(n+1,n+2-i); end; alfa; u; Q; eig-
val=roots(y)

Example 4.17 *Reduce the matrix*

$$A = \begin{pmatrix} 7 & 1 & 2 & 1 \\ 1 & 8 & 1 & -1 \\ 2 & 1 & 3 & 1 \\ 1 & -1 & 1 & 2 \end{pmatrix}$$

to symmetric tridiagonal form using Householder's method, and then find the approximation of the eigenvalues of A using the Strum sequence itera-

tion.

Solution. *Since the size of A is 4×4, we need two iterations to convert the given symmetric matrix into symmetric tridiagonal form. For the first iteration, we take $k = 1$, and we construct the elements of the vector $\mathbf{w}_1$ as follows:*

$$
\begin{aligned}
w_{11} &= 0 \\
w_{31} &= a_{31} = 2 \\
w_{41} &= a_{41} = 1 \\
w_{21} &= a_{21} \pm \sqrt{a_{21}^2 + a_{31}^2 + a_{41}^2} = 1 \pm \sqrt{1^2 + 2^2 + 1^1} = 1 \pm \sqrt{6}.
\end{aligned}
$$

Since the given coefficient $a_{21} > 0$, the positive sign must be used for w_{21}, and it gives

$$
w_{21} = 1 + 2.4495 = 3.4495.
$$

Thus, the vector $\mathbf{w}_1$ takes the form

$$
\mathbf{w}_1 = [0, 3.4495, 2, 1]^T
$$

and

$$
s_1 = \frac{2}{\mathbf{w}_1^T \mathbf{w}_1} = \frac{2}{(0)^2 + (3.4495)^2 + (2)^2 + 1^2} = \frac{2}{16.83} = 0.1183.
$$

Thus, the first transformation matrix Q_1 for the first iteration is

$$
Q_1 = I - s_1 \mathbf{w}_1 \mathbf{w}_1^T = \begin{pmatrix} 1 & 0 & 0 & 0 \\ 0 & 1 & 0 & 0 \\ 0 & 0 & 1 & 0 \\ 0 & 0 & 0 & 1 \end{pmatrix} - 0.1188 \begin{pmatrix} 0 \\ 3.4495 \\ 2 \\ 1 \end{pmatrix} \begin{pmatrix} 0 & 3.4495 & 2 & 1 \end{pmatrix},
$$

and it gives

$$
Q_1 = \begin{pmatrix} 1.0000 & 0 & 0 & 0 \\ 0 & -0.4082 & -0.8165 & -0.4082 \\ 0 & -0.8165 & 0.5266 & -0.2367 \\ 0 & -0.4082 & -0.2367 & 0.8816 \end{pmatrix}.
$$

Therefore,

$$A_2 = Q_1^T A_1 Q_1 = \begin{pmatrix} 7.0000 & -2.4495 & 0.0000 & 0 \\ -2.4495 & 4.6667 & 1.5700 & 1.1933 \\ 0.0000 & 1.5700 & 4.7816 & 2.9972 \\ 0 & 1.1933 & 2.9972 & 3.5518 \end{pmatrix}.$$

Now for $k = 2$, we construct the elements of the vector $\mathbf{w}_2$ as follows:

$$\begin{aligned}
w_{12} &= 0 \\
w_{22} &= 0 \\
w_{42} &= 1.1933 \\
w_{32} &= 1.5700 \pm \sqrt{(1.5700)^2 + (1.1933)^2} = 1.5700 \pm \sqrt{0000003.855}. \\
&= 1.5700 \pm 1.9721
\end{aligned}$$

Since the given coefficient $a_{32} > 0$, the positive sign must be used for w_{32}, and it gives

$$w_{32} = 1.5700 + 1.9721 = 3.5421.$$

Thus, the vector $\mathbf{w}_2$ takes the form

$$\mathbf{w}_2 = [0, 0, 3.5421, 1.1933]^T$$

and

$$s_2 = \frac{2}{\mathbf{w}_2^T \mathbf{w}_2} = \frac{2}{13.9704} = 0.1432.$$

Thus, the second transformation matrix Q_2 for the second iteration is

$$Q_2 = I - s_2 \mathbf{w}_2 \mathbf{w}_2^T = \begin{pmatrix} 1 & 0 & 0 & 0 \\ 0 & 1 & 0 & 0 \\ 0 & 0 & 1 & 0 \\ 0 & 0 & 0 & 1 \end{pmatrix} - 0.1432 \begin{pmatrix} 0 \\ 0 \\ 3.5421 \\ 1.1933 \end{pmatrix} \begin{pmatrix} 0 & 0 & 3.5421 & 1.1933 \end{pmatrix},$$

and it gives

$$Q_2 = \begin{pmatrix} 1.0000 & 0 & 0 & 0 \\ 0 & -0.4082 & 0.8971 & 0.1690 \\ 0 & -0.8165 & -0.2760 & -0.5071 \\ 0 & -0.4082 & -0.3450 & 0.8452 \end{pmatrix}.$$

Therefore,

$$A_3 = Q_2^T A_2 Q_2 = \begin{pmatrix} 7.0000 & -2.4495 & 0.0000 & 0.0000 \\ -2.4495 & 4.6667 & -1.9720 & 0.0000 \\ 0.0000 & -1.9720 & 7.2190 & -0.2100 \\ 0.0000 & 0.0000 & -0.2100 & 1.1143 \end{pmatrix} = T,$$

which is the symmetric tridiagonal form.

To find the eigenvalues of this symmetric tridiagonal matrix we use the Sturm sequence iteration

$$f_4(\lambda) = (a_4 - \lambda) f_3(\lambda) - b_4^2 f_2(\lambda),$$

where

$$f_3(\lambda) = (a_3 - \lambda) f_2(\lambda) - b_3^2 f_1(\lambda)$$

and

$$f_2(\lambda) = (a_2 - \lambda) f_1(\lambda) - b_2^2 f_0(\lambda),$$

with

$$f_1(\lambda) = (a_1 - \lambda) \quad and \quad f_0(\lambda) = 1.$$

Since

$$a_1 = 7.0000, \quad a_2 = 4.6667, \quad a_3 = 7.2190, \quad a_4 = 1.1143$$

and

$$b_2 = -2.4495, \quad b_3 = -1.9720, \quad b_4 = -0.2100.$$

Thus,

$$f_4(\lambda) = \lambda^4 - 20\lambda^3 + 128.0002\lambda^2 - 284.0021\lambda + 183.0027 = 0,$$

and solving this characteristic equation, we get

$$\lambda_1 = 9.2510, \quad \lambda_1 = 7.1342, \quad \lambda_3 = 2.5100, \quad \lambda_4 = 1.1047,$$

which are the eigenvalues of the symmetric tridiagonal matrix T and are also the eigenvalues of the given matrix A. Once the eigenvalues of A are obtained, then the corresponding eigenvectors of A can be obtained by using the shifted inverse power method. ●

4.6 Matrix Decomposition Methods

In the following we will discuss two matrix decomposition methods, called the QR method and the LR method, which help us to find the eigenvalues of a given general matrix.

4.6.1 QR Method

We know that the Jacobi, Given's, and Householder's methods are applicable only to symmetric matrices for finding all the eigenvalues of a matrix A. First, we describe the QR method, which can find all the eigenvalues of a general matrix. In this method we decompose an arbitrary real matrix A into a product QR, where Q is an orthogonal matrix and R is an upper-triangular matrix with nonnegative diagonal elements. Note that when A is nonsingular, this decomposition is unique.

Starting with $A_1 = A$, the QR method iteratively computes similar matrices A_i, $i = 2, 3, \ldots$, in two stages:

(1) Factor A_i into $Q_i R_i$, i.e., $A_i = Q_i R_i$.

(2) Define $A_{i+1} = R_i Q_i$.

Note that from stage (1), we have

$$R_i = Q_i^{-1} A_i,$$

and using this, stage (2) can be written as

$$A_{i+1} = Q_i^{-1} A_i Q_i = Q_i^T A_i Q_i,$$

where all A_i are similar to A, and thus have the same eigenvalues. It turns out that in the case where the eigenvalues of A all have different magnitude,

$$|\lambda_1| > |\lambda_2| > \cdots > |\lambda_n|,$$

then the QR iterates A_i approach an upper-triangular matrix, and thus the elements of the main diagonal approach the eigenvalues of a given matrix

A. When there are distinct eigenvalues of the same size, the iterates A_i may not approach an upper-triangular matrix; however, they do approach a matrix that is near enough to an upper-triangular matrix to allow us to find the eigenvalues of A.

If a given matrix A is symmetric and tridiagonal, since the QR transformation preserves symmetry, all subsequent matrices A_i will be symmetric and, hence, tridiagonal. Thus, the combined method of first reducing a symmetric matrix to a symmetric tridiagonal form by the Householder transformations and then applying the QR method is probably the most effective for evaluating all the eigenvalues of a symmetric matrix.

The simplest way of calculating the QR decomposition of an $n \times n$ matrix A is to premultiply A by a series of rotation matrices, and the values of $p, q,$ and θ are chosen to annihilate one of the lower-triangular elements. The value of θ, which is chosen to create zero in the (q, p) position, is defined as

$$\theta = -\arctan\left(\frac{a_{qp}}{a_{pp}}\right).$$

The first stage of the decomposition annihilates the element in position $(2,1)$ using the rotation matrix Q_{12}^T. The next two stages annihilate elements in positions $(3,1)$ and $(3,2)$ using the rotation matrices Q_{13}^T and Q_{23}^T, respectively. The process continues in this way, creating zeros in row order, until the rotation matrix Q_{n-1n}^T is used to annihilate the element in the position $(n, n-1)$. The zeros created are retained in a similar way as in Given's method, and an upper-triangular matrix R is produced after $\frac{n(n-1)}{2}$ premultiplications, i.e.,

$$Q_{n-1n}^T \cdots Q_{13}^T Q_{12}^T A = R,$$

which can be rearranged as

$$A = (Q_{12} Q_{13} \cdots Q_{n-1n})R = QR,$$

since $Q_{pq}^T = Q_{pq}^{-1}$.

Example 4.18 *Find the first QR iteration for the matrix*

$$A = \begin{pmatrix} 1 & 4 & 3 \\ 2 & 3 & 1 \\ 2 & 6 & 5 \end{pmatrix}.$$

Solution. Step I. *Create a zero in the $(2,1)$ position by using the first orthogonal transformation matrix*

$$Q_{12} = \begin{pmatrix} c & s & 0 \\ -s & c & 0 \\ 0 & 0 & 1 \end{pmatrix}$$

and

$$\theta = -\arctan\left(\frac{a_{21}}{a_{11}}\right) = -\arctan(2) = -70.4833,$$

$$\cos\theta \approx 0.4472 \quad and \quad \sin\theta \approx -0.8944.$$

Then

$$Q_{12} = \begin{pmatrix} 0.4472 & -0.8944 & 0 \\ 0.8944 & 0.4472 & 0 \\ 0 & 0 & 1 \end{pmatrix} \quad and \quad Q_{12}^T = \begin{pmatrix} 0.4472 & -0.8944 & 0 \\ 0.8944 & 0.4472 & 0 \\ 0 & 0 & 1 \end{pmatrix},$$

which gives

$$Q_{12}^T A = \begin{pmatrix} 2.2360 & 4.4720 & 2.2360 \\ 0 & -2.2360 & -2.2360 \\ 2.0000 & 6.0000 & 5.0000 \end{pmatrix}.$$

Step II. *Create a zero in the $(3,1)$ position by using the second orthogonal transformation matrix*

$$Q_{13} = \begin{pmatrix} c & 0 & s \\ 0 & 1 & 0 \\ -s & 0 & c \end{pmatrix},$$

with

$$\theta = -\arctan\left(\frac{a_{31}}{a_{11}}\right) = -\arctan\left(\frac{2}{2.2360}\right) \approx -46.4585$$

$$\cos\theta \approx 0.7453, \quad and \quad \sin\theta \approx -0.6667.$$

Then

$$Q_{13} = \begin{pmatrix} 0.7453 & 0 & -0.6667 \\ 0 & 1 & 0 \\ 0.6667 & 0 & 0.7453 \end{pmatrix} \quad and \quad Q_{13}^T = \begin{pmatrix} 0.7453 & 0 & 0.6667 \\ 0 & 1 & 0 \\ -0.6667 & 0 & 0.7453 \end{pmatrix},$$

which gives

$$Q_{13}^T(Q_{12}^T A) = \begin{pmatrix} 3.0001 & 7.3336 & 5.0002 \\ 0 & -2.2360 & -2.2360 \\ 0.0001 & 1.4909 & 2.2363 \end{pmatrix}.$$

Step III. *Create a zero in the $(3,2)$ position by using the third orthogonal transformation matrix*

$$Q_{23} = \begin{pmatrix} 1 & 0 & 0 \\ 0 & c & s \\ 0 & -s & c \end{pmatrix},$$

with

$$\theta = -\arctan\left(\frac{a_{32}}{a_{22}}\right) = -\arctan\left(\frac{1.4909}{-2.2360}\right) \approx 37.4393$$

$$\cos\theta \approx 0.8320, \quad and \quad \sin\theta \approx 0.5548.$$

Then

$$Q_{23} = \begin{pmatrix} 1 & 0 & 0 \\ 0 & 0.8320 & 0.5548 \\ 0 & -0.5548 & 0.8320 \end{pmatrix} \quad and \quad Q_{23}^T = \begin{pmatrix} 1 & 0 & 0 \\ 0 & 0.8320 & -0.5548 \\ 0 & 0.5548 & 0.8320 \end{pmatrix},$$

which gives

$$R_1 = Q_{23}^T(Q_{13}^T Q_{12}^T A) = \begin{pmatrix} 3 & 7.3333 & 5 \\ 0 & -2.6874 & -3.1009 \\ 0 & 0 & 0.6202 \end{pmatrix},$$

which is the required upper-triangular matrix R_1. The matrix Q_1 can be computed as

$$Q_1 = Q_{12}Q_{13}Q_{23} = \begin{pmatrix} 0.3333 & -0.5788 & -0.7442 \\ 0.6667 & 0.7029 & -0.2481 \\ 0.6667 & -0.4134 & 0.6202 \end{pmatrix}.$$

Hence, the original matrix A can be decomposed as

$$A_1 = Q_1 R_1 = \begin{pmatrix} 0.9999 & 3.9997 & 2.9997 \\ 2.0001 & 3.0001 & 1.0000 \\ 2.0001 & 6.0001 & 5.0001 \end{pmatrix},$$

and the new matrix can be obtained as

$$A_2 = R_1 Q_1 = \begin{pmatrix} 9.2222 & 1.3506 & -0.9509 \\ -3.8589 & -0.6068 & -1.2564 \\ 0.4134 & -0.2564 & 0.3846 \end{pmatrix},$$

which is the required first QR iteration for the given matrix.   •

Note that if we continue in the same way with the 21 iterations, the new matrix A_{21} becomes the upper-triangular matrix

$$A_{21} = R_{20} Q_{20} = \begin{pmatrix} 8.5826 & -4.9070 & -2.1450 \\ 0 & 1 & -1.1491 \\ 0 & 0 & -0.5825 \end{pmatrix},$$

and its diagonal elements are the eigenvalues, $\lambda = 8.5826, 1, -0.5825$, of the given matrix A. Once the eigenvalues have been determined, the corresponding eigenvectors can be computed by the shifted inverse power method.

To get the above results using the MATLAB Command Window, we do the following:

```
>> A = [1 4 3; 2 3 1; 2 6 5];
>> sol = QRM(A);
```

Program 4.9

MATLAB m-file for the QR method

```
function sol=QRM(A)
[n,n]=size(A); M=0; for i=1:n; for j=1:n
if j < i; M=M+1;end; end;end;
for i=1:n; I(i,i)=1;end
dd=1; while dd > 0.0001; Q=I; Qs=I; kk=1;
for i=2:n; for j=1:i-1
```
$t = -atan((A(i,j)/A(j,j))); Q(j,j) = \cos(t);$
$Q(j,i) = \sin(t); Q(i,j) = -\sin(t); Q(i,i) = \cos(t); Q;$
$A = Q' * A; Qs(:,:,kk) = Q; kk = kk + 1; Q = I; end; end;$
$Q = Qs(:,:,M); forc = M - 1 : -1 : 1$
$Q = Qs(:,:,c) * Q; end; R = A; Q; A = R * Q; k = 1;$
```
for i=1:n; for j=1:n
```
$if\ j < i; m(k) = A(i,j); k = k + 1; end; end; end;$
$m; dd = max(abs(m)); end; for i=1:n; eigvals(i)=A(i,i); end$

Example 4.19 *Find the first QR iteration for the matrix*

$$A = \begin{pmatrix} 5 & -2 \\ -2 & 8 \end{pmatrix},$$

and if $(Q_1 R_1)\mathbf{x} = \mathbf{b}$ *and* $R_1\mathbf{x} = \mathbf{c}$*, with* $\mathbf{c} = Q_1^T\mathbf{b}$*, then find the solution of the linear system* $A\mathbf{x} = [7,8]^T$*.*

Solution. *First, create a zero in the* $(2,1)$ *position with the help of the orthogonal transformation matrix*

$$Q_{12} = \begin{pmatrix} c & s \\ -s & c \end{pmatrix},$$

and then, for finding the value of the θ*,* c*, and* s*, we calculate the*

$$\theta = -\arctan\left(\frac{a_{21}}{a_{11}}\right) = -\arctan(-0.4) = 0.3805,$$

$$\cos\theta \approx 0.9285 \quad and \quad \sin\theta \approx 0.3714.$$

So,

$$Q_1 = Q_{12} = \begin{pmatrix} 0.9285 & 0.3714 \\ -0.3714 & 0.9285 \end{pmatrix}$$

and

$$R_1 = Q_{12}^T A = \begin{pmatrix} 5.3853 & -4.8282 \\ 0 & 6.6852 \end{pmatrix}.$$

Since

$$\mathbf{c} = Q_1^T \mathbf{b} = \begin{pmatrix} 0.9285 & -0.3714 \\ 0.3714 & 0.9285 \end{pmatrix} \begin{pmatrix} 7 \\ 8 \end{pmatrix} = \begin{pmatrix} 3.5283 \\ 10.0278 \end{pmatrix},$$

therefore, solving the system

$$R_1 \mathbf{x} = \begin{pmatrix} 5.3853 & -4.8282 \\ 0 & 6.6852 \end{pmatrix} \begin{pmatrix} x_1 \\ x_2 \end{pmatrix} = \begin{pmatrix} 3.5283 \\ 10.0278 \end{pmatrix} = \mathbf{c},$$

we get

$$\begin{pmatrix} x_1 \\ x_2 \end{pmatrix} = \begin{pmatrix} 2.0000 \\ 1.5000 \end{pmatrix},$$

which is the required solution of the given system.   ●

4.6.2   LR Method

Another method, which is very similar to the QR method, is Rutishauser's
LR method. This method is based upon the decomposition of a matrix A
into the product of lower-triangular matrix L (with unit diagonal elements)
and upper-triangular matrix R. Starting with $A_1 = A$, the LR method
iteratively computes similar matrices A_i, $i = 2, 3, \ldots$, in two stages.

(1) Factor A_i into $L_i R_i$, i.e., $A_i = L_i R_i$.

(2) Define $A_{i+1} = R_i L_i$.

Each complete step is a similarity transformation because

$$A_{i+1} = R_i L_i = L_i^{-1} A_i L_i,$$

and so all of the matrices A_i have the same eigenvalues. This triangular
decomposition-based method enables us to reduce a given nonsymmetric

matrix to an upper-triangular matrix whose diagonal elements are the possible eigenvalues of a given matrix A, in decreasing order of magnitude. The rate at which the lower-triangular elements $a_{jk}^{(i)}$ of A_i converge to zero is of order $(\frac{\lambda_j}{\lambda_k})^i$, $j > k$.

This implies, in particular, that the order of convergence of the elements along the first subdiagonal is $(\frac{\lambda_{j+1}}{\lambda_j})^i$, and so convergence will be slow whenever two or more real eigenvalues are close together. The situation is more complicated if any of the eigenvalues are complex.

Since we know that the triangular decomposition is not always possible, we will use decomposition by partial pivoting, starting with

$$P_i A_i = L_i R_i,$$

where P_i represents the row permutations used in the decomposition. In order to preserve eigenvalues it is necessary to calculate A_{i+1} from

$$A_{i+1} = (R_i P_i) L_i.$$

It is easy to see that this is a similarity transformation because

$$A_{i+1} = (R_i P_i) L_i = L_i^{-1} P_i A_i P_i L_i,$$

and $P_i^{-1} = P_i$.

The matrix P_i does not have to be computed explicitly; $R_i P_i$ is just a column permutation of R_i using interchanges corresponding to row interchanges used in the decomposition of A_i.

Example 4.20 *Use the LR method to find the eigenvalues of the matrix*

$$A = \begin{pmatrix} 2 & -2 & 3 \\ 0 & 3 & -2 \\ 0 & -1 & 2 \end{pmatrix}.$$

Solution. *The exact eigenvalues of the given matrix A are $\lambda = 1, 2, 4$. The first triangular decomposition of $A = A_1$ produces*

$$L_1 = \begin{pmatrix} 1.0000 & 0 & 0 \\ 0 & 1.0000 & 0 \\ 0 & -0.3333 & 1.0000 \end{pmatrix}, \quad R_1 = \begin{pmatrix} 2.0000 & -2.0000 & 3.0000 \\ 0 & 3.0000 & -2.0000 \\ 0 & 0 & 1.3333 \end{pmatrix},$$

and no rows are interchanged. Then

$$A_2 = R_1 L_1 = \begin{pmatrix} 2.0000 & -3.0000 & 3.0000 \\ 0 & 3.6667 & -2.0000 \\ 0 & -0.4444 & 1.3333 \end{pmatrix}.$$

The second triangular decomposition of A_2 produces

$$L_2 = \begin{pmatrix} 1.0000 & 0 & 0 \\ 0 & 1.0000 & 0 \\ 0 & -0.1212 & 1.0000 \end{pmatrix}, \quad R_2 = \begin{pmatrix} 2.0000 & -3.0000 & 3.0000 \\ 0 & 3.6667 & -2.0000 \\ 0 & 0 & 1.0909 \end{pmatrix},$$

and again, no rows are interchanged. Then

$$A_3 = R_2 L_2 = \begin{pmatrix} 2.0000 & -3.3636 & 3.0000 \\ 0 & 3.9091 & -2.0000 \\ 0 & -0.1322 & 1.0909 \end{pmatrix}.$$

In a similar way, the next matrices in the sequence are

$$A_4 = \begin{pmatrix} 2 & -3.3636 & 3.0000 \\ 0 & 3.9091 & -2.0000 \\ 0 & 0 & 1.0233 \end{pmatrix} \begin{pmatrix} 1 & 0 & 0 \\ 0 & 1.0000 & 0 \\ 0 & -0.0338 & 1 \end{pmatrix}$$

$$= \begin{pmatrix} 2 & -3.4651 & 3.0000 \\ 0 & 3.9767 & -2.0000 \\ 0 & -0.0346 & 1.0233 \end{pmatrix}$$

$$A_5 = \begin{pmatrix} 2 & -3.4651 & 3.0000 \\ 0 & 3.9767 & -2.0000 \\ 0 & 0 & 1.0058 \end{pmatrix} \begin{pmatrix} 1 & 0 & 0 \\ 0 & 1.0000 & 0 \\ 0 & -0.0087 & 1 \end{pmatrix}$$

$$= \begin{pmatrix} 2 & -3.4912 & 3.0000 \\ 0 & 3.9942 & -2.0000 \\ 0 & -0.0088 & 1.0058 \end{pmatrix}$$

$$A_6 = \begin{pmatrix} 2 & -3.4912 & 3.0000 \\ 0 & 3.9942 & -2.0000 \\ 0 & 0 & 1.0015 \end{pmatrix} \begin{pmatrix} 1 & 0 & 0 \\ 0 & 1.0000 & 0 \\ 0 & -0.0022 & 1 \end{pmatrix}$$

$$= \begin{pmatrix} 2 & -3.4978 & 3.0000 \\ 0 & 3.9985 & -2.0000 \\ 0 & -0.0022 & 1.0015 \end{pmatrix}$$

$$A_7 = \begin{pmatrix} 2 & -3.4978 & 3 \\ 0 & 3.9985 & -2 \\ 0 & 0 & 1 \end{pmatrix} \begin{pmatrix} 1 & 0 & 0 \\ 0 & 1.0000 & 0 \\ 0 & -0.0005 & 1 \end{pmatrix} = \begin{pmatrix} 2 & -3.4995 & 3.0000 \\ 0 & 3.9996 & -2.0000 \\ 0 & -0.0005 & 1.0004 \end{pmatrix}$$

$$A_8 = \begin{pmatrix} 2 & -3.4995 & 3 \\ 0 & 3.9996 & -2 \\ 0 & 0 & 1 \end{pmatrix} \begin{pmatrix} 1 & 0 & 0 \\ 0 & 1.0000 & 0 \\ 0 & -0.0001 & 1 \end{pmatrix} = \begin{pmatrix} 2 & -3.4999 & 3.0000 \\ 0 & 3.9999 & -2.0000 \\ 0 & -0.0001 & 1.0001 \end{pmatrix}$$

$$A_9 = \begin{pmatrix} 2 & -3.4999 & 3 \\ 0 & 3.9999 & -2 \\ 0 & 0 & 1 \end{pmatrix} \begin{pmatrix} 1 & 0 & 0 \\ 0 & 1 & 0 \\ 0 & 0 & 1 \end{pmatrix} = \begin{pmatrix} 2 & -3.5 & 3 \\ 0 & 4 & -2 \\ 0 & 0 & 1 \end{pmatrix}.$$

•

4.6.3   Upper Hessenberg Form

In employing the QR method or the LR method to find the eigenvalues of a nonsymmetric matrix A, it is preferable to first use similarity transformations to convert A to upper Hessenberg form and then go on to demonstrate its usefulness in the QR and the LR methods.

Definition 4.4 *A matrix A is in upper Hessenberg form if*
$a_{ij} = 0$, *for all i, j,   such that $i - j > 1$.*   •

For example, in the following 4×4 matrix case, the nonzero elements are

$$A = \begin{pmatrix} 3 & 2 & 1 & 5 \\ 4 & 6 & 7 & 3 \\ 0 & 8 & 9 & 5 \\ 0 & 0 & 7 & 8 \end{pmatrix}.$$

Note that one way to characterize upper Hessenberg form is that it is almost triangular. This is important, since the eigenvalues of the triangular matrix are the diagonal elements. The upper Hessenberg form of a matrix A can be achieved by a sequence of Householder transformations or the Gaussian elimination procedure. Here, we will use the Gaussian elimination procedure since it is about a factor of 2 more efficient than Householder's method. It is possible to construct matrices for which the Householder reduction, being orthogonal, is stable and elimination is not, but such matrices are extremely rare in practice.

A general $n \times n$ matrix A can be reduced to upper Hessenberg form in exactly $n - 2$ steps.

Consider a 5×5 matrix

$$A = \begin{pmatrix} a_{11} & a_{12} & a_{13} & a_{14} & a_{15} \\ a_{21} & a_{22} & a_{23} & a_{24} & a_{25} \\ a_{31} & a_{32} & a_{33} & a_{34} & a_{35} \\ a_{41} & a_{42} & a_{43} & a_{44} & a_{45} \\ a_{51} & a_{52} & a_{53} & a_{54} & a_{55} \end{pmatrix}.$$

The first step of reducing the given matrix $A = A_1$ into upper Hessenberg form is to eliminate the elements in the $(3, 1), (4, 1)$, and $(5, 1)$ positions. It can be done by subtracting multiples $m_{31} = \frac{a_{31}}{a_{21}}, m_{41} = \frac{a_{41}}{a_{21}}$ and $m_{51} = \frac{a_{51}}{a_{21}}$ of row 2 from rows 3, 4, and 5, respectively, and considering the matrix

$$M_1 = \begin{pmatrix} 1 & 0 & 0 & 0 & 0 \\ 0 & 1 & 0 & 0 & 0 \\ 0 & m_{31} & 1 & 0 & 0 \\ 0 & m_{41} & 0 & 1 & 0 \\ 0 & m_{51} & 0 & 0 & 1 \end{pmatrix}.$$

Since we wish to carry out a similarity transformation to preserve eigenvalues, it is necessary to find the inverse matrix M^{-1} and compute

$$A_2 = M_1^{-1} A_1 M_1.$$

The right-hand side multiplication gives us

$$
A_2 = \begin{pmatrix}
a_{11} & a_{12}^{(2)} & a_{13} & a_{14} & a_{15} \\
a_{21} & a_{22}^{(2)} & a_{23} & a_{24} & a_{25} \\
0 & a_{32}^{(2)} & a_{33}^{(2)} & a_{34}^{(2)} & a_{35}^{(2)} \\
0 & a_{42}^{(2)} & a_{43}^{(2)} & a_{44}^{(2)} & a_{45}^{(2)} \\
0 & a_{52}^{(2)} & a_{53}^{(2)} & a_{54}^{(2)} & a_{55}^{(2)}
\end{pmatrix},
$$

where $a_{ij}^{(2)}$ denotes the new element in (i,j). In the second step, we eliminate the elements in the $(4,2)$ and $(5,2)$ positions. This can be done by subtracting multiples $m_{42} = \frac{a_{42}^{(2)}}{a_{32}^{(2)}}$ and $m_{52} = \frac{a_{52}^{(2)}}{a_{32}^{(2)}}$ of row 3 from rows 4 and 5, respectively, and considering the matrix

$$
M_2 = \begin{pmatrix}
1 & 0 & 0 & 0 & 0 \\
0 & 1 & 0 & 0 & 0 \\
0 & 0 & 1 & 0 & 0 \\
0 & 0 & m_{42} & 1 & 0 \\
0 & 0 & m_{52} & 0 & 1
\end{pmatrix}.
$$

Hence,

$$
A_3 = M_2^{-1} A_2 M_2 = \begin{pmatrix}
a_{11} & a_{12}^{(2)} & a_{13}^{(3)} & a_{14} & a_{15} \\
a_{21} & a_{22}^{(2)} & a_{23}^{(3)} & a_{24} & a_{25} \\
0 & a_{32}^{(2)} & a_{33}^{(3)} & a_{34}^{(2)} & a_{35}^{(2)} \\
0 & 0 & a_{43}^{(3)} & a_{44}^{(3)} & a_{45}^{(3)} \\
0 & 0 & a_{53}^{(3)} & a_{54}^{(3)} & a_{55}^{(3)}
\end{pmatrix},
$$

where $a_{ij}^{(3)}$ denotes the new element in (i,j). In the third step, we eliminate the elements in the $(5,3)$ position. This can be done by subtracting multiples $m_{53} = \frac{a_{53}^{(3)}}{a_{43}^{(3)}}$ of row 4 from row 5, and considering the matrix

$$M_3 = \begin{pmatrix} 1 & 0 & 0 & 0 & 0 \\ 0 & 1 & 0 & 0 & 0 \\ 0 & 0 & 1 & 0 & 0 \\ 0 & 0 & 0 & 1 & 0 \\ 0 & 0 & 0 & m_{53} & 1 \end{pmatrix}.$$

Hence,

$$A_4 = M_3^{-1} A_3 M_3 = \begin{pmatrix} a_{11} & a_{12}^{(2)} & a_{13}^{(3)} & a_{14}^{(4)} & a_{15} \\ a_{21} & a_{22}^{(2)} & a_{23}^{(3)} & a_{24}^{(4)} & a_{25} \\ 0 & a_{32}^{(2)} & a_{33}^{(3)} & a_{34}^{(4)} & a_{35}^{(2)} \\ 0 & 0 & a_{43}^{(3)} & a_{44}^{(4)} & a_{45}^{(4)} \\ 0 & 0 & 0 & a_{54}^{(4)} & a_{55}^{(4)} \end{pmatrix},$$

which is in upper Hessenberg form.

Example 4.21 *Use the Gaussian elimination method to convert the matrix*

$$A_1 = \begin{pmatrix} 5 & 3 & 6 & 4 & 9 \\ 4 & 6 & 5 & 3 & 4 \\ 4 & 2 & 3 & 1 & 1 \\ 2 & 4 & 6 & 3 & 3 \\ 2 & 5 & 6 & 4 & 7 \end{pmatrix}$$

into upper Hessenberg form.

Solution. *In the first step, we eliminate the elements in the $(3,1), (4,1)$ and $(5,1)$ positions. It can be done by subtracting multiples $m_{31} = \frac{4}{4} = 1, m_{41} = \frac{2}{4} = 0.5$, and $m_{51} = \frac{2}{4} = 0.5$ of row 2 from rows 3, 4, and 5,* *respectively. The matrices M_1 and M_1^{-1} are as follows:*

$$M_1 = \begin{pmatrix} 1 & 0 & 0 & 0 & 0 \\ 0 & 1 & 0 & 0 & 0 \\ 0 & -1 & 1 & 0 & 0 \\ 0 & -0.5 & 0 & 1 & 0 \\ 0 & -0.5 & 0 & 0 & 1 \end{pmatrix} \quad and \quad M_1^{-1} = \begin{pmatrix} 1 & 0 & 0 & 0 & 0 \\ 0 & 1 & 0 & 0 & 0 \\ 0 & 1 & 1 & 0 & 0 \\ 0 & 0.5 & 0 & 1 & 0 \\ 0 & 0.5 & 0 & 0 & 1 \end{pmatrix}.$$

Then the transformation is

$$A_2 = M_1^{-1} A_1 M_1 = \begin{pmatrix} 5 & 15.50 & 6 & 4 & 9 \\ 4 & 14.50 & 5 & 3 & 4 \\ 0 & -8.50 & -2 & -2 & -3 \\ 0 & 5.75 & 3.50 & 1.50 & 1 \\ 0 & 9.25 & 3.50 & 2.50 & 5 \end{pmatrix}.$$

In the second step, we eliminate the elements in the $(4,2)$ and $(5,2)$ positions. This can be done by subtracting multiples $m_{42} = \frac{5.75}{-8.50} = -0.6765$ and $m_{52} = \frac{9.25}{-8.50} = -1.0882$ of row 3 from rows 4 and 5, respectively. The matrices M_2 and M_2^{-1} are as follows:

$$M_2 = \begin{pmatrix} 1 & 0 & 0 & 0 & 0 \\ 0 & 1 & 0 & 0 & 0 \\ 0 & 0 & 1 & 0 & 0 \\ 0 & 0 & 0.6765 & 1 & 0 \\ 0 & 0 & 1.0882 & 0 & 1 \end{pmatrix} \quad and \quad M_2^{-1} = \begin{pmatrix} 1 & 0 & 0 & 0 & 0 \\ 0 & 1 & 0 & 0 & 0 \\ 0 & 0 & 1 & 0 & 0 \\ 0 & 0 & -0.6765 & 1 & 0 \\ 0 & 0 & -1.0882 & 0 & 1 \end{pmatrix}.$$

Then the transformation is

$$A_3 = M_2^{-1} A_1 M_2 = \begin{pmatrix} 5 & 15.50 & -6.50 & 4 & 9 \\ 4 & 14.50 & -1.3824 & 3 & 4 \\ 0 & -8.50 & 2.6176 & -2 & -3 \\ 0 & 0 & 3.1678 & 0.1471 & -1.0294 \\ 0 & 0 & -0.7837 & 0.3235 & 1.7353 \end{pmatrix}.$$

In the last step, we eliminate the elements in the $(5,3)$ position. This can be done by subtracting multiples $m_{53} = \frac{-0.7837}{3.1678} = -0.2474$ of row 4 *from row 5. The matrices M_3 and M_3^{-1} are as follows:*

$$M_3 = \begin{pmatrix} 1 & 0 & 0 & 0 & 0 \\ 0 & 1 & 0 & 0 & 0 \\ 0 & 0 & 1 & 0 & 0 \\ 0 & 0 & 0 & 1 & 0 \\ 0 & 0 & 0 & 0.2474 & 1 \end{pmatrix} \quad and \quad M_3^{-1} = \begin{pmatrix} 1 & 0 & 0 & 0 & 0 \\ 0 & 1 & 0 & 0 & 0 \\ 0 & 0 & 1 & 0 & 0 \\ 0 & 0 & 0 & 1 & 0 \\ 0 & 0 & 0 & -0.2474 & 1 \end{pmatrix}.$$

Then the transformation is

$$A_4 = \begin{pmatrix} 5 & 15.50 & -6.50 & 1.7733 & 9 \\ 4 & 14.50 & -1.3824 & 2.0104 & 4 \\ 0 & -8.50 & 2.6176 & -1.2578 & -3 \\ 0 & 0 & 3.1678 & 0.4017 & -1.0294 \\ 0 & 0 & 0 & -0.0064 & 1.4806 \end{pmatrix},$$

which is in required upper Hessenberg form. •

To get the above results using the MATLAB Command Window, we do the following:

```
>>>> A = [5 3 6 4 9; 4 6 5 3 4; 4 2 3 1 1; 2 4 6 3 3; 2 5 6 4 7];
>> sol = hes(A);
```

Program 4.10
MATLAB m-file for the Upper Hessenberg Form
function sol=hes(A)
$n = length(A(1,:));$ for i = 1:n-1; $m = eye(n);$
$[wj] = max(abs(A(i+1:n,i)));$
$if \ j > i+1;$
$t = m(i+1,:); m(i+1,:) = m(j,:);$
$m(j,:) = t; A = m * A * m'; end;$
$m = eye(n); m(i+2:n,i+1) = -A(i+2:n,i)/(A(i+1,i));$
$mi = m; mi(i+2:n,i+1) = -m(i+2:n,i+1);$
$A = m * A * mi; mesh(abs(A)); end$

Note that the above reduction fails if any $a_{j+1,j}^{(j)} = 0$ and, as in Gaussian elimination, is unstable whenever $|m_{ij}| > 1$. Row and column interchanges are used to avoid these difficulties (i.e., Gaussian elimination with pivoting). At step j, the elements below the diagonal in column j are examined. If the element of the largest modulus occurs in row r_j, say, then rows $j+1$ and r_j are interchanged. Here, we perform the transformation

$$A_{j+1} = M_j^{-1}(I_{j+1,r_j}^{-1} A_j I_{j+1,r_j}) M_j,$$

where I_{j+1,r_j} denotes a matrix obtained from the identity matrix by interchanging rows $j+1$ and r_j, and the elements of M_j are all less than or equal to one in the modulus. Note that

$$I_{j+1,r_j}^{-1} = I_{j+1,r_j}.$$

Example 4.22 *Use Gaussian elimination with pivoting to convert the matrix*

$$A = \begin{pmatrix} 3 & 2 & 1 & -1 \\ 1 & 4 & 2 & 1 \\ 2 & 2 & 3 & -2 \\ 5 & 1 & 2 & 3 \end{pmatrix}$$

into upper Hessenberg form.

Solution. *The element of the largest modulus below the diagonal occurs in the fourth row, so we need to interchange rows 2 and 3 and columns 2 and 3 to get*

$$A_1 = \mathbf{I}_{24} A \mathbf{I}_{24} = \begin{pmatrix} 1 & 0 & 0 & 0 \\ 0 & 0 & 0 & 1 \\ 0 & 0 & 1 & 0 \\ 0 & 1 & 0 & 0 \end{pmatrix} \begin{pmatrix} 3 & 2 & 1 & -1 \\ 1 & 4 & 2 & 1 \\ 2 & 2 & 3 & -2 \\ 5 & 1 & 2 & 3 \end{pmatrix} \begin{pmatrix} 1 & 0 & 0 & 0 \\ 0 & 0 & 0 & 1 \\ 0 & 0 & 1 & 0 \\ 0 & 1 & 0 & 0 \end{pmatrix},$$

which gives

$$A_1 = \begin{pmatrix} 3 & -1 & 1 & 2 \\ 5 & 3 & 2 & 1 \\ 2 & -2 & 3 & 2 \\ 1 & 1 & 2 & 4 \end{pmatrix}.$$

Now we eliminate the elements in the $(3, 1)$ and $(4, 1)$ positions. It can be done by subtracting multiples $m_{31} = \frac{2}{5} = 0.4$ and $m_{41} = \frac{1}{5} = 0.2$ of row 2 from rows 3 and 4, respectively. Then the transformation

$$A_2 = M^{-1}A_1M = \begin{pmatrix} 1 & 0 & 0 & 0 \\ 0 & 1 & 0 & 0 \\ 0 & -0.4 & 1 & 0 \\ 0 & -0.2 & 0 & 1 \end{pmatrix} \begin{pmatrix} 3 & -1 & 1 & 2 \\ 5 & 3 & 2 & 1 \\ 2 & -2 & 3 & 2 \\ 1 & 1 & 2 & 4 \end{pmatrix} \begin{pmatrix} 1 & 0 & 0 & 0 \\ 0 & 1 & 0 & 0 \\ 0 & 0.4 & 1 & 0 \\ 0 & 0.2 & 0 & 1 \end{pmatrix}$$

gives

$$A_2 = \begin{pmatrix} 3 & -0.2 & 1 & 2 \\ 5 & 4 & 2 & 1 \\ 0 & -2 & 2.2 & 1.6 \\ 0 & 1.8 & 1.6 & 3.8 \end{pmatrix}.$$

The element of the largest modulus below the diagonal in the second column occurs in the third row, and so there is no need to interchange the row and column. Now we eliminate the elements in the $(4, 2)$ position. This can be done by subtracting multiples $m_{42} = \frac{1.8}{-2} = -0.9$ of row 3 from row 4. Then the transformation

$$A_3 = M_2^{-1}A_2M_2 = \begin{pmatrix} 1 & 0 & 0 & 0 \\ 0 & 1 & 0 & 0 \\ 0 & 0 & 1 & 0 \\ 0 & 0 & 0.9 & 1 \end{pmatrix} \begin{pmatrix} 3 & -0.2 & 1 & 2 \\ 5 & 4 & 2 & 1 \\ 0 & -2 & 2.2 & 1.6 \\ 0 & 1.8 & 1.6 & 3.8 \end{pmatrix} \begin{pmatrix} 1 & 0 & 0 & 0 \\ 0 & 1 & 0 & 0 \\ 0 & 0 & 1 & 0 \\ 0 & 0 & 0.9 & 1 \end{pmatrix}$$

gives

$$A_3 = \begin{pmatrix} 3 & -0.2 & -0.8 & 2 \\ 5 & 4 & 1.1 & 1 \\ 0 & -2 & 0.76 & 1.6 \\ 0 & 0 & -1.136 & 5.24 \end{pmatrix},$$

which is in upper Hessenberg form. ●

Example 4.23 *Convert the following matrix to upper Hessenberg form and then apply the QR method to find its eigenvalues:*

$$A = \begin{pmatrix} 1 & 4 & 3 \\ 2 & 3 & 1 \\ 2 & 6 & 5 \end{pmatrix}.$$

Solution. *Since the upper Hessenberg form of the given matrix is*

$$H_1 = \begin{pmatrix} 1 & 7 & 3 \\ 2 & 4 & 1 \\ 0 & 7 & 4 \end{pmatrix},$$

then applying the QR method on the upper Hessenberg matrix H_1 will result in transformation matrices after iterations $1, 10, 14,$ and 19 as follows:

$$H_2 = R_1 Q1 = \begin{pmatrix} 7.0000 & -1.3460 & 2.0466 \\ -7.4297 & 1.8551 & -5.5934 \\ -0.0000 & -0.2268 & 0.1449 \end{pmatrix}$$

$$H_{10} = R_9 Q_9 = \begin{pmatrix} 8.5826 & 4.4029 & 6.8555 \\ -0.0000 & 1.0069 & -1.8962 \\ 0.0000 & 0.0058 & -0.5895 \end{pmatrix}$$

$$H_{14} = R_{13} Q_{13} = \begin{pmatrix} 8.5826 & 4.4248 & 6.8413 \\ -0.0000 & 1.0008 & -1.9013 \\ 0.0000 & 0.0007 & -0.5834 \end{pmatrix}$$

$$H_{19} = R_{18} Q_{18} = \begin{pmatrix} 8.5826 & 4.4279 & 6.8393 \\ -0.0000 & 0.9999 & -1.9020 \\ -0.0000 & -0.0000 & -0.5825 \end{pmatrix}.$$

In this case the QR method converges in 19 iterations faster than the QR method applied on the original matrix A in Example 4.18.   •

Note that the calculation of the QR decomposition is simplified if a given matrix is converted to upper Hessenberg form. So instead of applying the decomposition to the original matrix $A = A_1$, the original matrix is first transformed to the Hessenberg form. When $A_1 = H_1$ is in the upper Hessenberg form, all the subsequent H_i are also in the same form. Unfortunately, although transformation to upper Hessenberg form reduces the number of calculations at each step, the method may still prove to be computationally inefficient if the number of steps required for convergence is too large. Therefore, we use the more efficient process called the *shifting QR* method. Here, the iterative procedure

$$\begin{aligned} H_i &= Q_i R_i \\ H_{i+1} &= R_i Q_i \end{aligned}$$

is changed to

$$
\begin{aligned}
H_i - \mu_i \mathbf{I} &= Q_i R_i \\
H_{i+1} &= R_i Q_i + \mu_i \mathbf{I}.
\end{aligned}
\tag{4.36}
$$

This change is called *shift* because subtracting $\mu_i \mathbf{I}$ from H_i shifts the eigenvalues of the right side by μ_i as well as the eigenvalues of $R_i Q_i$. Adding $\mu_i \mathbf{I}$ in the second equation in (4.36) shifts the eigenvalues of H_{i+1} back to the original values. However, the shifts accelerate convergence of the eigenvalues close to μ_i.

4.6.4 Singular Value Decomposition

We have considered two principal methods for the decomposition of the matrix, QR decomposition and LR decomposition. There is another important method for matrix decomposition called *Singular Value Decomposition (SVD)*.

Here, we show that every rectangular real matrix A can be decomposed into a product UDV^T of two orthogonal matrices U and V and a generalized diagonal matrix D. The construction of UDV^T is based on the fact that for all real matrices A, a matrix $A^T A$ is symmetric, and therefore there exists an orthogonal matrix Q and a diagonal matrix D for which

$$
A^T A = Q D Q^T.
$$

As we know, the diagonal entries of D are the eigenvalues of $A^T A$. Now we show that they are nonnegative in all cases and that their square roots, called the *singular values* of A, can be used to construct UDV^T.

Singular Values of a Matrix

For any $m \times n$ matrix A, an $n \times n$ matrix $A^T A$ is symmetric and hence can be orthogonally diagonalized. Not only are the eigenvalues of $A^T A$ all real, they are all nonnegative. To show this, let λ be an eigenvalue of $A^T A$, with the corresponding unit vector $\mathbf{v}$. Then

$$
\begin{aligned}
0 \le \|A\mathbf{v}\|^2 &= (A\mathbf{v}).(A\mathbf{v}) = (A\mathbf{v})^T A\mathbf{v} = \mathbf{v}^T A^T A\mathbf{v} \\
&= \mathbf{v}^T \lambda \mathbf{v} = \lambda(\mathbf{v}.\mathbf{v}) = \lambda \|\mathbf{v}\|^2 = \lambda.
\end{aligned}
$$

It therefore makes sense to take (positive) square roots of these eigenvalues.

Definition 4.5 (Singular Values of a Matrix)

If A is an $m \times n$ matrix, the singular values of A are the square roots of the eigenvalues of $A^T A$ and are denoted by $\sigma_1, \ldots, \sigma_n$. It is conventional to arrange the singular values so that $\sigma_1 \geq \sigma_2 \geq \cdots \sigma_n$. •

Example 4.24 *Find the singular values of*

$$A = \begin{pmatrix} 1 & 0 & 1 \\ 1 & 1 & 0 \end{pmatrix}.$$

Solution. *Since the singular values of A are the square roots of the eigenvalues of $A^T A$, we compute*

$$A^T A = \begin{pmatrix} 1 & 1 \\ 0 & 1 \\ 1 & 0 \end{pmatrix} \begin{pmatrix} 1 & 0 & 1 \\ 1 & 1 & 0 \end{pmatrix} = \begin{pmatrix} 2 & 1 & 1 \\ 1 & 1 & 0 \\ 1 & 0 & 1 \end{pmatrix}.$$

The matrix $A^T A$ has eigenvalues $\lambda_1 = 3, \lambda_2 = 1$, and $\lambda_3 = 0$. Consequently, the singular values of A are $\sigma_1 = \sqrt{3} = 1.7321$, $\sigma_2 = \sqrt{1} = 1$, and $\sigma_3 = \sqrt{0} = 0$. •

Note that the singular values of A are not the same as its eigenvalues, but there is a connection between them if A is a symmetric matrix.

Theorem 4.8 *If $A = A^T$ is a symmetric matrix, then its singular values are the absolute values of its nonzero eigenvalues, i.e.,*

$$\sigma_i = |\lambda_i| > 0.$$ •

Theorem 4.9 *The condition number of a nonsingular matrix is the ratio between its largest singular value σ_1(or dominant singular value) and the smallest singular value σ_n, i.e.,*

$$K(A) = \frac{\sigma_1}{\sigma_n}.$$ •

Singular Value Decomposition

The following are some of the *properties* that make singular value decompositions useful:

1. All real matrices have singular value decompositions.

2. A real square matrix is invertible, if and only if all its singular values are nonzero.

3. For any $m \times n$ real rectangular matrix A, the number of nonzero singular values of A is equal to the rank of A.

4. If $A = UDV^T$ is a singular value decomposition of an invertible matrix A, then $A^{-1} = VD^{-1}U^T$.

5. For positive-definite symmetric matrices, the orthogonal decomposition QDQ^T and the singular value decomposition UDV^T coincide.

Theorem 4.10 (Singular Value Decomposition Theorem)

Every $m \times n$ matrix A can be factored into the product of an $m \times m$ matrix U with orthonormal columns, so $U^TU = \mathbf{I}$, the $m \times n$ diagonal matrix $D = diag(\sigma_1, \ldots, \sigma_r)$ that has the singular values of A as its diagonal entries, and an $n \times n$ matrix V with orthonormal rows, so $V^TV = \mathbf{I}$, i.e.,

$$A = UDV^T = (\mathbf{u}_1\mathbf{u}_2 \cdots \mathbf{u}_r\mathbf{u}_{r+1} \cdots \mathbf{u}_n)
\begin{pmatrix}
\sigma_1 & & & & & & \mathbf{0} \\
& \sigma_2 & & & & & \\
& & \ddots & & & & \\
& & & \sigma_r & & & \\
& & & & 0 & & \\
& & & & & \ddots & \\
\mathbf{0} & & & & & & 0
\end{pmatrix}
\begin{pmatrix}
\mathbf{v}_1^T \\
\mathbf{v}_2^T \\
\vdots \\
\mathbf{v}_r^T \\
\mathbf{v}_{r+1}^T \\
\vdots \\
\mathbf{v}_n^T
\end{pmatrix}.$$

●

Note that the columns of U, $\mathbf{u}_1, \mathbf{u}_2, \ldots, \mathbf{u}_r$, are called *left singular vectors* of A, and the columns of V, $\mathbf{v}_1, \mathbf{v}_2, \ldots, \mathbf{v}_r$, are called *right singular vectors* of A. The matrices U and V are not uniquely determined by A, but a matrix D must contain the singular values, $\sigma_1, \sigma_2, \ldots, \sigma_r$, of A.

To construct the orthogonal matrix V, we must find an orthonormal basis $\{\mathbf{v}_1, \mathbf{v}_2, \ldots, \mathbf{v}_n\}$ for $\mathbb{R}^n$ consisting of eigenvectors of an $n \times n$ symmetric matrix $A^T A$. Then

$$V = [\mathbf{v}_1 \mathbf{v}_2 \cdots \mathbf{v}_n],$$

is an orthogonal $n \times n$ matrix.

For the orthogonal matrix U, we first note that $\{A\mathbf{v}_1, A\mathbf{v}_2, \ldots, A\mathbf{v}_n\}$ is an orthogonal set of vectors in $\mathbb{R}^m$. To see this, suppose that $\mathbf{v}_i$ is an eigenvector of $A^T A$ corresponding to an eigenvalue λ_i, then, for $i \neq j$, we have

$$(A\mathbf{v}_i).(A\mathbf{v}_j) = (A\mathbf{v}_i)^T A\mathbf{v}_j = \mathbf{v}_i{}^T A^T A\mathbf{v}_j = \mathbf{v}_i{}^T \lambda_j \mathbf{v}_j = \lambda_j(\mathbf{v}_i.\mathbf{v}_j) = 0,$$

since the eigenvectors $\mathbf{v}_i$ are orthogonal. Now recall that the singular values satisfy $\sigma_i = \|A\mathbf{v}_i\|$ and that the first r of these are nonzero. Therefore, we can normalize $A\mathbf{v}_1, \ldots, A\mathbf{v}_r$ by setting

$$\mathbf{u}_i = \frac{1}{\sigma_i} A\mathbf{v}_i, \qquad \text{for} \quad i = 1, 2, \ldots, r.$$

This guarantees that $\{\mathbf{u}_1, \mathbf{u}_2, \ldots, \mathbf{u}_r\}$ is an orthonormal set in $\mathbb{R}^m$, but if $r < m$, it will not be a basis for $\mathbb{R}^m$. In this case, we extend the set $\{\mathbf{u}_1, \mathbf{u}_2, \ldots, \mathbf{u}_r\}$ to an orthonormal basis $\{\mathbf{u}_1, \mathbf{u}_2, \ldots, \mathbf{u}_m\}$ for $\mathbb{R}^m$.

Example 4.25 *Find the singular value decomposition of the following matrix:*

$$A = \begin{pmatrix} 1 & 0 & 1 \\ 1 & 1 & 0 \end{pmatrix}.$$

Solution. *We compute*

$$A^T A = \begin{pmatrix} 1 & 1 \\ 0 & 1 \\ 1 & 0 \end{pmatrix} \begin{pmatrix} 1 & 0 & 1 \\ 1 & 1 & 0 \end{pmatrix} = \begin{pmatrix} 2 & 1 & 1 \\ 1 & 1 & 0 \\ 1 & 0 & 1 \end{pmatrix},$$

and find that its eigenvalues are

$$\lambda_1 = 3, \quad \lambda_2 = 1, \quad \lambda_3 = 0,$$

with the corresponding eigenvectors

$$\begin{pmatrix} 2 \\ 1 \\ 1 \end{pmatrix}, \quad \begin{pmatrix} 0 \\ -1 \\ 1 \end{pmatrix}, \quad \begin{pmatrix} -1 \\ 1 \\ 1 \end{pmatrix}.$$

These vectors are orthogonal, so we normalize them to obtain

$$\mathbf{v}_1 = \begin{pmatrix} \dfrac{2}{\sqrt{6}} \\ \dfrac{1}{\sqrt{6}} \\ \dfrac{1}{\sqrt{6}} \end{pmatrix}, \quad \mathbf{v}_2 = \begin{pmatrix} 0 \\ -\dfrac{1}{\sqrt{2}} \\ \dfrac{1}{\sqrt{2}} \end{pmatrix}, \quad \mathbf{v}_3 = \begin{pmatrix} -\dfrac{1}{\sqrt{3}} \\ \dfrac{1}{\sqrt{3}} \\ \dfrac{1}{\sqrt{3}} \end{pmatrix}.$$

The singular values of A are

$$\sigma_1 = \sqrt{\lambda_1} = \sqrt{3}, \quad \sigma_2 = \sqrt{\lambda_2} = \sqrt{1} = 1, \quad \sigma_3 = \sqrt{\lambda_3} = \sqrt{0} = 0.$$

Thus,

$$V = \begin{pmatrix} \dfrac{2}{\sqrt{6}} & 0 & -\dfrac{1}{\sqrt{3}} \\ \dfrac{1}{\sqrt{6}} & -\dfrac{1}{\sqrt{2}} & \dfrac{1}{\sqrt{3}} \\ \dfrac{1}{\sqrt{6}} & \dfrac{1}{\sqrt{2}} & \dfrac{1}{\sqrt{3}} \end{pmatrix}, \quad D = \begin{pmatrix} \sqrt{3} & 0 & 0 \\ 0 & 1 & 0 \end{pmatrix}.$$

To find U, we compute

$$\mathbf{u}_1 = \frac{1}{\sigma_1} A\mathbf{v}_1 = \frac{1}{\sqrt{3}} \begin{pmatrix} 1 & 0 & 1 \\ 1 & 1 & 0 \end{pmatrix} \begin{pmatrix} \dfrac{2}{\sqrt{6}} \\[2mm] \dfrac{1}{\sqrt{6}} \\[2mm] \dfrac{1}{\sqrt{6}} \end{pmatrix} = \begin{pmatrix} \dfrac{1}{\sqrt{2}} \\[2mm] \dfrac{1}{\sqrt{2}} \end{pmatrix}$$

and

$$\mathbf{u}_2 = \frac{1}{\sigma_2} A\mathbf{v}_2 = \frac{1}{1} \begin{pmatrix} 1 & 0 & 1 \\ 1 & 1 & 0 \end{pmatrix} \begin{pmatrix} 0 \\[2mm] -\dfrac{1}{\sqrt{2}} \\[2mm] \dfrac{1}{\sqrt{2}} \end{pmatrix} = \begin{pmatrix} \dfrac{1}{\sqrt{2}} \\[2mm] -\dfrac{1}{\sqrt{2}} \end{pmatrix}.$$

These vectors already form an orthonormal basis for $\mathbb{R}^2$, so we have

$$U = \begin{pmatrix} \dfrac{1}{\sqrt{2}} & \dfrac{1}{\sqrt{2}} \\[3mm] \dfrac{1}{\sqrt{2}} & -\dfrac{1}{\sqrt{2}} \end{pmatrix}.$$

This yields the SVD

$$A = \begin{pmatrix} \dfrac{1}{\sqrt{2}} & \dfrac{1}{\sqrt{2}} \\[3mm] \dfrac{1}{\sqrt{2}} & -\dfrac{1}{\sqrt{2}} \end{pmatrix} \begin{pmatrix} \sqrt{3} & 0 & 0 \\ 0 & 1 & 0 \end{pmatrix} \begin{pmatrix} \dfrac{2}{\sqrt{6}} & 0 & -\dfrac{1}{\sqrt{3}} \\[3mm] \dfrac{1}{\sqrt{6}} & -\dfrac{1}{\sqrt{2}} & \dfrac{1}{\sqrt{3}} \\[3mm] \dfrac{1}{\sqrt{6}} & \dfrac{1}{\sqrt{2}} & \dfrac{1}{\sqrt{3}} \end{pmatrix}.$$

$\bullet$

The MATLAB built-in function **svd** performs the *SVD* of a matrix. Thus, to reproduce the above results using the MATLAB Command Window, we do the following:

```
>> A = [1 0 1; 1 1 0];
>> [U, D, V] = svd(A);
```

The SVD occurs in many applications. For example, if we can compute the *SVD* accurately, then we can solve a linear system very efficiently. Since we know that the nonzero singular values of A are the square roots of the nonzero eigenvalues of a matrix AA^T, which are the same as the nonzero eigenvalues of $A^T A$, there are exactly $r = rank(A)$ positive singular values.

Suppose that A is square and has full rank. Then if $A\mathbf{x} = \mathbf{b}$, we have

$$
\begin{aligned}
UDV^T\mathbf{x} &= \mathbf{b} \\
U^T UDV^T\mathbf{x} &= U^T\mathbf{b} \\
DV^T\mathbf{x} &= U^T\mathbf{b} \\
V^T\mathbf{x} &= D^{-1}U^T\mathbf{b} \\
VV^T\mathbf{x} &= VD^{-1}U^T\mathbf{b} \\
\mathbf{x} &= VD^{-1}U^T\mathbf{b}
\end{aligned}
$$

(since $U^T U = 1, VV^T = 1$ by orthogonality).

Example 4.26 *Find the solution of the linear system $A\mathbf{x} = \mathbf{b}$ using SVD, where*

$$
A = \begin{pmatrix} -4 & -6 \\ 3 & -8 \end{pmatrix} \quad and \quad \mathbf{b} = \begin{pmatrix} 1 \\ 4 \end{pmatrix}.
$$

Solution. *First we have to compute the SVD of A. For this we have to compute*

$$
A^T A = \begin{pmatrix} -4 & 3 \\ -6 & -8 \end{pmatrix} \begin{pmatrix} -4 & -6 \\ 3 & -8 \end{pmatrix} = \begin{pmatrix} 25 & 0 \\ 0 & 100 \end{pmatrix}.
$$

The characteristic polynomial of $A^T A$ is

$$
\lambda^2 - 125\lambda + 2500 = (\lambda - 100)(\lambda - 25) = 0,
$$

and it gives the eigenvalues of $A^T A$:

$$\lambda_1 = 100 \quad and \quad \lambda_2 = 25.$$

Corresponding to the eigenvalues λ_1 and λ_2, we can have the eigenvectors

$$\begin{pmatrix} 0 \\ 1 \end{pmatrix} \quad and \quad \begin{pmatrix} 1 \\ 0 \end{pmatrix},$$

respectively. These vectors are orthogonal, so we normalize them to obtain

$$\mathbf{v}_1 = \begin{pmatrix} 0 \\ 1 \end{pmatrix} \quad and \quad \mathbf{v}_2 = \begin{pmatrix} 1 \\ 0 \end{pmatrix}.$$

The singular values of A are

$$\sigma_1 = \sqrt{\lambda_1} = \sqrt{100} = 10 \quad and \quad \sigma_2 = \sqrt{\lambda_2} = \sqrt{25} = 5.$$

Thus,

$$V = \begin{pmatrix} 0 & 1 \\ 1 & 0 \end{pmatrix} \quad and \quad D = \begin{pmatrix} 10 & 0 \\ 0 & 5 \end{pmatrix}.$$

To find U, we compute

$$\mathbf{u}_1 = \frac{1}{\sigma_1} A \mathbf{v}_1 = \frac{1}{10} \begin{pmatrix} -4 & -6 \\ 3 & -8 \end{pmatrix} \begin{pmatrix} 0 \\ 1 \end{pmatrix} = \begin{pmatrix} -0.6 \\ -0.8 \end{pmatrix}$$

and

$$\mathbf{u}_2 = \frac{1}{\sigma_2} A \mathbf{v}_2 = \frac{1}{5} \begin{pmatrix} -4 & -6 \\ 3 & -8 \end{pmatrix} \begin{pmatrix} 1 \\ 0 \end{pmatrix} = \begin{pmatrix} -0.8 \\ 0.6 \end{pmatrix}.$$

These vectors already form an orthonormal basis for $\mathbb{R}^2$, so we have

$$U = \begin{pmatrix} -0.6 & -0.8 \\ -0.8 & 0.6 \end{pmatrix}.$$

This yields the SVD

$$A = \begin{pmatrix} -0.6 & -0.8 \\ -0.8 & 0.6 \end{pmatrix} \begin{pmatrix} 10 & 0 \\ 0 & 5 \end{pmatrix} \begin{pmatrix} 0 & 1 \\ 1 & 0 \end{pmatrix}.$$

Now to find the solution of the given linear system, we solve

$$\mathbf{x} = V D^{-1} U^T \mathbf{b}$$

or

$$\begin{pmatrix} x_1 \\ x_2 \end{pmatrix} = \begin{pmatrix} 0 & 1 \\ 1 & 0 \end{pmatrix} \begin{pmatrix} 0.1 & 0 \\ 0 & 0.2 \end{pmatrix} \begin{pmatrix} -0.6 & -0.8 \\ -0.8 & 0.6 \end{pmatrix} = \begin{pmatrix} -0.04 \\ -0.14 \end{pmatrix}.$$

So

$$x_1 = -0.04 \quad and \quad x_2 = -0.14,$$

which is the solution of the given linear system. •

4.7 Summary

We discussed many numerical methods for finding eigenvalues and eigenvectors. Many eigenvalue problems do not require computation of all of the eigenvalues. The power method gives us a mechanism for computing the dominant eigenvalue along with its associated eigenvector for an arbitrary matrix. The convergence rate of the power method is poor when the two largest eigenvalues in magnitude are nearly equal. The technique of shifting the matrix by an amount $(-\mu \mathbf{I})$ can help us to overcome this disadvantage, and it can also be used to find intermediate eigenvalues by the power method. Also, if a matrix A is symmetric, then the power method gives faster convergence to the dominant eigenvalue and associated eigenvector. The inverse power method is used to estimate the least dominant eigenvalue of a nonsingular matrix. The inverse power method is guaranteed to converge if a matrix A is diagonalizable with the single least dominant nonzero eigenvalue. The inverse power method requires more computational effort than the power method, because a linear algebraic system must be solved at each iteration. The LU decomposition method (Chapter 1) can be used to efficiently accomplish this task. We also discussed the deflation method to obtain other eigenvalues once the dominant eigenvalue is known, and the Gerschgorin Circles theorem, which gives a crude approximation of the location of the eigenvalues of a matrix.

A technique for symmetric matrices, which occurs frequently, is the Jacobi method. It is an iterative method that uses orthogonal similarity transformations based on plane rotations to reduce a matrix to a diagonal form with diagonal elements as the eigenvalues of a matrix. The rotation matrices are used at the same time to form a matrix whose columns contain the eigenvectors of the matrix. The disadvantage of this method is that it may take many rotations to converge to a diagonal form. The rate of convergence of this method is increased by first preprocessing a matrix by Given's method and Householder transformations. These methods use the orthogonal similarity transformations to convert a given symmetric matrix to a symmetric tridiagonal matrix.

In the last section we discussed methods that depend on matrix decomposition. Methods such as the QR method and the LR method can be applied to a general matrix. To improve the computational efficiency of these methods, instead of applying the decomposition to an original matrix, an original matrix is first transformed to upper Hessenberg form. We discussed the singular values of a matrix and the singular value decomposition of a matrix in the last section of this chapter.

4.8 Problems

1. Find the first four iterations of the power method applied to each of the following matrices:

(a) $\begin{pmatrix} 2 & 3 & 1 \\ 1 & 4 & -1 \\ 3 & 1 & 2 \end{pmatrix}$, start with $\mathbf{x}_0 = [0, 1, 1]^T$.

(b) $\begin{pmatrix} 5 & 4 & 6 \\ 2 & 2 & -3 \\ 3 & 1 & 1 \end{pmatrix}$, start with $\mathbf{x}_0 = [1, 1, 1]^T$.

(c) $\begin{pmatrix} 1 & 1 & 1 \\ -2 & 2 & 1 \\ 5 & 1 & 1 \end{pmatrix}$, start with $\mathbf{x}_0 = [1, 1, 0]^T$.

(d) $\begin{pmatrix} 3 & 0 & 0 & 2 \\ 0 & 3 & 0 & -1 \\ 1 & 0 & 2 & 2 \\ 0 & 0 & 4 & 2 \end{pmatrix}$, start with $\mathbf{x}_0 = [1, 0, 0, 0]^T$.

2. Find the first four iterations of the power method with Euclidean scaling applied to each of the following matrices:

(a) $\begin{pmatrix} 2 & 1 & 2 \\ 1 & 4 & -1 \\ 2 & -1 & 2 \end{pmatrix}$, start with $\mathbf{x}_0 = [1, 0, 1]^T$.

(b) $\begin{pmatrix} 2 & 4 & -1 \\ 4 & 2 & -3 \\ -1 & -3 & 5 \end{pmatrix}$, start with $\mathbf{x}_0 = [1, 0, 0]^T$.

(c) $\begin{pmatrix} 3 & 2 & 4 \\ 2 & 3 & -1 \\ 4 & -1 & 3 \end{pmatrix}$, start with $\mathbf{x}_0 = [0, 1, 1]^T$.

(d) $\begin{pmatrix} 3 & 1 & 0 & 1 \\ 1 & 4 & -1 & 3 \\ 0 & -1 & 5 & 1 \\ 1 & 3 & 1 & 2 \end{pmatrix}$, start with $\mathbf{x}_0 = [1, 0, 1, 1]^T$.

3. Repeat Problem 2 using the power method with maximum entry scaling.

4. Repeat Problem 1 using the inverse power method.

5. Find the first four iterations of the following matrices by using the shifted inverse power method:

(a) $\begin{pmatrix} 2 & 3 & 3 \\ 1 & 4 & -1 \\ 3 & 1 & 2 \end{pmatrix}$, start with $\mathbf{x}_0 = [0, 1, 1]^T$, $\mu = 4.5$.

(b) $\begin{pmatrix} 1 & 1 & -1 \\ 2 & 1 & -3 \\ 2 & -4 & 1 \end{pmatrix}$, start with $\mathbf{x}_0 = [1, 1, 1]^T$, $\mu = 5$.

(c) $\begin{pmatrix} 1 & 1 & 1 \\ -2 & 2 & 1 \\ 3 & 3 & 3 \end{pmatrix}$,   start with $\mathbf{x}_0 = [1, 1, 0]^T$, $\mu = 4$.

(d) $\begin{pmatrix} 3 & 0 & 3 & 2 \\ 1 & 3 & 0 & -1 \\ 1 & 0 & 2 & 2 \\ 0 & 0 & 0 & 2 \end{pmatrix}$,   start with $\mathbf{x}_0 = [1, 0, 0, 0]^T$, $\mu = 3.5$.

6. Find the dominant eigenvalue and corresponding eigenvector by using the power method, with $\mathbf{x}^{(0)} = [1, 1, 1]^t$ (only four iterations):

$$A = \begin{pmatrix} 3 & 0 & 1 \\ 2 & 2 & 2 \\ 4 & 2 & 5 \end{pmatrix}.$$

Also, solve by using the inverse power method by taking the initial value of the eigenvalue by using the Rayleigh quotient theorem.

7. Find the dominant eigenvalue and corresponding eigenvector of the matrix A by using the power method. Start with $x^{(0)} = [2, 1, 0, -1]^T$ and $\epsilon = 0.0001$:

$$A = \begin{pmatrix} 3 & 1 & -2 & 1 \\ 1 & 8 & -1 & 0 \\ -2 & -1 & 3 & -1 \\ 1 & 0 & -1 & 8 \end{pmatrix}.$$

Also, use the shifted inverse power method with the same $x^{(0)}$ as given above to find the eigenvalue nearest to μ, which can be calculated by using the Rayleigh quotient.

8. Find the dominant eigenvalue and corresponding eigenvector by using the power method, with $u^0 = [1, 1, 1]^t$ (only four iterations):

$$A = \begin{pmatrix} 3 & 0 & 1 \\ 2 & 2 & 2 \\ 4 & 2 & 5 \end{pmatrix}.$$

Also, solve by using the inverse power method by taking the initial value of the eigenvalue by using the Rayleigh quotient.

9. Use the Gerschgorin Circles theorem to determine the bounds for the eigenvalues of each of the given matrices:

(a) $\begin{pmatrix} 3 & 2 & 1 \\ 2 & 3 & 0 \\ 1 & 0 & 3 \end{pmatrix}$, (b) $\begin{pmatrix} 1 & 1 & 1 \\ 1 & 1 & 0 \\ 1 & 0 & 1 \end{pmatrix}$, (c) $\begin{pmatrix} 2 & -2 & 1 \\ -2 & 1 & 1 \\ 1 & 1 & 2 \end{pmatrix}$.

10. Consider the matrix

$$A = \begin{pmatrix} 1 & 1 & -2 \\ -1 & 2 & 1 \\ 0 & 1 & -1 \end{pmatrix},$$

which has an eigenvalue 2 with eigenvector $[1, 3, 1]^T$. Use the deflation method to find the remaining eigenvalues and eigenvectors of A.

11. Consider the matrix

$$A = \begin{pmatrix} 2 & -3 & 6 \\ 0 & 3 & -4 \\ 0 & 2 & -3 \end{pmatrix},$$

which has an eigenvalue 2 with eigenvector $[1, 0, 0]^T$. Use the deflation method to find the remaining eigenvalues and eigenvectors of A.

12. Consider the matrix

$$A = \begin{pmatrix} 8 & -2 & -3 & 1 \\ 7 & -1 & -3 & 1 \\ 6 & -2 & -1 & 1 \\ 5 & -2 & -3 & 4 \end{pmatrix},$$

which has an eigenvalue 4 with eigenvector $[1, 1, 1, 1]^T$. Use the deflation method to find the remaining eigenvalues and eigenvectors of A.

13. Find the eigenvalues and corresponding eigenvectors of each of the following matrices by using the Jacobi method:

(a) $\begin{pmatrix} 5 & 3 & 7 \\ 3 & 1 & 6 \\ 7 & 6 & 2 \end{pmatrix}$, (b) $\begin{pmatrix} 2 & -1.5 & 0 \\ -1.5 & 2 & -0.5 \\ 0 & -0.5 & 2 \end{pmatrix}$, (c) $\begin{pmatrix} 4 & 6 & 7 \\ 6 & 5 & -3 \\ 7 & -3 & 2 \end{pmatrix}$,

$$\text{(d)} \begin{pmatrix} 0.4 & 0.3 & 0.1 \\ 0.3 & 0.5 & 0.2 \\ 0.1 & 0.2 & 0.6 \end{pmatrix}, \quad \text{(e)} \begin{pmatrix} 4 & 4 & 4 & 1 \\ 4 & 6 & 1 & 4 \\ 4 & 1 & 6 & 4 \\ 1 & 4 & 4 & 6 \end{pmatrix}, \quad \text{(f)} \begin{pmatrix} 2 & -1 & 3 & 2 \\ -1 & 3 & 1 & -2 \\ 3 & 1 & 4 & 1 \\ 2 & -2 & 1 & -3 \end{pmatrix}.$$

14. Use the Jacobi method to find all the eigenvalues and eigenvectors of the matrix

$$A = \begin{pmatrix} 5 & -2 & -0.5 & 1.5 \\ -2 & 5 & 1.5 & -0.5 \\ -0.5 & 1.5 & 5 & -2 \\ 1.5 & -0.5 & -2 & 5 \end{pmatrix}.$$

15. Use the Sturm sequence iteration to find the number of eigenvalues of the following matrices lying in the given intervals (a, b):

$$\text{(a)} \begin{pmatrix} 2 & -1 & 0 \\ -1 & 2 & -1 \\ 0 & -1 & 2 \end{pmatrix}, \quad (-1,3) \quad \text{(b)} \begin{pmatrix} 5 & -1 & 0 \\ -1 & 2 & 2 \\ 0 & 2 & 3 \end{pmatrix}, \quad (0,4).$$

16. Use the Sturm sequence iteration to find the eigenvalues of the following matrices:

$$\text{(a)} \begin{pmatrix} 1 & 4 & 0 \\ 4 & 1 & 4 \\ 0 & 1 & 1 \end{pmatrix}, \quad \text{(b)} \begin{pmatrix} 1 & 2 & 0 \\ 2 & 2 & 1 \\ 0 & 4 & 4 \end{pmatrix}, \quad \text{(c)} \begin{pmatrix} 1 & 2 & 0 \\ 2 & 1 & 2 \\ 0 & 2 & 1 \end{pmatrix}.$$

17. Find the eigenvalues and eigenvectors of the given symmetric matrix A by using the Jacobi method:

$$A = \begin{pmatrix} 0.6532 & 0.2165 & 0.0031 \\ & 0.4105 & 0.0052 \\ & & 0.2132 \end{pmatrix}.$$

Also, use Given's method to tridiagonalize the above matrix.

18. Use Given's method to convert the given matrix into tridiagonal form:

$$A = \begin{pmatrix} 2 & -1 & 3 & 2 \\ -1 & 3 & 1 & -2 \\ 3 & 1 & 4 & 1 \\ 2 & -2 & 1 & -3 \end{pmatrix}.$$

Also find the characteristic equation by using the Sturm sequence iteration.

19. Use Given's method to convert each matrix considered in Problem 9 into tridiagonal form.

20. Use Given's method to convert each matrix considered in Problem 5 into tridiagonal form and then use the Sturm sequence iteration to find the eigenvalues of each matrix.

21. Use Householder's method to convert each matrix considered in Problem 9 into tridiagonal form.

22. Use Householder's method to convert each matrix into tridiagonal form and then use the Sturm sequence iteration to find the eigenvalues of each matrix:

$$
\textbf{(a)} \begin{pmatrix} 2 & 3 & 4 \\ 3 & 4 & 5 \\ 4 & 5 & 6 \end{pmatrix}, \quad \textbf{(b)} \begin{pmatrix} 5 & -2 & 1 \\ -2 & 7 & 9 \\ 1 & 9 & 8 \end{pmatrix}, \quad \textbf{(c)} \begin{pmatrix} 4 & -2 & 1 & 4 \\ -2 & 5 & 0 & 3 \\ 1 & 0 & 6 & 2 \\ 4 & 3 & 2 & 7 \end{pmatrix}.
$$

23. Use Householder's method to place the following matrix in tridiagonal form:

$$
A = \begin{pmatrix} 7 & 1 & 2 & 1 \\ 1 & 8 & 1 & -1 \\ 2 & 1 & 3 & 1 \\ 1 & -1 & 1 & 2 \end{pmatrix}.
$$

Also, find the characteristic equation.

24. Find the first four QR iterations for each of the given matrices:

$$
\textbf{(a)} \begin{pmatrix} 1 & 0 & 2 \\ -2 & 1 & 1 \\ -2 & -5 & 1 \end{pmatrix}, \quad \textbf{(b)} \begin{pmatrix} 2 & -1 & 2 \\ 3 & 1 & 0 \\ 0 & 2 & 1 \end{pmatrix}, \quad \textbf{(c)} \begin{pmatrix} -21 & -9 & 12 \\ 0 & 6 & 0 \\ -24 & -8 & 15 \end{pmatrix}.
$$

25. Find the first 15 QR iterations for each of the matrices in Problem 9.

26. Use the QR method to find the eigenvalues of the matrix

$$A = \begin{pmatrix} 2 & -1 & 0 \\ -1 & -1 & -2 \\ 0 & -2 & 3 \end{pmatrix}.$$

27. Find the eigenvalues using the LR method for each of the given matrices:

(a) $\begin{pmatrix} 3 & 1 & 1 \\ 2 & 1 & 1 \\ 1 & 1 & 1 \end{pmatrix}$, (b) $\begin{pmatrix} 2 & 1 & 2 \\ 3 & 1 & 0 \\ 1 & 2 & 1 \end{pmatrix}$, (c) $\begin{pmatrix} 4 & 0 & 1 \\ -2 & 1 & 0 \\ -2 & 0 & 1 \end{pmatrix}.$

28. Find the eigenvalues using the LR method for each of the given matrices:

(a) $\begin{pmatrix} 1 & 2 & 4 \\ 5 & 1 & 1 \\ 2 & 1 & 1 \end{pmatrix}$, (b) $\begin{pmatrix} 3 & 3 & 3 \\ 3 & 3 & 3 \\ -3 & -3 & -3 \end{pmatrix}$, (c) $\begin{pmatrix} 15 & 13 & 20 \\ -21 & 12 & 15 \\ -8 & -8 & 11 \end{pmatrix}.$

29. Transform each of the given matrices into upper Hessenberg form:

(a) $\begin{pmatrix} 1 & 6 & 4 \\ 5 & 1 & 3 \\ 2 & 4 & 4 \end{pmatrix}$, (b) $\begin{pmatrix} 5 & 4 & 3 \\ 2 & 3 & 3 \\ -3 & -3 & 8 \end{pmatrix}$, (c) $\begin{pmatrix} 2 & 5 & 2 \\ 11 & 6 & 7 \\ 9 & 15 & 22 \end{pmatrix}$,

(d) $\begin{pmatrix} 2 & -1 & 4 & 2 \\ 3 & 2 & 3 & 2 \\ 1 & 2 & 2 & 2 \\ 2 & -3 & 4 & 4 \end{pmatrix}$, (e) $\begin{pmatrix} 2 & 1 & -2 & -3 \\ 2 & 2 & -3 & 2 \\ -3 & -3 & 4 & 5 \\ 7 & 8 & 3 & 2 \end{pmatrix}$,

(f) $\begin{pmatrix} 9 & 2 & 1 & -2 \\ 2 & 1 & 1 & -5 \\ -2 & 1 & 6 & -2 \\ -2 & -1 & 1 & -3 \end{pmatrix}.$

30. Transform each of the given matrices into upper Hessenberg form using Gaussian elimination with pivoting. Then use the QR method and the LR method to find their eigenvalues:

(a) $\begin{pmatrix} 11 & 33 & 45 \\ 12 & 21 & 23 \\ 18 & 22 & 31 \end{pmatrix}$, (b) $\begin{pmatrix} 4 & 3 & 3 \\ 2 & 5 & 4 \\ -3 & 2 & 1 \end{pmatrix}$, (c) $\begin{pmatrix} 14 & 22 & 2 & 1 \\ 5 & 1 & 5 & -2 \\ 6 & 1 & 6 & 1 \\ 7 & -2 & 7 & 4 \end{pmatrix}$.

31. Find the singular values for each of the given matrices:

(a) $\begin{pmatrix} 2 & 0 & 1 \\ 0 & 2 & 0 \end{pmatrix}$, (b) $\begin{pmatrix} 3 & 0 & 0 \\ -2 & 3 & -2 \\ 2 & 0 & 5 \end{pmatrix}$, (c) $\begin{pmatrix} 1 & 0 & 1 \\ 0 & 1 & 0 \\ 0 & 1 & 2 \end{pmatrix}$,

(d) $\begin{pmatrix} 4 & 0 & 1 \\ 0 & 1 & 0 \\ 2 & 1 & 1 \end{pmatrix}$, (e) $\begin{pmatrix} 2 & 0 & 1 \\ -4 & 6 & -2 \\ 2 & 0 & 7 \end{pmatrix}$, (f) $\begin{pmatrix} 2 & 0 & 1 & 2 \\ 0 & 1 & 1 & 3 \\ 0 & 3 & 2 & 1 \\ 1 & 0 & 3 & 1 \end{pmatrix}$.

32. Show that the singular values of the following matrices are the same as the eigenvalues of the matrices:

(a) $\begin{pmatrix} 4 & 2 & 1 \\ 2 & 8 & 0 \\ 1 & 0 & 8 \end{pmatrix}$, (b) $\begin{pmatrix} 3 & 0 & 1 \\ 0 & 5 & 0 \\ 1 & 0 & 5 \end{pmatrix}$, (c) $\begin{pmatrix} 2 & 0 & 0 \\ 0 & 6 & 0 \\ 0 & 0 & 7 \end{pmatrix}$.

33. Show that all singular values of an orthogonal matrix are 1.

34. Show that if A is a positive-definite matrix, then A has a singular value decomposition of the form QDQ^T.

35. Find an SVD for each of the given matrices:

(a) $\begin{pmatrix} 0 & -4 \\ -6 & 0 \end{pmatrix}$, (b) $\begin{pmatrix} 2 & 1 & 0 \\ 1 & 3 & 0 \end{pmatrix}$,

(c) $\begin{pmatrix} 1 & 0 \\ 1 & 2 \\ -1 & 3 \end{pmatrix}$, (d) $\begin{pmatrix} 1 & 2 & 1 \\ 1 & 2 & 1 \end{pmatrix}$.

36. Find an *SVD* for each of the given matrices:

(a) $\begin{pmatrix} 0 & -2 \\ -3 & 0 \end{pmatrix}$,   (b) $\begin{pmatrix} 2 & 0 & 0 \\ 0 & 3 & 0 \end{pmatrix}$,

(c) $\begin{pmatrix} 1 & 0 \\ 1 & 1 \\ -1 & 1 \end{pmatrix}$,   (d) $\begin{pmatrix} 1 & 1 & 1 \\ 1 & 1 & 1 \end{pmatrix}$.

37. Find an *SVD* for each of the given matrices:

(a) $\begin{pmatrix} 0 & -2 & 1 \\ -3 & 0 & 2 \\ 0 & 1 & 1 \end{pmatrix}$,   (b) $\begin{pmatrix} 2 & -4 & 3 \\ 6 & 6 & 3 \\ -4 & 2 & 4 \end{pmatrix}$,

(c) $\begin{pmatrix} 1 & 2 & 0 \\ 1 & 1 & 1 \\ -1 & 1 & 2 \end{pmatrix}$,   (d) $\begin{pmatrix} 8 & -3 & 7 & 1 \\ 3 & 11 & 3 & 2 \\ 1 & 2 & 5 & 2 \\ 2 & 0 & 7 & 2 \end{pmatrix}$,

(e) $\begin{pmatrix} 2 & 1 & -2 & -13 \\ 11 & 12 & -3 & 12 \\ 3 & 22 & 24 & 15 \\ 7 & 8 & 3 & 2 \end{pmatrix}$,   (f) $\begin{pmatrix} 10 & 1 & 0 & -2 \\ 2 & 1 & 1 & -5 \\ 4 & 6 & 7 & -2 \\ 5 & -1 & 1 & -3 \end{pmatrix}$.

38. Find the solution of each of the following linear systems, $A\mathbf{x} = \mathbf{b}$, using singular value decomposition:

(a)
$$A = \begin{pmatrix} 1 & -3 \\ 3 & -5 \end{pmatrix}, \quad \mathbf{x} = \begin{pmatrix} x_1 \\ x_2 \end{pmatrix}, \quad \mathbf{b} = \begin{pmatrix} 1 \\ 2 \end{pmatrix}.$$

(b)
$$A = \begin{pmatrix} 1 & -1 \\ 1 & 4 \end{pmatrix}, \quad \mathbf{x} = \begin{pmatrix} x_1 \\ x_2 \end{pmatrix}, \quad \mathbf{b} = \begin{pmatrix} 1.1 \\ 0.5 \end{pmatrix}.$$

(c)

$$A = \begin{pmatrix} 3 & -1 & 4 \\ -1 & 0 & 1 \\ 4 & 1 & 2 \end{pmatrix}, \quad \mathbf{x} = \begin{pmatrix} x_1 \\ x_2 \\ x_3 \end{pmatrix}, \quad \mathbf{b} = \begin{pmatrix} 1 \\ 2 \\ 3 \end{pmatrix}.$$

(d)

$$A = \begin{pmatrix} 4 & 3 & 2 \\ 1 & 2 & -1 \\ 1 & 3 & 2 \end{pmatrix}, \quad \mathbf{x} = \begin{pmatrix} x_1 \\ x_2 \\ x_3 \end{pmatrix}, \quad \mathbf{b} = \begin{pmatrix} 2.5 \\ 1.5 \\ 0.85 \end{pmatrix}.$$

39. Find the solution each of the following linear systems, $A\mathbf{x} = \mathbf{b}$, using singular value decomposition:

(a)

$$A = \begin{pmatrix} 2 & 2 \\ 1 & 3 \end{pmatrix}, \quad \mathbf{x} = \begin{pmatrix} x_1 \\ x_2 \end{pmatrix}, \quad \mathbf{b} = \begin{pmatrix} 1 \\ 0.9 \end{pmatrix}.$$

(b)

$$A = \begin{pmatrix} 1 & 0 \\ 3 & -2 \end{pmatrix}, \quad \mathbf{x} = \begin{pmatrix} x_1 \\ x_2 \end{pmatrix}, \quad \mathbf{b} = \begin{pmatrix} 1 \\ 2 \end{pmatrix}.$$

(c)

$$A = \begin{pmatrix} 1 & -1 & 0 \\ 2 & 0 & 1 \\ 3 & 0 & 2 \end{pmatrix}, \quad \mathbf{x} = \begin{pmatrix} x_1 \\ x_2 \\ x_3 \end{pmatrix}, \quad \mathbf{b} = \begin{pmatrix} 1 \\ 1 \\ 1 \end{pmatrix}.$$

(d)

$$A = \begin{pmatrix} 1 & 2 & 3 \\ 2 & 1 & 2 \\ 1 & 1 & 1 \end{pmatrix}, \quad \mathbf{x} = \begin{pmatrix} x_1 \\ x_2 \\ x_3 \end{pmatrix}, \quad \mathbf{b} = \begin{pmatrix} 1 \\ 0 \\ 1 \end{pmatrix}.$$

Chapter 5

Interpolation and Approximation

5.1 Introduction

In this chapter we describe the numerical methods for the approximation of functions other than elementary functions. The main purpose of these techniques is to replace a complicated function with one that is simpler and more manageable. We sometimes know the value of a function $f(x)$ at a set of points (say, $x_0 < x_1 < x_2 \cdots < x_n$), but we do not have an analytic expression for $f(x)$ that lets us calculate its value at an arbitrary point. We will concentrate on techniques that may be adapted if, for example, we have a table of values of functions that may have been obtained from some physical measurement or some experiments or long numerical calculations that cannot be cast into a simple functional form. The task now is to estimate $f(x)$ for an arbitrary point x by, in some sense, drawing a smooth curve through (and perhaps beyond) the data points x_i. If the desired x is between the largest and smallest of the data points, then the problem is called *interpolation*; and if x is outside that range, it is called *extrapolation*. Here, we shall restrict our attention to interpolation. It is a rational process generally used in estimating a missing functional value by taking a weighted average of known functional values at neighboring data points.

The interpolation scheme must model a function, in between or beyond the known data point, by some plausible functional form. The form should be sufficiently general to be able to approximate large classes of functions that might arise in practice. The functional forms are polynomials, trigonometric functions, rational functions, and exponential functions. However, we shall restrict our attention to *polynomials*. The polynomial functions are widely used in practice, since they are easy to determine, evaluate, differentiate, and are integrable. Polynomial interpolation provides some mathematical tools that can be used in developing methods for approximation theory, numerical differentiation, numerical integration, and numerical solutions of ordinary differential equations and partial differential equations. A set of data points we consider here may be *equally or unequally spaced* in the independent variable x. Several procedures can be used to fit approximation polynomials either individually or for both. For example, Lagrange interpolatory polynomials, Newton's divided difference interpolatory polynomials, and Aitken's interpolatory polynomials can be used for unequally spaced or equally spaced, and procedures based on differences can be used for equally spaced, including Newton forward and backward difference polynomials, Gauss forward and backward difference polynomials, Bessel difference polynomials and Stirling difference polynomials. These methods are quite easy to apply. But here, we discuss only the Lagrange interpolation method, Newton's divided differences interpolation method, and Aitken's interpolation method. We shall also discuss another polynomial interpolation known as the Chebyshev polynomial. This type of polynomial interpolates the given function over the interval $[-1, 1]$.

The other approach to approximate a function is called the *least squares approximation*. This approach is suitable if the given data points are experimental data. We shall discuss linear, nonlinear, plane, and trigonometric least squares approximation of a function. We shall also discuss the least squares solution of overdetermined and underdetermined linear systems. At the end of the chapter, we discuss least squares with QR decomposition and singular value decomposition.

5.2 Polynomial Approximation

The general form of an *nth*-degree polynomial is

$$p_n(x) = a_0 + a_1 x + a_2 x^2 + \cdots + a_n x^n, \tag{5.1}$$

where n denotes the degree of the polynomial and $a_0, a_1, \ldots, a_n$ are constant coefficients. Since there are $(n+1)$ coefficients, $(n+1)$ data points are required to obtain a unique value for the coefficients. The important property of polynomials that makes them suitable for approximating functions is due to the following *Weierstrass Approximation theorem.*

Theorem 5.1 (Weierstrass Approximation Theorem)

If $f(x)$ is a continuous function in the closed interval $[a, b]$, then for every $\epsilon > 0$ there exists a polynomial $p_n(x)$, where the value of n depends on the value of ϵ, such that for all x in $[a, b]$,

$$|f(x) - p_n(x)| < \epsilon. \tag{5.2}$$

Consequently, any continuous function can be approximated to any accuracy by a polynomial of high enough degree.   •

Suppose we have a given a set of $(n+1)$ data points relating dependent variables $f(x)$ to an independent variable x as follows:

$$\begin{array}{c|cccc} x & x_0 & x_1 & \cdots & x_n \\ \hline f(x) & f(x_0) & f(x_1) & \cdots & f(x_n) \end{array}.$$

Generally, the data points $x_0, x_1, \ldots, x_n$ are arbitrary, and assume the interval between the two adjacent points is not the same, and assume that the data points are organized in such a way that $x_0 < x_1 < x_2 < \cdots < x_{n-1} < x_n$.

When the data points in a given functional relationship are not equally spaced, the interpolation problem becomes more difficult to solve. The basis for this assertion lies in the fact that the interpolating polynomial coefficient will depend on the functional values as well as on the data points given in the table.

5.2.1   Lagrange Interpolating Polynomials

This is one of the most popular and well-known interpolation methods used to approximate the functions at an arbitrary point x. The Lagrange interpolation method provides a direct approach for determining interpolated values, regardless of the data points spacing, i.e., it can be fitted to unequally spaced or equally spaced data. To discuss the Lagrange interpolation method, we start with the simplest form of interpolation, i.e,, *linear interpolation*. The interpolated value is obtained from the equation of a straight line that passes through two tabulated values, one on each side of the required value. This straight line is a first-degree polynomial. The problem of determining a polynomial of degree one that passes through the distinct points (x_0, y_0) and (x_1, y_1) is the same as approximating the function $f(x)$ for which $f(x_0) = y_0$ and $f(x_1) = y_1$ by means of first degree polynomial interpolation. Let us consider the construction of a linear polynomial $p_1(x)$ passing through two data points $(x_0, f(x_0))$ and $(x_1, f(x_1))$, as shown in Figure 5.1. Let us consider a linear polynomial of the form

$$p_1(x) = a_0 + a_1 x. \tag{5.3}$$

Since a polynomial of degree one has two coefficients, one might expect to be able to choose two conditions that satisfy

$$p_1(x_k) = f(x_k); \qquad k = 0, 1.$$

When $p_1(x)$ passes through point $(x_0, f(x_0))$, we have

$$p_1(x_0) = a_0 + a_1 x_0 = y_0 = f(x_0),$$

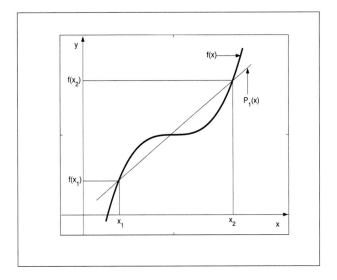

Figure 5.1: Linear Lagrange interpolation.

and if it passes through point $(x_1, f(x_1))$, we have

$$p_1(x_1) = a_0 + a_1 x_1 = y_1 = f(x_1).$$

Solving the last two equations gives the unique solution

$$a_0 = \frac{x_0 y_1 - x_1 y_0}{x_0 - x_1}; \quad a_1 = \frac{y_1 - y_0}{x_1 - x_0}. \tag{5.4}$$

Putting these values in (5.3), we have

$$p_1(x) = \left(\frac{x - x_1}{x_0 - x_1}\right) y_0 + \left(\frac{x - x_0}{x_1 - x_0}\right) y_1,$$

which can also be written as

$$p_1(x) = L_0(x) f(x_0) + L_1(x) f(x_1), \tag{5.5}$$

where

$$L_0(x) = \frac{x - x_1}{x_0 - x_1} \quad \text{and} \quad L_1(x) = \frac{x - x_0}{x_1 - x_0}. \tag{5.6}$$

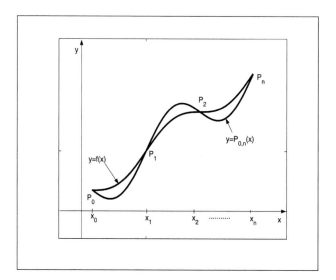

Figure 5.2: General Lagrange interpolation.

Note that when $x = x_0$, then $L_0(x_0) = 1$ and $L_1(x_0) = 0$. Similarly, when $x = x_1$, then $L_0(x_1) = 0$ and $L_1(x_1) = 1$. The polynomial (5.5) is known as the *linear Lagrange interpolating polynomial* and (5.6) are the *Lagrange coefficient polynomials*. To generalize the concept of linear interpolation, consider the construction of a polynomial $p_n(x)$ of degree at most n that passes through $(n + 1)$ distinct points $(x_0, f(x_0)), \ldots, (x_n, f(x_n))$ (Figure 5.2) and satisfies the interpolation conditions

$$p_n(x_k) = f(x_k); \qquad k = 0, 1, 2, \ldots, n. \qquad (5.7)$$

Assume that there exists polynomial $L_k(x)$ $(k = 0, 1, 2, \ldots, n)$ of degree n having the property

$$L_k(x_j) = \begin{cases} 0, & \text{for} \quad k \neq j \\ 1, & \text{for} \quad k = j \end{cases} \qquad (5.8)$$

and

$$\sum_{k=0}^{n} L_k(x) = 1. \qquad (5.9)$$

The polynomial $p_n(x)$ is given by

$$
\begin{aligned}
p_n(x) &= L_0(x)f(x_0) + L_1(x)f(x_1) + \cdots + L_{i-1}(x)f(x_{i-1}) \\
&\quad + L_i(x)f(x_i) + \cdots + L_n(x)f(x_n) \\
&= \sum_{k=0}^{n} L_k(x)f(x_k).
\end{aligned}
\tag{5.10}
$$

It is clearly a polynomial of degree at most n and satisfies the conditions (5.7) since

$$
\begin{aligned}
p_n(x_i) = L_0(x_i)f(x_0) &+ L_1(x_i)f(x_1) + \cdots + L_{i-1}(x_i)f(x_{i-1}) \\
&+ L_i(x_i)f(x_i) + \cdots + L_n(x_i)f(x_n),
\end{aligned}
$$

which implies that

$$
p_n(x_i) = f(x_i).
$$

It remains to be shown how the polynomial $L_i(x)$ can be constructed so that it satisfies (5.8). If $L_i(x)$ is to satisfy (5.8), then it must contain a factor

$$
(x - x_0)(x - x_1)\cdots(x - x_{i-1})(x - x_{i+1})\cdots(x - x_n).
\tag{5.11}
$$

Since this expression has exactly n terms and $L_i(x)$ is a polynomial of degree n, we can deduce that

$$
L_i(x) = A_i(x - x_0)(x - x_1)\cdots(x - x_{i-1})(x - x_{i+1})\cdots(x - x_n),
\tag{5.12}
$$

for some multiplicative constant A_i. Let $x = x_i$, then the value of A_i is chosen so that

$$
A_i = 1/(x_i - x_0)(x_i - x_1)\cdots(x_i - x_{i-1})(x_i - x_{i+1})\cdots(x_i - x_n),
\tag{5.13}
$$

where none of the terms in the denominator can be zero, from the assumption of distinct points. Hence,

$$
L_i(x) = \prod_{k=0}^{n}\left(\frac{x - x_k}{x_i - x_k}\right), \quad i \neq k.
\tag{5.14}
$$

The interpolating polynomial can now be readily evaluated by substituting (5.14) into (5.10) to give

$$f(x) \approx p_n(x) = \sum_{i=0}^{n} \prod_{k=0}^{n} \left(\frac{x - x_k}{x_i - x_k} \right) f(x_i), \qquad i \neq k. \qquad (5.15)$$

This formula is called the *Lagrange interpolation formula of degree n* and the terms in (5.14) are called the Lagrange coefficient polynomials.

To show the *uniqueness* of the interpolating polynomial $p_n(x)$, we suppose that in addition to the polynomial $p_n(x)$ the interpolation problem has another solution $q_n(x)$ of degree $\leq n$, whose graph passes through (x_i, y_i), $i = 0, 1, \ldots, n$. Then define

$$r_n(x) = p_n(x) - q_n(x)$$

of a degree not greater than n. Since

$$r_n(x_i) = p_n(x_i) - q_n(x_i) = f(x_i) - f(x_i) = 0,$$

the polynomial $r_n(x)$ vanishes at $n + 1$ point. But by using the following well-known result from the theory of equations:

"If a polynomial of degree n vanishes at $n + 1$ distinct points, then the polynomial is identically zero."

Hence, $r_n(x)$ vanishes identically, or equivalently, at $p_n(x) = q_n(x)$.

Example 5.1 *Let $f(x) = 0$ be defined by the three numbers $-h, 0, h$, where $h \neq 0$. Use the Lagrange interpolating polynomial to construct the polynomial $p(x)$, which interpolates $f(x)$ at the given numbers. Then show that this polynomial can be written in the following form:*

$$p(x) = \frac{1}{2h^2}[f(-h) - 2f(0) + f(h)]x^2 + \frac{1}{2h}[f(h) - f(-h)]x + f(0).$$

Solution. *Given three distinct points $x_0 = -h, x_1 = 0$, and $x_2 = h$ and using the quadratic Lagrange interpolating polynomial as*

$$p_2(x) = \frac{(x - x_1)(x - x_2)}{(x_0 - x_1)(x_0 - x_2)}f(x_0) + \frac{(x - x_0)(x - x_2)}{(x_1 - x_0)(x_1 - x_2)}f(x_1) + \frac{(x - x_0)(x - x_1)}{(x_2 - x_0)(x_2 - x_1)}f(x_2),$$

at these data points, we get

$$p_2(x) = \frac{(x-0)(x-h)}{(-h-0)(-h-h)}f(-h) + \frac{(x+h)(x-h)}{(0+h)(0-h)}f(0) + \frac{(x+h)(x-0)}{(h+h)(h-0)}f(h)$$

or

$$p_2(x) = \frac{(x^2-xh)}{(-h)(-2h)}f(-h) + \frac{(x^2-h^2)}{(h)(-h)}f(0) + \frac{(x^2+xh)}{(2h)(h)}f(h).$$

Separating the coefficients of x^2, x, and a constant term, we get

$$p_2(x) = \left(\frac{f(-h)}{2h^2} + \frac{f(0)}{-h^2} + \frac{f(h)}{2h^2}\right)x^2 + \left(\frac{-hf(-h)}{2h^2} + \frac{hf(h)}{2h^2} + \frac{f(h)}{2h^2}\right)x + \left(\frac{-h^2f(0)}{-h^2}\right).$$

Simplifying, we obtain

$$p(x) = \frac{1}{2h^2}[f(-h)-2f(0)+f(h)]x^2 + \frac{1}{2h}[f(h)-f(-h)]x + f(0).$$

Example 5.2 *Let $p_2(x)$ be the quadratic Lagrange interpolating polynomial for the data: $(1,2),(2,3),(3,\alpha)$. Find the value of α if the constant term in $p_2(x)$ is 5. Also, find the approximation of $f(2.5)$.*

Solution. *Consider the quadratic Lagrange interpolating polynomial as*

$$p_2(x) = L_0(x)f(x_0) + L_1(x)f(x_1) + L_2(x)f(x_2),$$

and using the given data points, we get

$$p_2(x) = L_0(x)(2) + L_1(x)(3) + L_2(x)(\alpha),$$

where the Lagrange coefficients can be calculated as follows:

$$L_0(x) = \frac{(x - x_1)(x - x_2)}{(x_0 - x_1)(x_0 - x_2)} = \frac{(x - 2)(x - 3)}{(1 - 2)(1 - 3)}$$

$$= \frac{1}{2}(x^2 - 5x + 6)$$

$$L_1(x) = \frac{(x - x_0)(x - x_2)}{(x_1 - x_0)(x_1 - x_2)} = \frac{(x - 1)(x - 3)}{(2 - 1)(2 - 3)}$$

$$= -(x^2 - 4x + 3)$$

$$L_2(x) = \frac{(x - x_0)(x - x_1)}{(x_2 - x_0)(x_2 - x_1)} = \frac{(x - 1)(x - 2)}{(3 - 1)(3 - 2)}$$

$$= \frac{1}{2}(x^2 - 3x + 2).$$

Thus,

$$p_2(x) = \frac{1}{2}(x^2 - 5x + 6)(2) - (x^2 - 4x + 3)(3) + \frac{1}{2}(x^2 - 3x + 2)(\alpha).$$

Separating the coefficients of x^2, x, and a constant term, we get

$$p_2(x) = (1 - 3 + \frac{\alpha}{2})x^2 + (-5 + 12 - \frac{3\alpha}{2})x + (6 - 9 + \alpha)$$

or

$$p_2(x) = (-2 + \frac{\alpha}{2})x^2 + (7 - \frac{3\alpha}{2})x + (-3 + \alpha).$$

Since the given value of the constant term is 5, using this, we get

$$(-3 + \alpha) = 5, \quad gives \quad \alpha = 8.$$

Now using this value of α, the approximation of $f(x)$ and given $x = 2.5$, we get

$$p_2(2.5) = (-2 + \left(\frac{8}{2}\right)(2.5)^2 + (7 - \left(\frac{24}{2}\right)(2.5) + (-3 + 8),$$

and it gives

$$f(2.5) \approx p_2(2.5) = 12.50 - 12.50 + 5 = 5.$$

Example 5.3 *Let $f(x) = x + 1/x$, with points $x_0 = 1, x_1 = 1.5, x_2 = 2.5$, and $x_3 = 3$. Find the quadratic Lagrange polynomial for the approximation of $f(2.7)$. Also, find the relative error.*

Solution. *Consider the quadratic Lagrange interpolating polynomial as*

$$p_2(x) = L_0(x)f(x_0) + L_1(x)f(x_1) + L_2(x)f(x_2),$$

where the Lagrange coefficients are as follows:

$$L_0(x) = \frac{(x - x_1)(x - x_2)}{(x_0 - x_1)(x_0 - x_2)}$$

$$L_1(x) = \frac{(x - x_0)(x - x_2)}{(x_1 - x_0)(x_1 - x_2)}$$

$$L_2(x) = \frac{(x - x_0)(x - x_1)}{(x_2 - x_0)(x_2 - x_1)}.$$

Since the given interpolating point is $x = 2.7$, the best three points for the quadratic polynomial should be

$$x_0 = 1.5, \quad x_1 = 2.5, \quad x_2 = 3,$$

and the function values at these points are

$$f(x_0) = 2.167, \quad f(x_1) = 2.9, \quad f(x_2) = 3.333.$$

So using these values, we have

$$p_2(x) = 2.167L_0(x) + 2.9L_1(x) + 3.333L_2(x),$$

where

$$L_0(x) = \frac{(x - 2.5)(x - 3)}{(1.5 - 2.5)(1.5 - 3)} = \frac{1}{1.5}(x^2 - 5.5x + 7.5)$$

$$L_1(x) = \frac{(x - 1.5)(x - 3)}{(2.5 - 1.5)(2.5 - 3)} = \frac{1}{-0.5}(x^2 - 4.5x + 4.5)$$

$$L_2(x) = \frac{(x - 1.5)(x - 2.5)}{(3 - 1.5)(3 - 2.5)} = \frac{1}{0.75}(x^2 - 4x + 3.75).$$

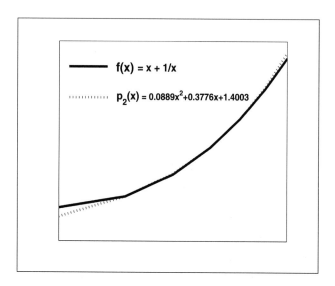

Figure 5.3: Quadratic approximation of a function.

Thus,

$$p_2(x) = \frac{2.167}{1.5}(x^2 - 5.5x + 7.5) = \frac{2.9}{-0.5}(x^2 - 4.5x + 4.5) = \frac{3.333}{0.75}(x^2 - 4x + 3.75),$$

and simplifying, we get

$$p_2(x) = 0.0889x^2 + 0.3776x + 1.4003,$$

which is the required quadratic polynomial. At $x = 2.7$, we have

$$f(2.7) \approx p_2(2.7) = 3.0679.$$

The relative error is

$$\frac{|f(2.7) - p_2(2.7)|}{|f(2.7)|} = \frac{|3.0704 - 3.0679|}{|3.0704|} = 0.0008.$$

●

Note that the sum of the Lagrange coefficients is equal to 1 as it should be:

$$L_0(2.7) + L_1(2.7) + L_2(2.7) = -0.0400 + 0.7200 + 0.3200 = 1.$$

Using MATLAB commands, the above results can be reproduced as follows:

```
>> x = [1.5 2.5 3];
>> y = x + 1/x;
>> x0 = 2.7;
>> sol = lint(x, y, x0);
```

Program 5.1

MATLAB m-file for the Lagrange Interpolation Method

```
function fi=lint(x,y,x0)
dxi=x0-x; m=length(x); L=zeros(size(y));
L(1) = prod(dxi(2 : m))/prod(x(1) − x(2 : m));
L(m) = prod(dxi(1 : m − 1))/prod(x(m) − x(1 : m − 1));
for j=2:m-1
num = prod(dxi(1 : j − 1)) ∗ prod(dxi(j + 1 : m));
dem = prod(x(j) − x(1 : j − 1)) ∗ prod(x(j) − x(j + 1 : m));
L(j)=num/dem; end; fi = sum(y. ∗ L);
```

Example 5.4 *Using the cubic Lagrange interpolation formula*

$$p(x) = \alpha_1 f(0) + \alpha_2 f(0.2) + \alpha_3 f(0.4) + \alpha_4 f(0.6),$$

for the approximation of $f(0.5)$, show that

$$\alpha_1 + \alpha_2 + \alpha_3 + \alpha_4 = 1.$$

Solution. *Consider the cubic Lagrange interpolating polynomial*

$$p_3(x) = \alpha_1 f(x_0) + \alpha_2 f(x_1) + \alpha_3 f(x_2) + \alpha_4 f(x_3),$$

where the values of $\alpha_1, \alpha_2, \alpha_3, \alpha_4$ can be defined as follows:

$$\alpha_1 = \frac{(x - x_1)(x - x_2)(x - x_3)}{(x_0 - x_1)(x_0 - x_2)(x_0 - x_3)}$$

$$\alpha_2 = \frac{(x - x_0)(x - x_2)(x - x_3)}{(x_1 - x_0)(x_1 - x_2)(x_1 - x_3)}$$

$$\alpha_3 = \frac{(x - x_0)(x - x_1)(x - x_3)}{(x_2 - x_0)(x_2 - x_1)(x_2 - x_3)}$$

$$\alpha_4 = \frac{(x - x_0)(x - x_1)(x - x_2)}{(x_3 - x_0)(x_3 - x_1)(x_3 - x_2)}.$$

Using the given values as $x_0 = 0, x_1 = 0.2, x_2 = 0.4, x_3 = 0.6$, and the interpolating point $x = 0.5$, we obtain

$$\alpha_1 = \frac{(0.5 - 0.2)(0.5 - 0.4)(0.5 - 0.6)}{(0 - 0.2)(0 - 0.4)(0 - 0.6)} = 0.0625$$

$$\alpha_2 = \frac{(0.5 - 0)(0.5 - 0.4)(0.5 - 0.6)}{(0.2 - 0)(0.2 - 0.4)(0.2 - 0.6)} = -0.3125$$

$$\alpha_3 = \frac{(0.5 - 0)(0.5 - 0.2)(0.5 - 0.6)}{(0.4 - 0)(0.4 - 0.2)(0.4 - 0.6)} = 0.9375$$

$$\alpha_4 = \frac{(0.5 - 0)(0.5 - 0.2)(0.5 - 0.4)}{(0.6 - 0)(0.6 - 0.2)(0.6 - 0.4)} = 0.3125.$$

Thus,

$$\alpha_1 + \alpha_2 + \alpha_3 + \alpha_4 = 0.0625 - 0.3125 + 0.9375 + 0.3125 = 1. \bullet$$

Error Formula

As with any numerical technique, it is important to obtain bounds for the errors involved. Now we discuss the error term when the Lagrange polynomial is used to approximate the continuous function $f(x)$. It is similar to the error term for the well-known Taylor polynomial, except that the factor $(x - x_0)^{n+1}$ is replaced with the product $(x - x_0)(x - x_1) \cdots (x - x_n)$. This is expected because interpolation is exact at each of the $(n + 1)$ data points x_k, where we have

$$f(x_k) - p_n(x_k) = y_k - y_k = 0, \quad \text{for} \quad k = 1, 2, \ldots, n. \tag{5.16}$$

Theorem 5.2 (Error Formula of the Lagrange Polynomial)

If $f(x)$ has $(n + 1)$ derivatives on interval I, and if it is approximated by a polynomial $p_n(x)$ passing through $(n + 1)$ data points on I, then the error E_n is given by

$$E_n = f(x) - p_n(x) = \frac{f^{(n+1)}(\eta(x))}{(n + 1)!}(x - x_0)(x - x_1) \cdots (x - x_n), \quad \eta(x) \in I,$$

$$\tag{5.17}$$

where $p_n(x)$ is the Lagrange interpolating polynomial (5.10) and an unknown point $\eta(x) \in (x_0, x_n)$. ●

The error formula (5.17) is an important theoretical result because Lagrange polynomials are used extensively for deriving numerical differentiation and integration methods. Error bounds for these techniques are obtained from the Lagrange error formula.

Example 5.5 *Find the linear Lagrange polynomial that passes through the points $(0, f(0)$ and $(\pi, f(\pi))$ to approximate the function $f(x) = 2 \cos x$. Also, find a bound for the error in the linear interpolation of $f(x)$.*

Solution. *Given $x_0 = 0$ and $x_1 = \pi$, then the linear Lagrange polynomial $p_1(x)$*

$$p_1(x) = \frac{(x - x_1)}{(x_0 - x_1)} f(x_0) + \frac{(x - x_0)}{(x_1 - x_0)} f(x_1)$$

interpolating $f(x)$ at these points is

$$p_1(x) = \frac{(x - \pi)}{(0 - \pi)} f(0) + \frac{(x - 0)}{(\pi - 0)} f(\pi).$$

By using the function values at the given data points, we get

$$f(x) \approx p_1(x) = \frac{(x - \pi)}{(0 - \pi)}(2) + \frac{(x - 0)}{(\pi - 0)}(-2) = 2 - \frac{4x}{\pi}.$$

To compute a bound of error in the linear interpolation of $f(x)$, we use the linear Lagrange error formula (5.17)

$$f(x) - p_1(x) = \frac{(x - x_0)(x - x_1)}{2!} f''(\eta(x)),$$

where $\eta(x)$ is a unknown point between $x_0 = 0$ and $x_1 = \pi$. Hence,

$$|f(x) - p_1(x)| = \left| \frac{(x - 0)(x - \pi)}{2} \right| |f''(\eta(x))|.$$

The value of $f''(\eta(x))$ cannot be computed exactly because $\eta(x)$ is not known. But we can bound the error by computing the largest possible value for $|f''(\eta(x))|$. So the bound $|f''(x)|$ on $[0, \pi]$ can be obtained as

$$M = \max_{0 \leq x \leq \pi} |f''(x)| = \max_{0 \leq x \leq \pi} |-2 \cos x| = 2,$$

and so for $|f''(\eta(x))| \leq M$, *we have*

$$|f(x) - p_1(x)| \leq \frac{2}{2}|(x - 0)(x - \pi)|.$$

Since the function $|(x-0)(x-\pi)|$ *attains its maximum in* $[0, \pi]$ *and occurs at* $x = \frac{(0+\pi)}{2}$, *the maximum value is* $\frac{(\pi-0)^2}{4}$.

This follows easily by noting that the function $(x - 0)(x - \pi)$ *is a quadratic and has two roots* 0 *and* π. *Hence, its maximum value occurs midway between these roots. Thus, for any* $x \in [0, \pi]$, *we have*

$$|f(x) - p_1(x)| \leq \frac{(\pi - 0)^2}{4} = \frac{\pi^2}{4},$$

which is the required bound of error in the linear interpolation of $f(x)$. •

Example 5.6 *Use the best Lagrange interpolating polynomial to find the approximation of* $f(1.5)$, *if* $f(-2) = 2, f(-1) = 1.5$, $f(1) = 3.5$, *and* $f(2) = 5$. *Estimate the error bound if the maximum value of* $|f^{(4)}(x)|$ *is* 0.025 *in the interval* $[-2, 2]$.

Solution. *Since the given number of points,* $x_0 = -2, x_1 = -1, x_2 = 1, x_3 = 2$, *are four, the best Lagrange interpolating polynomial to find the approximation of* $f(1.5)$ *will be the cubic. The cubic Lagrange interpolating polynomial for the approximation of the given function is*

$$f(x) \approx p_3(x) = L_0(x)f(x_0) + L_1(x)f(x_1) + L_2(x)f(x_2) + L_3(x)f(x_3),$$

and taking $f(-2) = 2, f(-1) = 1.5, f(1) = 3.5, f(2) = 5$, *and the interpolating point* $x = 1.5$, *we have*

$$f(1.5) \approx p_3(1.5) = L_0(1.5)f(-2) + L_1(1.5)f(-1) + L_2(1.5)f(1) + L_3(1.5)f(2)$$

or

$$f(1.5) \approx p_3(1.5) = 2L_0(1.5) + 1.5L_1(1.5) + 3.5L_2(1.5) + 5L_3(1.5).$$

The Lagrange coefficients can be calculated as follows:

$$L_0(1.5) = \frac{(1.5+1)(1.5-1)(1.5-2)}{(-2+1)(-2-1)(-2-2)} = 0.0521$$

$$L_1(1.5) = \frac{(1.5+2)(1.5-1)(1.5-2)}{(-1+2)(-1-1)(-1-2)} = -0.1458$$

$$L_2(1.5) = \frac{(1.5+2)(1.5+1)(1.5-2)}{(1+2)(1+1)(1-2)} = 0.7292$$

$$L_3(1.5) = \frac{(1.5+2)(1.5+1)(1.5-1)}{(2+2)(2+1)(2-1)} = 0.3646.$$

Putting these values of the Lagrange coefficients in the above equation, we get

$$f(1.5) \approx p_3(1.5) = 2(0.0521)+1.5(-0.1458)+3.5(0.7292)+5(0.3646) = 4.2607,$$

which is the required cubic interpolating polynomial approximation of the function at the given point $x = 1.5$.

To compute an error bound for the approximation of the given function in the interval $[-2, 2]$, we use the following cubic error formula:

$$|f(x) - p_3(x)| = \frac{|f^{(4)}(\eta(x))|}{4!}|(x - x_0)(x - x_1)(x - x_2)(x - x_3)|.$$

Since

$$|f^{(4)}(\eta(x))| \leq M = \max_{-2 \leq x \leq 2}|f^{(4)}(x)| = 0.025,$$

$$|f(1.5) - p_3(1.5)| \leq \frac{M}{4!}|(1.5 + 2)(1.5 + 1)(1.5 - 1)(1.5 - 2)|,$$

which gives

$$|f(1.5) - p_3(1.5)| \leq \frac{(0.025)(2.1875)}{24} = 0.0023,$$

the desired error bound. ●

Example 5.7 *Determine the spacing h in a table of equally spaced values of the function $f(x) = e^x$ between the smallest point $a = 1$ and the largest point $b = 2$, so that interpolation with a second-degree polynomial in this table will yield the desired accuracy.*

Solution. *Suppose that the given table contains the function values $f(x_i)$, for the points $x_i = 1 + ih$, $i = 0, 1, \ldots, n$, where $n = \frac{(2-1)}{h}$. If $x \in [x_{i-1}, x_{i+1}]$, then we approximate the function $f(x)$ by degree 2 polynomial $p_2(x)$, which interpolates $f(x)$ at x_{i-1}, x_i, x_{i+1}. Then the error formula (5.17) for these data points becomes*

$$|f(x) - p_2(x)| = \left| \frac{(x - x_{i-1})(x - x_i)(x - x_{i+1})}{3!} \right| \left| f'''(\eta(x)) \right|,$$

where $\eta(x) \in (x_{i-1}, x_{i+1})$. Since the point $\eta(x)$ is unknown, we cannot estimate $f'''(\eta(x))$, so we let

$$|f'''(\eta(x))| \leq M = \max_{1 \leq x \leq 2} |f'''(x)|.$$

Then

$$|f(x) - p_2(x)| \leq \frac{M}{6} |(x - x_{i-1})(x - x_i)(x - x_{i+1})|.$$

Since $f(x) = e^x$ and $f'''(x) = e^x$,

$$|f'''(\eta(x))| \leq M = e^2 = 7.3891.$$

Now to find the maximum value of $|(x - x_{i-1})(x - x_i)(x - x_{i+1})|$, we have

$$\max_{x \in [x_{i-1}, x_{i+1}]} |(x - x_{i-1})(x - x_i)(x - x_{i+1})| = \max_{t \in [-h,h]} |(t - h)t(t + h)|$$

$$= \max_{t \in [-h,h]} |t(t^2 - h^2)|,$$

using the linear change of variables $t = x - x_i$. As we can see, the function $H(t) = t^3 - th^2$ vanishes at $t = -h$ and $t = h$, so the maximum value of $|H(t)|$ on $[-h, h]$ must occur at one of the extremes of $H(t)$, which can be found by solving the equation

$$H'(t) = 3t^2 - h^2 = 0, \quad \text{gives}, \quad t = \pm h/\sqrt{3}.$$

Hence,

$$\max_{x\in[x_{i-1},x_{i+1}]}|(x-x_{i-1})(x-x_i)(x-x_{i+1})| = \frac{2h^3}{3\sqrt{3}}.$$

Thus, for any $x \in [1,2]$, we have

$$|f(x) - p_2(x)| \leq \frac{(2h^3/3\sqrt{3})e^2}{6} = \frac{h^3 e^2}{9\sqrt{3}},$$

if $p_2(x)$ is chosen as the polynomial of degree 2, which interpolates $f(x) = e^x$ at the three tabular points nearest x. If we wish to obtain six decimal place accuracy this way, we would have to choose h so that

$$\frac{h^3 e^2}{9\sqrt{3}} < 5 \times 10^{-7},$$

which implies that

$$h^3 < 10.5483 \times 10^{-7}$$

and gives $h = 0.01$. ●

While the Lagrange interpolation formula is at the heart of polynomial interpolation, it is not, by any stretch of the imagination, the most practical way to use it. Just consider for a moment that if we had to add an additional data point in the previous Example 5.6, in order to find the cubic polynomial $p_3(x)$, we would have to repeat the whole process again because we cannot use the solution of the quadratic polynomial $p_2(x)$ in the construction of the cubic polynomial $p_3(x)$. Therefore, one can note that the Lagrange method is not particularly efficient for large values of n, the degree of the polynomial. When n is large and the data for x is ordered, some improvement in efficiency can be obtained by considering only the data pairs in the vicinity of the x values for which $f(x)$ is sought.

One will be quickly convinced that there must be better techniques available. In the following, we discuss some of the more practical approaches to polynomial interpolation. They are Newton's, Aitken's, and Chebyshev's interpolation formulas. In using the first two schemes, the construction of the difference table plays an important role. It must be noted that in using the Lagrange interpolation scheme there was no need to construct a difference table.

5.2.2   Newton's General Interpolating Formula

We noted in the previous section that for a small number of data points
one can easily use the Lagrange formula for the interpolating polynomial.
However, for a large number of data points there will be many multiplications and, more significantly, whenever a new data point is added to an
existing set, the interpolating polynomial has to be completely recalculated. Here, we describe an efficient way of organizing the calculations to
overcome these disadvantages.

Let us consider the *nth*-degree polynomial $p_n(x)$ that agrees with the
function $f(x)$ at the distinct numbers $x_0, x_1, \ldots, x_n$. The divided differences of $f(x)$ with respect to $x_0, x_1, \ldots, x_n$ are derived to express $p_n(x)$ in
the form

$$
\begin{aligned}
p_n(x) = a_0 \; &+ \; a_1(x - x_0) + a_2(x - x_0)(x - x_1) + \cdots \\
&+ \; a_n(x - x_0)(x - x_1)\cdots(x - x_{n-1}),
\end{aligned}
\tag{5.18}
$$

for appropriate constants $a_0, a_1, \ldots, a_n$.

Now determine the constants first by evaluating $p_n(x)$ at x_0, and we
have

$$
p_n(x_0) = a_0 = f(x_0).
\tag{5.19}
$$

Similarly, when $p_n(x)$ is evaluated at x_1, then

$$
p_n(x_1) = a_0 + a_1(x_1 - x_0) = f(x_1),
$$

which implies that

$$
a_1 = \frac{f(x_1) - f(x_0)}{x_1 - x_0}.
\tag{5.20}
$$

Now we express the interpolating polynomial in terms of divided differences.

Divided Differences

First, we define the *zeroth divided difference* at the point x_i by

$$f[x_i] = f(x_i), \tag{5.21}$$

which is simply the value of the function $f(x)$ at x_i.

The *first-order* or *first divided difference* at the points x_i and x_{i+1} can be defined by

$$f[x_i, x_{i+1}] = \frac{f[x_{i+1}] - f[x_i]}{x_{i+1} - x_i} = \frac{f(x_{i+1}) - f(x_i)}{x_{i+1} - x_i}. \tag{5.22}$$

In general, the *nth divided difference* $f[x_i, x_{i+1}, \ldots, x_{i+n}]$ is defined by

$$f[x_i, x_{i+1}, \ldots, x_{i+n}] = \frac{f[x_{i+1}, x_{i+2}, \ldots, x_{i+n}] - f[x_i, x_{i+1}, \ldots, x_{i+n-1}]}{x_{i+n} - x_i}. \tag{5.23}$$

By using this definition, (5.19) and (5.20) can be written as

$$a_0 = f[x_0]; \qquad a_1 = f[x_0, x_1],$$

respectively. Similarly, one can have the values of other constants involved in (5.18) such as

$$\begin{aligned}
a_2 &= f[x_0, x_1, x_2] \\
a_3 &= f[x_0, x_1, x_2, x_3] \\
\cdots &= \cdots \\
\cdots &= \cdots \\
a_n &= f[x_0, x_1, \ldots, x_n].
\end{aligned}$$

Putting the values of these constants in (5.18), we get

$$\begin{aligned}
p_n(x) = f[x_0] &+ f[x_0, x_1](x - x_0) + f[x_0, x_1, x_2](x - x_0)(x - x_1) \\
&+ \cdots + f[x_0, x_1, \ldots, x_n](x - x_0)(x - x_1) \cdots (x - x_{n-1}), \tag{5.24}
\end{aligned}$$

which can also be written as

$$p_n(x) = f[x_0] + \sum_{k=1}^{n} f[x_0, x_1, \ldots, x_k](x - x_0)(x - x_1) \cdots (x - x_{k-1}). \tag{5.25}$$

This type of polynomial is known as *Newton's interpolatory divided difference polynomial*. Table 5.1 shows the divided differences for a function $f(x)$. One can easily show that (5.25) is simply a rearrangement of the La-

Table 5.1: Divided difference table for a function $y = f(x)$.

k	x_k	Zeroth Divided Difference	First Divided Difference	Second Divided Difference	Third Divided Difference
0	x_0	$f[x_0]$			
1	x_1	$f[x_1]$	$f[x_0, x_1]$		
2	x_2	$f[x_2]$	$f[x_1, x_2]$	$f[x_0, x_1, x_2]$	
3	x_3	$f[x_3]$	$f[x_2, x_3]$	$f[x_1, x_2, x_3]$	$f[x_0, x_1, x_2, x_3]$

grange form defined by (5.10). For example, the Newton divided difference interpolation polynomial of degree one is

$$p_1(x) = f[x_0] + f[x_0, x_1](x - x_0),$$

which implies that

$$
\begin{aligned}
p_1(x) &= f(x_0) + \left(\frac{f(x_1) - f(x_0)}{x_1 - x_0} \right)(x - x_0) \\
&= \frac{(x_1 - x_0)f(x_0) + (x - x_0)f(x_1) - f(x_0)(x - x_0)}{x_1 - x_0} \\
&= \left(\frac{x - x_1}{x_0 - x_1} \right) f(x_0) + \left(\frac{x - x_0}{x_1 - x_0} \right) f(x_1),
\end{aligned}
$$

which is the Lagrange interpolating polynomial of degree one. Similarly, one can show the equivalent for the *n*th-degree polynomial. ●

Example 5.8 *Construct the fourth divided differences table for the function* $f(x) = 4x^4 + 3x^3 + 2x^2 + 10$ *using the values* $x = 3, 4, 5, 6, 7, 8$.

Solution. *The results are listed in Table 5.5.*

From the results in Table 5.5, one can note that the nth divided difference for the nth polynomial equation is always constant and the (n+1)th divided difference is always zero for the nth polynomial equation.   ●

Using the following MATLAB commands one can construct Table 5.5 as follows:

```
>> x = [3 4 5 6 7 8];
>> y = 4 * x.^ 4+3 * x.^ 3+2 * x.^ 2+10;
>> D = divdiff(x, y);
```

Table 5.2: Divided differences table for $f(x) = e^x$ at the given points.

k	x_k	Zeroth Divided Difference	First Divided Difference	Second Divided Difference	Third Divided Difference	Fourth Divided Difference	Fifth Divided Difference
0	3	433					
1	4	1258	825				
2	5	2935	1677	426			
3	6	5914	2979	651	75		
4	7	10741	4827	924	91	4	
5	8	18058	7317	1245	107	4	0

Example 5.9 *Write Newton's interpolating polynomials in the form $a + bx + cx^2$ and show that $a + b + c = 2$ by using the following data points:*

$$\begin{array}{c|ccc} x & 0 & 1 & 3 \\ \hline f(x) & 1 & 2 & 3 \end{array}.$$

Solution. *First, we construct the divided differences table for the given data points. The result of the divided differences is listed in Table 5.3. Since Newton's interpolating polynomial of degree 2 can be written as*

$$p_2(x) = f[x_0] + f[x_0, x_1](x - x_0) + f[x_0, x_1, x_2](x - x_0)(x - x_1),$$

Table 5.3: Divided differences table for Example 5.9.

k	x_k	Zeroth Divided Difference	First Divided Difference	Second Divided Difference
0	0	1		
1	1	2	1	
2	3	3	$\dfrac{1}{2}$	$-\dfrac{1}{6}$

by using Table 5.3, we have

$$p_2(x) = 1 + (1)(x - 0) + \left(-\frac{1}{6}\right)(x - 0)(x - 1),$$

which gives

$$p_2(x) = \frac{(6 + 7x - x^2)}{6},$$

and from it, we have

$$a + b + c = \frac{6}{6} + \frac{7}{6} + \frac{-1}{6} = 2.$$

●

Example 5.10 *Show that Newton's interpolating polynomial $p_2(x)$ of degree 2 satisfies the interpolation conditions*

$$p_2(x_i) = f(x_i), \qquad i = 0, 1, 2.$$

Solution. *Since Newton's interpolating polynomial of degree 2 is*

$$p_2(x) = f[x_0] + f[x_0, x_1](x - x_0) + f[x_0, x_1, x_2](x - x_0)(x - x_1),$$

first, taking $x = x_0$, we have

$$p_2(x_0) = f[x_0] + 0 + 0 = f(x_0).$$

Now taking $x = x_1$, we have

$$p_2(x_1) = f[x_0] + f[x_0, x_1](x_1 - x_0) + 0 = f(x_0) + \frac{f(x_1) - f(x_0)}{x_1 - x_0}(x_1 - x_0),$$

and it gives

$$p_2(x_1) = f(x_0) + f(x_1) - f(x_0) = f(x_1).$$

Finally, taking $x = x_2$, we have

$$p_2(x_2) = f[x_0] + f[x_0, x_1](x_2 - x_0) + f[x_0, x_1, x_2](x_2 - x_0)(x_2 - x_1),$$

which can be written as

$$p_2(x_2) = f[x_0] + f[x_0, x_1](x_2 - x_0) + \frac{f[x_1, x_2] - f[x_0, x_1]}{x_2 - x_0}(x_2 - x_0)(x_2 - x_1),$$

which gives

$$p_2(x_2) = f[x_0] + f[x_0, x_1](x_1 - x_0) + f[x_1, x_2](x_2 - x_1).$$

From (5.22), we have

$$p_2(x_2) = f[x_0] + \frac{f(x_1) - f(x_0)}{x_1 - x_0}(x_1 - x_0) + \frac{f(x_2) - f(x_1)}{x_2 - x_1}(x_2 - x_1),$$

which gives

$$p_2(x_2) = f(x_0) + f(x_1) - f(x_0) + f(x_2) - f(x_1) = f(x_2). \qquad \bullet$$

Program 5.2
MATLAB m-file for the Divided Differences

```
function D=divdiff(x,y)
% Construct divided difference table
m = length(x); D = zeros(m, m); D(:, 1) = y(:);
for j=2:m; for i=j:m
D(i, j) = (D(i, j-1) - D(i-1, j-1))/(x(i) - x(i-j+1));
end; end
```

The main advantage of the Newton divided difference form over the Lagrange form is that polynomial $p_n(x)$ can be calculated from polynomial $p_{n-1}(x)$ by adding just one extra term, since it follows from (5.25) that

$$p_n(x) = p_{n-1}(x) + f[x_0, x_1, \ldots, x_n](x - x_0)(x - x_1) \cdots (x - x_{n-1}). \quad (5.26)$$

Example 5.11 (a) *Construct the divided difference table for the function* $f(x) = \ln(x + 2)$ *in the interval* $0 \le x \le 3$ *for the stepsize* $h = 1$.
(b) *Use Newtons's divided difference interpolation formula to construct the interpolating polynomials of degree 2 and degree 3 to approximate* $\ln(3.5)$.
(c) *Compute error bounds for the approximations in part (b).*

Solution. (a) *The results of the divided differences are listed in Table 5.4.*

(b) *First, we construct the second degree polynomial* $p_2(x)$ *by using the quadratic Newton interpolation formula as follows:*

$$p_2(x) = f[x_0] + f[x_0, x_1](x - x_0) + f[x_0, x_1, x_2](x - x_0)(x - x_1),$$

then with the help of the divided differences Table 5.4, we get

$$p_2(x) = 0.6932 + 0.4055(x - 0) - 0.0589(x - 0)(x - 1),$$

which implies that

$$p_2(x) = -0.0568x^2 + 0.4644x + 0.6932.$$

Then at $x = 1.5$, *we have*

$$p_2(1.5) = 1.2620,$$

with possible actual error

$$f(1.5) - p_2(1.5) = 1.2528 - 1.2620 = -0.0072.$$

Now to construct the cubic interpolatory polynomial $p_3(x)$ *that fits at all four points, we only have to add one more term to the polynomial* $p_2(x)$:

$$p_3(x) = p_2(x) + f[x_0, x_1, x_2, x_3](x - x_0)(x - x_1)(x - x_2)$$

or

$$\begin{aligned} p_3(x) &= p_2(x) + 0.0089(x - 0)(x - 1)(x - 2) \\ p_3(x) &= p_2(x) + 0.0089(x^3 - 3x^2 + 2x), \end{aligned}$$

then at $x = 1.5$, *we get*

$$p_3(1.5) = p_2(1.5) + 0.0089((1.5)^3 - 3(1.5)^2 + 2(1.5))$$
$$p_3(1.5) = 1.2620 - 0.0033 = 1.2587,$$

with possible actual error

$$f(1.5) - p_3(1.5) = 1.2528 - 1.2587 = -0.0059.$$

We note that the estimated value of $f(1.5)$ *by the cubic interpolating polynomial is closer to the exact solution than the quadratic polynomial.*

Table 5.4: Divide differences table for Example 5.11.

k	x_k	Zeroth Divided Difference	First Divided Difference	Second Divided Difference	Third Divided Difference
0	0	**0.6932**			
1	1	1.0986	**0.4055**		
2	2	1.3863	0.2877	**- 0.0589**	
3	3	1.6094	0.2232	- 0.0323	**0.0089**

(c) Now to compute the error bounds for the approximations in part (b), we use the error formula (5.17). For the polynomial $p_2(x)$, *we have*

$$|f(x) - p_2(x)| = \frac{|f'''(\eta(x))|}{3!}|(x - x_0)(x - x_1)(x - x_2)|.$$

Since the third derivative of the given function is

$$f'''(x) = \frac{2}{(x+2)^3}$$

and

$$|f'''(\eta(x))| = \left|\frac{2}{(\eta(x)+2)^3}\right|, \quad for \quad \eta(x) \in (0,2),$$

then

$$M = \max_{0 \le x \le 2}\left|\frac{2}{(x+2)^3}\right| = 0.25$$

and

$$|f(1.5) - p_2(1.5)| \le (0.375)\frac{(0.25)}{6} = 0.0156,$$

which is the required error bound for the approximation $p_2(1.5)$.

Since the error bound for the cubic polynomial $p_3(x)$ is

$$|f(x) - p_3(x)| = \frac{|f^{(4)}(\eta(x))|}{4!}|(x - x_0)(x - x_1)(x - x_2)(x - x_3)|,$$

taking the fourth derivative of the given function, we have

$$f^{(4)}(x) = \frac{-6}{(x+2)^4}$$

and

$$|f^{(4)}(\eta(x))| = \left|\frac{-6}{(\eta(x)+2)^4}\right|, \qquad for \quad \eta(x) \in (0,3).$$

Since

$$|f^{(4)}(0)| = 0.375$$
$$|f^{(4)}(3)| = 0.0096,$$

$$|f^{(4)}(\eta(x))| \le \max_{0 \le x \le 3}\left|\frac{-6}{(x+2)^4}\right| = 0.375$$

and

$$|f(1.5) - p_3(1.5)| \le (0.5625)\frac{(0.375)}{24} = 0.0088,$$

which is the required error bound for the approximation $p_3(1.5)$.   ●

    Note that in Example 5.11, we used the value of the quadratic polynomial $p_2(1.5)$ in calculating the cubic polynomial $p_3(1.5)$. It was possible because the initial value for both polynomials was the same as $x_0 = 0$. But the situation will be quite different if the initial point for both polynomials is different. For example, if we have to find the approximate value of $\ln(4.5)$, then the suitable data points for the quadratic polynomial will be $x_0 = 1, x_1 = 2, x_2 = 3$ and for the cubic polynomial will be $x_0 = 0, x_1 = 1, x_2 = 2, x_3 = 3$. So for getting the best approximation of

$\ln(4.5)$ by the cubic polynomial $p_3(2.5)$, we cannot use the value of the quadratic polynomial $p_2(2.5)$ in the cubic polynomial $p_3(2.5)$. The best way is to use the cubic polynomial form

$$
\begin{aligned}
p_3(2.5) &= f[0] + f[0,1](2.5-0) + f[0,1,2](2.5-0)(2.5-1) \\
&+ f[0,1,2,3](2.5-0)(2.5-1)(2.5-2),
\end{aligned}
$$

which gives

$$
p_3(2.5) = 0.6932 + 1.0137 - 0.2208 + 0.0166 = 1.5027.
$$

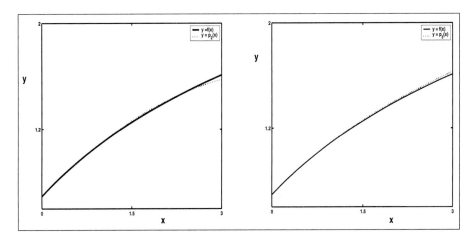

Figure 5.4: Quadratic and cubic polynomial approximations of the function.

MATLAB commands can reproduce the results of Example 5.11 as follows:

```
>> x = [0 1 2 3];
>> y = log(x + 2);
>> x0 = 1.5;
>> Y = Ndivf(x, y, x0);
```

Program 5.3

MATLAB m-file for Newton's Linear Interpolation Method

```
function Y=Ndivf(x,y,x0)
m = length(x); D = zeros(m, m); D(:, 1) = y(:);
for j=2:m; for i=j:m;
D(i, j) = (D(i, j−1)−D(i−1, j−1))/(x(i)−x(i−j+1)); end; end;
Y = D(m, m) * ones(size(x0));
for i = m − 1 : −1 : 1;
Y = D(i, i) + (x0 − x(i)) * Y; end
```

Example 5.12 *Let $x_0 = 0.5, x_1 = 0.7, x_2 = 0.9, x_3 = 1.1, x_4 = 1.3$, and $x_5 = 1.5$. Use Newton polynomial $p_5(x)$ of degree five to approximate the function $f(x) = e^x$ at $x = 0.6$, when $p_4(0.6) = 1.9112$. Also, compute an error bound for your approximation.*

Solution. *Since the fifth-degree Newton polynomial $p_5(x)$ is defined as*

$$p_5(x) = p_4(x) + (x−x_0)(x−x_1)(x−x_2)(x−x_3)(x−x_4)f[x_0, x_1, x_2, x_3, x_4, x_5],$$

and using the given data points, we have

$$p_5(0.6) = p_4(0.6) + (0.0010)f[0.5, 0.7, 0.9, 1.1, 1.3, 1.5].$$

Now we compute the fifth-order divided differences of the function as follows. Thus,

$$p_5(0.6) = 1.9112 + (0.0010)(0.0228) = 1.9112.$$

Since the error bound for the fifth-degree polynomial $p_5(x)$ is

$$|f(x)−p_5(x)| = \frac{|f^{(6)}(\eta(x))|}{6!}|(x−x_0)(x−x_1)(x−x_2)(x−x_3)(x−x_4)(x−x_5)|,$$

taking the sixth derivative of the given function, we have

$$f^{(6)}(x) = e^x$$

and

$$|f^{(6)}(\eta(x))| = e^{(\eta(x))}, \qquad for \quad \eta(x) \in (0.5, 1.5).$$

Table 5.5: Divided differences for $f(x) = e^x$ at the given points.

k	x_k	Zeroth Divided Difference	First Divided Difference	Second Divided Difference	Third Divided Difference	Fourth Divided Difference	Fifth Divided Difference
0	0.5	1.6487					
1	0.7	2.0138	1.8252				
2	0.9	2.4596	2.2293	1.0102			
3	1.1	3.0042	2.7228	1.2339	0.3728		
4	1.3	3.6693	3.3257	1.5071	0.4553	0.1032	
5	1.5	4.4817	4.0620	1.8408	0.4553	0.1260	0.0228

Since

$$|f^{(6)}(0.5)| = 1.6487$$
$$|f^{(6)}(1.5)| = 4.4817,$$

$$|f^{(6)}(\eta(x))| \leq \max_{0.5 \leq x \leq 1.5} |e^x| = 4.4817.$$

Thus, we get

$$|f(0.6) - p_5(0.6)| \leq (0.00095)(4.4817)/720 = 0.000006,$$

which is the required error bound for the approximation $p_5(0.6)$.   ●

Example 5.13 *Consider the points $x_0 = 0.5, x_1 = 1.5, x_2 = 2.5, x_3 = 3.0, x_4 = 4.5$, and for a function $f(x)$, the divided differences are*

$$f[x_2] = 73.8125, f[x_1, x_2] = 59.5, f[x_0, x_1, x_2] = 23.7,$$

$$f[x_1, x_2, x_3] = 47.25, f[x_0, x_1, x_2, x_3, x_4] = 1.$$

Using this information, construct the complete divided differences table for the given data points.

Solution. *Since we know the third divided difference is defined as*

$$f[x_0, x_1, x_2, x_3] = \frac{f[x_1, x_2, x_3] - f[x_0, x_1, x_2]}{x_3 - x_0},$$

using the given data points, we get

$$f[x_0, x_1, x_2, x_3] = \frac{47.25 - 23.5}{3.0 - 0.5} = 9.50.$$

Similarly, the other third divided difference $f[x_1, x_2, x_3, x_4]$ can be computed by using the fourth divided difference formula as follows:

$$f[x_0, x_1, x_2, x_3, x_4] = \frac{f[x_1, x_2, x_3, x_4] - f[x_0, x_1, x_2, x_3]}{x_4 - x_0}$$

$$1.0 = \frac{f[x_1, x_2, x_3, x_4] - 9.50}{4.5 - 0.5}$$

$$f[x_1, x_2, x_3, x_4] = 4.0 + 9.50 = 13.50.$$

Now finding the remaining second-order divided difference $f[x_2, x_3, x_4]$, we use the third-order divided difference as follows:

$$f[x_1, x_2, x_3, x_4] = \frac{f[x_2, x_3, x_4] - f[x_1, x_2, x_3]}{x_4 - x_1}$$

$$13.50 = \frac{f[x_2, x_3, x_4] - 47.25}{4.5 - 1.5}$$

$$f[x_2, x_3, x_4] = 40.50 + 47.25 = 87.75.$$

Finding the first-order divided difference $f[x_0, x_1]$, we use the second-order divided difference as follows:

$$f[x_0, x_1, x_2] = \frac{f[x_1, x_2] - f[x_0, x_1]}{x_2 - x_0}$$

$$23.50 = \frac{59.50 - f[x_0, x_1]}{2.5 - 0.5}$$

$$f[x_2, x_3, x_4] = 59.50 - 47.0 = 12.50.$$

Similarly, the other two first-order divided differences $f[x_2, x_3]$ and $f[x_3, x_4]$ can be calculated as follows:

$$f[x_1, x_2, x_3] = \frac{f[x_2, x_3] - f[x_1, x_2]}{x_3 - x_1}$$

$$47.25 = \frac{f[x_2, x_3] - 59.50}{3.0 - 1.5}$$

$$f[x_2, x_3] = 70.8750 + 59.50 = 130.375$$

and

$$f[x_2, x_3, x_4] = \frac{f[x_3, x_4] - f[x_2, x_3]}{x_4 - x_2}$$

$$87.75 = \frac{f[x_3, x_4] - 130.375}{4.5 - 2.5}$$

$$f[x_2, x_3, x_4] = 175.50 + 130.375 = 305.875.$$

Also, the remaining zeroth-order divided differences can be calculated as follows:

$$f[x_1, x_2] = \frac{f[x_2] - f[x_1]}{x_2 - x_1}$$

$$59.50 = \frac{73.8125 - f[x_1]}{2.5 - 1.5}$$

$$f[x_1] = 73.8125 - 59.50 = 14.3125$$

and

$$f[x_0, x_1] = \frac{f[x_1] - f[x_0]}{x_1 - x_0}$$

$$59.50 = \frac{73.8125 - f[x_1]}{2.5 - 1.5}$$

$$f[x_1] = 73.8125 - 59.50 = 14.3125.$$

Finally,

$$f[x_2, x_3] = \frac{f[x_3] - f[x_2]}{x_3 - x_2}$$

$$130.375 = \frac{f[x_3] - 73.8125}{3.0 - 2.5}$$

$$f[x_3] = 65.1875 + 73.8125 = 139.00$$

and

$$f[x_3, x_4] = \frac{f[x_4] - f[x_3]}{x_4 - x_3}$$

$$305.875 = \frac{f[x_4] - 139.00}{4.5 - 3.0}$$

$$f[x_4] = 458.8125 + 139.00 = 597.8125,$$

which completes the divided differences table as shown in Table 5.6.　　　●

Table 5.6: Complete divided differences table for the given points.

k	x_k	Zeroth Divided Difference	First Divided Difference	Second Divided Difference	Third Divided Difference	Fourth Divided Difference
0	0.5	1.8125				
1	1.5	14.3125	12.5000			
2	2.5	73.8125	59.5000	23.5000		
3	3.0	139.000	130.375	47.2500	9.5000	
4	4.5	597.8125	305.875	87.7500	13.500	1.0000

Example 5.14 *If $f(x) = p(x)q(x)$, then show that*

$$f[x_0, x_1] = p(x_1)q[x_0, x_1] + q(x_0)p[x_0, x_1].$$

Also, find the values of $p[0, 1]$ and $q[0, 1]$, when $f[0, 1] = 4, f(1) = 5, p(1) = q(0) = 2$.

Solution. *The first-order divided difference can be written as*

$$f[x_0, x_1] = \frac{f(x_1) - f(x_0)}{x_1 - x_0}.$$

Now using $f(x_1) = p(x_1)q(x_1)$ *and* $f(x_0) = p(x_0)q(x_0)$ *in the above formula, we have*

$$f[x_0, x_1] = \frac{p(x_1)q(x_1) - p(x_0)q(x_0)}{x_1 - x_0}.$$

Adding and subtracting the term $p(x_1)q(x_0)$, *we obtain*

$$f[x_0, x_1] = \frac{p(x_1)q(x_1) - p(x_1)q(x_0) + p(x_1)q(x_0) - p(x_0)q(x_0)}{x_1 - x_0},$$

which can be written as

$$f[x_0, x_1] = p(x_1)\frac{q(x_1) - q(x_0)}{x_1 - x_0} + q(x_0)\frac{p(x_1) - p(x_0)}{x_1 - x_0}.$$

Thus,

$$f[x_0, x_1] = p(x_1)q[x_0, x_1] + q(x_0)p[x_0, x_1].$$

Given $x_0 = 0, x_1 = 1, f(1) = 5,$ *and* $f[0, 1] = 4,$ *we obtain*

$$f[0, 1] = \frac{f(1) - f(0)}{1 - 0} = f(1) - f(0)$$

or

$$f[0, 1] = 4 = 5 - f(0), \quad gives \quad f(0) = 1.$$

Also,

$$f(1) = 5 = p(1)q(1) = 2q(1), \quad gives \quad q(1) = \frac{5}{2}$$

and

$$f(0) = 1 = p(0)q(0) = 2p(0), \quad gives \quad p(0) = \frac{1}{2}.$$

Hence,

$$p[0, 1] = \frac{p(1) - p(0)}{1 - 0} = p(1) - p(0) = 2 - \frac{1}{2} = \frac{3}{2}$$

and

$$q[0, 1] = \frac{q(1) - q(0)}{1 - 0} = q(1) - q(0) = \frac{5}{2} - 2 = \frac{1}{2}.$$

•

In the case of the Lagrange interpolating polynomial we derive an expression for the truncation error in the form given by (5.17), namely, that

$$R_{n+1}(x) = \frac{f^{(n+1)}(\eta(x))}{(n+1)!}, L_n(x),$$

where $L_n(x) = (x - x_0)(x - x_1) \cdots (x - x_n)$.

For Newton's divided difference formula, we obtain, following the same reasoning as above,

$$
\begin{aligned}
f(x) = \; & f[x_0] + f[x_0, x_1](x - x_0) + f[x_0, x_1, x_2](x - x_0)(x - x_1) + \cdots \\
+ \; & f[x_0, x_1, \ldots, x_n](x - x_0)(x - x_1) \cdots (x - x_{n-1}) \\
+ \; & f[x_0, x_1, \ldots, x_n, x](x - x_0)(x - x_1) \cdots (x - x_{n-1})(x - x_n),
\end{aligned}
$$

which can also be written as

$$f(x) = p_n(x) + f[x_0, x_1, \ldots, x_n, x](x - x_0)(x - x_1) \cdots (x - x_n) \qquad (5.27)$$

or

$$f(x) - p_n(x) = L_n(x) f[x_0, x_1, \ldots, x_n, x]. \qquad (5.28)$$

Since the interpolation polynomial agreeing with $f(x)$ at $x_0, x_1, \ldots, x_n$ is unique, it follows that these two error expressions must be equal.

Theorem 5.3 *Let $p_n(x)$ be the polynomial of degree at most n that interpolates a function $f(x)$ at a set of $n + 1$ distinct points $x_0, x_1, \ldots, x_n$. If x is a point different from the points $x_0, x_1, \ldots, x_n$, then*

$$f(x) - p_n(x) = f[x_0, x_1, \ldots, x_n, x] \prod_{j=0}^{n} (x - x_j). \qquad (5.29)$$

•

One can easily show the relationship between the divided differences and the derivative. From (5.23), we have

$$f[x_0, x_1] = \frac{f(x_1) - f(x_0)}{x_1 - x_0}.$$

Now applying the Mean Value theorem to the above equation implies that when the derivative f' exists, we have

$$f[x_0, x_1] = f'(\eta(x))$$

for the unknown point $\eta(x)$, which lies between x_0 and x_1. The following theorem generalizes this result.

Theorem 5.4 (Divided Differences and Derivatives)

Suppose that $f \in C^n[a, b]$ and $x_0, x_1, \ldots, x_n$ are distinct numbers in $[a, b]$. Then for some point $\eta(x)$ in the interval (a, b) spanned by $x_0, \ldots, x_n$ exists

$$f[x_0, x_1, \ldots, x_n] = \frac{f^{(n)}(\eta(x))}{n!}. \tag{5.30}$$

●

Example 5.15 *Let $f(x) = x \ln x$, and the points $x_0 = 1.1, x_1 = 1.2, x_2 = 1.3$. Find the best approximate value for the unknown point $\eta(x)$ by using the relation (5.30).*

Solution. *Given $f(x) = x \ln x$, then*

$$\begin{aligned}
f(1.1) &= 1.1 \ln(1.1) = 0.1048 \\
f(1.2) &= 1.2 \ln(1.2) = 0.2188 \\
f(1.3) &= 1.3 \ln(1.3) = 0.3411.
\end{aligned}$$

Since the relation (5.30) for the given data points is

$$f[x_0, x_1, x_2] = \frac{f''(\eta(x))}{2!}, \tag{5.31}$$

to compute the value of the left-hand side of the relation (5.31), we have to find the values of the first-order divided differences

$$f[x_0, x_1] = \frac{f(x_1) - f(x_0)}{x_1 - x_0} = \frac{0.2188 - 0.1048}{1.2 - 1.1} = 1.1400$$

and

$$f[x_1, x_2] = \frac{f(x_2) - f(x_1)}{x_2 - x_1} = \frac{0.3411 - 0.2188}{1.3 - 1.2} = 1.2230.$$

Using these values, we can compute the second-order divided difference as

$$f[x_0, x_1, x_2] = \frac{f[x_1, x_2] - f[x_0, x_1]}{x_2 - x_0} = \frac{1.2230 - 1.1400}{1.3 - 1.1} = 0.4150.$$

Now we calculate the right-hand side of the relation (5.31) for the given points, which gives us

$$\frac{f''(x_0)}{2} = \frac{1}{2x_0} = 0.4546$$

$$\frac{f''(x_1)}{2} = \frac{1}{2x_1} = 0.4167$$

$$\frac{f''(x_2)}{2} = \frac{1}{2x_2} = 0.3846.$$

We note that the left-hand side of (5.31) is nearly equal to the right-hand side when $x_1 = 1.2$. Hence, the best approximate value of $\eta(x)$ is $x_1 = 1.2$. •

Properties of Divided Differences

Now we discuss some of the properties of divided differences as follows:

1. If $p_n(x)$ is a polynomial of degree n, then the divided differences of order n is always constant and $(n+1), (n+2), \ldots$ are identically zero.

2. The divided difference is a symmetric function of its arguments. Thus, if $(t_0, \ldots, t_n)$ is a permutation of $(x_0, x_1, \ldots, x_n)$, then

$$f[t_0, t_1, \ldots, t_n] = f[x_0, x_1, \ldots, x_n].$$

This can be verified easily, since the divided differences on both sides of the above equation are the coefficients of x^n in the polynomial of degree at most n that interpolates $f(x)$ at the $n+1$ distinct points $t_0, t_1, \ldots, t_n$ and $x_0, x_1, \ldots, x_n$. These two polynomials are, of course, the same.

3. The interpolating polynomial of degree n can be obtained by adding a single term to the polynomial of degree $(n-1)$ expressed in the Newton form:

$$p_n(x) = p_{n-1}(x) + f[x_0, \ldots, x_n] \prod_{j=0}^{n-1} (x - x_j).$$

4. The divided difference $f[x_0, \ldots, x_{n-1}]$ is the coefficient of x^{n-1} in the polynomial that interpolates $(x_0, f_0), (x_1, f_1), \ldots, (x_{n-1}, f_{n-1})$.

5. A sequence of divided differences may be constructed recursively from the formula

$$f[x_0, \ldots, x_n] = \frac{f[x_1, \ldots, x_n] - f[x_0, \ldots, x_{n-1}]}{x_n - x_0},$$

and the zeroth-order divided difference is defined by

$$f[x_i] = f(x_i), \quad i = 0, 1, \ldots, n.$$

6. Another useful property of divided differences can be obtained by using the definitions of the divided differences (5.23) and (5.24), which can be extended to the case where some or all of the points x_i are coincident, provided that $f(x)$ is sufficiently differentiable. For example, define

$$f[x_0, x_0] = \lim_{x_1 \to x_0} f[x_0, x_1] = \lim_{x_1 \to x_0} \frac{f(x_1) - f(x_0)}{x_1 - x_0} = f'(x_0). \quad (5.32)$$

For an arbitrary $n \geq 1$, let all the points in Theorem 5.4 approach x_0. This leads to the definition

$$f[x_0, x_0, \ldots, x_0] = \frac{f^{(n)}(x_0)}{n!},$$

where the left-hand side denotes the *nth* divided difference, for which all points are x_0.

Example 5.16 *Let $f(x) = x^2 e^{2x} + \ln(x+1)$ and $x_0 = 0, x_1 = 1$. Using (5.30) and the above divide difference property 6, calculate the values of the divided differences $f[1,1,0,0]$ and $f[0,0,0,1,1,1]$.*

Solution. *Since*

$$
\begin{aligned}
f[1,1,0,0] &= f[0,0,1,1] = \frac{f[0,1,1] - f[0,0,1]}{1-0} \\
&= f[0,1,1] - f[0,0,1] \\
&= \left[\frac{f[1,1] - f[0,1]}{1-0}\right] - \left[\frac{f[0,1] - f[0,0]}{1-0}\right] \\
&= f[1,1] - 2f[0,1] + f[0,0] \\
&= f[1,1] - 2\left[\frac{f[1] - f[0]}{1-0}\right] + f[0,0] \\
&= f'(1) - 2f(1) + 2f(0) + f'(0),
\end{aligned}
$$

the given function is

$$ f(x) = x^2 e^{2x} + \ln(x+1), $$

and its first derivative can be calculated as

$$ f'(x) = 2x e^{2x} + 2x^2 e^{2x} + \frac{1}{(x+1)}, $$

so their values at the given points are

$$
\begin{aligned}
f(0) &= 0 \quad \text{and} \quad f(1) = e^2 + \ln 2 \\
f'(0) &= 1 \quad \text{and} \quad f'(1) = 4e^2 + 0.5.
\end{aligned}
$$

Thus, we have

$$ f[1,1,0,0] = f'(1) - 2f(1) + 2f(0) + f'(0) = 4e^2 + 0.5 - 2(e^2 + \ln 2) + 0 + 1, $$

and it gives

$$ f[1,1,0,0] = 2e^2 - 2\ln 2 + 1.5 = 14.8918. $$

Also,

$$f[0,0,0,1,1,1] = \frac{f[0,0,1,1,1] - f[0,0,0,1,1]}{1-0}$$

$$= f[0,0,1,1,1] - f[0,0,0,1,1]$$

$$= \left[\frac{f[0,1,1,1] - f[0,0,1,1]}{1-0}\right] - \left[\frac{f[0,0,1,1] - f[0,0,0,1]}{1-0}\right]$$

$$= f[0,1,1,1] - 2f[0,0,1,1] + f[0,0,0,1]$$

$$= f[1,1,1] - 3f[0,0,1,1] + f[0,0,0]$$

$$= \frac{f''(1)}{2!} - 3f[0,0,1,1] + \frac{f''(0)}{2!}.$$

Since the second derivative of the given function is

$$f''(x) = e^{2x}(2 + 8x + 4x^2) - \frac{1}{(x+1)^2},$$

and its values at the given points are

$$f''(0) = 2 - 1 = 1 \quad and \quad f''(1) = 14e^2 - 0.25,$$

using these values and $f[0,0,1,1] = 14.8918$, *we get*

$$f[0,0,0,1,1,1] = \frac{14e^2 - 0.25}{2!} - 3(14.8918) + \frac{1}{2}$$

$$= 51.60895 - 44.6754 - 0.5 = 6.43355,$$

which is the required value of the fifth-order divided difference of the function at the given points. ●

There are many schemes for the efficient implementation of divided difference interpolation, such as those due to the *Aitken's Method*, which is designed for the easy evaluation of the polynomial, taking the points closest to the one of interest first and computing only those divided differences that are actually necessary for the computation. The implementation is iterative in nature; additional data points are included one at a time until successive estimates $p_k(x)$ and $p_{k+1}(x)$ of $f(x)$ agree to some specified accuracy or until all data has been used.

5.2.3 Aitken's Method

This is an iterative interpolation process based on the repeated application of a simple interpolation method. This elegant method may be used to interpolate between both equal and unequal spaced data points. The basis of this method is equivalent to generating a sequence of the Lagrange polynomials, but it is a very efficient formulation. The method is used to compute an interpolated value using successive, higher degree polynomials until further increases in the degree of the polynomials give a negligible improvement on the interpolated value.

Suppose we want to fit a polynomial function, for the purpose of interpolation, to the following data points:

$$
\begin{array}{c|cccc}
x & x_0 & x_1 & \cdots & x_n \\
\hline
f(x) & f(x_0) & f(x_1) & \cdots & f(x_n)
\end{array}.
$$

In order to estimate the value of the function $f(x)$ corresponding to any given value of x, we consider the following expression:

$$
P_{01}(x) = \frac{1}{x_1 - x_0} \begin{vmatrix} x - x_0 & f(x_0) \\ x - x_1 & f(x_1) \end{vmatrix} = \frac{1}{x_1 - x_0}\left[(x-x_0)f(x_1)-(x-x_1)f(x_0)\right]
$$

and

$$
P_{02}(x) = \frac{1}{x_2 - x_0} \begin{vmatrix} x - x_0 & f(x_0) \\ x - x_2 & f(x_2) \end{vmatrix} = \frac{1}{x_2 - x_0}\left[(x-x_0)f(x_2)-(x-x_2)f(x_0)\right].
$$

In general,

$$
\begin{aligned}
P_{0m}(x) &= \frac{1}{x_m - x_0} \begin{vmatrix} x - x_0 & f(x_0) \\ x - x_m & f(x_m) \end{vmatrix} \\
&= \frac{1}{x_m - x_0}\left[(x - x_0)f(x_m) - (x - x_m)f(x_0)\right].
\end{aligned}
\tag{5.33}
$$

It represents a first-degree polynomial and is equivalent to a linear interpolation using the data points $(x_0, f(x_0))$ and $(x_m, f(x_m))$. One can easily verify that

$$
P_{0m}(x_0) = f(x_0) \qquad \text{and} \qquad P_{0m}(x_m) = f(x_m).
\tag{5.34}
$$

Similarly, the second-degree polynomials are generated as follows:

$$P_{012}(x) = \frac{1}{x_2 - x_1} \begin{vmatrix} x - x_1 & P_{01}(x) \\ x - x_2 & P_{02}(x) \end{vmatrix}$$

$$= \frac{1}{x_2 - x_1} \left[(x - x_1) P_{02}(x) - (x - x_2) P_{01}(x) \right]$$

and

$$P_{01m}(x) = \frac{1}{x_m - x_1} \begin{vmatrix} x - x_1 & P_{01}(x) \\ x - x_m & P_{0m}(x) \end{vmatrix}$$

$$= \frac{1}{x_m - x_1} \left[(x - x_1) P_{0m}(x) - (x - x_m) P_{01}(x) \right], \quad (5.35)$$

where m can now take any value from 2 to n, and P_{01m} denotes a polynomial of degree 2 that passes through three points $(x_0, f(x_0)), (x_1, f(x_1))$, and $(x_m, f(x_m))$. By repeated use of this procedure, higher degree polynomials can be generated. In general, one can define this procedure as follows:

$$P_{012\cdots n}(x) = \frac{1}{x_n - x_{n-1}} \begin{vmatrix} P_{01\cdots(n-1)}(x) & f(x_{n-1}) \\ P_{01\cdots n}(x) & f(x_n) \end{vmatrix}$$

$$= \frac{1}{x_n - x_{n-1}} \left[P_{01\cdots(n-1)}(x) f(x_n) - P_{01\cdots n}(x) f(x_{n-1}) \right]. \quad (5.36)$$

This is a polynomial of degree n and it fits all the data. Table 5.7 shows the construction of $P_{012\cdots n}(x)$. When using Aitken's method in practice, only the values of the polynomials for specified values of x are computed and coefficients of the polynomials are not determined explicitly. Furthermore, if for a specified x, the stage is reached when the difference in value between successive degree polynomials is negligible, then the procedure can be terminated. It is an advantage of this method compared with the Lagrange interpolation formula.

Table 5.7: Aitken's scheme to approximate a function.

k	x_k	$x - x_k$	$f(x_k)$	First Order	Second Order	Third Order		Nth Order
0	x_0	$x - x_0$	$f(x_0)$					
1	x_1	$x - x_1$	$f(x_1)$	$P_{01}(x)$				
2	x_2	$x - x_2$	$f(x_2)$	$P_{02}(x)$	$P_{012}(x)$			
3	x_3	$x - x_3$	$f(x_3)$	$P_{03}(x)$	$P_{013}(x)$			
...	...	...	...	...	...	...	...	...
n	x_n	$x - x_n$	$f(x_n)$	$P_{0n}(x)$	$P_{01n}(x)$	$P_{012n}(x)$	...	$P_{012\cdots n}(x)$

Example 5.17 *Apply Aitken's method to the approximate evaluation of* $\ln x$ *at* $x = 4.5$ *from the following data points:*

x	2	3	4	5
$f(x)$	0.6932	1.0986	1.3863	1.6094

Solution. *To find the estimate value of* $\ln(4.5)$*, using the given data points, we have to compute all the unknowns required in the given problem as follows:*

$$P_{01}(x) = \frac{1}{x_1 - x_0} \begin{vmatrix} x - x_0 & f(x_0) \\ x - x_1 & f(x_1) \end{vmatrix}$$

$$P_{01}(4.5) = \frac{1}{3 - 2} \begin{vmatrix} 4.5 - 2 & 0.6932 \\ 4.5 - 3 & 1.0986 \end{vmatrix}$$

$$= (2.5)(1.0986) - (1.5)(0.6932) = 1.7067$$

and

$$P_{02}(x) = \frac{1}{x_2 - x_0} \begin{vmatrix} x - x_0 & f(x_0) \\ x - x_2 & f(x_2) \end{vmatrix}$$

$$P_{02}(4.5) = \frac{1}{4 - 2} \begin{vmatrix} 4.5 - 2 & 0.6932 \\ 4.5 - 4 & 1.3863 \end{vmatrix}$$

$$= \frac{1}{2}\Big[(2.5)(1.3863) - (0.5)(0.6932)\Big] = 1.5596$$

and

$$P_{03}(x) \quad = \quad \frac{1}{x_3 - x_0} \begin{vmatrix} x - x_0 & f(x_0) \\ x - x_3 & f(x_3) \end{vmatrix}$$

$$P_{03}(4.5) \quad = \quad \frac{1}{5 - 2} \begin{vmatrix} 4.5 - 2 & 0.6932 \\ 4.5 - 5 & 1.6094 \end{vmatrix}$$

$$= \quad \frac{1}{3} \Big[(2.5)(1.6094) - (-0.5)(0.6932) \Big] = 1.4567.$$

Similarly, the values of second-degree polynomials can be generated as follows:

$$P_{012}(x) \quad = \quad \frac{1}{x_2 - x_1} \begin{vmatrix} x - x_1 & P_{01}(x) \\ x - x_2 & P_{02}(x) \end{vmatrix}$$

$$P_{012}(4.5) \quad = \quad \frac{1}{4 - 3} \begin{vmatrix} 4.5 - 3 & P_{01}(4.5) \\ 4.5 - 4 & P_{02}(4.5) \end{vmatrix}$$

$$= \quad (1.5)(1.5596) - (0.5)(1.7067) = 1.4860$$

and

$$P_{013}(x) \quad = \quad \frac{1}{x_3 - x_1} \begin{vmatrix} x - x_1 & P_{01}(x) \\ x - x_3 & P_{03}(x) \end{vmatrix}$$

$$P_{013}(4.5) \quad = \quad \frac{1}{5 - 3} \begin{vmatrix} 4.5 - 3 & P_{01}(4.5) \\ 4.5 - 5 & P_{03}(4.5) \end{vmatrix}$$

$$= \quad \frac{1}{2} \Big[(1.5)(1.4567) - (-0.5)(1.7067) \Big] = 1.5193.$$

Finally, the values of third-degree polynomials can be generated as follows:

$$P_{0123}(x) \quad = \quad \frac{1}{x_3 - x_2} \begin{vmatrix} x - x_2 & P_{012}(x) \\ x - x_3 & P_{013}(x) \end{vmatrix}$$

$$P_{0123}(4.5) \quad = \quad \frac{1}{5 - 4} \begin{vmatrix} 4.5 - 4 & P_{012}(4.5) \\ 4.5 - 4 & P_{013}(4.5) \end{vmatrix}$$

$$= \quad (0.5)(1.5193) - (-0.5)(1.4860) = 1.5027.$$

Table 5.8: Approximate solution for Example 5.17.

k	x_k	$f(x_k)$	$x - x_k$	First-Order	Second-Order	Third-Order
0	2.0	0.6932	2.5			
1	3.0	1.0986	1.5	1.7067		
2	4.0	1.3863	0.5	1.5596	1.4860	
3	5.0	1.6094	-0.5	1.4567	1.5193	1.5027

The results obtained are listed in Table 5.8. Note that the approximate value of $\ln(4.5)$ *is* $P_{0123}(4.5) = 1.5027$ *and its exact value is* 1.5048. •

To get the above results using the MATLAB Command Window, we do the following:

```
>> x = [2 3 4 5];
>> y = [0.6932 1.0986 1.3863 1.6094];
>> x0 = 4.5;
>> P = Aitken1(x, y, x0);
```

Program 5.4
MATLAB m-file for Aitken's Method
```
function P=Aitken1(x,y,x0)
n=size(x,1);
if n==1
n=size(x,2); end
for i=1:n
P(i,1)=y(i); end
for i=2:n
t=0;
for j=2:i
t=t+1;
P(i,j)=(P(i,j-1)*(x0-x(j-1))-P(t,t)*(x0-x(i)))/(x(i)-x(j-1));
end; end
```

5.2.4 **Chebyshev Polynomials**

Here, we discuss polynomial interpolation for $f(x)$ over the interval $[-1, 1]$ based on the points

$$-1 \leq x_0 < x_1 < x_2 < \ldots < x_n \leq 1.$$

This special type of polynomial is known as a *Chebyshev polynomial.*

Chebyshev polynomials are used in many parts of numerical analysis and more generally in mathematics and physics. Basically, Chebyshev polynomials are used to minimize approximation error. These polynomials are of the form

$$T_n(x) = \cos(n \cos^{-1}(x)), \quad \text{for} \quad x \in [-1, 1]. \tag{5.37}$$

The representation of (5.37) may not appear to be a polynomial, but we will show it is a polynomial of degree n. To simplify the manipulation of (5.37), we introduce

$$\theta = \cos^{-1}(x), \quad \text{or} \quad x = \cos\theta, \quad 0 \leq \theta \leq \pi. \tag{5.38}$$

Then

$$T_n(x) = \cos(n\theta), \quad \text{for} \quad x \in [-1, 1]. \tag{5.39}$$

For example, taking $n = 0$, then

$$T_0(x) = \cos(0.\theta) = \cos(0) = 1,$$

and for $n = 1$, gives

$$T_1(x) = \cos(1.\theta) = \cos(\theta) = x.$$

Also, by taking $n = 2$, we have

$$T_2(x) = \cos(2.\theta) = \cos(2\theta),$$

and using the standard identity, $\cos(2\theta) = 2\cos^2(\theta) - 1$, we get

$$T_2(x) = \cos(2\theta) = 2\cos^2(\theta) - 1 = 2x^2 - 1.$$

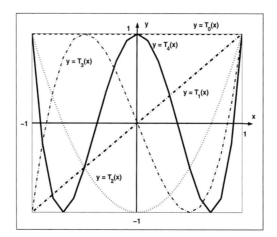

Figure 5.5: Graphs of $T_0(x), T_1(x), T_2(x), T_3(x), T_4(x)$.

The graphs of the Chebyshev polynomials $T_0(x), T_1(x), T_2(x), T_3(x)$, and $T_4(x)$ are given in Figure 5.5.

The first few Chebyshev polynomials are as follows:

$$
\begin{aligned}
T_0(x) &= \cos(0.\theta) = 1 \\
T_1(x) &= \cos(1.\theta) = x \\
T_2(x) &= \cos(2.\theta) = 2x^2 - 1 \\
T_3(x) &= \cos(3.\theta) = 4x^3 - 3x \\
T_4(x) &= \cos(4.\theta) = 8x^4 - 8x^2 + 1 \\
T_5(x) &= \cos(5.\theta) = 16x^5 - 20x^3 + 5x \\
T_6(x) &= \cos(6.\theta) = 32x^6 - 48x^4 + 18x^2 - 1 \\
T_7(x) &= \cos(7.\theta) = 64x^7 - 112x^5 + 56x^3 - 7x \\
T_8(x) &= \cos(8.\theta) = 128x^8 - 256x^6 + 160x^4 - 32x^2 + 1 \\
T_9(x) &= \cos(9.\theta) = 256x^9 - 576x^7 + 432x^5 - 120x^3 + 9x \\
T_10(x) &= \cos(10.\theta) = 512x^{10} - 1280x^8 + 1120x^6 - 600x^4 + 50x^2 - 1.
\end{aligned}
$$

To get the coefficients of the above Chebyshev polynomials using the MATLAB Command Window, we do the following:

```
>> n = 10;
>> T = CHEBP(n);
>> T =
512 0 − 1280 0 1120 0 − 400 0 50 0 − 1
```

Note that we got the coefficients of the Chebyshev polynomials in descending order of powers.

Program 5.5
MATLAB m-file for Computing Chebyshev
Polynomials
```
function T = CHEBP(n)
x0 = 1; x1 = [1 0];
if n == 0  T = x0;
elseif n == 1; T = x1;  else
for i=2:n
T = [2 * x1 0] − [0 0 x0];
x0 = x1; x1 = T; end  end
```

The higher order polynomials can be generated from the recursion relation called the *triple recursion relation*. This relation can be easily constructed with the help of the trigonometric addition formulas

$$\cos(A \pm B) = \cos(A)\cos(B) \mp \sin(A)\sin(B).$$

For any $n \geq 1$, apply these identities to get

$$
\begin{aligned}
T_{n+1}(x) &= \cos((n+1)\theta) = \cos(n\theta + \theta) = \cos(n\theta)\cos(\theta) - \sin(n\theta)\sin(\theta) \\
T_{n-1}(x) &= \cos((n-1)\theta) = \cos(n\theta - \theta) = \cos(n\theta)\cos(\theta) + \sin(n\theta)\sin(\theta).
\end{aligned}
$$

By adding $T_{n+1}(x)$ and $T_{n-1}(x)$, we get

$$T_{n+1}(x) + T_{n-1}(x) = 2\cos(n\theta)\cos(\theta) = 2xT_n(x),$$

because $\cos(n\theta) = T_n(x)$ and $\cos(\theta) = x$.

So the relation

$$T_{n+1}(x) = 2xT_n(x) - T_{n-1}(x), \quad n \geq 1 \tag{5.40}$$

is called the triple recursion relation for the Chebyshev polynomials.

Theorem 5.5 (Properties of Chebyshev Polynomials)

The functions $T_n(x)$ satisfy the following properties:

1. *Each $T_n(x)$ is a polynomial of degree n.*

2. *$T_{n+1}(x) = 2xT_n(x) - T_{n-1}(x), \quad n \geq 1$.*

3. *$T_n(x) = 2^{n-1}x^n + $ lower order terms.*

4. *When $n = 2m$, $T_{2m}(x)$ is an even function, i.e., $T_{2m}(-x) = T_{2m}(x)$.*

5. *When $n = 2m + 1$, $T_{2m+1}(x)$ is an odd function, i.e., $T_{2m+1}(-x) = -T_{2m+1}(x)$.*

6. *$T_n(x)$ has n distinct zeros x_k (called Chebyshev points) that lie on the interval $[-1, 1]$:*

$$x_k = \cos\left[\frac{(2k+1)\pi}{2(n+1)}\right], \quad for \quad k = 0, 1, \ldots, n.$$

7. *$|T_n(x)| \leq 1, \quad for \quad -1 \leq x \leq 1$.*

8. *Chebyshev polynomials have some unusual properties. They form an orthogonal set. To show the orthogonality of the Chebyshev polynomials, consider*

$$\int_{-1}^{1} \frac{T_n(x)T_m(x)}{\sqrt{1-x^2}}dx = \int_{-1}^{1} \frac{\cos(n\cos^{-1}x)\cos(m\cos^{-1}x)}{\sqrt{1-x^2}}dx.$$

Let $\theta = \cos^{-1}x$, then

$$d\theta = -\frac{1}{\sqrt{1-x^2}}dx$$

and

$$x = -1, \quad \theta = \pi, \quad \text{and} \quad x = 1, \quad \theta = 0.$$

Suppose that $n \neq m$, and since

$$\cos n\theta \cos m\theta = \frac{1}{2}\left[\cos(n+m)\theta + \cos(n-m)\theta\right],$$

then we get

$$\int_{-1}^{1} \frac{T_n(x)T_m(x)}{\sqrt{1-x^2}}dx = \frac{1}{2}\int_{0}^{\pi} \cos[(n+m)\theta]d\theta + \frac{1}{2}\int_{0}^{\pi} \cos[(n-m)\theta]d\theta.$$

Solving the right-hand side, we obtain

$$\int_{-1}^{1} \frac{T_n(x)T_m(x)}{\sqrt{1-x^2}}dx = \left[\frac{1}{2(n+m)}\sin[(n+m)\theta] + \frac{1}{2(n-m)}\sin[(n-m)\theta]\right]_{0}^{\pi}$$

or

$$\int_{-1}^{1} \frac{T_n(x)T_m(x)}{\sqrt{1-x^2}}dx = 0.$$

Now when $n = m$, we have

$$\begin{aligned}
\int_{-1}^{1} \frac{T_n(x)T_n(x)}{\sqrt{1-x^2}}dx &= \frac{1}{2}\int_{0}^{\pi} \cos[(n+n)\theta]d\theta + \frac{1}{2}\int_{0}^{\pi} \cos[(n-n)\theta]d\theta \\
&= \frac{1}{2}\int_{0}^{\pi} \cos(2n\theta)d\theta + \frac{1}{2}\int_{0}^{\pi} \cos(0)d\theta \\
&= \frac{1}{2}\left[\frac{\sin(2n\theta)}{2n}\right]_{0}^{\pi} + \frac{1}{2}[\theta]_{0}^{\pi} \\
&= 0,
\end{aligned}$$

for each $n \geq 1$. ●

In the following example, we shall find the Chebyshev points for the linear, quadratic, and cubic interpolations for the given function.

Example 5.18 *Let $f(x) = x^2 e^x$ on the interval $[-1, 1]$. Then the Chebyshev points for the linear interpolation $(n = 1)$ are given by*

$$x_0 = \cos\left[\frac{(2(0) + 1)\pi}{2(1 + 1)}\right] = \cos\frac{\pi}{4} = 0.7071$$

$$x_1 = \cos\left[\frac{(2(1) + 1)\pi}{2(1 + 1)}\right] = \cos\frac{3\pi}{4} = -0.7071.$$

Now using the linear Lagrange polynomial using these two Chebyshev points, we have

$$p_1(x) = L_0(x)f(x_0) + L_1(x)f(x_1),$$

where

$$L_0(x) = \frac{(x - x_1)}{(x_0 - x_1)} = \frac{(x + 0.7071)}{(0.7071 + 0.7071)} = \frac{(x + 0.7071)}{1.4142}$$

$$L_1(x) = \frac{(x - x_0)}{(x_1 - x_0)} = \frac{(x - 0.7071)}{(-0.7071 - 0.7071)} = \frac{(x - 0.7071)}{-1.4142},$$

and the function values at the Chebyshev points are

$$f(0.7071) = (0.7071)^2 e^{0.7071} = 1.0140$$
$$f(-0.7071) = (-0.7071)^2 e^{-0.7071} = 0.2465.$$

Thus,

$$p_1(x) = \frac{(x + 0.7071)}{1.4142}(1.0140) + \frac{(x - 0.7071)}{-1.4142}(0.2465)$$

gives

$$p_1(x) = 0.6302 + 0.5427x.$$

Now to find the quadratic Lagrange polynomial, we need to calculate three Chebyshev points as follows:

$$x_0 = \cos\left[\frac{(2(0) + 1)\pi}{2(2 + 1)}\right] = \cos\frac{\pi}{6} = 0.8660$$

$$x_1 = \cos\left[\frac{(2(1) + 1)\pi}{2(2 + 1)}\right] = \cos\frac{3\pi}{6} = 0.0$$

$$x_2 = \cos\left[\frac{(2(2) + 1)\pi}{2(2 + 1)}\right] = \cos\frac{5\pi}{6} = -0.8660.$$

For the quadratic polynomial, we have

$$p_2(x) = L_0(x)f(x_0) + L_1(x)f(x_1) + L_2(x)f(x_2),$$

where

$$L_0(x) = \frac{(x - 0.0)(x + 0.8660)}{(0.8660 - 0.0)(0.8660 + 0.8660)} = \frac{(x^2 + 0.8660x)}{1.4999}$$

$$L_1(x) = \frac{(x - 0.8660)(x + 0.8660)}{(0.0 - 0.8660)(0.0 + 0.8660)} = \frac{(x^2 + 0.7500)}{(-0.7500)}$$

$$L_2(x) = \frac{(x - 0.8660)(x - 0.0)}{(-0.8660 - 0.8660)(-0.8660 - 0.0)} = \frac{(x^2 - 0.7500)}{0.7500},$$

and the function values are

$$
\begin{aligned}
f(0.8660) &= (0.8660)^2 e^{0.8660} = 1.7829 \\
f(0.0) &= (0.0)^2 e^{0.0} = 0.0 \\
f(-0.8660) &= (-0.8660)^2 e^{-0.8660} = 0.3155.
\end{aligned}
$$

So,

$$p_2(x) = 0.8939x + 1.6101x^2.$$

Similarly, for the cubic polynomial, we need the following:

$$x_0 = \cos\left[\frac{(2(0) + 1)\pi}{2(3 + 1)}\right] = \cos\frac{\pi}{8} = 0.9239$$

$$x_1 = \cos\left[\frac{(2(1) + 1)\pi}{2(3 + 1)}\right] = \cos\frac{3\pi}{8} = 0.3827$$

$$x_2 = \cos\left[\frac{(2(2) + 1)\pi}{2(3 + 1)}\right] = \cos\frac{5\pi}{8} = -0.3827$$

$$x_3 = \cos\left[\frac{(2(3) + 1)\pi}{2(3 + 1)}\right] = \cos\frac{7\pi}{8} = -0.9239$$

and

$$L_0(x) = \frac{(x - 0.3827)(x + 0.3827)(x + 0.9239)}{(0.9239 - 0.3827)(0.9239 + 0.3827)(0.9239 + 0.9239)}$$

$$= \frac{(x^3 + 0.9239x^2 - 0.1464x - 0.1353)}{1.3066}$$

$$L_1(x) = \frac{(x - 0.9239)(x + 0.3827)(x + 0.9239)}{(0.3827 - 0.9239)(0.3827 + 0.3827)(0.3827 + 0.9239)}$$

$$= \frac{(x^3 + 0.3827x^2 - 0.8536x - 0.3267)}{-0.5412}$$

$$L_2(x) = \frac{(x - 0.9239)(x - 0.3827)(x + 0.9239)}{(-0.3827 - 0.9239)(-0.3827 - 0.3827)(-0.3827 + 0.9239)}$$

$$= \frac{(x^3 - 0.3827x^2 - 0.8536x - 0.3267)}{0.5412}$$

$$L_3(x) = \frac{(x - 0.9293)(x - 0.3827)(x + 0.3827)}{(-0.9239 - 0.9293)(-0.9239 - 0.3827)(0.9239 + 0.3827)}$$

$$= \frac{(x^3 - 0.9239x^2 - 0.1465x + 0.1354)}{3.1638}$$

and

$$
\begin{aligned}
f(0.8660) &= (0.9239)^2 e^{0.9239} = 2.1503 \\
f(0.3827) &= (0.3827)^2 e^{0.3827} = 0.2147 \\
f(-0.3827) &= (-0.3827)^2 e^{-0.3827} = 0.0999 \\
f(-0.9239) &= (-0.9239)^2 e^{-0.9239} = 0.3388.
\end{aligned}
$$

Thus,

$$p_3(x) = -0.1389 - 0.0756x + 1.7592x^2 + 1.5404x^3. \qquad \bullet$$

Note that because $T_n(x) = \cos(n\theta)$, Chebyshev polynomials have a succession of maximums and minimums of alternating signs, each of magnitude one. Also, $|\cos(n\theta)| = 1$, for $n\theta = 0, \pi, 2\pi, \ldots$, and because θ varies from 0 to π as x varies from 1 to -1, $T_n(x)$ assumes its maximum magnitude of unity $(n + 1)$ times on the interval $[-1, 1]$. An important result of Chebyshev polynomials is the fact that, of all polynomials of degree n where the coefficient of x^n is unity, the polynomial $(\frac{1}{2^{n-1}})T_n(x)$ has a smaller bound to its magnitude on the interval $[-1, 1]$ than any other. Because the maximum magnitude of $T_n(x)$ is one, the upper bound referred to is $\frac{1}{2^{n-1}}$.

Theorem 5.6 *Let $n \geq 1$ be an integer, and consider all possible monic polynomials (a polynomial whose highest-degree term has a coefficient of 1) of degree n. Then the degree n monic polynomial with the smallest maximum absolute value on $[-1, 1]$ is $\frac{1}{2^{n-1}}$.* ●

It is important to note that polynomial interpolation using equally spaced data points, whether expressed in the Lagrange interpolation formula or Newton's interpolation formula, is most accurate in the middle range of the interpolation domain, but the error of the interpolation increases toward the edges. While the spacings determined by a Chebyshev polynomial are largest at the center of the interpolation domain and decrease toward the edges, errors become more evenly distributed throughout the domain and their magnitudes become less than with equally spaced points. Since the error formula for Lagrange and Newton's polynomials satisfy

$$f(x) = p_n(x) + E_n(x),$$

where

$$E_n(x) = R(x)\frac{f^{(n+1)}}{(n+1)!},$$

and $R(x)$ is the polynomial of degree $(n+1)$

$$R(x) = (x - x_0)(x - x_1) \cdots (x - x_n),$$

using this relationship, we have

$$|E_n(x)| \leq |R(x)| \frac{\max\limits_{-1 \leq x \leq 1} \left| f^{(n+1)} \right|}{(n+1)!}.$$

Here, we are looking to get a minimum of $\max\limits_{-1 \leq x \leq 1} |R(x)|$.

The Russian mathematician Chebyshev studied how to minimize the upper bound for $|E_n(x)|$. One upper bound can be formed by taking the product of the maximum value of $|R(x)|$ over all x in $[-1, 1]$ and the maximum value $\left| \dfrac{f^{(n+1)}}{(n+1)!} \right|$ over all x in $[-1,1]$. To minimize the factor $\max |R(x)|$, Chebyshev found that $x_0, x_1, \ldots, x_n$ should be chosen so that

$$R(x) = \frac{1}{2^n} T_{n+1}(x).$$

Theorem 5.7 *Let $f \in C^{n+1}([-1,1])$ be given, and let $p_n(x)$ be the nth degree polynomial interpolated to f using the Chebyshev points. Then*

$$\max_{-1 \le x \le 1} |f(x) - p_n(x)| \le \frac{1}{2^n(n+1)!} \max_{-1 \le x \le 1} \left| f^{(n+1)} \right|. \tag{5.41}$$

Note that

$$\frac{1}{2^n} \le \max_{-1 \le x \le 1} |(x - x_0)(x - x_1) \cdots (x - x_n)|,$$

for any choice of $x_0, x_1, \ldots, x_n$ on the interval $[-1,1]$.   ●

Example 5.19 *Construct the Lagrange interpolating polynomials of degree 2 on the interval $[-1,1]$ to $f(x) = (x+2)e^x$ using equidistant and the Chebyshev points.*

Solution. *First, we construct the polynomial with the use of the three equidistant points*

$$x_0 = -1, \quad x_1 = 0, \quad x_2 = 1,$$

and their corresponding function values

$$f(-1) = e^{-1}, \quad f(0) = 2, \quad f(1) = 3e.$$

Then the Lagrange polynomial at equidistant points is

$$\begin{aligned}
p_2(x) &= \frac{(x-0)(x-1.0)}{(-1.0-0)(-1.0-1.0)}(e^{-1}) \\[2mm]
&= \frac{(x+1.0)(x-1.0)}{(0+1.0)(0-1.0)}(2) \\[2mm]
&= \frac{(x+1.0)(x-0.0)}{(1.0+1.0)(1.0-0)}(3e).
\end{aligned}$$

Simplifying this, we get

$$p_2(x) = 2.2614x^2 + 3.8935x + 2,$$

the required polynomial at equidistant points.

Similarly, we can obtain the polynomial using the Chebyshev points

$$x_0 = \cos\frac{\pi}{6} = 0.8660$$

$$x_1 = \cos\frac{3\pi}{6} = 0.0$$

$$x_2 = \cos\frac{5\pi}{6} = -0.8660,$$

with their corresponding function values

$$f(x_0) = 6.81384$$

$$f(x_1) = 2.0$$

$$f(x_2) = 0.4770,$$

as follows:

$$Q_2(x) = \frac{(x-0)(x+0.8660)}{(0.8660-0)(0.8660+0.8660)}(6.81384)$$

$$= \frac{(x-0.8660)(x+0.8660)}{(0-0.8660)(0+0.8660)}(2)$$

$$= \frac{(x-0.8660)(x-0)}{(-0.8660-0.8660)(-0.8660-0)}(0.4770).$$

Thus, the Lagrange polynomial at the Chebyshev points is

$$Q_2(x) = 2.1941x^2 + 3.6587x + 2.$$

Note that the coefficients $p_2(x)$ and $Q_2(x)$ are different because they use different points and function values. Also, the actual errors at $x = 0.5$ using both polynomials are

$$f(0.5) - p_2(0.5) = -0.3903$$

and

$$f(0.5) - Q_2(0.5) = -0.2561.$$

•

Changing Intervals: $[a, b]$ to $[-1, 1]$

The Chebyshev polynomial of interpolation can be applied to any range other than $[-1, 1]$ by mapping $[-1, 1]$ onto the range of interest. Writing the range of interpolation as $[a, b]$, the mapping is given by

$$x = \frac{b(1 + z) + a(1 - z)}{2} \quad \text{or} \quad z = \frac{2x - a - b}{b - a},$$

where $a \leq x \leq b$ and $-1 \leq z \leq 1$.

The required Chebyshev points on $T_{n+1}(z)$ on $[-1, 1]$ are

$$z_k = \cos\left((2n + 1 - 2k)\frac{\pi}{2(n + 1)}\right) \quad \text{or} \quad k = 0, 1, \ldots, n, \qquad (5.42)$$

and the interpolating points on $[a, b]$ are obtained as

$$x_k = \frac{b(1 + z_k) + a(1 - z_k)}{2} \quad \text{or} \quad k = 0, 1, \ldots, n. \qquad (5.43)$$

Theorem 5.8 (Lagrange–Chebyshev Approximation Polynomial)

Suppose that $p_n(x)$ is the Lagrange polynomial that is based on the Chebyshev points

$$x_k = \frac{b(1 + z_k) + a(1 - z_k)}{2} \quad \text{or} \quad k = 0, 1, \ldots, n.$$

If $f \in C^{n+1}[a, b]$, then the error bound formula is

$$|f(x) - p_n(x)| \leq \frac{2(b-a)^{n+1}}{4^{n+1}(n+1)} \max_{a \leq x \leq b} \left| f^{(n+1)}(x) \right|. \tag{5.44}$$

●

Example 5.20 *Find the three Chebyshev points in $1 \leq x \leq 3$ and then write the Lagrange interpolation to interpolate $\ln(x+1)$. Also, compute an error bound.*

Solution. *Given $a = 1, b = 3, n = 2$, and $k = 0, 1, 2$, the three Chebyshev points can be calculated as follows:*

$$z_0 = \cos\left((4+1-0)\frac{\pi}{6}\right) = \cos\frac{5\pi}{6} = -0.86603$$

$$z_1 = \cos\left((4+1-2)\frac{\pi}{6}\right) = \cos\frac{\pi}{2} = 0.0$$

$$z_2 = \cos\left((4+1-4)\frac{\pi}{6}\right) = \cos\frac{\pi}{6} = 0.5.$$

Now we compute the interpolating points on $[1, 3]$ as follows:

$$x_0 = \frac{3(1+z_0) + 1(1-z_0)}{2} = \frac{3(1-0.86603) + 1(1+0.86603)}{2} = 1.13398$$

$$x_1 = \frac{3(1+z_1) + 1(1-z_1)}{2} = \frac{3(1+0.0) + 1(1-0.0)}{2} = 2.0$$

$$x_0 = \frac{3(1+z_2) + 1(1-z_2)}{2} = \frac{3(1+0.5) + 1(1-0.5)}{2} = 2.5.$$

Now we compute the function values at these interpolating points as:

$$\begin{aligned}
f(x_0) &= \ln(1+x_0) = \ln(2.13398) = 0.75799 \\
f(x_1) &= \ln(1+x_1) = \ln(3) = 1.09861 \\
f(x_2) &= \ln(1+x_2) = \ln(3.5) = 1.25276.
\end{aligned}$$

Thus, the Lagrange interpolating polynomial becomes

$$p_2(x) \;=\; \frac{(x-2)(x-2.5)}{(1.13398-2)(1.13398-2.5)}(0.75799)$$

$$+\; \frac{(x-1.13398)(x-2.5)}{(2-1.13398)(2-2.5)}(1.09861)$$

$$+\; \frac{(x-1.13398)(x-2)}{(2.5-1.13398)(2.5-2)}(1.25276)$$

and simplifying it, we get

$$p_2(x) = -0.06223x^2 + 0.58834x + 0.17086.$$

To compute the error bound, we use formula (5.44) as

$$|f(x) - p_2(x)| \le \frac{2(3-1)^{2+1}}{4^{2+1}(2+1)} \max_{1 \le x \le 3} \left| f^{(2+1)}(x) \right|,$$

then using $f^{(3)}(x) = \dfrac{2}{(x+1)^3}$, *we get*

$$|f(x) - p_2(x)| \le \frac{2(3-1)^{2+1}}{4^{2+1}(2+1)} \max_{1 \le x \le 3} \left| \frac{2}{(x+1)^3} \right|$$

or

$$|f(x) - p_2(x)| \le 0.01042,$$

which is the required error bound.   ●

Theorem 5.9 (Chebyshev Approximation Polynomial)

The Chebyshev approximation polynomial $p_n(x)$ *of degree n for a function* $f(x)$ *over the interval* $[-1, 1]$ *can be written as*

$$f(x) \approx p_n(x) = \sum_{i=1}^{n} a_i T_i(x), \qquad\qquad (5.45)$$

where the coefficients of the polynomial can be calculated as

$$a_0 = \frac{1}{n+1}\sum_{j=0}^{n} f(x_j)T_0(x_j)$$

(5.46)

$$a_i = \frac{2}{n+1}\sum_{j=0}^{n} f(x_j)T_i(x_j), \quad for\ i = 1, 2, \ldots, n,$$

(5.47)

and the polynomial T_i is

$$T_i(x_j) = \cos\left(\frac{(2j+1)i\pi}{2n+2}\right).$$

(5.48)

Example 5.21 *Construct the Chebyshev polynomial of degree 4 and Lagrange interpolating polynomial of degree 4 (using equidistant points) on the interval $[-1, 1]$ to approximate the function $f(x) = (x+2)\ln(x+2)$.*

Solution. *The Chebyshev polynomial of degree 4 to approximate the given function can be written as*

$$p_4(x) = a_0 T_0(x) + a_1 T_1(x) + a_2 T_2(x) + a_3 T_3(x) + a_4 T_4(x).$$

First, we compute the coefficients a_0, a_1, a_2, a_3, and a_4 by using (5.46) and (5.48), and the points $x_j = \cos(\frac{(2j+1)\pi}{10})$, for $j = 0, 1, 2, 3, 4$, as follows:

$$a_0 = \frac{1}{5}\sum_{j=0}^{4}(x_j + 2)\ln(x_j + 2)T_0(x_j) = 1.5156$$

$$a_1 = \frac{2}{5}\sum_{j=0}^{4}(x_j + 2)\ln(x_j + 2)T_1(x_j) = 1.6597$$

$$a_2 = \frac{2}{5}\sum_{j=0}^{4}(x_j + 2)\ln(x_j + 2)T_2(x_j) = 0.1308$$

$$a_3 = \frac{2}{5}\sum_{j=0}^{4}(x_j + 2)\ln(x_j + 2)T_3(x_j) = -0.0115$$

$$a_4 = \frac{2}{5}\sum_{j=0}^{4}(x_j+2)\ln(x_j+2)T_4(x_j) = 0.0015.$$

Using these values, we have

$$p_4(x) = 1.5156T_0(x)+1.6597T_1(x)+0.1308T_2(x)-0.0115T_3(x)+0.0015T_4(x).$$

Since we know that

$$
\begin{aligned}
T_0(x) &= \cos(0.\theta) = 1\\
T_1(x) &= \cos(1.\theta) = x\\
T_2(x) &= \cos(2.\theta) = 2x^2 - 1\\
T_3(x) &= \cos(3.\theta) = 4x^3 - 3x\\
T_4(x) &= \cos(4.\theta) = 8x^4 - 8x^2 + 1,
\end{aligned}
$$

we have

$$p_4(x) = 1.5156(1)+1.6597(x)+0.1308(2x^2-1)-0.0115(4x^3-3x)+0.0015(8x^4-8x^2-$$

or

$$p_4(x) = 1.3863 + 1.9918x - 0.0120x^2 - 0.0460x^3 + 0.0120x^4,$$

which is the required Chebyshev approximation polynomial for the given function.

Now we construct the Lagrange polynomial of degree 4 using the equidistant points on the interval $[-1, 1]$,

$$x_0 = -1, \ x_1 = -0.5, \ x_2 = 0, \ x_3 = 0.5, \ x_4 = 1,$$

and the functional values at these points,

$$f(x_0) = 0.0, \ f(x_1) = 0.6082, \ f(x_2) = 1.3863, \ f(x_3) = 2.2907, \ f(x_4) = 3.2958,$$

as follows:

$$
\begin{aligned}
p_4(x) &= L_0(x)(0) + L_1(x)(0.6082) + L_2(x)(1.3863) + L_3(x)(2.2907) + L_4(x)(3.2958) \\
&= 0.6082L_1(x) + 1.3863L_2(x) + 2.2907L_3(x) + 3.2958L_4(x).
\end{aligned}
$$

The values of the unknown Lagrange coefficients are as follows:

$$
L_1(x) = \frac{(x+1)(x-0)(x-0.5)(x-1)}{(-0.5+1)(-0.5-0)(-0.5-0.5)(-0.5-1)} = \frac{x^4 - 0.5x^3 - x^2 + 0.5x}{0.375}
$$

$$
L_2(x) = \frac{(x+1)(x+0.5)(x-0.5)(x-1)}{(0+1)(0+0.5)(0-0.5)(0-1)} = \frac{x^4 - 1.25x^2 + 0.25}{0.25}
$$

$$
L_3(x) = \frac{(x+1)(x+0.5)(x-0)(x-1)}{(0.5+1)(0.5+0.5)(0.5-0)(0.5-1)} = \frac{x^4 + 0.5x^3 - x^2 - 0.5x}{-0.375}
$$

$$
L_4(x) = \frac{(x+1)(x+0.5)(x-0)(x-0.5)}{(1+1)(1+0.5)(1-0)(1-0.5)} = \frac{x^4 + x^3 - 0.25x^2 - 0.25x}{1.5}.
$$

Thus,

$$
p_4(x) = 1.3863 + 3.3159x - 2.9942x^2 - 1.6680x^3 + 3.2558x^4,
$$

which is the Lagrange interpolating polynomial of degree 4 to approximate the given function. ●

To get the coefficients of the above Chebyshev polynomial approximation we use the MATLAB Command Window as follows:

```
>> n = 4; a = -1; b = 1;
>> y =' (x + 2). * log(x + 2)';
>> A = CHEBPA(y, n, a, b);
>> A =
     1.5156 1.6597 0.1308  - 0.0115 0.0015
```

Program 5.6

MATLAB m-file for Computing Coefficients of Chebyshev
Polynomial Approximation

```
function C= CHEBPA(fn,n,a,b)
if nargin==2, a = -1; b = 1;  end
d = pi/(2 * n + 2);  A = zeros(1, n + 1);
for i=1:n+1
x(i) = cos((2 * i - 1) * d);  end
x = (b - a) * x/2 + (a + b)/2;  y = eval(fn);
for i=1:n+1
z = (2 * i - 1) * d;
for j=1:n+1
A(j) = A(j) + y(i) * cos((j - 1) * z); end  end
A = 2 * A/(n + 1);  A(1) = A(1)/2;
```

5.3   Least Squares Approximation

In fitting a curve to given data points, there are two basic approaches. One is to have the graph of the approximating function pass exactly through the given data points. The methods of polynomial interpolation approximation discussed in the previous sections have this special property. If the data values are experimental then they may contain errors or have a limited number of significant digits. In such cases, the polynomial interpolation methods may yield unsatisfactory results. The second approach, which is discussed here, is usually more satisfactory for experimental data and uses an approximating function that graphs a smooth curve having the general shape suggested by the data values but not, in general, passing exactly through all of the data points. Such an approach is known as *least squares* data fitting. The least squares method seeks to minimize the sum (over all data points) of the squares of the differences between the function value and the data value. The method is based on results from calculus that demonstrate that a function, in this case, the total squared error, attains a minimum value when its partial derivatives are zero.

The least squares method of evaluating empirical formulas has been used for many years. In engineering, curve fitting plays an important role in the analysis, interpretation, and correlation of experimental data with mathematical models formulated from fundamental engineering principles.

5.3.1 Linear Least Squares

To introduce the idea of linear least squares approximation, consider the experimental data shown in Figure 5.6.

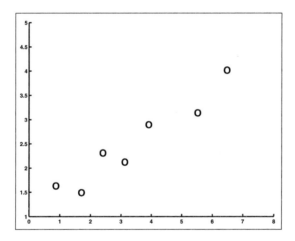

Figure 5.6: Least squares approximation.

A Lagrange interpolation of a polynomial of degree 6 could easily be constructed for this data. However, there is no justification for insisting that the data points be reproduced exactly, and such an approximation may well be very misleading since unwanted oscillations are likely. A more satisfactory approach would be to find a straight line that passes close to all seven points. One such possibility is shown in Figure 5.7. Here, we have to decide what criterion is to be adopted for constructing such an approximation. The most common approach for this curve is known as linear least squares data fitting. The linear least squares approach defines the correct straight line as the one that minimizes the sum of the squares of the distances between the data points and the line. The least squares straight line approximations are an extremely useful and common approximate fit. The

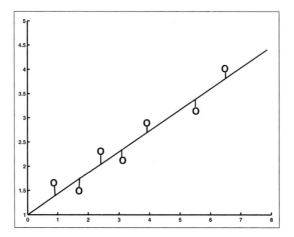

Figure 5.7: Least squares approximation.

solution to linear least squares approximation is an important application of the solution of systems of linear equations and leads to other interesting ideas of numerical linear algebra. The least squares approximation is not restricted to a straight line. However, in order to motivate the general case we consider this first. The straight line

$$p_1(x) = a + bx \qquad (5.49)$$

should be fitted through the given points $(x_1, y_1), \ldots, (x_n, y_n)$ so that the sum of the squares of the distances of these points from the straight line is minimum, where the distance is measured in the vertical direction (the y-direction). Hence, it will suffice to minimize the function

$$E(a, b) = \sum_{j=1}^{n} (y_j - a - bx_j)^2. \qquad (5.50)$$

The minimum of E occurs if the partial derivatives of E with respect to a and b become zero. Note that $\{x_j\}$ and $\{y_j\}$ are constant in (5.50) and unknown parameters a and b are variables. Now differentiate E with respect to variable a by making the other variable b fixed and then put it equal to zero, which gives

$$\frac{\partial E}{\partial a} = -2 \sum_{j=1}^{n} (y_j - a - bx_j) = 0. \tag{5.51}$$

Now hold variable a and differentiate E with respect to variable b and then put it equal to zero, and we obtain

$$\frac{\partial E}{\partial b} = -2 \sum_{j=1}^{n} x_j(y_j - a - bx_j) = 0. \tag{5.52}$$

Equations (5.51) and (5.52) may be rewritten after dividing by –2 as follows:

$$\sum_{j=1}^{n} y_j \quad - \quad \sum_{j=1}^{n} a - b \sum_{j=1}^{n} x_j = 0$$

$$\sum_{j=1}^{n} x_j y_j \quad - \quad a \sum_{j=1}^{n} x_j - b \sum_{j=1}^{n} x_j^2 = 0,$$

which can be arranged to form a 2×2 system that is known as the *normal equations*

$$na \quad + \quad b \sum_{j=1}^{n} x_j = \sum_{j=1}^{n} y_j$$

$$a \sum_{j=1}^{n} x_j \quad + \quad b \sum_{j=1}^{n} x_j^2 = \sum_{j=1}^{n} x_j y_j.$$

Now writing in matrix form, we have

$$\begin{pmatrix} n & S1 \\ S1 & S3 \end{pmatrix} \begin{pmatrix} a \\ b \end{pmatrix} = \begin{pmatrix} S2 \\ S4 \end{pmatrix}, \tag{5.53}$$

where

$$\begin{array}{rcl} S1 & = & \sum x_j \\ S2 & = & \sum y_j \\ S3 & = & \sum x_j^2 \\ S4 & = & \sum x_j y_j. \end{array}$$

Table 5.9: Find the coefficients of (5.53).

i	x_i	y_i	x_i^2	$x_i y_i$
1	1.0000	1.0000	1.0000	1.0000
2	2.0000	2.0000	4.0000	4.0000
3	3.0000	2.0000	9.0000	6.0000
4	4.0000	3.0000	16.0000	12.0000
n=4	S1=10	S2=8	S3=30	S4=23

In the foregoing equations the summation is over j from 1 to n.
The solution of the above system (5.53) can be obtained easily as

$$a = \frac{S3S2 - S1S4}{nS3 - (S1)^2} \quad \text{and} \quad b = \frac{nS4 - S1S2}{nS3 - (S1)^2}. \tag{5.54}$$

The formula (5.53) reduces the problem of finding the parameters for a least squares linear fit to simple matrix multiplication.

We shall call a and b the least squares linear parameters for the data and the linear guess function with parameters, i.e.,

$$p_1(x) = a + bx$$

will be called the least squares line (or regression line) for the data.

Example 5.22 *Using the method of least squares, fit a straight line to the four points* $(1, 1), (2, 2), (3, 2),$ *and* $(4, 3)$.

Solution. *The sums required for the normal equation (5.53) are easily obtained using the values in Table 5.9. The linear system involving a and b in (5.53) form is*

$$\begin{pmatrix} 4 & 10 \\ 10 & 30 \end{pmatrix} \begin{pmatrix} a \\ b \end{pmatrix} = \begin{pmatrix} 8 \\ 23 \end{pmatrix}.$$

Then solving the above linear system using LU decomposition by the Cholesky method discussed in Chapter 1, the solution of the linear system is

$$a = 0.5 \quad \text{and} \quad b = 0.6.$$

Thus, the least squares line is

$$p_1(x) = 0.5 + 0.6x.$$

•

Clearly, $p_1(x)$ replaces the tabulated functional relationship given by $y = f(x)$. The original data along with the approximating polynomials are shown graphically in Figure 5.8.

Use the MATLAB Command Window as follows:

```
>> x = [1 2 3 4];
>> y = [1 2 2 3];
>> [a, b] = linefit(x, y);
```

To plot Figure 5.7 one can use the MATLAB Command Window:

```
>> xfit = 0 : 0.1 : 5;
>> yfit = 0.6 * xfit + 0.5;
>> plot(x, y,' o', xfit, yfit,' −');
```

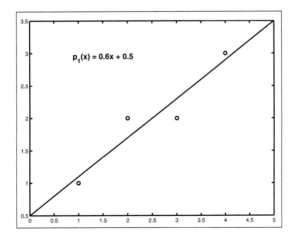

Figure 5.8: Least squares fit of four data points to a line.

Table 5.10: Error analysis of the linear fit.

i	x_i	y_i	$p_1(x_i)$	$abs(y_i - p_1(x_i))$
1	1.0000	1.0000	1.1000	0.1000
2	2.0000	2.0000	1.7000	0.3000
3	3.0000	2.0000	2.3000	0.3000
4	4.0000	3.0000	2.9000	0.1000

Table 5.10 shows the error analysis of the straight line using least squares approximation.
Hence, we have

$$E(a, b) = \sum_{i=1}^{4} (y_i - p_1(x_i))^2 = 0.2000.$$

•

Program 5.7

MATLAB m-file for the Linear Least Squares Fit
function [a,b]=linefit(x,y)
n=length(x);
$S1 = sum(x);$
$S2 = sum(y);$
$S3 = sum(x.*x);$
$S4 = sum(x.*y);$
$a = (n*S4 - S1*S2)/(n*S3 - (S1)\,\hat{}\,2);$
$b = (S3*S2 - S4*S1)/(n*S3 - (S1)\,\hat{}\,2);$
$for\ k = 1:n$
$p_1 = a + b*x(k);$
$Error(k) = abs(p_1 - y(k));$
end
$Error = sum(Error.*Error);$

5.3.2 Polynomial Least Squares

In the previous section we discussed a procedure to derive the equation of a straight line using least squares, which works very well if the measured data are intrinsically linear. But in many cases, data from experimental results are not linear. Therefore, now we show how to find the least squares parabola, and the extension to a polynomial of higher degree is easily made. The general problem of approximating a set of data $\{(x_i, y_i), i = 0, 1, \ldots, m\}$ with a polynomial of degree $n < m - 1$ is

$$p_n(x) = b_0 + b_1 x + b_2 x^2 + \cdots + b_n x^n. \tag{5.55}$$

Then the error E takes the form

$$
\begin{aligned}
E \;&=\; \sum_{j=1}^{m}(y_j - p_n(x_j))^2 \\[2mm]
&=\; \sum_{j=1}^{m} y_j^2 - 2\sum_{j=1}^{m} p_n(x_j)y_j + \sum_{j=1}^{m}(p_n(x_j))^2 \\[2mm]
&=\; \sum_{j=1}^{m} y_j^2 - 2\sum_{j=1}^{m}\left(\sum_{i=0}^{n} b_i x_j^i\right) y_j + \sum_{j=1}^{m}\left(\sum_{i=0}^{n} b_i x_j^i\right)^2 \\[2mm]
&=\; \sum_{j=1}^{m} y_j^2 - 2\sum_{i=0}^{n} b_i \left(\sum_{j=1}^{m} y_j x_j^i\right) + \sum_{i=0}^{n}\sum_{k=0}^{n} b_i b_k \left(\sum_{j=1}^{m} x_j^{i+k}\right).
\end{aligned}
$$

Like in linear least squares, for E to be minimized, it is necessary that $\partial E/\partial b_i = 0$, for each $i = 0, 1, 2, \ldots, n$. Thus, for each i,

$$0 = \frac{\partial E}{\partial b_i} = -2\sum_{j=1}^{m} y_j x_j^i + 2\sum_{k=1}^{n} b_k \sum_{j=1}^{m} x_j^{i+k}. \tag{5.56}$$

This gives $(n+1)$ *normal equations* in the $(n+1)$ unknowns b_i,

$$\sum_{k=1}^{n} b_k \sum_{j=1}^{m} x_j^{i+k} = \sum_{j=1}^{m} y_j x_j^i, \quad i = 0, 1, 2, \ldots, n. \tag{5.57}$$

It is helpful to write the equations as follows:

$$b_0 \sum_{j=1}^{m} x_j^0 + b_1 \sum_{j=1}^{m} x_j^1 + b_2 \sum_{j=1}^{m} x_j^2 + \cdots + b_n \sum_{j=1}^{m} x_j^n = \sum_{j=1}^{m} y_j x_j^0$$

$$b_0 \sum_{j=1}^{m} x_j^1 + b_1 \sum_{j=1}^{m} x_j^2 + b_2 \sum_{j=1}^{m} x_j^3 + \cdots + b_n \sum_{j=1}^{m} x_j^{n+1} = \sum_{j=1}^{m} y_j x_j^1$$

$$\vdots$$

$$b_0 \sum_{j=1}^{m} x_j^n + b_1 \sum_{j=1}^{m} x_j^{n+1} + b_2 \sum_{j=1}^{m} x_j^{n+2} + \cdots + b_n \sum_{j=1}^{m} x_j^{2n} = \sum_{j=1}^{m} y_j x_j^n.$$

Note that the coefficients matrix of this system is symmetric and positive-definite. Hence, the normal equations possess a unique solution.

Example 5.23 *Find the least squares polynomial approximation of degree 2 to the following data:*

$$\frac{x_j \quad 0 \quad 1 \quad 2 \quad 4 \quad 6}{y_j \quad 3 \quad 1 \quad 0 \quad 1 \quad 4}.$$

Solution. *The coefficients of the least squares polynomial approximation of degree 2,*

$$p_2(x) = b_0 + b_1 x + b_2 x^2,$$

are the solution values b_0, b_1, and b_2 of the linear system

$$\left. \begin{array}{r} b_0 m \quad + b_1 \sum x_j^1 + b_2 \sum x_j^2 = \sum y_j x_j^0 \\[2mm] b_0 \sum x_j^1 + b_1 \sum x_j^2 + b_2 \sum x_j^3 = \sum y_j x_j^1 \\[2mm] b_0 \sum x_j^2 + b_1 \sum x_j^3 + b_2 \sum x_j^4 = \sum y_j x_j^2 \end{array} \right\}. \qquad (5.58)$$

The sums required for the normal equation (5.58) are easily obtained using the values in Table 5.11. The linear system involving unknown coefficients

Table 5.11: Find the coefficients of (5.58).

i	x_i	y_i	x_i^2	x_i^3	x_i^4	$x_i y_i$	$x_i^2 y_i$
1	0.00	3.00	0.00	0.00	0.00	0.00	0.00
2	1.00	1.00	1.00	1.00	1.00	1.00	1.00
3	2.00	0.00	4.00	8.00	16.00	0.00	0.00
4	4.00	1.00	16.00	64.00	256.00	4.00	16.00
5	6.00	4.00	36.00	216.00	1296.00	24.00	144.00
m=5	13.00	9.00	57.00	289.00	1569.00	29.00	161.00

$b_0, b_1,$ and b_2 is

$$5b_0 + 13b_1 + 57b_2 = 9$$
$$13b_0 + 57b_1 + 289b_2 = 29$$
$$57b_0 + 289b_1 + 1569b_2 = 161.$$

Then solving the above linear system, the solution of the linear system is

$$b_0 = 2.8252, \qquad b_1 = -2.0490, \qquad b_2 = 0.3774.$$

Hence, the parabola equation becomes

$$p_2(x) = 2.8252 - 2.0490x + 0.3774x^2.$$

●

Use the MATLAB Command Window as follows:

```
>> x = [0 1 2 4 6];
>> y = [3 1 0 1 4];
>> n = 2;
>> C = polyfit(x, y, n);
```

Clearly, $p_2(x)$ replaces the tabulated functional relationship given by $y = f(x)$. The original data along with the approximating polynomials are shown graphically in Figure 5.9. To plot Figure 5.9 one can use the MATLAB Command Window as follows:

```
>> xfit = -1 : 0.1 : 7;
>> yfit = 2.8252 - 2.0490. * xfit + 0.3774. * xfit. * xfit;
>> plot(x, y,' o', xfit, yfit,' -');
```

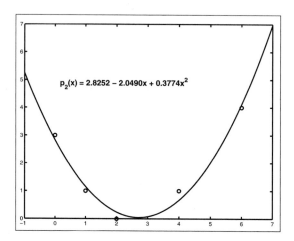

Figure 5.9: Least squares fit of five data points to a parabola.

Program 5.8

MATLAB m-file for the Polynomial Least Squares Fit
function C=polyfit(x,y,n)
m=length(x); for $i = 1 : 2 * n + 1$
$a(i) = sum(x.\hat{\ } (i-1))$;end
$for\ i = 1 : n + 1$ % coefficients of vector b
$b(i) = sum(y. * x.\hat{\ } (i-1))$;end
for i=1:n+1; for j=1:n+1
$A(i, j) = a(j + i - 1)$; end;end
$C = A \backslash b'$; % Solving linear system
$for\ k = 1 : m; S = C(1); for\ i = 2 : m + 1$
$S = S + C(i). * x(k).\hat{\ } (i-1)$;end;
$p_2(k) = S; Error(k) = abs(y(k) - p_2(k))$; end
$Error = sum(Error. * Error)$;

Table 5.12: Error analysis of the polynomial Fit.

i	x_i	y_i	$p_2(x_i)$	$abs(y_i - p_2(x_i))$
1	0.0000	3.0000	2.8252	0.1748
2	1.0000	1.0000	1.1535	0.1535
3	2.0000	0.0000	0.2367	0.2367
4	4.0000	1.0000	0.6674	0.3326
5	6.0000	4.0000	4.1173	0.1173
	13.000	9.0000	9.0001	1.0148

Table 5.12 shows the error analysis of the parabola using least squares approximation. Hence, the error associated with the least squares polynomial approximation of degree 2 is

$$E(b_0, b_1, b_2) = \sum_{j=1}^{5} (y_i - p(x_i))^2 = 0.2345.$$

5.3.3 Nonlinear Least Squares

Although polynomials are frequently used as the approximating function, they are by no means the only possibilities. The most popular forms of nonlinear curves are the exponential forms

$$y(x) = ax^b \tag{5.59}$$

or

$$y(x) = ae^{bx}. \tag{5.60}$$

We can develop the normal equations for these analogously to the previous development for least squares. The least squares error for (5.59) is given by

$$E(a, b) = \sum_{j=1}^{n} (y_j - ax_j^b)^2, \tag{5.61}$$

with associated normal equations

$$\frac{\partial E}{\partial a} = -2\sum_{j=1}^{n}(y_j - ax_j^b)x_j^b = 0$$

$$\frac{\partial E}{\partial b} = -2\sum_{j=1}^{n}(y_j - ax_j^b)(abx_j^{b-1}) = 0 \qquad\qquad (5.62)$$

Then the set of normal equations (5.62) represents the system of two equations in the two unknowns a and b. Such nonlinear simultaneous equations can be solved using Newton's method for nonlinear systems. The details of this method of nonlinear systems will be discussed in Chapter 7.

Example 5.24 *Find the best-fit of the form $y = ax^b$ by using the data*

x	1	2	4	10
y	2.87	4.51	6.11	9.43

by Newton's method, starting with the initial approximation $(a_0, b_0) = (2,1)$ and taking a desired accuracy within $\epsilon = 10^{-5}$.

Solution. *The normal equation is*

$$\sum_{j=1}^{4} y_j x_j^b - a\sum_{j=1}^{n} x_j^{2b} = 0$$

$$\sum_{j=1}^{4} y_j x_j^b \ln x_j - a\sum_{j=1}^{4} x_j^{2b} \ln x_j = 0 \qquad\qquad (5.63)$$

By using the given data points, the nonlinear system (5.63) gives

$$2.87 - a(1 + 2^{2b} + 4^{2b} + 10^{2b}) + 4.5(2^b) + 6.11(4^b) + 9.43(10^b) = 0$$
$$-a(0.69(2^{2b}) + 1.39(4^{2b}) + 2.30(10^{2b})) + 3.12(2^b) + 8.47(4^b) + 21.72(10^b) = 0.$$

Let us consider the two functions

$$f_1(a,b) = 2.87 - a(1 + 2^{2b} + 4^{2b} + 10^{2b}) + 4.5(2^b) + 6.11(4^b) + 9.43(10^b)$$
$$f_2(a,b) = -a(0.69(2^{2b}) + 1.39(4^{2b}) + 2.30(10^{2b})) + 3.12(2^b) + 8.47(4^b)$$
$$+21.72(10^b),$$

and their derivatives with respect to unknown variables a and b:

$$\frac{\partial f_1}{\partial a} = -(1 + 2^{2b} + 4^{2b} + 10^{2b})$$

$$\frac{\partial f_1}{\partial b} = -a(1.39(2^{2b}) + 2.77(4^{2b}) + 4.61(10^{2b}) + 3.12(2^b) + 8.47(4^b) + 21.71(10^b))$$

$$\frac{\partial f_2}{\partial a} = -(0.69(2^{2b}) + 1.39(4^{2b}) + 2.30(10^{2b}))$$

$$\frac{\partial f_2}{\partial b} = -a(0.96(2^{2b}) + 3.84(4^{2b}) + 10.61(10^{2b}) + 2.16(2^b) + 11.74(4^b) + 50.01(10^b)).$$

Since Newton's formula for the system of two nonlinear equations is

$$\left(\begin{array}{c} a_{k+1} \\ b_{k+1} \end{array} \right) = \left(\begin{array}{c} a_k \\ b_{k+1} \end{array} \right) - J^{-1}(a_k, b_k) \left(\begin{array}{c} f_1(a_k, b_k) \\ f_2(a_k, b_k) \end{array} \right),$$

where

$$J = \left(\begin{array}{cc} \dfrac{\partial f_1}{\partial a} & \dfrac{\partial f_1}{\partial b} \\[2mm] \dfrac{\partial f_2}{\partial a} & \dfrac{\partial f_2}{\partial b} \end{array} \right),$$

let us start with the initial approximation $(a_0, b_0) = (2, 1)$, and the values of the functions at this initial approximation are as follows:

$$f_1(2, 1) = -111.39$$
$$f_2(2, 1) = -253.216.$$

The Jacobian matrix J and its inverse J^{-1} at the given initial approximation can be calculated as

$$J(2, 1) = \left(\begin{array}{cc} -121 & -763.576 \\ -255.248 & -1700.534 \end{array} \right)$$

and

$$J^{-1}(2, 1) = \left(\begin{array}{cc} -0.1565 & 0.0703 \\ 0.0235 & -0.0111 \end{array} \right).$$

Substituting all these values in the above Newton's formula, we get the first approximation as

$$\begin{pmatrix} a_1 \\ b_1 \end{pmatrix} = \begin{pmatrix} 2.0 \\ 1.0 \end{pmatrix} - \begin{pmatrix} -0.1565 & 0.0703 \\ 0.0235 & -0.0111 \end{pmatrix} \begin{pmatrix} -111.39 \\ -253.216 \end{pmatrix} = \begin{pmatrix} 2.3615 \\ 0.7968 \end{pmatrix}.$$

Similarly, the second iteration using $(a_1, b_1) = (2.3615, 0.7968)$ *gives*

$$\begin{pmatrix} a_2 \\ b_2 \end{pmatrix} = \begin{pmatrix} 2.3615 \\ 0.7968 \end{pmatrix} - \begin{pmatrix} -0.2323 & 0.1063 \\ 0.0339 & -0.0169 \end{pmatrix} \begin{pmatrix} -35.4457 \\ -81.1019 \end{pmatrix} = \begin{pmatrix} 2.7444 \\ 0.6282 \end{pmatrix}.$$

The first two and the further steps of the method are listed in Table 7.4 by taking the desired accuracy within $\epsilon = 10^{-5}$.

Table 5.13: Solution of a system of two nonlinear equations.

n	a-approx. a_n	b-approx. b_n
00	2.00000	1.00000
01	2.36151	0.79684
02	2.74443	0.62824
03	2.99448	0.52535
04	3.07548	0.49095
05	3.08306	0.48754
06	3.08314	0.48751

Hence, the best nonlinear fit is

$$y(x) = 3.08314x^{0.48751}.$$

●

But remember that nonlinear simultaneous equations are more difficult to solve than linear equations. Because of this difficulty, the exponential forms are usually linearized by taking logarithms before determining the required parameters. Therefore, taking logarithms of both sides of (5.59), we get

$$\ln y = \ln a + b \ln x,$$

which may be written as

$$Y = A + BX, \tag{5.64}$$

with $A = \ln a$, $B = b$, $X = \ln x$, and $Y = \ln y$. The values of A and B can be chosen to minimize

$$E(A, B) = \sum_{j=1}^{n} (Y_j - (A + BX_j))^2, \tag{5.65}$$

where $X_j = \ln x_j$ and $Y_j = \ln y_j$. After differentiating E with respect to A and B and then putting the results equal to zero, we get the normal equations in linear form as

$$nA \quad + \quad B\sum_{j=1}^{n} X_j = \sum_{j=1}^{n} Y_j$$

$$A\sum_{j=1}^{n} X_j \quad + \quad B\sum_{j=1}^{n} X_j^2 = \sum_{j=1}^{n} X_j Y_j.$$

Then writing the above equations in matrix form, we have

$$\begin{pmatrix} n & S1 \\ S1 & S3 \end{pmatrix} \begin{pmatrix} A \\ B \end{pmatrix} = \begin{pmatrix} S2 \\ S4 \end{pmatrix}, \tag{5.66}$$

where

$$\begin{aligned} S1 &= \sum X_j \\ S2 &= \sum Y_j \\ S3 &= \sum X_j^2 \\ S4 &= \sum X_j Y_j. \end{aligned}$$

Table 5.14: Find the coefficients of (5.67).

i	X_i	Y_i	X_i^2	X_iY_i
1	0.0000	1.0543	0.0000	0.0000
2	0.6932	1.5063	0.4805	1.0442
3	1.3863	1.8099	1.9218	2.5091
4	2.3026	2.2439	5.3020	5.1668
n=4	S1=4.3821	S2=6.6144	S3=7.7043	S4=8.7201

In the foregoing equations the summation is over j from 1 to n. The solution of the above system can be obtained easily as

$$
\left.
\begin{aligned}
A &= \frac{S3S2 - S1S4}{nS3 - (S1)^2} \\[2mm]
B &= \frac{nS4 - S1S2}{nS3 - (S1)^2}
\end{aligned}
\right\}. \tag{5.67}
$$

Now the data set may be transformed to $(\ln x_j, \ln y_j)$ and determining a and b is a linear least squares problem. The values of unknowns a and b can be deduced from the relations

$$
a = e^A \quad \text{and} \quad b = B. \tag{5.68}
$$

Thus, the nonlinear guess function with parameters a and b

$$
y(x) = ax^b
$$

will be called the nonlinear least squares approximation for the data.

Example 5.25 *Find the best-fit of the form* $y = ax^b$ *by using the following data:*

x	1	2	4	10
y	2.87	4.51	6.11	9.43

Solution. *The sums required for the normal equation (5.66) are easily obtained using the values in Table 5.14. The linear system involving A and*

B in (5.66) form is

$$\begin{pmatrix} 4 & 4.3821 \\ 4.3821 & 7.7043 \end{pmatrix} \begin{pmatrix} A \\ B \end{pmatrix} = \begin{pmatrix} 6.6144 \\ 8.7201 \end{pmatrix}.$$

Then solving the above linear system, the solution of the linear system is

$$A = 1.0975 \quad and \quad B = 0.5076.$$

Using the values of A and B in (5.68), we have the values of the parameters a and b as

$$a = e^A = 2.9969 \quad and \quad b = B = 0.5076.$$

Hence, the best nonlinear fit is

$$y(x) = 2.9969 x^{0.5076}.$$

•

Use the MATLAB Command Window as follows:

```
>> x = [1 2 4 10];
>> y = [2.87 4.51 6.11 9.43];
>> [A, B] = exp1fit(x, y);
```

Program 5.9

MATLAB m-file for the Nonlinear Least Squares Fit
function [A,B]=exp1fit(x,y)% Least square fit $y = ax^b$
%Transform the data from (x,y) to (X,Y), $X = log(x), Y = log(y)$;
$n = length(x); X = log(x); Y = log(y)$;
$S1 = sum(X); S2 = sum(Y); S3 = sum(X.*X); S4 = sum(W.*Z)$;
$B = (n*S4 - S1*S2)/(n*S3 - (S1)\,\hat{}\,2)$;
$A = (S3*S2 - S4*S1)/(n*S3 - (S1)\hat{}\,2)$;
$b = B; a = exp(A)$;
for k=1:n
$y = a*X(k).\hat{}\,b; Error(k) = abs(y(k) - y)$; end
$Error = sum(Error.*Error)$;

Clearly, $y(x)$ replaces the tabulated functional relationship given by $y = f(x)$. The original data along with the approximating polynomials are shown graphically in Figure 5.10. To plot Figure 5.10, one can use the MATLAB Command Window as follows:

```
>> xfit = 0 : 0.1 : 11;
>> yfit = 2.9969 * xfit. ^ 0.5076;
>> plot(x, y,' o', xfit, yfit,' −');
```

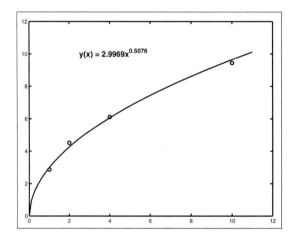

Figure 5.10: Nonlinear least squares fit.

Table 5.15 shows the error analysis of the nonlinear least squares approximation.

Table 5.15: Error analysis of the nonlinear fit.

i	x_i	y_i	$y(x_i)$	$abs(y_i - y(x_i))$
1	1.0000	2.870	2.9969	0.1269
2	2.000	4.510	4.2605	0.2495
3	4.000	6.110	6.0569	0.0531
4	10.000	9.430	9.6435	0.2135

Hence, the error associated with the nonlinear least squares approximation
is

$$E(a,b) = \sum_{i=1}^{4} (y_i - ax_i^{bx})^2 = 0.1267.$$

•

Similarly, for the other nonlinear curve $y(x) = ae^{bx}$, the least squares error
is defined as

$$E(a,b) = \sum_{j=1}^{n} (y_j - ae^{bx_j})^2, \tag{5.69}$$

which gives the associated normal equations as

$$\left. \begin{array}{ll} \dfrac{\partial E}{\partial a} & = -2\sum_{j=1}^{n}(y_j - ae^{bx_j})e^{bx_j} \quad = 0 \\[4mm] \dfrac{\partial E}{\partial a} & = -2\sum_{j=1}^{n}(y_j - ae^{bx_j})ax_j e^{bx_j} = 0 \end{array} \right\}. \tag{5.70}$$

Then the set of normal equations (5.70) represents the nonlinear simultaneous system.

Example 5.26 *Find the best-fit of the form $y = ae^{bx}$ by using the data*

x	0	0.25	0.4	0.5
y	9.532	7.983	4.826	5.503

*by Newton's method, starting with the initial approximation $(a_0, b_0) = (8,0)$
and taking the desired accuracy within $\epsilon = 10^{-5}$.*

Solution. *The normal equation is*

$$\left. \begin{array}{l} \displaystyle\sum_{j=1}^{4} y_j e^{bx_j} - a\sum_{j=1}^{4} e^{2bx_j} = 0 \\[6mm] \displaystyle\sum_{j=1}^{4} y_j x_j e^{bx_j} - a\sum_{j=1}^{4} x_j e^{2bx_j} = 0 \end{array} \right\}. \tag{5.71}$$

By using the given data points, the nonlinear system (5.71) gives

$$9.53 - a(1 + e^{0.5b} + e^{0.8b} + e^b) + 7.98e^{0.25b} + 4.83e^{0.4b} + 5.503e^{0.5b} = 0$$
$$-a(0.25e^{0.5b} + 0.4e^{0.8b} + 0.5e^b) + 1.996e^{0.25b} + 1.93e^{0.4b} + 2.752e^{0.5b} = 0.$$

Let us consider the two functions

$$f_1(a, b) = 9.53 - a(1 + e^{0.5b} + e^{0.8b} + e^b) + 7.98e^{0.25b} + 4.83e^{0.4b} + 5.503e^{0.5b}$$
$$f_2(a, b) = -a(0.25e^{0.5b} + 0.4e^{0.8b} + 0.5e^b) + 1.996e^{0.25b} + 1.93e^{0.4b} + 2.752e^{0.5b}$$

and their derivatives with respect to unknown variables a and b:

$$\frac{\partial f_1}{\partial a} = -(1 + e^{0.5b} + e^{0.8b} + e^b)$$

$$\frac{\partial f_1}{\partial b} = -a(0.5e^{0.5b} + 0.8e^{0.8b} + e^b) + 1.996e^{0.25b} + 1.93e^{0.4b} + 2.752e^{0.5b}$$

$$\frac{\partial f_2}{\partial a} = -(0.25e^{0.5b} + 0.4e^{0.8b} + 0.5e^b)$$

$$\frac{\partial f_2}{\partial b} = -a(0.125e^{0.5b} + 0.32e^{0.8b} + 0.5e^b) + 0.499e^{0.25b} + 0.772e^{0.4b} + 1.376e^{0.5b}.$$

Since Newton's formula for the system of two nonlinear equations is

$$\begin{pmatrix} a_{k+1} \\ b_{k+1} \end{pmatrix} = \begin{pmatrix} a_k \\ b_{k+1} \end{pmatrix} - J^{-1}(a_k, b_k) \begin{pmatrix} f_1(a_k, b_k) \\ f_2(a_k, b_k) \end{pmatrix},$$

where

$$J = \begin{pmatrix} \dfrac{\partial f_1}{\partial a} & \dfrac{\partial f_1}{\partial b} \\ \dfrac{\partial f_2}{\partial a} & \dfrac{\partial f_2}{\partial b} \end{pmatrix},$$

let us start with the initial approximation $(a_0, b_0) = (8, 0)$, and the values of the functions at this initial approximation are as follows:

$$f_1(2, 1) = -4.156$$
$$f_2(2, 1) = -2.522.$$

*The Jacobian matrix J and its inverse J^{-1} at the given initial approxima-
tion can be computed as*

$$J(8,0) = \begin{pmatrix} -4 & -11.722 \\ -1.15 & -4.913 \end{pmatrix}$$

and

$$J^{-1}(8,0) = \begin{pmatrix} -0.7961 & 1.8993 \\ 0.1863 & -0.6481 \end{pmatrix}.$$

*Substituting all these values in the above Newton's formula, we get the first
approximation as*

$$\begin{pmatrix} a_1 \\ b_1 \end{pmatrix} = \begin{pmatrix} 8.0 \\ 0.0 \end{pmatrix} - \begin{pmatrix} -0.7961 & 1.8993 \\ 0.1863 & -0.6481 \end{pmatrix} \begin{pmatrix} -4.156 \\ -2.522 \end{pmatrix} = \begin{pmatrix} 9.48168 \\ -0.86015 \end{pmatrix}.$$

Similarly, the second iteration using $(a_1, b_1) = (9.48168, -0.86015)$ gives

$$\begin{pmatrix} a_2 \\ b_2 \end{pmatrix} = \begin{pmatrix} 9.48168 \\ -0.86015 \end{pmatrix} - \begin{pmatrix} -0.87813 & 2.19437 \\ 0.20559 & -0.92078 \end{pmatrix} \begin{pmatrix} -1.4546 \\ -0.6856 \end{pmatrix} = \begin{pmatrix} 9.70881 \\ -1.19239 \end{pmatrix}.$$

*The first two and the further steps of the method are listed in Table 7.4
by taking the desired accuracy within $\epsilon = 10^{-5}$.*

Table 5.16: Solution of nonlinear system.

Iteration	a-approximation	b-approximation
n	a_n	b_n
00	8.00000	0.00000
01	9.48168	-0.86015
02	9.70881	-1.19239
03	9.72991	-1.26193
04	9.73060	-1.26492
05	9.73060	-1.26492

Hence, the best nonlinear fit is

$$y(x) = 9.73060e^{-1.26492x}.$$

Once again, to make this exponential form a linearized form, we take the logarithms of both sides of (5.60), and we get

$$\ln y = \ln a + bx,$$

which may be written as

$$Y = A + BX, \tag{5.72}$$

with $A = \ln a$, $B = b$, $X = x$, and $Y = \ln y$. The values of A and B can be chosen to minimize

$$E(A, B) = \sum_{j=1}^{n}(Y_j - (A + BX_j))^2, \tag{5.73}$$

where $X_j = x_j$ and $Y_j = \ln y_j$. By solving the linear normal equations of the form

$$\left. \begin{array}{c} nA \quad + \quad B\sum\limits_{j=1}^{n} X_j = \sum\limits_{j=1}^{n} Y_j \\[2em] A\sum\limits_{j=1}^{n} X_j \quad + \quad B\sum\limits_{j=1}^{n} X_j^2 = \sum\limits_{j=1}^{n} X_j Y_j \end{array} \right\} \tag{5.74}$$

to get the values of A and B, the data set may be transformed to $(x_j, \ln y_j)$ and determining a and b is a linear least squares problem. The values of unknowns a and b are deduced from the relations

$$a = e^A \qquad \text{and} \quad b = B. \tag{5.75}$$

Thus, the nonlinear guess function with parameters a and b

$$y(x) = ae^{bx}$$

will be called the nonlinear least squares approximation for the data.

Example 5.27 *Find the best-fit of the form $y = ae^{bx}$ by using the following data:*

x	0	0.25	0.4	0.5
y	9.532	7.983	4.826	5.503

Solution. *The sums required for the normal equation (5.74) are easily obtained using the values in Table 5.17.*

Table 5.17: Find the coefficients of (5.74).

i	X_i	Y_i	X_i^2	$X_i Y_i$
1	0.0000	2.2546	0.0000	0.0000
2	0.2500	2.0773	0.0625	0.5193
3	0.4000	1.5740	0.1600	0.6296
4	0.5000	1.7053	0.2500	0.8527
n=4	1.1500	7.6112	0.4725	2.0016

The linear system involving unknown coefficients A and B is

$$4A \quad + \quad 1.1500B = 7.6112$$
$$1.1500A + 0.4725B = 2.0016.$$

Then solving the above linear system, the solution of the linear system is

$$A = 2.2811 \quad and \quad B = -1.3157.$$

Using the values in (5.75), we have the values of the unknown parameters a and b as

$$a = e^A = 9.7874, \quad and \quad b = B = -1.3157.$$

Hence, the best nonlinear fit is

$$y(x) = 9.7874x^{-1.3157}.$$

Use the MATLAB Command Window as follows:

```
>> x = [0 0.25 0.4 0.5];
>> y = [9.532 7.983 4.826 5.503];
>> [A, B] = exp2fit(x, y);
```

Clearly, $y(x)$ replaces the tabulated functional relationship given by $y = f(x)$. The original data along with the approximating polynomials are shown graphically in Figure 5.14. To plot Figure 5.14, one can use the MATLAB Command Window as follows:

```
>> xfit = -0.1 : 0.1 : 0.6;
>> yfit = 9.7874 * exp(-1.3157. * xfit);
>> plot(x, y,' o', xfit, yfit,' -');
```

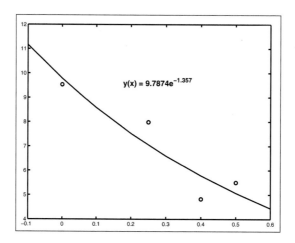

Figure 5.11: Nonlinear least squares fit.

Note that the value of a and b calculated for the linearized problem will not necessarily be the same as the values obtained for the original least squares problem. In this example, the nonlinear system becomes

$$9.532 - a + 7.983e^{0.25b} - ae^{0.5b} + 4.826e^{0.4b} - ae^{0.8b} + 5.503e^{0.5b} - ae^{b} = 0$$

$$1.996e^{0.25b} - 0.25a^2e^{0.5b} + 1.930e^{0.4b} - 0.4a^2e^{0.8b} + 2.752ae^{0.5b} - 0.5a^2e^{b} = 0.$$

Table 5.18: Error analysis of the nonlinear fit.

i	x_i	y_i	$y(x_i)$	$abs(y_i - y(x_i))$
1	0.000	9.532	9.7872	0.2552
2	0.250	7.983	7.0439	0.9391
3	0.400	4.826	5.7823	0.9563
4	0.500	5.503	5.0695	0.4335

Now Newton's method for nonlinear systems can be applied to this system, and we get the values of a and b as

$$a = 9.731 \qquad \text{and} \quad b = -1.265.$$

Table 5.18 shows the error analysis of the nonlinear least squares approximation.

Hence, the error associated with the nonlinear least squares approximation is

$$E(a, b) = \sum_{i=1}^{4} (y_i - ae^{bx_i})^2 = 2.0496.$$

●

Table 5.19 shows the conversion of nonlinear forms into linear forms by using a change of variables and constants.

Example 5.28 *Find the best-fit of the form* $y = axe^{-bx}$ *by using the change of variables to linearize the following data points:*

x	1.5	2.5	4.0	5.5
y	3.0	4.3	6.5	7.0

Solution. *Write the given form* $y = axe^{-bx}$ *into the form*

$$\frac{y}{x} = ae^{-bx},$$

and taking the logarithms of both sides of the above equation, we get

$$\ln\left(\frac{y}{x}\right) = \ln a + (-bx),$$

Table 5.19: Conversion of nonlinear forms into linear forms.

No.	Nonlinear Form	Linear Form	Change of Variables and Constants
	$y = f(x)$	$Y = A + BX$	
1	$y = ax + b/x^2$	$Y = A + BX$	$Y = y/x, X = 1/x, A = a, B = b$
2	$y = 1/(a + bx)^2$	$Y = A + BX$	$Y = (y)^{1/2}, X = x, A = a, B = b$
3	$y = 1/(a + bx^2)$	$Y = A + BX$	$Y = 1/y, X = x^2, A = a, B = b$
4	$y = 1/(2 + ae^{bx})$	$Y = A + BX$	$Y = \ln(1/y - 2), X = x, A = a, B = b$
5	$y = axe^{-bx}$	$Y = A + BX$	$Y = \ln(y/x), X = x, A = \ln(a), B = -b$
6	$y = a + b/\ln x$	$Y = A + BX$	$Y = y, X = 1/\ln x, A = a, B = b$
7	$y = (a + bx)^{-3}$	$Y = A + BX$	$Y = 1/y^{1/3}, X = x, A = a, B = b$

Table 5.20: Find the coefficients of (5.67).

i	x_i	y_i	$X_i = x_i$	$Y_i = \ln(y_i/x_i)$	X_i^2	$X_i Y_i$
1	1.50	3.0	1.50	0.6932	2.25	1.0398
2	2.50	4.3	2.50	0.5423	6.25	1.3558
3	4.00	6.5	4.00	0.4855	16.00	1.9420
4	5.50	7.0	5.50	0.2412	3.25	1.3266
n=4			S1=13.50	S2=1.9622	S3=54.750	S4=5.6642

which may be written as

$$Y = A + BX,$$

with

$$A = \ln a, \quad B = -b, \quad X = x, \quad Y = \ln\left(\frac{y}{x}\right).$$

Then the sums required for the normal equation (5.66) are easily obtained using the values in Table 5.20. The linear system involving A and B in (5.66) form is

$$\begin{pmatrix} 4 & 13.50 \\ 13.50 & 54.750 \end{pmatrix} \begin{pmatrix} A \\ B \end{pmatrix} = \begin{pmatrix} 1.9622 \\ 5.6642 \end{pmatrix}.$$

Then solving the above linear system, the solution of the linear system is

$$A = 0.8426 \quad and \quad B = -0.1043.$$

Using these values of A and B, we have the values of the parameters a and b as

$$a = e^A = 2.3224 \quad and \quad b = -B = 0.1043.$$

Hence,

$$y(x) = 2.3224e^{-0.1043x}$$

is the best nonlinear fit. ●

Program 5.10

MATLAB m-file for the Nonlinear Least Squares Fit
function [A,B]=exp2fit(x,y) % Least square fit $y = ae^{bx}$
% Transform the data from (x,y) to (x,Y), $Y = log(Y)$;
$n = length(x); Y = log(y); S1 = sum(x); S2 = sum(Y);$
$S3 = sum(x.*x); S4 = sum(x.*Y);$
$B = (n*S4 - S1*S2)/(n*S3 - (S1)\^{} 2);$
$A = (S3*S2 - S4*S1)/(n*S3 - (S1)\^{} 2);$
$b = B; a = exp(A)$
for k=1:n
$y = a*exp(b*x(k)); Error(k) = abs(y(k) - y);$ end
$Error = sum(Error.*Error);$

5.3.4 Least Squares Plane

Many problems arise in engineering and science where the dependent variable is a function of two or more variables. For example, $z = f(x, y)$ is a two-variables function. Consider the least squares plane

$$z = ax + by + c, \tag{5.76}$$

for the n points $(x_1, y_1, z_1), \ldots,$ and (x_n, y_n, z_n) is obtained by minimizing

$$E(a, b, c) = \sum_{j=1}^{n}(z_j - ax_j - by_j - c)^2, \tag{5.77}$$

and the function $E(a, b, c)$ is minimum when

$$\frac{\partial E}{\partial a} = 2\sum_{j=1}^{n}(z_j - ax_j - bx_j - c)(-x_j) = 0$$

$$\frac{\partial E}{\partial b} = 2\sum_{j=1}^{n}(z_j - ax_j - bx_j - c)(-y_j) = 0$$

$$\frac{\partial E}{\partial c} = 2\sum_{j=1}^{n}(z_j - ax_j - bx_j - c)(-1) = 0.$$

Dividing by 2 and rearranging gives the normal equations

$$\left.\begin{array}{l}(\sum x_j^2)a + (\sum x_j y_j)b + (\sum x_j)c = \sum z_j x_j \\ (\sum x_j y_j)a + (\sum y_j^2)b + (\sum y_j)c = \sum z_j y_j \\ (\sum x_j)a + (\sum y_j)b + nc = \sum z_j \end{array}\right\} . \qquad (5.78)$$

The above linear system can be solved for unknowns a, b, and c.

Example 5.29 *Find the least squares plane $z = ax + by + c$ by using the following data:*

x_j	1	1	2	2	2
y_j	1	2	1	2	3
z_j	7	9	10	11	12

Solution. *The sums required for the normal equation (5.78) are easily obtained using the values in Table 5.23.*

The linear system (5.78) involving unknown coefficients a, b, and c is

$$\begin{array}{rrrrrl}14a &+& 15b &+& 8c &= 82 \\ 15a &+& 19b &+& 9c &= 93 \\ 8a &+& 9b &+& 5c &= 49.\end{array}$$

Then solving the above linear system, the solution of the linear system is

$$a = 2.400, \quad b = 1.200, \quad c = 3.800.$$

Table 5.21: Find the coefficients of (5.78).

i	x_i	y_i	z_i	x_i^2	$x_i y_i$	y_i^2	$x_i z_i$	$y_i z_i$
1	1.000	1.000	7.000	1.000	1.000	1.000	7.000	7.000
2	1.000	2.000	9.000	1.000	2.000	4.000	9.000	18.000
3	2.000	1.000	10.000	4.000	2.000	1.000	20.000	10.000
4	2.000	2.000	11.000	4.000	4.000	4.000	22.000	22.000
5	2.000	3.000	12.000	4.000	6.000	9.000	24.000	36.000
n=5	8.000	9.000	49.000	14.000	15.000	19.000	82.000	93.000

Table 5.22: Error Analysis of the Plane fit.

i	x_i	y_i	z_i	$z(x_i, y_i)$	$abs(z_i - z)$
1	1.0000	1.0000	7.0000	7.4000	0.4000
2	1.0000	2.0000	9.0000	8.6000	0.4000
3	2.0000	1.0000	10.0000	9.8000	0.2000
4	2.0000	2.0000	11.0000	11.0000	0.0000
5	2.0000	3.0000	12.0000	12.2000	0.2000

Hence, the least squares plane fit is

$$z = 2.400x + 1.200y + 3.800.$$

Use the MATLAB Command Window as follows:

```
>> x = [1 1 2 2 2];
>> y = [1 2 1 2 3];
>> z = [7 9 10 11 12];
>> sol = planefit(x, y, z);
```

Table 5.22 shows the error analysis of the least squares plane approximation. Hence, the error associated with the least squares plane approximation is

$$E(a, b, c) = \sum_{i=1}^{4} (z_i - ax_i + by_i + c)^2 = 0.4000. \qquad \bullet$$

Program 5.11
MATLAB m-file for the Least Squares Plane Fit
function Sol=planefit(x,y,z)
$n = length(x); S1 = sum(x); S2 = sum(y); S3 = sum(z);$
$S4 = sum(x.*x); S5 = sum(x.*y); S6 = sum(y.*y);$
$S7 = sum(z.*x); S8 = sum(z.*y);$
$A = [S4\ S5\ S1; S5\ S6\ S2; S1\ S2\ n]; B = [S7\ S8\ S3]'; C = A\backslash B;$
for k=1:n
$z = C(1)*x(k) + C(2)*y(k) + C(3);$
$Error(k) = abs(z(k) - z);$ end
$Error = sum(Error.*Error);$

5.3.5 Trigonometric Least Squares Polynomial

This is another popular form of the polynomial frequently used as the approximating function. Since we know that a series of the form

$$p(x) = \frac{a_0}{2} + \sum_{i=1}^{m}[a_i\cos(kx) + b_i\sin(kx)] \tag{5.79}$$

is called a *trigonometric polynomial* of order m, here we shall approximate the given data points with the function (5.79) using the least squares method. The least squares error for (5.79) is given by

$$E(a_0, a_1, \ldots, a_n, b_1, b_2, \ldots, b_n) = \sum_{j=1}^{n}[y_j - p(x_j)]^2, \tag{5.80}$$

and with the associated normal equations

$$\left.\begin{aligned}\frac{\partial E}{\partial a_k} &= 0, \quad \text{for} \quad k = 0,1,\ldots,m \\ \frac{\partial E}{\partial b_k} &= 0, \quad \text{for} \quad k = 1,2,\ldots,m\end{aligned}\right\}, \tag{5.81}$$

gives

$$
\left.
\begin{aligned}
\sum_{j=1}^{n}[\frac{a_0}{2}\sum_{i=1}^{m}a_i\cos(kx_j)+\sum_{i=1}^{m}b_i\sin(kx_j)] &= \sum_{j=1}^{n}y_j \\[2em]
\sum_{j=1}^{n}[\frac{a_0}{2}\sum_{i=1}^{m}a_i\cos(kx_j)+\sum_{i=1}^{m}b_i\sin(kx_j)]\cos(kx_j) &= \sum_{j=1}^{n}\cos(kx_j)y_j \\[2em]
\sum_{j=1}^{n}[\frac{a_0}{2}\sum_{i=1}^{m}a_i\cos(kx_j)+\sum_{i=1}^{m}b_i\sin(kx_j)]\sin(kx_j) &= \sum_{j=1}^{n}\sin(kx_j)y_j \\
\text{for} & \qquad k=1,2,\ldots,m
\end{aligned}
\right\}.
$$

$$(5.82)$$

Then the set of these normal equations (5.83) represents the system of $(2m+1)$ equations in $(2m+1)$ unknowns and can be solved using any numerical method discussed in Chapter 2. Note that the derivation of the coefficients a_k and b_k is usually called *discrete Fourier analysis*.

For $m=1$, we can write the normal equations (5.83) in the form

$$
\left.
\begin{aligned}
\frac{a_0}{2}+a_1\sum_{j=1}^{n}\cos(x_j)+b_1\sum_{j=1}^{n}\sin(x_j) &= \sum_{j=1}^{n}y_j \\[2em]
a_0\sum_{j=1}^{n}\cos(x_j)+a_1\sum_{j=1}^{n}\cos^2(x_j)+b_1\sum_{j=1}^{n}\cos(x_j)\sin(x_j) &= \sum_{j=1}^{n}\cos(x_j)y_j \\[2em]
a_0\sum_{j=1}^{n}\sin(x_j)+a_1\sum_{j=1}^{n}\cos(x_j)\sin(x_j)+b_1\sum_{j=1}^{n}\sin^2(x_j) &= \sum_{j=1}^{n}\sin(x_j)y_j
\end{aligned}
\right\},
$$

$$(5.83)$$

where j is the number of data points. By writing the above equations in matrix form, we have

$$
\begin{pmatrix} n/2 & S1 & S2 \\ S1 & S3 & S4 \\ S2 & S4 & S5 \end{pmatrix}\begin{pmatrix} a_0 \\ a_1 \\ b_1 \end{pmatrix}=\begin{pmatrix} S6 \\ S7 \\ S8 \end{pmatrix}, \qquad (5.84)
$$

where

$$S1 = \sum_{j=1}^{n} \cos(x_j), \quad S2 = \sum_{j=1}^{n} \sin(x_j)$$

$$S3 = \sum_{j=1}^{n} \cos^2(x_j), \quad S4 = \sum_{j=1}^{n} \cos(x_j)\sin(x_j)$$

$$S5 = \sum_{j=1}^{n} \sin^2(x_j), \quad S6 = \sum_{j=1}^{n} y_j$$

$$S7 = \sum_{j=1}^{n} \cos(x_j)y_j, \quad S8 = \sum_{j=1}^{n} \sin(x_j)y_j,$$

which represents a linear system of three equations in three unknowns a_0, a_1, and b_1. Note that the coefficients matrix of this system is symmetric and positive-definite. Hence, the normal equations possess a unique solution.

Example 5.30 *Find the trigonometric least squares polynomial* $p_1(x) = a_0 + a_1 \cos x + b_1 \sin x$ *that approximates the following data:*

x_j	0.1	0.2	0.3	0.4	0.5
y_j	0.9	0.75	0.64	0.52	0.40

.

Solution. *To find the trigonometric least squares polynomial*

$$p_1(x) = a_0 + a_1 \cos x + b_1 \sin x,$$

we have to solve the system

$$\begin{pmatrix} n/2 & S1 & S2 \\ S1 & S3 & S4 \\ S2 & S4 & S5 \end{pmatrix} \begin{pmatrix} a_0 \\ a_1 \\ b_1 \end{pmatrix} = \begin{pmatrix} S6 \\ S7 \\ S8 \end{pmatrix},$$

where

$$S1 = \sum_{j=1}^{n} \cos(x_j) = 4.7291, \quad S2 = \sum_{j=1}^{n} \sin(x_j) = 1.4629$$

$$S3 = \sum_{j=1}^{n} \cos^2(x_j) = 4.4817, \quad S4 = \sum_{j=1}^{n} \cos(x_j)\sin(x_j) = 1.3558$$

$$S5 = \sum_{j=1}^{n} \sin^2(x_j) = 0.5183, \quad S6 = \sum_{j=1}^{n} y_j = 3.2100$$

$$S7 = \sum_{j=1}^{n} \cos(x_j)y_j = 3.0720, \quad S8 = \sum_{j=1}^{n} \sin(x_j)y_j = 0.8223.$$

So,

$$\begin{pmatrix} 2.5 & 4.7291 & 1.4629 \\ 4.7291 & 4.4817 & 1.3558 \\ 1.4629 & 1.3558 & 0.5183 \end{pmatrix} \begin{pmatrix} a_0 \\ a_1 \\ b_1 \end{pmatrix} = \begin{pmatrix} 3.2100 \\ 3.0720 \\ 0.8223 \end{pmatrix},$$

and using the Gauss elimination method, we get the values of unknown as

$$a_0 = -0.0001, \quad a_1 = 0.9850, \quad b_1 = -0.9897.$$

Thus, we get the best trigonometric least squares polynomial

$$p(x) = -0.0001 + 0.9850 \cos x - 0.9897 \sin x,$$

which approximates the given data.

Note that $C = \cos(x_j)$ and $S = \sin(x_j)$. ●

The original data along with the approximating polynomial are shown graphically in Figure 5.12. To plot Figure 5.12, one can use the MATLAB Command Window as follows:

```
>> x = [0.1 0.2 0.3 0.4 0.5];
>> y = [0.9 0.75 0.64 0.52 0.40];
>> xfit = -0.1 : 0.1 : 0.6;
>> yfit = -0.0001 + 0.9850. * cos(xfit) - 0.9897. * sin(xfit);
>> plot(x, y,'o', xfit, yfit,'-');
```

Table 5.23: Find the coefficients of (5.83).

j	x_j	y_j	C	S	C^2	S^2	CS	Cy_j	Sy_j
1	0.1	0.90	0.9950	0.0998	0.9900	0.0100	0.0993	0.8955	0.0899
2	0.2	0.75	0.9801	0.1987	0.9605	0.0395	0.1947	0.7350	0.1490
3	0.3	0.64	0.9553	0.2955	2.000	0.0873	0.2823	0.6114	0.1891
4	0.4	0.52	0.9211	0.3894	4.000	0.1516	0.3587	0.4790	0.2025
5	0.5	0.40	0.8776	0.4794	6.000	0.2298	0.4207	0.3510	0.1918
n=5	1.5	3.21	4.7291	1.4629	4.4817	0.5183	1.3558	3.0720	0.8223

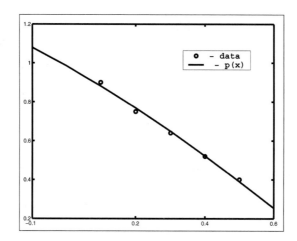

Figure 5.12: Trigonometric least squares fit.

5.3.6 Least Squares Solution of an Overdetermined System

In Chapter 2 we discussed methods for computing the solution **x** to a linear system $A\mathbf{x} = \mathbf{b}$ when the coefficient matrix A is square (number of rows and columns are equal). For square matrix A, the linear system usually has a unique solution. Now we consider linear systems where the coefficient matrix is *rectangular* (number of rows and columns are not equal). If A has m rows and n columns, then **x** is a vector with n components and **b** is a vector with m components. If the number of rows is greater than the number of columns ($m > n$), then the linear system is called an *overdetermined system*. Typically, an overdetermined system has no solution. This type of system generally arises when dealing with experimental data. It is also common in optimization-related problems.

Consider the following overdetermined linear system of two equations in one variable:

$$2x_1 = 3$$
$$4x_1 = 1. \tag{5.85}$$

Now using Gauss elimination to solve this system, we obtain

$$0 = -5,$$

which is impossible and hence, the given system (5.85) is inconsistent. Writing the given system in vector form, we get

$$\begin{pmatrix} 2 \\ 4 \end{pmatrix} x_1 = \begin{pmatrix} 2 \\ 4 \end{pmatrix}. \tag{5.86}$$

The left-hand side of (5.86) is $[0,0]^T$ when $x_1 = 0$, and is $[2,4]^T$ when $x_1 = 1$. Note that as x_1 takes on all possible values, the left-hand side of (5.86) generates the line connecting the origin and the point $(2,4)$ (Figure 5.13). On the other hand, the right-hand side of (5.86) is the vector $[3,1]^T$. Since the point $(3,1)$ does not lie on the line, the left-hand side and the right-hand side of (5.86) are never equal. The given system (5.86) is only consistent when the point corresponding to the right-hand side is contained in the line corresponding to the left-hand side. Thus, the least squares solution to (5.86) is the value of x_1 for which the point on the line is closest to the point $(3,1)$. In Figure 5.13, we see that the point $(1,2)$ on the line is closest to $(3,1)$, which we got when $x_1 = \frac{1}{2}$. So the least squares solution to (5.85) is $x_1 = \frac{1}{2}$. Now consider the following linear system of three equations in two variables:

$$\left.\begin{array}{ccccc} a_{11}x_1 & + & a_{12}x_2 & = & b_1 \\ a_{21}x_1 & + & a_{22}x_2 & = & b_2 \\ a_{31}x_1 & + & a_{32}x_2 & = & b_3 \end{array}\right\}. \tag{5.87}$$

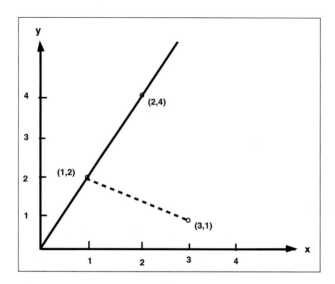

Figure 5.13: Least squares solution to an overdetermined system.

Again, it is impossible to find a solution that can satisfy all of the equations unless two of the three equations are dependent. That is, if only two out of the three equations are unique, then a solution is possible. Otherwise, our best hope is to find a solution that minimizes the error, i.e., the least squares solution. Now, we discuss the method for finding the least squares solution to the overdetermined system.

In the least squares method, $\hat{\mathbf{x}}$ is chosen so that the Euclidean norm of residual $\mathbf{r} = \mathbf{b} - A\hat{\mathbf{x}}$ is as small as possible. The residual corresponding to system (5.87) is

$$\mathbf{r} = \begin{pmatrix} b_1 - a_{11}x_1 - a_{12}x_2 \\ b_2 - a_{21}x_1 - a_{22}x_2 \\ b_3 - a_{31}x_1 - a_{32}x_2 \end{pmatrix}.$$

The l_2-norm of the residual is the square root of the sum of each component squared:

$$\|\mathbf{r}\|_2 = \sqrt{r_1^2 + r_2^2 + r_3^2} \ .$$

Since minimizing $\|\mathbf{r}\|_2$ is equivalent to minimizing $(\|\mathbf{r}\|_2)^2$, the least squares solution to (5.87) is the values for x_1 and x_2, which minimize the expression

$$(b_1 - a_{11}x_1 - a_{12}x_2)^2 + (b_2 - a_{21}x_1 - a_{22}x_2)^2 + (b_3 - a_{31}x_1 - a_{32}x_2)^2. \quad (5.88)$$

Minimizing x_1 and x_2 is done by differentiating (5.88), with respect to x_1 and x_2, and setting the derivatives to zero. Then solving for x_1 and x_2, we obtain the least squares solution $\hat{\mathbf{x}} = [x_1, x_2]^T$ to the system (5.87).

For a general overdetermined linear system $A\mathbf{x} = \mathbf{b}$, the residual is $\mathbf{r} = \mathbf{b} - A\hat{\mathbf{x}}$ and the l_2-norm of the residual is the square root of $\mathbf{r}^T\mathbf{r}$. The least squares solution to the linear system minimizes

$$\mathbf{r^T r} = (\|\mathbf{r}\|_2)^2 = (\mathbf{b} - A\hat{\mathbf{x}})^T(\mathbf{b} - A\hat{\mathbf{x}}). \quad (5.89)$$

The above equation (5.89) attains minimum when the partial derivative with respect to each of the variables $x_1, x_2, \ldots, x_n$ is zero. Since

$$\mathbf{r^T r} = r_1^2 + r_2^2 + \cdots + r_m^2 , \quad (5.90)$$

and the *ith* component of the residual $\mathbf{r}$ is

$$r_i = b_i - a_{i1}x_1 - a_{i2}x_2 - \cdots - a_{in}x_n ,$$

the partial derivative of $\mathbf{r}^T\mathbf{r}$ with respect to x_j is given by

$$\frac{\partial}{\partial x_j}\mathbf{r^T r} = -2r_1a_{1j} - 2r_2a_{2j} - \cdots - 2r_ma_{mj}. \quad (5.91)$$

From the right side of (5.91) we see that the partial derivative of $\mathbf{r}^T\mathbf{r}$ with respect to x_j is -2 times the product between the *jth* column of A and $\mathbf{r}$. Note that the *jth* column of A is the *jth* row of A^T. Since the *jth* component of $A^T\mathbf{r}$ is equal to the jth column of A times $\mathbf{r}$, the partial derivative of $\mathbf{r}^T\mathbf{r}$ with respect to x_j is the *jth* component of the vector $-2A^T\mathbf{r}$. The l_2-norm of the residual is minimized at the point $\mathbf{x}$ where all the partial derivatives vanish, i.e.,

$$\frac{\partial}{\partial x_1}\mathbf{r^T r} = \frac{\partial}{\partial x_2}\mathbf{r^T r} = \cdots = \frac{\partial}{\partial x_n}\mathbf{r^T r} = 0. \quad (5.92)$$

Since each of these partial derivatives is -2 times the corresponding component of $A^T\mathbf{r}$, we conclude that

$$A^T\mathbf{r} = 0. \quad (5.93)$$

Replacing $\mathbf{r}$ by $\mathbf{b} - A\mathbf{x}$ gives

$$A^T(\mathbf{b} - A\hat{\mathbf{x}}) = 0 \qquad (5.94)$$

or

$$A^T A\hat{\mathbf{x}} = A^T\mathbf{b}, \qquad (5.95)$$

which is called the *normal equation*.

Any $\hat{\mathbf{x}}$ that minimizes the l_2-norm of the residual $\mathbf{r} = \mathbf{b} - A\hat{\mathbf{x}}$ is a solution to the normal equation (5.95). Conversely, any solution to the normal equation (5.95) is a least squares solution to the overdetermined linear system.

Example 5.31 *Solve the following overdetermined linear system of three equations in two unknowns:*

$$\begin{array}{rrrrrl}
2x_1 & + & 5x_2 & + & x_3 & = & 1 \\
3x_1 & - & 4x_2 & + & 2x_3 & = & 3 \\
4x_1 & + & 3x_2 & + & 3x_3 & = & 5 \\
5x_1 & - & 2x_2 & + & 4x_3 & = & 7.
\end{array}$$

Solution. *The matrix form of the given system is*

$$\begin{pmatrix} 2 & 5 & 1 \\ 3 & -4 & 2 \\ 4 & 3 & 3 \\ 5 & -2 & 4 \end{pmatrix} \begin{pmatrix} x_1 \\ x_2 \end{pmatrix} = \begin{pmatrix} 1 \\ 3 \\ 5 \\ 7 \end{pmatrix}.$$

Then using the normal equation (5.95), we obtain

$$\begin{pmatrix} 2 & 3 & 4 & 5 \\ 5 & -4 & 3 & -2 \\ 1 & 2 & 3 & 4 \end{pmatrix} \begin{pmatrix} 2 & 5 & 1 \\ 3 & -4 & 2 \\ 4 & 3 & 3 \\ 5 & -2 & 4 \end{pmatrix} \begin{pmatrix} x_1 \\ x_2 \end{pmatrix} = \begin{pmatrix} 2 & 3 & 4 & 5 \\ 5 & -4 & 3 & -2 \\ 1 & 2 & 3 & 4 \end{pmatrix} \begin{pmatrix} 1 \\ 3 \\ 5 \\ 7 \end{pmatrix},$$

which reduces the given system as

$$\begin{pmatrix} 54 & 0 & 40 \\ 0 & 54 & -2 \\ 40 & -2 & 30 \end{pmatrix} \begin{pmatrix} x_1 \\ x_2 \end{pmatrix} = \begin{pmatrix} 66 \\ -6 \\ 50 \end{pmatrix}.$$

Solving this simultaneous linear system, the values of unknowns are

$$x_1 = -1.00, \quad x_2 = 0.00, \quad x_3 = 3.00,$$

and they are the least squares solution of the given overdetermined system.

●

Using the MATLAB Command Window, the above result can be reproduced as follows:

```
>> A = [2 5 1; 3 −4 2; 4 3 3; 5 −2 4];
>> b = [1; 3; 5; 7];
>> x = overD(A, b);
```

Program 5.12
MATLAB m-file for the Overdetermined Linear System
function sol=overD(A,b)
$x = (A' * A) \backslash (A' * b);$ % Solve the normal equations
sol=x;

5.3.7 Least Squares Solution of an Underdetermined System

We consider again such linear systems where the coefficient matrix is **rectangular** (number of rows and columns are not equal). If A has m rows and n columns, then **x** is a vector with n components and **b** is a vector with m components. If the number of rows is smaller than the number of columns ($m < n$), then the linear system is called an *underdetermined system*. Typically, an underdetermined system has infinitely many solutions. This type of system generally arises in optimization theory and in economic modeling.

In general, the coefficient in row i and column j for the matrix AA^T is the dot product between row i and row j from A.

Notice that the coefficient matrix AA^T is symmetric, so when forming the matrix AA^T, we just evaluate the coefficients that are on the diagonal or above the diagonal, whereas the coefficients below the diagonal are determined from the symmetry property. Consider the equation

$$4x_1 + 3x_2 = 15. \tag{5.96}$$

We want to find the least squares solution to (5.96). The set of all points (x_1, x_2) that satisfy (5.96) forms a line with slope $-\frac{4}{3}$ and the distance from the origin to the point (x_1, x_2) is $(x_1^2 + x_2^2)^{\frac{1}{2}}$.

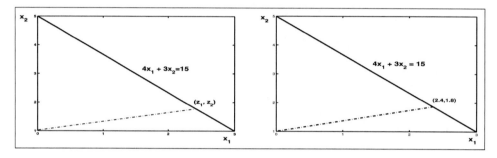

Figure 5.14: Least squares solution of underdetermined system.

To find the least squares solution to (5.96), we choose the point (x_1, x_2) that is as close to the origin as possible. The point (z_1, z_2) in Figure 5.14, which is closest to the origin, is the least squares solution to (5.96). We see in Figure 5.14 that the vector from the origin to (z_1, z_2) is orthogonal to the line $4x_1 + 3x_2 = 15$.

The collection of points forming this perpendicular have the form

$$t \begin{pmatrix} 4 \\ 3 \end{pmatrix},$$

where t is an arbitrary scalar. Since (z_1, z_2) lies on this perpendicular, there exists a value of t such that

$$\begin{pmatrix} z_1 \\ z_2 \end{pmatrix} = t \begin{pmatrix} 4 \\ 3 \end{pmatrix},$$

and this implies that
$$z_1 = 4t \quad \text{and} \quad z_2 = 3t.$$

Since $x_1 = z_1$ and $x_2 = z_2$ must satisfy (5.96), we have
$$4z_1 + 3z_2 = 15. \tag{5.97}$$

Substituting $z_1 = 4t$ and $z_2 = 3t$ into (5.97) gives
$$25t = 15 \quad \text{or} \quad t = 3/5 = 0.6.$$

Thus, the least squares solution to (5.96) is
$$\begin{pmatrix} z_1 \\ z_2 \end{pmatrix} = 0.6 \begin{pmatrix} 4 \\ 3 \end{pmatrix} = \begin{pmatrix} 2.4 \\ 1.8 \end{pmatrix}.$$

Now, let us consider a general underdetermined linear system $A\mathbf{x} = \mathbf{b}$ and suppose that $\mathbf{p}$ is any solution to the linear system and $\mathbf{q}$ is any vector for which
$$A\mathbf{q} = \mathbf{0}.$$

Since
$$A(\mathbf{p} + q) = A\mathbf{p} + A\mathbf{q} = A\mathbf{p} = \mathbf{b},$$

we see that $\mathbf{p} + q$ is a solution to $A\mathbf{x} = \mathbf{b}$, whenever $A\mathbf{q} = \mathbf{0}$. Conversely, it is also true because, if $A\mathbf{x} = \mathbf{b}$, then
$$\mathbf{x} = \mathbf{p} + (\mathbf{x} - p) = \mathbf{p} + \mathbf{q},$$

where $\mathbf{q} = \mathbf{x} - p$.

Since
$$A(\mathbf{x} - p) = \mathbf{b} - \mathbf{b} = \mathbf{0},$$

then $\mathbf{x}$ can be expressed as $\mathbf{x} = \mathbf{p} + \mathbf{q}$ and $A\mathbf{z} = \mathbf{0}$.

The set of all $\mathbf{q}$ such that $A\mathbf{q} = \mathbf{0}$ is called the *null space* of A (*kernel of A*) letting
$$N = \{\mathbf{q} : A\mathbf{q} = \mathbf{0}\}.$$

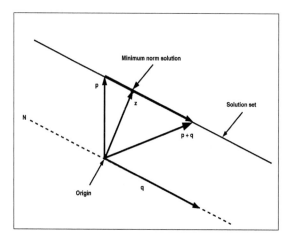

Figure 5.15: Least squares solution of underdetermined system.

Any solutions $\mathbf{x}$ of the underdetermined linear system $A\mathbf{x} = \mathbf{b}$ are sketched in Figure 5.15, for $\mathbf{x} = \mathbf{p} + q$ and $\mathbf{q} \in N$.

From Figure 5.15, the solution closest to the origin is perpendicular to N.

In linear algebra, the set of vectors perpendicular to the null space of A are linear combinations of the rows of A, so if

$$
A = \begin{pmatrix}
a_{11} & a_{12} & \cdots & a_{1n} \\
a_{21} & a_{22} & \cdots & a_{2n} \\
\vdots & \vdots & \vdots & \vdots \\
a_{m1} & a_{m2} & \cdots & a_{mn}
\end{pmatrix}, \quad m < n,
$$

then

$$
\mathbf{z} = \begin{pmatrix} z_1 \\ z_2 \\ \vdots \\ z_n \end{pmatrix} = t_1 \begin{pmatrix} a_{11} \\ a_{12} \\ \vdots \\ a_{1n} \end{pmatrix} + t_2 \begin{pmatrix} a_{21} \\ a_{22} \\ \vdots \\ a_{2n} \end{pmatrix} + \cdots + t_m \begin{pmatrix} a_{m1} \\ a_{m2} \\ \vdots \\ a_{mn} \end{pmatrix}
$$

or

$$\mathbf{s} = \begin{pmatrix} a_{11}t_1 & + & a_{21}t_2 & + & \cdots & + & a_{m1}t_m \\ a_{12}t_1 & + & a_{22}t_2 & + & \cdots & + & a_{m2}t_m \\ \vdots & & \vdots & & \vdots & & \vdots \\ a_{1n}t_1 & + & a_{2n}t_2 & + & \cdots & + & a_{mn}t_m \end{pmatrix} = \begin{pmatrix} a_{11} & a_{21} & \cdots & a_{m1} \\ a_{12} & a_{22} & \cdots & a_{m2} \\ \vdots & \vdots & \vdots & \vdots \\ a_{1n} & a_{2n} & \cdots & a_{mn} \end{pmatrix} \begin{pmatrix} t_1 \\ t_2 \\ \vdots \\ t_m \end{pmatrix}.$$

So,

$$\mathbf{z} = A^T \mathbf{t},$$

where

$$\mathbf{t} = \begin{pmatrix} t_1 \\ t_2 \\ \vdots \\ t_m \end{pmatrix}.$$

Substituting $\mathbf{x} = \mathbf{z} = A^T\mathbf{t}$ into the linear system $A\mathbf{x} = \mathbf{b}$, we have

$$AA^T\mathbf{t} = \mathbf{b}, \tag{5.98}$$

and solving this equation yields $\mathbf{t}$, i.e.,

$$\mathbf{t} = (AA^T)^{-1}\mathbf{b},$$

while the least squares solution $\mathbf{s}$ to the underdetermined system is

$$\mathbf{z} = A^T\mathbf{t} = A^T(AA^T)^{-1}\mathbf{b}. \tag{5.99}$$

Now solving the underdetermined equation (5.96),

$$(\begin{matrix} 4 & 3 \end{matrix}) \begin{pmatrix} x_1 \\ x_2 \end{pmatrix} = 15,$$

we first use (5.98) as follows:

$$(\begin{matrix} 4 & 3 \end{matrix}) \begin{pmatrix} 4 \\ 3 \end{pmatrix} t = 15,$$

which gives $t = \frac{15}{25} = 0.6$. Now using (5.99), we have

$$\mathbf{z} = A^T\mathbf{t} = \begin{pmatrix} 4 \\ 3 \end{pmatrix} (0.6) = \begin{pmatrix} 2.4 \\ 1.8 \end{pmatrix},$$

the required least squares solution of the given underdetermined equation.

●

Example 5.32 *Solve the following underdetermined linear system of two equations in three unknowns:*

$$
\begin{array}{rrrrr}
x_1 & + & 2x_2 & - & 3x_3 & = & 42 \\
5x_1 & - & x_2 & + & x_3 & = & 54.
\end{array}
$$

Solution. *The matrix form of the given system is*

$$
\begin{pmatrix} 1 & 2 & -3 \\ 5 & -1 & 1 \end{pmatrix} \begin{pmatrix} x_1 \\ x_2 \\ x_3 \end{pmatrix} = \begin{pmatrix} 42 \\ 54 \end{pmatrix}.
$$

Then using the normal equation (5.99)

$$
AA^T \mathbf{t} = \mathbf{b},
$$

we obtain

$$
\begin{pmatrix} 1 & 2 & -3 \\ 5 & -1 & 1 \end{pmatrix} \begin{pmatrix} 1 & 5 \\ 2 & -1 \\ -3 & 1 \end{pmatrix} \begin{pmatrix} t_1 \\ t_2 \end{pmatrix} = \begin{pmatrix} 42 \\ 54 \end{pmatrix},
$$

which reduces the given system to

$$
\begin{pmatrix} 14 & 0 \\ 0 & 27 \end{pmatrix} \begin{pmatrix} t_1 \\ t_2 \end{pmatrix} = \begin{pmatrix} 42 \\ 54 \end{pmatrix}.
$$

Solving the above linear system, the values of unknowns are

$$
t_1 = 3 \quad t_2 = 2.
$$

Since the best least squares solution $\mathbf{z}$ *to the given linear system is* $\mathbf{z} = A^T \mathbf{t}$, *i.e.,*

$$
\begin{pmatrix} z_1 \\ z_2 \\ z_3 \end{pmatrix} = \begin{pmatrix} 1 & 5 \\ 2 & -1 \\ -3 & 1 \end{pmatrix} \begin{pmatrix} 3 \\ 2 \end{pmatrix} = \begin{pmatrix} 13 \\ 4 \\ -7 \end{pmatrix},
$$

it is called the least squares solution of the given underdetermined system.

●

Using the MATLAB Command Window, the above result can be reproduced as:

```
>> A = [1 2 − 3; 5 − 1 1];
>> b = [42; 54];
>> x = underD(A, b);
```

Program 5.13

MATLAB m-file for the Underdetermined Linear System
function sol=underD(A,b)
$t = (A * A') \backslash (b);$ % Solve the normal equations
$x = A' * t;$ % Solve the normal equations
sol=x;

5.3.8 The Pseudoinverse of a Matrix

If A is an $n \times n$ matrix with linearly independent columns, then it is invertible, and the unique solution to the linear system $A\mathbf{x} = \mathbf{b}$ is $\mathbf{x} = A^{-1}\mathbf{b}$. If $m > n$ and A is an $m \times n$ matrix with linearly independent columns, then a system $A\mathbf{x} = \mathbf{b}$ has no exact solution, but the best approximation is given by the unique least squares solution $\hat{\mathbf{x}} = (A^T A)^{-1} A^T \mathbf{b}$. The matrix $(A^T A)^{-1} A^T$, therefore, plays the role of an inverse of A in this situation.

Definition 5.1 (Pseudoinverse of a Matrix)

If A is a matrix with linearly independent columns, then the pseudoinverse of a matrix is the matrix A^+ defined by

$$A^+ = (A^T A)^{-1} A^T. \tag{5.100}$$

For example, consider the matrix

$$A = \begin{pmatrix} 1 & 2 \\ 2 & 3 \\ 3 & 4 \end{pmatrix},$$

then we have

$$A^T A = \begin{pmatrix} 1 & 2 & 3 \\ 2 & 3 & 4 \end{pmatrix} \begin{pmatrix} 1 & 2 \\ 2 & 3 \\ 3 & 4 \end{pmatrix} = \begin{pmatrix} 14 & 20 \\ 20 & 29 \end{pmatrix},$$

and its inverse will be of the form

$$(A^T A)^{-1} = \begin{pmatrix} \dfrac{29}{6} & -\dfrac{10}{3} \\[2ex] -\dfrac{10}{3} & \dfrac{7}{3} \end{pmatrix}.$$

Thus, the pseudoinverse of the matrix is

$$(A^T A)^{-1} A^T = A^+ = \begin{pmatrix} \dfrac{29}{6} & -\dfrac{10}{3} \\[2ex] -\dfrac{10}{3} & \dfrac{7}{3} \end{pmatrix} \begin{pmatrix} 1 & 2 & 3 \\ 2 & 3 & 4 \end{pmatrix}$$

$$= \begin{pmatrix} -\dfrac{11}{6} & -\dfrac{1}{3} & \dfrac{7}{6} \\[2ex] \dfrac{4}{3} & \dfrac{1}{3} & -\dfrac{2}{3} \end{pmatrix}.$$

●

The pseudoinverse of a matrix can be obtained using the MATLAB Command Window as follows:

```
>> A = [1 2; 2 3; 3 4];
>> pinv(A);
```

Note that if A is a square matrix, then $A^+ = A^{-1}$ and in such a case, the least squares solution of a linear system $A\mathbf{x} = \mathbf{b}$ is the exact solution, since

$$\hat{\mathbf{x}} = A^+ \mathbf{b} = A^{-1} \mathbf{b}.$$

Example 5.33 *Find the pseudoinverse of the matrix of the following linear system, and then use it to compute the least squares solution of the system:*

$$\begin{aligned} x_1 &+ 2x_2 = 3 \\ 2x_1 &- 3x_2 = 4. \end{aligned}$$

Solution. *The matrix form of the given system is*

$$\begin{pmatrix} 1 & 2 \\ 2 & -3 \end{pmatrix} \begin{pmatrix} x_1 \\ x_2 \end{pmatrix} = \begin{pmatrix} 3 \\ 4 \end{pmatrix},$$

and so

$$A^T A = \begin{pmatrix} 1 & 2 \\ 2 & -3 \end{pmatrix} \begin{pmatrix} 1 & 2 \\ 2 & -3 \end{pmatrix} = \begin{pmatrix} 5 & -4 \\ -4 & 13 \end{pmatrix}.$$

The inverse of the matrix $A^T A$ can be computed as

$$(A^T A)^{-1} = \begin{pmatrix} \dfrac{13}{49} & \dfrac{4}{49} \\[2mm] \dfrac{4}{49} & \dfrac{5}{49} \end{pmatrix}.$$

The pseudoinverse of the matrix of the given system is

$$(A^T A)^{-1} A^T = A^+ = \begin{pmatrix} \dfrac{13}{49} & \dfrac{4}{49} \\[2mm] \dfrac{4}{49} & \dfrac{5}{49} \end{pmatrix} \begin{pmatrix} 1 & 2 \\ 2 & -3 \end{pmatrix} = \begin{pmatrix} \dfrac{3}{7} & \dfrac{2}{7} \\[2mm] \dfrac{2}{7} & -\dfrac{1}{7} \end{pmatrix}.$$

Now we compute the least squares solution of the system as

$$\hat{x} = A^+ b = \begin{pmatrix} \dfrac{3}{7} & \dfrac{2}{7} \\[2mm] \dfrac{2}{7} & -\dfrac{1}{7} \end{pmatrix} \begin{pmatrix} 3 \\ 4 \end{pmatrix},$$

which gives

$$\hat{x} = \begin{pmatrix} \dfrac{17}{7} \\[2mm] \dfrac{2}{7} \end{pmatrix},$$

and this is the least squares solution of the given system.   •

The least squares solution to the linear system by the pseudoinverse of a matrix can be obtained using the MATLAB Command Window as follows:

```
>> A = [1 2; 2 − 3];
>> b = [3 4]';
>> x = pinv(A) * b;
```

Theorem 5.10 *Let A be a matrix with linearly independent columns, then A^+ of A satisfies:*

1. $AA^+A = A$.

2. $A^+AA^+ = A^+$.

3. $(A^T)^+ = (A^+)^T$.

                                                                    •

Theorem 5.11 *If A is a square matrix with linearly independent columns, then:*

1. $A^+ = A^{-1}$.

2. $(A^+)^+ = A$.

3. $(A^T)^+ = (A^+)^T$.

                                                                    •

5.3.9   Least Squares with QR Decomposition

The least squares solutions discussed previously suffer from a frequent problem. The matrix $A^T A$ of the normal equation is usually ill-conditioned, therefore, a small numerical error in performing the Gauss elimination will result in a large error in the least squares.

Usually, the Gauss elimination for $A^T A$ of size $n \geq 5$ does not yield any good approximate solutions. It turns out that the QR decomposition of A (discussed in Chapter 4) yields a more reliable way of computing the least squares approximation of linear system $A\mathbf{x} = \mathbf{b}$. The idea behind this approach is that because orthogonal matrices preserve length, they should preserve the length of the error as well.

Let A have linearly independent columns and let $A = QR$ be a QR decomposition. In this decomposition, we express a matrix as the product of an orthogonal matrix Q and an upper triangular matrix R.

For $\hat{\mathbf{x}}$, a least squares solution of $A\mathbf{x} = \mathbf{b}$, we have

$$\begin{aligned} A^T A\hat{\mathbf{x}} &= A^T \mathbf{b} \\ (QR)^T (QR)\hat{\mathbf{x}} &= (QR)^T \mathbf{b} \\ R^T Q^T QR\hat{\mathbf{x}} &= R^T Q^T \mathbf{b} \\ R^T R\hat{\mathbf{x}} &= R^T Q^T \mathbf{b} \quad (\text{because } Q^T Q = \mathbf{I}). \end{aligned}$$

Since R is invertible, so is R^T, and hence

$$R\hat{\mathbf{x}} = Q^T \mathbf{b},$$

or equivalently,

$$\hat{\mathbf{x}} = R^{-1} Q^T \mathbf{b}.$$

Since R is an upper triangular matrix, in practice it is easier to solve $R\hat{\mathbf{x}} = Q^T \mathbf{b}$ directly (using backward substitution) than to invert R and compute $R^{-1} Q^T \mathbf{b}$.

Theorem 5.12 *If A is an $m \times n$ matrix with linearly independent columns and if $A = QR$ is a QR decomposition, then the unique least squares solutions $\hat{\mathbf{x}}$ of $A\mathbf{x} = \mathbf{b}$ is, theoretically, given by*

$$\hat{\mathbf{x}} = R^{-1} Q^T \mathbf{b}, \tag{5.101}$$

and it is usually computed by solving the system

$$R\hat{\mathbf{x}} = Q^T \mathbf{b}. \tag{5.102}$$

Example 5.34 *A QR decomposition of A is given. Use it to find a least squares solution of the linear system* $A\mathbf{x} = \mathbf{b}$, *where*

$$A = \begin{pmatrix} 2 & 2 & 6 \\ 1 & 4 & -3 \\ 2 & -4 & 9 \end{pmatrix}, \quad \mathbf{b} = \begin{pmatrix} 1 \\ -1 \\ 4 \end{pmatrix}$$

and

$$Q = \begin{pmatrix} \dfrac{2}{3} & \dfrac{1}{3} & -\dfrac{2}{3} \\[2mm] \dfrac{1}{3} & \dfrac{2}{3} & \dfrac{2}{3} \\[2mm] -\dfrac{2}{3} & -\dfrac{2}{3} & \dfrac{1}{3} \end{pmatrix}, \quad R = \begin{pmatrix} 3 & 0 & 9 \\ 0 & 6 & -6 \\ 0 & 0 & -3 \end{pmatrix}.$$

Solution. *For the right-hand side of (5.102), we obtain*

$$Q^T\mathbf{b} = \begin{pmatrix} \dfrac{2}{3} & \dfrac{1}{3} & -\dfrac{2}{3} \\[2mm] \dfrac{1}{3} & \dfrac{2}{3} & -\dfrac{2}{3} \\[2mm] -\dfrac{2}{3} & \dfrac{2}{3} & \dfrac{1}{3} \end{pmatrix} \begin{pmatrix} 1 \\ -1 \\ 4 \end{pmatrix} = \begin{pmatrix} 3 \\ -3 \\ 0 \end{pmatrix}.$$

Hence, (5.102) can be written as

$$\begin{pmatrix} 3 & 0 & 9 \\ 0 & 6 & -6 \\ 0 & 0 & -3 \end{pmatrix} \begin{pmatrix} x_1 \\ x_2 \\ x_3 \end{pmatrix} = \begin{pmatrix} 3 \\ -3 \\ 0 \end{pmatrix}$$

or

$$\begin{aligned} 3x_1 \quad\quad + \;\; 9x_3 &= \;\; 3 \\ 6x_2 \;-\;\; 6x_3 &= -3 \\ -\;\; 3x_3 &= \;\; 0. \end{aligned}$$

Now using backward substitution, we obtain

$$\hat{\mathbf{x}} = [x_1, x_2, x_3]^T = [1, -\frac{1}{2}, 0]^T,$$

which is called the least squares solution of the given system. •

So we conclude that
$$R\hat{\mathbf{x}} = Q^T\mathbf{b}$$

must be satisfied by the solution of $A^T A\hat{\mathbf{x}} = A^T\mathbf{b}$, but because, in general, R is not an even square, we cannot use multiplication by $(R^T)^{-1}$ to arrive at this conclusion. In fact, it is not true, in general, that the solution of

$$R\hat{\mathbf{x}} = Q^T\mathbf{b}$$

even exists; after all, $A\mathbf{x} = \mathbf{b}$ is equivalent to $QR\mathbf{x} = \mathbf{b}$, i.e., to $R\mathbf{x} = Q^T\mathbf{b}$, so $R\mathbf{x} = Q^T\mathbf{b}$ can have an actual solution $\mathbf{x}$ only if $A\mathbf{x} = \mathbf{b}$ does. However, we are getting close to finding the least squares solution. Here, we need to find a way to simplify the expression

$$R^T R\hat{\mathbf{x}} = R^T Q^T\mathbf{b}. \tag{5.103}$$

The matrix R is upper triangular, and because we have restricted ourselves to the case where $m \geq n$, we may write the $m \times n$ matrix R as

$$R = \begin{pmatrix} R_1 \\ 0 \end{pmatrix}, \tag{5.104}$$

in partitioned (block) form, where R_1 is an upper triangular $n \times n$ matrix and 0 represents an $(m - n) \times n$ zero matrix. Since $rank(R) = n$, R_1 is nonsingular. Hence, every diagonal element of R_1 must be nonzero. Now we may rewrite

$$R^T R\hat{\mathbf{x}} = R^T Q^T\mathbf{b}$$

as

$$\begin{pmatrix} R_1 \\ 0 \end{pmatrix}^T \begin{pmatrix} R_1 \\ 0 \end{pmatrix} \hat{\mathbf{x}} = \begin{pmatrix} R_1 \\ 0 \end{pmatrix}^T Q^T\mathbf{b}$$

$$\begin{pmatrix} R_1^T & 0^T \end{pmatrix} \begin{pmatrix} R_1 \\ 0 \end{pmatrix} \hat{\mathbf{x}} = \begin{pmatrix} R_1^T & 0^T \end{pmatrix} Q^T\mathbf{b}$$

$$R_1^T R_1\hat{\mathbf{x}} = \begin{pmatrix} R_1^T & 0^T \end{pmatrix} (Q^T\mathbf{b}).$$

Note that multiplying by the block 0^T (an $n \times (m-n)$ zero matrix) on the right-hand side simply means that the last $(m - n)$ components of $Q^T\mathbf{b}$ do not affect the computation. Since R_1 is nonsingular, then we have

$$R_1 \hat{\mathbf{x}} = (R_1^T)^{-1} \begin{pmatrix} R_1^T & 0^T \end{pmatrix} (Q^T \mathbf{b})$$

$$= \begin{pmatrix} \mathbf{I}_n & 0^T \end{pmatrix} (Q^T \mathbf{b}).$$

The left-hand side, $R_1 \hat{\mathbf{x}}$, is $(n \times n) \times (n \times 1) \longrightarrow n \times 1$, and the right-hand side is $(n \times (n + (m-n))) \times (m \times m) \times (m \times 1) \longrightarrow n \times 1$. If we define the vector $\mathbf{q}$ to be equal to the first n components of $Q^T \mathbf{b}$, then this becomes

$$R_1 \hat{\mathbf{x}} = \mathbf{q}, \qquad (5.105)$$

which is a square linear system involving a nonsingular upper triangular $n \times n$ matrix. So (5.105) is called the *least squares solution* of the overdetermined system $A\mathbf{x} = \mathbf{b}$, with QR decomposition by backward substitution, where $A = QR$ is the QR decomposition of A and $\mathbf{q}$ is essentially $Q^T \mathbf{b}$.

Note that the last $(m - n)$ columns of Q are not needed to solve the least squares solution of the linear system with QR decomposition. The block-matrix representation of Q corresponding to R (by (5.104)) is

$$Q = [Q_1, Q_2],$$

where Q_1 is the matrix composed of the first m columns of Q and Q_2 is a matrix composed of the remaining columns of Q. Note that only the first n columns of Q are needed to create A using the coefficients in R, and we can save effort and memory in the process of creating the QR decomposition. The so-called short QR decomposition of A is

$$A = Q_1 R_1. \qquad (5.106)$$

The only difference between the full QR decomposition and the short decomposition is that the full QR decomposition contains the additional $(m - n)$ columns of Q.

Example 5.35 *Find the least squares solution of the following linear system* $A\mathbf{x} = \mathbf{b}$ *using* QR *decomposition, where*

$$A = \begin{pmatrix} 2 & 1 \\ 1 & 0 \\ 3 & 1 \end{pmatrix}, \quad \mathbf{x} = \begin{pmatrix} x_1 \\ x_2 \end{pmatrix}, \quad \mathbf{b} = \begin{pmatrix} 1.9 \\ 0.9 \\ 2.8 \end{pmatrix}.$$

Solution. *First, we find the QR decomposition, and we will get*

$$
Q = \begin{pmatrix} -0.5345 & 0.6172 & -0.5774 \\ -0.2673 & -0.7715 & -0.5774 \\ -0.8018 & -0.1543 & 0.5774 \end{pmatrix} \quad and \quad R = \begin{pmatrix} -3.7417 & -1.3363 \\ 0 & 0.4629 \\ 0 & 0 \end{pmatrix}
$$

and

$$
Q^T \mathbf{b} = \begin{pmatrix} -0.5345 & -0.2673 & -0.8018 \\ 0.6172 & -0.7715 & -0.1543 \\ -0.5774 & -0.5774 & 0.5774 \end{pmatrix} \begin{pmatrix} 1.9 \\ 0.9 \\ 2.8 \end{pmatrix} = \begin{pmatrix} -3.5011 \\ 0.0463 \\ 0.0000 \end{pmatrix},
$$

so that

$$
R_1 = \begin{pmatrix} -3.7417 & -1.3363 \\ 0 & 0.4629 \end{pmatrix} \quad and \quad \mathbf{q} = \begin{pmatrix} -3.5011 \\ 0.0463 \end{pmatrix}.
$$

Hence, we must solve (5.105), i.e.,

$$
R_1 \hat{\mathbf{x}} = \mathbf{q},
$$

i.e.,

$$
\begin{pmatrix} -3.7417 & -1.3363 \\ 0 & 0.4629 \end{pmatrix} \begin{pmatrix} x_1 \\ x_2 \end{pmatrix} = \begin{pmatrix} -3.5011 \\ 0.0463 \end{pmatrix}.
$$

Using backward substitution, we obtain

$$
\hat{\mathbf{x}} = [x_1, x_2]^T = [0.9000, 0.1000]^T,
$$

the least squares solution of the given system. ●

The MATLAB built-in function **qr** returns the QR decomposition of a matrix. There are two ways of calling **qr**, which are

```
>> [Q, R] = qr(A);
>> [Q_1, R_1] = qr(A, 0);
```

where Q and Q_1 are orthogonal matrices and R and R_1 are upper triangular matrices. The above first form returns the full QR decomposition (i.e., if A is $(m \times n)$, then Q is $(m \times m)$ and R is $(m \times n)$). The second form returns the short QR decomposition, where Q_1 and R_1 are the matrices in (5.106).

In Example 5.35, we apply the full QR decomposition of A using the first form of the built-in function **qr** as

```
>> A = [2 1; 1 0; 3 1];
>> [Q, R] = qr(A);
```

The short QR decomposition of A can be obtained by using the second form of the built-in function **qr** as

```
>> A = [2 1; 1 0; 3 1];
>> [Q_1, R_1] = qr(A, 0);
```

$$Q_1 =$$
$$\begin{matrix} -0.5345 & 0.6172 \\ -0.2673 & -0.7715 \\ -0.8018 & -0.1543 \end{matrix}$$
$$R_1 =$$
$$\begin{matrix} -3.7417 & -1.3363 \\ 0 & 0.4629 \end{matrix}$$

As expected, Q_1 and the first two columns of Q are identical, as are R_1 and the first two rows of R. The short QR decomposition of A possesses all the necessary information in the columns of Q_1 and R_1 to reconstruct A.

5.3.10 Least Squares with Singular Value Decomposition

One of the advantages of the Singular Value Decomposition (SVD) method is that we can efficiently compute the least squares solution. Consider the problem of finding the least squares solution of the overdetermined linear system $A\mathbf{x} = \mathbf{b}$. We discussed previously that the least squares solution of $A\mathbf{x} = \mathbf{b}$ is the solution of $A^T A\hat{\mathbf{x}} = A^T\mathbf{b}$, i.e., the solution of

$$(UDV^T)^T UDV^T\hat{\mathbf{x}} = (UDV^T)^T\mathbf{b}$$

$$
\begin{aligned}
VD^T U^T U D V^T \hat{\mathbf{x}} &= VD^T U^T \mathbf{b} \\
VD^T D V^T \hat{\mathbf{x}} &= VD^T U^T \mathbf{b} \\
V^T V D^T D V^T \hat{\mathbf{x}} &= V^T V D^T U^T \mathbf{b} \\
D^T D V^T \hat{\mathbf{x}} &= D^T U^T \mathbf{b} \\
D V^T \hat{\mathbf{x}} &= U^T \mathbf{b} \\
V^T \hat{\mathbf{x}} &= D^{-1} U^T \mathbf{b} \\
\hat{\mathbf{x}} &= V D^{-1} U^T \mathbf{b}.
\end{aligned}
$$

This is the same formal solution that we found for the linear system $A\mathbf{x} = \mathbf{b}$ (see Chapter 6), but recall that A is no longer a square matrix.

Note that in exact arithmetic, the solution to a least squares problem via normal equations QR and SVD is exactly the same. The main difference between these two approaches is the numerical stability of the methods. To find the least squares solution of the overdetermined linear system with SVD, we will find D_1 as

$$
D = \begin{pmatrix} D_1 \\ 0 \end{pmatrix}
$$

in partitioned (block) form, where D_1 is an $n \times n$ matrix and 0 represents an $(m-n) \times n$ zero matrix. If we define the right-hand vector $\mathbf{q}$ to be equal to the first n components of $U^T \mathbf{b}$, then the least squares solution of the overdetermined linear system is to solve the system

$$
D_1 V^T \hat{\mathbf{x}} = \mathbf{q} \tag{5.107}
$$

or

$$
\hat{\mathbf{x}} = V D_1^{-1} \mathbf{q}. \tag{5.108}
$$

Example 5.36 *Find the least squares solution of the following linear system $A\mathbf{x} = \mathbf{b}$ using SVD, where*

$$
A = \begin{pmatrix} 1 & 1 \\ 0 & 1 \\ 1 & 0 \end{pmatrix}, \quad \mathbf{x} = \begin{pmatrix} x_1 \\ x_2 \end{pmatrix}, \quad \mathbf{b} = \begin{pmatrix} 1 \\ 1 \\ 1 \end{pmatrix}.
$$

Solution. *First, we find the SVD of the given matrix. The first step is to find the eigenvalues of the following matrix:*

$$A^T A = \begin{pmatrix} 1 & 0 & 1 \\ 1 & 1 & 0 \end{pmatrix} \begin{pmatrix} 1 & 1 \\ 0 & 1 \\ 1 & 0 \end{pmatrix} = \begin{pmatrix} 2 & 1 \\ 1 & 2 \end{pmatrix}.$$

The characteristic polynomial of $A^T A$ is

$$p(\lambda) = \lambda^2 - 4\lambda + 3 = (\lambda - 3)(\lambda - 1) = 0,$$

which gives

$$\lambda_1 = 3, \quad \lambda_2 = 1,$$

and the eigenvalues of $A^T A$ and the corresponding eigenvectors are

$$\begin{pmatrix} 1 \\ 1 \end{pmatrix} \quad and \quad \begin{pmatrix} -1 \\ 1 \end{pmatrix}.$$

These vectors are orthogonal, so we normalize them to obtain

$$\mathbf{v}_1 = \begin{pmatrix} \dfrac{\sqrt{2}}{2} \\ \dfrac{\sqrt{2}}{2} \end{pmatrix} \quad and \quad \mathbf{v}_2 = \begin{pmatrix} -\dfrac{\sqrt{2}}{2} \\ \dfrac{\sqrt{2}}{2} \end{pmatrix}.$$

The singular values of A are

$$\sigma_1 = \sqrt{\lambda_1} = \sqrt{3} \quad and \quad \sigma_2 = \sqrt{\lambda_2} = \sqrt{1} = 1.$$

Thus,

$$V = \begin{pmatrix} \dfrac{\sqrt{2}}{2} & -\dfrac{\sqrt{2}}{2} \\ \dfrac{\sqrt{2}}{2} & \dfrac{\sqrt{2}}{2} \end{pmatrix} = \begin{pmatrix} 0.7071 & -0.7071 \\ 0.7071 & 0.7071 \end{pmatrix}$$

and

$$D = \begin{pmatrix} \sqrt{3} & 0 \\ 0 & 1 \\ 0 & 0 \end{pmatrix} = \begin{pmatrix} 1.7321 & 0 \\ 0 & 1.0000 \\ 0 & 0 \end{pmatrix}.$$

To find U, we first compute

$$\mathbf{u}_1 = \frac{1}{\sigma_1} A \mathbf{v}_1 = \frac{\sqrt{3}}{3} \begin{pmatrix} 1 & 1 \\ 0 & 1 \\ 1 & 0 \end{pmatrix} \begin{pmatrix} \dfrac{\sqrt{2}}{2} \\[2mm] \dfrac{\sqrt{6}}{2} \end{pmatrix} = \begin{pmatrix} \dfrac{\sqrt{6}}{3} \\[2mm] \dfrac{\sqrt{6}}{6} \\[2mm] \dfrac{\sqrt{6}}{2} \end{pmatrix},$$

and similarly,

$$\mathbf{u}_2 = \frac{1}{\sigma_2} A \mathbf{v}_2 = \begin{pmatrix} 0 \\[2mm] \dfrac{\sqrt{2}}{2} \\[2mm] -\dfrac{\sqrt{2}}{2} \end{pmatrix}.$$

These are two of the three column vectors of U, and they already form an orthonormal basis for $\mathbb{R}^2$. Now to find the third column vector $\mathbf{u}_3$ of U, we will look for a unit vector $\mathbf{u}_3$ that is orthogonal to

$$\sqrt{6}\,\mathbf{u}_1 = \begin{pmatrix} 2 \\ 1 \\ 1 \end{pmatrix} \quad and \quad \sqrt{2}\,\mathbf{u}_2 = \begin{pmatrix} 0 \\ 1 \\ -1 \end{pmatrix}.$$

To satisfy these two orthogonality conditions, the vector $\mathbf{u}_3$ must be a solution of the homogeneous linear system

$$\begin{pmatrix} 2 & 1 & 1 \\ 0 & 1 & -1 \end{pmatrix} \begin{pmatrix} x_1 \\ x_2 \\ x_3 \end{pmatrix} = \begin{pmatrix} 0 \\ 0 \\ 0 \end{pmatrix},$$

which gives the general solution of the system

$$\begin{pmatrix} x_1 \\ x_2 \\ x_3 \end{pmatrix} = \alpha \begin{pmatrix} -1 \\ 1 \\ 1 \end{pmatrix}, \quad \alpha \in \mathbb{R}.$$

By normalizing the vector on the right-hand side, we get

$$
\mathbf{u}_3 = \begin{pmatrix} -\dfrac{1}{\sqrt{3}} \\[2mm] \dfrac{1}{\sqrt{3}} \\[2mm] -\dfrac{1}{\sqrt{3}} \end{pmatrix}.
$$

So we have

$$
U = \begin{pmatrix} \dfrac{\sqrt{6}}{3} & 0 & -\dfrac{1}{\sqrt{3}} \\[2mm] \dfrac{\sqrt{6}}{6} & \dfrac{\sqrt{2}}{2} & \dfrac{1}{\sqrt{3}} \\[2mm] \dfrac{\sqrt{6}}{2} & -\dfrac{\sqrt{2}}{2} & \dfrac{1}{\sqrt{3}} \end{pmatrix} = \begin{pmatrix} 0.8165 & 0.0000 & -0.5774 \\ 0.4082 & 0.7071 & 0.5774 \\ 0.4082 & -0.7071 & 0.5774 \end{pmatrix}.
$$

This yields the SVD

$$
A = \begin{pmatrix} \dfrac{\sqrt{6}}{3} & 0 & -\dfrac{1}{\sqrt{3}} \\[2mm] \dfrac{\sqrt{6}}{6} & -\dfrac{\sqrt{2}}{2} & \dfrac{1}{\sqrt{3}} \\[2mm] \dfrac{\sqrt{6}}{2} & \dfrac{\sqrt{2}}{2} & \dfrac{1}{\sqrt{3}} \end{pmatrix} \begin{pmatrix} \sqrt{3} & 0 \\ 0 & 1 \\ 0 & 0 \end{pmatrix} \begin{pmatrix} \dfrac{\sqrt{2}}{2} & \dfrac{\sqrt{2}}{2} \\[2mm] \dfrac{\sqrt{2}}{2} & -\dfrac{\sqrt{2}}{2} \end{pmatrix}.
$$

Hence,

$$
D_1 = \begin{pmatrix} 1.7321 & 0 \\ 0 & 1.0000 \end{pmatrix} \quad and \quad D_1^{-1} = \begin{pmatrix} 0.5774 & 0 \\ 0 & 1.0000 \end{pmatrix}.
$$

Also,

$$U^T \mathbf{b} = \begin{pmatrix} 0.8165 & 0.4082 & 0.4082 \\ 0.0000 & 0.7071 & -0.7071 \\ -0.5774 & 0.5774 & 0.5774 \end{pmatrix} \begin{pmatrix} 1 \\ 1 \\ 1 \end{pmatrix} = \begin{pmatrix} 1.6330 \\ 0.0000 \\ 0.5774 \end{pmatrix},$$

and from it, we obtain

$$\mathbf{q} = \begin{pmatrix} 1.6330 \\ 0.0000 \end{pmatrix}.$$

Thus, we must solve (5.108), i.e.,

$$\hat{\mathbf{x}} = V D_1^{-1} \mathbf{q},$$

which gives

$$\begin{pmatrix} x_1 \\ x_2 \end{pmatrix} = \begin{pmatrix} 0.7071 & -0.7071 \\ 0.7071 & 0.7071 \end{pmatrix} \begin{pmatrix} 0.5774 & 0 \\ 0 & 1.0000 \end{pmatrix} \begin{pmatrix} 1.6330 \\ 0.0000 \end{pmatrix} = \begin{pmatrix} 0.6667 \\ 0.6667 \end{pmatrix},$$

the least squares solution of the given system. ●

Like QR decomposition, the MATLAB built-in function **svd** returns the SVD of a matrix. There are two ways of calling **svd**:

```
>> [U, D, V] = svd(A);
>> [U_1, D_1, V] = svd(A, 0);
```

Here, A is any matrix, D is a diagonal matrix having singular values of A in the diagonal, and U and V are orthogonal matrices. The first form returns the full SVD and the second form returns the short SVD. The second decomposition is useful when A is an $m \times n$ matrix with $m > n$. The second form of SVD gives U_1, the first n columns of U, and square $(n \times n)$ D_1. When $m > n$, the full SVD of A gives a D matrix with only zeros in the last $(m - n)$ rows. Note that there is no change in V in both forms.

In Example 5.36, we apply the full SVD of A using the first form of the built-in function **svd**:

> >> $A = [1\ 1; 0\ 1; 1\ 0]$;
> >> $[U, D, V] = svd(A)$;

The short *SVD* of A can be obtained by using the second form of the built-in function **svd**:

> >> $A = [1\ 1; 0\ 1; 1\ 0]$;
> >> $[U_1, D_1, V] = svd(A, 0)$;

$$
U_1 =
$$
$$
\begin{array}{rr}
0.8165 & 0.0000 \\
0.4082 & 0.7071 \\
0.4082 & -0.7071
\end{array}
$$
$$
D_1 =
$$
$$
\begin{array}{rr}
1.7321 & 0 \\
0 & 1.0000
\end{array}
$$
$$
V =
$$
$$
\begin{array}{rr}
0.7071 & -0.7071 \\
0.7071 & 0.7071
\end{array}
$$

As expected, U_1 and the first two columns of U are identical, as are D_1 and the first two rows of D (no change in V in either form). The short *SVD* of A possesses all the necessary information in the columns of U_1 and D_1 (with V also) to reconstruct A.

Note that when m and n are similar in size, *SVD* is significantly more expensive to compute than QR decomposition. If m and n are equal, then solving a least squares problem by *SVD* is about an order of magnitude more costly than using QR decomposition. So for least squares problems it is generally advisable to use QR decomposition. When a least squares problem is known to be a difficult one, using *SVD* is probably justified.

5.4   Summary

In this chapter, we discussed the procedures for developing approximating polynomials for discrete data. First, we discussed the Lagrange and Newton divided differences polynomials and both yield the same interpolation for a given set of n data pairs $(x, f(x))$. The pairs are not required to be ordered, nor is the independent variable required to be equally spaced. The dependent variable is approximated as a single-valued function. The Lagrange polynomial works well for small data points. The Newton divided differences polynomial is generally more efficient than the Lagrange polynomial, and it can be adjusted easily for additional data. For efficient implementation of divided difference interpolation, we used Aitken's method, which is designed for the easy evaluation of the polynomial. We also discussed the Chebyshev polynomial interpolation of the function over the interval $[-1, 1]$. These types of polynomials are used to minimize approximation error.

Procedures for developing least squares approximation for discrete data were also discussed. Least squares approximations are useful for large sets of data and sets of rough data. Least squares polynomial approximation is straightforward for one independent variable and for two independent variables. The least squares normal equations corresponding to polynomials approximating functions are linear, which leads to very efficient solution procedures. For nonlinear approximating functions, the least squares normal equations are nonlinear, which leads to complicated solution procedures. We discussed the trigonometric least squares polynomial for approximating the given function. We also discussed the least squares solutions to overdetermined linear systems and the underdetermined linear systems. In the last section, we discussed the least squares solutions of linear systems with the pseudoinverse of matrices, QR decomposition, and SVD.

5.5 Problems

1. Use the Lagrange interpolation formula based on the points $x_0 = 0, x_1 = 1$ and $x_2 = 2$ to find the equation of the quadratic polynomial to approximate $f(x) = \frac{x+1}{x+5}$ at $x = 1.5$. Compute the absolute error.

2. Let $f(x) = x^2 \sin(\frac{x\pi}{4})$, where x is in radian. Find the quadratic Lagrange interpolation polynomial by using the best of the points $x_0 = 0, x_1 = 1, x_2 = 2$, and $x_3 = 4$ to find the approximation of the

function $f(x)$ at $x = 0.5$ and $x = 3.5$. Compute the error bounds for each case.

3. Use the quadratic Lagrange interpolation formula to show that $A + B = 1 - C$, such that $p_2(1.4) = Af(0) + Bf(1) + Cf(2)$.

4. Let $f(x) = x + 2\ln(x + 2)$. Use the quadratic Lagrange interpolation formula based on the points $x_0 = 0, x_1 = 1, x_2 = 2$, and $x_3 = 3$ to approximate $f(0.5)$ and $f(2.8)$. Also, compute the error bounds for your approximations.

5. Let $p_2(x)$ be the quadratic Lagrange interpolating polynomial for the data: $(0,0)$, $(1, \alpha)$, $(2, 3)$. Find the value of α, if the coefficient of x^2 in $p_2(x)$ is $\frac{1}{2}$.

6. Consider the function $f(x) = e^{x^2} \ln(x + 1)$ and $x = 0, 0.25, 0.5, 1$. Then use the suitable Lagrange interpolating polynomial to approximate $f(0.75)$. Also, compute an error bound for your approximation.

7. Consider the following table having the data for $f(x) = e^{3x} \cos 2x$:

x	0.1	0.2	0.4	0.5
$f(x)$	1.32295	1.67828	2.31315	2.42147

Find the cubic Lagrange polynomial $p_3(x)$ and use it to approximate $f(0.3)$. Also, estimate the actual error and the error bound for the approximation.

8. Construct the divided differences table for the function $f(x) = x^4 + 4x^3 + 2x^2 + 11x + 21$, for the values $x = 1.5, 2.5, 3.5, 4.5, 5.5, 6.5$.

9. Construct the divided differences table for the function $f(x) = (x + 2)e^{x-3}$, for the values $x = 2.1, 3.2, 4.3, 5.4, 6.5, 7.6$.

10. Consider the following table:

x	0.5	1.5	2.5	3.5	4.5	5.5
$f(x)$	2.70	6.97	11.89	21.06	36.34	41.45

(a) Construct the divided differences table for the tabulated function.

(b) Compute the Newton interpolating polynomials $p_2(x)$ and $p_3(x)$ at $x = 3.7$.

11. Consider the following table of $f(x) = \sqrt{x}$:

x	4	5	6	7	8
$f(x)$	2.0000	2.2361	2.4495	2.6458	2.8284

(a) Construct the divided differences table for the tabulated function.

(b) Find the Newton interpolating polynomials $p_3(x)$ and $p_4(x)$ at $x = 5.9$.

(c) Compute error bounds for your approximations in part (b).

12. Let $f(x) = \ln(x+3) \sin x$, with $x_0 = 0, x_1 = 2, x_2 = 2.5, x_3 = 4, x_4 = 4.5$. Then:

(a) Construct the divided differences table for the given data points.

(b) Find the Newton divided difference polynomials $p_2(x), p_3(x)$ and $p_4(x)$ at $x = 2.4$.

(c) Compute error bounds for your approximations in part (b).

(d) Compute the actual error.

13. Show that if $x_0, x_1,$ and x_2 are distinct, then

$$f[x_0, x_1, x_2] = f[x_1, x_2, x_0] = f[x_2, x_0, x_1].$$

14. The divided differences form of the interpolating polynomial $p_3(x)$ is

$$\begin{aligned} p_3(x) &= f[x_0] + (x - x_0)f[x_0, x_1] + (x - x_0)(x - x_1)f[x_0, x_1, x_2, x_0] \\ &+ (x - x_0)(x - x_1)(x - x_2)f[x_0, x_1, x_2, x_3]. \end{aligned}$$

By expressing these divided differences in terms of the function values $f(x_i)(i = 0, 1, 2, 3)$, verify that $p_3(x)$ does pass through the points $(x_i, f(x_i))(i = 0, 1, 2, 3)$.

15. Let $f(x) = x^2 + e^x$ and $x_0 = 0, x_1 = 1$. Use the divided differences to find the value of the second divided difference $f[x_0, x_1, x_0]$.

16. Let $f(x) = \ln(x+2)e^{x^2}$ and $x_0 = 0.5, x_1 = 1.5$. Use the divided differences to find the value of the third divided difference $f[x_0, x_1, x_0, x_1]$.

17. Use property 1 of the Chebyshev polynomial to construct $T_4(x)$ using $T(3)$ and $T_2(x)$.

18. Find the Chebyshev polynomial $p_3(x)$ that approximates the function $f(x) = e^{2x+1}$ over $[-1, 1]$.

19. Find the Lagrange–Chebyshev polynomial approximation $p_3(x)$ of $f(x) = ln(x+2)$ on $[-1, 1]$. Also, compute the error bound.

20. Find the four Chebyshev points in $2 \le x \le 4$ and write the Lagrange interpolation to interpolate $e^x(x+2)$. Compute the error bound.

21. Apply Aitken's method to approximate $f(1.5)$ by using the following data points:

x	1	2	3	4
$f(x)$	1	4	10	17

22. Consider the following table:

x	1.0	1.1	1.2	1.3	1.4	1.5
$f(x)$	0.8415	0.8912	0.9320	0.9636	0.9854	0.9975

Use Aitken's method to find the estimated value of $f(1.21)$.

23. Let $f(x) = \cos(x-2)e^{-x}$, with points $x_0 = 0$, $x_1 = 1$, $x_2 = 2, x_3 = 3$, and $x_4 = 4$. Use Aitken's method to find the estimated value of $f(2.5)$.

24. **(a)** Let $f(x) = \frac{xe^x}{x+1}$, with points $x_0 = 0.2, x_1 = 1.1, x_2 = 2.3$, and $x_3 = 3.5$. Use Aitken's method to find the estimated value of $f(2.5)$.
    (b) Let $f(x) = x + \frac{3}{x^2+1}$, with points $x_0 = 1.5$, $x_1 = 2.5$, $x_2 = 3.5$, and $x_3 = 4.5$. Use Aitken's method to find the estimated value of $f(3.9)$.

25. Find the least squares line fit $y = ax + b$ for the following data:

(a) $(-2, 1), (-1, 2), (0, 3), (1, 4)$.
(b) $(1.5, 1.4), (2, 2.4), (3, 3.9), (4, 4.7)$.

(c) $(2, 0), (3, 4), (4, 10), (5, 16)$.
(d) $(2, 2.6), (3, 3.4), (4, 4.9), (5, 5.4), (8, 4.6)$.
(e) $(-4, 1.2), (-2, 2.8), (0, 6.2), (2, 7.8), (4, 13.2)$.

26. Repeat Problem 25 to find the least squares parabolic fit $y = a + bx + cx^2$.

27. Find the least squares parabolic fit $y = a + bx + cx^2$ for the following data:

(a) $(-1, 0), (0, -2), (0.5, -1), (1, 0)$.
(b) $(-3, 15), (-1, 5), (1, 1), (3, 5)$.
(c) $(-3, -1), (-1, 25), (1, 25), (3, 1)$.
(d) $(0, 3), (1, 1), (2, 0), (4, 1), (6, 4)$.
(e) $(-2, 10), (-1, 1), (0, 0), (1, 2), (2, 9)$.

28. Repeat Problem 27 to find the best fit of the form $y = ax^b$.

29. Find the best fit of the form $y = ae^{bx}$ for the following data:

(a) $(5, 5.7), (10, 7.5), (15, 8.9), (20, 9.9)$.
(b) $(-1, 0.1), (1, 2.3), (2, 10), (3, 45)$.
(c) $(3, 4), (4, 10), (5, 16), (6, 20)$.
(d) $(-2, 1), (-1, 2), (0, 3), (1, 3), (2, 4)$.
(e) $(-1, 6.62), (0, 3.94), (1, 2.17), (2, 1.35), (3, 0.89)$.

30. Use a change of variable to linearize each of the following data points:

(a) For the given data $(1, 10), (3, 20), (5, 35), (7, 55), (9, 70)$, find the least squares curve $f(x) = \frac{4}{(1 + ce^{ax})}$.
(b) For the given data $(0.5, 2), (1.5, 5), (2.5, 6.5), (4, 7), (6.5, 11)$, find the least squares curve $f(x) = \frac{1}{(a + bx^2)}$.
(c) For the given data $(1.3, \frac{1}{4}), (1.8, \frac{1}{9}), (2.5, \frac{1}{16}), (3.6, \frac{1}{25})$, $(4.2, \frac{1}{36})$, find the least squares curve $f(x) = \frac{1}{(a + bx)^2}$.
(d) For the given data $(4, 5.6), (7, 7.2), (9, 11.5), (12, 15.5), (17, 18.7)$, find the least squares curve $f(x) = \frac{a \ln x + b}{\ln x}$.

31. Find the least squares planes for the following data:

   (a) $(0, 1, 2), (1, 0, 3), (2, 1, 4), (0, 2, 1)$.
   (b) $(1, 7, 1), (2, 5, 6), (3, 1, -2), (2, 1, 0)$.
   (c) $(3, 1, -3), (2, 1, -1), (2, 2, 0), (1, 1, 1), (1, 2, 3)$.
   (d) $(5, 4, 3), (3, 7, 9), (3, 2, 3), (4, 4, 4), (5, 7, 8)$.

32. Find the plane $z = ax + by + c$ that best fits the following data:

   (a) $(1, 2, 3), (1, -2, 1), (2, 1, 3), (2, 2, 1)$.
   (b) $(2, 4, -1), (2, 2, 5), (1, 3, 1), (7, 8, 2)$.
   (c) $(3, -1, 1), (2, 3, -2), (9, 6, -2), (7, 1, 2)$.
   (d) $(1, 2, 2), (3, 1, 6), (1, 2, 2), (2, 5, 1)$.

33. Find the trigonometric least squares polynomial fit $y = a_0 + a_1 \cos x + b_1 \sin x$ for each of the following data:

   (a) $(1.5, 7.5), (2.5, 11.4), (3.5, 15.3), (4.5, 19.2), (5.5, 23.5)$.
   (b) $(0.2, 3.0), (0.4, 5.0), (0.6, 7.0), (0.8, 9.0), (1.0, 11.0)$.
   (d) $(-2.0, 1.5), (-1.0, 2.5), (0.0, 3.5), (1.0, 4.5), (2.0, 5.5)$.
   (c) $(1.1, 6.5), (1.3, 8.3), (1.5, 10.4), (1.7, 12.9), (1.9, 14.6)$.

34. Repeat Problem 25 to find the trigonometric least squares polynomial fit $y = a_0 + a_1 \cos x + b_1 \sin x$.

35. Find the least squares solution for each of the following overdetermined systems:

   (a)
$$\begin{aligned}
3x_1 &+ x_2 &= 1 \\
2x_1 &+ 5x_2 &= 2 \\
x_1 &- 4x_2 &= 3
\end{aligned}$$

   (b)
$$\begin{aligned}
7x_1 &+ 6x_2 &= 5 \\
3x_1 &+ 5x_2 &= 1 \\
2x_1 &+ 6x_2 &= 2
\end{aligned}$$

(c)

$$
\begin{aligned}
2x_1 - 3x_2 + 4x_3 &= 13 \\
x_1 + 5x_2 + 3x_3 &= 7 \\
3x_1 + 2x_2 + x_3 &= 11 \\
4x_1 + x_2 + 5x_3 &= 10
\end{aligned}
$$

(d)

$$
\begin{aligned}
4x_1 - 3x_2 + 4x_3 &= 9 \\
3x_1 + 2x_2 - 5x_3 &= 3 \\
2x_1 + 5x_2 - 9x_3 &= 5 \\
4x_1 + 12x_2 + 3x_3 &= 7
\end{aligned}
$$

36. Find the least squares solution for each of the following overdetermined systems:

(a)

$$
\begin{aligned}
12x_1 - 9x_2 &= 7 \\
11x_1 + 21x_2 &= 13 \\
17x_1 - 22x_2 &= 24
\end{aligned}
$$

(b)

$$
\begin{aligned}
x_1 + 5x_2 + 19x_3 &= 13 \\
4x_1 - 2x_2 + 3x_3 &= 14 \\
3x_1 + x_2 - x_3 &= 12 \\
5x_1 - 4x_2 + 4x_3 &= 19
\end{aligned}
$$

(c)

$$
\begin{aligned}
2x_1 - 4x_2 + 11x_3 &= 9 \\
3x_1 - 5x_2 + 5x_3 &= 3 \\
11x_1 + 11x_2 - 7x_3 &= 2 \\
x_1 - 8x_2 + 3x_3 &= 7
\end{aligned}
$$

(d)

$$
\begin{aligned}
2x_1 + 5x_2 + 2x_3 + 2x_4 &= 12 \\
x_1 + 4x_2 + 6x_3 + x_4 &= 14 \\
3x_1 + 7x_2 + 2x_3 - 2x_4 &= 23 \\
5x_1 - 2x_2 + 11x_3 + 7x_4 &= 11 \\
x_1 + 4x_2 - 7x_3 + 13x_4 &= 19
\end{aligned}
$$

37. Find the least squares solution for each of the following overdetermined systems:

(a)

$$
\begin{aligned}
7x_1 + 6x_2 - 4x_3 &= -3 \\
8x_1 + 5x_2 + 3x_3 &= -5 \\
9x_1 + 3x_2 + 5x_3 &= 6 \\
-3x_1 + 2x_2 + 6x_3 &= 7
\end{aligned}
$$

(b)

$$
\begin{aligned}
x_1 + 3x_2 + 9x_3 &= 23 \\
2x_1 - 2x_2 + 6x_3 &= 24 \\
3x_1 - 7x_2 + 5x_3 &= 11 \\
4x_1 - 4x_2 + 9x_3 &= 22
\end{aligned}
$$

(c)

$$
\begin{aligned}
2x_1 - 5x_2 + 4x_3 + 3x_4 &= 15 \\
3x_1 + 2x_2 + x_3 + 5x_4 &= 14 \\
7x_1 - 3x_2 + 4x_3 + 9x_4 &= 13 \\
11x_1 + 8x_2 + 5x_3 + 7x_4 &= 12 \\
3x_1 + x_2 - 2x_3 + 6x_4 &= 11
\end{aligned}
$$

(d)

$$
\begin{aligned}
x_1 + 7x_2 + 7x_3 - 3x_4 &= 5 \\
3x_1 - 2x_2 - 5x_3 + 2x_4 &= 6 \\
3x_1 + 3x_2 - 5x_3 - 2x_4 &= 7 \\
x_1 - 3x_2 + 12x_3 + 11x_4 &= 8 \\
2x_1 + 4x_2 - 15x_3 + 3x_4 &= 9
\end{aligned}
$$

38. Find the least squares solution for each of the following overdetermined systems:

(a)

$$
\begin{aligned}
2x_1 - 9x_2 &= 5 \\
4x_1 + 8x_2 &= 1 \\
12x_1 + 13x_2 &= 2
\end{aligned}
$$

(b)

$$
\begin{aligned}
3x_1 &+ 4x_2 + 9x_3 &= 3 \\
4x_1 &- 2x_2 + 7x_3 &= 4 \\
2x_1 &- 8x_2 + 4x_3 &= 2 \\
x_1 &- 4x_2 + 4x_3 &= 9
\end{aligned}
$$

(c)

$$
\begin{aligned}
2x_1 &- 24x_2 + 9x_3 &= 3 \\
11x_1 &- 15x_2 + 14x_3 &= 1 \\
13x_1 &+ 21x_2 - 6x_3 &= 2 \\
x_1 &- 8x_2 + 3x_3 &= 3
\end{aligned}
$$

(d)

$$
\begin{aligned}
x_1 &+ 5x_2 + 2x_3 - 2x_4 &= 1 \\
3x_1 &- 2x_2 + 6x_3 + x_4 &= 1 \\
3x_1 &+ 7x_2 - 5x_3 - 2x_4 &= 2 \\
5x_1 &- 2x_2 + 12x_3 + 11x_4 &= 1 \\
x_1 &+ 4x_2 - 15x_3 + 3x_4 &= 1
\end{aligned}
$$

39. Find the least squares solution for each of the following underdetermined systems:

(a)

$$
\begin{aligned}
2x_1 &+ x_2 + x_3 &= 4 \\
x_1 &+ 3x_2 + 4x_3 &= 5
\end{aligned}
$$

(b)

$$
\begin{aligned}
x_1 &- 3x_2 + 4x_3 &= 2 \\
-x_1 &+ 2x_2 + 5x_3 &= 11
\end{aligned}
$$

(c)

$$
\begin{aligned}
x_1 &+ 5x_2 + 3x_3 - 5x_4 &= 15 \\
2x_1 &+ 5x_2 + 6x_3 + x_4 &= 18 \\
3x_1 &+ 2x_2 + 5x_3 - 3x_4 &= 10
\end{aligned}
$$

(d)

$$
\begin{aligned}
2x_1 &+ 5x_2 + 2x_3 + x_4 &= -5 \\
4x_1 &+ 3x_2 - 2x_3 + 9x_4 &= 6 \\
x_1 &+ x_2 + 3x_3 + 8x_4 &= 12
\end{aligned}
$$

40. Find the least squares solution for each of the following underdetermined systems:

(a)
$$2x_1 + 11x_2 + 7x_3 = 13$$
$$5x_1 + 13x_2 + 3x_3 = 11$$

(b)
$$12x_1 + 8x_2 - 9x_3 = 4$$
$$15x_1 - 13x_2 + 14x_3 = 5$$

(c)
$$x_1 + 5x_2 + 4x_3 + 3x_4 = 22$$
$$7x_1 + 15x_2 + 6x_3 + x_4 = 15$$
$$-3x_1 + 7x_2 + 3x_3 + 6x_4 = 19$$

(d)
$$x_1 + 5x_2 + x_3 + 13x_4 = 3$$
$$2x_1 + 15x_2 + 12x_3 + 9x_4 = 8$$
$$x_1 - 11x_2 + 17x_3 + 22x_4 = -2$$

41. Find the least squares solution for each of the following underdetermined systems:

(a)
$$3x_1 - 9x_2 + 5x_3 = 21$$
$$4x_1 + 17x_2 + 15x_3 = 23$$

(b)
$$9x_1 - 5x_2 - 8x_3 = 14$$
$$7x_1 - 3x_2 + 4x_3 = 11$$

(c)
$$2x_1 - 6x_2 + 7x_3 + 9x_4 = 9$$
$$5x_1 + 11x_2 + 9x_3 - 4x_4 = 8$$
$$2x_1 + 5x_2 + 6x_3 + 8x_4 = 7$$

(d)
$$3x_1 + 5x_2 + 7x_3 + 3x_4 = 33$$
$$2x_1 + 21x_2 + 2x_3 + 29x_4 = 18$$
$$5x_1 - 31x_2 + 19x_3 + 12x_4 = 22$$

42. Find the least squares solution for each of the following underdetermined systems:

(a)
$$\begin{aligned} 3x_1 &+ 23x_2 &+ 14x_3 &= 51 \\ 4x_1 &- 37x_2 &+ 35x_3 &= 13 \end{aligned}$$

(b)
$$\begin{aligned} 5x_1 &- 16x_2 &- 18x_3 &= 44 \\ 2x_1 &- 23x_2 &+ 34x_3 &= 51 \end{aligned}$$

(c)
$$\begin{aligned} 2x_1 &+ 7x_2 &- 9x_3 &+ 5x_4 &= 19 \\ -x_1 &+ 11x_2 &+ 18x_3 &- 24x_4 &= 27 \\ -3x_1 &+ 15x_2 &+ 6x_3 &+ 8x_4 &= 39 \end{aligned}$$

(d)
$$\begin{aligned} 13x_1 &+ 5x_2 &- 13x_3 &+ 23x_4 &= 17 \\ 17x_1 &- 22x_2 &- 12x_3 &+ 29x_4 &= 28 \\ 15x_1 &+ 11x_2 &- 19x_3 &+ 22x_4 &= 32 \end{aligned}$$

43. Find the pseudoinverse of each of the matrices:

(a) $\quad A = \begin{pmatrix} 1 & 3 \\ 1 & 4 \\ 2 & 1 \end{pmatrix}$ 　(b) $\quad A = \begin{pmatrix} 1 & -2 \\ 2 & -3 \\ 5 & 1 \end{pmatrix}$

(c) $\quad A = \begin{pmatrix} 2 & 3 \\ 3 & 4 \\ 5 & 2 \end{pmatrix}$ 　(d) $\quad A = \begin{pmatrix} 1 & 2 \\ 4 & -2 \\ -1 & 1 \end{pmatrix}$

44. Find the pseudoinverse of each of the matrices:

(a) $\quad A = \begin{pmatrix} 2 & 2 \\ 3 & 0 \\ 1 & 4 \end{pmatrix}$ 　(b) $\quad A = \begin{pmatrix} 3 & -1 \\ 2 & 3 \\ 6 & 5 \end{pmatrix}$

(c) $\quad A = \begin{pmatrix} 3 & 0 \\ 2 & 1 \\ 1 & -1 \\ -1 & 2 \end{pmatrix}$ $\qquad$ (d) $\quad A = \begin{pmatrix} 1 & -1 \\ 1 & 1 \\ 2 & 3 \\ 3 & 5 \end{pmatrix}$

45. Find the least squares solution for each of the following linear systems by using the pseudoinverse of the matrices:

(a)
$$2x_1 + 5x_2 = 4$$
$$x_1 + 11x_2 = 5$$

(b)
$$3x_1 - 7x_2 = 2$$
$$5x_1 + 2x_2 = 5$$

(c)
$$x_1 + 3x_2 + 2x_3 = 1$$
$$2x_1 - 7x_2 + x_3 = 1$$
$$5x_1 + 3x_2 + 5x_3 = 1$$

(d)
$$2x_1 + 7x_2 + 12x_3 = 3$$
$$x_1 + 13x_2 - 2x_3 = 2$$
$$7x_1 + 11x_2 + 3x_3 = 9$$

46. Find the least squares solution for each of the following linear systems by using the pseudoinverse of the matrices:

(a)
$$7x_1 + 13x_2 = 3$$
$$8x_1 + 15x_2 = 1$$

(b)
$$2x_1 + 5x_2 = 14$$
$$5x_1 - 3x_2 = 11$$

(c)
$$3x_1 + 5x_2 + 3x_3 = 2$$
$$x_1 + 5x_2 + 6x_3 = 5$$
$$-3x_1 + 2x_2 + 3x_3 = 9$$

(d)

$$
\begin{array}{rcrcrcl}
x_1 & + & 3x_2 & + & 4x_3 & = & 13 \\
x_1 & - & 6x_2 & + & 17x_3 & = & 9 \\
4x_1 & - & 15x_2 & + & 9x_3 & = & 2
\end{array}
$$

47. A QR decomposition of A is given. Use it to find a least squares solution of $A\mathbf{x} = \mathbf{b}$, where

$$
A = \begin{pmatrix} 0 & 3 \\ 0 & 4 \\ 5 & 10 \end{pmatrix}, \qquad \mathbf{b} = \begin{pmatrix} 3 \\ 0 \\ -4 \end{pmatrix},
$$

$$
Q = \begin{pmatrix} 0 & -0.6000 & -0.8000 \\ 0 & -0.8000 & 0.6000 \\ -1.0000 & 0 & 0 \end{pmatrix}, \qquad R = \begin{pmatrix} 5 & -10 \\ 0 & -5 \\ 0 & 0 \end{pmatrix}.
$$

48. A QR decomposition of A is given. Use it to find a least squares solution of $A\mathbf{x} = \mathbf{b}$, where

$$
A = \begin{pmatrix} 1 & 0 \\ 2 & -1 \\ -1 & 1 \end{pmatrix}, \qquad \mathbf{b} = \begin{pmatrix} 1 \\ 1 \\ 1 \end{pmatrix},
$$

$$
Q = \begin{pmatrix} -0.4082 & 0.7071 & -0.5774 \\ -0.8165 & -0.0000 & 0.5774 \\ 0.4082 & 0.7071 & 0.5774 \end{pmatrix}, \qquad R = \begin{pmatrix} -2.4495 & 1.2247 \\ 0 & 0.7071 \\ 0 & 0 \end{pmatrix}.
$$

49. A QR decomposition of A is given. Use it to find a least squares solution of $A\mathbf{x} = \mathbf{b}$, where

$$
A = \begin{pmatrix} 2 & 1 \\ 2 & 0 \\ 1 & 1 \end{pmatrix}, \qquad \mathbf{b} = \begin{pmatrix} 2 \\ 3 \\ -1 \end{pmatrix},
$$

$$
Q = \begin{pmatrix} -0.6667 & 0.3333 & -0.6667 \\ -0.6667 & -0.6667 & 0.3333 \\ -0.3333 & 0.6667 & 0.6667 \end{pmatrix}, \qquad R = \begin{pmatrix} -3.0000 & -1.0000 \\ 0 & 1.0000 \\ 0 & 0 \end{pmatrix}.
$$

50. A QR decomposition of A is given. Use it to find a least squares solution of $A\mathbf{x} = \mathbf{b}$, where

$$A = \begin{pmatrix} 1 & 1 \\ -1 & 1 \\ 1 & 1 \\ 1 & -1 \end{pmatrix}, \qquad \mathbf{b} = \begin{pmatrix} 1 \\ 2 \\ 4 \\ -1 \end{pmatrix},$$

$$Q = \begin{pmatrix} -0.5000 & -0.5000 & -0.7000 & -0.1000 \\ 0.5000 & -0.5000 & -0.1000 & 0.7000 \\ -0.5000 & -0.5000 & 0.7000 & 0.1000 \\ -0.5000 & 0.5000 & -0.1000 & 0.7000 \end{pmatrix}, \qquad R = \begin{pmatrix} -2 & 0 \\ 0 & -2 \\ 0 & 0 \\ 0 & 0 \end{pmatrix}.$$

51. A QR decomposition of A is given. Use it to find a least squares solution of $A\mathbf{x} = \mathbf{b}$, where

$$A = \begin{pmatrix} 1 & 2 \\ 1 & -1 \end{pmatrix}, \qquad \mathbf{b} = \begin{pmatrix} 1 \\ 1 \end{pmatrix}.$$

$$Q = \begin{pmatrix} 0.7071 & -0.7071 \\ 0.7071 & 0.7071 \end{pmatrix}, \qquad R = \begin{pmatrix} 1.4142 & 0.7071 \\ 0.0000 & -2.1213 \end{pmatrix}.$$

52. A QR decomposition of A is given. Use it to find a least squares solution of $A\mathbf{x} = \mathbf{b}$, where

$$A = \begin{pmatrix} 1 & 2 \\ -1 & 3 \end{pmatrix}, \qquad \mathbf{b} = \begin{pmatrix} 2 \\ 1 \end{pmatrix},$$

$$Q = \begin{pmatrix} 0.7071 & 0.7071 \\ -0.7071 & 0.7071 \end{pmatrix}, \qquad R = \begin{pmatrix} 1.4142 & -0.7071 \\ 0.0000 & 3.5356 \end{pmatrix}.$$

53. A QR decomposition of A is given. Use it to find a least squares solution of $A\mathbf{x} = \mathbf{b}$, where

$$A = \begin{pmatrix} 1 & 0 \\ 1 & 1 \end{pmatrix}, \qquad \mathbf{b} = \begin{pmatrix} 1 \\ -1 \end{pmatrix},$$

$$Q = \begin{pmatrix} 0.7071 & -0.7071 \\ 0.7071 & 0.7071 \end{pmatrix}, \qquad R = \begin{pmatrix} 1.4142 & 0.7071 \\ 0.0000 & 0.7071 \end{pmatrix}.$$

54. A QR decomposition of A is given. Use it to find a least squares solution of $A\mathbf{x} = \mathbf{b}$, where

$$A = \begin{pmatrix} 1 & 3 & 0 \\ 0 & -1 & 8 \\ 1 & 2 & 4 \end{pmatrix}, \qquad \mathbf{b} = \begin{pmatrix} 1 \\ 2 \\ 3 \end{pmatrix},$$

$$Q = \begin{pmatrix} 0.7071 & -0.4082 & -0.5774 \\ 0 & 0.8165 & -0.5774 \\ 0.7071 & 0.4082 & 0.5774 \end{pmatrix}, \quad R = \begin{pmatrix} 1.4142 & 3.5355 & 2.8284 \\ 0 & -1.2247 & 8.1650 \\ 0 & 0 & -2.3094 \end{pmatrix}.$$

55. A QR decomposition of A is given. Use it to find a least squares solution of $A\mathbf{x} = \mathbf{b}$, where

$$A = \begin{pmatrix} 1 & 0 & 2 \\ -1 & 2 & 0 \\ -1 & -2 & 2 \end{pmatrix}, \qquad \mathbf{b} = \begin{pmatrix} 1 \\ 1 \\ 1 \end{pmatrix},$$

$$Q = \begin{pmatrix} 0.7071 & -0.4082 & -0.5774 \\ 0 & 0.8165 & -0.5774 \\ 0.7071 & 0.4082 & 0.5774 \end{pmatrix}, \quad R = \begin{pmatrix} 1.4142 & 3.5355 & 2.8284 \\ 0 & -1.2247 & 8.1650 \\ 0 & 0 & -2.3094 \end{pmatrix}.$$

56. Find the least squares solution of each of the following linear systems $A\mathbf{x} = \mathbf{b}$ using QR decomposition:

(a)

$$A = \begin{pmatrix} 5 & 3 \\ 1 & 3 \\ 2 & 1 \end{pmatrix}, \quad \mathbf{x} = \begin{pmatrix} x_1 \\ x_2 \end{pmatrix}, \quad \mathbf{b} = \begin{pmatrix} 4.9 \\ 0.8 \\ 1.7 \end{pmatrix}.$$

(b)

$$A = \begin{pmatrix} 1 & -1 & 4 \\ 0 & 2 & 1 \\ 1 & 1 & 0 \\ 2 & -1 & 1 \end{pmatrix}, \quad \mathbf{x} = \begin{pmatrix} x_1 \\ x_2 \\ x_3 \end{pmatrix}, \quad \mathbf{b} = \begin{pmatrix} 1.1 \\ 0.2 \\ 0.9 \\ 1.7 \end{pmatrix}.$$

(c)

$$A = \begin{pmatrix} 1 & -1 & 1 \\ -1 & 4 & 2 \\ -2 & 1 & 2 \\ 1 & 4 & 2 \end{pmatrix}, \quad \mathbf{x} = \begin{pmatrix} x_1 \\ x_2 \\ x_3 \end{pmatrix}, \quad \mathbf{b} = \begin{pmatrix} 0.7 \\ -0.8 \\ -1.5 \\ 1.02 \end{pmatrix}.$$

(d)

$$A = \begin{pmatrix} 3 & 2 & 1 \\ 1 & 2 & 2 \\ 1 & 0 & -1 \\ 2 & 1 & -2 \end{pmatrix}, \quad \mathbf{x} = \begin{pmatrix} x_1 \\ x_2 \\ x_3 \end{pmatrix}, \quad \mathbf{b} = \begin{pmatrix} 2.5 \\ 1.1 \\ 0.8 \\ 1.9 \end{pmatrix}.$$

57. Find the least squares solution of each of the following linear systems $A\mathbf{x} = \mathbf{b}$ using QR decomposition:

(a)

$$A = \begin{pmatrix} 2 & 1 \\ 1 & 1 \\ 2 & 2 \end{pmatrix}, \quad \mathbf{x} = \begin{pmatrix} x_1 \\ x_2 \end{pmatrix}, \quad \mathbf{b} = \begin{pmatrix} 1 \\ 1 \\ 1 \end{pmatrix}.$$

(b)

$$A = \begin{pmatrix} 1 & -1 \\ 0 & 2 \\ 1 & 1 \end{pmatrix}, \quad \mathbf{x} = \begin{pmatrix} x_1 \\ x_2 \end{pmatrix}, \quad \mathbf{b} = \begin{pmatrix} 0.1 \\ 1.7 \\ 0.9 \end{pmatrix}.$$

(c)

$$A = \begin{pmatrix} 3 & -1 \\ -1 & 4 \\ -2 & 1 \end{pmatrix}, \quad \mathbf{x} = \begin{pmatrix} x_1 \\ x_2 \end{pmatrix}, \quad \mathbf{b} = \begin{pmatrix} 2.7 \\ -0.8 \\ -1.5 \end{pmatrix}.$$

(d)

$$A = \begin{pmatrix} 4 & 2 \\ 1 & 2 \\ 1 & 0 \end{pmatrix}, \quad \mathbf{x} = \begin{pmatrix} x_1 \\ x_2 \end{pmatrix}, \quad \mathbf{b} = \begin{pmatrix} 3.5 \\ 1.1 \\ 0.8 \end{pmatrix}.$$

58. Solve Problem 55 using singular value decomposition.

59. Find the least squares solution of each of the following linear systems $A\mathbf{x} = \mathbf{b}$ using singular value decomposition:

(a)

$$A = \begin{pmatrix} -2 & 2 \\ -1 & 1 \\ 2 & 2 \end{pmatrix}, \quad \mathbf{x} = \begin{pmatrix} x_1 \\ x_2 \end{pmatrix}, \quad \mathbf{b} = \begin{pmatrix} -1.8 \\ -0.9 \\ 1.9 \end{pmatrix}.$$

(b)

$$A = \begin{pmatrix} 1 & -1 \\ 1 & 2 \\ 1 & 1 \end{pmatrix}, \quad \mathbf{x} = \begin{pmatrix} x_1 \\ x_2 \end{pmatrix}, \quad \mathbf{b} = \begin{pmatrix} 1.1 \\ 0.7 \\ 0.9 \end{pmatrix}.$$

(c)

$$A = \begin{pmatrix} 1 & 0 \\ 1 & 1 \\ -1 & 1 \end{pmatrix}, \quad \mathbf{x} = \begin{pmatrix} x_1 \\ x_2 \end{pmatrix}, \quad \mathbf{b} = \begin{pmatrix} 0.9 \\ 0.85 \\ -0.9 \end{pmatrix}.$$

(d)

$$A = \begin{pmatrix} 3 & 2 \\ 1 & -1 \\ 1 & 3 \end{pmatrix}, \quad \mathbf{x} = \begin{pmatrix} x_1 \\ x_2 \end{pmatrix}, \quad \mathbf{b} = \begin{pmatrix} 2.5 \\ 1.05 \\ 0.85 \end{pmatrix}.$$

60. Solve Problem 56 using singular value decomposition.

Chapter 6

Linear Programming

6.1 Introduction

In this chapter, we give an introduction to linear programming. *Linear Programming (LP)* is a mathematical method for finding optimal solutions to problems. It deals with the problem of optimizing (maximizing or minimizing) a linear function, subject to the constraints imposed by a system of linear inequalities. It is widely used in industry and in government. Historically, linear programming was first developed and applied in 1947 by George Dantzig, Marshall Wood, and their associates in the U.S. Air Force; the early applications of *LP* were thus in the military field. However, the emphasis in applications has now moved to the general industrial area. *LP* today is concerned with the efficient use or allocation of limited resources to meet desired objectives.

Before formally defining a *LP* problem, we define the concepts *linear function* and *linear inequality*.

Definition 6.1 (Linear Function)

A function $Z(x_1, x_2, \ldots, x_N)$ of $x_1, x_2, \ldots, x_N$ is a linear function, if and only if for some set of constants $c_1, c_2, \ldots, c_N$,

$$Z(x_1, x_2, \ldots, x_N) = c_1 x_1 + c_2 x_2 + \cdots + c_N x_N.$$

For example, $Z(x_1, x_2) = 30x_1 + 50x_2$ is a linear function of x_1 and x_2, but $Z(x_1, x_2) = 2x_1^2 x_2$ is not a linear function of x_1 and x_2. •

Definition 6.2 (Linear Inequality)

For any function $Z(x_1, x_2, \ldots, x_N)$ and any real number b, the inequalities

$$Z(x_1, x_2, \ldots, x_N) \leq b$$

and

$$Z(x_1, x_2, \ldots, x_N) \geq b$$

are linear inequalities. For example, $3x_1 + 2x_2 \leq 11$ and $10x_1 + 15x_2 \geq 17$ are linear inequalities, but $2x_1^2 x_2 \geq 3$ is not a linear inequality. •

Definition 6.3 (Linear Programming Problem)

An LP problem is an optimization problem for which we do the following:

1. *We attempt to maximize or to minimize a linear function of the decision variables. The function that is to be maximized or minimized is called the objective function.*

2. *The values of the decision variables must satisfy a set of constraints. Each constraint must be a linear equation or linear inequality.*

3. *A sign restriction is associated with each variable. For any variable x_i, the sign restriction specifies that x_i must be either a nonnegative $(x_i \geq 0)$ or* **unrestricted sign.**

In the following, we discuss some LP problems involving linear functions, inequality constraints, equality constraints, and sign restriction. •

6.2 General Formulation

Let $x_1, x_2, \ldots, x_N$ be N variables in an *LP* problem. The problem is to find the values of the variables $x_1, x_2, \ldots, x_N$ to maximize (or minimize) a given linear function of the variables, subject to a given set of constraints that are linear in the variables.

The general formulation for an *LP* problem is

$$\text{maximize} \quad Z = c_1 x_1 + c_2 x_2 + \cdots + c_N x_N, \tag{6.1}$$

subject to the constraints

$$
\begin{array}{ccccccc}
a_{11} x_1 & + & a_{12} x_2 & + & \cdots & + & a_{1N} x_N & \leq & b_1 \\
a_{21} x_1 & + & a_{22} x_2 & + & \cdots & + & a_{2N} x_N & \leq & b_2 \\
\vdots & & \vdots & & \cdots & & \vdots & & \vdots \\
a_{M1} x_1 & + & a_{M2} x_2 & + & \cdots & + & a_{MN} x_N & \leq & b_M
\end{array} \tag{6.2}
$$

and

$$x_1 \geq 0, x_2 \geq 0, \ldots, x_N \geq 0, \tag{6.3}$$

where $a_{ij}(i = 1, 2, \ldots, M; j = 1, 2, \ldots, N)$ are constants (called *constraint coefficients*), $b_i(i = 1, 2, \ldots, M)$ are constants (called *resources values*), and $c_j(j = 1, 2, \ldots, N)$ are constants (called *cost coefficients*). We call Z the objective function.

In matrix form, the general formulation can be written as

$$\text{maximize} \quad Z = \mathbf{c}^T \mathbf{x}, \tag{6.4}$$

subject to the constraints

$$A\mathbf{x} \leq \mathbf{b} \tag{6.5}$$

and

$$\mathbf{x} \geq \mathbf{0}, \tag{6.6}$$

where

$$A = \begin{pmatrix} a_{11} & a_{12} & \cdots & a_{1N} \\ a_{21} & a_{22} & \cdots & a_{2N} \\ \vdots & \vdots & \vdots & \vdots \\ a_{M1} & a_{M2} & \cdots & a_{MN} \end{pmatrix} \tag{6.7}$$

$$\mathbf{b} = \begin{pmatrix} b_1 \\ b_2 \\ \vdots \\ b_M \end{pmatrix}, \quad \mathbf{c} = \begin{pmatrix} c_1 \\ c_2 \\ \vdots \\ c_N \end{pmatrix}, \quad \mathbf{x} = \begin{pmatrix} x_1 \\ x_2 \\ \vdots \\ x_N \end{pmatrix}, \quad \mathbf{0} = \begin{pmatrix} 0 \\ 0 \\ \vdots \\ 0 \end{pmatrix} \qquad (6.8)$$

and $\mathbf{c}^T$ denotes the transpose of the vector $\mathbf{c}$.

6.3   Terminology

The following terms are commonly used in LP:

- Decision variables: variables $x_1, x_2, \ldots, x_N$ in (6.1).

- Objective function: function Z given by (6.1).

- Objective function coefficients: constants $c_1, \ldots, c_N$ in (6.1).

- Constraints coefficients: constants a_{ij} in (6.2).

- Nonnegativity constraints: constraints given by (6.3).

- Feasible solution: set of $x_1, x_2, \ldots, x_N$ values that satisfy all the constraints.

- Feasible region: collection of all feasible solutions.

- Optimal solution: feasible solution that gives an optimal value of the objective function (i.e., the maximum value of Z in (6.1)).

6.4   Linear Programming Problems

Example 6.1 *(Product-Mix Problem)*

The Handy-Dandy Company wishes to schedule the production of a kitchen appliance that requires two resources—labor and material. The company is considering three different models and its production engineering department has furnished the following data: the supply of raw material is restricted to 360 pounds per day. The daily availability of labor is 250 hours. Formulate a linear programming model to determine the daily production rate of the various models in order to maximize the total profit.

	Model A	Model B	Model C
Labor (hours per unit)	7	3	6
Material (hours per unit)	6	7	8
Profit ($ per unit)	6	6	3

6.4.1 Formulation of Mathematical Model

Step I. *Identify the Decision Variables: the unknown activities to be determined are the daily rate of production for the three models.*
Representing them by algebraic symbols:
x_A: *Daily production of model A*
x_B: *Daily production of model B*
x_C: *Daily production of model C*

Step II. *Identify the Constraints: in this problem, the constraints are the limited availability of the two resources—labor and material. Model A requires 7 hours of labor for each unit, and its production quantity is x_A. Hence, the labor requirement for model A alone will be $7x_A$ hours (assuming a linear relationship). Similarly, models B and C will require $3x_B$ and $6x_C$ hours, respectively. Thus, the total requirement of labor will be $7x_A + 3x_B + 6x_C$, which should not exceed the available 250 hours. So the labor constraint becomes*

$$7x_A + 3x_B + 6x_C \leq 250.$$

Similarly, the raw material requirements will be $6x_A$ pounds for model A, $7x_B$ pounds for model B, and $8x_C$ pounds for model C. Thus, the raw material constraint is given by

$$6x_A + 7x_B + 8x_C \leq 360.$$

In addition, we restrict the variables x_A, x_B, x_C to have only nonnegative values, i.e.,

$$x_A \geq 0, \quad x_B \geq 0, \quad x_C \geq 0.$$

These are called the nonnegativity constraints, which the variables must satisfy. Most practical linear programming problems will have this nonnegative restriction on the decision variables. However, the general framework of linear programming is not restricted to nonnegative values.

Step III. *Identify the Objective: the objective is to maximize the total profit for sales. Assuming that a perfect market exists for the product such that all that is produced can be sold, the total profit from sales becomes*

$$Z = 6x_A + 6x_B + 3x_C.$$

Thus, the complete mathematical model for the product mix problem may now be summarized as follows:

Find numbers x_A, x_B, x_C, which will maximize

$$Z = 6x_A + 6x_B + 3x_C,$$

subject to the constraints

$$
\begin{array}{rcl}
7x_A + 3x_B + 6x_C & \leq & 250 \\
6x_A + 7x_B + 8x_C & \leq & 360
\end{array}
$$

$$x_A \geq 0, \quad x_B \geq 0, \quad x_C \geq 0.$$

●

Example 6.2 (An Inspection Problem)

A company has two grades of inspectors, 1 and 2, who are to be assigned for a quality control inspection. It is requires that at least 1800 pieces be inspected per 8-hour day. Grade 1 inspectors can check pieces at the rate of 25 per hour, with an accuracy of 98%. Grade 2 inspectors check at the rate of 15 pieces per hour, with an accuracy of 95%. The wage rate of a Grade 1 inspector is $4.00 per hour, that of a Grade 2 inspector is $3.00 per *hour. Each time an error is made by an inspector, the cost to the company is $2.00. The company has available for the inspection job eight Grade 1 inspectors and ten Grade 2 inspectors. The company wants to determine the optimal assignment of inspectors, which will minimize the total cost of the inspection.*

6.4.2   Formulation of Mathematical Model

Let x_1 and x_2 denote the number of Grade 1 and Grade 2 inspectors assigned for inspection. Since the number of available inspectors in each grade is limited, we have the following constraints:

$$x_1 \leq 8 \quad (\textit{Grade 1})$$
$$x_1 \leq 10 \quad (\textit{Grade 2}).$$

The company requires at least 1800 pieces to be inspected daily. Thus, we get

$$8(25)x_1 + 8(15)x_2 \geq 1800$$

or

$$200x_1 + 120x_2 \geq 1800,$$

which can also be written as

$$5x_1 + 3x_2 \geq 45.$$

To develop the objective function, we note that the company incurs two types of costs during inspections: wages paid to the inspector, and the cost of the inspection error. The hourly cost of each Grade 1 inspector is

$$\$4 + 2(25)(0.02) = \$5 \textit{ per hour.}$$

Similarly, for each Grade 2 inspector the cost is

$$\$3 + 2(15)(0.05) = \$4.50 \textit{ per hour.}$$

Thus, the objective function is to minimize the daily cost of inspection given by

$$Z = 8(5x_1 + 4.5x_2) = 40x_1 + 36x_2.$$

The complete formulation of the linear programming problem thus becomes

$$minimize \quad Z = 40x_1 + 36x_2,$$

subject to the constraints

$$
\begin{aligned}
x_1 \quad & \quad \le \quad 8 \\
x_2 & \le \quad 10 \\
5x_1 + 3x_2 & \ge \quad 45
\end{aligned}
$$

$$x_1 \ge 0, \qquad x_2 \ge 0.$$

6.5   Graphical Solution of LP Models

In the last section, two examples were presented to illustrate how practical problems can be formulated mathematically as *LP* problems. The next step after formulation is to solve the problem mathematically to obtain the best possible solution. In this section, a graphical procedure to solve *LP* problems involving only two variables is discussed. Though in practice such small problems are usually not encountered, the graphical procedure is presented to illustrate some of the basic concepts used in solving large *LP* problems.

Example 6.3 *A company manufactures two types of mobile phones, model A and model B. It takes 5 hours and 2 hours to manufacture A and B, respectively. The company has 900 hours available per week for the production of mobile phones. The manufacturing cost of each model A is \$8 and the manufacturing cost of a model B is \$10. The total funds available per week for production is \$2800. The profit on each model A is \$3, and the profit on each model B is \$2. How many of each type of mobile phone should be manufactured weekly to obtain the maximum profit?*

Solution. *We first find the inequalities that describe the time and monetary constraints. Let the company manufacture x_1 of model A and x_2 of model B. Then the total manufacturing time is $(5x_1 + 2x_2)$ hours. There are 900 hours available. Therefore,*

$$5x_1 + 2x_2 \le 900,$$

Now the cost of manufacturing x_1 of model A at \$8 each is \$8x_1, and the cost of manufacturing x_2 of model B at \$10 each is \$10x_2. Thus, the total production cost is $(8x_1 + 10x_2)$. There is \$2800 available for production of mobile phones. Therefore,

$$8x_1 + 10x_2 \leq 2800.$$

Furthermore, x_1 and x_2 represent the numbers of mobile phones manufactured. These numbers cannot be negative. Therefore, we get two more constraints

$$x_1 \geq 0, \quad x_2 \geq 0.$$

Next, we find a mathematical expression for profit. Since the weekly profit on x_1 mobile phones at \$3 per mobile phone is \$3x_1 and the weekly profit on x_2 mobile phones at \$2 per mobile phone is \$2x_2, the total weekly profit is \$$(3x_1 + 2x_2)$. Let the profit function Z be defined as

$$Z = 3x_1 + 2x_2.$$

Thus, the mathematical model for the given LP problem with the profit function and the system of linear inequalities may be written as

$$maximize \quad Z = 3x_1 + 2x_2,$$

subject to the constraints

$$
\begin{aligned}
5x_1 + 2x_2 &\leq 900 \\
8x_1 + 10x_2 &\leq 2800
\end{aligned}
$$

$$x_1 \geq 0, \quad x_2 \geq 0.$$

In this problem, we are interested in determining the values of the variables x_1 and x_2 that will satisfy all the restrictions and give the maximum value of the objective function. As a first step in solving this problem, we want to identify all possible values of x_1 and x_2 that are nonnegative and satisfy the constraints. The solution of an LP problem is merely finding the best feasible solution (optimal solution) in the feasible region (set of all feasible solutions). In our example, an optimal solution is a feasible solution which maximizes the objective function $3x_1 + 2x_2$. The value of the *objective function corresponding to an optimal solution is called the **optimal value** of the LP problem.*

To represent the feasible region in a graph, every constraint is plotted, and all values of x_1, x_2 that will satisfy these constraints are identified. The nonnegativity constraints imply that all feasible values of the two variables will be in the first quadrant. It can be shown that the graph of the constraint $5x_1 + 2x_2 \le 900$ consists of points on and below the straight line $5x_1 + 2x_2 = 900$. Similarly, the points that satisfy the inequality $8x_1 + 10x_2 \le 2800$ are on and below the straight line $8x_1 + 10x_2 = 2800$.

The feasible region is given by the shaded region $ABCO$ as shown in Figure 6.1. Obviously, there is an infinite number of feasible points in this

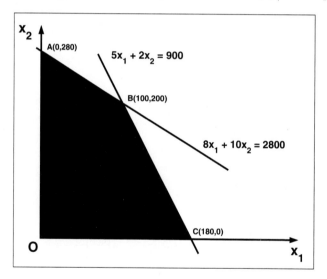

Figure 6.1: Feasible region of Example 6.3.

region. Our objective is to identify the feasible point with the largest value of the objective function Z.

It has been proved that the maximum value of Z will occur at a vertex of the feasible region, namely, at one of the points $A, B, C,$ or O; if there is more than one point at which the maximum occurs, it will be along one *edge of the region, such as AB or BC. Hence, we only have to examine the value of the objective function $Z = 3x_1 + 2x_2$ at the vertices A, B, C, and O. These vertices are found by determining the points of intersection of the lines. We obtain:*

$$
\begin{aligned}
A(0, 280): &\quad Z_A = 3(0) + 2(280) &&= 560 \\
B(100, 200): &\quad Z_B = 3(100) + 2(200) &&= 700 \\
C(180, 0): &\quad Z_C = 3(180) + 2(0) &&= 540 \\
O(0, 0): &\quad Z_O = 3(0) + 2(0) &&= 0.
\end{aligned}
$$

The maximum value of Z is 700 at B. Thus, the maximum value of $Z = 3x_1 + 2x_2$ is 700, when $x_1 = 100$ and $x_2 = 200$. The interpretation of these results is that the maximum weekly profit is $700 and this occurs when 100 model A mobile phones and 200 model B mobile phones are manufactured.   ●

In using the Optimization Toolbox, **linprog** solves LP problems:

$$
\min_x Z' * x \quad \text{subject to:} \qquad A * x <= b
$$
$$
Aeq * x == beq
$$
$$
lb <= x <= ub.
$$

In solving LP problems using **linprog**, we use the following:

Syntax:

```
>> x = linprog(Z, A, b)
```

solves $\min Z' * x$ such that $A * x <= b$.

```
>> x = linprog(Z, A, b, Aeq, beq)
```

solves $\min Z' * x$ such that $A * x <= b$, $Aeq * x == beq$. If no inequalities exist, then set $A = [\,]$ and $b = [\,]$.

```
>> x = linprog(Z, A, b, Aeq, beq, lb, ub)
```

defines a set of lower and upper bounds on the design variables, x, so that the solution is always in the range $lb <= x <= ub$. If no equalities exist, then set $Aeq = [\ ]$ and $beq = [\ ]$.

$$>> [x, Fval] = linprog(Z, A, b)$$

returns the value of the objective function at x, $Fval = Z' * x$.

$$>> [x, Fval, exitflag] = linprog(Z, A, b)$$

returns a value *exitflag* that describes the exit condition.

$$>> [x, Fval, exitflag, output] = linprog(Z, A, b)$$

returns a structure containing information about the optimization.

Input Parameters:

Z is the objective function coefficients.

A is a matrix of inequality constraint coefficients.

b is the right-hand side in inequality constraints.

$Aeqs$ the matrix of equality constraints.

beq is the right-hand side in equality constraints.

lb is the lower bounds of the desgin. values; -Inf == unbounded below. Embty lb ==> -Inf on all variables.

ub is the upper bounds on the desgin. values; Inf == unbounded above. Embty ub ==> Inf on all variables.

Output Parameters:

x optimal design parameters.

$Fval$ optimal design parameters.

exitflag   exit conditions of **linprog**. If *exitflag* is:

     > 0 then **linprog** converged with a solution x.

     $= 0$ then **linprog** reached the maximum number of iterations without converging to a solution x.

     < 0 then the problem was infeasible or **linprog** failed. For example, if *exitflag* = −2, then no feasible point was found.

if *exitflag* = −3, then the problem is unbounded.

if *exitflag* = −4, then the NaN value was encountered during execution of the algorithm.

if *exitflag* = −5, then both the primal and dual problems are infeasible.

     output structure, and the fields of the structure are:

          the number of iterations taken,

          the type of algorithm used,

          the number of conjugate gradient iterations (if used).

Now to reproduce the above results (Example 6.3) using MATLAB, we do the following:

First, enter the coefficients:

```
>> Z = [3 2];
>> A = [5 2; 8 10];
>> b = [900; 2800];
```

Second, evaluate **linprog**:

```
>> [x, Fval, exitflag, output] = linprog(−Z, A, b);
```

MATLAB's answer is:
optimization terminated successfully.

Now evaluate x and Z as follows:

```
>> x =
     100.0000
     200.0000
```

and

```
>> Objective = −Fval =
   700.0000
```

Note that the optimization functions in the toolbox minimize the objective function. To maximize a function Z, apply an optimization function to minimize $-Z$. The resulting point where the maximum of Z occurs is also the point where the minimum of $-Z$ occurs.

We graph the regions specified by the constraints. Let's put in the two constraint inequalities:

```
>> X = 0 : 200;
>> Y1 = max((900 − 5. ∗ X)./2, 0);
>> Y2 = max((2800 − 8. ∗ X)./10, 0);
>> Ytop = min([Y1; Y2]);
>> area(X, Ytop);
```

Like the optimization toolbox function **linprog**, MATLAB's Simulink Toolbox has a built-in function, **simlp**, that implements the solution of an LP problem. We will now use **simlp** on the above problem as follows:

```
>> Z = [3 2];
>> A = [5 2; 8 10];
>> b = [900; 2800];
>> simlp(−Z, A, b);
>> ans
    100.0000
    200.0000
```

This is the same answer we obtained before. Note that we entered the negative of the coefficient vector for the objective function Z because **simlp** also searches for a minimum rather than a maximum.

Example 6.4 *Find the maximum value of*

$$Z = 2x_1 + x_2,$$

subject to the constraints

$$3x_1 + 5x_2 \leq 20$$
$$4x_1 + x_2 \leq 15$$

and

$$x_1 \geq 0, \qquad x_2 \geq 0.$$

Solution. *The constraints are represented graphically by the shaded region ACBO in Figure 6.2. The vertices of the feasible region are found by*

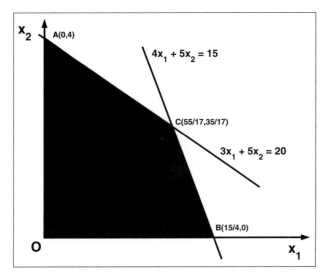

Figure 6.2: Feasible region of Example 6.4.

determining the points of intersections of the lines. We get

$$A(0,4): \qquad Z_A = 2(0) + 4 \qquad = 4$$

$$B\left(\frac{15}{4},0\right): \qquad Z_B = 2\left(\frac{15}{4}\right) + 0 \qquad = \frac{15}{2}$$

$$C\left(\frac{55}{17},\frac{35}{17}\right): \qquad Z_C = 2\left(\frac{55}{17}\right) + \frac{35}{17} = \frac{145}{17}$$

$$O(0,0): \qquad Z_O = 0 + 0 \qquad = 0$$

Thus, the maximum value of the objective function Z is $\frac{145}{17}$; namely, when $x_1 = \frac{55}{17}$ and $x_2 = \frac{35}{17}$.   •

To get the above results using MATLAB's built-in function, **simlp**, we do the following:

$$>> Z = [2\ 1];$$
$$>> A = [3\ 5; 4\ 1];$$
$$>> b = [20; 15];$$
$$>> simlp(-Z, A, b);$$
$$>> ans$$
$$\qquad 3.2353$$
$$\qquad 2.0588$$

In fact, we can compute

$$>> [x, y] = solve('3 * x + 5 * y = 20', '4 * x + y = 15');$$
$$>> x$$
$$\qquad 3.2353$$
$$>> y$$
$$\qquad 2.0588$$

6.5.1   Reversed Inequality Constraints

For cases where some or all of the constraints contain inequalities with the sign reversed ($\geq$ rather than $\leq$), the $\geq$ signs can be converted to $\leq$ signs by multiplying both sides of the constraints by -1. Thus, the constraint

$$a_{i1}x_1 + a_{i2}x_2 + \cdots + a_{iN}x_N \geq b_i$$

is equivalent to

$$-a_{i1}x_1 - a_{i2}x_2 - \cdots - a_{iN}x_N \leq -b_i.$$

6.5.2   Equality Constraints

For cases where some or all of the constraints contain equalities, the problem can be reformulated by expressing an equality as two inequalities with opposite signs. Thus, the constraint

$$a_{i1}x_1 + a_{i2}x_2 + \cdots + a_{iN}x_N = b_i$$

is equivalent to

$$a_{i1}x_1 + a_{i2}x_2 + \cdots + a_{iN}x_N \leq b_i$$
$$a_{i1}x_1 + a_{i2}x_2 + \cdots + a_{iN}x_N \geq b_i.$$

6.5.3   Minimum Value of a Function

An *LP* problem that involves determining the minimum value of an objective function Z can be solved by looking for the maximum value of $-Z$, the negative of Z, over the same feasible region. Thus, the problem

$$\text{minimize}\quad Z = c_1x_1 + c_2x_2 + \cdots + c_Nx_N$$

is equivalent to

$$\text{maximize}\quad -Z = (-c_1)x_1 + (-c_2)x_2 + \cdots + (-c_N)x_N.$$

Theorem 6.1 (Minimum Value of a Function)

The minimum value of a function Z over a region S occurs at the point(s) of the maximum value of $-Z$ and is the negative of that maximum value.

Proof. *Let Z have a minimum value Z_A at the point A in the region. Then if B is any other point in the region,*

$$Z_A \leq Z_B.$$

Multiply both sides of this inequality by -1 to get

$$-Z_A \geq -Z_B.$$

This implies that A is a point of maximum value $-Z$. Furthermore, the minimum value is Z_A, the negative of the maximum value $-Z_A$. The above steps can be reversed, proving that the converse holds, and this verifies the result.   ●

In summary, a minimization *LP* problem in which

1. the objective function is to be minimized (rather than maximized) or;

2. the constraints contain equalities (= rather than ≤) or;

3. the constraints contain inequalities with the sign reversed (≥ rather than ≤);

can be reformulated in terms of the general solution formulation given by (6.1), (6.2), and (6.3).

The following diet problem is an example of a general *LP* problem, in which the objective function is to be minimized and the constraints contain ≥ signs.

Example 6.5 (Diet Problem)

The diet problem arises in the choice of foods for a healthy diet. The problem is to determine the foods in a diet that minimize the total cost per day, subject to constraints that ensure minimum daily nutritional requirements. Let

- M = *number of nutrients*

- N = *number of types of food*

- a_{ij} = *number of units of nutrient i in food ($i = 1, 2, \ldots, M$; $j = 1, 2, \ldots, N$)*

- b_i = *number of units of nutrient i required per day ($i = 1, 2, \ldots, M$)*

- c_i = *cost per unit of food j ($j = 1, 2, \ldots, N$) x_j = number of units of food j in the diet per day ($j = 1, 2, \ldots, N$)*

The objective is to find the values of the N variables $x_1, x_2, \ldots, x_N$ to minimize the total cost per day, Z. The LP formulation for the diet problem is

$$\text{minimze} \qquad Z = c_1 x_1 + c_2 x_2 + \cdots + c_N x_N,$$

subject to the constraints

$$
\begin{array}{ccccccc}
a_{11}x_1 & + & a_{12}x_2 & + & \cdots & + & a_{1N}x_N & \geq & b_1 \\
a_{21}x_1 & + & a_{22}x_2 & + & \cdots & + & a_{2N}x_N & \geq & b_2 \\
\vdots & & \vdots & & \cdots & & \vdots & & \vdots \\
a_{M1}x_1 & + & a_{M2}x_2 & + & \cdots & + & a_{MN}x_N & \geq & b_M
\end{array}
$$

and

$$x_1 \geq 0, x_2 \geq 0, \ldots, x_N \geq 0,$$

where $a_{ij}, b_i, c_j (i = 1, 2, \ldots, M; j = 1, 2, \ldots, N)$ are constants. •

Example 6.6 *Consider the inspection problem given by Example 6.2:*

$$\text{minimize} \qquad Z = 40x_1 + 36x_2,$$

subject to the constraints

$$
\begin{array}{rcrcl}
x_1 & & & \leq & 8 \\
& & x_2 & \leq & 10 \\
5x_1 & + & 3x_2 & \geq & 45
\end{array}
$$

$$x_1 \geq 0, \qquad x_2 \geq 0.$$

In this problem, we are interested in determining the values of the variables x_1 and x_2 that will satisfy all the restrictions and give the least value of the objective function. As a first step in solving this problem, we want to identify all possible values of x_1 and x_2 that are nonnegative and satisfy the constraints. For example, a solution $x_1 = 8$ and $x_2 = 10$ is positive and satisfies all the constraints. In our example, an optimal solution is a feasible solution which minimizes the objective function $40x_1 + 36x_2$.

To represent the feasible region in a graph, every constraint is plotted, and all values of x_1, x_2 that will satisfy these constraints are identified. The nonnegativity constraints imply that all feasible values of the two variables will be in the first quadrant. The constraint $5x_1 + 3x_2 \geq 45$ requires that any feasible solution (x_1, x_2) to the problem should be on one side of the straight line $5x_1 + 3x_2 = 45$. The proper side is found by testing whether the origin satisfies the constraint or not. The line $5x_1 + 3x_2 = 45$ is first plotted by taking two convenient points (for example, $x_1 = 0, x_2 = 15$ and $x_1 = 9, x_2 = 0$).

Similarly, the constraints $x_1 \leq 8$ and $x_2 \leq 10$ are plotted. The feasible region is given by the region $ACBO$ as shown in Figure 6.3. Obviously

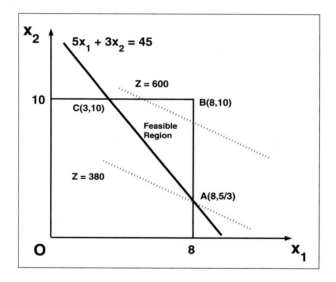

Figure 6.3: Feasible region of Example 6.6.

there is an infinite number of feasible points in this region. Our objective is to identify the feasible point with the lowest value of the objective function Z.

Observe that the objective function, given by $Z = 40x_1 + 36x_2$, represents a straight line if the value of Z is fixed a priori. Changing the value of Z essentially translates the entire line to another straight line parallel to itself. In order to determine an optimal solution, the objective function line is drawn for a convenient value of Z such that it passes through one or more *points in the feasible region. Initially, Z is chosen as 600. By moving this line closer to the origin, the value of Z is further decreased (Figure 6.3). The only limitation on this decrease is that the straight line $Z = 40x_1 + 36x_2$ contains at least one point in the feasible region ABC. This clearly occurs at the corner point A given by $x_1 = 8, x_2 = \frac{5}{3}$. This is the best feasible point giving the lowest value of Z as 380. Hence, $x_1 = 8$, $x_2 = \frac{5}{3}$ is an optimal solution, and $Z = 380$ is the optimal value for the LP problem.*

Thus, for the inspection problem the optimal utilization is achieved by using eight Grade 1 inspectors and 1.67 Grade 2 inspectors. The fractional value $x_2 = \frac{5}{3}$ suggests that one of the Grade 2 inspectors is utilized only 67% of the time. If this is not possible, the normal practice is to round off the fractional values to get an optimal integer solution as $x_1 = 8, x_2 = 2$. (In general, rounding off the fractional values will not produce an optimal integer solution.) •

Unique Optimal Solution

In Example 6.6, the solution $x_1 = 8, x_2 = \frac{5}{3}$ is the only feasible point with the lowest value of Z. In other words, the values of Z corresponding to the other feasible solution in Figure 6.3 exceed the optimal value of 380. Hence, for this problem, the solution $x_1 = 8, x_2 = \frac{5}{3}$ is the *unique optimal solution*.

Alternative Optimal Solutions

In some *LP* problems, there may exist more than one feasible solution such that their objective values are equal to the optimal values of the linear program. In such cases, all of these feasible solutions are optimal solutions, and the *LP* problem is said to have *alternative* or *multiple optimal solutions*. To illustrate this, consider the following *LP* problem.

Example 6.7 *Find the minimum value of*

$$Z = x_1 + 2x_2,$$

subject to the constraints

$$
\begin{aligned}
x_1 + 2x_2 &\leq 10 \\
x_1 + x_2 &\geq 1 \\
x_2 &\leq 4
\end{aligned}
$$

and

$$x_1 \geq 0, \qquad x_2 \geq 0.$$

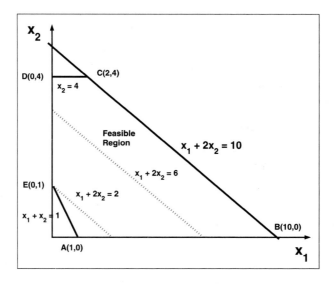

Figure 6.4: Feasible region of Example 6.7.

Solution. *The feasible region is shown in Figure 6.4. The objective function lines are drawn for $Z = 2, 6$, and 10. The optimal value for the LP problem is 10, and the corresponding objective function line $x_1 + 2x_2 = 10$ coincides with side BC of the feasible region. Thus, the corner point feasible solutions $x_1 = 10, x_2 = 0(B)$, and $x_1 = 2, x_2 = 4(C)$, and all other points on the line BC are optimal solutions.* ●

Unbounded Solution

Some *LP* problems may not have an optimal solution. In other words, it is possible to find better feasible solutions continuously improving the objective function values. This would have been the case if the constraint $x_1 + 2x_2 \leq 10$ were not given in Example 6.7. In this case, moving farther away from the origin increases the objective function $x_1 + 2x_2$ and maximized Z would be $+\infty$. When there exists no finite optimum, the *LP* problem is said to have an *unbounded solution*.

It is inconceivable for a practical problem to have an unbounded solution, since this implies that one can make infinite profit from a finite amount of resources. If such a solution is obtained in a practical problem it generally means that one or more constraints have been omitted inadvertently during the initial formulation of the problem. These constraints would have prevented the objective function from assuming infinite values.

Theorem 6.2 *If there exists an optimal solution to an LP problem, then at least one of the corner points of the feasible region will always qualify to be an optimal solution.* •

Notice that each feasible region we have discussed is such that the whole of the segment of a straight line joining two points within the region lies within that region. Such a region is called a *convex*. A theorem states that the feasible region in an *LP* problem is a convex (see Figure 6.5).

In the following section, we will use an iterative procedure called the *simplex method* for solving an *LP* problem based on Theorem 6.2. Even though the feasible region of an *LP* problem contains an infinite number of points, an optimal solution can be determined by merely examining the finite number of corner points in the feasible region. Before we discuss the simplex method, we discuss the canonical and standard forms of a linear program.

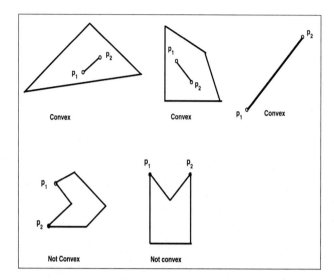

Figure 6.5: Convex and nonconvex sets in $\mathbb{R}^2$.

6.5.4   LP Problem in Canonical Form

The general LP problem can always be put in the following form, which is referred to as the *canonical form*:

$$\text{maximize} \quad Z = \sum_{i=1}^{M} c_i x_i,$$

subject to the constraints

$$\sum_{i=1}^{M} a_{ij} x_j \leq b_i, \quad j = 1, 2, \ldots, N$$

$$x_i \geq 0, \quad i = 1, 2, \ldots, M.$$

The characteristics of this form are:

1. All decision variables are nonnegative.

2. All the constraints are of the $\leq$ form.

3. The objective function is to maximize.

4. All the right-hand sides $b_i \geq 0, \quad i = 1, 2, \ldots, M$.

5. The matrix A contains M identity columns of an $M \times M$ identity matrix $\mathbf{I}$.

6. The objective function coefficients corresponding to those M identity columns are zero.

Note that the variables corresponding to the M identity columns are called *basic variables* and the remaining variables are called *nonbasic variables*. The feasible solution obtained by setting the nonbasic variables equal to zero and using the constraint equations to solve for the basic variables is called the *basic feasible solution*.

6.5.5   LP Problem in Standard Form

The standard form of an LP problem with M constraints and N variables can be represented as follows:

$$\text{maximize } (minimize) \quad Z = c_1 x_1 + c_2 x_2 + \cdots + c_N x_N, \qquad (6.9)$$

subject to the constraints

$$
\begin{aligned}
a_{11} x_1 &+ a_{12} x_2 + \cdots + a_{1N} x_N = b_1 \\
a_{21} x_1 &+ a_{22} x_2 + \cdots + a_{2N} x_N = b_2 \\
&\vdots \\
a_{M1} x_1 &+ a_{M2} x_2 + \cdots + a_{MN} x_N = b_M
\end{aligned}
\qquad (6.10)
$$

and

$$
\begin{aligned}
x_1 \geq 0, \ x_2 \geq 0, \ \ldots, x_N &\geq 0 \\
b_1 \geq 0, \ b_2 \geq 0, \ \ldots, b_M &\geq 0.
\end{aligned}
\qquad (6.11)
$$

The main features of the standard form are:

- The objective function is of the maximization or minimization type.

- All constraints are expressed as equations.

- All variables are restricted to be nonnegative.

- The right-hand side constant of each constraint is nonnegative.

In matrix-vector notation, the standard *LP* problem can be expressed as:

$$\text{maximize } (minimize) \quad Z = \mathbf{cx} \tag{6.12}$$

subject to the constraints

$$A\mathbf{x} = \mathbf{b} \tag{6.13}$$

and

$$\mathbf{x} \geq \mathbf{0}$$
$$\mathbf{b} \geq \mathbf{0}, \tag{6.14}$$

where A is an $M \times N$ matrix, $\mathbf{x}$ is an $N \times 1$ column vector, $\mathbf{b}$ is an $M \times 1$ column vector, and $\mathbf{x}$ is an $1 \times N$ row vector. In other words,

$$A = \begin{pmatrix} a_{11} & a_{12} & \cdots & a_{1N} \\ a_{21} & a_{22} & \cdots & a_{2N} \\ \vdots & \vdots & \vdots & \vdots \\ a_{M1} & a_{M2} & \cdots & a_{MN} \end{pmatrix}, \tag{6.15}$$

$$\mathbf{b} = \begin{pmatrix} b_1 \\ b_2 \\ \vdots \\ b_M \end{pmatrix}, \quad \mathbf{c} = \begin{pmatrix} c_1 & c_2 & \cdots & c_N \end{pmatrix}, \quad \mathbf{x} = \begin{pmatrix} x_1 \\ x_2 \\ \vdots \\ x_N \end{pmatrix}. \tag{6.16}$$

In practice, A is called a coefficient matrix, $\mathbf{x}$ is the decision vector, $\mathbf{b}$ is the requirement vector, and $\mathbf{c}$ is the profit (cost) vector of an *LP* problem.

Note that to convert an *LP* problem into standard form, each inequality constraint must be replaced by an equation constraint by introducing new variables that are *slack variables* or *surplus variables*. We illustrate this procedure using the following problem.

Example 6.8 (Leather Limited)

Leather Limited manufactures two types of belts: the deluxe model and the regular model. Each type requires 3 square yards of leather. A regular belt requires 5 hours of skilled labor and a deluxe belt requires 4 hours. Each week, 55 square yards of leather and 75 hours of skilled labor are available. Each regular belt contributes \$10 to profit and each deluxe belt, \$15. Formulate the LP problem.

Solution. *Let x_1 be the number of deluxe belts and x_2 be the regular belts that are produced weekly. Then the appropriate LP problem sets the form:*

$$maximize \quad Z = 10x_1 + 15x_2,$$

subject to the constraints

$$3x_1 + 3x_2 \leq 55$$
$$4x_1 + 5x_2 \leq 75$$

$$x_1 \geq 0, \quad x_2 \geq 0.$$

To convert the above inequality constraints to equality constraints, we define for each $\leq$ constraint a slack variable u_i (u_i = slack variable for ith constraint), which is the amount of the resource unused in the ith constraint. Because $x_1 + x_2$ square yards of leather are being used, and 40 square yards are available, we define u_1 by

$$u_1 = 55 - 3x_1 - 3x_2 \quad or \quad 3x_1 + 3x_2 + u_1 = 55.$$

Similarly, we define u_2 by

$$u_2 = 75 - 4x_1 - 5x_2 \quad or \quad 4x_1 + 5x_2 + u_2 = 75.$$

Observe that a point (x_1, x_2) satisfies the ith constraint, if and only if $u_i \geq 0$. Thus, the converted LP problem

$$maximize \quad Z = 10x_1 + 15x_2,$$

subject to the constraints

$$3x_1 + 3x_2 + u_1 \qquad = 55$$
$$4x_1 + 5x_2 \qquad + u_2 = 75$$

$$x_1 \geq 0, \; x_2 \geq 0, \; u_1 \geq 0, \; u_2 \geq 0$$

is in standard form.

In summary, if constraint i of an LP problem is a $\leq$ constraint, then we convert it to an equality constraint by adding the slack variable u_i to the ith constraint and adding the sign restriction $u_i \geq 0$. $\bullet$

Now we illustrate how a $\geq$ constraint can be converted to an equality constraint. Let us consider the diet problem discussed in Example 6.5. To convert the *ith* $\geq$ constraint to an equality constraint, we define an **excess variable** (surplus variable) v_i (v_i will always be the excess variable for the *ith* constraint). We define v_i to be the amount by which the *ith* constraint is oversatisfied. Thus, for the diet problem,

$$v_1 = a_{11}x_1 + a_{12}x_2 + \cdots + a_{1N}x_N - b_1$$

or

$$a_{11}x_1 + a_{12}x_2 + \cdots + a_{1N}x_N - v_1 = b_1.$$

We do the same for the other remaining $\geq$ constraints; the converted standard form of the diet problem after adding the sign restrictions $v_i \geq 0 (i = 1, 2, \ldots, M)$ may be written as

$$\text{minimize} \quad Z = c_1x_1 + c_2x_2 + \cdots + c_Nx_N,$$

subject to the constraints

$$
\begin{array}{llllllll}
a_{11}x_1 & + & \cdots & + & a_{1N}x_N & - & v_1 & & & = & b_1 \\
a_{21}x_1 & + & \cdots & + & a_{2N}x_N & & & - & v_2 & = & b_2 \\
\vdots & & \cdots & & \vdots & & \vdots & \cdots & & & \vdots \\
a_{M1}x_1 & + & \cdots & + & a_{MN}x_N & & & & - \; v_M & = & b_M
\end{array}
$$

and

$$x_i \geq 0, \; v_i \geq 0 \qquad (i = 1, 2, \ldots, M).$$

A point $(x_1, x_2, \ldots, x_N)$ satisfies the *ith* $\geq$ constraint, if and only if v_i is nonnegative.

In summary, if the *ith* constraint of an *LP* problem is a $\geq$ constraint, then it can be converted to an equality constraint by subtracting an excess variable v_i from the *ith* constraint and adding the sign restriction $v_i \geq 0$.

If an *LP* problem has both $\leq$ and $\geq$ constraints, then simply apply the procedures we have described to the individual constraints. For example, the *LP* problem

$$\text{maximize} \quad Z = 55x_1 + 60x_2,$$

subject to the constraints

$$
\begin{array}{rcrcl}
x_1 & & & \leq & 30 \\
& & x_2 & \leq & 45 \\
15x_1 & + & 25x_2 & \leq & 70 \\
30x_1 & + & 35x_2 & \geq & 90
\end{array}
$$

$$x_1 \geq 0, \quad x_2 \geq 0$$

can be easily transformed into standard form by adding slack variables u_1, u_2, and u_3, respectively, to the first three constraints and <u>subtracting</u> <u>an excess variable v_4</u> from the fourth constraint. Then we add the sign restrictions

$$u_1 \geq 0, \quad u_2 \geq 0, \quad u_3 \geq 0, \quad v_4 \geq 0.$$

This yields the following *LP* problem in standard form

$$\text{maximize} \quad Z = 55x_1 + 60x_2,$$

subject to the constraints

$$
\begin{array}{rcrcrcrcrcl}
x_1 & & & + & u_1 & & & & & = & 30 \\
& & x_2 & & & + & u_2 & & & = & 45 \\
15x_1 & + & 25x_2 & & & & & + & u_3 & & = & 70 \\
30x_1 & + & 35x_2 & & & & & & & - & v_4 & = & 90
\end{array}
$$

$$x_1 \geq 0, \ x_2 \geq 0, \ u_1 \geq 0, \ u_2 \geq 0, u_3 \geq 0, \ v_4 \geq 0.$$

6.5.6 **Some Important Definitions**

Let us review the basic definitions using the standard form of an *LP* problem given by

$$\text{maximize} \quad Z = \mathbf{cx},$$

subject to the constraints

$$A\mathbf{x} = \mathbf{b}$$

$$\mathbf{x} \geq 0.$$

1. *Feasible Solution.* A feasible solution is a nonnegative vector $\mathbf{x}$ satisfying the constraints $A\mathbf{x} = \mathbf{b}$.

2. *Feasible Region.* The feasible region, denoted by S, is the set of all feasible solutions. Mathematically,

$$S = \{\mathbf{x} | A\mathbf{x} = \mathbf{b}, \quad \mathbf{x} > 0\}.$$

   If the feasible set S is empty, then the linear program is said to be *infeasible.*

3. *Optimal Solution.* An optimal solution is a vector $\mathbf{x^0}$ such that it is feasible and its value of the objective function $(\mathbf{cx^0})$ is larger than that of any other feasible solution. Mathematically, $\mathbf{x^0}$ is optimal, if and only if $\mathbf{x^0} \in S$ and $\mathbf{cx^0} \geq \mathbf{cx}$ for all $\mathbf{x} \in S$.

4. *Optimal Value.* The optimal value of a linear program is the value of the objective function corresponding to the optimal solution. If Z^0 is the optimal value, then $Z^0 = \mathbf{cx^0}$.

5. *Alternate Optimum.* When a linear program has more than one optimal solution, it is said to have an alternate optimal solution. In this case, there exist more than one feasible solution having the same optimal value (Z^0) for their objective functions.

6. *Unique Optimum.* The optimal solution of a linear program is said to be unique when there exists no other optimal solution.

7. *Unbounded Solution.* When a linear program does not pose a finite optimum (i.e., $Z_{max} \to \infty$), it is said to have an unbounded solution.

6.6 The Simplex Method

The graphical method of solving an *LP* problem introduced in the last section has its limitations. The method demonstrated for two variables can be extended to *LP* problems involving three variables, but for problems involving more than two variables, the graphical approach becomes impractical. Here, we introduce the other approach called the *simplex method*, which is an algebraic method that can be used for any number of variables. This method was developed by George B. Dantzig in 1947.

It can be used to solve maximization or minimization problems with any standard constraints.

Before proceeding further with our discussion with the simplex algorithm, we must define the concept of a basic solution to a linear system (6.13).

6.6.1 Basic and Nonbasic Variables

Consider a linear system $A\mathbf{x} = \mathbf{b}$ of M linear equations in N variables (assume $N \geq M$).

Definition 6.4 (Basic Solution)

A basic solution to $A\mathbf{x} = \mathbf{b}$ is obtained by setting $N - M$ variables equal to 0 and solving for the values of the remaining M variables. This assumes that setting the $N - M$ variables equal to 0 yields unique values for the remaining M variables or, equivalently, the columns for the remaining M variables are linearly independent. ●

To find a basic solution to $A\mathbf{x} = \mathbf{b}$, we choose a set of $N - M$ variables (the *nonbasic variables*) and set each of these variables equal to 0. Then we solve for the values of the remaining $N - (N - M)M$ variables (the *basic variables*) that satisfy $A\mathbf{x} = \mathbf{b}$.

Definition 6.5 (Basic Feasible Solution)

Any basic solution to a linear system (6.13) in which all variables are nonnegative is a **basic feasible solution**. •

The simplex method deals only with basic feasible solutions in the sense that it moves from one basic solution to another. Each basic solution is associated with an iteration. As a result, the maximum number of iterations in the simplex method cannot exceed the number of basic solutions of the standard form. We can thus conclude that the maximum number of iterations cannot exceed

$$\binom{N}{M} = \frac{N!}{(N-M)!M!}.$$

The basic–nonbasic swap gives rise to two suggestive concepts: The *entering variable* is a current nonbasic variable that will "enter" the set of basic variables at the next iteration. The *leaving variable* is a current basic variable that will "leave" the basic solution in the next iteration.

Definition 6.6 (Adjacent Basic Feasible Solution)

For any LP problem with M constraints, two basic feasible solutions are said to be adjacent if their sets of basic variables have $M-1$ basic variables in common. In other words, an adjacent feasible solution differs from the present basic feasible solution in exactly one basic variable. •

We now give a general description of how the simplex algorithm solves *LP* problems.

6.6.2   The Simplex Algorithm

1. Set up the *initial simplex tableau*.

2. Locate the negative element in the last row, other than the last element, that is largest in magnitude (if two or more entries share this property, any one of these can be selected). If all such entries are nonnegative, the tableau is in final form.

3. Divide each positive element in the column defined by this negative entry into the corresponding element of the last column.

4. Select the divisor that yields the smallest quotient. This element is called the pivot element (if two or more elements share this property, any one of these can be selected as the pivot).

5. Now use operations to create a 1 in the pivot location and zeros elsewhere in the pivot column.

6. Repeat steps 2-5 until all such negative elements have been eliminated from the last row. The final matrix is called the *final simplex tableau* and it leads to the optimal solution.

Example 6.9 *Determine the maximum value of the function*

$$Z = 3x_1 + 5x_2 + 8x_3,$$

subject to the constraints

$$
\begin{aligned}
x_1 + x_2 + x_3 &\le 100 \\
3x_1 + 2x_2 + 4x_3 &\le 200 \\
x_1 + 2x_2 + x_3 &\le 150
\end{aligned}
$$

$$x_1 \ge 0, \quad x_2 \ge 0, \quad x_3 \ge 0.$$

Solution. *Take the three slack variables u_1, u_2, and u_3, which must be added to the given 3 constraints to get the standard constraints, which may be written in the LP problem as*

$$
\begin{aligned}
x_1 + x_2 + x_3 + u_1 &= 100 \\
3x_1 + 2x_2 + 4x_3 + u_2 &= 200 \\
x_1 + 2x_2 + x_3 + u_3 &= 150
\end{aligned}
$$

$$x_1 \ge 0, \ x_2 \ge 0, \ x_3 \ge 0, \ u_1 \ge 0, \ u_2 \ge 0, \ u_3 \ge 0.$$

The objective function $Z = 3x_1 + 5x_2 + 8x_3$ is rewritten in the form

$$-3x_1 - 5x_2 - 8x_3 + Z = 0.$$

Thus, the entire problem now becomes that of determining the solution to the following system of equations:

$$
\begin{aligned}
x_1 &+ x_2 + x_3 + u_1 &&&&&&= 100 \\
3x_1 &+ 2x_2 + 4x_3 &&+ u_2 &&&&= 200 \\
x_1 &+ 2x_2 + x_3 &&&+ u_3 &&= 150 \\
-3x_1 &- 5x_2 - 8x_3 &&&&+ Z &= 0.
\end{aligned}
$$

Since we know that the simplex algorithm starts with an initial basic feasible solution, by inspection, we see that if we set nonbasic variables $x_1 = x_2 = x_3 = 0$, we can solve for the values of the basic variables u_1, u_2, u_3. So the basic feasible solution for the basic variables is

$$u_1 = 100, \quad u_2 = 200, \quad u_3 = 150, \quad x_1 = x_2 = x_3 = 0.$$

It is important to observe that each basic variable may be associated with the row of the canonical form in which the basic variable has a coefficient of 1. Thus, for the initial canonical form, u_1 may be thought of as the basic variable for row 1, as may u_2 for row 2, and u_3 for row 3. To perform the simplex algorithm, we also need a basic (although not necessarily nonnegative) variable for the last row. Since Z appears in the last row with a coefficient of 1, and Z does not appear in any other row, we use Z as its basic variable. With this convention, the basic feasible solution for our initial canonical form has

basic variables $\{u_1, u_2, u_3, Z\}$ and   nonbasic variables $\{x_1, x_2, x_3\}$.

For this basic feasible solution

$$u_1 = 100, \quad u_2 = 200, \quad u_3 = 150, \quad Z = 0, \quad x_1 = x_2 = x_3 = 0$$

Note that a slack variable can be used as a basic variable for an equation if the right-hand side of the constraint is nonnegative.

Thus, the simplex tableaus are as follows:

basis	x_1	x_2	x_3	u_1	u_2	u_3	Z	constants
u_1	1	1	1	1	0	0	0	100
u_2	3	2	④	0	1	0	0	200
u_3	1	2	0	0	0	1	0	150
Z	-3	-5	-8	0	0	0	1	0

basis	x_1	x_2	x_3	u_1	u_2	u_3	Z	constants
u_1	1	1	1	1	0	0	0	100
u_2	$\frac{3}{4}$	$\frac{1}{2}$	1	0	$\frac{1}{4}$	0	0	50
u_3	1	2	0	0	0	1	0	150
Z	-3	-5	-8	0	0	0	1	0

basis	x_1	x_2	x_3	u_1	u_2	u_3	Z	constants
u_1	$\frac{1}{4}$	$\frac{1}{2}$	0	1	$-\frac{1}{4}$	0	0	50
x_3	$\frac{3}{4}$	$\frac{1}{2}$	1	0	$\frac{1}{4}$	0	0	50
u_3	1	2	0	0	0	1	0	150
Z	3	-1	0	0	2	0	1	400

basis	x_1	x_2	x_3	u_1	u_2	u_3	Z	constants
u_1	$\frac{1}{4}$	$\frac{1}{2}$	0	1	$-\frac{1}{4}$	0	0	50
x_3	$\frac{3}{4}$	$\frac{1}{2}$	1	0	$\frac{1}{4}$	0	0	50
u_3	1	②	0	0	0	1	0	150
Z	3	-1	0	0	2	0	1	400

basis	x_1	x_2	x_3	u_1	u_2	u_3	Z	constants
u_1	0	0	0	1	$-\frac{1}{4}$	$-\frac{1}{4}$	0	$\frac{25}{2}$
x_3	$\frac{1}{2}$	0	1	0	$\frac{1}{4}$	$-\frac{1}{4}$	0	$\frac{25}{2}$
x_2	$\frac{1}{2}$	1	0	0	0	$\frac{1}{2}$	0	75
Z	$\frac{7}{2}$	0	0	0	2	$\frac{1}{2}$	1	475

Since all negative elements have been eliminated from the last row, the final tableau gives the following system of equations:

$$u_1 \;-\; \frac{1}{4}u_2 \;-\; \frac{1}{4}u_3 \;=\; \frac{25}{2}$$

$$\frac{1}{2}x_1 \;+\; x_3 \;+\; \frac{1}{4}u_2 \;-\; \frac{1}{4}u_3 \;=\; \frac{25}{2}$$

$$\frac{1}{2}x_1 \;+\; x_2 \;+\; \frac{1}{2}u_3 \;=\; 75$$

$$\frac{7}{2}x_1 \;+\; 2u_2 \;+\; \frac{1}{2}u_3 \;+\; Z \;=\; 475.$$

The constraints are

$$x_1 \geq 0, \ x_2 \geq 0, \ x_3 \geq 0, \ u_1 \geq 0, \ u_2 \geq 0, \ u_3 \geq 0.$$

The final equation, under the constraints, implies that Z has a maximum value of 475 when $x_1 = 0, u_2 = 0, u_3 = 0$. On substituting these values back into the equations, we get

$$x_2 = 75, \quad x_3 = \frac{25}{2}, \quad u_1 = \frac{25}{2}.$$

Thus, $Z = 3x_1 + 5x_2 + 8x_3$ has a maximum value of 475 at

$$x_1 = 0, \quad x_2 = 75, \quad x_3 = \frac{25}{2}.$$

Note that the reasoning at this maximum value of Z implies that the element in the last row and the last column of the final tableau will always correspond to the maximum value of the objective function Z. ●

In the following example, we illustrate the application of the simplex method when there are many optimal solutions.

Example 6.10 *Determine the maximum value of the function*

$$Z = 8x_1 + 2x_2,$$

subject to the constraints

$$4x_1 + x_2 \ \leq \ 32$$
$$4x_1 + 3x_2 \ \leq \ 48$$

$$x_1 \geq 0, \quad x_2 \geq 0.$$

Solution. *Take the two slack variables u_1 and u_2, which must be added to the given 2 constraints to get the standard constraints, which may be written in the LP problem as*

$$4x_1 \ + \ x_2 \ + \ u_1 \ \qquad = \ 32$$
$$4x_1 \ + \ 3x_2 \ \qquad + \ u_2 \ = \ 48$$

$$x_1 \geq 0, \ x_2 \geq 0, \ u_1 \geq 0, \ u_2 \geq 0.$$

The objective function $Z = 8x_1 + 2x_2$ is rewritten in the form

$$-8x_1 - 2x_2 + Z = 0.$$

Thus, the entire problem now becomes that of determining the solution to the following system of equations:

$$
\begin{aligned}
4x_1 &+ x_2 + u_1 && && = 32 \\
4x_1 &+ 3x_2 && + u_2 && = 48 \\
-8x_1 &- 2x_2 && && + Z = 0.
\end{aligned}
$$

The simplex tableaus are as follows:

basis	x_1	x_2	u_1	u_2	Z	constants
u_1	④	1	1	0	0	32
u_2	4	3	0	1	0	48
Z	-8	-2	0	0	1	0

basis	x_1	x_2	u_1	u_2	Z	constants
u_1	1	$\frac{1}{4}$	$\frac{1}{4}$	0	0	8
u_2	4	3	0	1	0	48
Z	-8	-2	0	0	1	0

basis	x_1	x_2	u_1	u_2	Z	constants
x_1	1	$\frac{1}{4}$	$\frac{1}{4}$	0	0	8
u_2	0	2	-1	1	0	16
Z	0	0	2	0	1	64

Since all negative elements have been eliminated from the last row, the final tableau gives the following system of equations:

$$x_1 + \frac{1}{4}x_2 + \frac{1}{4}u_1 = 8$$

$$2x_2 - u_1 + u_2 = 16$$

$$2u_1 + Z = 64,$$

with the constraints

$$x_1 \geq 0, \ x_2 \geq 0, \ u_1 \geq 0, \ u_2 \geq 0.$$

The last equation implies that Z has a maximum value of 64 when $u_1 = 0$. On substituting these values back into the equations, we get

$$x_1 \ + \ \tfrac{1}{4}x_2 \qquad\qquad\qquad = \ 8$$

$$2x_2 \quad + \ u_2 \ = \ 16,$$

with the constraints

$$x_1 \geq 0, \quad x_2 \geq 0, \quad u_2 \geq 0.$$

Any point (x_1, x_2) that satisfies these conditions is an optimal solution. Thus, $Z = 8x_1 + 2x_2$ has a maximum value of 64. This is achieved at a point on the line $x_1 + \tfrac{1}{4}x_2 = 8$ between $(6,8)$ and $(8,0)$. ●

To use the simplex method set 'LargeScale' to 'off' and 'simplex' to 'on' in options:

```
>> option = optimset('Large','off','simplex','on');
```

Then call the function **linprog** with the options input argument:

```
>> Z = [8 : 2];
>> A = [4 1; 4 3];
>> b = [32; 48];
>> lb = [0; 0]; ub = [20; 20];
>> [x, Fval, exitflag, output] = linprog(-Z, A, b, [ ], [ ], lb, ub);
```

6.6.3   Simplex Method for Minimization Problem

In the last two examples, we used the simplex method for finding the maximum value of the objective function Z. In the following, we will apply the method for the minimization problem.

Example 6.11 *Determine the minimum value of the function*

$$Z = -2x_1 + x_2,$$

subject to the constraints

$$
\begin{aligned}
2x_1 + x_2 &\leq 20 \\
x_1 - x_2 &\leq 4 \\
-x_1 + x_2 &\leq 5
\end{aligned}
$$

$$x_1 \geq 0, \quad x_2 \geq 0.$$

Solution. *We can solve this LP problem using two different approaches.*

First Approach: *Put* $Z1 = -Z$, *then minimizing* Z *is equivalent to maximizing* $Z1$. *For this first approach, find* $Z1_{max}$, *then* $Z_{min} = -Z1_{max}$. *Let*

$$Z1 = -Z = 2x_1 - x_2,$$

then the problem reduces to maximize $Z1 = 2x_1 - x_2$ *under the same constraints. Introducing slack variables, we have*

$$
\begin{aligned}
2x_1 &+ x_2 &+ u_1 & & & & &= 20 \\
x_1 &- x_2 & &+ u_2 & & & &= 4 \\
-x_1 &+ x_2 & & &+ u_3 & & &= 5 \\
-2x_1 &+ x_2 & & & &+ Z1 &= 0
\end{aligned}
$$

$$x_1 \geq 0, \ x_2 \geq 0, \ u_1 \geq 0, \ u_2 \geq 0, \ u_3 \geq 0.$$

The simplex tableaus are as follows:

basis	x_1	x_2	u_1	u_2	u_3	Z_1	constants
u_1	2	1	1	0	0	0	20
u_2	①	-1	0	1	0	0	4
u_3	-1	1	0	0	1	0	5
Z_1	-2	1	0	0	0	1	0

basis	x_1	x_2	u_1	u_2	u_3	Z_1	constants
u_1	0	③	1	-2	0	0	12
x_1	1	-1	0	1	0	0	4
u_3	0	0	0	1	1	0	9
Z_1	0	-1	0	2	0	1	8

basis	x_1	x_2	u_1	u_2	u_3	Z_1	constants
x_2	0	1	$\frac{1}{3}$	$-\frac{2}{3}$	0	0	4
x_1	1	0	$\frac{1}{3}$	$\frac{1}{3}$	0	0	8
u_3	0	0	0	1	1	0	9
Z_1	0	0	$\frac{1}{3}$	$\frac{4}{3}$	0	1	12

Thus, the final tableau gives the following system of equations:

$$x_2 + \frac{1}{3}u_1 - \frac{2}{3}u_2 \qquad\qquad = 4$$

$$x_1 \qquad + \frac{1}{3}u_1 + \frac{1}{3}u_2 \qquad\qquad = 8$$

$$u_2 + u_3 \qquad\qquad = 9$$

$$\frac{1}{3}u_1 + \frac{4}{3}u_2 \qquad + Z1 = 12,$$

with the constraints

$$x_1 \geq 0, \ x_2 \geq 0, \ u_1 \geq 0, \ u_2 \geq 0, u_3 \geq 0.$$

The last equation implies that Z_1 has a maximum value of 12 when $u_1 = u_2 = 0$. Thus, $Z1 = 2x_1 - x_2$ has a maximum value of 12 at $x_1 = 8$ and $x_2 = 4$. Since $Z = -Z1 = -12$, the minimum value of the objective function $Z = -2x_1 + x_2$ is -12.

Second Approach: *To decrease Z, we have to pick out the largest positive entry in the bottom row to find a pivotal column. Thus, the problem now becomes that of determining the solution to the following system of equations:*

$$
\begin{aligned}
2x_1 + x_2 + u_1 \qquad\qquad\qquad &= 20 \\
x_1 - 2x_2 \qquad + u_2 \qquad\qquad &= 4 \\
-x_1 + x_2 \qquad\qquad + u_3 \qquad &= 5 \\
2x_1 - x_2 \qquad\qquad\qquad + Z &= 0.
\end{aligned}
$$

The simplex tableaus are as follows:

basis	x_1	x_2	u_1	u_2	u_3	Z	constants
u_1	2	1	1	0	0	0	20
u_2	①	-1	0	1	0	0	4
u_3	-1	1	0	0	1	0	5
Z	2	-1	0	0	0	1	0

basis	x_1	x_2	u_1	u_2	u_3	Z	constants
u_1	0	③	1	-2	0	0	12
x_1	1	-1	0	1	0	0	4
u_3	0	0	0	1	1	0	9
Z	0	1	0	-2	0	1	-8

basis	x_1	x_2	u_1	u_2	u_3	Z	constants
x_2	0	1	$\frac{1}{3}$	$-\frac{2}{3}$	0	0	4
x_1	1	0	$\frac{1}{3}$	$\frac{1}{3}$	0	0	8
u_3	0	0	0	1	1	0	9
Z	0	0	$-\frac{1}{3}$	$-\frac{4}{3}$	0	1	-12

Thus, $Z = -2x_1 + x_2$ has a minimum value of -12 at $x_1 = 8$ and $x_2 = 4$. •

6.7 Unrestricted in Sign Variables

In solving an LP problem with the simplex method, we used the ratio test to determine the row in which an entering variable becomes a basic variable. Recall that the ratio test depends on the fact that any feasible point requires all variables to be nonnegative. Thus, if some variables are allowed to be unrestricted in sign, the ratio test and therefore the simplex method are no longer valid. Here, we show how an LP problem with restricted in sign variables can be transformed into an LP problem in which all variables are required to be nonnegative.

For each unrestricted in sign variable x_i, we begin by defining two new variables x_i' and x_i''. Then substitute $x_i' - x_i''$ for x_i in each constraint and in the objective function. Also, add the sign restrictions $x_i' \geq 0$ and $x_i'' \geq 0$. Now all the variables are nonnegative, therefore, we can use the simplex method. Note that each basic feasible solution can have either $x_i' > 0$ (and $x_i'' = 0$), or $x_i'' > 0$ (and $x_i' = 0$), or $x_i' = x_i'' = 0$ ($x_i = 0$).

Example 6.12 *Consider the following LP problem:*

$$maximize \quad Z = 30x_1 - 4x_2,$$

subject to the constraints

$$
\begin{aligned}
5x_1 \quad - \quad x_2 &\leq 30 \\
x_1 \qquad\qquad &\leq 5
\end{aligned}
$$

$$x_1 \geq 0, \quad x_2 \text{ unrestricted.}$$

Solution. *Since x_2 is unrestricted in sign, we replace x_2 by $x_2' - x_2''$ in the objective function, and in the first constraint we obtain*

$$maximize \quad Z = 30x_1 - 4x_2' + 4x_2'',$$

subject to the constraints

$$
\begin{aligned}
5x_1 \quad - \quad x_2' \quad + \quad x_2'' &\leq 30 \\
x_1 \qquad\qquad\qquad &\leq 5
\end{aligned}
$$

$$x_1 \geq 0, \quad x_2' \geq 0, \quad x_2'' \geq 0.$$

Now convert the problem into standard form by adding two slack variables, u_1 and u_2, in the first and second constraints, respectively, and we get

$$maximize \quad Z = 30x_1 - 4x_2' + 4x_2'',$$

subject to the constraints

$$
\begin{aligned}
5x_1 \quad - \quad x_2' \quad + \quad x_2'' \quad + \quad u_1 \qquad\qquad &= 30 \\
x_1 \qquad\qquad\qquad\qquad + \quad u_2 &= 5
\end{aligned}
$$

$$x_1 \geq 0, \ x_2' \geq 0, \ x_2'' \geq 0, \ u_1 \geq 0, \ u_2 \geq 0.$$

The simplex tableaus are as follows:

basis	x_1	x_2'	x_2''	u_1	u_2	Z	constants
u_1	5	-1	1	1	0	0	30
u_2	①	0	0	0	1	0	5
Z	-30	4	-4	0	0	1	0

basis	x_1	x_2'	x_2''	u_1	u_2	Z	constants
u_1	0	-1	①	1	-5	0	5
x_1	1	0	0	0	1	0	5
Z	0	4	-4	0	30	1	150

basis	x_1	x_2'	x_2''	u_1	u_2	Z	constants
x_2''	0	-1	1	1	-5	0	5
x_1	1	0	0	0	1	0	5
Z	0	0	0	4	10	1	170

We now have an optimal solution to the LP problem given by $x_1 = 5, x_1' = 0, x_2'' = 5, u_1 = u_2 = 0$, and maximum $Z = 170$.

Note that the variables x_2' and x_2'' will never both be basic variables in the same tableau. ●

6.8   Finding a Feasible Basis

A major requirement of the simplex method is the availability of an initial basic feasible solution in canonical form. Without it, the initial simplex tableau cannot be found. There are two basic approaches to finding an initial basic feasible solution.

6.8.1   By Trial and Error

Here, a basic variable is chosen arbitrarily for each constraint, and the system is reduced to canonical form with respect to those basic variables. If the resulting canonical system gives a basic feasible solution (i.e., the right-hand side constants are nonnegative), then the initial tableau can be set up to start the simplex method. It is also possible that during the canonical reduction some of the right-hand side constants may become negative. In that case, the basic solution obtained will be infeasible, and the simplex method cannot be started. Of course, one can repeat the process by trying a different set of basic variables for the canonical reduction and hope for a basic feasible solution. Now it is clearly obvious that the trial and error method is very inefficient and expensive. In addition, if a problem does not possess a feasible solution, it will take a long time to realize this.

6.8.2   Use of Artificial Variables

This is a systematic way of getting a canonical form with a basic feasible solution when none is available by inspection. First, an *LP* problem is converted to standard form such that all the variables are nonnegative, the constraints are equations, and all the right-hand side constants are nonnegative. Then each constraint is examined for the existence of a basic variable. If none is available, a new variable is added to act as the basic variable in that constraint. In the end, all the constraints will have a basic variable, and by definition we have a canonical system. Since the right-hand side elements are nonnegative, an initial simplex tableau can be formed readily. Of course, the additional variables have no meaning to the original problem. These are merely added so that we will have a ready canonical system to start the simplex method. Hence, these variables are termed *artificial variables* as opposed to the real decision variables in the problem. Eventually they will be forced to zero lest they unbalance the equations. To illustrate the use of artificial variables, consider the following *LP* problem:

Example 6.13 *Consider the minimization problem*

$$minimize \quad Z = -3x_1 + x_2 + x_3,$$

subject to the constraints

$$
\begin{aligned}
x_1 &- 2x_2 + x_3 &\le 11 \\
-4x_1 &+ x_2 + 2x_3 &\ge 3 \\
2x_1 &\quad\quad - x_3 &= -1
\end{aligned}
$$

$$x_1 \ge 0, \quad x_2 \ge 0, \quad x_3 \ge 0.$$

First, the problem is converted to the standard form as follows:

$$minimize \quad Z = -3x_1 + x_2 + x_3,$$

subject to the constraints

$$
\begin{aligned}
x_1 &- 2x_2 + x_3 + u_1 &= 11 \\
-4x_1 &+ x_2 + 2x_3 &- v_2 &= 3 \\
-2x_1 &\quad\quad + x_3 &= 1
\end{aligned}
$$

$$x_1 \geq 0, \ x_2 \geq 0, \ x_3 \geq 0, \ u_1 \geq 0, \ v_2 \geq 0.$$

The slack variable u_1 in the first constraint equation is a basic variable. Since there are no basic variables in the other constraint equations, we add artificial variables, w_3 and w_4, to the second and third constraint equations, respectively. To retain the standard form, w_3 and w_4 will be restricted to be nonnegative. Thus, we now have an **artificial system** *given by:*

$$
\begin{array}{rcrcrcrcrcrcrcr}
x_1 & - & 2x_2 & + & x_3 & + & u_1 & & & & & & & = & 11 \\
-4x_1 & + & x_2 & + & 2x_3 & & & - & v_2 & + & w_3 & & & = & 3 \\
-2x_1 & & & + & x_3 & & & & & & & + & w_4 & = & 1
\end{array}
$$

$$x_1 \geq 0, \ x_2 \geq 0, \ x_3 \geq 0, \ u_1 \geq 0, \ v_2 \geq 0, \ w_3 \geq 0, \ w_4 \geq 0.$$

The artificial system has a basic feasible solution in canonical form given by

$$x_1 = x_2 = x_3 = 0, u_1 = 11, v_2 = 0, w_3 = 3, w_4 = 1.$$

But this is not a feasible solution to the original problem due to the presence of the artificial variables w_3 and w_4 at positive values. ●

On the other hand, it is easy to see that any basic feasible solution to the artificial system in which the artificial variables (w_3 and w_4 in the above example) are zero is automatically a basic feasible solution to the original problem. Hence, the object is to reduce the artificial variables to zero as soon as possible. This can be accomplished in two ways, and each one gives rise to a variant of the simplex method, the *Big M simplex method* and the *Two-Phase simplex method*.

6.9 Big M Simplex Method

In this approach, the artificial variables are assigned a very large cost in the objective function. The simplex method, while trying to improve the objective function, will find the artificial variables uneconomical to maintain as basic variables with positive values. Hence, they will be quickly replaced in the basis by the real variables with smaller costs. For hand calculations it is not necessary to assign a specific cost value to the artificial variables. The general practice is to assign the letter M as the cost in a minimization problem, and $-M$ as the profit in a maximization problem, with the assumption that M is a very large positive number.

The following steps describe the Big M simplex method:

1. Modify the constraints so that the right-hand side of each constraint is nonnegative. This requires that each constraint with a negative right-hand side be multiplied through by -1.

2. Convert each inequality constraint to standard form. This means that if constraint i is a $\leq$ constraint, we add a slack variable u_i, and if i is a $\geq$ constraint, we subtract a surplus variable v_i.

3. If (after step 1 has been completed) constraint i is a $\geq$ or $=$ constraint, add an artificial variable w_i. Also, add the sign restriction $w_i \geq 0$.

4. Let M denote a very large positive number. If an LP problem is a minimization problem, add (for each artificial variable) Mw_i to the objective function. If an LP problem is a maximization problem, add (for each artificial variable) $-Mw_i$ to the objective function.

5. Since each artificial variable will be in the starting basis, all artificial variables must be eliminated from the last row before beginning the simplex method. This ensures that we begin with the canonical form. In choosing the entering variable, remember that M is a very large positive number. Now solve the transformed problem by the simplex method. If all artificial variables are equal to zero in the optimal solution, we have found the optimal solution to the original problem. If any artificial variables are positive in the optimal solution, the original problem is infeasible.

Example 6.14 *To illustrate the Big M simplex method, let us consider the standard form of Example 6.13:*

$$minimize \quad Z = -3x_1 + x_2 + x_3,$$

subject to the constraints

$$
\begin{array}{rrrrrrr}
x_1 & - & 2x_2 & + & x_3 & + & u_1 & & & = & 11 \\
-4x_1 & + & x_2 & + & 2x_3 & & & - & v_2 & = & 3 \\
-2x_1 & & & + & x_3 & & & & & = & 1
\end{array}
$$

$$x_1 \geq 0, \; x_2 \geq 0, \; x_3 \geq 0, \; u_1 \geq 0, \; v_2 \geq 0.$$

Solution. *In order to derive the artificial variables to zero, a large cost will be assigned to w_3 and w_4 so that the objective function becomes:*

$$minimize \quad Z = -3x_1 + x_2 + x_3 + Mw_3 + Mw_4,$$

where M is a very large positive number. Thus, the LP problem with its artificial variables becomes:

$$minimize \quad Z = -3x_1 + x_2 + x_3 + Mw_3 + Mw_4,$$

subject to the constraints

$$
\begin{aligned}
x_1 \;\; - \;\; 2x_2 \;\; + \;\; x_3 \;\; + \;\; u_1 \qquad\qquad\qquad\qquad &= 11 \\
-4x_1 \;\; + \;\; x_2 \;\; + \;\; 2x_3 \qquad\quad - \;\; v_2 \;\; + \;\; w_3 \qquad &= 3 \\
-2x_1 \qquad\qquad + \;\; x_3 \qquad\qquad\qquad\quad + \;\; w_4 &= 1
\end{aligned}
$$

$$x_1 \geq 0, \; x_2 \geq 0, \; x_3 \geq 0, \; u_1 \geq 0, \; v_2 \geq 0, \; w_3 \geq 0, \; w_4 \geq 0.$$

Note the reason behind the use of the artificial variables. We have three equations and seven unknowns. Hence, the starting basic solution must include $7 - 3 = 4$ zero variables. If we put $x_1, x_2, x_3,$ and v_2 at the zero level, we immediately obtain the solution $u_1 = 11, w_3 = 3,$ and $w_4 = 1$, which is the required starting feasible solution. Having constructed a starting feasible solution, we must "condition" the problem so that when we put it in tabular form, the right-hand side column will render the starting solution directly. This is done by using the constraint equations to substitute out w_3 and w_4 in the objective function. Thus,

$$
\begin{aligned}
w_3 &= 3 \;\; + \;\; 4x_1 \;\; - \;\; x_2 \;\; - \;\; 2x_3 \qquad + \;\; v_2 \\
w_4 &= 1 \;\; + \;\; 2x_1 \qquad\qquad - \;\; x_3.
\end{aligned}
$$

The objective function thus becomes

$$Z = -3x_1 + x_2 + x_3 + M(3 + 4x_1 - x_2 - 2x_3 + v_2) + M(1 + 2x_1 - x_3)$$

or

$$Z = (-3 + 6M)x_1 + (1 - M)x_2 + (1 - 3M)x_3 + Mv_2 + 4M,$$

and the Z-equation now appears in the tableau as

$$Z - (-3 + 6M)x_1 - (1 - M)x_2 - (1 - 3M)x_3 - Mv_2 = 4M.$$

Now we can see that at the starting solution, given $x_1 = x_2 = x_3 = v_2 = 0$, the value of Z is $4M$, as it should be when $u_1 = 11, w_3 = 3$, and $w_4 = 1$.

The sequence of tableaus leading to the optimum solution is shown in the following:

basis	x_1	x_2	x_3	u_1	v_2	w_3	w_4	Z	constants
u_1	1	-2	1	1	0	0	0	0	11
w_3	-4	1	2	0	-1	1	0	0	3
w_4	-2	0	①	0	0	0	1	0	1
Z	$3 - 6M$	$-1 + M$	$-1 + 3M$	0	$-M$	0	0	1	$4M$

basis	x_1	x_2	x_3	u_1	v_2	w_3	w_4	Z	constants
u_1	3	-2	0	1	0	0	-1	0	10
w_3	0	①	0	0	-1	1	-2	0	1
x_3	-2	0	1	0	0	0	1	0	1
Z	1	$-1 + M$	0	0	$-M$	0	$1 - 3M$	1	$M + 1$

basis	x_1	x_2	x_3	u_1	v_2	w_3	w_4	Z	constants
u_1	③	0	0	1	-2	2	-5	0	12
x_2	0	1	0	0	-1	1	-2	0	1
x_3	-2	0	1	0	0	0	1	0	1
Z	1	0	0	0	-1	$1 - M$	$-1 - M$	1	2

Now both the artificial variables w_3 and w_4 have been reduced to zero. Thus, Tableau 3 represents a basic feasible solution to the original problem. Of course, this is not an optimal solution since x_1 can reduce the objective function further by replacing u_1 in the basis.

basis	x_1	x_2	x_3	u_1	v_2	w_3	w_4	Z	constants
x_1	1	0	0	$\frac{1}{3}$	$-\frac{2}{3}$	$\frac{2}{3}$	$-\frac{5}{3}$	0	4
x_2	0	1	0	0	-1	1	-2	0	1
x_3	0	0	1	0	0	0	1	0	9
Z	0	0	0	$-\frac{1}{3}$	$-\frac{1}{3}$	$\frac{1}{3} - M$	$\frac{2}{3} - M$	1	-2

Tableau 4 is optimal, and the unique optimal solution is given by $x_1 = 4, x_2 = 1, x_3 = 9, u_1 = 0, v_2 = 0$, and the minimum $z = -2$.   •

Note that an artificial variable is added merely to act as a basic variable in a particular equation. Once it is replaced by a real (decision) variable, there is no need to retain the artificial variable in the simplex tableaus. In other words, we could have omitted the column corresponding to the artificial variable w_4 in Tableaus 2, 3, and 4. Similarly, the column corresponding to w_3 could have been dropped from Tableaus 3 and 4.

When the Big M simplex method terminates with an optimal tableau, it is sometimes possible for one or more artificial variables to remain as basic variables at positive values. This implies that the original problem is infeasible, since no basic feasible solution is possible to the original system if it includes even one artificial variable at a positive value. In other words, the original problem without artificial variables does not have a feasible solution. Infeasibility is due to the presence of inconsistent constraints in the formulation of the problem. In economic terms, this means that the resources of the system are not sufficient to meet the expected demands.

Also, note that for computer solutions, M has to be assigned a specific value. Usually the largest value that can be represented in the computer solution is assumed.

6.10   Two-Phase Simplex Method

A drawback of the Big M simplex method is that assigning a very large value to the constant M can sometimes create computational problems in a digital computer. The Two-Phase method is designed to alleviate this difficulty. Although the artificial variables are added in the same manner employed in the Big M simplex method, the use of the constant M is eliminated by solving the problem in two phases (hence, the name "Two-Phase" method). These two phases are outlined as follows:

Phase 1. This phase consists of finding an initial basic feasible solution to the original problem. In other words, the removal of the artificial variables is taken up first. For this an artificial objective function is created, which is the sum of all the artificial variables. The artificial objective function is then minimized using the simplex method. If the minimum value of the artificial problem is zero, then all the artificial variables have been reduced to zero, and we have a basic feasible solution to the original problem. Go to Phase 2. Otherwise, if the minimum is positive, the problem has no feasible solution. Stop.

Phase 2. The basic feasible solution found is optimized with respect to the original objective function. In other words, the final tableau of Phase 1 becomes the initial tableau for Phase 2 after changing the objective function. The simplex method is once again applied to determine the optimal solution.

The following steps describe the Two-Phase simplex method. Note that steps $1 - 3$ for the Two-Phase simplex method are similar to steps $1 - 3$ for the Big M simplex method.

1. Modify the constraints so that the right-hand side of each constraint is nonnegative. This requires that each constraint with a negative right-hand side be multiplied through by -1.

2. Convert each inequality constraint to standard form. This means that if constraint i is a $\leq$ constraint, we add a slack variable u_i, and if i is a $\geq$ constraint, we subtract a surplus variable v_i.

3. If (after step 1 has been completed) constraint i is a $\geq$ or $=$ constraint, add an artificial variable w_i. Also, add the sign restriction $w_i \geq 0$.

4. For now, ignore the original LP's objective function. Instead solve an LP problem whose objective function is *minimize* $W = $ (sum of all the artificial variables). This is called the Phase 1 LP problem. The act of solving the Phase 1 LP problem will force the artificial variables to be zero.

   Note that:
   If the optimal value of W is equal to zero, and no artificial variables

are in the optimal Phase 1 basis, then we drop all columns in the optimal Phase 1 tableau that correspond to the artificial variables. We now combine the original objective function with the constraints from the optimal Phase 1 tableau. This yields the Phase 2 *LP* problem. The optimal solution to the Phase 2 *LP* problem is the optimal solution to the original *LP* problem.

If the optimal value W is greater than zero, then the original *LP* problem has no feasible solution.

If the optimal value of W is equal to zero and at least one artificial variable is in the optimal Phase 1 basis, then we can find the optimal solution to the original *LP* problem if, at the end of Phase 1, we drop from the optimal Phase 1 tableau all nonbasic artificial variables and any variable from the original problem that has a negative coefficient in the last row of the optimal Phase 1 tableau.

Example 6.15 *To illustrate the Two-Phase simplex method, let us consider again the standard form of Example 6.13:*

$$minimize \quad Z = -3x_1 + x_2 + x_3,$$

subject to the constraints

$$
\begin{aligned}
x_1 - 2x_2 + x_3 + u_1 &= 11 \\
-4x_1 + x_2 + 2x_3 \quad - v_2 &= 3 \\
-2x_1 \quad + x_3 &= 1
\end{aligned}
$$

$$x_1 \geq 0, \ x_2 \geq 0, \ x_3 \geq 0, \ u_1 \geq 0, \ v_2 \geq 0.$$

Solution.
Phase 1 Problem:
Since we need artificial variables w_3 and w_4 in the second and third equations, the Phase 1 problem reads as

$$minimize \quad W = w_3 + w_4,$$

subject to the constraints

$$
\begin{aligned}
x_1 - 2x_2 + x_3 + u_1 \quad &= 11 \\
-4x_1 + x_2 + 2x_3 \quad - v_2 + w_3 \quad &= 3 \\
-2x_1 \quad + x_3 \quad + w_4 &= 1
\end{aligned}
$$

$$x_1 \geq 0, \ x_2 \geq 0, \ x_3 \geq 0, \ u_1 \geq 0, \ v_2 \geq 0, \ w_3 \geq 0, \ w_4 \geq 0.$$

Because w_3 and w_4 are in the starting solution, they must be substituted out in the objective function as follows:

$$
\begin{aligned}
W &= w_3 + w_4 \\
&= (3 + 4x_1 - x_2 - 2x_3 + v_2) + (1 + 2x_1 - x_3) \\
&= 4 + 6x_1 - x_2 - 3x_3 + v_2,
\end{aligned}
$$

and the W equation now appears in the tableau as

$$W - 6x_1 + x_2 + 3x_3 - v_2 = 4.$$

The initial basic feasible solution for the Phase 1 problem is given below:

basis	x_1	x_2	x_3	u_1	v_2	w_3	w_4	W	constants
u_1	1	-2	1	1	0	0	0	0	11
w_3	-4	1	2	0	-1	1	0	0	3
w_4	-2	0	①	0	0	0	1	0	1
W	-6	1	3	0	-1	0	0	1	4

basis	x_1	x_2	x_3	u_1	v_2	w_3	w_4	W	constants
u_1	3	-2	0	1	0	0	-1	0	10
w_3	0	①	0	0	-1	1	-2	0	1
x_3	-2	0	1	0	0	0	1	0	1
W	0	$+1$	0	0	-1	0	-3	1	1

basis	x_1	x_2	x_3	u_1	v_2	w_3	w_4	W	constants
u_1	3	0	0	1	-2	2	-5	0	12
x_2	0	1	0	0	-1	1	-2	0	1
x_3	-2	0	1	0	0	0	1	0	1
W	0	0	0	0	0	-1	1	1	0

We now have an optimal solution to the Phase 1 LP problem, given by $x_1 = 0, x_2 = 1, x_3 = 1, u_1 = 12, v_2 = 0, w_3 = 0, w_4 = 0$, and minimum $W = 0$. Since the artificial variables $w_3 = 0$ and $w_4 = 0$, Tableau 3 represents a basic feasible solution to the original problem.

Phase 2 Problem: *The artificial variables have now served their purpose and must be dispensed with in all subsequent computations. This means that the equations of the optimum tableau in Phase 1 can be written as*

$$
\begin{array}{rcrcrcr}
3x_1 & & & + & u_1 & - & 2v_2 & = & 12 \\
& x_2 & & & & - & v_2 & = & 1 \\
-2x_1 & & + & x_3 & & & & = & 1.
\end{array}
$$

These equations are exactly equivalent to those in the standard form of the original problem (before artificial variables are added). Thus, the original problem can be written as

$$
minimize \quad Z = -3x_1 + x_2 + x_3,
$$

subject to the constraints

$$
\begin{array}{rcrcrcr}
3x_1 & & & + & u_1 & - & 2v_2 & = & 12 \\
& x_2 & & & & - & v_2 & = & 1 \\
-2x_1 & & + & x_3 & & & & = & 1
\end{array}
$$

$$
x_1 \geq 0, \ x_2 \geq 0, \ x_3 \geq 0, \ u_1 \geq 0, \ v_2 \geq 0.
$$

As we can see, the principal contribution of the Phase 1 computations is to provide a ready starting solution to the original problem. Since the problem has three equations and five variables, by putting $5 - 3 = 2$ variables, namely, $x_1 = v_2 = 0$, we immediately obtain the starting basic feasible solution $u_1 = 12, x_2 = 1$, and $x_3 = 1$.

To solve the problem, we need to substitute the basic variables x_1, x_2, and x_3 in the objective function. This is accomplished by using the constraint equations as follows:

$$
\begin{aligned}
Z &= -3x_1 + x_2 + x_3 \\
&= -3(4 - \tfrac{1}{3}u_1 + \tfrac{2}{3}v_2) + (1 + v_2) + 2(4 - \tfrac{1}{3}u_1 + \tfrac{2}{3}v_2) \\
&= -2 + \tfrac{1}{3}u_1 + \tfrac{1}{3}v_2.
\end{aligned}
$$

Thus, the starting tableau for Phase 2 becomes:

basis	x_1	x_2	x_3	u_1	v_2	Z	constants
u_1	③	0	0	1	-2	0	12
x_2	0	1	0	0	-1	0	1
x_3	-2	0	1	0	0	0	1
Z	0	0	0	$-\frac{1}{3}$	$-\frac{1}{3}$	1	-2

basis	x_1	x_2	x_3	u_1	v_2	Z	constants
x_1	1	0	0	$\frac{1}{3}$	$-\frac{2}{3}$	0	4
x_2	0	1	0	0	-1	0	1
x_3	0	0	1	$\frac{2}{3}$	$-\frac{4}{3}$	0	9
Z	0	0	0	$-\frac{1}{3}$	$-\frac{1}{3}$	1	-2

An optimal solution has been reached, and it is given by $x_1 = 4, x_2 = 1, x_3 = 9, u_1 = 0, v_2 = 0$, *and minimum* $Z = -2$. •

Comparing the Big M simplex method and the Two-Phase simplex method, we observe the following:

- The basic approach to both methods is the same. Both add the artificial variables to get the initial canonical system and then derive them to zero as soon as possible.

- The sequence of tableaus and the basis changes are identical.

- The number of iterations are the same.

- The Big M simplex method solves the linear problem in one pass, while the Two-Phase simplex method solves it in two stages as two linear programs.

6.11 Duality

From both the theoretical and practical points of view, the theory of *duality* is one of the most important and interesting concepts in linear programming. Each *LP* problem has a related *LP* problem called the *dual problem*.

The original *LP* problem is called the *primal problem*. For the primal problem defined by (6.1)–(6.3) above, the corresponding dual problem is to find the values of the M variables $y_1, y_2, \ldots, y_M$ to solve the following:

$$\text{minimize} \quad V = b_1 y_1 + b_2 y_2 + \cdots + b_M y_M, \tag{6.17}$$

subject to the constraints

$$
\begin{array}{ccccccc}
a_{11} y_1 & + & a_{21} y_2 & + & \cdots & + & a_{M1} y_M & \geq & c_1 \\
a_{12} y_1 & + & a_{22} y_2 & + & \cdots & + & a_{M2} y_M & \geq & c_2 \\
\vdots & & \vdots & & \cdots & & \vdots & & \vdots \\
a_{1N} y_1 & + & a_{2N} y_2 & + & \cdots & + & a_{MN} y_M & \geq & c_N
\end{array}
\tag{6.18}
$$

and

$$y_1 \geq 0, \ y_2 \geq 0, \ \ldots, \ y_M \geq 0. \tag{6.19}$$

In matrix notation, the primal and the dual problems are formulated as

Primal

Maximize $Z = \mathbf{c}^T \mathbf{x}$
subject to the constraints

$$A\mathbf{x} \leq \mathbf{b}$$

$$\mathbf{x} \geq \mathbf{0}$$

Dual

Minimize $V = \mathbf{b}^T \mathbf{y}$
subject to the constraints

$$A^T \mathbf{y} \geq \mathbf{c}$$

$$\mathbf{y} \geq \mathbf{0},$$

where

$$
A = \begin{pmatrix}
a_{11} & a_{12} & \cdots & a_{1N} \\
a_{21} & a_{22} & \cdots & a_{2N} \\
\vdots & \vdots & \vdots & \vdots \\
a_{M1} & a_{M2} & \cdots & a_{MN}
\end{pmatrix}
$$

$$
\mathbf{b} = \begin{pmatrix} b_1 \\ b_2 \\ \vdots \\ b_M \end{pmatrix}, \quad
\mathbf{c} = \begin{pmatrix} c_1 \\ c_2 \\ \vdots \\ c_N \end{pmatrix}, \quad
\mathbf{x} = \begin{pmatrix} x_1 \\ x_2 \\ \vdots \\ x_N \end{pmatrix}, \quad
\mathbf{y} = \begin{pmatrix} y_1 \\ y_2 \\ \vdots \\ y_M \end{pmatrix}
$$

and $\mathbf{c}^T$ denotes the transpose of the vector $\mathbf{c}$.

The concept of a dual can be introduced with the help of the following *LP* problem.

Example 6.16 *Write the dual of the following linear problem:*

Primal Problem:

$$maximize \quad Z = x_1 + 2x_2 - 3x_3 + 4x_4,$$

subject to the following constraints

$$x_1 + 2x_2 + 2x_3 - 3x_4 \leq 25$$
$$2x_1 + x_2 - 3x_3 + 2x_4 \leq 15$$

$$x_1 \geq 0, \quad x_2 \geq 0, \quad x_3 \geq 0, \quad x_4 \geq 0.$$

The above linear problem has two constraints and four variables. The dual of this primal problem is written as:

Dual Problem:

$$minimize \quad V = 25y_1 + 15y_2,$$

subject to the constraints

$$y_1 + 2y_2 \geq 1$$
$$2y_1 + 2y_2 \geq 2$$
$$2y_1 - 3y_2 \geq -3$$
$$-3y_1 + 2y_2 \geq 4$$

$$y_1 \geq 0, \quad y_2 \geq 0,$$

where y_1 and y_2 are called the dual variables. ●

6.11.1 Comparison of Primal and Dual Problems

Comparing the primal and the dual problems, we observe the following relationships:

1. The objective function coefficients of the primal problem have became the right-hand side constants of the dual. Similarly, the right-hand side constants of the primal have become the cost coefficients of the dual.

2. The inequalities have been reversed in the constraints.

3. The objective function is changed from maximization in primal to minimization in dual.

4. Each column in the primal corresponds to a constraint (row) in the dual. Thus, the number of dual constraints is equal to the number of primal variables.

5. Each constraint (row) in the primal corresponds to a column in the dual. Hence, there is one dual variable for every primal constraint.

6. The dual of the dual is the primal problem.

In both the primal and the dual problems, the variables are nonnegative and the constraints are inequalities. Such problems are called *symmetric dual linear programs*.

Definition 6.7 (Symmetric Form)

A linear program is said to be in symmetric form, if all the variables are restricted to be nonnegative, and all the constraints are inequalities (in a maximization problem the inequalities must be in "less than or equal to" form, while in a minimization problem they must be "greater than or equal to"). •

The general rules for writing the dual of a linear program in symmetric form are summarized below:

1. Define one (nonnegative) dual variable for each primal constraint.

2. Make the cost vector of the primal the right-hand side constants of the dual.

3. Make the right-hand side vector of the primal the cost vector of the dual.

4. The transpose of the coefficient matrix of the primal becomes the constraint matrix of the dual.

5. Reverse the direction of the constraint inequalities.

6. Reverse the optimization direction, i.e., change minimizing to maximizing and vice versa.

Example 6.17 *Write the following linear problem in symmetric form and then find its dual:*

$$minimize \quad Z = 2x_1 + 4x_2 + 3x_3 + 5x_4 + 3x_5 + 4x_6,$$

subject to the constraints

$$
\begin{array}{rcl}
x_1 + x_2 + x_3 & \leq & 300 \\
x_4 + x_5 + x_6 & \leq & 600 \\
x_1 + x_4 & \geq & 200 \\
x_2 + x_5 & \geq & 300 \\
x_3 + x_6 & \geq & 400
\end{array}
$$

$$x_1 \geq 0, \ x_2 \geq 0, \ x_3 \geq 0, \ x_4 \geq 0, \ x_5 \geq 0, \ x_6 \geq 0.$$

Solution. *For the above linear program (minimization) to be in symmetric form, all the constraints must be in "greater than or equal to" form. Hence, we multiply the first two constraints by* -1, *then we have the primal problem as*

$$minimize \quad Z = 2x_1 + 4x_2 + 3x_3 + 5x_4 + 3x_5 + 4x_6,$$

subject to the constraints

$$
\begin{array}{rcl}
-x_1 - x_2 - x_3 & \geq & -300 \\
-x_4 - x_5 - x_6 & \geq & -600 \\
x_1 + x_4 & \geq & 200 \\
x_2 + x_5 & \geq & 300 \\
x_3 + x_6 & \geq & 400
\end{array}
$$

$$x_1 \geq 0, \ x_2 \geq 0, \ x_3 \geq 0, \ x_4 \geq 0, \ x_5 \geq 0, \ x_6 \geq 0.$$

The dual of the above primal problem becomes

$$maximize \quad V = -300y_1 - 600y_2 + 200y_3 + 300y_4 + 400y_5,$$

subject to the constraints

$$
\begin{array}{rrrrrr}
-y_1 & & + \ y_3 & & & \leq \ 2 \\
-y_1 & & & + \ y_4 & & \leq \ 4 \\
-y_1 & & & & + \ y_5 & \leq \ 3 \\
& - \ y_2 & + \ y_3 & & & \leq \ 5 \\
& - \ y_2 & & + \ y_4 & & \leq \ 3 \\
& - \ y_2 & & & + \ y_5 & \leq \ 4
\end{array}
$$

$$y_1 \geq 0, \ y_2 \geq 0, \ y_3 \geq 0, \ y_4 \geq 0, \ y_5 \geq 0.$$

●

6.11.2   Primal-Dual Problems in Standard Form

In most *LP* problems, the dual is defined for various forms of the primal depending on the types of the constraints, the signs of the variables, and the sense of optimization. Now we introduce a definition of the dual that automatically accounts for all forms of the primal. It is based on the fact that any *LP* problem must be put in the standard form before the model is solved by the simplex method. Since all the primal-dual computations are obtained directly from the simplex tableau, it is logical to define the dual in a way that is consistent with the standard form of the primal.

Example 6.18 *Write the standard form of the primal-dual problem of the following linear problem:*

$$maximize \quad Z = 5x_1 + 12x_2 + 4x_3,$$

subject to the constraints

$$
\begin{array}{rrrrl}
x_1 & + \ 2x_2 & + \ x_3 & \leq & 10 \\
2x_1 & - \ x_2 & + \ x_3 & = & 8
\end{array}
$$

$$x_1 \geq 0, \quad x_2 \geq 0, \quad x_3 \geq 0.$$

Solution. *The given primal can be put in the standard primal as*

$$\text{maximize} \quad Z = 5x_1 + 12x_2 + 4x_3,$$

subject to the constraints

$$
\begin{aligned}
x_1 &+& 2x_2 &+& x_3 &+& u_1 &=& 10 \\
2x_1 &-& x_2 &+& x_3 & & &=& 8
\end{aligned}
$$

$$x_1 \geq 0, \quad x_2 \geq 0, \quad x_3 \geq 0, \quad u_1 \geq 0.$$

Notice that u_1 is a slack in the first constraint. Now its dual form can be written as

$$\text{minimize} \quad V = 10y_1 + 8y_2,$$

subject to the constraints

$$
\begin{aligned}
y_1 + 2y_2 &\geq& 5 \\
2y_1 - y_2 &\geq& 12 \\
y_1 + y_2 &\geq& 4
\end{aligned}
$$

$$y_1 \geq 0, \quad y_2 \text{ unrestricted.}$$

●

Example 6.19 *Write the standard form of the primal-dual problem of the following linear problem:*

$$\text{minimize} \quad Z = 5x_1 - 2x_2,$$

subject to the constraints

$$
\begin{aligned}
-x_1 &+& x_2 &\geq& -3 \\
2x_1 &+& 3x_2 &\leq& 5
\end{aligned}
$$

$$x_1 \geq 0, \quad x_2 \geq 0.$$

Solution. *The given primal can be put in the standard primal form as*

$$\text{minimize} \quad Z = 5x_1 - 2x_2,$$

subject to the constraints

$$
\begin{array}{rrrrrrl}
x_1 & - & x_2 & + & u_1 & & = & 3 \\
2x_1 & + & 3x_2 & & & + u_2 & = & 5
\end{array}
$$

$$
x_1 \geq 0, \quad x_2 \geq 0, \quad u_1 \geq 0, \quad u_2 \geq 0.
$$

Notice that u_1 and u_2 are slack in the first and second constraints. Their dual form is

$$
maximize \quad V = 3y_1 + 5y_2,
$$

subject to the constraints

$$
\begin{array}{rrrcr}
y_1 & + & 2y_2 & \leq & 5 \\
-y_1 & + & 3y_2 & \leq & -2 \\
y_1 & & & \leq & 0 \\
& & y_2 & \leq & 0
\end{array}
$$

y_1, y_2 *unrestricted.*

●

Theorem 6.3 (Duality Theorem)

If the primal problem has an optimal solution, then the dual problem also has an optimal solution, and the optimal values of their objective functions are equal, i.e.,

$$
Maximize\ Z = Minimize\ V.
$$

●

It can be shown that when a primal problem is solved by the simplex method, the final tableau contains the optimal solution to the dual problem in the objective row under the columns of the slack variables, i.e., the first dual variable is found in the objective row under the first slack variable, the second is found under the second slack variable, and so on.

Example 6.20 *Find the dual of the following linear problem:*

$$\text{maximize} \quad Z = 12x_1 + 9x_2 + 15x_3,$$

subject to the constraints

$$
\begin{array}{rcrcrcl}
2x_1 & + & x_2 & + & x_3 & \leq & 30 \\
x_1 & + & x_2 & + & 3x_3 & \leq & 40
\end{array}
$$

$$x_1 \geq 0, \quad x_2 \geq 0, \quad x_3 \geq 0,$$

and then find its optimal solution.

Solution. *The dual of this problem is*

$$\text{minimize} \quad V = 30y_1 + 40y_2,$$

subject to the constraints

$$
\begin{array}{rcrcl}
2y_1 & + & y_2 & \geq & 12 \\
y_1 & + & y_2 & \geq & 9 \\
y_1 & + & 3y_2 & \geq & 15
\end{array}
$$

$$y_1 \geq 0, \quad y_2 \geq 0.$$

Introducing the slack variables u_1 and u_2 in order to convert the given linear problem to the standard form, we obtain

$$\text{maximize} \quad Z = 12x_1 + 9x_2 + 15x_3$$

subject to the following constraints

$$
\begin{array}{rcrcrcrcrcl}
2x_1 & + & x_2 & + & x_3 & + & u_1 & & & = & 30 \\
x_1 & + & x_2 & + & 3x_3 & & & + & u_2 & = & 40
\end{array}
$$

$$x_1 \geq 0, \ x_2 \geq 0, \ x_3 \geq 0, \ u_1 \geq 0, \ u_2 \geq 0.$$

We now apply the simplex method and obtain the following tableaus:

basis	x_1	x_2	x_3	u_1	u_2	Z	constants
u_1	2	1	1	1	0	0	30
u_2	1	1	③	0	1	0	40
Z	-12	-9	-15	0	0	1	0

basis	x_1	x_2	x_3	u_1	u_2	Z	constants
u_1	$\frac{5}{3}$	$\frac{2}{3}$	0	1	$-\frac{1}{3}$	0	$\frac{50}{3}$
x_3	$\frac{1}{3}$	$\frac{1}{3}$	1	0	$\frac{1}{3}$	0	$\frac{40}{3}$
Z	-7	-4	0	0	5	1	200

basis	x_1	x_2	x_3	u_1	u_2	Z	constants
x_1	1	$\frac{2}{5}$	0	$\frac{3}{5}$	$-\frac{1}{5}$	0	10
x_3	0	$\frac{1}{5}$	1	$-\frac{1}{5}$	$\frac{2}{5}$	0	10
Z	0	$-\frac{6}{5}$	0	$\frac{21}{5}$	$\frac{18}{5}$	1	270

basis	x_1	x_2	x_3	u_1	u_2	Z	constants
x_2	$\frac{5}{2}$	1	0	$\frac{3}{2}$	$-\frac{1}{2}$	0	25
x_3	$-\frac{1}{2}$	0	1	$-\frac{1}{2}$	$\frac{1}{2}$	0	5
Z	3	0	0	6	3	1	300

Thus, the optimal solution to the given primal problem is

$$x_1 = 0, \quad x_2 = 25, \quad x_3 = 5,$$

and the optimal value of the objective function Z is 300.

The optimal solution to the dual problem is found in the objective row under the slack variables u_1 and u_2 columns as

$$y_1 = 6 \quad and \quad y_2 = 3.$$

Thus, the optimal value of the dual objective function is

$$V = 30(6) + 40(3) = 300,$$

which we expect from the Duality theorem 6.3.   ●

In the following we give another important duality theorem, which gives the relationship between the primal and dual solutions.

Theorem 6.4 (Weak Duality Theorem)

Consider the symmetric primal-dual linear problems:

Primal	**Dual**
Maximize $Z = \mathbf{c}^T\mathbf{x}$	*Minimize* $V = \mathbf{b}^T\mathbf{y}$
subject to the constraints	*subject to the constraints*
$A\mathbf{x} \leq \mathbf{b}$	$A^T\mathbf{y} \geq \mathbf{c}$
$\mathbf{x} \geq 0$	$\mathbf{x} \geq 0$

The value of the objective function of the minimization problem (dual) for any feasible solution is always greater than or equal to that of the maximization problem (primal). •

Example 6.21 *Consider the following LP problem:*

Primal:

$$maximize \quad Z = x_1 + 2x_2 + 3x_3 + 4x_4,$$

subject to the constraints

$$x_1 + 2x_2 + 2x_3 + 3x_4 \leq 20$$
$$2x_1 + x_2 + 3x_3 + 2x_4 \leq 20$$
$$x_1 \geq 0,\ x_2 \geq 0,\ x_3 \geq 0,\ x_4 \geq 0.$$

Its dual form is:

Dual:

$$minimize \quad V = 20y_1 + 20y_2,$$

subject to the constraints

$$y_1 + 2y_2 \geq 1$$
$$2y_1 + y_2 \geq 2$$
$$2y_1 + 3y_2 \geq 3$$
$$3y_1 + 2y_2 \geq 4$$

$$y_1 \geq 0, \quad y_2 \geq 0.$$

The feasible solution for the primal is $x_1 = x_2 = x_3 = x_4 = 1$, *and* $y_1 = y_2 = 1$ *is feasible for the dual. The value of the primal objective is*

$$Z = \mathbf{c}^{\mathbf{T}}\mathbf{x} = 10,$$

and the value of the dual objective is

$$V = \mathbf{b}^{\mathbf{T}}\mathbf{y} = 40.$$

Note that

$$\mathbf{c}^{\mathbf{T}}\mathbf{x} < \mathbf{b}^{\mathbf{T}}\mathbf{y},$$

which satisfies the Weak Duality Theorem 6.4. •

6.12 Sensitivity Analysis in Linear Programming

Sensitivity analysis refers to the study of the changes in the optimal solution and optimal value of objective function Z due to the input data coefficients. The need for such an analysis arises in various circumstances. Often management is not entirely sure about the values of the constants and wants to know the effects of changes. There may be different kinds of modifications:

1. Changes in the right-hand side constants b_i.

2. Changes in the objective function coefficients c_i.

3. Changes in the elements a_{ij} of the coefficient matrix A.

4. Introducing additional constraints or deleting some of the existing constraints.

5. Adding or deleting decision variables.

We will discuss here only changes in the right-hand side constants b_i, which are the most common in sensitivity analysis.

Example 6.22 *A small towel company makes two types of towels, standard and deluxe. Both types have to go through two processing departments, cutting and sewing. Each standard towel needs 1 minute in the cutting department and 3 minutes in the sewing department. The total available time in cutting is 160 minutes for a production run. Each deluxe towel needs 2 minutes in the cutting department and 2 minutes in the sewing department. The total available time in sewing is 240 minutes for a production run. The profit on each standard towel is \$1.00, whereas the profit on each deluxe towel is \$1.50. Determine the number of towels of each type to produce to maximize profit.*

Solution. *Let x_1 and x_2 be the number of standard towels and deluxe towels, respectively. Then the LP problem is*

$$\text{maximize} \quad Z = x_1 + 1.5x_2,$$

subject to the constraints

$$
\begin{array}{rrrll}
x_1 & + & 2x_2 & \leq & 160 \quad \text{(cutting dept.)} \\
3x_1 & + & 2x_2 & \leq & 240 \quad \text{(sewing dept.)}
\end{array}
$$

$$x_1 \geq 0, \quad x_2 \geq 0.$$

After converting the problem to the standard form and then applying the simplex method, one can easily get the final tableau as follows:

basis	x_1	x_2	u_1	u_2	Z	constants
x_2	0	1	$\frac{3}{4}$	$-\frac{1}{4}$	0	60
x_1	1	0	$-\frac{1}{2}$	$\frac{1}{2}$	0	40
Z	0	0	$\frac{5}{8}$	$\frac{1}{8}$	1	130

The optimal solution is

$$x_1 = 40, x_2 = 60, u_1 = u_2 = 0, \quad Z_{max} = 130.$$

Now let us ask a typical sensitivity analysis question:

*Suppose we increase the maximum number of minutes at the cutting depart-
ment by 1 minute, i.e., if the maximum minutes at the cutting department
is 161 instead of 160, what would be the optimal solution?*

Then the revised LP problem will be

$$\text{maximize} \quad Z = x_1 + 1.5x_2,$$

subject to the constraints

$$
\begin{aligned}
x_1 + 2x_2 &\leq 161 \qquad \text{(cutting dept.)} \\
3x_1 + 2x_2 &\leq 240 \qquad \text{(sewing dept.)}
\end{aligned}
$$

$$x_1 \geq 0, \quad x_2 \geq 0.$$

*Of course, we can again solve this revised problem using the simplex method.
However, since the modification is not drastic, we would wonder whether
there is an easy way to utilize the final tableau for the original problem instead of going through all the iteration steps for the revised problem. There
is a way, and this way is the key idea of the sensitivity analysis.*

1. *Since the slack variable for the cutting department is u_1, then use the
   u_1-column.*

2. *Modify the right most column (constants) using the u_1-column as
   subsequently shown, giving the final tableau for the revised problem
   as follows:*

basis	x_1	x_2	u_1	u_2	Z	constants
x_2	0	1	$\frac{3}{4}$	$-\frac{1}{4}$	0	$60 + 1 \times \frac{3}{4}$
x_1	1	0	$-\frac{1}{2}$	$\frac{1}{2}$	0	$40 + 1 \times \left(-\frac{1}{2}\right)$
Z	0	0	$\frac{5}{8}$	$\frac{1}{8}$	1	$130 + 1 \times \frac{5}{8}$

*(where in the last column, the first entry is the original entry, the second
one is one unit (minutes) increased, and the final one is the u_1-column
entry), i.e.,*

basis	x_1	x_2	u_1	u_2	Z	constants
x_2	0	1	$\frac{3}{4}$	$-\frac{1}{4}$	0	$60\frac{3}{4}$
x_1	1	0	$-\frac{1}{2}$	$\frac{1}{2}$	0	$39\frac{1}{2}$
Z	0	0	$\frac{5}{8}$	$\frac{1}{8}$	1	$130\frac{5}{8}$

then the optimal solution for the revised problem is

$$x_1 = 39\frac{1}{2}, x_2 = 60\frac{3}{4}, u_1 = u_2 = 0, \quad Z_{max} = 130\frac{5}{8}.$$

Let us try one more revised problem:

Assume that the maximum number of minutes at the sewing department is reduced by 8, making the maximum minutes $240 - 8 = 232$. The final tableau for this revised problem will be given as follows:

basis	x_1	x_2	u_1	u_2	Z	constants
x_2	0	1	$\frac{3}{4}$	$-\frac{1}{4}$	0	$60 + (-8) \times (-\frac{1}{4}) = 62$
x_1	1	0	$-\frac{1}{2}$	$\frac{1}{2}$	0	$40 + (-8) \times (\frac{1}{2}) = 36$
Z	0	0	$\frac{5}{8}$	$\frac{1}{8}$	1	$130 + (-8) \times (\frac{1}{8}) = 129$

then the optimal solution for the revised problem is

$$x_1 = 36, x_2 = 62, u_1 = u_2 = 0, \quad Z_{max} = 129.$$

The bottom-row entry, $\frac{5}{8}$, represents the net profit increase for a one unit (minute) increase of the available time at the cutting department. It is called the shadow price at the cutting department. Similarly, another bottom-row entry, $\frac{1}{8}$, is called the shadow price at the sewing department.

In general, the shadow price for a constraint is defined as the change in the optimal value of the objective function when one unit is increased in the right-hand side of the constraint.

A negative entry in the bottom row represents the net profit increase when one unit of the variable in that column is introduced. For example, if a negative entry in the x_1 column is $-\frac{1}{4}$, then introducing one unit of x_1 will result in $\$(\frac{1}{4}) = 25$ cents net profit gain. Therefore, the bottom-row entry, $\frac{5}{8}$, in the preceding tableau represents that the net profit loss is $\$(\frac{5}{8})$ when one unit of u_1 is introduced, keeping the constraint, 160, the same.

Now, suppose the constraint at the cutting department is changed from 160 to 161. If this increment of 1 minute is credited to u_1 as a slack, or unused time at the cutting department, the total profit will remain the same because the unused time will not contribute to a profit increase. However, if this $u_1 = 1$ is given up, or reduced (which is the opposite of introduced), it will yield a net profit gain of $\$(\frac{5}{8})$. •

6.13 **Summary**

In this chapter we gave a brief introduction to linear programming. Problems were described by systems of linear inequalities. One can see that small systems can be solved in a graphical manner, but that large systems are solved using row operations on matrices by means of the simplex method. For finding the basic feasible solution to artificial systems, we discussed the Big M simplex method and the Two-Phase simplex method.

In this chapter we also discussed the concept of duality in linear programming. Since the optimal primal solution can be obtained directly from the optimal dual tableau (and vice-versa), it is advantageous computationally to solve the dual when it has fewer constraints than the primal. Duality provides an economic interpretation that sheds light on the unit worth or shadow price of the different resources. It also explains the condition of optimality by introducing the new economic definition of inputted costs for each activity. We closed this chapter with a presentation of the important technique of sensitivity analysis, which gives linear programming the dynamic characteristic of modifying the optimum solution to reflect changes in the model.

6.14 **Problems**

1. The Oakwood Furniture Company has 12.5 units of wood on hand from which to manufacture tables and chairs. Making a table uses two units of wood and making a chair uses one unit. Oakwood's distributor will pay $20 for each table and $15 for each chair, but they will not accept more than eight chairs, and they want at least twice as many chairs as tables. How many tables and chairs should the company produce to maximize its revenue? Formulate this as a linear programming problem.

2. The Mighty Silver Ball Company manufactures three kinds of pinball machines, each requiring a different manufacturing technique. The Super Deluxe Machine requires 17 hours of labor, 8 hours of testing, and yields a profit of $300. The Silver Ball Special requires 10 hours of labor, 4 hours of testing, and yields a profit of $200. The Bumper King requires 2 hours of labor, 2 hours of testing, and yields a profit of $100. There are 1000 hours of labor and 500 hours of testing available.

   In addition, a marketing forecast has shown that the demand for the Super Deluxe is no more that 50 machines, demand for the Silver Ball Special is no more than 80, and demand for the Bumper King is no more than 150. The manufacturer wants to determine the optimal production schedule that will maximize the total profit. Formulate this as a linear programming problem.

3. Consider a diet problem in which a college student is interested in finding a minimum cost diet that provides at least 21 units of Vitamin A and 12 units of Vitamin B from five foods with the following properties:

Food	1	2	3	4	5
Vitamin A content	1	0	1	1	2
Vitamin B content	0	1	2	1	1
Cost per unit (cents)	20	20	31	11	12

Formulate this as a linear programming problem.

4. Consider a problem of scheduling the weekly production of a certain item for the next 4 weeks. The production cost of the item is $10 for the first two weeks, and $15 for the last two weeks. The weekly demands are 300, 700, 800, and 900 units, which must be met. The plant can produce a maximum of 700 units each week. In addition, the company can employ overtime during the second and third weeks. This increases weekly production by an additional 200 units, but the cost of production increases by $5 per unit. Excess production can be stored at a cost of $3 an item per week. How should the production be scheduled to minimize the total cost? Formulate this as a linear programming problem.

5. An oil refinery can blend three grades of crude oil to produce regular and super gasoline. Two possible blending processes are available. For each production run the older process uses 5 units of crude A, 7 units of crude B, and 2 units of crude C to produce 9 units of regular and 7 units of super gasoline. The newer process uses 3 units of crude A, 9 units of crude B, and 4 units of crude C to produce 5 units of regular and 9 units of super gasoline for each production run. Because of prior contract commitments, the refinery must produce at least 500 units of regular gasoline and at least 300 units of super for the next month. It has available 1500 units of crude A, 1900 units of crude B, and 1000 units of crude C. For each unit of regular gasoline produced the refinery receives $6, and for each unit of super it receives $9. Determine how to use the resources of crude oil and the two blending processes to meet the contract commitments and, at the same time, maximize revenue. Formulate this as a linear programming problem.

6. A tailor has 80 square yards of cotton material and 120 square yards of woolen material. A suit requires 2 square yards of cotton and 1

square yard of wool. A dress requires 1 square yard of cotton and 3 square yards of wool. How many of each garment should the tailor make to maximize income if a suit and a dress each sell for $90? What is the maximum income? Formulate this as a linear programming problem.

7. A trucking firm ships the containers of two companies, A and B. Each container from company A weighs 40 pounds and is 2 cubic feet in volume. Each container from company B weighs 50 pounds and is 3 cubic feet in volume. The trucking firm charges company A $2.20 for each container shipped and charges company B $3.00 for each container shipped. If one of the firm's trucks cannot carry more than $37,000$ pounds and cannot hold more than 2000 cubic feet, how many containers from companies A and B should a truck carry to maximize the shipping charges?

8. A company produces two types of cowboy hats. Each hat of the first type requires twice as much labor time as does each hat of the second type. If all hats are of the second type only, the company can produce a total of 500 hats a day. The market limits daily sales of the first and second types to 150 and 200 hats, respectively. Assume that the profit per hat is $8 for type 1 and $5 for type 2. Determine the number of hats of each type to produce to maximize profit.

9. A company manufactures two types of hand calculators, of model A and model B. It takes 1 hour and 4 hours in labor time to manufacture each A and B, respectively. The cost of manufacturing A is $30 and that of manufacturing B is $20. The company has 1600 hours of labor time available and $18,000$ in running costs. The profit on each A is $10 and on each B is $8. What should the production schedule be to ensure maximum profit?

10. A clothing manufacturer has 10 square yards of cotton material, 10 square yards of wool material, and 6 square yards of silk material. A pair of slacks requires 1 square yard of cotton, 2 square yards of wool, and 1 square yard of silk. A skirt requires 2 square yards of cotton, 1 square yard of wool, and 1 square yard of silk. The net profit on

a pair of slacks is \$3 and the net profit on a skirt is \$4. How many skirts and how many slacks should be made to maximize profit?

11. A manufacturer produces sacks of chicken feed from two ingredients, A and B. Each sack is to contain at least 10 ounces of nutrient N_1, at least 8 ounces of nutrient N_2, and at least 12 ounces of nutrient N_3. Each pound of ingredient A contains 2 ounces of nutrient N_1, 2 ounces of nutrient N_2, and 6 ounces of nutrient N_3. Each pound of ingredient B contains 5 ounces of nutrient N_1, 3 ounces of nutrient N_2, and 4 ounces of nutrient N_3. If ingredient A costs 8 cents per pound and ingredient B costs 9 cents per pound, how much of each ingredient should the manufacturer use in each sack of feed to minimize the cost?

12. The Apple Company has a contract with the government to supply 1200 microcomputers this year and 2500 next year. The company has the production capacity to make 1400 microcomputers each year, and it has already committed its production line for this level. Labor and management have agreed that the production line can be used for at most 80 overtime shifts each year, each shift costing the company an additional \$20,000. In each overtime shift, 50 microcomputers can be manufactured this year and used to meet next year's demand, but must be stored at a cost of \$100 per unit. How should the production be scheduled to minimize cost?

13. Solve each of the following linear programming problems using the graphical method:

    (a)   Maximize: $Z = 2x_1 + x_2$
               Subject to: $4x_1 + x_2 \le 36$
                             $4x_1 + 3x_2 \le 60$
                             $x_1 \ge 0, x_2 \ge 0$

    (b)   Maximize: $Z = 2x_1 + x_2$
               Subject to: $4x_1 + x_2 \le 16$
                               $x_1 + x_2 \le 7$

$$x_1 \geq 0, x_2 \geq 0$$

(c) Maximize: $Z = 4x_1 + x_2$
Subject to: $2x_1 + x_2 \leq 4$
$6x_1 + x_2 \leq 8$
$x_1 \geq 0, x_2 \geq 0$

(d) Minimize: $Z = 2x_1 - 7x_2 - 3x_3$
Subject to: $x_1 + 2x_2 + x_3 \leq 5$
$x_1 + x_3 \leq 10$
$x_1 \geq 0, \quad x_2 \geq 0, \quad x_3 \geq 0$

14. Solve each of the following linear programming problems using the graphical method:

(a) Maximize: $Z = 6x_1 - 2x_2$
Subject to: $x_1 - x_2 \leq 1$
$3x_1 - x_2 \leq 6$
$x_1 \geq 0, x_2 \geq 0$

(b) Maximize: $Z = 4x_1 + 4x_2$
Subject to: $2x_1 + 7x_2 \leq 21$
$7x_1 + 2x_2 \leq 49$
$x_1 \geq 0, x_2 \geq 0$

(c) Maximize: $Z = 3x_1 + 2x_2$
Subject to: $2x_1 + x_2 \leq 2$
$3x_1 + 4x_2 \geq 12$
$x_1 \geq 0, \quad x_2 \geq 0$

(d) Maximize: $Z = -2x_1 - x_2 + 4x_3$
Subject to: $3x_1 - x_2 + 2x_3 \leq 25$
$-x_1 - x_2 + 2x_3 \leq 20$
$-x_1 - x_2 + x_3 \leq 5$
$x_1 \geq 0, x_2 \geq 0, x_3 \geq 0$

15. Solve each of the following linear programming problems using the graphical method:

(a) Minimize: $Z = -3x_1 - x_2$
Subject to: $x_1 + x_2 \leq 150$
$4x_1 + x_2 \leq 450$
$x_1 \geq 0, x_2 \geq 0$

(b) Minimize: $Z = -2x_1 + x_2$
Subject to: $2x_1 + 3x_2 \leq 14$
$4x_1 + 5x_2 \leq 16$
$x_1 \geq 0, x_2 \geq 0$

(c) Minimize: $Z = -2x_1 + x_2$
Subject to: $2x_1 + x_2 \leq 440$
$4x_1 + x_2 \leq 680$
$x_1 \geq 0, x_2 \geq 0$

(d) Maximize: $Z = x_1 - x_2$
Subject to: $x_1 + x_2 \leq 6$
$x_1 - x_2 \geq 0$
$-x_1 - x_2 \geq 3$
$x_1 \geq 0, x_2 \geq 0$

16. Solve each of the following linear programming problems using the graphical method:

(a) Minimize: $Z = 3x_1 + 5x_2$
Subject to: $3x_1 + 2x_2 \geq 36$
$3x_1 + 5x_2 \geq 45$
$x_1 \geq 0, x_2 \geq 0$

(b) Minimize: $Z = 3x_1 - 8x_2$
Subject to: $2x_1 - x_2 \leq 4$

$$3x_1 + 11x_2 \leq 33$$
$$3x_1 + 4x_2 \geq 24$$
$$x_1 \geq 0, x_2 \geq 0$$

(c)   Minimize: $Z = 3x_1 - 5x_2$
Subject to: $2x_1 - x_2 \leq -2$
$$4x_1 - x_2 \geq 0$$
$$x_2 \leq 3$$
$$x_1 \geq 0, x_2 \geq 0$$

(d)   Minimize: $Z = -3x_1 + 2x_2$
Subject to: $3x_1 - x_2 \geq -5$
$$-x_1 + x_2 \geq 1$$
$$2x_1 + 4x_2 \geq 12$$
$$x_1 \geq 0, x_2 \geq 0$$

17. Solve each of the following linear programming problems using the simplex method:

(a)   Maximize: $Z = 3x_1 + 2x_2$
Subject to: $-x_1 + 2x_2 \leq 4$
$$3x_1 + 2x_2 \leq 14$$
$$x_1 - x_2 \leq 3$$
$$x_1 \geq 0, x_2 \geq 0$$

(b)   Maximize: $Z = x_1 + 3x_2$
Subject to: $4x_1 \leq 5$
$$x_1 + 2x_2 \leq 10$$
$$x_2 \leq 4$$
$$x_1 \geq 0, x_2 \geq 0$$

(c)   Maximize: $Z = 10x_1 + 5x_2$
Subject to: $x_1 + x_2 \leq 180$
$$3x_1 + 2x_2 \leq 480$$

$$x_1 \geq 0, x_2 \geq 0$$

(d)   Maximize: $Z = x_1 - 4x_2$
     Subject to: $x_1 + 2x_2 \leq 5$
             $x_1 + 6x_2 \leq 7$
             $x_1 \geq 0, x_2 \geq 0$

18. Solve Problem 13 using the simplex method.

19. Solve each of the following linear programming problems using the simplex method:

(a)   Maximize: $Z = 2x_1 + 4x_2 + x_3$
     Subject to: $-x_1 + 2x_2 + 3x_3 \leq 6$
             $-x_1 + 4x_2 + 5x_3 \leq 5$
             $-x_1 + 5x_2 + 7x_3 \leq 7$
             $x_1 \geq 0, x_2 \geq 0, x_3 \geq 0$

(b)   Minimize: $Z = 2x_1 - x_2 + 2x_3$
     Subject to: $-x_1 + x_2 + x_3 \leq 4$
             $x_1 + 2x_2 + 3x_3 \leq 3$
             $-x_1 + x_2 - x_3 \leq 6$
             $x_1 \geq 0, x_2 \geq 0, x_3 \geq 0$

(c)   Maximize: $Z = 100x_1 + 200x_2 + 50x_3$
     Subject to: $5x_1 + 5x_2 + 10x_3 \leq 1000$
             $10x_1 + 8x_2 + 5x_3 \leq 2000$
             $10x_1 + 5x_2 \leq 500$
             $x_1 \geq 0, x_2 \geq 0, x_3 \geq 0$

(d)   Minimize: $Z = 6x_1 + 3x_2 + 4x_3$
     Subject to: $x_1 \geq 30$
             $2x_2 \leq 50$
             $x_3 \geq 20$

$$x_1 + x_2 + x_3 \geq 120$$
$$x_1 \geq 0, x_2 \geq 0, x_3 \geq 0$$

20. Solve each of the following linear programming problems using the simplex method:

(a) Minimize: $Z = x_1 + 2x_2 + 3x_3 + 4x_4$
     Subject to: $x_1 + 2x_2 + 2x_3 + 3x_4 \leq 20$
                  $2x_1 + x_2 + 3x_3 + 2x_4 \leq 20$
                  $x_1 \geq 0, x_2 \geq 0, x_3 \geq 0, x_4 \geq 0$

(b) Maximize: $Z = x_1 + 2x_2 + 4x_3 - x_4$
     Subject to: $5x_1 + 4x_3 + 6x_4 \leq 20$
                  $4x_1 + 2x_2 + 2x_3 + 8x_4 \leq 40$
                  $x_1 \geq 0, x_2 \geq 0, x_3 \geq 0, x_4 \geq 0$

(c) Minimize: $Z = 3x_1 + x_2 + x_3 + x_4$
     Subject to: $2x_1 + 2x_2 + x_3 + 2x_4 \leq 4$
                  $3x_1 + x_2 + 2x_3 + 4x_4 \leq 6$
                  $x_1 \geq 0, x_2 \geq 0, x_3 \geq 0, x_4 \geq 0$

(d) Minimize: $Z = x_1 + 2x_2 - x_3 + 3x_4$
     Subject to: $2x_1 + 4x_2 + 5x_3 + 6x_4 \leq 24$
                  $4x_1 + 4x_2 + 2x_3 + 2x_4 \leq 4$
                  $x_1 \geq 0, x_2 \geq 0, x_3 \geq 0, x_4 \geq 0$

21. Use Big M simplex method to solve each of the following linear programming problems:

(a) Minimize: $Z = 4x_1 + 4x_2 + x_3$
     Subject to: $x_1 + x_2 + x_3 \leq 2$
                  $2x_1 + x_2 + 0x_3 \leq 3$
                  $2x_1 + x_2 + 3x_3 \geq 3$
                  $x_1 \geq 0, x_2 \geq 0, x_3 \geq 0$

(b)   Minimize: $Z = 2x_1 + 3x_2$
Subject to: $0.5x_1 + 0.25x_2 \leq 4$
$x_1 + 3x_2 \geq 36$
$x_1 + x_2 = 10$
$x_1 \geq 0, x_2 \geq 0$

(c)   Minimize: $Z = -3x_1 - 2x_2$
Subject to: $x_1 + x_2 = 10$
$x_1 + 0x_2 \geq 4$
$x_1 \geq 0, x_2 \geq 0$

(d)   Minimize: $Z = 2x_1 + 3x_2$
Subject to: $0.5x_1 + 0.25x_2 \leq 4$
$x_1 + 3x_2 \geq 20$
$x_1 \geq 0, \quad x_2 \geq 0$

22. Use the Two-Phase simplex method to solve each of the following linear programming problems:

(a)   Minimize: $Z = 2x_1 + 3x_2$
Subject to: $0.5x_1 + 0.25x_2 \leq 4$
$x_1 + 3x_2 \geq 36$
$x_1 + x_2 = 10$
$x_1 \geq 0, x_2 \geq 0$

(b)   Maximize: $Z = 2x_1 + 3x_2 - 5x_3$
Subject to: $x_1 + x_2 + x_3 = 7$
$2x_1 - 5x_2 + x_3 \geq 10$
$x_1 \geq 0, x_2 \geq 0, x_3 \geq 0$

(c)   Minimize: $Z = 4x_1 + x_2$
Subject to: $3x_1 + x_2 = 3$
$4x_1 + 3x_2 - x_3 = 6$
$4x_1 + 2x_2 + x_4 = 4$

$$x_1 \geq 0, x_2 \geq 0, x_3 \geq 0, x_4 \geq 0$$

(d)   Minimize: $Z = 40x_1 + 10x_2 + 7x_5 + 14x_6$

Subject to:
$$x_1 - x_2 + 2x_5 = 0$$
$$-2x_1 + x_2 - 2x_5 = 0$$
$$x_1 + x_3 + x_5 - x_6 = 3$$
$$2x_2 + x_3 + x_4 + 2x_5 + x_6 = 4$$
$$x_i \geq 0, i = 1, \ldots, 6$$

23. Write the duals of each of the following linear programming problems:

(a)   Maximize: $Z = -5x_1 + 2x_2$

Subject to:
$$-x_1 + x_2 \leq -3$$
$$2x_1 + 3x_2 \leq 5$$
$$x_1 \geq 0, x_2 \geq 0$$

(b)   Maximize: $Z = 5x_1 + 6x_2 + 4x_3$

Subject to:
$$x_1 + 4x_2 + 6x_3 \leq 12$$
$$2x_1 + x_2 + 2x_3 \leq 11$$
$$x_1 \geq 0, x_2 \geq 0, x_3 \geq 0$$

(c)   Minimize: $Z = 6x_1 + 3x_2$

Subject to:
$$6x_1 - 3x_2 + x_3 \geq 2$$
$$3x_1 + 4x_2 + x_3 \geq 5$$
$$x_1 \geq 0, x_2 \geq 0, x_3 \geq 0$$

(d)   Minimize: $Z = 2x_1 + x_2 + 2x_3$

Subject to:
$$-x_1 + x_2 + x_3 = 4$$
$$-x_1 + x_2 - x_3 \leq 6$$
$$x_1 \geq 0, x_2 \geq 0, x_3 \geq 0$$

24. Write the duals of each of the following linear programming problems:

(a)   Minimize: $Z = 3x_1 + 4x_2 + 6x - 3$

Subject to: $x_1 + x_2 \geq 10$

$$x_1 \geq 0, x_3 \geq 0, x_2 \leq 0$$

(b)   Maximize: $Z = 5x_1 + 2x_2 + 3x_3$
Subject to: $x_1 + 5x_2 + 2x_3 = 30$
$x_1 - x_2 - 6x_3 \leq 40$
$x_1 \geq 0, x_2 \geq 0, x_3 \geq 0$

(c)   Minimize: $Z = x_1 + 2x_2 + 3x_3 + 4x_4$
Subject to: $2x_1 + 2x_2 + 2x_3 + 3x_4 \geq 30$
$2x_1 + x_2 + 3x_3 + 2x_4 \geq 20$
$x_1 \geq 0, \ x_2 \geq 0, \ x_3 \geq 0, \ x_4 \geq 0$

(d)   Maximize: $Z = x_1 + x_2$
Subject to: $2x_1 + x_2 = 5$
$3x_1 - x_2 = 6$
$x_1 \geq 0, x_2 \geq 0$

(e)   Maximize: $Z = 2x_1 + 4x_2$
Subject to: $4x_1 + 3x_2 = 50$
$2x_1 + 5x_2 = 75$
$x_1 \geq 0, x_2 \geq 0$

(f)   Maximize: $Z = x_1 + 2x_2 + 3x - 3$
Subject to: $2x_1 + x_2 + x_3 = 5$
$x_1 + 3x_2 + 4x_3 = 3$
$x_1 \geq 0, x_2 \geq 0$

Chapter 7

Nonlinear Programming

7.1 Introduction

In the previous chapter, we studied linear programming problems in some
detail. For such cases, our goal was to maximize or minimize a linear function subject to linear constraints. But in many interesting maximization
and minimization problems, the objective function may not be a linear
function, or some of the constraints may not be linear constraints. Such
an optimization problem is called a Nonlinear Programming (*NLP*) problem.

An *NLP* problem is characterized by terms or groups of terms that
involve intrinsically nonlinear functions. For example, the transcendental
functions such as $\sin(x), \cos(x)$, or exponential functions and logarithmic
functions such as $e^x, \ln(x+1)$, etc. Nonlinearities also arise as a result of
interactions between two or more variables, such as $x \ln y, xy, x^y$, and so on.

Remember that in studying linear programming solution techniques, there was a basic underlying structure that was exploited in solving those problems. This structure found that an optimal solution could be achieved by solving linear systems of equations. It was also known that an optimal solution would always be found at an extreme point of the feasible solution space. Though in solving *NLP* problems, an optimal solution might be found at an extreme point, or at a point of discontinuity, in addition, algorithmic techniques might involve the solution of simultaneous linear systems of equations, simultaneous nonlinear equations, or both. Before formally defining an *NLP* problem, we begin with a review of material from differential calculus, Taylor's series approximations, and definitions of the gradient vector and the Hessian matrix of functions of n variables, which will be needed for our study of nonlinear programming. A discussion of quadratic functions and convex functions and sets is also included.

7.2   Review of Differential Calculus

We begin with a review of material from differential calculus, which will be needed for the discussion of nonlinear programming.

7.2.1   Limits of Functions

The concept of the limit of a function f is one of the fundamental ideas that distinguishes calculus from algebra and trigonometry. One of the important things to know about a function f is how its outputs will change when the inputs change. If the inputs get closer and closer to some specific value a, for example, will the outputs get closer to some specific value L? If they do, we want to know that because it means we can control the outputs by controlling the inputs.

Definition 7.1 (Limits)

The equation

$$\lim_{x \to a} f(x) = L$$

means that as x gets closer to a (real number), the value of $f(x)$ gets arbitrarily closer to L. Note that the limit of a function may or may not exist. ●

Example 7.1 *Find the limit, if it exists:*

$$\lim_{x \to 3} \frac{2x^3 - 6x^2 + x - 3}{x - 3}.$$

Solution. *The domain of the given function*

$$f(x) = \frac{2x^3 - 6x^2 + x - 3}{x - 3}$$

is all the real numbers except the number 3. *To find the limit we shall change the form of* $f(x)$ *by factoring it as follows:*

$$\lim_{x \to 3} \frac{(2x^2 + 1)(x - 3)}{x - 3}.$$

When we investigate the limit as $x \to 3$, *we assume that* $x \neq 3$. *Hence,* $x - 3 \neq 0$, *and we can now cancel the factor* $x - 3$. *Thus,*

$$\lim_{x \to 3} 2x^2 + 1 = 2(3)^2 + 1 = 19,$$

and the given limit exists. ●

Example 7.2 *Find the limit, if it exists:*

$$\lim_{x \to 3} \sqrt{x - 3}.$$

Solution. *The given function is*

$$f(x) = \sqrt{x - 3}.$$

To find the limit of $f(x)$, *we have to find the one-side limits, i.e., the right-hand limit*

$$\lim_{x \to 3^+} f(x),$$

and the left-hand limit

$$\lim_{x \to 3^-} f(x).$$

First, we find

$$\lim_{x \to 3^+} \sqrt{x - 3}.$$

If $x > 3$, then $x - 3 > 0$, and hence $f(x) = \sqrt{x - 3}$ is a real number; i.e., $f(x)$ is defined. Thus,

$$\lim_{x \to 3^-} \sqrt{x - 3} = \sqrt{3 - 3} = 0.$$

Now we find the left-hand limit

$$\lim_{x \to 3^-} \sqrt{x - 3}.$$

But this limit does not exist because $f(x) = \sqrt{x - 3}$ is not a real number, if $x < 3$. Thus, the limit of the function

$$\lim_{x \to 3} \sqrt{x - 3},$$

does not exist because $f(x)$ is not defined throughout an open interval containing 3. •

The relationship between one-sided limits and limits is described in the following theorem.

Theorem 7.1

$$\lim_{x \to a} f(x) = L, \quad \text{if and only if} \quad \lim_{x \to a^+} f(x) = L = \lim_{x \to a^-} f(x).$$

•

7.2.2   Continuity of a Function

In mathematics and science, we use the word continuous to describe a process that goes on without abrupt changes. In fact, our experience leads us to assume that this is an essential feature of many natural processes. The issue of continuity has become one of practical as well as theoretical importance. As scientists, we need to know when continuity is called for, what it is, and how to test for it.

Definition 7.2 (Continuity)

A function $f(x)$ is continuous at a point a, if the following conditions are satisfied:

(i) *$f(a)$ must be defined (real number).*

(ii) *$\lim\limits_{x \to a} f(x)$ must exist.*

(iii) *$\lim\limits_{x \to a} f(x) = f(a)$.*

Note that if $f(x)$ is not a continuous function at $x = a$, then we say that $f(x)$ is discontinuous (or has a discontinuity) at a. ●

Example 7.3 *Show that the function*

$$f(x) = \begin{cases} \dfrac{16 - x^2}{4 - x}, & \text{if } x \neq 4 \\[2mm] 8, & \text{if } x = 4 \end{cases}$$

is continuous at $x = 4$.

Solution. *The given function is continuous at $x = 4$ because*

(i) *$f(4) = 8$ is a real number.*

(ii) $\lim\limits_{x \to 4} \dfrac{16 - x^2}{4 - x} = \lim\limits_{x \to 4} \dfrac{(4 - x)(4 + x)}{4 - x} = \lim\limits_{x \to 4}(x + 4) = 8$ *exists.*

(iii) $\lim\limits_{x \to 4} \dfrac{16 - x^2}{4 - x} = 8 = f(4)$. ●

7.2.3 Derivative of a Function

Derivatives are the functions that measure the rates at which things change. We use them to calculate velocities and accelerations, to predict the effect of flight maneuvers on the heart, and to describe how sensitive formulas are to errors in measurement.

Definition 7.3 (Differentiation)

The derivative of a function $f(x)$ at an arbitrary point x can be denoted as $f'(x)$ and is defined as

$$f'(x) = \lim_{x \to a} \frac{f(x+h) - f(x)}{h}, \quad \text{provided limits exist.}$$

The process of finding derivatives is called differentiation. ●

If a limit does not exist, then the function is not differentiable. Remember that the derivative of $f(x)$ at $x = a$, i.e., $f'(a)$ is called the *slope* of $f(x)$. If $f'(a) > 0$, then $f(x)$ is *increasing* at $x = a$, whereas if $f'(a) < 0$, then $f(x)$ is *decreasing* at $x = a$.

Basic Rules of Differentiation

Functions	Derivatives of the Functions
$f(x) = c$	$f'(x) = 0$
$f(x) = x$	$f'(x) = 1$
$f(x) = ax + b$	$f'(x) = a$
$f(x) = x^n$	$f'(x) = nx^{n-1}$
$f(x) = e^{bx}$	$f'(x) = be^{bx}$
$f(x) = a^x$	$f'(x) = a^x \ln a$
$f(x) = \ln x$	$f'(x) = \frac{1}{x}$
$f(x) = [g(x)]^n$	$f'(x) = n[g(x)]^{n-1}g'(x)$
$f(x) = [f_1(x) + f_2(x)]$	$f'(x) = f_1'(x) + f_2'(x)$
$f(x) = [f_1(x)f_2(x)]$	$f'(x) = f_1'(x)f_2(x) + f_1(x)f_2'(x)$
$f(x) = \dfrac{f_1(x)}{f_2(x)}$	$f'(x) = \dfrac{f_2(x)f_1'(x) - f_1(x)f_2'(x)}{[f_2(x)]^2}$

Higher Derivatives

Sometimes we have to find the derivatives of derivatives. For this we can take sequential derivatives to form second derivatives, third derivatives, and so on. As we have seen, if we differentiate a function f, we obtain another function denoted f'. If f' has a derivative, it is denoted f'' and is called the *second derivative* of f. The *third derivative* of f, denoted f''', is the derivative of the second derivative. In general, if n is a positive integer, then f^n denotes the *nth derivative* of f and is found by starting with f and differentiating, successively, n times.

Example 7.4 *Find the first three derivatives of the function*

$$f(x) = 2x^4 - 5x^3 + x^2 - 4x + 1.$$

Solution. *By using the differentiation rule, we find the first derivative of the given function as follows:*

$$f'(x) = 8x^3 - 15x^2 + 2x - 4.$$

Similarly, we can find the second and the third derivatives of the function as follows:

$$
\begin{aligned}
f''(x) &= 24x^2 - 30x + 2 \\
f'''(x) &= 48x - 30,
\end{aligned}
$$

which are the required first three derivatives of the function.   ●

To plot the above function $f(x)$ and its first three derivatives $f'(x)$, $f''(x)$, $f'''(x)$, we use the following MATLAB commands:

```
>> x = [-3 : 0.01 : 3];
>> y = 2 * x.^ 4-5 * x.^ 3+x.^ 2-4 * x + 1;
>> plot(x, y)
>> holdon
>> dy = 8 * x.^ 3-15 * x.^ 2+2 * x - 4;
>> plot(x, dy)
>> ddy = 24 * x.^ 2-30 * x + 2;
>> plot(x, ddy)
>> dddy = 48 * x - 30;
>> plot(x, dddy)
```

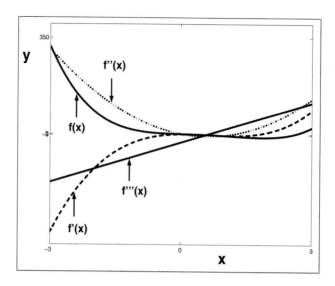

Figure 7.1: Higher-order derivatives of the function.

7.2.4 Local Extrema of a Function

One of the principal applications of derivatives is to find the *local maximum* and *local minimum* values (local extrema) of a function in an interval. Points at which the first derivative of the function is zero ($f'(x) = 0$) are called *critical points* of $f(x)$. Although the condition $f'(c) = 0$ (c is called the critical point of $f(x)$) is used to find extrema, it does not guarantee that $f(x)$ have a local extremum there. For example, $f(x) = x^3$; $f'(0) = 0$, but $f(x)$ has no extreme value at $x = 0$.

Let $f(x)$ be continuous on the open interval (a, b) and let $f'(x)$ exist and be continuous on (a, b). If $f'(x) > 0$ in (a, c) and $f'(x) < 0$ in (c, b), then $f(x)$ is *concave downward* at $x = c$. On the other hand, if $f'(x) < 0$ in (a, c) and $f'(x) > 0$ in (c, b), then $f(x)$ is *concave upward* at $x = c$. The type of concavity is related to the sign of the second derivative, and so we have the second derivative test to determine if a critical point is local extremum or not.

Theorem 7.2 (Second Derivative Test)

If $f'(c) = 0$ and $f''(x)$ exists, then:

(i) *if $f''(x) < 0$, then $f(x)$ has a local maximum at $x = c$;*

(ii) *if $f''(x) > 0$, then $f(x)$ has a local minimum at $x = c$;*

(iii) *if $f''(x) = 0$, then there is no information at $x = c$.*

A point $x = D$ is called an inflection point, if $f(x)$ is concave downward on one side of D and concave on the other side. Consequently, $f''(x) = 0$ at an inflection point. It is not necessary that $f'(x) = 0$ at an inflection point. ●

Example 7.5 *Find the local extrema and inflection points of the function*

$$f(x) = x^3 - 2x^2 + x + 1$$

over the entire x-axis.

Solution. *The first derivative of the function is*

$$f'(x) = 3x^2 - 4x + 1,$$

and the equation

$$f'(x) = 3x^2 - 4x + 1 = (3x - 1)(x - 1) = 0$$

shows that there are two critical points, $x = \frac{1}{3}$ and $x = 1$. The second derivative of the function is

$$f''(x) = 6x - 4.$$

The fact that $f''(\frac{1}{3}) = -2 < 0$ and $f''(1) = 2 > 0$ tells us that the critical point $x = \frac{1}{3}$ is the local maximum $(f(\frac{1}{3}) = \frac{31}{27})$ and $x = 1$ is the local minimum $(f(1) = 1)$ of $f(x)$. The inflection point is given by $f''(x) = 0$, or at $x = \frac{2}{3}$, i.e., $(\frac{2}{3}, -\frac{1}{3})$. ●

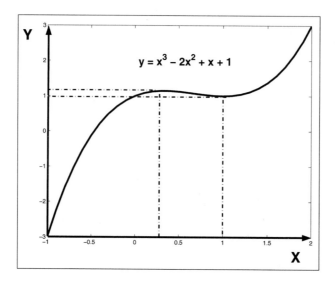

Figure 7.2: Local extrema of the function.

Example 7.6 *Find the maximum and minimum values of the function*

$$f(x) = x^3 - 6x^2 - 15x + 1$$

on the closed interval $[-2, 6]$.

Solution. *First, we find the critical points of the function by differentiating the function, which gives*

$$f'(x) = 3x^2 - 12x - 15 = 3(x + 1)(x - 5).$$

Since the derivative exists for every x, *the only critical points are those for which the derivative is zero—i.e.,* -1 *and* 5. *As* $f(x)$ *is continuous on* $[-2, 6]$, *it follows that the maximum and minimum values are among the numbers* $f(-2), f(-1), f(5)$, *and* $f(6)$. *Calculating these values, we obtain the following:*

$$f(-2) = -1, \quad f(-1) = 9, \quad f(5) = -99, \quad f(6) = -89.$$

Thus, the minimum value of $f(x)$ *on* $[-2, 6]$ *is the smallest function value* $f(5) = -99$, *and the maximum value is the largest value* $f(-1) = 9$ *on* $[-2, 6]$.

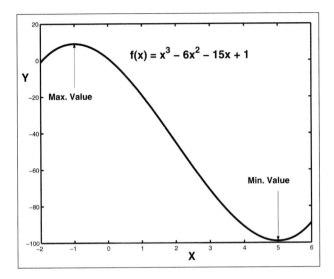

Figure 7.3: Absolute extrema of the function.

Note that the extrema of a function on the closed interval $[a, b]$ is also called the absolute extrema of a function.   •

The MATLAB command **fminbnd** can be used to find the minimum of a function of a single variable within the interval $[a, b]$. It has the form:

$$x = fminbnd('function', a, b)$$

Note that the function can be entered as a string, as the name of a function file, or as the name of an inline function, i.e.,

$$
\begin{aligned}
&>> f = inline('x\hat{\ } 3 - 6 * x\hat{\ } 2 - 15 * x + 1'); \\
&>> a = -2; b = 6; \\
&>> x = fminbnd('f', a, b) \\
&x = \\
&\quad 5.0000
\end{aligned}
$$

The value of the function at the minimum can be added to the output by using the following command:

```
>> [x fvalue] = fminbnd('x^3 - 6 * x^2 - 15 * x + 1', -2, 6)
x =
    5.0000
fvalue =
    -99.0000
```

Also, the **fminbnd** command can be used to find the maximum of a function, which can be done by multiplying the function by -1 and then finding the minimum. For example:

```
>> [x fvalue] = fminbnd('-x^3 + 6 * x^2 + 15 * x - 1', -2, 6)
x =
    -1.0000
fvalue =
    -9.0000
```

Note that the maximum of the function is at $x = -1$, and the value of the function at this point is 9.

Definition 7.4 (Partial Differentiation)

Let f be a function of two variables. The first partial derivative of f with respect to x_1 and x_2 are the functions f_{x_1} and f_{x_2}, such that

$$f_{x_1}(x_1, x_2) = \lim_{h \to 0} \frac{f(x_1 + h, x_2) - f(x_1, x_2)}{h}$$

$$f_{x_2}(x_1, x_2) = \lim_{h \to 0} \frac{f(x_1, x_2 + h) - f(x_1, x_2)}{h}.$$

In this definition, x_1 and x_2 are fixed (but arbitrary) and h is the only variable; hence, we use the notation for limits of functions of one variable instead of the $(x_1, x_2) \to (a, b)$ notation introduced previously. We can find partial derivatives without using limits, as follows. If we let $y = b$ and define a function g of one variable by $g(x_1) = f(x_1, x_2)$, then $g'(x_1) = f_{x_1}(x_1, b) = f_{x_1}(x_1, x_2)$. Thus, to find $f_{x_1}(x_1, x_2)$, we regard x_2 as a constant and differentiate $f(x_1, x_2)$ with respect to x. Similarly, to find $f_{x_2}(x_1, x_2)$, we regard the variable x_1 as a constant and differentiate $f(x_1, x_2)$ with respect to x_2. •

Notations for Partial Derivatives

If $z = f(x_1, x_2)$, then the first partial derivative of a function is defined as

$$f_{x_1} = \frac{\partial f}{\partial x_1}, \qquad f_{x_2} = \frac{\partial f}{\partial x_2}$$

$$f_{x_1}(x_1, x_2) = \frac{\partial}{\partial x_1} f(x_1, x_2) = \frac{\partial z}{\partial x_1} = z_{x_1}$$

$$f_{x_2}(x_1, x_2) = \frac{\partial}{\partial x_2} f(x_1, x_2) = \frac{\partial z}{\partial x_2} = z_{x_2}.$$

Second Partial Derivatives

If f is a function of two variables x_1 and x_2, then f_{x_1} and f_{x_2} are also functions of two variables, and we may consider their first partial derivatives. These are called the *second partial derivatives* of f and are denoted as follows:

$$\frac{\partial}{\partial x_1} f_{x_1} = (f_{x_1})_{x_1} = f_{x_1 x_1} = \frac{\partial}{\partial x_1} \left(\frac{\partial f}{\partial x_1} \right) = \frac{\partial^2 f}{\partial x_1^2}$$

$$\frac{\partial}{\partial x_2} f_{x_1} = (f_{x_1})_{x_2} = f_{x_1 x_2} = \frac{\partial}{\partial x_2} \left(\frac{\partial f}{\partial x_1} \right) = \frac{\partial^2 f}{\partial x_2 \partial x_1}$$

$$\frac{\partial}{\partial x_1} f_{x_2} = (f_{x_2})_{x_1} = f_{x_2 x_1} = \frac{\partial}{\partial x_1} \left(\frac{\partial f}{\partial x_2} \right) = \frac{\partial^2 f}{\partial x_1 \partial x_2}$$

$$\frac{\partial}{\partial x_2} f_{x_2} = (f_{x_2})_{x_2} = f_{x_2 x_2} = \frac{\partial}{\partial x_2} \left(\frac{\partial f}{\partial x_2} \right) = \frac{\partial^2 f}{\partial x_2^2}.$$

Theorem 7.3 *Let f be a function of two variables x_1 and x_2. If f, f_{x_1}, $f_{x_2}, f_{x_1 x_2}$, and $f_{x_2 x_1}$ are continuous on an open region $\mathbf{R}$, then*

$$f_{x_1 x_2} = f_{x_2 x_1},$$

throughout $\mathbf{R}$.   ●

Example 7.7 *Find the first partial derivatives of the function*

$$f(x_1, x_2) = x_1^2 + \sqrt{x_1^2 + x_2^2 + 4},$$

and also compute the value of $f_{x_1 x_2}(1, 2)$.

Solution. *The first partial derivatives of the given function are as follows:*

$$f_{x_1} = \frac{\partial f}{\partial x_1} = 2x_1 + \frac{x_1}{\sqrt{x_1^2 + x_2^2 + 4}}$$

$$f_{x_2} = \frac{\partial f}{\partial x_2} = \frac{x_2}{\sqrt{x_1^2 + x_2^2 + 4}}.$$

Similarly, the second derivative is

$$f_{x_1 x_2}(x_1, x_2) = \frac{\partial f_{x_1}}{\partial x_2} = -\frac{x_1 x_2}{(x_1^2 + x_2^2 + 4)^{3/2}},$$

and its value at $(1, 2)$ is

$$f_{x_1 x_2}(1, 2) = -\frac{(1)(2)}{(1^2 + 2^2 + 4)^{3/2}} = -0.0741.$$

●

To plot the above function $f(x_1, x_2)$ and its partial derivatives f_{x_1}, f_{x_2}, $f_{x_1 x_2}$, we use the following MATLAB commands

```
>> syms x₁ x₂;
>> f = x₁ˆ 2+sqrt(x₁ˆ 2+x₂ˆ 2+4);
>> fx₁ = diff(f, x₁); fx₂ = diff(f, x₂);
>> fx₁x₂ = (fx₁, x₂);
>> simplify(fx₁x₂);
>> subs(fx₁x₂, {x₁, x₂}, {1, 2});
>> ezmesh(f);
>> ezmesh(fx₁);
>> ezmesh(fx₁x₂);
```

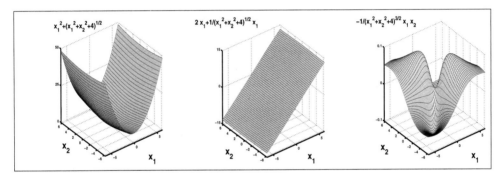

Figure 7.4: Partial derivatives of the function.

Example 7.8 *Find the second partial derivatives of the function*

$$f(x_1, x_2, x_3) = 3x_1^3 + 4x_1^2 x_2 + 2x_2^3 - 2x_2 x_3^2 + 4x_3^3.$$

Solution. *The first partial derivatives of the given function are as follows:*

$$f_{x_1} = \frac{\partial f}{\partial x_1} = 9x_1^2 + 8x_1 x_2$$

$$f_{x_2} = \frac{\partial f}{\partial x_2} = 4x_1^2 + 6x_2^2 - 2x_3^2$$

$$f_{x_3} = \frac{\partial f}{\partial x_3} = -4x_2 x_3 + 12x_3^2.$$

Similarly, the second derivatives of the functions are as follows:

$$(f_{x_1})_{x_1} = \frac{\partial f_{x_1}}{\partial x_1{}^2} = \frac{\partial^2 f}{\partial x_1{}^2} = 18x_1 + 8x_2$$

$$(f_{x_2})_{x_2} = \frac{\partial f_{x_2}}{\partial x_2{}^2} = \frac{\partial^2 f}{\partial x_2{}^2} = 12x_2$$

$$(f_{x_3})_{x_3} = \frac{\partial f_{x_3}}{\partial x_3{}^2} = \frac{\partial^2 f}{\partial x_3{}^2} = -4x_2 + 24x_3$$

$$(f_{x_1})_{x_2} = \frac{\partial f_{x_1}}{\partial x_2} = \frac{\partial^2 f}{\partial x_2 \partial x_1} = 8x_1 = \frac{\partial^2 f}{\partial x_1 \partial x_2} = (f_{x_2})_{x_1}$$

$$(f_{x_1})_{x_3} = \frac{\partial f_{x_1}}{\partial x_3} = \frac{\partial^2 f}{\partial x_3 \partial x_1} = 0 = \frac{\partial f_{x_3}}{\partial x_1} = \frac{\partial^2 f}{\partial x_1 \partial x_3} = (f_{x_3})_{x_1}$$

$$(f_{x_2})_{x_3} = \frac{\partial f_{x_2}}{\partial x_3} = \frac{\partial^2 f}{\partial x_3 \partial x_2} = -4x_3 = \frac{\partial f_{x_3}}{\partial x_2} = \frac{\partial^2 f}{\partial x_2 \partial x_3} = (f_{x_3})_{x_2}.$$

●

In the following we give a theorem that is analogous to the Second Derivative Test for functions of one variable.

Theorem 7.4 (Second Partials Test)

Suppose that $f(x_1, x_2)$ has continuous partial derivatives in a neighborhood of a point (x_{10}, x_{20}) and that $\nabla f(x_{10}, x_{20}) = \mathbf{0}$. Let

$$D = D(x_{10}, x_{20}) = f_{x_1 x_1}(x_{10}, x_{20}) f_{x_2 x_2}(x_{10}, x_{20}) - f_{x_1 x_2}^2(x_{10}, x_{20}).$$

Then:

(i) *if $D > 0$ and $f_{x_1 x_1}(x_{10}, x_{20}) < 0$, $f(x_{10}, x_{20})$ is a local maximum value;*

(ii) *if $D > 0$ and $f_{x_1 x_1}(x_{10}, x_{20}) > 0$, $f(x_{10}, x_{20})$ is a local minimum value;*

(iii) *if $D < 0$, $f(x_{10}, x_{20})$ is not an extreme value ((x_{10}, x_{20}) is a saddle point);*

(iv) if $D = 0$, the test is inconclusive. •

Example 7.9 *Find the extrema, if any, of the function*

$$f(x_1, x_2) = 4x_1^3 + 2x_2^2 - 12x_1 + 8x_2 + 2.$$

Solution. *Since the first derivatives of the function with respect to x_1 and x_2 are*

$$f_{x_1}(x_1, x_2) = 12x_1^2 - 12 \quad and \quad f_{x_2}(x_1, x_2) = 4x_2 + 8,$$

the critical points obtained by solving the simultaneous equations $f_{x_1}(x_1, x_2) = f_{x_2}(x_1, x_2) = 0$, are $(1, -2)$ and $(-1, -2)$.

To find the critical points for the given function $f(x_1, x_2)$ using MAT-LAB commands we do the following:

```
>> syms x₁ x₂;
>> f = 4 * x₁ ^ 3 + 2 * x₂ ^ 2 - 12 * x₁ + 8 * x₂ + 2;
>> fx₁ = diff(f, x₁);
>> fx₂ = diff(f, x₂);
>> [x₁c, x₂c] = solve(fx₁, fx₂);
x₁c =
    [ 1]
    [-1]
x₂c =
    [-2]
    [-2]
```

Similarly, the second partial derivatives of the function are

$$f_{x_1 x_1}(x_1, x_2) = 24x_1, \quad f_{x_2 x_2}(x_1, x_2) = 4, \quad and \quad f_{x_1 x_2}(x_1, x_2) = 0.$$

Thus, at the critical point $(1, -2)$, we get

$$D = f_{x_1 x_1}(1, -2) f_{x_2 x_2}(1, -2) - f_{x_1 x_2}^2(1, -2) = 24(4) - 0 = 96 > 0.$$

Furthermore, $f_{x_1 x_1}(1, -2) = 24 > 0$ and so, by (ii), $f(1, -2) = -14$ is a local minimum value of the given function $f(x_1, x_2)$.

Now testing the given function at the other critical point, $(-1, -2)$, we find that

$$D = f_{x_1 x_1}(-1, -2) f_{x_2 x_2}(-1, -2) - f_{x_1 x_2}^2(-1, -2) = (-24)(4) - 0 = -96 < 0.$$

Thus, by (iii), $(-1, -2)$ is a saddle point and $f(1, -2)$ is not an extremum.
●

To get the above results we use the following MATLAB commands:

```
>> syms x₁ x₂
>> f = 4 * x₁ ^ 3 + 2 * x₂ ^ 2 - 12 * x₁ + 8 * x₂ + 2;
>> fx₁ = diff(f, x₁);
>> fx₁x₁ = diff(fx₁, x₁);
>> fx₂ = diff(f, x₂);
>> fx₂x₂ = diff(fx₂, x₂);
>> fx₁x₂ = diff(fx₁, x₂);
>> D = fx₁x₁ * fx₂x₂ - fx₁x₂ ^ 2;
>> [ax₁ ax₂] = solve(fx₁, fx₂);
>> T = [ax₁ ax₂ subs(D, x₁ x₂, ax₁ ax₂) subs(fx₁x₁, x₁ x₂, ax₁ ax₂)];
>> double(T);
ans =
        1   -2    96    24
       -1   -2   -96   -24
```

To plot a symbolic expression f that contains two variables x_1 and x_2, we use the **ezplot** command as follows:

```
>> ezplot(f);
```

7.2.5   Directional Derivatives and the Gradient Vector

Here, we introduce a type of derivative, called a *directional derivative*, that enables us to find the rate of change of a function of two or more variables in any direction.

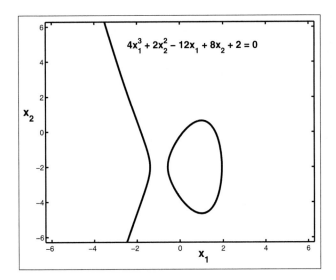

Figure 7.5: Local extrema of the function.

Definition 7.5 (Directional Derivatives)

Let $z = f(x_1, x_2)$ be a function, and the directional derivative of $f(x_1, x_2)$ at the point (x_{10}, x_{20}) in the direction of a unit vector $\mathbf{u} = (u_1, u_2)$ is given by

$$D_u f(x_{10}, x_{20}) = \lim_{h \to 0} \frac{f(x_{10} + hu_1, x_{20} + hu_2) - f(x_{10}, x_{20})}{h}$$

provided the limit exists.   ●

Notice that this definition is similar to the definition of a partial derivative, except that in this case, both variables may be changed. Further, one can observe that the directional derivative in the direction of the positive x_1-axis (i.e., in the direction of the unit vector $\mathbf{u} = (1, 0)$) is

$$D_u f(x_{10}, x_{20}) = \lim_{h \to 0} \frac{f(x_{10} + h, x_{20}) - f(x_{10}, x_{20})}{h},$$

which is the partial derivative $\frac{\partial f}{\partial x_1}$. Likewise, the directional derivative in the direction of the positive x_2-axis (i.e., in the direction of the unit vector

$\mathbf{u} = (0, 1))$ is

$$D_u f(x_{10}, x_{20}) = \lim_{h \to 0} \frac{f(x_{10}, x_{20} + h) - f(x_{10}, x_{20})}{h},$$

which is the partial derivative $\frac{\partial f}{\partial x_2}$. So this means that any directional derivative can be calculated simply in terms of the first derivatives.

Theorem 7.5 *If $f(x_1, x_2)$ is a differentiable function of x_1 and x_2, then $f(x_1, x_2)$ has a directional derivative in the direction of any unit vector $\mathbf{u} = (u_1, u_2)$ and can be written as*

$$D_u f(x_1, x_2) = \frac{\partial f}{\partial x_1} u_1 + \frac{\partial f}{\partial x_2} u_2.$$

•

Example 7.10 *Find the directional derivative of $f(x, y) = x_1^2 x_2 + 3x_2^2$ at the point $(1, 2)$ in the direction of the unit vector $\mathbf{u} = (3, -2)$.*

Solution. *Since we know that*

$$D_u f(x_1, x_2) = \frac{\partial f}{\partial x_1} u_1 + \frac{\partial f}{\partial y} u_2,$$

we can easily compute the first partial derivatives of the given function as

$$f_{x_1} = \frac{\partial f}{\partial x_1} = 2x_1 x_2 \quad \text{and} \quad f_{x_2} = \frac{\partial f}{\partial x_2} = x_1^2 + 6x_2,$$

and their values at the given point $(1, 2)$ are

$$f_{x_1}(1, 2) = 4 \quad \text{and} \quad f_{x_2}(1, 2) = 13.$$

Thus, for the given unit vector $\mathbf{u} = (3, -2)$, and using Theorem 7.5, we have

$$D_u f(1, 2) = 4(3) + 13(-2) = 12 - 26 = -14,$$

which is the required solution.

•

To get the results of Example 7.10, we the following MATLAB commands:

```
>> syms x₁ x₂;
>> u1 = 3; u2 = -2;
>> f = x₁^ 2x₂ + 3 * x₂^ 2;
>> fx₁ = (f, x₁); fx₂ = (f, x₂);
>> dᵤf = subs(fx₁, {x₁, x₂}, {1, 2}) * u1 + subs(fx₂, {x₁, x₂}, {1, 2}) * u2;
```

It is useful to combine the first partial derivatives of a function into a single vector function called a *gradient*. We denote the gradient of a function f by *grad f* or ∇f.

Definition 7.6 (The Gradient Vector)

Let $z = f(x_1, x_2)$ be a function, then the gradient of $f(x_1, x_2)$ is the vector function ∇f defined by

$$\nabla f(x_1, x_2) = \left(\frac{\partial f}{\partial x_1}, \frac{\partial f}{\partial x_2} \right) = \frac{\partial f}{\partial x_1}\mathbf{i} + \frac{\partial f}{\partial x_2}\mathbf{j},$$

provided that both partial derivatives exist. ●

Similarly, the vector of partial derivatives of a function $f(x_1, x_2, \ldots, x_n)$ with respect to the point $\mathbf{x} = (x_1, x_2, \ldots, x_n)$ is defined as

$$\nabla f(\mathbf{x}) = \begin{pmatrix} \dfrac{\partial f(\mathbf{x})}{\partial x_1} \\[2ex] \dfrac{\partial f(\mathbf{x})}{\partial x_2} \\[1ex] \vdots \\[1ex] \dfrac{\partial f(\mathbf{x})}{\partial x_n} \end{pmatrix}.$$

One can easily compute the directional derivatives by using the following theorem.

Theorem 7.6 *If $f(x_1, x_2)$ is a differentiable function of x_1 and x_2, and $\mathbf{u}$ is any unit vector, then*

$$D_u f(x_1, x_2) = \nabla f(x_1, x_2).\mathbf{u}.$$

•

Example 7.11 *Find the gradient of the function $f(x_1, x_2, x_3) = x_1 \cos(x_2 x_3)$ at the point $(2, 0, 2)$.*

Solution. *The gradient of the function is*

$$\nabla f(x_1, x_2, x_3) = (f_{x_1}, f_{x_2}, f_{x_3}) = (\cos(x_2 x_3), -x_1 x_3 \sin(x_2 x_3), -x_1 x_2 \sin(x_2 x_3)).$$

At the given point $(2, 0, 2)$, we have

$$\nabla f(2, 0, 2) = (1, 0, 0).$$

•

The MATLAB command **jacobian** can be used to get the gradient of the function as follows:

```
>> syms x1 x2 x3
>> f = x1 * cos(x2 * x3);
>> gradf = jacobian(f, [x1, x2, x3]);
ans =

[cos(x2 * x3), -x1 * sin(x2 * x3) * x3, -x1 * sin(x2 * x3) * x2]
```

Theorem 7.7 *The point $\bar{\mathbf{x}}$ is a stationary point of $f(\mathbf{x})$, if and only if*

$$\nabla f(\bar{\mathbf{x}}) = \mathbf{0}.$$

•

Example 7.12 *Consider the function*

$$f(x_1, x_2) = x_1^2 + (x_2 - 1)^2 + 3,$$

then

$$\nabla f(x_1, x_2) = \begin{pmatrix} 2x_1 \\ 2(x_2 - 1) \end{pmatrix},$$

and at the stationary point $\bar{\mathbf{x}} = [0, \frac{1}{2}]^T$, *the gradient vector of the function is*

$$\nabla f\left(0, \frac{1}{2}\right) = \begin{pmatrix} 0 \\ -1 \end{pmatrix}.$$

●

7.2.6 Hessian Matrix

If a function f is twice continuous differentiable, then there exists a matrix H of second partial derivatives, called the *Hessian* matrix, whose entries are given by

$$h_{ij} = \frac{\partial^2 f}{\partial x_i \partial x_j}, \quad i, j = 1, 2, \ldots, n.$$

For example, the Hessian matrix of size 2×2 can be written as

$$H(x_1, x_2) = \nabla^2 f(x_1, x_2) = \begin{pmatrix} \dfrac{\partial^2 f}{\partial x_1^2} & \dfrac{\partial^2 f}{\partial x_1 \partial x_2} \\[3mm] \dfrac{\partial^2 f}{\partial x_2 \partial x_1} & \dfrac{\partial^2 f}{\partial x_2^2} \end{pmatrix}.$$

This matrix is formally referred to as the Hessian of f. Note that the Hessian matrix is square and symmetric.

Example 7.13 *Find the Hessian matrix of the function*

$$f(x_1, x_2) = x_1^2 + 2(x_2 - 1)^2 + 3x_1 x_2 + 4.$$

Solution. *The first-order partial derivatives of the given function are*

$$\frac{\partial f}{\partial x_1} = 2x_1 + 3x_2, \quad \frac{\partial f}{\partial x_2} = 4(x_2 - 1) + 3x_1,$$

and the second-order partial derivatives of the given functions are

$$\frac{\partial^2 f}{\partial x_1^2} = 2, \quad \frac{\partial^2 f}{\partial x_2^2} = 4.$$

Also, the mixed partial derivatives are

$$\frac{\partial^2 f}{\partial x_1 \partial x_2} = 3, \quad \frac{\partial^2 f}{\partial x_2 \partial x_1} = 3.$$

Thus,

$$H(x_1, x_2) = \begin{pmatrix} 2 & 3 \\ 3 & 4 \end{pmatrix}$$

is the Hessian matrix of the given function.   •

To get the above results we use the following MATLAB commands:

```
>> syms x₁ x₂
>> fun = x₁^2+2*(x₂ − 1)^2+3*x₁*x₂ + 4;
>> T = Hessian(fun, [x₁, x₂])
>> double(T)
          2  3
ans =
          3  4
```

Program 7.1
MATLAB m-file for the Hessian Matrix
```
function H = Hessian(fun,Vars)
n = numel(Vars);
Hess = vpa(ones(n,n));
for j = 1:n;
for i = 1:n;
Hess(j,i) = diff(diff(fun,Vars(i),1),Vars(j),1);
end  end
H=Hess;
```

For the n-dimensional case, the Hessian matrix $H(\mathbf{x})$ is defined as follows:

$$
H(\mathbf{x}) =
\begin{pmatrix}
\dfrac{\partial^2 f(\mathbf{x})}{\partial x_1^2} & \dfrac{\partial^2 f(\mathbf{x})}{\partial x_1 \partial x_2} & \cdots & \dfrac{\partial^2 f(\mathbf{x})}{\partial x_1 \partial x_n} \\[2.5ex]
\dfrac{\partial^2 f(\mathbf{x})}{\partial x_2 \partial x_1} & \dfrac{\partial^2 f(\mathbf{x})}{\partial x_2^2} & \cdots & \dfrac{\partial^2 f(\mathbf{x})}{\partial x_2 \partial x_n} \\[2.5ex]
\vdots & \vdots & \vdots & \vdots \\[1.5ex]
\dfrac{\partial^2 f(\mathbf{x})}{\partial x_n \partial x_1} & \dfrac{\partial^2 f(\mathbf{x})}{\partial x_n \partial x_2} & \cdots & \dfrac{\partial^2 f(\mathbf{x})}{\partial x_n^2}
\end{pmatrix}.
$$

Checking the sign of the second derivative when $n = 1$ corresponds to checking the definition of the Hessian matrix when $n > 1$. Let us consider a constant matrix H of size $n \times n$ and a nonzero n-dimensional vector $\mathbf{z}$, then:

1. H is positive-definite, if and only if $\mathbf{z}^T H \mathbf{z} > 0$.

2. H is positive-semidefinite, if and only if $\mathbf{z}^T H \mathbf{z} \geq 0$.

3. H is negative-definite, if and only if $\mathbf{z}^T H \mathbf{z} < 0$.

4. H is negative-semidefinite, if and only if $\mathbf{z}^T H \mathbf{z} \leq 0$.

Note that if H is the zero matrix (so that it is both positive-semidefinite and negative-semidefinite) or if the sign of $\mathbf{z}^T H \mathbf{z}$ varies with the choice of $\mathbf{z}$, we shall say that H is *indefinite*.

The relationships between the Hessian matrix definiteness and the classification of stationary points are discussed in the following two theorems.

Theorem 7.8 (Minima of a Function)

If $\bar{\mathbf{x}}$ is a stationary point of $f(\mathbf{x})$, then the following conditions are satisfied:

1. $H(\bar{\mathbf{x}})$ is positive-definite implies that $\bar{\mathbf{x}}$ is a strict minimizing point.

2. $H(\bar{\mathbf{x}})$ *is positive-semidefinite for all* $\bar{\mathbf{x}}$, *in some neighborhood of* $\bar{\mathbf{x}}$, *implies that* $\bar{\mathbf{x}}$ *is a minimizing point.*

3. $\bar{\mathbf{x}}$ *is a minimizing point implies that* $H(\bar{\mathbf{x}})$ *is positive-semidefinite.* •

Theorem 7.9 (Maxima of a Function)

If $\bar{\mathbf{x}}$ *is a stationary point of* $f(\mathbf{x})$, *then the following conditions are satisfied:*

1. $H(\bar{\mathbf{x}})$ *is negative-definite implies that* $\bar{\mathbf{x}}$ *is a strict maximizing point.*

2. $H(\bar{\mathbf{x}})$ *is negative-semidefinite for all* $\bar{\mathbf{x}}$, *in some neighborhood of* $\bar{\mathbf{x}}$, *implies that* $\bar{\mathbf{x}}$ *is a maximizing point.*

3. $\bar{\mathbf{x}}$ *is a maximizing point implies that* $H(\bar{\mathbf{x}})$ *is negative-semidefinite.*•

Since we know that the second derivative test for a function of one variable gives no information when the second derivative of a function is zero, similarly, if $H(\bar{\mathbf{x}})$ is indefinite or if $H(\bar{\mathbf{x}})$ is positive-semidefinite at $\bar{\mathbf{x}}$ but not all points are in a neighborhood of $\bar{\mathbf{x}}$, then the function might have a maximum, or a minimum, or neither at $\bar{\mathbf{x}}$.

Example 7.14 *Consider the function*

$$f(x_1, x_2) = 2x_1^2 + 2x_2^2 + 4,$$

then the Hessian matrix of the given function can be found as

$$H(x_1, x_2) = \begin{pmatrix} 4 & 0 \\ 0 & 4 \end{pmatrix}.$$

To check the definiteness of H, take

$$\mathbf{z}^T H \mathbf{z} = \begin{pmatrix} z_1 & z_2 \end{pmatrix} \begin{pmatrix} 4 & 0 \\ 0 & 4 \end{pmatrix} \begin{pmatrix} z_1 \\ z_2 \end{pmatrix},$$

which gives

$$\mathbf{z}^T H \mathbf{z} = \begin{pmatrix} z_1 & z_2 \end{pmatrix} \begin{pmatrix} 4z_1 \\ 4z_2 \end{pmatrix} = 4z_1^2 + 4z_2^2.$$

Note that

$$\mathbf{z}^T H \mathbf{z} = 4(z_1^2 + z_2^2) > 0,$$

for $\mathbf{z} \neq \mathbf{0}$, *so the Hessian matrix is positive-definite and the stationary point* $\bar{\mathbf{x}} = [0, 0]^T$ *is a strict local minimizing point.* ●

Example 7.15 *Consider the function*

$$f(x_1, x_2) = (x_1 - 1)^2 - \frac{1}{2} x_2^2 + 1,$$

then the gradient vector of the given function is

$$\nabla f(x_1, x_2) = \begin{pmatrix} 2(x_1 - 1) \\ -x_2 \end{pmatrix}.$$

The stationary point can be found as

$$\nabla f(x_1, x_2) = \begin{pmatrix} 2(x_1 - 1) \\ -x_2 \end{pmatrix} = \mathbf{0},$$

which gives

$$\bar{\mathbf{x}} = [1, 0]^T.$$

The Hessian matrix for the given function is

$$H(x_1, x_2) = \begin{pmatrix} 2 & 0 \\ 0 & -1 \end{pmatrix}.$$

To check the definiteness of H, *take*

$$\mathbf{z}^T H \mathbf{z} = \begin{pmatrix} z_1 & z_2 \end{pmatrix} \begin{pmatrix} 2 & 0 \\ 0 & -1 \end{pmatrix} \begin{pmatrix} z_1 \\ z_2 \end{pmatrix},$$

and it gives

$$\mathbf{z}^T H \mathbf{z} = \begin{pmatrix} z_1 & z_2 \end{pmatrix} \begin{pmatrix} 2z_1 \\ -z_2 \end{pmatrix} = 2z_1^2 - z_2^2.$$

The sign of $\mathbf{z}^T H \mathbf{z}$ *clearly depends on the particular values taken on by* z_1 *and* z_2, *so the Hessian matrix is indefinite and the stationary point* $\mathbf{x}$ *cannot be classified on the basis of this test.* ●

7.2.7   Taylor's Series Expansion

Let $f(x)$ be a function of a single variable x, and if $f(x)$ has continuous derivatives $f'(x), f''(x), \ldots$, then Taylor's series can be used to approximate this function about x_0 as follows:

$$f(x) = f(x_0) + (x - x_0)f'(x_0) + \frac{(x - x_0)^2}{2!}f''(x_0) + \frac{(x - x_0)^3}{3!}f'''(x_0) + \cdots.$$

A linear approximation of the function can be obtained by truncating the above series after the second term, i.e.,

$$f(x) \approx T_1(x) = f(x_0) + (x - x_0)f'(x_0), \qquad (7.1)$$

whereas the quadratic approximation of the function can be computed as

$$f(x) \approx T_2(x) = f(x_0) + (x - x_0)f'(x_0) + \frac{(x - x_0^2)}{2!}f''(x_0). \qquad (7.2)$$

Example 7.16 *Find the cubic approximation of the function $f(x) = e^x \cos x$ expanded about $x_0 = 0$.*

Solution. *The cubic approximation of the function about x_0 is*

$$T_3(x) = f(x_0) + \frac{(x - x_0)}{1!}f'(x_0) + \frac{(x - x_0)^2}{2!}f''(x_0) + \frac{(x - x_0)^3}{3!}f'''(x_0). \ (7.3)$$

Since $f(x) = e^x \cos x$, then $f(x_0) = f(0) = 1$, and calculating the derivatives required for the desired polynomial $T_3(x)$, we get

$$
\begin{array}{llll}
f'(x) & = & e^x(\cos x - \sin x), & f'(x_0) = f'(0) = 1 \\
f''(x) & = & -2e^x \sin x, & f''(x_0) = f''(0) = 0 \\
f'''(x) & = & -2e^x(\cos x + \sin x), & f'''(x_0) = f'''(0) = -2.
\end{array}
$$

Putting all these values in (7.3), we get

$$T_3(x) = 1 + x + 0 - 2\frac{x^3}{6}.$$

Thus,

$$f(x) \approx T_3(x) = 1 + x - \frac{x^3}{3}$$

is the cubic approximation of the given function about $x_0 = 0$.   •

The MATLAB command for a Taylor polynomial is **taylor**$(f, n+1, a)$, where f is the function, n is the order of the polynomial, and a is the point about which the expansion is made. We can use the following MATLAB commands to get the above results:

```
>> syms x
>> f = inline('exp(x) * cos(x)');
>> taylor(f(x), 4, 0)
ans =      1 + x − 1/3 * x^ 3
```

Now consider a function $f(x, y)$ of two variables and all of its partial derivatives are continuous, then we can approximate this function about a given point (x_0, y_0) using Taylor's series as follows:

$$f(x, y) = f(x_0, y_0) + \left[\frac{\partial f(x_0, y_0)}{\partial x}(x - x_0) + \frac{\partial f(x_0, y_0)}{\partial y}(y - y_0) \right]$$

$$+ \frac{1}{2} \left[\frac{\partial^2 f(x_0, y_0)}{\partial x^2}(x - x_0)^2 + 2\frac{\partial^2 f(x_0, y_0)}{\partial x \partial y}(x - x_0)(y - y_0) \right.$$

$$+ \left. \frac{\partial^2 f(x_0, y_0)}{\partial y^2}(y - y_0)^2 \right] + \cdots. \tag{7.4}$$

Writing the above expression in more compact form by using matrix notation gives

$$f(x, y) = f(x_0, y_0) + \begin{bmatrix} \dfrac{\partial f(x_0, y_0)}{\partial x} & \dfrac{\partial f(x_0, y_0)}{\partial y} \end{bmatrix} \begin{bmatrix} x - x_0 \\ y - y_0 \end{bmatrix}$$

$$+ \frac{1}{2}(x - x_0)(y - y_0) \begin{bmatrix} \dfrac{\partial^2 f(x_0, y_0)}{\partial x^2} & \dfrac{\partial^2 f(x_0, y_0)}{\partial x \partial y} \\ \dfrac{\partial^2 f(x_0, y_0)}{\partial x \partial y} & \dfrac{\partial^2 f(x_0, y_0)}{\partial y^2} \end{bmatrix} \begin{bmatrix} x - x_0 \\ y - y_0 \end{bmatrix} + \cdots.$$

Denote the first derivative of f by ∇f and the second derivative by $\nabla^2 f$, then a 2×1 vector called a *gradient vector* is defined as

$$\nabla f(x_0, y_0) = \begin{bmatrix} \dfrac{\partial f(x_0, y_0)}{\partial x} \\[2ex] \dfrac{\partial f(x_0, y_0)}{\partial y} \end{bmatrix},$$

and a 2×2 matrix called the *Hessian* matrix is defined as

$$\nabla^2 f(x_0, y_0) = \begin{bmatrix} \dfrac{\partial^2 f(x_0, y_0)}{\partial x^2} & \dfrac{\partial^2 f(x_0, y_0)}{\partial x \partial y} \\[2ex] \dfrac{\partial^2 f(x_0, y_0)}{\partial y \partial x} & \dfrac{\partial^2 f(x_0, y_0)}{\partial y^2} \end{bmatrix}.$$

Also, note that

$$\nabla f(x_0, y_0)^T = \begin{bmatrix} \dfrac{\partial f(x_0, y_0)}{\partial x} & \dfrac{\partial f(x_0, y_0)}{\partial y} \end{bmatrix}$$

and

$$\begin{bmatrix} x - x_0 \\ y - y_0 \end{bmatrix} = \begin{bmatrix} x \\ y \end{bmatrix} - \begin{bmatrix} x_0 \\ y_0 \end{bmatrix} = (\mathbf{x} - \mathbf{x}^*) \equiv \Delta \mathbf{x}.$$

Thus, Taylor's series for a function of two variables can be written as

$$f(x, y) = f(x_0, y_0) + \nabla f(x_0, y_0)^T \Delta \mathbf{x} + \frac{1}{2} \Delta \mathbf{x}^T \nabla^2 f(x_0, y_0) \Delta \mathbf{x} + \cdots. \quad (7.5)$$

Similarly, for a function of n variables, $\mathbf{x} = [x_1, x_2, \ldots, x_n]$, an $n \times 1$ gradient vector and an $n \times n$ Hessian matrix can be used in Taylor's series for n variables defined as

$$\nabla f(\mathbf{x}) = \begin{bmatrix} \dfrac{\partial f}{\partial x_1} \\[2ex] \dfrac{\partial f}{\partial x_2} \\[2ex] \vdots \\[2ex] \dfrac{\partial f}{\partial x_n} \end{bmatrix}$$

and

$$\nabla^2 f(\mathbf{x}) = \begin{bmatrix} \dfrac{\partial^2 f(\mathbf{x})}{\partial x_1^2} & \dfrac{\partial^2 f(\mathbf{x})}{\partial x_1 \partial x_2} & \cdots & \dfrac{\partial^2 f(\mathbf{x})}{\partial x_1 \partial x_n} \\[2ex] \dfrac{\partial^2 f(\mathbf{x})}{\partial x_2 \partial x_1} & \dfrac{\partial^2 f(\mathbf{x})}{\partial x_2^2} & \cdots & \dfrac{\partial^2 f(\mathbf{x})}{\partial x_2 \partial x_n} \\[2ex] \vdots & \vdots & \vdots & \vdots \\[2ex] \dfrac{\partial^2 f(\mathbf{x})}{\partial x_n \partial x_1} & \dfrac{\partial^2 f(\mathbf{x})}{\partial x_n \partial x_2} & \cdots & \dfrac{\partial^2 f(\mathbf{x})}{\partial x_n^2} \end{bmatrix}.$$

For a continuously differentiable function, the mixed second partial derivatives are symmetric, i.e.,

$$\frac{\partial^2 f(\mathbf{x})}{\partial x_i \partial x_j} = \frac{\partial^2 f(\mathbf{x})}{\partial x_j \partial x_i}, \qquad i,j = 1,2,\ldots,n,$$

which implies that the Hessian matrix is always symmetric. Thus, Taylor's series approximation for a function of several variables about given point $\mathbf{x}^*$ can be written as

$$f(\mathbf{x}) = f(\mathbf{x}^*) + \nabla f(\mathbf{x}^*)^T \Delta \mathbf{x} + \frac{1}{2} \Delta \mathbf{x}^T \nabla^2 f(\mathbf{x}^*) \Delta \mathbf{x} + \cdots.$$

Example 7.17 *Find the linear and quadratic approximations of the function*

$$f(x_1, x_2) = x_1^2 + 5x_1 x_2^2 + 3x_2^2 + 4x_1^3 x_2^3$$

at the given point $(a,b) = (1,1)$ *using Taylor's series formulas.*

Solution. *The first-order partial derivatives of the function are*

$$\frac{\partial f}{\partial x_1} = 2x_1 + 5x_2^2 + 12x_1^2 x_2^3,$$

and

$$\frac{\partial f}{\partial x_2} = 10x_1 x_2 + 6x_2 + 12x_1^3 x_2^2.$$

Thus, the gradient of the function is

$$\nabla f(x_1, x_2)^T = [2x_1 + 5x_2^2 + 12x_1^2 x_2^3 \quad 10x_1 x_2 + 6x_2 + 12x_1^3 x_2^2]$$

and

$$\Delta \mathbf{x} = \begin{bmatrix} x_1 - a \\ x_2 - b \end{bmatrix}.$$

Linear Approximation Formula

The linear approximation formula for two variables is

$$f(x_1, x_2) \approx T_1(x_1, x_2) = f(a, b) + \nabla f(a, b)^T \Delta \mathbf{x}.$$

Given $(a, b) = (1, 1)$, we get the following values:

$$f(a, b) \quad = \quad f(1, 1) = 13$$

$$\nabla f(1, 1)^T \quad = \quad [19 \quad 28]$$

$$\Delta \mathbf{x} \quad = \quad \begin{bmatrix} x_1 - 1 \\ x_2 - 1 \end{bmatrix}.$$

Putting these values in the above linear approximation formula, we get

$$f(x_1, x_2) \approx T_1(x_1, x_2) = 13 + [19 \quad 28] \begin{bmatrix} x_1 - 1 \\ x_2 - 1 \end{bmatrix}$$

or

$$f(x_1, x_2) \approx T_1(x_1, x_2) = 19x_1 + 28x_2 - 34,$$

which is the required linear approximation for the given function.

Quadratic Approximation Formula

The quadratic approximation formula for two variables is defined as

$$f(x_1, x_2) \approx T_2(x_1, x_2) = f(a, b) + \nabla f(a, b)^T \Delta \mathbf{x} + \frac{1}{2} \Delta \mathbf{x}^T \nabla^2 f(a, b) \Delta \mathbf{x}.$$

Now we compute the second-order partial derivatives as follows:

$$\frac{\partial^2 f}{\partial x_1^2} = 2 + 24x_1 x_2^3$$

$$\frac{\partial^2 f}{\partial x_2^2} = 20x_2 + x_2 + 24x_1^3 x_2$$

$$\frac{\partial^2 f}{\partial x_1 \partial x_2} = 10x_2 + 36x_1^2 x_2^2$$

$$\frac{\partial^2 f}{\partial x_2 \partial x_1} = 10x_2 + 36x_1^2 x_2^2,$$

and so

$$\nabla^2 f(x_1, x_2) = \begin{bmatrix} \dfrac{\partial^2 f(x_1,x_2)}{\partial x_1^2} & \dfrac{\partial^2 f(x_1,x_2)}{\partial x_1 \partial x_2} \\ \dfrac{\partial^2 f(x_1,x_2)}{\partial x_2 \partial x_1} & \dfrac{\partial^2 f(x_1,x_2)}{\partial x_2^2} \end{bmatrix} = \begin{bmatrix} 2 + 24x_1 x_2^3 & 10x_2 + 36x_1^2 x_2^2 \\ 10x_2 + 36x_1^2 x_2^2 & 20x_2 + x_2 + 24x_1^3 x_2 \end{bmatrix}.$$

Thus,

$$\nabla^2 f(1,1)^T = \begin{bmatrix} 72 & 46 \\ 46 & 45 \end{bmatrix}.$$

So using the quadratic approximation formula

$$f(x_1, x_2) \approx T_2(x_1, x_2) = T_1(x_1, x_2) + \frac{1}{2}\Delta\mathbf{x}^T \nabla^2 f(a,b)\Delta\mathbf{x},$$

we get

$$f(x_1, x_2) \approx T_2(x_1, x_2) = T_1(x_1, x_2) + \frac{1}{2}\begin{bmatrix} (x_1 - 1) & (x_2 - 1) \end{bmatrix}\begin{bmatrix} 72 & 46 \\ 46 & 45 \end{bmatrix}\begin{bmatrix} x_1 - 1 \\ x_2 - 1 \end{bmatrix},$$

which gives

$$\begin{aligned} f(x_1, x_2) \approx T_2(x_1, x_2) = {} & 19x_1 + 28x_2 - 34 \\ & + \frac{1}{2}\left[72x_1^2 + 45x_2^2 - 236x_1 - 182x_2 + 92x_1 x_2 + 117\right]. \end{aligned}$$

So

$$f(x_1, x_2) \approx T_2(x_1, x_2) = 36x_1^2 + 22.5x_2^2 - 97x_1 - 63x_2 + 46x_1x_2 + 24.5,$$

which is the required quadratic approximation of the function.   ●

7.2.8   Quadratic Forms

Quadratic forms play an important role in geometry. Given

$$\mathbf{x} = (x_1, x_2, \ldots, x_n)^T$$

and

$$A = \begin{pmatrix} a_{11} & a_{12} & \cdots & a_{1n} \\ a_{21} & a_{22} & \cdots & a_{2n} \\ a_{31} & a_{32} & \cdots & a_{3n} \\ \vdots & \vdots & \vdots & \vdots \\ a_{n1} & a_{n2} & \cdots & a_{nn} \end{pmatrix},$$

then the function

$$q(\mathbf{x}) = \mathbf{x}^T A \mathbf{x} = \begin{pmatrix} x_1 & x_2 & \cdots & x_n \end{pmatrix} \begin{pmatrix} a_{11} & a_{12} & \cdots & a_{1n} \\ a_{21} & a_{22} & \cdots & a_{2n} \\ a_{31} & a_{32} & \cdots & a_{3n} \\ \vdots & \vdots & \vdots & \vdots \\ a_{n1} & a_{n2} & \cdots & a_{nn} \end{pmatrix} \begin{pmatrix} x_1 \\ x_2 \\ \vdots \\ x_n \end{pmatrix}$$

$$= \sum_{i=1}^{n} \sum_{j=1}^{n} a_{ij} x_i x_j,$$

can be used to represent any quadratic polynomial in the variables $x_1, x_2, \ldots, x_n$ and is called a *quadratic form*. The matrix A of a quadratic form can always be forced to be symmetric because

$$\mathbf{x}^T A \mathbf{x} = \mathbf{x}^T \left[\frac{(A + A^T)}{2} \right] \mathbf{x},$$

and the matrix $\frac{(A+A^T)}{2}$ is always symmetric. The symmetric matrix A associated with the quadratic form is called the *matrix of the quadratic form*.

Example 7.18 *What is the quadratic form of the associated matrix*

$$A = \begin{pmatrix} 2 & 3 & -1 \\ 3 & 6 & 0 \\ -1 & 0 & 4 \end{pmatrix}?$$

Solution. *If*

$$\mathbf{x} = \begin{pmatrix} x_1 \\ x_2 \\ x_3 \end{pmatrix},$$

then

$$\mathbf{x}^T A \mathbf{x} = \begin{pmatrix} x_1 & x_2 & x_3 \end{pmatrix} \begin{pmatrix} 2 & 3 & -1 \\ 3 & 6 & 0 \\ -1 & 0 & 4 \end{pmatrix} \begin{pmatrix} x_1 \\ x_2 \\ x_3 \end{pmatrix}$$

or

$$\mathbf{x}^T A \mathbf{x} = \begin{pmatrix} x_1 & x_2 & x_3 \end{pmatrix} \begin{pmatrix} 2x_1 & + & 2x_2 & - & x_3 \\ 2x_1 & + & 6x_2 & & \\ -x_1 & & & + & 4x_3 \end{pmatrix}.$$

Thus,

$$\mathbf{x}^T A \mathbf{x} = 2x_1^2 + 2x_1x_2 - x_1x_3 + 2x_1x_2 + 3x_2^2 - x_1x_3 + 4x_3^2$$

or

$$\mathbf{x}^T A \mathbf{x} = 2x_1^2 + 4x_1x_2 - 2x_1x_3 + 6x_2^2 + 4x_3^2.$$

After rearranging the terms, we have

$$\mathbf{x}^T A \mathbf{x} = (x_1 + 2x_2)^2 + 2x_2^2 + (x_1 - x_3)^2 + 3x_3^2.$$

Hence,

$$(x_1 + 2x_2)^2 + 2x_2^2 + (x_1 - x_3)^2 + 3x_3^2 > 0,$$

unless

$$x_1 = x_2 = x_3 = 0.$$

●

Example 7.19 *Find the matrix associated with the quadratic form*

$$q(x_1, x_2, x_3) = 4x_1^2 + 2x_2^2 + 2x_3^2 + 2x_1x_2 - 4x_1x_3 + 4x_2x_3.$$

Solution. *The coefficients of the squared terms x_i^2 go on the diagonal as a_{ii}, and the product terms $x_i x_j$ are split between a_{ij} and a_{ji}. The elements a_{ij} can be computed as $a_{ij} = \frac{(a_{ij}+a_{ji})}{2}$, which gives*

$$A = \begin{pmatrix} 4 & 1 & -2 \\ 1 & 2 & 2 \\ -2 & 2 & 2 \end{pmatrix}.$$

Thus,

$$\mathbf{x}^T A \mathbf{x} = \begin{pmatrix} x_1 & x_2 & x_3 \end{pmatrix} \begin{pmatrix} 4 & 1 & -2 \\ 1 & 2 & 2 \\ -2 & 2 & 2 \end{pmatrix} \begin{pmatrix} x_1 \\ x_2 \\ x_3 \end{pmatrix}.$$

                                                      ●

The quadratic form is said to be:

1. positive-definite, if $q(\mathbf{x}) > 0$ for all $\mathbf{x} \neq \mathbf{0}$;

2. positive-semidefinite, if $q(\mathbf{x}) \geq 0$ for all $\mathbf{x}$ and there exists $\mathbf{x} \neq \mathbf{0}$, such that $q(\mathbf{x}) = 0$;

3. negative-definite, if $-q(\mathbf{x})$ is positive-definite;

4. negative-semidefinite, if $-q(\mathbf{x})$ is positive-semidefinite;

5. indefinite in all other cases.

It can be proved that the necessary and sufficient conditions for the realization of the preceding cases are:

Theorem 7.10 *Let $q(x) = \mathbf{x}^T A \mathbf{x}$, then:*

1. *$q(\mathbf{x})$ is positive-definite (-semidefinite), if the values of the principal minor determinants of A are positive (nonnegative). In this case, A is said to be positive-definite (-semidefinite).*

2. $q(\mathbf{x})$ is negative-definite, if the value of the kth principal minor determinant of A has the sign of $(-1)^k$, $k = 1, 2, \ldots, n$. In this case,
A is called negative-definite.

3. $q(\mathbf{x})$ is negative-semidefinite, if the kth principal minor determinant
of A either is zero or has the sign of $(-1)^k$, $k = 1, 2, \ldots, n$.   •

Theorem 7.11 *Let A be an $n \times n$ symmetric matrix. The quadratic form
$q(\mathbf{x}) = \mathbf{x}^T A \mathbf{x}$ is:*

1. *positive-definite, if and only if all of the eigenvalues of A are positive;*

2. *positive-semidefinite, if and only if all of the eigenvalues of A are
nonnegative;*

3. *negative-definite, if and only if all of the eigenvalues of A are negative;*

4. *negative-semidefinite, if and only if all of the eigenvalues of A are
nonpositive;*

5. *indefinite, if and only if A has both positive and negative eigenvalues.*
•

Example 7.20 *Classify*

$$q(x_1, x_2, x_3) = 7x_1^2 + 7x_2^2 + 7x_3^2 - 2x_1x_2 - 2x_1x_3 - 2x_2x_3$$

as positive-definite, negative-definite, indefinite, or none of these.

Solution. *The matrix of the quadratic form is*

$$A = \begin{pmatrix} 7 & -1 & -1 \\ -1 & 7 & -1 \\ -1 & -1 & 7 \end{pmatrix}.$$

*One can easily compute the eigenvalues of the above matrix, which are $8, 8$,
and 5. Since all of these eigenvalues are positive, $q(x_1, x_2, x_3)$ is a positive,
definite quadratic form.*   •

Theorem 7.12 *If A and B are n × n real matrices, with*

$$B = \frac{1}{2}(A + A^T),$$

then the corresponding quadratic forms of A and B are identical, and B is symmetric.

Proof. *Since* $\mathbf{x}^T A \mathbf{x}$ *is a* (1×1) *matrix (a real number), we have*

$$\mathbf{x}^T A^T \mathbf{x} = (\mathbf{x}^T A \mathbf{x})^T$$

and

$$
\begin{aligned}
\mathbf{x}^T B^T \mathbf{x} &= \frac{1}{2}\mathbf{x}^T(A + A^T)\mathbf{x} \\
&= \frac{1}{2}\mathbf{x}^T A \mathbf{x} + \frac{1}{2}\mathbf{x}^T A^T \mathbf{x} \\
&= \frac{1}{2}\mathbf{x}^T A \mathbf{x} + \frac{1}{2}(\mathbf{x}^T A \mathbf{x})^T \\
&= \frac{1}{2}\mathbf{x}^T A \mathbf{x} + \frac{1}{2}\mathbf{x}^T A \mathbf{x} \\
&= \mathbf{x}^T A \mathbf{x}.
\end{aligned}
$$

Also,

$$B^T = \frac{1}{2}(A + A^T)^T = \frac{1}{2}(A^T + A) = \frac{1}{2}(A + A^T) = B.$$

Note that the quadratic forms of A and B are the same but the matrices A and B are not, unless A is a symmetric. For example, for the matrix

$$A = \begin{pmatrix} 4 & 4 \\ 2 & 6 \end{pmatrix},$$

we have

$$B = \frac{1}{2}\left\{ \begin{pmatrix} 4 & 4 \\ 2 & 6 \end{pmatrix} + \begin{pmatrix} 4 & 2 \\ 4 & 6 \end{pmatrix} \right\},$$

and it gives

$$B = \frac{1}{2}\left\{ \begin{pmatrix} 8 & 6 \\ 6 & 12 \end{pmatrix} \right\} = \begin{pmatrix} 4 & 3 \\ 3 & 6 \end{pmatrix}.$$

Now for the symmetric matrix

$$A = \begin{pmatrix} 4 & 4 \\ 4 & 6 \end{pmatrix},$$

we have

$$B = \frac{1}{2} \left\{ \begin{pmatrix} 4 & 4 \\ 4 & 6 \end{pmatrix} + \begin{pmatrix} 4 & 4 \\ 4 & 6 \end{pmatrix} \right\},$$

which gives

$$B = \frac{1}{2} \left\{ \begin{pmatrix} 8 & 8 \\ 8 & 12 \end{pmatrix} \right\} = \begin{pmatrix} 4 & 4 \\ 4 & 6 \end{pmatrix}.$$

Also, the quadratic forms

$$\mathbf{x}^T A \mathbf{x} = \begin{pmatrix} x_1 & x_2 \end{pmatrix} \begin{pmatrix} 4 & 4 \\ 2 & 6 \end{pmatrix} \begin{pmatrix} x_1 \\ x_2 \end{pmatrix} = 4x_1^2 + 6x_1x_2 + 6x_2^2$$

and

$$\mathbf{x}^T B \mathbf{x} = \begin{pmatrix} x_1 & x_2 \end{pmatrix} \begin{pmatrix} 4 & 3 \\ 3 & 6 \end{pmatrix} \begin{pmatrix} x_1 \\ x_2 \end{pmatrix} = 4x_1^2 + 6x_1x_2 + 6x_2^2$$

are the same. ●

Example 7.21 *Classify*

$$q(x_1, x_2, x_3) = -9x_1^2 - 5.5x_2^2 - 9x_3^2 + 6x_1x_2 + 6x_1x_3 + 6x_2x_3$$

as positive-definite, negative-definite, indefinite, or none of these.

Solution. *The matrix of the quadratic form is*

$$A = \begin{pmatrix} -9 & 3 & 3 \\ 3 & -5.5 & 3 \\ 3 & 3 & -9 \end{pmatrix}.$$

The eigenvalues of the above matrix A are $-12, -10$, and -1.5, and since all of these eigenvalues are negative, $q(x_1, x_2, x_3)$ is a negative-definite quadratic form. ●

7.3   Nonlinear Equations and Systems

Here we study one of the fundamental problems of numerical analysis, namely, the numerical solution of nonlinear equations. Most equations, in practice, are nonlinear and are rarely of a form that allows the roots to be determined exactly. A nonlinear equation may be considered any one of the following types:

1. An equation may be an *algebraic equation (a polynomial equation of degree n)* expressible in the form:

$$a_n x^n + a_{n-1} x^{n-1} + \cdots + a_1 x + a_0 = 0, \quad a_n \neq 0, \quad n > 1,$$

where $a_n, a_{n-1}, \ldots, a_1$, and a_0 are constants. For example, the following equations are nonlinear:

$$x^2 + 5x + 6 = 0; \quad x^3 = 2x + 1; \quad x^{100} + x^2 + 1 = 0.$$

2. The power of the unknown variable involved in the equation must be difficult to manipulate. For example, the following equations are nonlinear:

$$x^{-1} + 2x = 1; \quad \sqrt{x} + x = 0; \quad x^{2/3} + \frac{2}{x} + 4 = 0.$$

3. An equation may be a *transcendental equation* that involves trigonometric functions, exponential functions, and logarithmic functions. For example, the following equations are nonlinear:

$$x = \cos(x); \quad e^x + x - 10 = 0; \quad x + \ln x = 10.$$

Definition 7.7 (Root of an Equation)

Assume that $f(x)$ is a continuous function. A number α for which $f(\alpha) = 0$ is called a root of the equation $f(x) = 0$ or a zero of the function $f(x)$. ●

There may be many roots of the given nonlinear equation, but we will seek the approximation of only one of its roots, which lies on the given interval $[a, b]$. This root may be *simple* (not repeating) or *multiple* (repeating).

Now, we shall discuss the methods for nonlinear equations in a *single variable*. The problem here can be written down simply as

$$f(x) = 0. \tag{7.6}$$

We seek the values of x called the *roots* of (7.6) or the *zeros* of the function $f(x)$ such that (7.6) is true. The roots of (7.6) may be real or complex. Here, we will look for the approximation of the real root of (7.6). There are many methods that will give us information about the real roots of (7.6). The methods we will discuss are all iterative methods. They are the bisection method, fixed-point method, and Newton's method.

7.3.1 Bisection Method

This is one of the simplest iterative techniques for determining the roots of (7.6), and it needs two initial approximations to start. It is based on the *Intermediate Value theorem*. This method is also called the *interval-halving method* because the strategy is to bisect, or halve, the interval from one endpoint of the interval to the other endpoint and then retain the half-interval whose ends still bracket the root. It is also referred to as a *bracketing method* or sometimes is called *Bolzano's method*. The fact that the function is required to change sign only once gives us a way to determine which half of the interval to retain; we keep the half on which $f(x)$ changes signs or becomes zero. The basis for this method can be easily illustrated by considering the function

$$y = f(x).$$

Our object is to find an x value for which y is zero. Using this method, we begin by supposing $f(x)$ is a continuous function defined on the interval $[a, b]$ and then by evaluating the function at two x values, say, a and b, such that

$$f(a).f(b) < 0.$$

The implication is that one of the values is negative and the other is positive. These conditions can be easily satisfied by sketching the function (Figure 7.6).

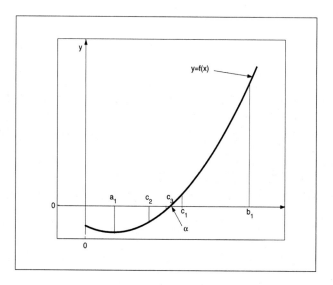

Figure 7.6: Bisection method.

Obviously, the function is negative at one endpoint a of the interval and positive at the other endpoint b and is continuous on $a \leq x \leq b$. Therefore, the root must lie between a and b (by the Intermediate Value theorem), and a new approximation to the root α can be calculated as

$$c = \frac{a+b}{2},$$

and, in general,

$$c_n = \frac{a_n + b_n}{2}, \qquad n \geq 1. \tag{7.7}$$

The iterative formula (7.7) is known as the *bisection method*.

If $f(c) \approx 0$, then $c \approx \alpha$ is the desired root, and, if not, then there are two possibilities. First, if $f(a).f(c) < 0$, then $f(x)$ has a zero between point a and point c. The process can then be repeated on the new interval $[a, c]$. Second, if $f(a).f(c) > 0$, it follows that $f(b).f(c) < 0$ since it is known that $f(b)$ and $f(c)$ have opposite signs. Hence, $f(x)$ has a zero between point c and point b and the process can be repeated with $[c, b]$. We see that after one step of the process, we have found either a zero or a new bracketing interval which is precisely half the length of the original one. The process continues until the desired accuracy is achieved. We use the bisection process in the following example.

Example 7.22 *Use the bisection method to find the approximation to the root of the equation*

$$x^3 = 4x - 2$$

that is located on the interval $[1.0, 2.0]$ accurate to within 10^{-4}.

Solution. *Since the given function $f(x) = x^3 - 4x + 2$ is a cubic polynomial function and is continuous on $[1.0, 2.0]$, starting with $a_1 = 1.0$ and $b_1 = 2.0$, we compute*

$$a_1 = 1.0: \quad f(a_1) = -2.0$$
$$b_1 = 2.0: \quad f(b_1) = 1.0,$$

and since $f(1.0).f(2.0) = -2 < 0$, so that a root of $f(x) = 0$ lies on the interval $[1.0, 2.0]$, using formula (7.7) (when $n = 1$), we get

$$c_1 = \frac{a_1 + b_1}{2} = \frac{1.0 + 2.0}{2} = 1.5; \quad f(c_1) = f(1.5) = 0.859375.$$

Hence, the function changes sign on $[a_1, c_1] = [1.5, 1.75]$. To continue, we squeeze from the right and set $a_2 = a_1$ and $b_2 = c_1$. Then the midpoint is

$$c_2 = \frac{a_2 + b_2}{2} = 1.625; \quad f(c_2) = 0.041056.$$

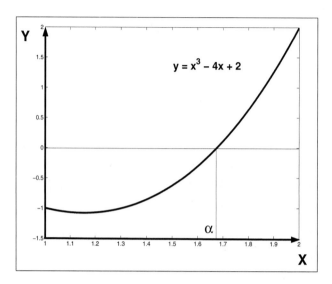

Figure 7.7: Bisection method.

Continuing in this manner we obtain a sequence $\{c_k\}$ of approximation shown by Table 7.1.

We see that the functional values approach zero as the number of iterations increases. We got the desired approximation to the root $\alpha = 1.6751309$ of the given equation $x^3 = 4x - 2$, i.e., $c_{17} = 1.675133$, which was obtained after 17 iterations, with accuracy $\epsilon = 10^{-4}$. ●

To use MATLAB commands for the bisection method, first we define a function m-file as fn.m for the equation as follows:

$$function\ y = fn(x)$$
$$y = x.\hat{}\ 3 - 4 * x + 2;$$

then use the single command:

$$>> s = bisect('fn', 1.0, 2.0, 1e-4)$$
$$s =$$
$$1.675133$$

Table 7.1: Solution of $x^3 = 4x - 2$ by the bisection method.

n	Left Endpoint a_n	Midpoint c_n	Right Endpoint b_n	Function Value $f(c_n)$
01	1.000000	1.500000	2.000000	−0.625000
02	1.500000	1.750000	2.000000	0.359375
03	1.500000	1.625000	1.750000	−0.208984
04	1.625000	1.687500	1.750000	0.055420
05	1.625000	1.656250	1.687500	−0.081635
06	1.656250	1.671875	1.687500	−0.014332
07	1.671875	1.679688	1.687500	0.020239
08	1.671875	1.675781	1.679688	0.002875
09	1.671875	1.673828	1.675781	−0.005748
10	1.673828	1.674805	1.675781	−0.001439
11	1.674805	1.675293	1.675781	0.000717
12	1.674805	1.675049	1.675293	−0.000361
13	1.674805	1.675171	1.675293	0.000177
14	1.674805	1.675110	1.675171	−0.000092
15	1.675110	1.675140	1.675171	0.000040
16	1.675110	1.675125	1.675140	−0.000025
17	1.675125	1.675133	1.675140	0.000009

Program 7.2

MATLAB m-file for the Bisection Method

```
function sol=bisect(fn,a,b,tol)
fa = feval(fn, a); fb = feval(fn, b);
if fa * fb > 0; fprintf('Endpoints have same sign') return
end
while abs (b − a) > tol
c = (a + b)/2; fc = feval(fn, c);
if fa * fc < 0; b = c; else a = c; end
end; sol=(a + b)/2;
```

Theorem 7.13 (Bisection Convergence and Error Theorem)

Let $f(x)$ be a continuous function defined on the initial interval $[a_0, b_0] = [a, b]$ and suppose $f(a).f(b) < 0$. Then the bisection method (7.7) generates a sequence $\{c_n\}_{n=1}^{\infty}$ approximating $\alpha \in (a, b)$, with the property

$$|\alpha - c_n| \leq \frac{b - a}{2^n}, \quad n \geq 1. \tag{7.8}$$

Moreover, to obtain an accuracy of

$$|\alpha - c_n| \leq \epsilon$$

(for $\epsilon = 10^{-k}$), it suffices to take

$$n \geq \frac{\ln\{10^k(b - a)\}}{\ln 2}, \tag{7.9}$$

where k is a nonnegative integer.   ●

The above Theorem 7.13 gives us information about bounds for errors in approximation and the number of bisections needed to obtain any given accuracy.

One drawback of the bisection method is that the convergence rate is raster slow. However, the rate of convergence is guaranteed. So for this reason it is often used as a start for a more efficient method used to find the roots of nonlinear equations. The method may give a false root, if $f(x)$ is discontinuous on the given interval $[a, b]$.

Procedure 7.1 (Bisection Method)

1. *Establish an interval $a \leq x \leq b$ such that $f(a)$ and $f(b)$ are opposite signs, i.e., $f(a).f(b) < 0$.*

2. *Choose an error tolerance ($\epsilon > 0$) value for the function.*

3. *Compute a new approximation for the root*

$$c_n = \frac{(a_n + b_n)}{2}; \quad n = 1, 2, 3, \ldots.$$

4. *Check the tolerance. If $|f(c_n)| \leq \epsilon$, then use $c_n (n = 1, 2, 3, \ldots)$ for the desired root; otherwise, continue.*

5. *Check; if $f(a_n).f(c_n) < 0$, then set $b_n = c_n$; otherwise, set $a_n = c_n$.*

6. *Go back to step 3 and repeat the process.*

7.3.2 Fixed-Point Method

This is another iterative method used to solve the nonlinear equation (7.6), and it needs one initial approximation to start. This is a very general method for finding the root of (7.6), and it provides us with a theoretical framework within which the convergence properties of subsequent methods can be evaluated. The basic idea of this method, which is also called the successive approximation method or function iteration, is to rearrange the original equation

$$f(x) = 0 \tag{7.10}$$

into an equivalent expression of the form

$$x = g(x). \tag{7.11}$$

Any solution of (7.11) is called a fixed point for the iteration function $g(x)$ and, hence, a root of (7.10).

Definition 7.8 (Fixed Point)

A fixed point of a function $g(x)$ is a real number α such that $\alpha = g(\alpha)$. ●

The task of solving (7.10) is therefore reduced to that of finding a point satisfying the fixed-point condition (7.11). The fixed-point method essentially solves two functions simultaneously; $y = x$ and $y = g(x)$. The point of intersection of these two functions is the solution to $x = g(x)$ and thus to $f(x) = 0$ (Figure 7.8).

This method is conceptually very simple. Since $g(x)$ is also nonlinear, the solution must be obtained iteratively. An initial approximation to the solution, say, x_0, must be determined. For choosing the best initial value

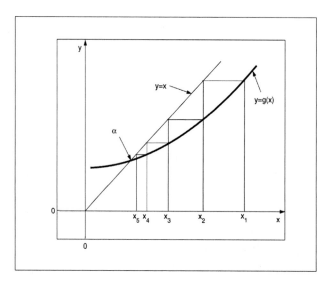

Figure 7.8: Fixed-point method.

x_0 for using this iterative method, we have to find an interval $[a, b]$ on which the original function $f(x)$ satisfies the sign property and then use the midpoint $\frac{a+b}{2}$ as the initial approximation x_0. Then this initial value x_0 is substituted in the function $g(x)$ to determine the next approximation x_1, and so on.

Definition 7.9 (Fixed-Point Method)

The iteration defined in

$$x_{n+1} = g(x_n); \qquad n \geq 0 \tag{7.12}$$

is called the fixed-point method or the fixed-point iteration.   ●

The value of the initial approximation x_0 is chosen arbitrarily and the hope is that the sequence $\{x_n\}_{n=0}^{\infty}$ converges to a number α which will automatically satisfy (7.10). Moreover, since (7.10) is a rearrangement of (7.11), α is guaranteed to be a zero of $f(x)$. In general, there are many different ways of rearranging (7.11) in (7.10) form. However, only some of these are likely to give rise to successful iterations; but sometimes we don't have successful iterations. To describe such behavior, we discuss the following example.

Example 7.23 *One of the possible rearrangements of the nonlinear equation $x^3 = 4x + 2$, which has the root on $[1, 2]$, is*

$$x_{n+1} = g(x_n) = \sqrt{\frac{4x_n - 2}{x_n}}; \qquad n \geq 0.$$

Then use the fixed-point iteration formula (7.12) to compute the approximation of the root of the equation accurate to within 10^{-4}, starting with $x_0 = 1.5$.

Solution. *Since $x_0 = 1.5$ is given, we have*

$$x_1 = g(x_0) = \sqrt{\frac{4x_0 - 2}{x_0}} = \sqrt{\frac{4(1.5) - 2}{1.5}} = 1.632993.$$

This and the further iterates are shown in Table 7.2.

Table 7.2: Solution of Example 7.23 .

n	$x_{n+1} = g(x_n) = \sqrt{\dfrac{4x_n - 2}{x_n}}$	$f(x_n)$
00	1.500000	−0.625000
01	1.632993	−0.177324
02	1.665910	−0.040313
03	1.673157	−0.008701
04	1.674710	−0.0018586
05	1.675041	−0.000397
06	1.675112	−0.000083
07	1.675127	−0.000017
08	1.675130	−0.000004

We note that the considered sequence converges, and it converged faster than the bisection method. The desired approximation to the root of the

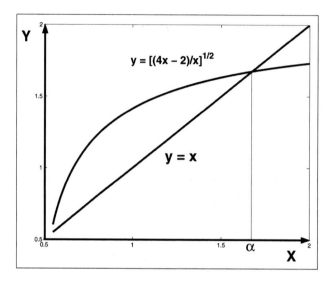

Figure 7.9: Fixed-point method.

given equation is $x_8 = 1.675130$, which we obtained after 8 iterations, with accuracy $\epsilon = 10^{-4}$. •

Theorem 7.14 (Fixed-Point Theorem)

If g is continuously differentiable on the interval $[a, b]$ and $g(x) \in [a, b]$ for all $x \in [a, b]$, then

(a) *g has at least one fixed point on the given interval $[a, b]$.*

Moreover, if the derivative $g'(x)$ of the function $g(x)$ exists on an interval $[a, b]$, which contains the starting value x_0, with

$$|g'(x)| \leq k < 1, \qquad for \ all \qquad x \in [a, b] \tag{7.13}$$

then:

(b) *The sequence (7.12) will converge to the attractive (unique) fixed-point α in $[a, b]$.*

(c) *The iteration (7.12) will converge to α for any initial approximation.*

(d) We have the error estimate

$$|\alpha - x_n| \le \frac{k^n}{1-k}|x_1 - x_0|, \qquad \text{for all} \quad n \ge 1. \qquad (7.14)$$

(e) The limit holds:

$$\lim_{n\to\infty} \frac{\alpha - x_{n+1}}{\alpha - x_n} = g'(\alpha). \qquad (7.15)$$

MATLAB commands for the above given rearrangement $x = g(x)$ of $f(x) = x^3 - 4x + 2$ by using the initial approximation $x_0 = 1.5$ can be written as follows:

$$
\begin{aligned}
&function\ y = fn(x) \\
&y = sqrt(4*x - 2)/x; \\
&>> x0 = 1.5; tol = 0.00001 \\
&>> sol = fixpt('fn', x0, tol);
\end{aligned}
$$

Program 7.3
MATLAB m-file for the Fixed-Point Method
function sol=fixpt(fn,x0,tol)
old= x0+1;
while abs(x0-old) > tol; old=x0;
$x0 = feval(fn, old)$; end; sol=$x0$;

Procedure 7.2 (Fixed-Point Method)

1. *Choose an initial approximation x_0 such that $x_0 \in [a, b]$.*

2. *Choose a convergence parameter $\epsilon > 0$.*

3. *Compute the new approximation x_{new} by using the iterative formula (7.12).*

4. *Check if $|x_{new} - x_0| < \epsilon$ then x_{new} is the desire approximate root; otherwise, set $x_0 = x_{new}$ and go to step 3.*

7.3.3 Newton's Method

This is one of the most popular and powerful iterative methods for finding roots of the nonlinear equation (7.6). It is also known as the method of tangents because after estimating the actual root, the zero of the tangent to the function at that point is determined. It always converges if the initial approximation is sufficiently close to the exact solution. This method is distinguished from the methods given in the previous sections, by the fact that it requires the evaluation of both the function $f(x)$ and the derivative of the function $f'(x)$ at arbitrary point x. Newton's method consists of geometrically expanding the tangent line at a current point x_i until it crosses zero, then setting the next guess x_{i+1} to the abscissa of that zero crossing (Figure 7.10). This method is also called the *Newton–Raphson method.*

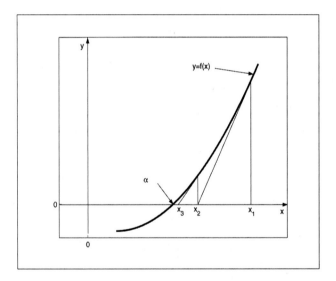

Figure 7.10: Newton's method.

There are many descriptions of Newton's method. We shall derive the method from the familiar Taylor's series expansion of a function in the neighborhood of a point.

Let $f \in C^2[a, b]$ and let x_n be the *nth* approximation to the root α such that $f'(x_n) \neq 0$, and $|\alpha - x_n|$ is small. Consider the first Taylor polynomial for $f(x)$ expanded about x_n, so we have

$$f(x) = f(x_n) + (x - x_n)f'(x_n) + \frac{(x - x_n)^2}{2}f''(\eta(x)), \qquad (7.16)$$

where $\eta(x)$ lies between x and x_n. Since $f(\alpha) = 0$, then (7.16), with $x = \alpha$, gives

$$f(\alpha) = 0 = f(x_n) + (\alpha - x_n)f'(x_n) + \frac{(\alpha - x_n)^2}{2}f''(\eta(\alpha)).$$

Since $|\alpha - x_n|$ is small, we neglect the term involving $(\alpha - x_n)^2$, and so

$$0 \approx f(x_n) + (\alpha - x_n)f'(x_n).$$

Solving for α, we get

$$\alpha \approx x_n - \frac{f(x_n)}{f'(x_n)}, \qquad (7.17)$$

which should be a better approximation to α than x_n. We call this approximation x_{n+1}, and we get

$$x_{n+1} = x_n - \frac{f(x_n)}{f'(x_n)}, \quad f'(x_n) \neq 0, \qquad \text{for all} \quad n \geq 0. \qquad (7.18)$$

The iterative method (7.18) is called Newton's method. Usually Newton's method converges well and quickly; its convergence cannot, however, be guaranteed, and it may sometimes converge to a different root from the one expected. In particular, there may be difficulties if the initial approximation is not sufficiently close to the actual root. The most serious problem of Newton's method is that some functions are difficult to differentiate analytically, and some functions cannot be differentiated analytically at all. Newton's method is not restricted to one dimension only. The method readily generalizes to multiple dimensions. It should be noted that this method is suitable for finding real as well as imaginary roots of polynomials.

Example 7.24 *Use Newton's method to find the root of the equation $x^3 - 4x + 2 = 0$ that is located on the interval $[1.0, 2.0]$ accurate to 10^{-4}, taking an initial approximation $x_0 = 1.5$.*

Solution. *Given*

$$f(x) = x^3 - 4x + 2,$$

since Newton's method requires that the value of the derivative of the function be found, the derivative of the function is

$$f'(x) = 3x^2 - 4.$$

Now evaluating $f(x)$ and $f'(x)$ at the give approximation $x_0 = 1.5$ gives

$$x_0 = 1.5, \qquad f(1.5) = -0.625, \qquad f'(1.5) = 2.750.$$

Using Newton's iterative formula (7.18), we get

$$x_1 = x_0 - \frac{f(x_0)}{f'(x_0)} = 1.5 - \frac{(-0.625)}{2.75} = 1.727273.$$

Now evaluating $f(x)$ and $f'(x)$ at the new approximation x_1, we get

$$x_1 = 1.727273, \qquad f(1.727273) = 0.244177, \qquad f'(1.727273) = 4.950413.$$

Using the iterative formula (7.18) again, we get the other new approximation as follows:

$$x_2 = x_1 - \frac{f(x_1)}{f'(x_1)} = 1.727273 - \frac{(0.244177)}{4.950413} = 1.677948.$$

Thus, the successive iterates are shown in Table 7.3. Just after the third iteration, the root is approximated to be $x_4 = 1.67513087056$ and the functional value is reduced to 4.05×10^{-10}. Since the exact solution is 1.67513087057, the actual error is 1×10^{-10}. We see that the convergence is faster than the methods considered previously. ●

To get the above results using MATLAB commands, first the function $x^3 - 4x + 2$ and its derivative $3x^2 - 4$ are saved in m-files called fn.m and dfn.m, respectively, written as follows:

Table 7.3: Solution of $x^3 = 4x - 2$ by Newton's method.

n	x_n	$f(x_n)$	$f'(x_n)$	Error $\alpha - x_n$
00	1.500000	-0.625000	2.750000	0.175131
01	1.727273	0.244177	4.950413	-0.052142
02	1.677948	0.012487	4.446529	-0.002817
03	1.675140	0.000040	4.418281	-0.000009
04	1.675131	4.05e-010	4.418190	-0.000000

$$function\ y = fn(x)$$
$$y = x.\hat{\ } 3 - 4 * x + 2;$$

and

$$function\ dy = dfn(x)$$
$$dy = 3 * x.\hat{\ } 2 - 4;$$

Then we do the following:

```
>> x0 = 1.5; tol = 0.00001;
>> sol = newton('fn','dfn',x0,tol);
```

Program 7.4

MATLAB m-file for Newton's Method

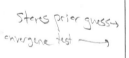

```
function sol=newton(fn,dfn,x0,tol)
old = x0+1;
while abs (x0 - old) > tol; old = x0;
x0 = old - feval(fn,old)/feval(dfn,old);
end; sol=x0;
```

7.3.4 System of Nonlinear Equations

A system of nonlinear algebraic equations may arise when one is dealing
with problems involving optimization and numerical integration (Gauss

quadratures). Generally, the system of equations may not be of the polynomial variety. Therefore, a system of n equations in n unknowns is called nonlinear, if one or more of the equations in the systems is/are nonlinear.

The numerical methods we discussed so far have been concerned with finding a root of a nonlinear algebraic equation with one independent variable. We now consider two numerical methods for solving systems of nonlinear algebraic equations in which each equation is a function of a specified number of variables.

Consider the system of two nonlinear equations with the two variables

$$f_1(x, y) = 0 \qquad (7.19)$$

and

$$f_2(x, y) = 0. \qquad (7.20)$$

The problem can be stated as follows:
Given the continuous functions $f_1(x, y)$ and $f_2(x, y)$, find the values $x = \alpha$ and $y = \beta$ such that

$$\begin{aligned} f_1(\alpha, \beta) &= 0 \\ f_2(\alpha, \beta) &= 0. \end{aligned} \qquad (7.21)$$

The functions $f_1(x, y)$ and $f_2(x, y)$ may be algebraic equations, transcendental, or any nonlinear relationship between the input x and y and the output $f_1(x, y)$ and $f_2(x, y)$. The solutions to (7.19) and (7.20) are the intersections of $f_1(x, y) = f_2(x, y) = 0$ (Figure 7.11). This problem is considerably more complicated than the solution of a single nonlinear equation. The one-point iterative method discussed above for the solution of a single equation may be extended to the system. So to solve the system of nonlinear equations we have many methods to choose from, but we will use Newton's method.

Newton's Method for the System

Consider the two nonlinear equations specified by the equations (7.19) and (7.20). Suppose that (x_n, y_n) is an approximation to a root (α, β); then

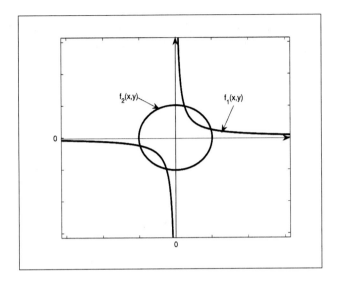

Figure 7.11: Nonlinear equation in two variables.

by using Taylor's theorem for functions of two variables for $f_1(x, y)$ and $f_2(x, y)$ expanding about (x_n, y_n), we have

$$
\begin{aligned}
f_1(x, y) &= f_1(x_n + (x - x_n), y_n + (y - y_n)) \\
&= f_1(x_n, y_n) + (x - x_n)\frac{\partial f_1(x_n, y_n)}{\partial x} + (y - y_n)\frac{\partial f_1(x_n, y_n)}{\partial y} + \cdots
\end{aligned}
$$

and

$$
\begin{aligned}
f_2(x, y) &= f_2(x_n + (x - x_n), y_n + (y - y_n)) \\
&= f_2(x_n, y_n) + (x - x_n)\frac{\partial f_2(x_n, y_n)}{\partial x} + (y - y_n)\frac{\partial f_2(x_n, y_n)}{\partial y} + \cdots.
\end{aligned}
$$

Since $f_1(\alpha, \beta) = 0$ and $f_2(\alpha, \beta) = 0$, these equations, with $x = \alpha$ and $y = \beta$, give

$$
\begin{aligned}
0 &= f_1(x_n, y_n) + (\alpha - x_n)\frac{\partial f_1(x_n, y_n)}{\partial x} + (\beta - y_n)\frac{\partial f_1(x_n, y_n)}{\partial y} + \cdots \\
0 &= f_2(x_n, y_n) + (\alpha - x_n)\frac{\partial f_2(x_n, y_n)}{\partial x} + (\beta - y_n)\frac{\partial f_2(x_n, y_n)}{\partial y} + \cdots.
\end{aligned}
$$

Newton's method has a condition that initial approximation (x_n, y_n) should be sufficiently close to the exact root (α, β), therefore, the higher order terms may be neglected to obtain

$$
\begin{aligned}
0 &\approx f_1(x_n, y_n) + (\alpha - x_n)\frac{\partial f_1(x_n, y_n)}{\partial x} + (\beta - y_n)\frac{\partial f_1(x_n, y_n)}{\partial y} \\
0 &\approx f_2(x_n, y_n) + (\alpha - x_n)\frac{\partial f_2(x_n, y_n)}{\partial x} + (\beta - y_n)\frac{\partial f_2(x_n, y_n)}{\partial y}.
\end{aligned}
$$

$$(7.22)$$

We see that this represents a system of two linear algebraic equations for α and β. Of course, since the higher order terms are omitted in the derivation of these equations, their solution (α, β) is no longer an exact root of (7.21) and (7.22). However, it will usually be a better approximation than (x_n, y_n), so replacing (α, β) by (x_{n+1}, y_{n+1}) in (7.21) and (7.22) gives the iterative scheme

$$
\begin{aligned}
0 &= f_1(x_n, y_n) + (x_{n+1} - x_n)\frac{\partial f_1(x_n, y_n)}{\partial x} + (y_{n+1} - y_n)\frac{\partial f_1(x_n, y_n)}{\partial y} \\
0 &= f_2(x_n, y_n) + (x_{n+1} - x_n)\frac{\partial f_2(x_n, y_n)}{\partial x} + (y_{n+1} - y_n)\frac{\partial f_2(x_n, y_n)}{\partial y}.
\end{aligned}
$$

Then writing in the matrix form, we have

$$
\begin{pmatrix} \dfrac{\partial f_1}{\partial x} & \dfrac{\partial f_1}{\partial y} \\[2mm] \dfrac{\partial f_2}{\partial x} & \dfrac{\partial f_2}{\partial y} \end{pmatrix}
\begin{pmatrix} x_{n+1} - x_n \\[2mm] y_{n+1} - y_n \end{pmatrix}
= - \begin{pmatrix} f_1 \\[2mm] f_2 \end{pmatrix},
\qquad (7.23)
$$

where f_1, f_2, and their partial derivatives f_{1x}, f_{1y} are evaluated at (x_n, y_n). Hence,

$$
\begin{pmatrix} x_{n+1} \\[2mm] y_{n+1} \end{pmatrix}
= \begin{pmatrix} x_n \\[2mm] y_n \end{pmatrix}
- \begin{pmatrix} \dfrac{\partial f_1}{\partial x} & \dfrac{\partial f_1}{\partial y} \\[2mm] \dfrac{\partial f_2}{\partial x} & \dfrac{\partial f_2}{\partial y} \end{pmatrix}^{-1}
\begin{pmatrix} f_1 \\[2mm] f_2 \end{pmatrix}.
\qquad (7.24)
$$

We call the matrix

$$J = \begin{pmatrix} \dfrac{\partial f_1}{\partial x} & \dfrac{\partial f_1}{\partial y} \\[2.5ex] \dfrac{\partial f_2}{\partial x} & \dfrac{\partial f_2}{\partial y} \end{pmatrix} \tag{7.25}$$

the *Jacobian matrix*.

Note that (7.23) can be written in the simplified form as

$$\begin{pmatrix} \dfrac{\partial f_1}{\partial x} & \dfrac{\partial f_1}{\partial y} \\[2.5ex] \dfrac{\partial f_2}{\partial x} & \dfrac{\partial f_2}{\partial y} \end{pmatrix} \begin{pmatrix} h \\ k \end{pmatrix} = - \begin{pmatrix} f_1 \\ f_2 \end{pmatrix},$$

where h and k can be evaluated as

$$\begin{aligned} h &= \dfrac{\left(-f_1 \dfrac{\partial f_2}{\partial y} + f_2 \dfrac{\partial f_1}{\partial y} \right)}{\left(\dfrac{\partial f_1}{\partial x} \dfrac{\partial f_2}{\partial y} - \dfrac{\partial f_1}{\partial y} \dfrac{\partial f_2}{\partial x} \right)} \\[3ex] k &= \dfrac{\left(f_1 \dfrac{\partial f_2}{\partial x} - f_2 \dfrac{\partial f_1}{\partial x} \right)}{\left(\dfrac{\partial f_1}{\partial x} \dfrac{\partial f_2}{\partial y} - \dfrac{\partial f_1}{\partial y} \dfrac{\partial f_2}{\partial x} \right)}, \end{aligned} \tag{7.26}$$

where all functions are to be evaluated at (x, y). Newton's method for a pair of equations in two unknowns is therefore

$$\begin{pmatrix} x_{n+1} \\ y_{n+1} \end{pmatrix} = \begin{pmatrix} x_n \\ y_n \end{pmatrix} + \begin{pmatrix} h \\ k \end{pmatrix}, \quad n = 0, 1, 2, \ldots, \tag{7.27}$$

where (h, k) are given by (7.26) and evaluated at (x_n, y_n).

At a starting approximation (x_0, y_0), the functions $f_1, f_{1x}, f_{1y}, f_2, f_{2x}$ and f_{2y} are evaluated. The linear equations are then solved for (x_1, y_1) and the whole process is repeated until convergence is obtained. Comparison of (7.18) and (7.24) shows that the above procedure is indeed an extension of Newton's method in one variable, where division by f' is generalized to premultiplication by J^{-1}.

Example 7.25 *Solve the following system of two equations using Newton's method with accuracy $\epsilon = 10^{-5}$:*

$$
\begin{aligned}
4x^3 + y &= 6 \\
x^2 y &= 1.
\end{aligned}
$$

Assume $x_0 = 1.0$ and $y_0 = 0.5$ as starting values.

Solution. *Obviously, this system of nonlinear equations has an exact solution of $x = 1.088282$ and $y = 0.844340$. Let us look at how Newton's method is used to approximate these roots. The first partial derivatives are as follows:*

$$
\begin{aligned}
f_1(x,y) &= 4x^3 + y - 6, & f_{1x} &= 12x^2, & f_{1y} &= 1 \\
f_2(x,y) &= x^2 y - 1, & f_{2x} &= 2xy, & f_{2y} &= x^2.
\end{aligned}
$$

At the given initial approximations $x_0 = 1.0$ and $y_0 = 0.5$, we get

$$
f_1(1.0, 0.5) = -1.5, \quad \frac{\partial f_1}{\partial x} = f_{1x} = 12, \quad \frac{\partial f_1}{\partial y} = f_{1y} = 1.0
$$

$$
f_2(1.0, 0.5) = -0.5, \quad \frac{\partial f_1}{\partial x} = f_{2x} = 1.0, \quad \frac{\partial f_2}{\partial y} = f_{2y} = 1.0.
$$

The Jacobian matrix J and its inverse J^{-1} at the given initial approximation can be calculated as

$$
J = \begin{pmatrix} \dfrac{\partial f_1}{\partial x} & \dfrac{\partial f_1}{\partial y} \\[2mm] \dfrac{\partial f_2}{\partial x} & \dfrac{\partial f_2}{\partial y} \end{pmatrix} = \begin{pmatrix} 12.0 & 1.0 \\ 1.0 & 1.0 \end{pmatrix}
$$

and

$$
J^{-1} = \frac{1}{11.0} \begin{pmatrix} 1.0 & -1.0 \\ -1.0 & 12.0 \end{pmatrix}.
$$

The Jacobian matrix can be found by using MATLAB commands as follows:

```
>> syms x y
>> fun = [4 * x ^ 3+y − 6, x ^ 2*y − 1];
>> var = [x, y];
>> R = jacobian(f, var);
```

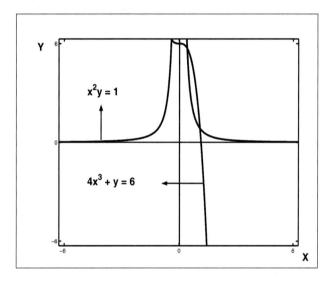

Figure 7.12: Graphical solution of the given nonlinear system.

Substituting all these values into (7.25), we get the first approximation as

$$\begin{pmatrix} x_1 \\ y_1 \end{pmatrix} = \begin{pmatrix} 1.0 \\ 0.5 \end{pmatrix} - \frac{1}{11.0} \begin{pmatrix} 1.0 & -1.0 \\ -1.0 & 12.0 \end{pmatrix} \begin{pmatrix} -1.5 \\ -0.5 \end{pmatrix} = \begin{pmatrix} 1.090909 \\ 0.909091 \end{pmatrix}.$$

Similarly, the second iteration gives

$$\begin{pmatrix} x_2 \\ y_2 \end{pmatrix} = \begin{pmatrix} 1.090909 \\ 0.909091 \end{pmatrix} - \frac{1}{15.012077} \begin{pmatrix} 1.190082 & -1.0 \\ -1.983471 & 14.280989 \end{pmatrix} \begin{pmatrix} 0.102178 \\ 0.081893 \end{pmatrix}$$

$$= \begin{pmatrix} 1.088264 \\ 0.844686 \end{pmatrix}.$$

The first two and the further steps of the method are listed in Table 7.4.

•

Table 7.4: Solution of a system of two nonlinear equations.

n	x-approx. x_n	y-approx. y_n	1st. func. $f_1(x_n, y_n)$	2nd. func. $f_2(x_n, y_n)$
00	1.000000	0.500000	-1.50000	-0.500000
01	1.090909	0.909091	0.102178	0.081893
02	1.088264	0.844686	0.000091	0.000377
03	1.088282	0.844340	0.000001	0.000001

Note that a typical iteration of this method for this pair of equations can be implemented in the MATLAB Command Window using:

```
>> f1 = 4 * x0^ 3 + y0 − 6; f2 = x0^ 2 * y0 − 1;
>> f1x = 12 * x0^ 2; f1y = 1; f2x = 2 * x0 * y0; f2y = x0^ 2;
>> D = f1x * f2y − f1y * f2x;
>> h = (f2 * f1y − f1 * f2y)/D; k = (f1 * f2x − f2 * f1x)/D;
>> x0 = x0 + h; y0 = y0 + k;
```

Using the starting value $(1.0, 0.5)$, the possible approximations are shown in Table 7.4.   •

We see that the values of both the functionals approach zero as the number of iterations is increased. We got the desired approximations to the roots after 3 iterations, with accuracy $\epsilon = 10^{-5}$.

Newton's method is fairly easy to implement for the case of two equations in two unknowns. We first need the function m-files for the equations and the partial derivatives. For the equations in Example 7.25, we do the following:

$$
\begin{array}{|l|}
\hline
function\ f = fn2(v) \\
\%Here\ f\ and\ v\ are\ vector\ quantities \\
x = v(1); y = v(2); \\
f(1) = 4 * x.\hat{\ }\, 2 + y - 6; \\
f(2) = x.\hat{\ }\, 2 * y - 1; \\
\hline
\end{array}
$$

$$
\begin{array}{|l|}
\hline
function\ J = dfn2(v) \\
\%Jacobian\ matrix\ for\ fn2.m \\
x = v(1); y = v(2); \\
J(1,1) = 12 * x.\hat{\ }\, 2; J(1,2) = 1; \\
J(2,1) = 2 * x * y; J(2,2) = x.\hat{\ }\, 2; \\
\hline
\end{array}
$$

Then the following MATLAB commands can be used to generate the solution of Example 7.25:

$$
\begin{array}{|l|}
\hline
>> s = newton2('fn2','dfn2',[1.0, 0.5], 1e-5) \\
s = \\
\quad 1.088282 \quad 0.844340 \\
\hline
\end{array}
$$

The m-file Newton2.m will need both the function and its partial derivatives as well as a starting vector and a tolerance. The following code can be used:

Program 7.5

MATLAB m-file for Newton's Method for a Nonlinear System

```
function sol=newton2(fn2,dfn2,x0,tol)
old=x0+1; while max(abs(x0-old))>tol; old=x0;
f = feval(fn2, old); f1 = f(1); f2 = f(2);
J=feval(dfn2,old);
f1x = J(1,1); f1y = J(1,2); f2x = J(2,1); f2y = J(2,2);
D = f1x * f2y - f1y * f2x;
h = (f2 * f1y - f1 * f2y)/D; k = (f1 * f2x - f2 * f1x)/D;
x0 = old + [h, k]; end; sol=x0;
```

Similarly, for a large system of equations it is convenient to use vector notation. Consider the system

$$
\mathbf{f}(\mathbf{x}) = \mathbf{0},
$$

where $\mathbf{f} = (f_1, f_2, \ldots, f_n)^T$ and $\mathbf{x} = (x_1, x_2, \ldots, x_n)^T$. Denoting the *nth* iterate by $\mathbf{x}^{[n]} = (x_1^{[n]}, x_2^{[n]}, x_3^{[n]}, \ldots, x_n^{[n]})^T$, then Newton's method is defined by

$$\mathbf{x}^{[n+1]} = \mathbf{x}^{[n]} - \left[J(\mathbf{x}^{[n]})\right]^{-1} \mathbf{f}(\mathbf{x}^{[n]}), \tag{7.28}$$

where the Jacobian matrix J is defined as

$$J = \begin{pmatrix} \dfrac{\partial f_1}{\partial x_1} & \dfrac{\partial f_1}{\partial x_2} & \cdots & \dfrac{\partial f_1}{\partial x_n} \\ & & & \\ \cdot & & & \cdot \\ \cdot & & & \cdot \\ \cdot & & & \cdot \\ \dfrac{\partial f_n}{\partial x_1} & \dfrac{\partial f_n}{\partial x_2} & \cdots & \dfrac{\partial f_n}{\partial x_n} \end{pmatrix}.$$

Since the iterative formula (7.28) involves the inverse of Jacobian J, in practice we do not attempt to find this explicitly. Instead of using the form of (7.28), we use the form

$$J(\mathbf{x}^{[n]})\mathbf{Z}^{[n]} = -\mathbf{f}(\mathbf{x}^{[n]}), \tag{7.29}$$

where $\mathbf{Z}^{[n]} = \mathbf{x}^{[n+1]} - \mathbf{x}^{[n]}$.

This represents a system of linear equations for $\mathbf{Z}^{[n]}$ and can be solved by any of the methods described in Chapter 3. Once $\mathbf{Z}^{[n]}$ has been found, the next iterate is calculated from

$$\mathbf{x}^{[n+1]} = \mathbf{Z}^{[n]} + \mathbf{x}^{[n]}. \tag{7.30}$$

There are two major disadvantages with this method:

1. The method may not converge unless the initial approximation is a good one. Unfortunately, there are no general means by which an initial solution can be obtained. One can assume such values for which $\det(J) \neq 0$. This does not guarantee convergence, but it does provide some guidance as to the appropriateness of one's initial approximation.

2. The method requires the user to provide the derivatives of each function with respect to each variable. Therefore, one must evaluate the n functions and the n^2 derivatives at each iteration. So solving systems of nonlinear equations is a difficult task. For systems of nonlinear equations that have analytical partial derivatives, Newton's method can be used; otherwise, multidimensional minimization techniques should be used.

Procedure 7.3 (Newton's Method for Two Nonlinear Equations)

1. *Choose the initial guess for the roots of the system so that the determinant of the Jacobian matrix is not zero.*

2. *Establish the tolerance $\epsilon (> 0)$.*

3. *Evaluate the Jacobian at initial approximations and then find the inverse of the Jacobian.*

4. *Compute a new approximation to the roots by using the iterative formula (7.30).*

5. *Check the tolerance limit. If $\|(x_n, y_n) - (x_{n-1}, y_{n-1})\| \leq \epsilon$, for $n \geq 0$, then end; otherwise, go back to step 3 and repeat the process.*

Fixed-Point Method for a System

It is sometimes convenient to solve a system of nonlinear equations by an iterative method that does not require the computation of partial derivatives. An example of the use of a *fixed-point* iteration for finding the zero of a nonlinear function of a single variable was discussed previously. Now we extend this idea to systems. The conditions that guarantee a fixed point for the vector function $\mathbf{g}(\mathbf{x})$ are similar for a fixed point of a function of a single variable.

The fixed-point iteration formula

$$x_{n+1} = g(x_n), \qquad n \geq 0$$

can be modified to solve the two simultaneous nonlinear equations

$$f_1(x, y) = 0$$
$$f_2(x, y) = 0.$$

These two nonlinear equations can be expressed in an equivalent form

$$x = g_1(x, y)$$
$$y = g_2(x, y),$$

and the iterative method to generate the sequences $\{x_n\}$ and $\{y_n\}$ is defined by the recurrence formulas

$$x_{n+1} = g_1(x_n, y_n), \quad n \geq 0$$
$$y_{n+1} = g_2(x_n, y_n), \quad n \geq 0 \tag{7.31}$$

for the given starting values x_0 and y_0.

The sufficient condition guaranteeing the convergence of the iterations defined by (7.31) or the convergence of $\{x_n\}$ to α and the convergence of $\{y_n\}$ to β, where the numbers α and β are such that

$$\alpha = g_1(\alpha, \beta)$$
$$\beta = g_2(\alpha, \beta)$$

are

$$\left|\frac{\partial g_1}{\partial x}\right| + \left|\frac{\partial g_2}{\partial x}\right| < 1, \quad \left|\frac{\partial g_1}{\partial y}\right| + \left|\frac{\partial g_2}{\partial y}\right| < 1 \tag{7.32}$$

or

$$\left(\frac{\partial g_1}{\partial x}\right)_n \left(\frac{\partial g_2}{\partial x}\right)_n - \left(\frac{\partial g_1}{\partial y}\right)_n \left(\frac{\partial g_2}{\partial y}\right)_n < 1. \tag{7.33}$$

Note that the fixed-point iteration may fail to converge even though the condition (7.32) is satisfied, unless the process is started with an initial guess (x_0, y_0) sufficiently close to (α, β).

Example 7.26 *Solve the following system of two equations using the fixed-point iteration, with accuracy $\epsilon = 10^{-5}$:*

$$4x^3 + y = 6$$
$$x^2 y = 1.$$

Assume $x_0 = 1.0$ and $y_0 = 0.5$ as starting values.

Solution. *Given the two functions*

$$\begin{aligned} f_1(x, y) &= 4x^3 + y - 6 \\ f_2(x, y) &= x^2 y - 1, \end{aligned}$$

let us consider the possible rearrangements of the given system of equations as follows:

$$x = g_1(x, y) = \left(\frac{6 - y}{4}\right)^{1/3}$$

$$y = g_2(x, y) = \left(\frac{1}{x^2}\right).$$

Thus,

$$x_{n+1} = g_1(x_n, y_n) = \left(\frac{6 - y_n}{4}\right)^{1/3} ; \quad n \geq 0$$

$$y_{n+1} = g_2(x_n, y_n) = \left(\frac{1}{x_n^2}\right) ; \quad n \geq 0,$$

and using the given initial approximation $(x_0 = 1.0, y_0 = 0.5)$, we get

$$x_1 = g_1(x_0, y_0) = \left(\frac{6 - y_0}{4}\right)^{1/3} = \left(\frac{6 - 0.5}{4}\right)^{1/3} = 1.111990$$

$$y_1 = g_2(x_0, y_0) = \left(\frac{1}{x_0^2}\right) = \left(\frac{1}{1^2}\right) = 1.$$

The first and the further iterations of the method, starting with the initial approximation $(x_0, y_0) = (1.0, 0.5)$ with accuracy 10^{-5}, are listed in Table 7.5.

Similarly, for a large system of equations it is convenient to use vector notation as follows:

$$\mathbf{x}^{(k+1)} = \mathbf{g}(\mathbf{x}^{(k)}), \quad k \geq 0,$$

where $\mathbf{g} = (g_1, g_2, \ldots, g_n)^T$ and $\mathbf{x} = (x_1, x_2, \ldots, x_n)^T$.   ●

Table 7.5: Solution of the given system by fixed-point iteration.

n	$x_{n+1} = g_1(x_n, y_n)$ $= \left(\dfrac{6 - y_n}{4}\right)^{1/3}$	$y_{n+1} = g_2(x_n, y_n)$ $= \left(\dfrac{1}{x_n^2}\right)$	1st func. $f_1(x_n, y_n)$	2nd func. $f_2(x_n, y_n)$
00	1.000000	0.500000	-1.50000	-0.500000
01	1.111990	1.000000	0.499999	0.236522
02	1.077217	0.808720	-0.191285	-0.061564
03	1.090782	0.861774	0.053047	0.025343
04	1.087054	0.840474	-0.021298	-0.006823
05	1.088554	0.846248	0.005775	0.002761
06	1.088148	0.843918	-0.002326	-0.000745
07	1.088312	0.844547	0.000634	0.000301
08	1.088267	0.844293	-0.000595	-0.000083
09	1.088285	0.844363	0.000066	0.000033
10	1.088280	0.844335	-0.000033	-0.000009
11	1.088282	0.844343	0.000004	0.000004

7.4   Convex and Concave Functions

Convex and concave functions play an extremely important role in the study of nonlinear programming problems.

Definition 7.10 (Convex Set)

A set S is convex if $x' \in S$ and $x'' \in S$ implies that all points on the line segment joining x' and x'' are members of S. This ensures that

$$cx' + (1 - c)x'', \quad for \quad 0 < c < 1$$

will be a member of S. For example, a vector subspace is convex, a ball in a normed vector space is convex (apply the triangle inequality), a hyperplane is a convex set, and a half-space is a convex set.

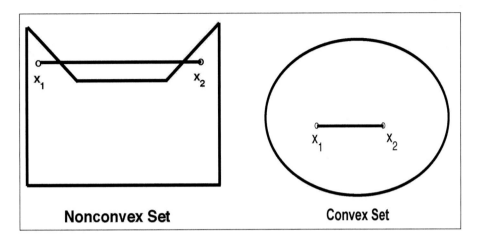

Figure 7.13: Convex and nonconvex sets.

Note that the intersection of convex sets is a convex set, but the union of convex sets is not necessarily a convex set.   •

Definition 7.11 (Convex Function)

A function $f(x_1, x_2, \ldots, x_n)$ that is defined for all points $(x_1, x_2, \ldots, x_n)$ in a convex set S is called a convex function on S, if for any $x' \in S$ and $x'' \in S$,

$$f(cx' + (1-c)x'') \leq cf(x') + (1-c)f(x'')$$

holds for $0 \leq c \leq 1$.
For example, the functions $f(x) = x^2$ and $f(x) = e^x$ are the convex functions.   •

Definition 7.12 (Concave Function)

A function $f(x_1, x_2, \ldots, x_n)$ that is defined for all points $(x_1, x_2, \ldots, x_n)$ in a convex set S is called a concave function on S, if for any $x' \in S$ and $x'' \in S$,

$$f(cx' + (1-c)x'') \geq cf(x') + (1-c)f(x'')$$

holds for $0 \leq c \leq 1$.
For example, the function $f(x) = \sqrt{x}$ is the concave function.   •

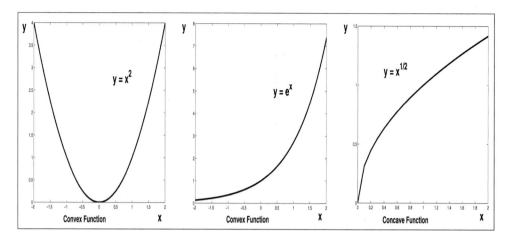

Figure 7.14: Convex and concave functions.

Let $y = f(x)$ be a function of a single variable. From Figure 7.15, we find that a function $f(x)$ is a convex function, if and only if the straight line joining any two points on the curve $y = f(x)$ is never below the curve $y = f(x)$. From the above figure we have:

$$
\begin{aligned}
\text{Point}\quad A &= (x', f(x')) \\
\text{Point}\quad B &= (cx' + (1 - c)x'', f(cx' + (1 - c)x'')) \\
\text{Point}\quad C &= (cx' + (1 - c)x'', cf(x') + (1 - c)f(x'')) \\
\text{Point}\quad D &= (x'', f(x'')),
\end{aligned}
$$

and from this, we get

$$
f(cx' + (1 - c)x'') \le cf(x') + (1 - c)f(x''),
$$

which implies that a function is convex.

A function $f(x)$ is a concave function, if and only if the straight line joining any two points on the curve $y = f(x)$ is never above the curve $y = f(x)$. From Figure 7.16, we have:

$$
\begin{aligned}
\text{Point}\quad A &= (x', f(x')) \\
\text{Point}\quad B &= (cx' + (1 - c)x'', cf(x') + (1 - c)f(x''))
\end{aligned}
$$

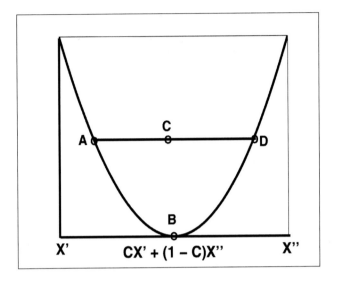

Figure 7.15: Convex function.

$$\text{Point} \quad C \; = \; (cx' + (1 - c)x'', f(cx' + (1 - c)x''))$$
$$\text{Point} \quad D \; = \; (x'', f(x'')),$$

which gives

$$f(cx' + (1 - c)x'') \geq cf(x') + (1 - c)f(x''),$$

which means that the function is concave.

Example 7.27 *Show that $f(x) = x^2$ is a convex function.*

Solution. *Let a function $f(x)$ be a convex function, then*

$$f(cx' + (1 - c)x'') \leq cf(x') + (1 - c)f(x'').$$

Given $f(x) = x^2$, then the left-hand side of the above inequality can be written as

$$f(cx' + (1 - c)x'') = (cx' + (1 - c)x'')^2 = c^2x'^2 + (1 - c)^2x''^2 + 2c(1 - c)x'x''.$$

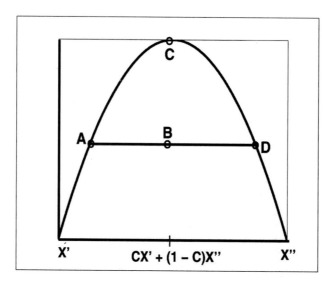

Figure 7.16: Concave function.

Also, the right-hand side of the inequality gives

$$cf(x') + (1 - c)f(x'') = cx'^2 + (1 - c)x''^2.$$

So using these values in the above inequality, we get

$$c^2 x'^2 + (1 - c)^2 x''^2 + 2c(1 - c)x'x'' \leq cx'^2 + (1 - c)x''^2,$$

or, it can be written as

$$x'^2(c^2 - c) + x''^2[(1 - c^2) - (1 - c)] + x'x''(2c(1 - c)) \leq 0.$$

Thus,

$$(c^2 - c)[x'^2 + x''^2 - 2x'x''] \leq 0$$

or

$$(c^2 - c)(x' - x'')^2 \leq 0.$$

For $c = 0$ and $c = 1$, this inequality holds with the equality, and for $c \in (0, 1)$, we have

$$c^2 - c < 0.$$

Also,

$$(x' - x'')^2 \geq 0,$$

which also holds. Hence, the given $f(x)$ is a convex function. ●

Example 7.28 *A linear function $f(x) = ax + b$ is both a convex and a concave function because it follows from*

$$\begin{aligned} f(cx' + (1 - c)x'') &= a(cx' + (1 - c)x'') + b \\ &= c(ax' + b) + (1 - c)(ax'' + b) \\ &= cf(x') + (1 - c)f(x''). \end{aligned}$$

●

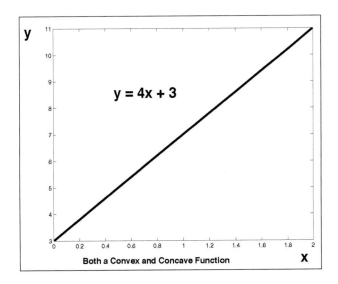

Figure 7.17: Both a convex and concave Function.

From the above definitions of convex and concave functions, we see that $f(x_1, x_2, \ldots, x_n)$ is a convex function, if and only if $-f(x_1, x_2, \ldots, x_n)$ is a concave function, and vice-versa.

From the following Figure 7.18, we see a function that is neither convex nor concave because the line segment AB lies below $y = f(x)$ and the line

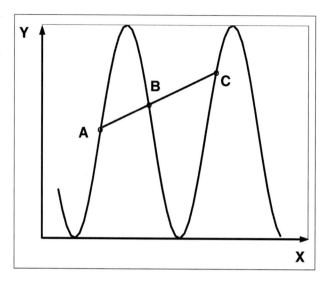

Figure 7.18: Neither a convex nor a concave function.

segment BC lies above $y = f(x)$.

A function $f(\mathbf{x})$ is said to be *strictly convex* if, for two distinct points x' and x'',
$$f(cx' + (1 - c)x'') < cf(x') + (1 - c)f(x''),$$
where $0 < c < 1$. Conversely, a function $f(\mathbf{x})$ is *strictly concave* if $-f(\mathbf{x})$ is strictly convex.

A special case of the convex (concave) function is the quadratic form
$$f(\mathbf{x}) = K\mathbf{x} + \mathbf{x}^T A\mathbf{x},$$
where K is a constant vector and A is a symmetric matrix. It can be proved that $f(\mathbf{x})$ is strictly convex, if A is positive-definite, and $f(\mathbf{x})$ is strictly concave, if A is negative-definite.

Properties of Convex Functions

1. If $f(x)$ is a convex function, then $af(x)$ is also a convex function, for any $a > 0$.

2. The *sum* of convex functions is also a convex function. For example, $f(x) = x^2$ and $g(x) = e^x$ are convex functions, so $h(x) = x^2 + e^x$ is also a convex function.

3. If $f(x)$ is a convex function, and $g(y)$ is another convex function whose value continuously increases, then the composite function $g(f(x))$ is also a convex function.

To check the convexity of a given function of a single variable, we can use the following theorem:

Theorem 7.15 (Convex Function)

Suppose that the second derivative of a function $f(x)$ exists for all x in a convex set S. Then $f(x)$ is a convex function on S, if and only if

$$f''(x) \geq 0, \qquad \text{for all} \quad x \in S.$$

For example, the function $f(x) = x^2$ is a convex function on $S = \mathbf{R}^1$ because

$$f'(x) = 2x, \qquad f''(x) = 2 \geq 0.$$

Theorem 7.16 (Concave Function)

Suppose that the second derivative of a function $f(x)$ exists for all x in a convex set S. Then $f(x)$ is a concave function on S, if and only if

$$f''(x) \leq 0, \qquad \text{for all} \quad x \in S.$$

For example, the function $f(x) = x^{1/2}$ is a concave function on $S = \mathbf{R}^1$ because

$$f'(x) = \frac{1}{2} x^{-1/2}, \qquad f''(x) = -\frac{1}{4} x^{-3/2} \leq 0.$$

Also, the function $f(x) = 3x + 2$ is both a convex and concave function on $S = \mathbf{R}^1$ because

$$f'(x) = 3, \qquad f''(x) = 0.$$

Using the definitions and the above two theorems, it is difficult to check for convexity of a given function because it would require consideration of infinite many points. However, using the sign of the Hessian matrix of the function, we can determine the convexity of a function.

Theorem 7.17

1. *A function $f(x_1, x_2, \ldots, x_n)$ is a convex function, if its Hessian matrix $H(x_1, x_2, \ldots, x_n)$ is at least positive-semidefinite.*

2. *A function $f(x_1, x_2, \ldots, x_n)$ is a concave function, if its Hessian matrix $H(x_1, x_2, \ldots, x_n)$ is at least negative-semidefinite.*

3. *A function $f(x_1, x_2, \ldots, x_n)$ is a nonconvex function, if its Hessian matrix $H(x_1, x_2, \ldots, x_n)$ is indefinite.* ●

Example 7.29 *Show that the function*

$$f(x_1, x_2, x_3) = 3x_1^2 + 2x_1x_2 + 2x_2^2 - 2x_2x_3 + 2x_3^2$$

is a convex function.

Solution. *First, we find the first and second partial derivatives of the given function as follows:*

$$
\begin{aligned}
f_{x_1} &= \frac{\partial f}{\partial x_1} = 6x_1 + 2x_2 \\[2mm]
f_{x_2} &= \frac{\partial f}{\partial x_2} = 2x_1 + 4x_2 - 2x_3 \\[2mm]
f_{x_3} &= \frac{\partial f}{\partial x_3} = -2x_2 + 4x_3,
\end{aligned}
$$

and the second derivatives of the functions are as follows:

$$
\begin{aligned}
(f_{x_1})_{x_1} &= \frac{\partial^2 f}{\partial x_1{}^2} = 6 \\[2mm]
(f_{x_2})_{x_2} &= \frac{\partial^2 f}{\partial x_2{}^2} = 4 \\[2mm]
(f_{x_3})_{x_3} &= \frac{\partial^2 f}{\partial x_3{}^2} = 2
\end{aligned}
$$

$$(f_{x_1})_{x_2} = \frac{\partial f_{x_1}}{\partial x_2} = 2 = \frac{\partial f_{x_2}}{\partial x_1} = (f_{x_2})_{x_1}$$

$$(f_{x_1})_{x_3} = \frac{\partial f_{x_1}}{\partial x_3} = 0 = \frac{\partial f_{x_3}}{\partial x_1} = (f_{x_3})_{x_1}$$

$$(f_{x_2})_{x_3} = \frac{\partial f_{x_2}}{\partial x_3} = -2 = \frac{\partial f_{x_3}}{\partial x_2} = (f_{x_3})_{x_2}.$$

Hence, the Hessian matrix for the given function can be found as

$$H(x_1, x_2, x_3) = \begin{pmatrix} 6 & 2 & 0 \\ 2 & 4 & -2 \\ 0 & -2 & 4 \end{pmatrix}.$$

To check the definiteness of H, take

$$\mathbf{z}^T H \mathbf{z} = (z_1, z_2, z_3) \begin{pmatrix} 6 & 2 & 0 \\ 2 & 4 & -2 \\ 0 & -2 & 4 \end{pmatrix} \begin{pmatrix} z_1 \\ z_2 \\ z_3 \end{pmatrix},$$

which gives

$$\mathbf{z}^T H \mathbf{z} = (z_1, z_2, z_3) \begin{pmatrix} 6z_1 + 2z_2 + 0z_3 \\ 2z_1 + 4z_2 - 2z_3 \\ 0z_1 - 2z_2 + 4z_3 \end{pmatrix}$$

$$= z_1(6z_1 + 2z_2) + z_2(2z_1 + 4z_2 - 2z_3) + z_3(-2z_2 + 4z_3).$$

Thus,

$$\mathbf{z}^T H \mathbf{z} = 6z_1^2 + 4z_1 z_2 + 4z_2^2 - 4z_2 z_3 + 4z_3^2.$$

Note that

$$\mathbf{z}^T H \mathbf{z} = 2(2z_1^2 + (z_1 + z_2)^2 + (z_2 - z_3)^2 + z_3^2) > 0,$$

for $\mathbf{z} \neq \mathbf{0}$, so the Hessian matrix is positive-definite. Hence, the function $f(x_1, x_2, x_3)$ is a convex function. ●

Another way to determine whether a function $f(x_1, x_2, \ldots, x_n)$ is a convex or concave function is to use the *principal minor test*, which helps us to determine the sign of the Hessian matrix. In the following, we discuss two definitions.

Definition 7.13 (Principal Minor)

An ith principal minor of an $n \times n$ matrix is the determinant of an $i \times i$ matrix obtained by deleting $(n-i)$ rows and the corresponding $(n-i)$ columns of a matrix.

For example, the matrix

$$A = \begin{pmatrix} -3 & -2 \\ -2 & -5 \end{pmatrix}$$

has -3 and -5 as the first principal minors, and the second principal minor is

$$-3(-5) - (-2)(-2) = 15 - 4 = 11,$$

which is the determinant of the given matrix.   •

Note that for an $n \times n$ square matrix, there are, in all, $2^n - 1$ principal minors (or determinants). Also, the first principal minors of a given matrix are just the diagonal entries of a matrix.

Definition 7.14 (Leading Principal Minor)

A kth leading principal minor of an $n \times n$ matrix is the determinant of a $k \times k$ matrix obtained by deleting $(n-k)$ rows and columns of a matrix. •

Let $H_k(x_1, x_2, \ldots, x_n)$ be the *kth* leading principal minor of the Hessian matrix evaluated at the point $(x_1, x_2, \ldots, x_n)$. Thus, if

$$f(x_1, x_2) = 3x_1^3 + 4x_1x_2 + 2x_2^2,$$

then

$$H_1(x_1, x_2) = 18x_1, \quad H_1(x_1, x_2) = 18x_1(4) - 4(4) = 72x_1 - 16.$$

Theorem 7.18 (Convex Function)

Suppose that a function $f(x_1, x_2, \ldots, x_n)$ has continuous second-order partial derivatives for each point $\mathbf{x} = (x_1, x_2, \ldots, x_n) \in S$. Then function $f(\mathbf{x})$ is a convex function on S, if and only if for each $\mathbf{x} \in S$ all principal minors are nonnegative.   •

Example 7.30 *Show that the function $f(x_1, x_2) = 3x_1^2 + 4x_1x_2 + 2x_2^2$ is a convex function on $S = \mathbf{R}^2$.*

Solution. *First, we find the Hessian matrix, which is of the form*

$$H(x_1, x_2) = \begin{pmatrix} 6 & 4 \\ 4 & 4 \end{pmatrix}.$$

The first principal minors of the Hessian matrix are the diagonal entries, both $6 > 0$ and $4 > 0$. The second principal minor of the Hessian matrix is the determinant of the Hessian matrix, which is

$$6(4) - (4)(4) = 24 - 16 = 8 > 0.$$

So for any point, all principal minors of the Hessian matrix $H(x_1, x_2)$ are nonnegative, therefore, Theorem 7.18 shows that $f(x_1, x_2)$ is a convex function on $\mathbf{R}^2$. ●

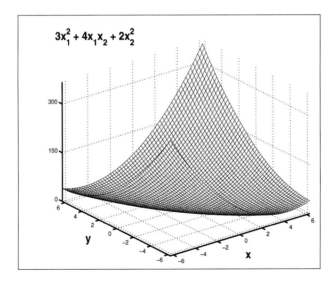

Figure 7.19: Convex function.

Theorem 7.19 (Concave Function)

Suppose that a function $f(x_1, x_2, \ldots, x_n)$ has continuous second-order partial derivatives for each point $\mathbf{x} = (x_1, x_2, \ldots, x_n) \in S$. Then function $f(\mathbf{x})$ is a concave function on S, if and only if for each $\mathbf{x} \in S$, $k = 1, 2, \ldots, n$, and all nonzero principal minors have the same sign as $(-1)^k$. ●

Example 7.31 *Show that the function $f(x_1, x_2) = -x_1^2 - 2x_1x_2 - 3x_2^2$ is a concave function on $S = \mathbf{R}^2$.*

Solution. *The Hessian matrix of the given function has the form*

$$H(x_1, x_2) = \begin{pmatrix} -2 & -2 \\ -2 & -6 \end{pmatrix}.$$

The first principal minors of the Hessian matrix are the diagonal entries (-2 and -6). These are both negative (nonpositive). The second principal minor is the determinant of the Hessian matrix $H(x_1, x_2)$ and equals

$$-2(-6) - (-2)(-2) = 12 - 4 = 8 > 0.$$

Thus, from Theorem 7.19, $f(x_1, x_2)$ is a concave function on $\mathbf{R}^2$. ●

Example 7.32 *Show that the function $f(x_1, x_2) = 2x_1^2 - 4x_1x_2 + 3x_2^2$ is not a convex nor a concave function on $S = \mathbf{R}^2$.*

Solution. *The Hessian matrix of the given function has the form*

$$H(x_1, x_2) = \begin{pmatrix} 2 & -4 \\ -4 & 3 \end{pmatrix}.$$

The first principal minors of the Hessian matrix are 2 and 3. Because both principal minors are positive, $f(x_1, x_2)$ cannot be concave. The second principal minor is the determinant of the Hessian matrix $H(x_1, x_2)$ and it is equal to

$$2(3) - (-4)(-4) = 6 - 16 = -10 < 0.$$

Thus, $f(x_1, x_2)$ cannot be a convex function on $\mathbf{R}^2$. Together, these facts show that $f(x_1, x_2)$ cannot be a convex nor a concave function. ●

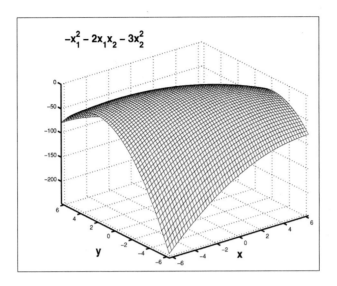

$-x_1^2 - 2x_1x_2 - 3x_2^2$

Figure 7.20: Concave function.

Example 7.33 *Show that the function*

$$f(x_1, x_2, x_3) = 2x_1^2 + 2x_2^2 + 2x_3^2 - 2x_1x_2 - 2x_2x_3 - 2x_1x_3$$

is a convex function on $S = \mathbf{R}^3$.

Solution. *The Hessian matrix of the given function has the form*

$$H(x_1, x_2, x_3) = \begin{pmatrix} 4 & -2 & -2 \\ -2 & 4 & -2 \\ -2 & -2 & 4 \end{pmatrix}.$$

By deleting rows (and columns) 1 and 2 of the Hessian matrix, we obtain the first-order principal minor $4 > 0$. By deleting rows (and columns) 1 and 3 of the Hessian matrix, we obtain the first-order principal minor $4 > 0$. By deleting rows (and columns) 2 and 3 of the Hessian matrix, we obtain the first-order principal minor $4 > 0$. By deleting row 1 and column 1 of the Hessian matrix, we find the second-order principal minor

$$|H_2| = \begin{vmatrix} 4 & -2 \\ -2 & 4 \end{vmatrix} = 16 - 4 = 12 > 0.$$

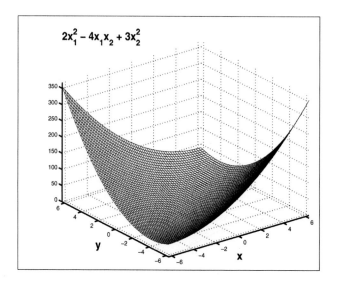

Figure 7.21: Neither a convex nor a concave function.

By deleting row 2 and column 2 of the Hessian matrix, we find the second-order principal minor

$$|H_2| = \begin{vmatrix} 4 & -2 \\ -2 & 4 \end{vmatrix} = 16 - 4 = 12 > 0.$$

By deleting row 3 and column 3 of the Hessian matrix, we find the second-order principal minor

$$|H_2| = \begin{vmatrix} 4 & -2 \\ -2 & 4 \end{vmatrix} = 16 - 4 = 12 > 0.$$

The third-order principal minor is simply the determinant of the Hessian matrix itself. Expanding by row 1 cofactors, we find the third-order principal minor as follows:

$$|H_3| = 4[(4)(4) - (-2)(-2)] - (-2)[(-2)(4) - (-2)(-2)]$$
$$+ (-2)[(-2)(-2) - (-2)(4)] = 0.$$

Because for all (x_1, x_2, x_3) all principal minors of the Hessian matrix are nonnegative, we have shown that $f(x_1, x_2, x_3)$ is a convex function on $\mathbf{R}^3$.

●

Example 7.34 *For what values of a, b, and c will the function*

$$f(x_1, x_2) = ax_1^2 + bx_1x_2 + cx_2^2$$

be a concave function on $\mathbf{R}^2$*?*

Solution. *The first-order partial derivatives are*

$$\frac{\partial f}{\partial x_1} = 2ax_1 + bx_2$$

and

$$\frac{\partial f}{\partial x_2} = bx_1 + 2cx_2.$$

Thus, the gradient of the function is

$$\nabla f(x_1, x_2) = \begin{pmatrix} 2ax_1 + bx_2 \\ bx_1 + 2cx_2 \end{pmatrix}.$$

The second-order partial derivatives are:

$$\frac{\partial^2 f}{\partial x_1^2} = 2a$$

$$\frac{\partial^2 f}{\partial x_2^2} = 2c$$

$$\frac{\partial^2 f}{\partial x_1 \partial x_2} = b = \frac{\partial^2 f}{\partial x_2 \partial x_1},$$

and so the Hessian matrix for the function is

$$\nabla^2 f(x_1, x_2) = H(x_1, x_2) = \begin{pmatrix} 2a & b \\ b & 2c \end{pmatrix}.$$

The first principal minors are

$$H_1 = 2a \quad and \quad H_1 = 2c,$$

and the second principal is the determinant of the Hessian matrix and is equal to

$$H_2 = |H| = 4ac - b^2.$$

If the given function is a concave function on $\mathbf{R}^2$, *then* a, b, *and* c *must satisfy the conditions*

$$a \leq 0, \ c \leq 0 \quad and \quad 4ac - b^2 \geq 0,$$

whereas $4ac - b^2 \geq 0$ *implies that*

$$|b| \leq \sqrt{4ac} \quad or \quad -\sqrt{4ac} \leq b \leq \sqrt{4ac}.$$

●

7.5 Standard Form of a Nonlinear Programming Problem

In solving *NLP* problems, we have to do the following:

Find an optimal solution $\mathbf{x} = (x_1, x_2, \ldots, x_n)^T$ in order to *minimize* or *maximize* an objective function, $f(\mathbf{x})$, subject to the constraint function $g_i(\mathbf{x})$, $i = 1, 2, \ldots, m$, which will be either equality constraints or inequality constraints type. Thus, the standard form of an *NLP* problem will be of the form:

$$\text{maximize}(minimize) \quad Z = f(\mathbf{x})$$

subject to   (7.34)

$$g_j(\mathbf{x}) \ (\leq, =, \geq) \ \mathbf{0}, \quad j = 1, 2, \ldots, m.$$

In the following we give two very important theorems that illustrate the importance of convex and concave functions in *NLP* problems.

Theorem 7.20 (Concave Function)

Consider the NLP problem (7.34) and assume it is a maximization problem. Suppose the feasible region S for NLP problem (7.34) is a convex set. If $f(x)$ *is concave on S, then any local maximum for NLP problem (7.34) is an optimal solution to this NLP problem.*   ●

Theorem 7.21 (Convex Function)

Consider the NLP problem (7.34) and assume it is a minimization problem. Suppose the feasible region S for NLP problem (7.34) is a convex set. If $f(x)$ is convex on S, then any local minimum for NLP problem (7.34) is an optimal solution to this NLP problem. •

The above two theorems demonstrate that if we are maximizing a concave function or minimizing a convex function over a convex feasible region S, then any local maximum or local minimum will solve *NLP* problem (7.34). As we solve *NLP* problems, we will repeatedly apply these two theorems. •

7.6 One-Dimensional Unconstrained Optimization

Optimization and root finding are related in the sense that both involve guessing and searching for a point on a function. In root finding, we look for zeros of a function or functions. While in optimization, we search for an extremum of a function, i.e., either the maximum or the minimum value of a function.

Here, we will discuss the *NLP* problem that consists of only an objective function, i.e., $z = f(\mathbf{x})$ and no constraints. Note that if the given objective function is convex (*concave*), then a unique solution will be found at an interior point to the feasible region where all derivatives are zero or at a point. We will discuss three one-dimensional optimization methods in the following sections.

7.6.1 Golden-Section Search

This is the first method we discuss for the single variable optimization that has a goal of finding the value of x that yields an extremum; either a maximum or minimum of a function $f(x)$. It is a simple, general-purpose, single-variable search method. It is similar to the bisection method for

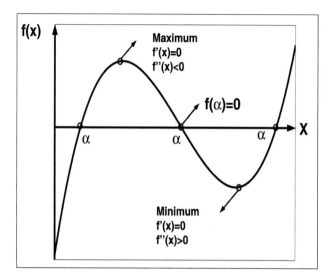

Figure 7.22: Relationship between optimization and root finding.

nonlinear equations.

This method is an iterative method and starts with two initial guesses, x_L and x_u, that bracket one local extremum of $f(x)$ (considered a maximum) and hence is called a *unimodel*. Next, we look for two interior points, x_1 and x_2, which can be chosen according to the golden ratio

$$d = \frac{\sqrt{5}-1}{2}(x_u - x_L),$$

which gives

$$x_1 = x_L + d$$
$$x_2 = x_u - d.$$

After finding the two interior points, the given function is evaluated at these points and two results can occur:

1. If $f(x_1) > f(x_2)$, then the domain of x to the left of x_2, from x_L to x_2, can be eliminated because it does not contain the maximum. Since the optimum lies on the interval (x_1, x_u), we set $x_2 = x_1$ for the next iteration.

2. If $f(x_1) < f(x_2)$, then the domain of x to the right of x_1, from x_1 to x_u, would have been eliminated. In this case, the optimum lies on the interval (x_L, x_2), and we set $x_2 = x_1$ for the next iteration.

3. If $f(x_1) = f(x_2)$, then the optimum lies on the interval (x_1, x_2).

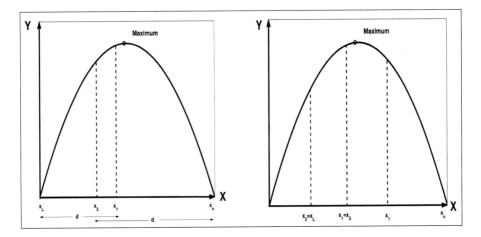

Figure 7.23: Graphical interpretation of golden-section search.

Remember that we do not have to recalculate all the function values for the next iteration, and we need only one new function value. For example, when the optimum is on the interval (x_1, x_u), then we set $x_2 = x_1$, i.e., the old x_1 becomes the new x_2 and $f(x_2) = f(x_1)$. After this, we have to find only the new x_1 for the next iteration, and it can be obtained as

$$x_1 = x_L + \frac{\sqrt{5} - 1}{2}(x_u - x_L).$$

A similar approach would be used for the other possible case, when the optimum is on the interval (x_L, x_2) by setting $x_2 = x_1$, i.e., the old x_2 becomes the new x_1 and $f(x_1) = f(x_2)$. Then we need to find only the new x_2 for the next iteration, which can be obtained as

$$x_2 = x_u - \frac{\sqrt{5} - 1}{2}(x_u - x_L).$$

As the iterations are repeated, the interval containing the optimum is reduced rapidly. In fact, with each iteration the interval is reduced by a factor of the golden ratio (about 61.8%). This means that after 10 iterations the interval is shrunk to about 0.008 or 0.8% of the initial interval.

Example 7.35 *Use golden-section search to find the approximation of the maximum of the function*

$$f(x) = 2x - 1.75x^2 + 1.1x^3 - 0.25x^4,$$

with initial guesses $x_L = 1$ and $x_u = 2.25$.

Solution. *To find the two interior points x_1 and x_2, first we compute the value of the golden ratio as*

$$d = \frac{\sqrt{5}-1}{2}(x_u - x_L) = \frac{\sqrt{5}-1}{2}(2.25 - 1) = 0.7725,$$

and with this value, we have the values of the interior points as follows:

$$\begin{aligned} x_1 &= x_L + d = 1 + 0.7725 = 1.7725, \\ x_2 &= x_u - d = 2.25 - 0.7725 = 1.4775. \end{aligned}$$

Next, we have to compute the function values at these interior points, which are:

$$\begin{aligned} f(x_1) &= f(1.7725) = 1.7049 \\ f(x_2) &= f(1.4775) = 1.4913. \end{aligned}$$

Since $f(x_1) > f(x_2)$, the maximum is on the interval defined by x_2, x_1, and x_u, i.e., (x_2, x_u). For this, we set the following scheme:

$$\begin{aligned} x_L &= x_2 = 1.4775 \\ x_2 &= x_1 = 1.7725 \\ x_u &= x_u = 2.25. \end{aligned}$$

So we have to find the new value of x_1, only for the second iteration, and it can be computed with the help of the new value of the golden ratio as follows:

$$d = \frac{\sqrt{5}-1}{2}(x_u - x_L) = \frac{\sqrt{5}-1}{2}(2.25 - 1.4775) = 0.4774$$

and

$$x_1 = x_L + d = 1.4775 + 0.4774 = 1.9549.$$

The function values at these new interior points are:

$$f(x_1) = f(1.9549) = 1.7887$$
$$f(x_2) = f(1.7725) = 1.7049.$$

Again, $f(x_1) > f(x_2)$, so the maximum is on the interval defined by x_2, x_1, and x_u. For this, we set the following scheme:

$$x_L = x_2 = 1.7725$$
$$x_2 = x_1 = 1.9549$$
$$x_u = x_u = 2.25.$$

The new value of x_1 and d can be computed as follows:

$$d = 0.2951 \quad and \quad x_1 = 2.0676.$$

Repeat the process, and the numerical results for the corresponding iterations, starting with the initial approximations $x_L = 1.0$ and $x_u = 2.25$ with accuracy 5×10^{-4}, are given in Table 7.6. From Table 7.6, we can see that within, 14 iterations (very slow), the result converges rapidly on the true value of 1.8082 at $x = 2.0793$. ●

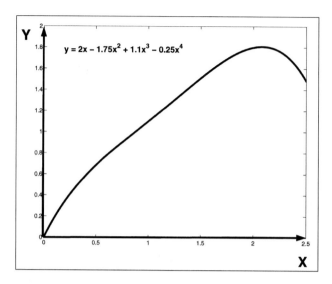

Figure 7.24: Graph of the given function.

Program 7.6

MATLAB m-file for the Golden-Section Search Method for Optimization

```
function sol=golden(fn,xL,xu,tol)
disp(' k  xL  f(xL)  x2  f(x2)  x1  f(x1)  xu  f(xu)  d ')
d=((sqrt(5)-1)/2)*(xu-xL); x1=xL+d; x2=xu-d;
fL=feval(fn,xL);fu=feval(fn,xu);
f1=feval(fn,x1);f2=feval(fn,x2); k = 0;
[k  xL  fL  x2  f2  x1  f1  xu  fu  d]
while abs(x1-x2)>tol; f1=feval(fn,x1);f2=feval(fn,x2);
k=k+1; if f1>f2; xL=x2; x2=x1; xu=xu;
d=((sqrt(5)-1)/2)*(xu-xL); x1=xL+d; f1=feval(fn,x1);
E=abs(x1-x2); sol=x2; f2=feval(fn,x2);
fL=feval(fn,xL); fu=feval(fn,xu);
[k  xL  fL  x2  f2  x1  f1  xu  fu  d  sol]
else; xu=x1; x1=x2; xL=xL; d=((sqrt(5)-1)/2)*(xu-xL);
x2=xu-d; f2=feval(fn,x2); E=abs(x1-x2); sol=x1;
f1=feval(fn,x1);fL=feval(fn,xL);fu=feval(fn,xu);
[k  xL  fL  x2  f2  x1  f1  xu  fu  d  sol]
end;end;
```

Table 7.6: Solution by the golden-section search.

n	x_L	$f(x_L)$	x_2	$f(x_2)$	x_1	$f(x_1)$	x_u	$f(x_u)$	d
00	1.0000	1.1000	1.4775	1.4913	1.7725	1.7049	2.2500	1.7631	0.7725
01	1.4775	1.4913	1.7725	1.7049	1.9549	1.7887	2.2500	1.7631	0.4774
02	1.7725	1.7049	1.9549	1.7887	2.0676	1.8080	2.2500	1.7631	0.2951
03	1.9549	1.7887	2.0676	1.8080	2.1373	1.8034	2.2500	1.7631	0.1824
04	1.9549	1.7887	2.0246	1.8042	2.0676	1.8080	2.1373	1.8034	0.1127
05	2.0246	1.8042	2.0676	1.8080	2.0942	1.8079	2.1373	1.8034	0.0697
06	2.0246	1.8042	2.0512	1.8071	2.0676	1.8080	2.0942	1.8079	0.0431
07	2.0512	1.8071	2.0676	1.8080	2.0778	1.8082	2.0942	1.8079	0.0266
08	2.0676	1.8080	2.0778	1.8082	2.0841	1.8081	2.0942	1.8079	0.0164
09	2.0676	1.8080	2.0739	1.8081	2.0778	1.8082	2.0841	1.8081	0.0102
10	2.0739	1.8081	2.0778	1.8082	2.0802	1.8082	2.0841	1.8081	0.0063
11	2.0778	1.8082	2.0802	1.8082	2.0817	1.8082	2.0841	1.8081	0.0039
12	2.0778	1.8082	2.0793	1.8082	2.0802	1.8082	2.0817	1.8082	0.0024
13	2.0778	1.8082	2.0787	1.8082	2.0793	1.8082	2.0802	1.8082	0.0015
14	2.0787	1.8082	2.0793	1.8082	2.0796	1.8082	2.0802	1.8082	0.0009

To use MATLAB commands for the golden-section search method, first we define a function m-file as fn.m for the equation as follows:

$$function \ y = fn(x)$$
$$y = 2*x - 1.75*x.\hat{\ }2 + 1.1*x.\hat{\ }3 - 0.25*x.\hat{\ }4;$$

then use the single command:

$$>> sol = golden('fn', 1.0, 2.25, 5e-4)$$
$$sol =$$
$$2.0793$$

7.6.2 Quadratic Interpolation

This iterative method is based on fitting a polynomial function through a given number of points. As the name indicates, the quadratic interpolation method uses three distinct points and fits a quadratic function through these points. The minimum of this quadratic function is computed using necessary points.

Since the method is iterative, a new set of three points is selected by comparing function values at this minimum point with three initial guesses. The process is repeated with the three new points until the interval on which the minimum lies becomes fairly small.

Similar to the previously discussed method, the golden-section search, this method also requires only one new function evaluation at each iteration. As the interval becomes small, the quadratic approximation becomes closer to the actual function, which speeds up convergence.

Derivation of the Formula

Just as there is only one straight line connecting two points, there is only one quadratic or parabola connecting three points. Suppose that we are given three distinct points x_0, x_1, and x_2 and a quadratic function $p(x)$ passing through the corresponding function values $f(x_0), f(x_1)$, and $f(x_2)$. Thus, if these three points jointly bracket an optimum, we can fit a quadratic function to the points as follows:

$$p(x) = \frac{(x-x_1)(x-x_2)}{(x_0-x_1)(x_0-x_2)}f(x_0) + \frac{(x-x_0)(x-x_2)}{(x_1-x_0)(x_1-x_2)}f(x_1)$$

$$+ \frac{(x-x_0)(x-x_1)}{(x_2-x_0)(x_2-x_1)}f(x_2).$$

The necessary condition for the minimum of this quadratic function can be obtained by differentiating it with respect to x. Set the result equal to zero, and solve the equation for an estimate of optimal x, i.e.,

$$p'(x) = \frac{(2x-x_1-x_2)}{(x_0-x_1)(x_0-x_2)}f(x_0) + \frac{(2x-x_0-x_2)}{(x_1-x_0)(x_1-x_2)}f(x_1)$$

$$+ \frac{(2x-x_0-x_1)}{(x_2-x_0)(x_2-x_1)}f(x_2)$$

$$= 0.$$

It can be shown by some algebraic manipulations that the minimum point (or optimal point) denoted x_{opt} is

$$x = x_{opt} = \frac{1}{2}\left(\frac{(x_1^2 - x_2^2)f(x_0) + (x_2^2 - x_0^2)f(x_1) + (x_0^2 - x_1^2)f(x_2)}{(x_1 - x_2)f(x_0) + (x_2 - x_0)f(x_1) + (x_0 - x_1)f(x_2)}\right), \quad (7.35)$$

which is called the *quadratic interpolation formula*.

After finding the new point (optimum point), the next job is to determine which one of the given three points is discarded before repeating the process. To discard a point we check the following:

1. $x_{opt} \leq x_1$:

   (i) If $f(x_1) \geq f(x_{opt})$, then the minimum of the actual function is on the interval (x_0, x_1), therefore, we will use the new three points x_0, x_{opt}, and x_1 for the next iteration.
   (ii) If $f(x_1) < f(x_{opt})$, then the minimum of the actual function is on the interval (x_{opt}, x_2), therefore, in this case we will use the new three points x_{opt}, x_1, and x_2 for the next iteration.

2. $x_{opt} > x_1$:
   (i) If $f(x_1) \geq f(x_{opt})$, then the minimum of the actual function is on the interval (x_1, x_2), therefore, we will use the new three points x_1, x_{opt}, and x_2 for the next iteration.
   (ii) If $f(x_1) < f(x_{opt})$, then the minimum of the actual function is on the interval (x_0, x_{opt}), therefore, in this case we will use the new three points x_0, x_1, and x_{opt} for the next iteration.

Example 7.36 *Use quadratic interpolation to find the approximation of the maximum of*

$$f(x) = 2x - 1.75x^2 + 1.1x^3 - 0.25x^4,$$

with initial guesses $x_0 = 1.75, x_1 = 2$, and $x_2 = 2.25$.

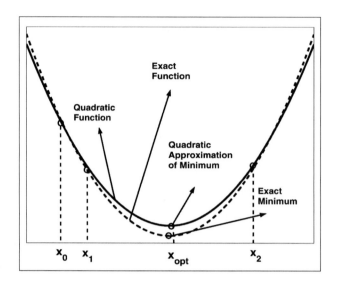

Figure 7.25: Graphical interpretation of quadratic interpolation.

Solution. *Using the three initial guesses, the corresponding functional values are:*

$$
\begin{aligned}
x_0 &= 1.75 & f(1.75) &= 1.6912 \\
x_1 &= 2.0 & f(2.0) &= 1.8000 \\
x_2 &= 2.25 & f(2.25) &= 1.7631.
\end{aligned}
$$

Using formula (7.35), we get

$$
x_{opt} = \frac{((2)^2 - (2.25)^2)(1.691) + ((2.25)^2 - (1.75)^2)(1.8) + ((1.75)^2 - (2)^2)(1.763)}{2(2 - 2.25)(1.691) + 2(2.25 - 1.75)(1.8) + 2(1.75 - 2)(1.763)}
$$

$$
= 2.0617,
$$

and the function value at this optimal point is

$$
f(2.0617) = 2(2.0617) - 1.75(2.0617)^2 - 1.1(2.0617)^3 - 0.25(2.0617)^4 = 1.8077.
$$

To perform the second iteration, we have to discard one point by using the same strategy as in the previous golden-section search. Since the function value at x_{opt} is greater than the intermediate point x_1 and the x_{opt} value is to the right of x_1, the first initial guess x_0 is discarded. So for the second *iteration, we will start from the following initial guesses:*

$$
\begin{aligned}
x_0 &= 2.0 & f(2.0) &= 1.8000 \\
x_1 &= 2.0617 & f(2.0617) &= 1.8077 \\
x_2 &= 2.25 & f(2.25) &= 1.7631.
\end{aligned}
$$

Using formula (7.35) again gives

$$
x_{opt} = \frac{((2.062)^2 - (2.25)^2)(1.8) + ((2.25)^2 - (2)^2)(1.808) + ((2)^2 - (2.062)^2)(1.763)}{2(2.062 - 2.25)(1.8) + 2(2.25 - 2)(1.808) + 2(2 - 2.062)(1.763)}
$$

$$
= 2.0741,
$$

and the function value at this optimal point is

$$
f(2.0741) = 2(2.0741) - 1.75(2.0741)^2 + 1.1(2.0741)^3 - 0.25(2.0741)^4 = 1.8081.
$$

Repeat the process and the numerical results for the corresponding iterations, starting with the initial approximations $x_0 = 1.75, x_1 = 2.0$, and $x_2 = 2.25$, with accuracy 5×10^{-2}, which are given in Table 7.7.

Table 7.7: Solution by quadratic interpolation method.

n	x_0	$f(x_0)$	x_1	$f(x_1)$	x_2	$f(x_2)$	x_3	$f(x_3)$
00	1.7500	1.6912	2.0000	1.8000	2.2500	1.7631	2.0617	1.8077
01	2.0000	1.8000	2.0617	1.8077	2.2500	1.7631	2.0741	1.8081
02	2.0741	1.8081	2.0781	1.8082	2.2500	1.7631	2.0781	1.8082
03	2.0781	1.8082	2.0790	1.8082	2.2500	1.7631	2.0790	1.8082
04	2.0790	1.8082	2.0793	1.8082	2.2500	1.7631	2.0793	1.8082
05	2.0793	1.8082	2.0793	1.8082	2.2500	1.7631	2.0793	1.8082

From Table 7.7, we can see that within five iterations, the result converges rapidly on the true value of 1.8082 at $x = 2.0793$. Also, note that for this problem the quadratic interpolation method converges only on one end of the interval, and sometimes the convergence can be slow for this reason.

•

To use MATLAB commands for the quadratic interpolation method, first we define a function m-file as fn.m for the equation as follows:

$$function\ y = fn(x)$$
$$y = 2*x - 1.75*x.\^\ 2 + 1.1*x.\^\ 3 - 0.25*x.\^\ 4;$$

then use the single command:

$$>> sol = Quadratic2('fn', 1.75, 2, 2.25, 5e - 2)$$
$$sol =$$
$$\qquad 2.0793$$

Remember that the procedure is essentially complete except for the choice of three initial points. Choosing three arbitrary values of x may cause problems if the denominator of the x_{opt} equation is zero. Assume that the three points are chosen as $0, \epsilon$, and 2ϵ, where a positive number ϵ is a chosen parameter (say, $\epsilon = 1$). In such case, the expression for x_{opt} takes the form

$$x_{opt} = \frac{(3f(x_0) - 4f(x_1) + f(x_2))}{(2f(x_0) - 4f(x_1) + 2f(x_2))},$$

and for the denominator to be greater than zero, we must have

$$2f(x_0) - 4f(x_1) + 2f(x_2) > 0,$$

i.e.,

$$\frac{f(x_0) + f(x_2)}{2} > f(x_1).$$

In the case of the convergence of the method, the interval on which the minimum lies becomes smaller, the quadratic function becomes closer to the actual function, and the process is terminated when

$$\left| \frac{(f(x_{opt}) - p(x_{opt}))}{(f(x_{opt}))} \right| \le tol,$$

where *tol* is a smaller convergence tolerance.

Program 7.7

MATLAB m-file for Quadratic Interpolation Method
function sol=Quadratic2(fn,x0,x1,x2,tol)
disp('k x0 $f(x0)$ x1 $f(x1)$ x2 $f(x2)$ x3 $f(x3)$')
f0=feval(fn,x0);f1=feval(fn,x1);f2=feval(fn,x2); k = 0;
$A = (f0 * (x1.\hat{} \ 2 - x2.\hat{} \ 2) + f1 * (x2.\hat{} \ 2 - x0.^2) + f2 * (x0.\hat{} \ 2 - x1.\hat{} \ 2));$
$B = (2 * f0 * (x1 - x2) + 2 * f1 * (x2 - x0) + 2 * f2 * (x0 - x1));$
$x3 = A/B$;f3=feval(fn,x3);
while abs(x2-x0)¿tol; k=k+1; if f3¿f1;
x0=x1;f0=f1;x1=x3;f1=f3; else; x2=x1;f2=f1;x1;x3;f1=f3;
disp('k x0 $f(x0)$ x1 $f(x1)$ x2 $f(x2)$ x3 $f(x3)$')
end;end; sol=x3;

7.6.3 Newton's Method

This is one of the best one-dimensional iterative methods for single variable optimization. Unlike other methods for one-dimensional optimization, this method requires only a single initial approximation.

Since for finding the root of nonlinear equation $f(x) = 0$, this method can be written as

$$x_{n+1} = x_n - \frac{f(x_n)}{f'(x_n)}, \quad n \geq 0,$$

a similar open approach can be used to find an optimum of $f(x)$ by defining a new function, $F(x) = f'(x)$. Thus, because the same optimal value $\mathbf{x}^*$ satisfies both the functions

$$F'(\mathbf{x}^*) = f'(\mathbf{x}^*) = 0,$$

Newton's method for optimization can be written as

$$x_{n+1} = x_n - \frac{f'(x_n)}{f''(x_n)}, \quad n \geq 0, \tag{7.36}$$

which can be used to find the minimum or maximum of $f(x)$, if $f(x)$ is twice continuously differentiable.

It should be noted that formula (7.36) can be obtained by using second-order Taylor's series for the single variable function $f(x)$ and setting the derivative of the series equal to zero, i.e., using

$$f(x) = f(x_0) + (x - x_0)f'(x_0) + \frac{(x - x_0)^2}{2!}f''(x_0) + \text{higher-order terms.}$$

Taking the derivative with respect to x and ignoring the higher-order term, we get

$$f'(x) \approx f'(x_0) + (x - x_0)f''(x_0).$$

Setting $f'(x) = 0$ and simplifying the expression for x, we obtain

$$x \approx x_0 - \frac{f'(x_0)}{f''(x_0)}.$$

It is an improved approximation and can be written as

$$x_1 = x_0 - \frac{f'(x_0)}{f''(x_0)},$$

or, in general, we have formula (7.36).

Example 7.37 *Use Newton's method to find the local maximum of the function*

$$f(x) = 2x - 1.75x^2 + 1.1x^3 - 0.25x^4,$$

with an initial guess $x_0 = 2.5$.

Solution. *To use formula (7.36), first we compute the first and second derivative of the given function as follows:*

$$f'(x) = 2 - 3.5x + 3.3x^2 - x^3$$

$$f''(x) = -3.5 - 6.6x - 3x^2.$$

Then using formula (7.36), we have

$$x_{n+1} = x_n - \frac{2 - 1.5x_n + 3.3x_n^2 - x_n^3}{-1.5 + 6.6x_n - 3x_n^2}, \quad n \geq 0.$$

Taking n = 0 and $x_0 = 2.5$, we get

$$x_1 = 2.1957 - \frac{2 - 1.5(2.1957) + 3.3(2.1957)^2 - (2.1957)^3}{-1.5 + 6.6(2.1957) - 3(2.1957)^2} = 2.0957,$$

which gives the function value $f(2.1957) = 1.7880$. Similarly, the second iteration can be obtained as

$$x_2 = 2.5 - \frac{2 - 1.5(2.5) + 3.3(2.5)^2 - (2.5)^3}{-1.5 + 6.6(2.5) - 3(2.5)^2} = 2.1917,$$

and the corresponding function value $f(2.0917) = 1.8080$.

Repeat the process; the numerical results for the corresponding iterations, starting with the initial approximation $x_0 = 2.5$ with accuracy 5×10^{-2}, are given in Table 7.8.

Table 7.8: Solution by Newton's method.

n	x_n	$f(x_n)$	$f'(x_n)$	$f''(x_n)$
00	2.5	1.4844	−1.7500	−5.7500
01	2.1957	1.7880	−0.3608	−3.4714
02	2.0917	1.8080	−0.0344	−2.8204
03	2.0795	1.8082	−0.00044	−2.7483
04	2.0793	1.8082	−7.5522e-008	−2.7474

From Table 7.8, we can see that within four iterations, the result converges rapidly on the true value of 1.8082 at $x = 2.0793$. Also, note that this method does not require initial guesses that bracket the optimum. In addition, this method also shares the disadvantage that it may be divergent. For confirming the convergence of the method we must check the correct sign of the second derivative of the function. For maximizing the function, the second derivative of the function should be less than zero, and it should be greater than zero for the minimizing problem. In both cases, the first derivative of the function should be close to zero as much as possible because optimum here means the same as the root of $f'(x) = 0$. Note that if the second derivative of the function equals zero at the given initial guess, then change the initial guess. ●

To get the above results using MATLAB commands, first the function $2x - 1.75x^2 + 1.1x^3 - 0.25x^4$ and its first and second derivatives $2 - 1.5x + 3.3x^2 - x^3$ and $-1.5 + 6.6x - 3x^2$ were saved in three m-files called fn.m, dfn.m, and ddfn.m, respectively, written as follows:

$$function \ y = fn(x)$$
$$y = 2 * x - 1.75 * x.\hat{\ } 2 + 1.1 * x.\hat{\ } 3 - 0.25 * x.\hat{\ } 4;$$

first derivative of the function

$$function \ dy = dfn(x)$$
$$dy = 2 - 1.5 * x + 3.3 * x.\hat{\ } 2 - x.\hat{\ } 3;$$

and the second derivative of the function,

$$function \ ddy = ddfn(x)$$
$$ddy = -1.5 + 6.6 * x - 3 * x.\hat{\ } 2;$$

after which we do the following.

$$\gg x0 = 2.5; tol = 5e - 6;$$
$$\gg sol = newtonO('fn', 'dfn', 'ddfn', x0, tol);$$

Program 7.8

MATLAB m-file for Newton's Method for Optimization
function sol=newtonO(fn,dfn,ddfn,x0,tol)
old = x0+1; k = 0;
while abs $(x0 - old) > tol$; old = x0;
$x0 = old - feval(dfn, old)/feval(ddfn, old);$
end; sol=x0;

7.7 Multidimensional Unconstrained Optimization

Just as the theory of linear programming is based on linear algebra with several variables, the theory of *NLP* is based on calculus with several variables. A convenient and familiar place to begin is therefore with the problem of finding the minimum or maximum of a nonlinear function in the absence of constraints.

Here, we will discuss the procedure to find an optimal solution (if it exists) or a local extremum for the following *NLP* problem:

$$\left. \begin{array}{rl} \text{maximize}(or\ minimize)\ z\ =\ & f(x_1, x_2, \ldots, x_n) \\ \text{subject to} & \\ \mathbf{x}\ =\ & (x_1, x_2, \ldots, x_n) \in \mathbf{R}^n \end{array} \right\}. \qquad (7.37)$$

We assume that the first and second partial derivatives of $f(x_1, x_2, \ldots, x_n)$ exist and are continuous at all points.

Theorem 7.22 (Local Extremum)

If $\bar{\mathbf{x}}$ is a local extremum for NLP problem (7.37), then

$$\frac{\partial f}{\partial x_i}(\bar{\mathbf{x}}) = \mathbf{0}, \quad i = 1, 2, \ldots, n, \qquad (7.38)$$

where the point $\bar{\mathbf{x}}$ is called the stationary point of a function $f(x)$. ●

The following theorems give conditions (involving the Hessian matrix of f) under which a stationary point is a local minimum, or a local maximum and not a local extremum.

Theorem 7.23 (Local Minimum)

If $H_k(\bar{\mathbf{x}}) > 0$, for $k = 1, 2, \ldots, n$, then the stationary point $\bar{\mathbf{x}}$ is a local minimum for NLP problem (7.37). ●

Theorem 7.24 (Local Maximum)

If $H_k(\bar{\mathbf{x}})$ is nonzero, for $k = 1, 2, \ldots, n$ and has the same sign as $(-1)^k$, then the stationary point $\bar{\mathbf{x}}$ is a local maximum for NLP problem (7.37). ●

Theorem 7.25 (Saddle Point)

If $H_k(\bar{\mathbf{x}}) \neq 0$, for $k = 1, 2, \ldots, n$, and the conditions of Theorem 7.23 and Theorem 7.24 do not hold, then the stationary point $\bar{\mathbf{x}}$ is not a local minimum for NLP problem (7.37). ●

Theorem 7.26 *If $H_k(\bar{\mathbf{x}}) = 0$, for $k = 1, 2, \ldots, n$, then the stationary point $\bar{\mathbf{x}}$ may be a local minimum, or a maximum, or a saddle point for NLP problem (7.37).* ●

Example 7.38 *Find all local minimum, local maximum, and saddle points for the function*

$$f(x_1, x_2) = \frac{1}{3}x_1^3 - \frac{2}{3}x_2^3 + \frac{1}{2}x_1^2 - 6x_1 + 32x_2 + 4.$$

Solution. *The first partial derivatives of the function are*

$$\frac{\partial f(x_1, x_2)}{\partial x_1} = x_1^2 + x_1 - 6$$

and

$$\frac{\partial f(x_1, x_2)}{\partial x_2} = -2x_2^2 + 32.$$

Since $\frac{\partial f}{\partial x_1}$ and $\frac{\partial f}{\partial x_2}$ exist for every (x_1, x_2), the only stationary points are the solutions of the system

$$\begin{cases} x_1^2 + x_1 - 6 &= 0 \\ -2x_2^2 + 32 &= 0. \end{cases}$$

Solving this system, we obtain the four stationary points

$$(-3, -4), (-3, 4), (2, -4), (2, 4).$$

The second partial derivatives of f are

$$\frac{\partial^2 f}{\partial x_1} = 2x_1 + 1, \quad \frac{\partial^2 f}{\partial x_2^2} = -4x_2$$

and

$$\frac{\partial^2 f}{\partial x_1 \partial x_2} = 0 = \frac{\partial^2 f}{\partial x_2 \partial x_1}.$$

Hence, the Hessian matrix for the function f(x) is

$$H(x_1, x_2) = \begin{pmatrix} 2x_1 + 1 & 0 \\ 0 & -4x_2 \end{pmatrix}.$$

At $(-3, -4)$ the Hessian matrix is

$$H(-3, -4) = \begin{pmatrix} -5 & 0 \\ 0 & 16 \end{pmatrix}.$$

Since

$$H_1(-3, -4) = -5 < 0$$

and

$$H_2(-3, -4) = \begin{vmatrix} -5 & 0 \\ 0 & 16 \end{vmatrix} = -80 \neq 0,$$

the conditions of Theorem 7.23 and Theorem 7.24 cannot be satisfied, therefore, the stationary point $\bar{\mathbf{x}} = (-3, -4)$ is not a local extremum for the given function. But Theorem 7.25 now implies that $\bar{\mathbf{x}} = (-3, -4)$ is a saddle point, i.e.,

$$(-3, -4, f(-3, -4)) = (-3, -4, -\frac{407}{6}).$$

At $(-3, 4)$ the Hessian matrix is

$$H(-3, 4) = \begin{pmatrix} -5 & 0 \\ 0 & -16 \end{pmatrix}.$$

Since

$$H_1(-3, 4) = -5 < 0$$

2,22.22222.2222222.

and

$$H_2(-3,4) = \begin{vmatrix} -5 & 0 \\ 0 & 16 \end{vmatrix} = -80 \neq 0,$$

the conditions of Theorem 7.24 are satisfied, and it shows that the stationary point $\bar{\mathbf{x}} = (-3,4)$ is a local maximum for the given function, i.e.,

$$f(-3,4) = \frac{617}{6}.$$

At $(2,-4)$ the Hessian matrix is

$$H(2,-4) = \begin{pmatrix} 5 & 0 \\ 0 & 16 \end{pmatrix}.$$

Since

$$H_1(2,-4) = 5 > 0$$

and

$$H_2(2,-4) = \begin{vmatrix} 5 & 0 \\ 0 & 16 \end{vmatrix} = 80 > 0,$$

the conditions of Theorem 7.23 are satisfied, and it shows that the stationary point $\bar{\mathbf{x}} = (2,-4)$ is a local minimum for the given function, i.e.,

$$f(2,-4) = -\frac{266}{3}.$$

Finally, at $(2,4)$ the Hessian matrix is

$$H(2,4) = \begin{pmatrix} 5 & 0 \\ 0 & -16 \end{pmatrix}.$$

Since

$$H_1(2,4) = 5 > 0$$

and

$$H_2(-3,-4) = \begin{vmatrix} -5 & 0 \\ 0 & 16 \end{vmatrix} = -80 \neq 0,$$

the conditions of Theorem 7.23 and Theorem 7.24 cannot be satisfied, therefore, the stationary point $\bar{\mathbf{x}} = (2,4)$ is not a local extremum for the given function. From Theorem 7.25, we see that $\bar{\mathbf{x}} = (2,4)$ is a saddle point, i.e.,

$$(2,4,f(2,4)) = (2,4,82).$$

Example 7.39 *Find all local minimum, local maximum, and saddle points for the function*

$$f(x_1, x_2) = -x_1^2 + x_2^2.$$

Solution. *The first partial derivatives of the function are*

$$\frac{\partial f(x_1, x_2)}{\partial x_1} = -2x_1, \quad \frac{\partial f(x_1, x_2)}{\partial x_2} = 2x_2.$$

Since $\frac{\partial f}{\partial x_1}$ and $\frac{\partial f}{\partial x_2}$ exist for every (x_1, x_2), the only stationary points are the solutions of the system

$$\begin{cases} -2x_1 &= 0 \\ 2x_2 &= 0. \end{cases}$$

Solving this system, we obtain the only stationary point $(0,0)$.
The second partial derivatives of f are

$$\frac{\partial^2 f}{\partial x_1^2} = -2, \quad \frac{\partial^2 f}{\partial x_2^2} = 2$$

and

$$\frac{\partial^2 f}{\partial x_1 \partial x_2} = 0 = \frac{\partial^2 f}{\partial x_2 \partial x_1}$$

Hence, the Hessian matrix for the function $f(x)$ is:

$$H(x_1, x_2) = \begin{pmatrix} -2 & 0 \\ 0 & 2 \end{pmatrix}.$$

At $(0,0)$ the Hessian matrix is

$$H(0,0) = \begin{pmatrix} -2 & 0 \\ 0 & 2 \end{pmatrix}.$$

Since

$$H_1(0,0) = -2 < 0$$

and

$$H_2(0,0) = \begin{vmatrix} -2 & 0 \\ 0 & 2 \end{vmatrix} = -4 \neq 0,$$

the conditions of Theorem 7.23 and Theorem 7.24 cannot be satisfied, therefore, the stationary point $\bar{\mathbf{x}} = (0,0)$ is not a local extremum for the given function. From Theorem 7.25, we conclude that $\bar{\mathbf{x}} = (0,0)$ is a saddle point, i.e.,

$$(0, 0, f(0,0)) = (0, 0, 0).$$

•

7.7.1   Gradient Methods

There are a number of techniques available for multidimensional unconstrained optimization. The techniques we discuss here require derivatives and therefore are called *gradient methods*. As the name implies, gradient methods explicitly use derivative information to generate efficient algorithms. Two methods will be discussed here, and they are called the *steepest ascent* and *steepest descent* methods.

Consider the following *NLP* problem:

$$\left.\begin{array}{rcl} \text{maximize } (or \ minimize) \ z & = & f(x_1, x_2, \ldots, x_n) \\ \text{subject to} & & \\ \mathbf{x} & = & (x_1, x_2, \ldots, x_n) \in \mathbf{R}^n \end{array}\right\}. \qquad (7.39)$$

We know that if $f(x_1, x_2, \ldots, x_n)$ is a *concave function*, then the optimal solution to the problem (7.39) (if it exists) will occur at a stationary point $\bar{\mathbf{x}}$ having

$$\frac{\partial f(\bar{\mathbf{x}})}{\partial x_1} = \frac{\partial f(\bar{\mathbf{x}})}{\partial x_2} = \cdots = \frac{\partial f(\bar{\mathbf{x}})}{\partial x_n} = 0.$$

Sometimes it is very easy to compute a stationary point of a function, but in many problems, it may be very difficult. Here, we discuss a method that can be used to approximate the stationary point of a function.

Definition 7.15 (Length of a Vector)

Given a vector $\mathbf{x} = (x_1, x_2, \ldots, x_n) \in \mathbf{R}^n$, the length of $\mathbf{x}$ is denoted by $\|\mathbf{x}\|$ and is defined as

$$\|\mathbf{x}\| = (x_1^2 + x_2^2 \cdots + x_n^2)^{1/2}. \qquad (7.40)$$

Note that any n-dimensional vector represents a direction in $\mathbf{R}^n$. Also, for any direction there are an infinite number of vectors representing that direction. For any vector $\mathbf{x}$, the vector

$$\|\mathbf{u}\| = \frac{\mathbf{x}}{\|\mathbf{x}\|} \tag{7.41}$$

is called a *unit vector* and will have a length of 1 and will define the same direction as $\mathbf{x}$.

Definition 7.16 (Gradient Vector)

Let $f(x_1, x_2, \ldots, x_n)$ be a function of n variables $\mathbf{x} = (x_1, x_2, \ldots, x_n)$, then the gradient vector for $f\mathbf{x}$ is denoted $\nabla f(\mathbf{x})$ and is defined as

$$\nabla f(\mathbf{x}) = \left[\frac{\partial f}{\partial x_1}(\bar{\mathbf{x}}), \frac{\partial f}{\partial x_2}(\bar{\mathbf{x}}), \ldots, \frac{\partial f}{\partial x_n}(\bar{\mathbf{x}}) \right]^T. \tag{7.42}$$

Also, $\nabla f(\mathbf{x})$ defines the direction $\dfrac{\nabla f(\mathbf{x})}{\|\nabla f(\mathbf{x})\mathbf{x}\|}$. ●

For example, if
$$f(x_1, x_2) = x_1^2 + x_2^2,$$

then
$$\nabla f(x_1, x_2) = [2x_1, 2x_2]^T.$$

Thus, at $(2, 3)$, the gradient vector of the function is

$$\nabla f(2, 3) = [2(2), 2(3)]^T = [4, 6]^T$$

and
$$\|\nabla f(2, 3)\| = \sqrt{4^2 + 6^2} = \sqrt{16 + 36} = \sqrt{52}.$$

So the gradient vector $\nabla f(2, 3)$ defines the direction

$$\frac{\nabla f(2, 3)}{\|\nabla f(2, 3)\|} = (\frac{4}{\sqrt{52}}, \frac{6}{\sqrt{52}}) = (0.5547, 0.8321).$$

Note that if any point $\bar{\mathbf{x}}$ lies on the curve $f(\mathbf{x})$, then the vector $\dfrac{\nabla f(\mathbf{x})}{\|\nabla f(\mathbf{x})\mathbf{x}\|}$ will be perpendicular to the curve $f(\mathbf{x})$.

For example, if $f(x_1, x_2) = x_1^2 + x_2^2$, then at $(2, 3)$

$$\frac{\nabla f(2, 3)}{\|\nabla f(2, 3)\|} = (0.5547, 0.8321)$$

is perpendicular to

$$x_1^2 + x_2^2 = 13.$$

Note that if at any point $\mathbf{x} = (x_1, x_2, \ldots, x_n)$ the gradient vector $\nabla f(\mathbf{x})$ points in the direction in which the function $f(\mathbf{x})$ is increasing most rapidly, it is called the *direction of steepest ascent*. So it follows that $-\nabla f(\mathbf{x})$ points in the direction in which $f(\mathbf{x})$ is decreasing more rapidly, and it is called the *direction of steepest descent*. In other words, we can say that if we are looking for a maximum of $f(\mathbf{x})$ using the initial point v_0, it seems sensible to look in the direction of steepest ascent, and for a minimum of $f(\mathbf{x})$ we look in the direction of steepest descent.

Also, moving from v_0 in the direction of $\nabla f(\mathbf{x})$ to get the local maximum, we have to find the new point v_1 as

$$v_1 = v_0 + \alpha_0 \nabla f(v_0),$$

for some $\alpha_0 > 0$. Since we desire v_1 to be as close as possible to the maximum, we need to find the unknown variable $\alpha_0 > 0$ such that

$$f(v_1) = f(v_0 + \alpha_0 \nabla f(v_0))$$

is as large as possible.

Since $f(v_0 + \alpha_0 \nabla f(v_0))$ is a function of the one variable α_0, α_0 can be found by using one-dimensional search. Since the steepest ascent method (also called the gradient method) is an iterative method, we need at each iteration a new value of the variable α_k, which helps us to maximize the function $f(v_k + \alpha_k \nabla f(v_k))$ at each step. The value of α_k can be computed by using the following form:

$$v_{k+1} = v_k + \alpha_k \nabla f(v_0), \quad \text{for} \quad k \geq 0.$$

Theorem 7.27 *Suppose we have a point v, and we move from v a small distance δ in a direction* **d**. *Then for a given δ, the maximum increase in the value of $f(\mathbf{x})$ will occur if we choose*

$$\mathbf{d} = \frac{\nabla f(\mathbf{x})}{\|\nabla f(\mathbf{x})\|}.$$

In short, if we move a small distance from v and we want $f(\mathbf{x})$ to increase as quickly as possible, then we should move in the direction of $\nabla f(v)$. ●

Beginning at any point v_0 and moving in the direction of $\nabla f(v_0)$ will result in a maximum rate of increase for f. So we begin by moving away from v_0 in the direction of $\nabla f(v_0)$. For some nonnegative value of α_0, we move to a point v_1, which can be written as

$$v_1 = v_0 + \alpha_0 \nabla f(v_0),$$

where α_0 solves the following one-dimensional optimization problem:

$$\left.\begin{array}{rl} z(\alpha) & = \; maximize \; f(v_0 + \alpha_0 \nabla f(v_0)) \\ \text{subject to} & \\ \alpha_0 & \geq \; 0 \end{array}\right\}.$$

If $\|\nabla f(v_1)\|$ is small, i.e.,

$$\|\nabla f(v_1)\| \leq \epsilon, \quad \text{given small} \quad \epsilon > 0,$$

we may terminate the process with the knowledge that v_1 is a good approximation of the stationary point $\bar{v}$ with $\nabla f(\bar{v}) = 0$.

But if $\|\nabla f(v_1)\|$ is not sufficiently small, then we move away from v_1 a distance α_1 in the direction of $\|\nabla f(v_1)\|$. As before, we discuss α_1 by solving

$$\left.\begin{array}{rl} z(\alpha) & = \; maximize \; f(v_1 + \alpha_1 \nabla f(v_1)) \\ \text{subject to} & \\ \alpha_1 & \geq \; 0 \end{array}\right\}.$$

We are at point v_2, which can be written as

$$v_2 = v_1 + \alpha_1 \nabla f(v_1).$$

If $\|\nabla f(v_2)\|$ is sufficiently small, then we terminate the process with the knowledge that v_2 is a good approximation of the stationary point $\bar{v}$ of the given function $f(\mathbf{x})$, with $\nabla f(\bar{v}) = 0$.

This process is called the *steepest ascent* method because, to generate points, we always move in the direction that maximizes the rate at which f increases (at least locally).

Example 7.40 *Use the steepest ascent method to approximate the solution to*

$$\begin{aligned} maximize\ z \quad &= \quad -(x_1 - 3)^2 - (x_2 - 2)^2 \\ subject\ to \quad & \\ (x_1, x_2) \quad &\in \quad \mathbf{R}^2, \end{aligned}$$

by starting with $v_0 = (1, 1)$.

Solution. *Given*

$$f(x_1, x_2) = -(x_1 - 3)^2 - (x_2 - 2)^2 \quad and \quad v_0 = (1, 1),$$

and the gradient vector of the given function, which is

$$\nabla f(x_1, x_2) = [-2(x_1 - 3), -2(x_2 - 2)]^T,$$

we choose to begin at the point $v_0 = (1, 1)$, so

$$\nabla f(1, 1) = [-2(1 - 3), -2(1 - 2)]^T = [4, 2]^T.$$

Thus, we must choose α_0 to maximize

$$f(\alpha_0) = f[(1, 1) + \alpha_0(4, 2)] = f(1 + 4\alpha_0, 1 + 2\alpha_0),$$

which can be simplified as

$$f(\alpha_0) = f(1 + 4\alpha_0, 1 + 2\alpha_0) = -(4\alpha_0 - 2)^2 - (2\alpha_0 - 1)^2.$$

Thus,

$$f(\alpha_0) = -20\alpha_0^2 + 20\alpha_0 - 5.$$

So solving the one-dimensional optimization problem, we need to find the value of α_0 which can be obtained as

$$f'(\alpha_0) = 0 = -40\alpha_0 + 20,$$

which gives

$$-40\alpha_0 + 20 = 0, \qquad \alpha_0 = 0.5.$$

Thus our new point can be found as

$$v_1 = v_0 + \alpha_0 \nabla f(v_0) = (1,1) + 0.5(4,2) = (3,2).$$

Since

$$\nabla f(3,2) = [0,0]^T,$$

we terminate the process. Thus, $(3,2)$ is the optimal solution to the given NLP problem because $f(x_1, x_2)$ is a concave function:

$$H(3,2) = \begin{pmatrix} -2 & 0 \\ 0 & -2 \end{pmatrix}.$$

The first principal minors of the Hessian matrix are the diagonal entries (-2 and -2). These are both negative (nonpositive). The second principal minor is the determinant of the Hessian matrix H and equals

$$-2(-2) - 0 = 4 \geq 0.$$

●

Procedure 7.4 (The Method of Steepest Ascent)

1. *Start with initial point $\mathbf{x}^{(0)}$ and initial (given) function $f_0(\mathbf{x})$.*

2. *Find the search direction; $d^0 = \nabla f_0(\mathbf{x}^{(0)})$; and if $d^0 = 0$, stop; $\mathbf{x}^{(0)}$ is the maximum.*

3. *Search the line $\mathbf{x} = \mathbf{x}^{(0)} + \alpha d^0$ for a maximum.*

4. *Find the approximation of α at which $f_0(\alpha)$ is maximized.*

5. *Update the estimate of the maximum:*

$$\mathbf{x}^{(k+1)} = \mathbf{x}^{(k)} + \alpha d^k.$$

6. *If* $\|\mathbf{x}^{(k+1)} - \mathbf{x}^{(k)}\| < \epsilon$ $(\epsilon > 0)$, *stop;* $\mathbf{x}^{(k+1)}$ *is the maximum; otherwise, repeat all steps.*

Example 7.41 *Use the steepest ascent method to approximate the solution to the problem*

$$\begin{aligned} maximize\ z \quad &= \quad x_1 + x_2 - x_1^2 - 2x_1 x_2 - 2x_2^2 \\ subject\ to \\ (x_1, x_2) \quad &\in \quad \mathbf{R}^2, \end{aligned}$$

by starting with $v_0 = (1, 1)$.

Solution. *Given*

$$f(x_1, x_2) = x_1 + x_2 - x_1^2 - 2x_1 x_2 - 2x_2^2 \quad and \quad v_0 = (1, 1),$$

the gradient vector of the given function can be evaluated as

$$\nabla f(x_1, x_2) = [1 - 2x_1 - 2x_2, 1 - 2x_1 - 2x_2]^T,$$

and it is at the given point $v_0 = (1, 1)$, *that we get*

$$\nabla f(1, 1) = [1 - 2(1) - 2(1), 1 - 2(1) - 2(1)]^T = [-3, -3]^T.$$

Thus, we must choose α_0 *to maximize*

$$f(\alpha_0) = f[(1, 1) + \alpha_0(-3, -3)] = f(1 + 4\alpha_0, 1 + 2\alpha_0),$$

which can be simplified as

$$f(\alpha_0) = f(1 + 4\alpha_0, 1 + 2\alpha_0) = -(4\alpha_0 - 2)^2 - (2\alpha_0 - 1)^2.$$

Thus,

$$f(\alpha_0) = -20\alpha_0^2 + 20\alpha_0 - 5.$$

So solving the one-dimensional optimization problem, we need to find the value of α_0, which can be obtained as follows:

$$f'(\alpha_0) = 0 = -40\alpha_0 + 20,$$

which gives

$$-40\alpha_0 + 20 = 0, \qquad \alpha_0 = 0.5.$$

Thus, our new point can be found as

$$v_1 = v_0 + \alpha_0 \nabla f(v_0) = (1,1) + 0.5(4,2) = (3,2).$$

Since

$$\nabla f(3,2) = [0,0]^T,$$

it is important to note that this method is a very slow convergent (linearly) method, therefore, it can mostly be used for providing the best initial guess for the approximation of an extreme value of a function for the other iterative methods. •

Example 7.42 *Use the steepest descent method to approximate the solution to the following problem*

$$\begin{aligned} minimize\ z \quad &= \quad 2x_1^2 + x_2^2 + x_1 - x_2 - x_1 x_2 \\ subject\ to \quad & \\ (x_1, x_2) \quad &\in \quad \mathbf{R}^2, \end{aligned}$$

by starting with $v_0 = (1,1)$.

Solution. *Given*

$$f(x_1, x_2) = 2x_1^2 + x_2^2 + x_1 - x_2 - x_1 x_2 \quad and \quad v_0 = (1,1),$$

the gradient vector of the given function can be computed as

$$\nabla f(x_1, x_2) = [4x_1 + 1 - x_2, 2x_2 - 1 - x_1]^T,$$

and its value at the given point $v_0 = (1,1)$ is

$$\nabla f(1,1) = [4,0]^T.$$

The new point is defined as

$$v_1 = v_0 + \alpha_0 d^0, \quad where \quad d^0 = -\nabla f(1,1) = [-4,0]^T.$$

Thus, we must choose α_0 to minimize

$$f(\alpha_0) = f[(1,1) + \alpha_0(-4,0)] = f(1 - 4\alpha_0, 1),$$

and simplifying it gives

$$f(\alpha_0) = 2(1 - 4\alpha_0)^2 + (1)^2 + (1 - 4\alpha_0) - 1 - (1 - 4\alpha_0)(1) = 32\alpha_0^2 - 16\alpha_0.$$

Thus, solving the one-dimensional optimization problem for finding the value of α_0, we do the following:

$$f'(\alpha_0) = 0 = 64\alpha_0 - 16,$$

which gives

$$64\alpha_0 = 16, \quad or \quad \alpha_0 = \frac{16}{64} = \frac{1}{4}.$$

Notice that

$$f''(\alpha_0) = f''\left(\frac{1}{4}\right) = 64 > 0,$$

so $\alpha_0 = \frac{1}{4}$ is the minimizing point.
Thus, our new point can be found as

$$v_1 = v_0 + \alpha_0 d^0 = (1,1) + \frac{1}{4}(-4,0) = (0,1)$$

and

$$\nabla f(v_1) = \nabla f(0,1) = [0,1]^T,$$

which gives

$$d^1 = -\nabla f(v_1) = -\nabla f(0,1) = [0,-1]^T.$$

Now we find α_1 to minimize

$$f(\alpha_1) = f[(0,1) + \alpha_1(0,-1)] = f(0, 1 - \alpha_1),$$

and it gives

$$f(\alpha_1) = 0 + (1 - \alpha_1)^2 + 0 - (1 - \alpha_1) = \alpha_1^2 - \alpha_1.$$

Again, solving the one-dimensional optimization problem for finding the value of α_1, we use the equation

$$f'(\alpha_1) = 0 = 2\alpha_1 - 1,$$

which gives $\alpha_1 = \frac{1}{2}$.
So the new point can be found as

$$v_2 = v_1 + \alpha_1 d^1 = (0, 1) + \frac{1}{2}(0, -1) = \left(0, \frac{1}{2}\right)$$

and

$$\nabla f(v_2) = \nabla f\left(0, \frac{1}{2}\right) = \left[\frac{1}{2}, 0\right]^T,$$

which gives

$$d^2 = -\nabla f(v_2) = -\nabla f\left(0, \frac{1}{2}\right) = \left[-\frac{1}{2}, 0\right]^T.$$

Similarly, we have the other iterations as follows:

$$\alpha_2 = \frac{1}{4}, \quad v_3 = \left(-\frac{1}{8}, \frac{1}{2}\right), \quad \nabla f(v_3) = \left[0, \frac{1}{8}\right]^T$$

$$\alpha_3 = \frac{1}{2}, \quad v_4 = \left(-\frac{1}{8}, \frac{7}{16}\right), \quad \nabla f(v_4) = \left[\frac{1}{16}, 0\right]^T.$$

Since $\nabla f(v_4) \approx 0$, the process can be terminated at this point. The approximate minimum point is given by $v_4 = (-\frac{1}{8}, \frac{7}{16})$. Notice that the gradients at points v_3 and v_4,

$$\left(0, \frac{1}{8}\right)\left(\frac{1}{16}, 0\right)^T = 0,$$

are orthogonal.   ●

7.7.2 Newton's Method

Newton's method can also be used for multidimensional maximization or minimization. This form of Newton's method can be obtained by using Taylor's series of several variables as follows:

$$f(\mathbf{x}) = f(\mathbf{x_0}) + \nabla f(\mathbf{x_0})(\mathbf{x} - \mathbf{x_0}) + \frac{1}{2}(\mathbf{x} - \mathbf{x_0})^T H(\mathbf{x_0})(\mathbf{x} - \mathbf{x_0}) + \cdots, \quad (7.43)$$

where $H(\mathbf{x_0})$ is called the Hessian matrix or, simply, the Hessian of $f(\mathbf{x})$. Take $\mathbf{x_0} = \mathbf{x}^*$ (for example, a minimum of f) and, ignoring the higher-order terms, we get

$$f(\mathbf{x}) \approx f(\mathbf{x}^*) + \nabla f(\mathbf{x}^*)^T (\mathbf{x} - \mathbf{x}^*) + \frac{1}{2}(\mathbf{x} - \mathbf{x}^*)^T H(\mathbf{x}^*)(\mathbf{x} - \mathbf{x}^*).$$

Since $\nabla f(\mathbf{x}^*) = 0$ because $\mathbf{x}^*$ is minimum of $f(\mathbf{x})$,

$$f(\mathbf{x}) \approx f(\mathbf{x}^*) + \frac{1}{2}(\mathbf{x} - \mathbf{x}^*)^T H(\mathbf{x}^*)(\mathbf{x} - \mathbf{x}^*).$$

Note that $\mathbf{x}^*$ is the local minimum value of $f(\mathbf{x}^*)$, so

$$(\mathbf{x} - \mathbf{x}^*)^T H(\mathbf{x}^*)(\mathbf{x} - \mathbf{x}^*) \geq 0,$$

at least for $\mathbf{x}$ near $\mathbf{x}^*$; if the minimum is strict local minimum, then

$$(\mathbf{x} - \mathbf{x}^*)^T H(\mathbf{x}^*)(\mathbf{x} - \mathbf{x}^*) > 0, \quad \text{for} \quad \mathbf{x} \neq \mathbf{x}^*,$$

showing that H is positive-definite.

From (7.43), we have $\nabla f(\mathbf{x}) = 0$, which gives

$$0 \approx 0 + \nabla f(\mathbf{x_0})^T + (\mathbf{x} - \mathbf{x_0})H(\mathbf{x_0}).$$

Also, it can be written as

$$(\mathbf{x} - \mathbf{x_0}) \approx -H^{-1}(\mathbf{x_0})\nabla f(\mathbf{x_0})^T$$

or

$$\mathbf{x} \approx \mathbf{x_0} - H^{-1}(\mathbf{x_0})\nabla f(\mathbf{x_0})^T,$$

which is a better approximation of $\mathbf{x}^*$. Hence, Newton's method for the extremum value of $f(\mathbf{x})$ (of several variables) is

$$\mathbf{x}_{k+1} = \mathbf{x}_k - H^{-1}(\mathbf{x}_k)\nabla f(\mathbf{x}_k)^T, \quad k \geq 0. \tag{7.44}$$

Note that if H is positive-definite, then it is nonsingular and the inverse of it exists. For example, if a given function is of two variables, then (7.44) can be written as

$$\begin{pmatrix} x_{k+1} \\ y_{k+1} \end{pmatrix} = \begin{pmatrix} x_k \\ y_k \end{pmatrix} - \begin{pmatrix} \dfrac{\partial^2 f}{\partial x^2} & \dfrac{\partial^2 f}{\partial x \partial y} \\[2ex] \dfrac{\partial^2 f}{\partial y \partial x} & \dfrac{\partial^2 f}{\partial y^2} \end{pmatrix}^{-1} \begin{pmatrix} \dfrac{\partial f}{\partial x} \\[2ex] \dfrac{\partial f}{\partial y} \end{pmatrix} \tag{7.45}$$

or for three variables

$$\begin{pmatrix} x_{k+1} \\ y_{k+1} \\ z_{k+1} \end{pmatrix} = \begin{pmatrix} x_k \\ y_k \\ z_k \end{pmatrix} - \begin{pmatrix} \dfrac{\partial^2 f}{\partial x^2} & \dfrac{\partial^2 f}{\partial x \partial y} & \dfrac{\partial^2 f}{\partial x \partial z} \\[2ex] \dfrac{\partial^2 f}{\partial y \partial x} & \dfrac{\partial^2 f}{\partial y^2} & \dfrac{\partial^2 f}{\partial y \partial z} \\[2ex] \dfrac{\partial^2 f}{\partial z \partial x} & \dfrac{\partial^2 f}{\partial z \partial y} & \dfrac{\partial^2 f}{\partial z^2} \end{pmatrix}^{-1} \begin{pmatrix} \dfrac{\partial f}{\partial x} \\[2ex] \dfrac{\partial f}{\partial y} \\[2ex] \dfrac{\partial f}{\partial z} \end{pmatrix}, \tag{7.46}$$

and so on. Note that in both formulas, (7.45) and (7.46), the Hessian matrix and gradient vector on the right-hand side are evaluated at $(x, y) = (x_k, y_k)$ and $(x, y, z) = (x_k, y_k, z_k)$, respectively.

Example 7.43 *Use Newton's method to find the local minimum of the given function*

$$f(x, y, z) = x + 2z + yz - x^2 - y^2 - z^2,$$

taking the starting values $(x_0, y_0, z_0)^T = (1, 1, 1)^T$.

Solution. *First, we compute the partial derivatives of the given function as*

$$\frac{\partial f}{\partial x} = 1 - 2x$$

$$\frac{\partial f}{\partial y} = z - 2y$$

$$\frac{\partial f}{\partial z} = 2 + y - 2z,$$

so the gradient of the function can be written as

$$\nabla f(x, y, z) = [1 - 2x, z - 2y, 2 + y - 2z]^T.$$

Also, the second partial derivatives are

$$\frac{\partial^2 f}{\partial x^2} = -2$$

$$\frac{\partial^2 f}{\partial y^2} = -2$$

$$\frac{\partial^2 f}{\partial z^2} = -2$$

$$\frac{\partial^2 f}{\partial x \partial y} = \frac{\partial^2 f}{\partial y \partial x} = 0$$

$$\frac{\partial^2 f}{\partial x \partial z} = \frac{\partial^2 f}{\partial z \partial x} = 0$$

$$\frac{\partial^2 f}{\partial y \partial z} = \frac{\partial^2 f}{\partial z \partial y} = 1,$$

so the Hessian of f is

$$H(x, y, z) = \begin{pmatrix} -2 & 0 & 0 \\ 0 & -2 & 1 \\ 0 & 1 & -2 \end{pmatrix}.$$

Thus, Newton's method for the given function is

$$
\begin{pmatrix} x_{k+1} \\ y_{k+1} \\ z_{k+1} \end{pmatrix} = \begin{pmatrix} x_k \\ y_k \\ z_k \end{pmatrix} - \begin{pmatrix} -2 & 0 & 0 \\ 0 & -2 & 1 \\ 0 & 1 & -2 \end{pmatrix}^{-1} \begin{pmatrix} 1 - 2x_k \\ z_k - 2y_k \\ 2 + y_k - 2z_k \end{pmatrix}.
$$

Starting with the initial approximation $(x_0, y_0, z_0) = (1, 1, 1)$ and $k = 0$ in the above formula, we have

$$
\begin{pmatrix} x_1 \\ y_1 \\ z_1 \end{pmatrix} = \begin{pmatrix} 1 \\ 1 \\ 1 \end{pmatrix} - \begin{pmatrix} -2 & 0 & 0 \\ 0 & -2 & 1 \\ 0 & 1 & -2 \end{pmatrix}^{-1} \begin{pmatrix} -1 \\ 1 \\ 1 \end{pmatrix}.
$$

Since the inverse of the Hessian matrix is

$$
H^{-1} = \begin{pmatrix} -2 & 0 & 0 \\ 0 & -2 & 1 \\ 0 & 1 & -2 \end{pmatrix}^{-1} = \begin{pmatrix} -\dfrac{1}{2} & 0 & 0 \\ 0 & -\dfrac{2}{3} & -\dfrac{1}{3} \\ 0 & -\dfrac{1}{3} & -\dfrac{2}{3} \end{pmatrix},
$$

using this value we have the first iteration as

$$
\begin{pmatrix} x_1 \\ y_1 \\ z_1 \end{pmatrix} = \begin{pmatrix} 1 \\ 1 \\ 1 \end{pmatrix} - \begin{pmatrix} -\dfrac{1}{2} & 0 & 0 \\ 0 & -\dfrac{2}{3} & -\dfrac{1}{3} \\ 0 & -\dfrac{1}{3} & -\dfrac{2}{3} \end{pmatrix} \begin{pmatrix} -1 \\ 1 \\ 1 \end{pmatrix} = \begin{pmatrix} \dfrac{1}{2} \\ 2 \\ 2 \end{pmatrix}.
$$

The norm of the gradient vector of the function at the new approximation can be calculated as

$$
\|\nabla f(x_1, y_1, z_1)\| = \left\| \left(\frac{1}{2}, 1, 1 \right)^T \right\| = \sqrt{\frac{1}{4} + 1 + 1} = \frac{3}{2}.
$$

Similarly, we have the other possible iterations as follows.

Second iteration:

$$
\begin{pmatrix} x_2 \\ y_2 \\ z_2 \end{pmatrix} = \begin{pmatrix} \dfrac{1}{2} \\ 2 \\ 2 \end{pmatrix} - \begin{pmatrix} -\dfrac{1}{2} & 0 & 0 \\ 0 & -\dfrac{2}{3} & -\dfrac{1}{3} \\ 0 & -\dfrac{1}{3} & -\dfrac{2}{3} \end{pmatrix} \begin{pmatrix} \dfrac{1}{2} \\ -1 \\ -1 \end{pmatrix} = \begin{pmatrix} \dfrac{1}{2} \\ \dfrac{2}{3} \\ \dfrac{2}{3} \end{pmatrix},
$$

and the norm

$$
\| \nabla f(x_2, y_2, z_2) \| = \left\| \left(0, -\frac{2}{3}, \frac{4}{3} \right)^T \right\| = \sqrt{0 + \frac{4}{9} + \frac{16}{9}} = \frac{2\sqrt{5}}{3}.
$$

Third iteration:

$$
\begin{pmatrix} x_3 \\ y_3 \\ z_3 \end{pmatrix} = \begin{pmatrix} \dfrac{1}{2} \\ \dfrac{2}{3} \\ \dfrac{2}{3} \end{pmatrix} - \begin{pmatrix} -\dfrac{1}{2} & 0 & 0 \\ 0 & -\dfrac{2}{3} & -\dfrac{1}{3} \\ 0 & -\dfrac{1}{3} & -\dfrac{2}{3} \end{pmatrix} \begin{pmatrix} 0 \\ -\dfrac{2}{3} \\ \dfrac{4}{3} \end{pmatrix} = \begin{pmatrix} \dfrac{1}{2} \\ \dfrac{2}{3} \\ \dfrac{4}{3} \end{pmatrix},
$$

and the norm

$$
\| \nabla f(x_3, y_3, z_3) \| = (0, 0, 0)^T = 0.
$$

We noted that the convergence is very fast because we start sufficiently close to the optimal solution, which can be easily computed analytically as

$$
\frac{\partial f}{\partial x} = 1 - 2x = 0
$$

$$
\frac{\partial f}{\partial y} = z - 2y = 0
$$

$$
\frac{\partial f}{\partial z} = 2 + y - 2z = 0,
$$

and solving the above system gives

$$(\bar{\mathbf{x}}, \bar{\mathbf{y}}, \bar{\mathbf{z}}) = \left(\frac{1}{2}, \frac{2}{3}, \frac{4}{3} \right).$$

•

7.8   Constrained Optimization

Here, we will discuss an *NLP* problem which that of both an objective function and constraints. The uniqueness of an optimal solution of the given *NLP* problem depends on the nature of both the objective function and the constraints. If the given objective function is *concave*, and the constraint set forms a convex region, then there will be only one maximization solution to the problem, and any stationary point must be a global maximization solution. But if the given objective function is *convex* and the constraint set also forms a convex region, then any stationary point will be a global minimum solution of the given *NLP* problem.

7.8.1   Lagrange Multipliers

Here, we will discuss a general, rather powerful method to maximize or minimize a function with one or more constraints. This method is due to Lagrange and is called the method of *Lagrange multipliers*.

This method can be used to solve the *NLP* problem in which all the constraints are equality constraints. We consider an *NLP* problem of the following type:

$$\text{maximize } (or \ minimize) \quad z = f(x_1, x_2, \ldots, x_n)$$

subject to

$$g_1(x_1, x_2, \ldots, x_n) = 0$$
$$g_2(x_1, x_2, \ldots, x_n) = 0$$
$$\vdots \qquad\qquad\qquad (7.47)$$
$$g_m(x_1, x_2, \ldots, x_n) = 0.$$

To solve problem (7.47), we associate a *multiplier* λ_i, for $i = 1, 2, \ldots, m$, with the *ith* constraints in (7.47) and form the *Lagrangian* as follows:

$$L(x_1, \ldots, x_n, \lambda_1, \ldots, \lambda_m) = f(x_1, \ldots, x_n) + \sum_{i=1}^{i=m} \lambda_i \left[b_i - g_i(x_1, \ldots, x_n) \right]. \quad (7.48)$$

Then we attempt to find an optimal point $(\bar{x}_1, \ldots, \bar{x}_n, \bar{\lambda}_1, \ldots, \bar{\lambda}_m)$ that maximizes (or minimizes) $L(x_1, \ldots, x_n, \lambda_1, \ldots, \lambda_m)$. If $(\bar{x}_1, \ldots, \bar{x}_n, \bar{\lambda}_1, \ldots, \bar{\lambda}_m)$ maximizes the Lagrangian L, then at $(\bar{x}_1, \ldots, \bar{x}_n, \bar{\lambda}_1, \ldots, \bar{\lambda}_m)$ we have

$$\frac{\partial L}{\partial \lambda_i} = b_i - g_i(x_1, \ldots, x_n) = 0,$$

where $\frac{\partial L}{\partial \lambda_i}$ is the partial derivative of L with respect to λ_i.

This shows that $(\bar{x}_1, \bar{x}_2, \ldots, \bar{x}_n)$ will satisfy the constraints in (7.47). To show that $(\bar{x}_1, \bar{x}_2, \ldots, \bar{x}_n)$ solves (7.47), let $(x_1', x_2', \ldots, x_n')$ be any point that is in (7.47)'s feasible region. Since $(\bar{x}_1, \bar{x}_2, \ldots, \bar{x}_n, \bar{\lambda}_1, \bar{\lambda}_2, \ldots, \bar{\lambda}_m)$ maximizes L, for any numbers $\lambda_1', \lambda_2', \ldots, \lambda_m')$, we have

$$L(\bar{x}_1, \bar{x}_2, \ldots, \bar{x}_n, \bar{\lambda}_1, \ldots, \bar{\lambda}_m) \geq L(x_1', x_2', \ldots, x_n', \lambda_1', \ldots, \lambda_m'). \quad (7.49)$$

Since $(\bar{x}_1, \ldots, \bar{x}_n)$ and $(x_1', \ldots, x_n')$ are both feasible in (7.47), the terms in (7.48) involving the λs are all zero, and (7.49) becomes

$$f(\bar{x}_1, \ldots, \bar{x}_n) \geq f(x_1', \ldots, x_n'). \quad (7.50)$$

Thus, $(\bar{x}_1, \ldots, \bar{x}_n)$ does solve problem (7.47). In short, if $(\bar{x}_1, \ldots, \bar{x}_n, \bar{\lambda}_1, \ldots, \bar{\lambda}_m)$ solves the unconstraint maximization problem

$$maximize \ L(x_1, \ldots, x_n, \lambda_1, \ldots, \lambda_m), \quad (7.51)$$

then $(\bar{x}_1, \ldots, \bar{x}_n)$ solves (7.47). Since we know that for $(\bar{x}_1, \bar{x}_2, \ldots, \bar{x}_n, \bar{\lambda}_1, \ldots, \bar{\lambda}_m)$ to solve (7.51), it is necessary that at $(\bar{x}_1, \ldots, \bar{x}_n, \bar{\lambda}_1, \ldots, \bar{\lambda}_m)$

$$\frac{\partial L}{\partial x_1} = \cdots = \frac{\partial L}{\partial x_n} = \frac{\partial L}{\partial \lambda_1} = \cdots = \frac{\partial L}{\partial \lambda_m} = 0. \quad (7.52)$$

The following theorems give conditions that any point $(\bar{x}_1, \ldots, \bar{x}_n, \bar{\lambda}_1, \ldots, \bar{\lambda}_m)$ that satisfies (7.52) will yield an optimal solution $(\bar{x}_1, \ldots, \bar{x}_n)$ to (7.47).

Theorem 7.28 *Suppose that NLP problem (7.47) is a maximization problem. If $f(x_1, x_2, \ldots, x_n)$ is a convex function and each $g_i(x_1, x_2, \ldots, x_n)$ for $i = 1, 2, \ldots, m$ is a linear function, then any point $(\bar{x}_1, \bar{x}_2, \ldots, \bar{x}_n, \bar{\lambda}_1, \bar{\lambda}_2, \ldots, \bar{\lambda}_m)$ satisfying (7.52) will yield an optimal solution $(\bar{x}_1, \bar{x}_2, \ldots, \bar{x}_n)$ to (7.47).* ●

Theorem 7.29 *Suppose that the NLP problem (7.47) is a minimization problem. If $f(x_1, x_2, \ldots, x_n)$ is a concave function and each $g_i(x_1, x_2, \ldots, x_n)$ for $i = 1, 2, \ldots, m$ is a linear function, then any point $(\bar{x}_1, \bar{x}_2, \ldots, \bar{x}_n, \bar{\lambda}_1, \bar{\lambda}_2, \ldots, \bar{\lambda}_m)$ satisfying (7.52) will yield an optimal solution $(\bar{x}_1, \bar{x}_2, \ldots, \bar{x}_n)$ to (7.47).* ●

Example 7.44 *A company is planning to spend $\$10,000$ advertising its product. It costs $\$3,000$ per minute to advertise on the internet, $\$2,000$ per minute by television, and $\$1,000$ per minute on radio. If the firm buys x_1 minutes of internet advertising, x_2 minutes of television advertising, and x_3 minutes of radio advertising, then its revenue in thousands of dollars is given by*

$$f(x_1, x_2, x_3) = x_1 + x_2 + x_3 - x_1^2 - x_2^2 - x_3^2.$$

How can the firm maximize its revenue?

Solution. *Given the following NLP problem:*

$$\text{maximize} \ \ w = f(x_1, x_2, x_3) = x_1 + x_2 + x_3 - x_1^2 - x_2^2 - x_3^2$$

subject to

$$3,000x_1 + 2,000x_2 + 1,000x_3 = 10,000,$$

the Lagrangian $L(x_1, x_2, x_3)$ is defined as

$$L(x_1, x_2, x_3) = x_1 + x_2 + x_3 - x_1^2 - x_2^2 - x_3^2 + \lambda(3x_1 + 2x_2 + x_3 - 10),$$

and we set

$$\frac{\partial L}{\partial x_1} = 0 = 1 - 2x_1 + 3\lambda$$

$$\frac{\partial L}{\partial x_2} = 0 = 1 - 2x_2 + 2\lambda$$

$$\frac{\partial L}{\partial x_3} = 0 = 1 - 2x_3 + \lambda$$

$$\frac{\partial L}{\partial \lambda} = 0 = 3x_1 + 2x_2 + x_3 - 10.$$

From the first equation of the above system, we have

$$1 - 2x_1 + 3\lambda = 0,$$

and it gives

$$x_1 = \frac{1 + 3\lambda}{2}.$$

The second equation of the above system is

$$1 - 2x_2 + 2\lambda = 0,$$

which gives

$$x_2 = \frac{1 + 2\lambda}{2}.$$

Also, the third equation of the system is

$$1 - 2x_3 + \lambda = 0,$$

which gives

$$x_3 = \frac{1 + \lambda}{2}.$$

Finally, the last equation of the system is simply the given constraint

$$3x_1 + 2x_2 + x_3 = 10,$$

and using the values of x_1, x_2, and x_3 in this equation, we get

$$3\left(\frac{1 + 3\lambda}{2}\right) + 2\left(\frac{1 + 2\lambda}{2}\right) + \left(\frac{1 + \lambda}{2}\right) = 10.$$

Simplifying this expression, we get

$$14\lambda = 14, \qquad \text{which gives} \quad \lambda = 1.$$

Using this value of $\lambda = 1$, we obtain

$$\mathbf{x}_1 = \frac{1 + 3(1)}{2} = 2.0$$

$$\mathbf{x}_2 = \frac{1 + 2(1)}{2} = 1.5$$

$$\mathbf{x}_3 = \frac{1 + 1}{2} = 1.0.$$

Thus, we get

$$\bar{\mathbf{x}} = [2.0, 1.5, 1.0]^T.$$

Now we compute the Hessian matrix of the function, which can help us show that the given function is concave. The Hessian for the given function is

$$H(2, 1.5, 1) = \begin{pmatrix} -2 & 0 & 0 \\ 0 & -2 & 0 \\ 0 & 0 & -2 \end{pmatrix}.$$

Since we know that the first principal minors are simply the diagonal elements of the Hessian,

$$H_1 = -2, \quad H_1 = -2, \quad H_1 = -2,$$

which are all negative, to find the second-order principal minors, we have to find the determinant of the matrices

$$\begin{pmatrix} -2 & 0 \\ 0 & -2 \end{pmatrix}, \quad \begin{pmatrix} -2 & 0 \\ 0 & -2 \end{pmatrix}, \quad \begin{pmatrix} -2 & 0 \\ 0 & -2 \end{pmatrix},$$

which can be obtained by just deleting row 1 and column 1, row 2 and column 2, and row 3 and column 3 of the Hessian matrix, respectively.

So the second-order principal minors are

$$H_2 = 4, \quad H_2 = 4, \quad H_2 = 4,$$

and all are nonnegative. The third-order principal minor is simply the determinant of the Hessian itself. So the determinant of the Hessian can be obtained by expanding row 1 cofactors as follows:

$$H_3 = -2[(-2)(-2) - 0] - 0 + 0 = -8,$$

which is the third-order minor and it is negative. So by Theorem 7.19 the given function is concave. Also, since the given constraint is linear, Theorem 7.28 shows that the Lagrange multiplier method does yield the optimal solution to the given NLP problem. Thus, the firm should purchase 2 minutes of television time, 1.5 minutes of radio time, and 1 minute of internet time, since $\lambda = 1$ and spending an extra δ (thousand) (for small) would increase the firm's revenues by approximately $\$1\delta$ (thousand).   ●

Example 7.45 *Find the maximum and minimum of the function*

$$f(x_1, x_2, x_3) = (x_1 - 1)^2 + (x_2 - 2)^2 + (x_3 - 2)^2$$

in the region

$$x_1^2 + x_2^2 + x_3^2 = 36$$

by using Lagrange multipliers.

Solution. *Given*

$$f(x_1, x_2, x_3) = (x_1 - 1)^2 + (x_2 - 2)^2 + (x_3 - 2)^2,$$

with the constraints

$$g_1(x_1, x_2, x_3) = x_1^2 + x_2^2 + x_3^2 - 36,$$

the Lagrangian $L(x, y, z)$ is defined as

$$L(x_1, x_2, x_3) = (x_1 - 1)^2 + (x_2 - 2)^2 + (x_3 - 2)^2 + \lambda(x_1^2 + x_2^2 + x_3^2 - 36),$$

which leads to the equations

$$\frac{\partial L}{\partial x_1} = 0 = 2(x_1 - 1) + 2x_1\lambda$$

$$\frac{\partial L}{\partial x_2} = 0 = 2(x_2 - 2) + 2x_2\lambda$$

$$\frac{\partial L}{\partial x_3} = 0 = 2(x_3 - 2) + 2x_3\lambda$$

$$\frac{\partial L}{\partial \lambda} = 0 = x_1^2 + x_2^2 + x_3^2 - 36.$$

Assume that $x_1 = x_2 = x_3 \neq 0$

$$\frac{1 - x_1}{x_1} = \lambda, \quad \frac{2 - x_2}{x_2} = \lambda, \quad \frac{2 - x_3}{x_3} = \lambda.$$

The first two equation, imply that

$$\frac{1 - x_1}{x_1} = \frac{2 - x_2}{x_2},$$

from which it follows that

$$x_2 = 2x_1.$$

Similarly, the first and third equations imply that

$$x_3 = 2x_1.$$

Putting the values of x_2 and x_3 in the constraint equation

$$x_1^2 + x_2^2 + x_3^2 - 36 = 0,$$

we obtain

$$9x^2 = 36 \quad or \quad x \pm 2.$$

Using these values of x_1, we get the two points

$$(2, 4, 4) \quad and \quad (-2, -4 - 4).$$

Since $f(2, 4, 4) = 9$ and $f(-2, -4, -4) = 81$, the function has a minimum value at the point $(2, 4, 4)$ and the maximum value at the other point $(-2, -4, -4)$. •

The above results can be reproduced using the following MATLAB commands:

```
>> syms x₁ x₂ x₃ lambda
>> f = (x₁ - 1)² + (x₂ - 2)² + (x₃ - 2)²;
>> g = x₁² + x₂² + x₃² - 36;
>> fx₁ = diff(f, x₁), fx₂ = diff(f, x₂), fx₃ = diff(f, x₃);
>> gx₁ = diff(g, x₁), gx₂ = diff(g, x₂), gx₃ = diff(g, x₃);
>> [alambda ax₁ ax₂ ax₃] = solve(fx₁ - lambda * gx₁, ...
fx₂ - lambda * gx₂, fx₃ - lambda * gx₃, g - 1)
>> [bx₁ bx₂ bx₃] = solve(fx₁, fx₂, fx₃);
>> [cx₁ cx₂ cx₃] = solve(gx₁, gx₂, gx₃);
>> T = [ax₁ ax₂ ax₃ subs(g, x₁ x₂ x₃, ax₁ ax₂ ax₃)...
subs(f, x₁ x₂ x₃, ax₁ ax₂ ax₃) bx₁ bx₂ bx₃...
subs(g, x₁ x₂ x₃, bx₁ bx₂ bx₃)...
subs(f, x₁ x₂ x₃, bx₁ bx₂ bx₃)...
cx₁ cx₂ cx₃ subs(g, x₁ x₂ x₃, cx₁ cx₂ cx₃)...
subs(f, x₁ x₂ x₃, cx₁ cx₂ cx₃)];
>> double(T);
```

Procedure 7.5 (The Method of Lagrange Multipliers)

1. *Verify that $n > m$ and each g_i has continuous first partials.*

2. *Form the Lagrangian function*

$$L(\mathbf{x}, \lambda) = f(\mathbf{x}) + \sum_{i=1}^{m} \lambda_i g_i(\mathbf{x}).$$

3. *Find all of the solutions $(\bar{\mathbf{x}}, \bar{\lambda})$ to the following system of nonlinear algebraic equations:*

$$\nabla L(\mathbf{x}, \lambda) = \nabla f(\mathbf{x}) + \sum_{i=1}^{m} \lambda_i \nabla g_i(\mathbf{x}) = 0$$

$$\frac{\partial L(\mathbf{x}, \lambda)}{\partial \lambda_i} = g_i(\mathbf{x}) = 0.$$

These equations are called the Lagrange conditions and $(\bar{\mathbf{x}}, \bar{\lambda})$ are the Lagrange points.

4. Examine each solution $(\bar{\mathbf{x}}, \bar{\lambda})$ to see if it is a minimizing point.

Example 7.46 *Maximize the function*

$$z = x_1^2 + x_2^2 + x_3^2,$$

subject to the constraints

$$2x_1 + x_2 + x_3 = 2$$
$$x_1 - x_2 - 3x_3 = 4.$$

Solution. *Given*

$$f(x_1, x_2, x_3) = x_1^2 + x_2^2 + x_3^2,$$

with constraints

$$g_1(x_1, x_2, x_3) = 2x_1 + x_2 + x_3 - 2$$
$$g_2(x_1, x_2, x_3) = x_1 - x_2 - 3x_3 - 4,$$

and $m = 2, n = 3$, $n > m$ as required, i.e., the method can be used for the given problem.

The gradients of the constraints are

$$\nabla g_1(\mathbf{x}) = \begin{pmatrix} 2 \\ 1 \\ 1 \end{pmatrix} \quad and \quad \nabla g_1(\mathbf{x}) = \begin{pmatrix} 1 \\ -1 \\ -3 \end{pmatrix},$$

so the gradients $\frac{\partial g_i}{\partial x_j}$ are continuous functions.

Now form the Lagrangian function

$$L(\mathbf{x}, \lambda) = x_1^2 + x_2^2 + x_3^2 + \lambda_1(2x_1 + x_2 + x_3 - 2) + \lambda_2(x_1 - x_2 - 3x_3 - 4).$$

Compute the derivatives of $L(\mathbf{x}, \lambda)$ and then set the derivatives equal to zero, which gives

$$\frac{\partial L}{\partial x_1} = 2x_1 + 2\lambda_1 + \lambda_2 = 0$$

$$\frac{\partial L}{\partial x_2} = 2x_2 + \lambda_1 - \lambda_2 = 0$$

$$\frac{\partial L}{\partial x_3} = 2x_3 + \lambda_1 - 3\lambda_2 = 0$$

$$\frac{\partial L}{\partial \lambda_1} = 2x_1 + x_2 + x_3 - 2 = 0$$

$$\frac{\partial L}{\partial \lambda_2} = x_1 - x_2 - 3x_3 - 4 = 0.$$

Note that this system is linear and quite easy to solve, but in many problems the systems are nonlinear, and in such cases the Lagrange conditions cannot be solved analytically, on account of the particular nonlinearities they contain. While in other cases an analytical solution is possible, some ingenuity might be needed to find it.

Thus, solving the above linear system, one can get the values of x_1, x_2, and x_3 as follows:

$$x_1 = -\frac{(2\lambda_1 + \lambda_2)}{2}$$

$$x_2 = -\frac{(\lambda_1 - \lambda_2)}{2}$$

$$x_3 = -\frac{(\lambda_1 - 3\lambda_2)}{2}.$$

Putting the values of x_1, x_2, and x_3 in the last two equations (constraints) of the above system, we get

$$2\left(\frac{(2\lambda_1 + \lambda_2)}{2}\right) + \left(\frac{(\lambda_1 - \lambda_2)}{2}\right) + \left(\frac{(\lambda_1 - 3\lambda_2)}{2}\right) + 2 = 0$$

$$\left(\frac{(2\lambda_1 + \lambda_2)}{2}\right) - \left(\frac{(\lambda_1 - \lambda_2)}{2}\right) - 3\left(\frac{(\lambda_1 - 3\lambda_2)}{2}\right) + 4 = 0.$$

After simplifying, we obtain

$$3\lambda_1 - \lambda_2 + 2 = 0$$
$$-2\lambda_1 + 11\lambda_2 + 8 = 0.$$

Solving this system, we get

$$\lambda_1 = -\frac{30}{31} \quad and \quad \lambda_2 = -\frac{28}{31}.$$

Using these values of the multipliers, we have

$$x_1 = \frac{44}{31}, \quad x_2 = \frac{1}{31}, \quad x_3 = -\frac{27}{31}.$$

Thus, one solution to the Lagrange conditions is

$$\bar{\mathbf{x}} = \left[\frac{44}{31}, \frac{1}{31}, -\frac{27}{31}\right]^T \quad and \quad \bar{\lambda} = \left[-\frac{30}{31}, -\frac{28}{31}\right]^T.$$

Note that this solution is unique, but in many problems it is not, and multiple solutions must be sought.

One can easily examine that the solution is a minimizing point as the gradients of the constraints

$$\nabla g_1(\mathbf{x}) = \begin{pmatrix} 2 \\ 1 \\ 1 \end{pmatrix} \quad and \quad \nabla g_1(\mathbf{x}) = \begin{pmatrix} 1 \\ -1 \\ -3 \end{pmatrix}$$

are linearly independent vectors.   ●

Some applications may involve more than two constraints. In particular, consider the problem of finding the extremum of $f(x_1, x_2, x_3, x_4)$ subject to the constraints $g_i(\mathbf{x}) = 0$ (for $i = 1, 2, 3$). If $f(\mathbf{x})$ has an extremum subject to these constraints, then the following conditions must be satisfied for some real numbers $\lambda_1, \lambda_2,$ and λ_3:

$$\nabla f + \lambda_1 \nabla g_1 + \lambda_2 \nabla g_2 + \lambda_3 \nabla g_3 = 0.$$

By equating components and using constraints, we obtain a system of seven equations in the seven unknowns $x_1, x_2, x_3, x_4, \lambda_1, \lambda_2, \lambda_3$. This method can also be extended to functions of more than 4 variables and to more than 3 constraints.

Theorem 7.30 (Lagrange Multipliers Theorem)
Given an NLP problem:

$$minimize \ \ z = f(x_1, x_2, \ldots, x_n)$$

subject to

$$g_1(x_1, x_2, \ldots, x_n) = 0$$
$$g_2(x_1, x_2, \ldots, x_n) = 0$$
$$\vdots \qquad\qquad\qquad (7.53)$$
$$g_m(x_1, x_2, \ldots, x_n) = 0.$$

If $\bar{\mathbf{x}}$ is a local minimizing point for the NLP problem (7.53), and $n > m$ (there are more variables than constraints) and the constraints $g_i(i = 1, , 2, \ldots, m)$ have continuous first derivatives with respect to $x_j(j = 1, 2, \ldots, n)$ and the gradient $\nabla g_i(\bar{\mathbf{x}})$ are linearly independent vectors, then there is a vector $\lambda = [\lambda_1, \lambda_2, \ldots, \lambda_m]^T$ such that

$$\nabla f(\bar{\mathbf{x}}) + \sum_{i=1}^{m} \lambda_i \nabla g_i(\bar{\mathbf{x}}) = \mathbf{0}, \qquad\qquad (7.54)$$

where the numbers λ_i are called Lagrange multipliers.   •

Note that, in general, a Lagrange multiplier for an equality-constraint problem can be of either sign. The requirement that the $\nabla g_i(\bar{\lambda})$ be linearly independent is called a *constraint qualification*.

The above Theorem 7.30 gives a condition that must necessarily be satisfied by any minimizing point $\bar{\mathbf{x}}$, namely,

$$\nabla f(\bar{\mathbf{x}}) + \sum_{i=1}^{m} \lambda_i \nabla g_i(\bar{\mathbf{x}}) = \mathbf{0}, \ \text{for} \ \ \lambda \in \mathbf{R}^m.$$

For fixed $\bar{\mathbf{x}}$, this vector equation is simply a system of linear equations in the variables λ_i $(i = 1, 2, \ldots, m)$. If the assumptions of Theorem 7.30 hold, and if there is no λ such that the preceding gradient equation holds at $\bar{\mathbf{x}}$, then the point $\bar{\mathbf{x}}$ cannot be a minimizing point.

Example 7.47 *Consider the following problem*

$$minimize \ \ z = f(x_1, x_2, x_3) = 20 + 2x_1 + 2x_2 + x_2^3$$

subject to

$$g_1(x_1, x_2, x_3) = x_1^2 + x_2^2 + x_3^2 - 11 = 0$$
$$g_2(x_1, x_2, x_3) = x_1 + x_2 + x_3 - 3 = 0.$$

Solution. *Given the objective function*

$$f(x_1, x_2, x_3) = 20 + 2x_1 + 2x_2 + x_3^2,$$

and the constraints

$$g_1(x_1, x_2, x_3) = x_1^2 + x_2^2 + x_3^2 - 11 = 0$$
$$g_2(x_1, x_2, x_3) = x_1 + x_2 + x_3 - 3 = 0,$$

the Lagrangian $L(x_1, x_2, x_3)$ *is defined as*

$$L(x_1, x_2, x_3) = 20 + 2x_1 + 2x_2 + x_3^2 + \lambda(x_1^2 + x_2^2 + x_3^2 - 11) + \mu(x_1 + x_2 + x_3 - 3).$$

Now calculate the gradients of $L, f, g_1,$ *and* $g_2,$ *and then set*

$$\nabla L(\mathbf{x}) = \nabla f(\mathbf{x}) + \lambda \nabla g_1(\mathbf{x}) + \mu \nabla g_2(\mathbf{x}) = 0,$$

which gives

$$\frac{\partial L}{\partial x_1} = 0 = 2 + 2\lambda x_1 + \mu$$

$$\frac{\partial L}{\partial x_2} = 0 = 2 + 2\lambda x_2 + \mu$$

$$\frac{\partial L}{\partial x_3} = 0 = 2x_3 + 2\lambda x_3 + \mu.$$

Writing the above system in matrix form, we have

$$\begin{pmatrix} 0 \\ 0 \\ 0 \end{pmatrix} = \begin{pmatrix} 2 \\ 2 \\ 2x_3 \end{pmatrix} + \lambda \begin{pmatrix} 2x_1 \\ 2x_2 \\ 2x_3 \end{pmatrix} + \mu \begin{pmatrix} 1 \\ 1 \\ 1 \end{pmatrix}.$$

Suppose that the feasible point $\bar{\mathbf{x}} = [-1, 3, 1]^T$ *is the minimizing point, then*

$$
\begin{pmatrix} 0 \\ 0 \\ 0 \end{pmatrix} = \begin{pmatrix} 2 \\ 2 \\ 2 \end{pmatrix} + \lambda \begin{pmatrix} -2 \\ 6 \\ 2 \end{pmatrix} + \mu \begin{pmatrix} 1 \\ 1 \\ 1 \end{pmatrix}
$$

or

$$
\begin{aligned}
0 &= 2 - 2\lambda + \mu \\
0 &= 2 + 6\lambda + \mu \\
0 &= 2 + 2\lambda + \mu.
\end{aligned}
$$

By adding the first and third equations, we get

$$
0 = 4 + 2\mu \qquad gives \quad \mu = -\frac{1}{2},
$$

and using this, we get

$$
\lambda = \frac{3}{4}.
$$

Putting the values of λ *and* μ *in the second equation, we see that it does not satisfy. Thus, there is no* (λ, μ) *such that this system has a solution. So* $\bar{\mathbf{x}}$ *cannot be the minimization point. Remember that these equations are only necessary conditions for a minimizing point. If a solution* λ *does exist for a given* $\bar{\mathbf{x}}$*, then that point* $\bar{\mathbf{x}}$ *may be or may not be a minimizing point.*

●

7.8.2 The Kuhn–Tucker Conditions

The KT conditions play an important role in the general theory of non-linear programming, and in particular, they are the conditions that must be used in solving problems with inequality constraints. In the Lagrange method, we found that Lagrangian multipliers could be utilized in solving equality-constrained optimization problems. Kuhn and Tucker have extended this theory to include the general *NLP* problem with both equality and inequality constraints. In the Lagrange method we used the gradient condition and the original equality constraint equations to find the stationary points of an equality-constrained problem. In a similar way, here we can use the gradient condition, the orthogonality condition, and the original constraint inequalities to find the stationary points for an inequality-constrained problem.

The KT conditions are first-order necessary conditions for a general constrained minimization problem written in the standard form

$$\text{minimize} \ \ z = f(\mathbf{x})$$

subject to

$$g_i(\mathbf{x}) \ \leq \ 0, \quad i = 1, 2, \ldots, m$$
$$h_i(\mathbf{x}) \ = \ 0, \quad i = 1, 2, \ldots, p$$

In the above standard form, the objective function is the minimization type, all constraint right-hand sides are zero, and the inequality constraints are the less-than type. Before applying the KT conditions, it is necessary to note that the given problem has been converted to this standard form.

The KT conditions represent a set of equations that must be satisfied for all local minimizing points of a considered minimization problem. Since they are only necessary conditions, the points that satisfy these conditions are only candidates for being a local minimum and are usually known as *KT points*. Then one can also check sufficient conditions to determine if a given KT point actually is a local minimum or not.

It is important to make sure that to apply the results of this section, all the *NLP* constraints must be $\leq$ constraints. A constraint of the form

$$h(x_1, x_2, \ldots, x_n) \geq 0$$

must be written as

$$-h(x_1, x_2, \ldots, x_n) \leq 0.$$

Also, a constraint of the form

$$h(x_1, x_2, \ldots, x_n) = 0$$

can be replaced by

$$h(x_1, x_2, \ldots, x_n) \leq 0$$

and

$$-h(x_1, x_2, \ldots, x_n) \leq 0.$$

For example,

$$x_1 + 3x_2 = 4$$

can be replaced by

$$x_1 + 3x_2 \leq 4$$

and

$$-x_1 - 3x_2 \leq -4.$$

7.8.3   Karush–Kuhn–Tucker Conditions

1. Form the Lagrangian function

$$L(\mathbf{x}, \lambda) = f(\mathbf{x}) + \sum_{i=1}^{m} \lambda_i g_i(\mathbf{x}).$$

2. Find all of the solutions $(\bar{\mathbf{x}}, \bar{\lambda})$ to the following system of nonlinear algebraic equations and inequalities:

$$\frac{\partial L}{\partial x_j} = 0, \quad j = 1, 2, \ldots, n \quad (gradient\ condition)$$

$$g_i(\mathbf{x}) \leq 0, \quad i = 1, 2, \ldots, m \quad (feasibility\ condition)$$

$$\lambda_i g_i(\mathbf{x}) = 0, \quad i = 1, 2, \ldots, m \quad (orthogonality\ condition)$$

$$\lambda_i \geq 0, \quad i = 1, 2, \ldots, m \quad (nonnegativity\ condition)$$

3. If the functions $g_i(\mathbf{x})$ are all convex, the point $(\bar{\mathbf{x}})$ is the global minimizing point. Otherwise, examine each solution $(\bar{\mathbf{x}}, \bar{\lambda})$ to see if $(\bar{\mathbf{x}})$ is a minimizing point.

Note that for each inequality constraint, we need to consider the following two possibilities.

Inactive Inequality Constraints: such constraints for which $g(\bar{\mathbf{x}}) < 0$. These constraints do not determine the optimum and hence are not needed in developing optimality conditions.

Active Inequality Constraints: such constraints for which $g(\bar{\mathbf{x}}) = 0$.

Now, we discuss necessary and sufficient conditions for $\bar{\mathbf{x}} = (\bar{x}_1, \bar{x}_2, \ldots, \bar{x}_n)$ to be an optimal solution for the following *NLP* problem:

$$\text{maximize } (or\ minimize)\ \ z = f(x_1, x_2, \ldots, x_n)$$

subject to

$$g_1(x_1, x_2, \ldots, x_n) \leq 0 \tag{7.55}$$
$$g_2(x_1, x_2, \ldots, x_n) \leq 0 \tag{7.56}$$
$$\vdots$$
$$g_m(x_1, x_2, \ldots, x_n) \leq 0$$

The following theorems give conditions (KT conditions) that are necessary for a point

$$\bar{\mathbf{x}} = (\bar{x}_1, \bar{x}_2, \ldots, \bar{x}_n)$$

to solve (7.55).

Theorem 7.31 (Necessary Conditions, Maximization Problem)

Suppose (7.55) is a maximum problem. If $\bar{\mathbf{x}} = (\bar{x}_1, \bar{x}_2, \ldots, \bar{x}_n)$ is an optimal solution to (7.55), then $\bar{\mathbf{x}} = (\bar{x}_1, \bar{x}_2, \ldots, \bar{x}_n)$ must satisfy the m constraints in (7.55), and there exist multipliers $\bar{\lambda} = (\bar{\lambda}_1, \bar{\lambda}_2, \ldots, \bar{\lambda}_m)$ satisfying

$$\frac{\partial f(\bar{\mathbf{x}})}{\partial x_j} - \sum_{i=1}^{m} \bar{\lambda}_i \frac{\partial g_i(\bar{\mathbf{x}})}{\partial x_j} = 0, \ \ j = 1, 2, \ldots, n \tag{7.57}$$

$$\bar{\lambda}_i[g_i(\bar{\mathbf{x}})] = 0, \ \ i = 1, 2, \ldots, m \tag{7.58}$$

$$\bar{\lambda}_i \geq 0, \ \ i = 1, 2, \ldots, m. \tag{7.59}$$

•

Theorem 7.32 (Necessary Conditions, Minimization Problem)

Suppose (7.55) is a minimization problem. If $\bar{\mathbf{x}} = (\bar{x}_1, \bar{x}_2, \ldots, \bar{x}_n)$ is an optimal solution to (7.55), then $\bar{\mathbf{x}} = (\bar{x}_1, \bar{x}_2, \ldots, \bar{x}_n)$ must satisfy the m constraints in (7.55), and there exist multipliers $\bar{\lambda} = (\bar{\lambda}_1, \bar{\lambda}_2, \ldots, \bar{\lambda}_m)$ satisfying

$$\frac{\partial f(\bar{\mathbf{x}})}{\partial x_j} + \sum_{i=1}^{m} \bar{\lambda}_i \frac{\partial g_i(\bar{\mathbf{x}})}{\partial x_j} = 0, \quad j = 1, 2, \ldots, n \quad (7.60)$$

$$\bar{\lambda}_i[g_i(\bar{\mathbf{x}})] = 0, \quad i = 1, 2, \ldots, m \quad (7.61)$$

$$\bar{\lambda}_i \geq 0, \quad i = 1, 2, \ldots, m. \quad (7.62)$$

●

In many situations, the KT conditions are applied to the *NLP* problem in which the variables must be nonnegative. For example, the KT-conditions can be used to find the optimal solution to the following *NLP* problem:

$$\text{maximize } (or \ minimize) \ \ z = f(x_1, x_2, \ldots, x_n)$$

subject to

$$g_1(x_1, x_2, \ldots, x_n) \leq 0$$
$$g_2(x_1, x_2, \ldots, x_n) \leq 0$$
$$\vdots$$
$$g_m(x_1, x_2, \ldots, x_n) \leq 0$$
$$-x_1 \leq 0$$
$$-x_2 \leq 0$$
$$\vdots$$
$$-x_n \leq 0.$$

If we associate multipliers $\mu_1, \mu_2, \ldots, \mu_n$ with these nonnegative constraints, Theorem 7.31 and Theorem 7.32 can be written as follows.

Theorem 7.33 (Necessary Conditions, Maximization Problem)

Consider a maximum problem:

$$\text{maximize (or minimize)} \quad z = f(x_1, x_2, \ldots, x_n)$$

subject to

$$g_1(x_1, x_2, \ldots, x_n) \leq 0$$
$$g_2(x_1, x_2, \ldots, x_n) \leq 0$$
$$\vdots$$
$$g_m(x_1, x_2, \ldots, x_n) \leq 0 \qquad (7.63)$$
$$-x_1 \leq 0$$
$$-x_2 \leq 0$$
$$\vdots$$
$$-x_n \leq 0.$$

If $\bar{\mathbf{x}} = (\bar{x}_1, \bar{x}_2, \ldots, \bar{x}_n)$ is a an optimal solution to (7.63), then $\bar{\mathbf{x}} = (\bar{x}_1, \bar{x}_2, \ldots, \bar{x}_n)$ must satisfy the m constraints in (7.63), and there exist multipliers $\bar{\lambda} = (\bar{\lambda}_1, \bar{\lambda}_2, \ldots, \bar{\lambda}_m, \bar{\mu}_1, \bar{\mu}_2, \ldots, \bar{\mu}_n)$ satisfying

$$\frac{\partial f(\bar{\mathbf{x}})}{\partial x_j} - \sum_{i=1}^{m} \bar{\lambda}_i \frac{\partial g_i(\bar{\mathbf{x}})}{\partial x_j} + \mu_j = 0, \quad j = 1, 2, \ldots, n \qquad (7.64)$$

$$\bar{\lambda}_i [g_i(\bar{\mathbf{x}})] = 0, \quad i = 1, 2, \ldots, m \qquad (7.65)$$

$$\left[\frac{\partial f(\bar{\mathbf{x}})}{\partial x_j} - \sum_{i=1}^{m} \bar{\lambda}_i \frac{\partial g_i(\bar{\mathbf{x}})}{\partial x_j} \right] \bar{\mathbf{x}} = 0 \qquad (7.66)$$

$$\bar{\lambda}_i \geq 0, \quad i = 1, 2, \ldots, m \qquad (7.67)$$
$$\bar{\mu}_j \geq 0, \quad j = 1, 2, \ldots, n. \qquad (7.68)$$

●

Since $\bar{\mu}_j \geq 0$, the first equation in the above system is equivalent to

$$\frac{\partial f(\bar{\mathbf{x}})}{\partial x_j} - \sum_{i=1}^{m} \bar{\lambda}_i \frac{\partial g_i(\bar{\mathbf{x}})}{\partial x_j} \leq 0, \quad j = 1, 2, \ldots, n.$$

Thus, the KT conditions for the above maximization problem with non-negative constraints may be written as

$$\frac{\partial f(\bar{\mathbf{x}})}{\partial x_j} - \sum_{i=1}^{m} \bar{\lambda}_i \frac{\partial g_i(\bar{\mathbf{x}})}{\partial x_j} \leq 0, \quad j = 1, 2, \ldots, n \qquad (7.69)$$

$$\bar{\lambda}_i[g_i(\bar{\mathbf{x}})] = 0, \quad i = 1, 2, \ldots, m \qquad (7.70)$$

$$\left[\frac{\partial f(\bar{\mathbf{x}})}{\partial x_j} - \sum_{i=1}^{m} \bar{\lambda}_i \frac{\partial g_i(\bar{\mathbf{x}})}{\partial x_j}\right] \bar{\mathbf{x}} = 0 \qquad (7.71)$$

$$\bar{\lambda}_i \geq 0, \quad i = 1, 2, \ldots, m. \qquad (7.72)$$

Theorem 7.34 (Necessary Conditions, Minimization Problem)

Suppose (7.69) is a minimization problem. If $\bar{\mathbf{x}} = (\bar{x}_1, \ldots, \bar{x}_n)$ is an optimal solution to (7.63), then $\bar{\mathbf{x}} = (\bar{x}_1, \ldots, \bar{x}_n)$ must satisfy the m constraints in (7.63), and there exist multipliers $\bar{\lambda} = (\bar{\lambda}_1, \ldots, \bar{\lambda}_m, \bar{\mu}_1, \bar{\mu}_2, \ldots, \bar{\mu}_n)$ satisfying

$$\frac{\partial f(\bar{\mathbf{x}})}{\partial x_j} - \sum_{i=1}^{m} \bar{\lambda}_i \frac{\partial g_i(\bar{\mathbf{x}})}{\partial x_j} - \mu_j = 0, \quad j = 1, 2, \ldots, n \qquad (7.73)$$

$$\bar{\lambda}_i[g_i(\bar{\mathbf{x}})] = 0, \quad i = 1, 2, \ldots, m \qquad (7.74)$$

$$\left[\frac{\partial f(\bar{\mathbf{x}})}{\partial x_j} + \sum_{i=1}^{m} \bar{\lambda}_i \frac{\partial g_i(\bar{\mathbf{x}})}{\partial x_j}(\bar{\mathbf{x}})\right] \bar{\mathbf{x}} = 0 \qquad (7.75)$$

$$\bar{\lambda}_i \geq 0, \quad i = 1, 2, \ldots, m \qquad (7.76)$$
$$\bar{\mu}_j \geq 0, \quad j = 1, 2, \ldots, n. \qquad (7.77)$$

•

Since $\bar{\mu}_j \geq 0$, the first equation in the above system is equivalent to

$$\frac{\partial f}{\partial x_j}(\bar{\mathbf{x}}) + \sum_{i=1}^{m} \bar{\lambda}_i \frac{\partial g_i}{\partial x_j}(\bar{\mathbf{x}}) \geq 0, \quad j = 1, 2, \ldots, n.$$

Thus, the KT conditions for the above maximization problem with non-negative constraints may be written as

$$\frac{\partial f(\bar{\mathbf{x}})}{\partial x_j} + \sum_{i=1}^{m} \bar{\lambda}_i \frac{\partial g_i(\bar{\mathbf{x}})}{\partial x_j}(\bar{\mathbf{x}}) \ \geq \ 0, \quad j = 1, 2, \ldots, n \qquad (7.78)$$

$$\bar{\lambda}_i [g_i(\bar{\mathbf{x}})] \ = \ 0, \quad i = 1, 2, \ldots, m \qquad (7.79)$$

$$\left[\frac{\partial f}{\partial x_j}(\bar{\mathbf{x}}) + \sum_{i=1}^{m} \bar{\lambda}_i \frac{\partial g_i}{\partial x_j}(\bar{\mathbf{x}}) \right] \bar{\mathbf{x}} \ = \ 0 \qquad (7.80)$$

$$\bar{\lambda}_i \ \geq \ 0, \quad i = 1, 2, \ldots, m. \qquad (7.81)$$

Theorems 7.31, 7.32, 7.33, and 7.34 give the necessary conditions for a point $\bar{\mathbf{x}} = (\bar{x}_1, \bar{x}_2, \ldots, \bar{x}_n)$ to be an optimal solution to (7.55) and (7.69). The following theorems give the sufficient conditions for a point $\bar{\mathbf{x}} = (\bar{x}_1, \bar{x}_2, \ldots, \bar{x}_n)$ to be an optimal solution to (7.55) and (7.69).

Theorem 7.35 (Sufficient Conditions, Maximization Problem)

Suppose (7.55) is a maximization problem. If $f(x_1, \ldots, x_n)$ is a concave function and $g_1(x_1, \ldots, x_n), \ldots, g_m(x_1, \ldots, x_n)$ are convex functions, then any point $\bar{\mathbf{x}} = (\bar{x}_1, \ldots, \bar{x}_n)$ satisfying the hypothesis of Theorem 7.31 is an optimal solution to (7.55). Also, if (7.63) is a maximization problem, $f(x_1, \ldots, x_n)$ is a concave function, and $g_1(x_1, \ldots, x_n), \ldots, g_m(x_1, \ldots, x_n)$ are convex functions, then any point $\bar{\mathbf{x}} = (\bar{x}_1, \ldots, \bar{x}_n)$ satisfying the hypothesis of Theorem 7.33 is an optimal solution to (7.63). ●

Theorem 7.36 (Sufficient Conditions, Minimization Problem)

Suppose (7.55) is a minimization problem. If $f(x_1, \ldots, x_n)$ is a convex function and $g_1(x_1, \ldots, x_n), \ldots, g_m(x_1, \ldots, x_n)$ are convex functions, then any point $\bar{\mathbf{x}} = (\bar{x}_1, \ldots, \bar{x}_n)$ satisfying the hypothesis of Theorem 7.32 is an optimal solution to (7.55). Also, if (7.63) is a minimization problem, $f(x_1, \ldots, x_n)$ is a convex function, and $g_1(x_1, \ldots, x_n), \ldots, g_m(x_1, \ldots, x_n)$ are convex functions, then any point $\bar{\mathbf{x}} = (\bar{x}_1, \ldots, \bar{x}_n)$ satisfying the hypothesis of Theorem 7.34 is an optimal solution to (7.63). ●

Example 7.48 *Minimize the function*

$$z = (x_1 - 2)^2 + (x_2 - 3)^2 + (x_3 - 3)^2,$$

subject to the inequality constraints

$$
\begin{aligned}
x_1 + x_2 - x_3 &\leq 1 \\
2x_1 - x_2 - 2x_3 &\leq 2.
\end{aligned}
$$

(a) *Write down the KT conditions for the problem.*
(b) *Find all the solutions of the KT conditions for the problem.*
(c) *Find all the local minimizing points.*

Solution. (a) *Given*

$$f(x_1, x_2, x_3) = (x_1 - 2)^2 + (x_2 - 3)^2 + (x_3 - 3)^2,$$

with constraints

$$
\begin{aligned}
g_1(x_1, x_2, x_3) &= x_1 + x_2 - x_3 - 1 \leq 0 \\
g_2(x_1, x_2, x_3) &= 2x_1 - x_2 - 2x_3 - 2 \leq 0,
\end{aligned}
$$

first, we form the Lagrangian function as follows:

$$L(\mathbf{x}, \lambda) = (x_1-2)^2+(x_2-3)^2+(x_3-3)^2+\lambda_1[x_1+x_2-x_3-1]+\lambda_2[2x_1-x_2-2x_3-2].$$

Write down the KT conditions:

$$\frac{\partial L}{\partial x_1} = 2(x_1 - 2) + \lambda_1 + 2\lambda_2 = 0$$

$$\frac{\partial L}{\partial x_2} = 2(x_2 - 3) + \lambda_1 - \lambda_2 = 0$$

$$\frac{\partial L}{\partial x_3} = 2(x_3 - 3) - \lambda_1 - 2\lambda_2 = 0$$

$$\frac{\partial L}{\partial \lambda_1} = g_1(\mathbf{x}) = x_1 + x_2 - x_3 - 1 \leq 0$$

$$\frac{\partial L}{\partial \lambda_2} = g_1(\mathbf{x}) = 2x_1 - x_2 - 2x_3 - 2 \leq 0$$

$$\lambda_1[x_1 + x_2 - x_3 - 1] = 0$$

$$\lambda_2[2x_1 - x_2 - 2x_3 - 2] = 0$$

$$\lambda_1 \geq 0, \quad \lambda_2 \geq 0.$$

(b) *Consider the four possible cases:*

$$\lambda_1 = 0, \qquad \lambda_2 = 0$$
$$\lambda_1 \neq 0, \qquad \lambda_2 \neq 0$$
$$\lambda_1 = 0, \qquad \lambda_2 \neq 0$$
$$\lambda_1 \neq 0, \qquad \lambda_2 = 0.$$

First Case: *When* $\lambda_1 = 0$, $\lambda_2 = 0$, *then using the set of equations we got from the gradient condition, we get*

$$2(x_1 - 2) = 0 \quad gives \quad x_1 = 2$$
$$2(x_2 - 3) = 0 \quad gives \quad x_2 = 3$$
$$2(x_3 - 3) = 0 \quad gives \quad x_3 = 3.$$

Putting these values of $x_1, x_2,$ and x_3 in the given first constraint, we have

$$x_1 + x_2 - x_3 - 1 = 2 + 3 - 3 - 1 = 1 \nleq 0.$$

Hence, this case does not hold.

Second Case: *When* $\lambda_1 \neq 0$, $\lambda_2 \neq 0$, *then again using the set of equations we got from the gradient condition, we get*

$$2(x_1 - 2) + \lambda_1 + 2\lambda_2 = 0 \quad gives \quad x_1 = \frac{1}{2}[4 - \lambda_1 - 2\lambda_2]$$

$$2(x_2 - 3) + \lambda_1 - \lambda_2 = 0 \quad gives \quad x_2 = \frac{1}{2}[6 - \lambda_1 + \lambda_2]$$

$$2(x_3 - 3) - \lambda_1 - 2\lambda_2 = 0 \quad gives \quad x_3 = \frac{1}{2}[6 + \lambda_1 + 2\lambda_2].$$

Since $\lambda_1 \neq 0$, $\lambda_2 \neq 0$, from the orthogonality condition, we get

$$\begin{aligned} x_1 + x_2 - x_3 - 1 &= 0 \\ 2x_1 - x_2 - 2x_3 - 2 &= 0. \end{aligned}$$

Now putting the values of x_1, x_2, and x_3 in this system, we get

$$\begin{aligned} 3\lambda_1 + 9\lambda_2 &= -14 \\ 3\lambda_1 + 3\lambda_2 &= 2. \end{aligned}$$

Solving this system gives

$$\lambda_1 = \frac{10}{3} \quad and \quad \lambda_2 = -\frac{8}{3}.$$

But from the nonnegativity condition, $\lambda_2 \geq 0$. So this case also does not hold.

Third Case: *When $\lambda_1 = 0$, $\lambda_2 \neq 0$, then using the gradient condition, we get*

$$2(x_1 - 2) + 2\lambda_2 = 0 \quad gives \quad x_1 = 2 - \lambda_2$$

$$2(x_2 - 3) - \lambda_2 = 0 \quad gives \quad x_2 = \frac{1}{2}[6 + \lambda_2]$$

$$2(x_3 - 3) - 2\lambda_2 = 0 \quad gives \quad x_3 = 3 + \lambda_2.$$

Since $\lambda_2 \neq 0$, then from the orthogonality condition, we get

$$2x_1 - x_2 - 2x_3 - 2 = 0.$$

Now using the values of x_1, x_2, and x_3, we get

$$2(2 - \lambda_2) - \frac{1}{2}(6 + \lambda_2) - 2(3 + \lambda_2) - 2 = 0, \quad \lambda_2 = -\frac{8}{7}.$$

So this case also does not hold.

Fourth Case: *When* $\lambda_1 \neq 0, \quad \lambda_2 = 0$ *then using the gradient condition, we get*

$$2(x_1 - 2) + \lambda_1 \;=\; 0 \quad gives \quad x_1 = \frac{1}{2}[4 - \lambda_1]$$

$$2(x_2 - 3) + \lambda_1 \;=\; 0 \quad gives \quad x_2 = \frac{1}{2}[6 - \lambda_1]$$

$$2(x_3 - 3) - \lambda_1 \;=\; 0 \quad gives \quad x_3 = \frac{1}{2}[6 + \lambda_1].$$

Since $\lambda_1 \neq 0$, *then from the orthogonality condition, we get*

$$x_1 + x_2 - x_3 - 1 = 0.$$

Now using the values of x_1, x_2, *and* x_3, *we get*

$$\frac{1}{2}(4 - \lambda_1) + \frac{1}{2}(6 - \lambda_1) - \frac{1}{2}(6 + \lambda_1) - 1 = 0, \quad \lambda_1 = \frac{2}{3}.$$

So this case holds and

$$\bar{\mathbf{x}} = \left[\frac{5}{3}, \frac{8}{3}, \frac{10}{3}\right]^T \quad and \quad \bar{\lambda} = \left[\frac{2}{3}, 0\right]^T$$

is the only KT point.

(c) *Now we will check whether the functions* f, g_1, g_2 *are convex or not. First, we check the convexity of the objective function as follows:*

$$f(x_1, x_2, x_3) = (x_1 - 2)^2 + (x_2 - 3)^2 + (x_3 - 3)^2$$

$$\frac{\partial f}{\partial x_1} \;=\; 2(x_1 - 2); \quad \frac{\partial f^2}{\partial x_1^2} = 2$$

$$\frac{\partial f}{\partial x_2} \;=\; 2(x_2 - 3); \quad \frac{\partial f^2}{\partial x_2^2} = 2$$

$$\frac{\partial f}{\partial x_3} \;=\; 2(x_3 - 3); \quad \frac{\partial f^2}{\partial x_3^2} = 2$$

$$\frac{\partial f^2}{\partial x_1 \partial x_2} \;=\; \frac{\partial f^2}{\partial x_1 \partial x_3} = \frac{\partial f^2}{\partial x_2 \partial x_3} = 0.$$

Hence, the Hessian matrix for the objective function is

$$H = \begin{pmatrix} 2 & 0 & 0 \\ 0 & 2 & 0 \\ 0 & 0 & 2 \end{pmatrix}.$$

By deleting rows (and columns) 1 and 2 of the Hessian matrix, we obtain the first-order principal minor $2 > 0$. By deleting rows (and columns) 1 and 3 of the Hessian matrix, we obtain the first-order principal minor $2 > 0$. By deleting rows (and columns) 2 and 3 of the Hessian matrix, we obtain the first-order principal minor $2 > 0$. By deleting row 1 and column 1 of the Hessian matrix, we find the second-order principal minor

$$|H_2| = \left| \begin{pmatrix} 2 & 0 \\ 0 & 2 \end{pmatrix} \right| = 4 - 0 = 4 > 0.$$

By deleting row 2 and column 2 of the Hessian matrix, we find the second-order principal minor

$$|H_2| = \left| \begin{pmatrix} 2 & 0 \\ 0 & 2 \end{pmatrix} \right| = 4 - 0 = 4 > 0.$$

By deleting row 3 and column 3 of the Hessian matrix, we find the second-order principal minor

$$|H_2| = \left| \begin{pmatrix} 2 & 0 \\ 0 & 2 \end{pmatrix} \right| = 4 - 0 = 4 > 0.$$

The third-order principal minor is simply the determinant of the Hessian matrix itself, which is

$$|H_3| = 8 > 0.$$

Because for all (x_1, x_2, x_3) all principal minors of the Hessian matrix are nonnegative, we have shown that $f(x_1, x_2, x_3)$ is a convex function.

Also, since the functions g_1 and g_2 are linear and thus convex by the definition of a convex function, hence all the functions are convex and the point $\bar{\mathbf{x}} = [\frac{5}{3}, \frac{8}{3}, \frac{10}{3}]^T$ is the global minimum.

Note for this problem there is one active constraint, $g_1(\bar{\mathbf{x}}) = 0$, so $\lambda_1 \neq 0$, and one inactive constraint, $g_2(\bar{\mathbf{x}}) < 0$, so $\lambda_2 \neq 0$. ●

7.9 Generalized Reduced-Gradient Method

Here, we will consider the equality-constrained problems and how they can be converted from the inequality-constrained problems. The equality-constrained problems have much importance in nonlinear optimization. As in the case of linear programming, many nonlinear optimizations are easily formulated in a such way that they contain equality constraints. Also, the theory of equality-constrained nonlinear programming leads naturally to a more complete theory that encompasses both equality and inequality constraints. The advantage of using equality constraints here is that it helps us to eliminate variables by the use of constraint equations from some variables in terms of the others. After eliminating the variables, we could easily solve the resulting reduced unconstrained problem by using a simple calculus method. The elimination of the variables will be very simple if the constraints are linear equations. But for the nonlinear case we can use Taylor's method to convert the given nonlinear constraint equations into linear constraint equations.

First, we consider the *NLP* problem having equality constraints in linear form.

Example 7.49 *Minimize the following function*

$$z = x_1^2 + x_2^2 + x_3^2 + x_4^2,$$

subject to the linear constraints

$$x_1 + 2x_2 + 3x_3 + 5x_4 = 10$$
$$x_1 + 2x_2 + 5x_3 + 6x_4 = 15.$$

Solution. *Given the objective function*

$$f(\mathbf{x}) = x_1^2 + x_2^2 + x_3^2 + x_4^2$$

and the linear constraints

$$g_1(\mathbf{x}) = x_1 + 2x_2 + 3x_3 + 5x_4 - 10$$
$$g_2(\mathbf{x}) = x_1 + 2x_2 + 5x_3 + 6x_4 - 15,$$

we will solve the given constraints for two of the variables in terms of the other two. Solving for x_1 and x_3 in terms of x_2 and x_4, we multiply the first constraint equation by 5 and the second constraint equation by 3 and subtract the results, which gives

$$x_1 = \frac{5}{2} - 2x_2 - \frac{7}{2}x_4.$$

Next, subtracting the two given constraint equations, we get

$$x_3 = \frac{5}{2} - \frac{1}{2}x_4.$$

Putting these two expressions for x_1 and x_3 into the given objective function, we obtain the new problem (called the reduced problem):

$$minimize \quad f(x_2, x_4) = \left(\frac{5}{2} - 2x_2 - \frac{7}{2}x_4\right)^2 + x_2^2 + \left(\frac{5}{2} - \frac{1}{2}x_4\right)^2 + x_4^2,$$

or it can be written as

$$minimize \quad f(x_2, x_4) = \frac{25}{2} - 10x_2 + 5x_2^2 - 20x_4 + 14x_2x_4 + \frac{27}{2}x_4^2.$$

One can note that this is an unconstraint problem now, and it can be solved by setting the first partial derivatives with respect to x_2 and x_4 equal to zero, i.e.,

$$\frac{\partial f}{\partial x_2} = -10 + 10x_2 + 14x_4 = 0$$

and

$$\frac{\partial f}{\partial x_4} = -20 + 14x_2 + 27x_4 = 0.$$

Thus, we have a linear system of the form

$$\begin{aligned} 10x_2 + 14x_4 &= 10 \\ 14x_2 + 27x_4 &= 20. \end{aligned}$$

Solving this system by taking $x_2 = (1 - \frac{14}{10}x_4)$ from the first equation and then putting this value in the second equation we get

$$14\left(1 - \frac{14}{10}\right) + 27x_4 = 20,$$

which gives $x_4 = \frac{30}{37}$, and it implies that $x_2 = -\frac{5}{37}$.

Now using these values of x_2 and x_4, we obtain the other two variables x_1 and x_3 as follows:

$$x_1 = \frac{5}{2} + 2\frac{5}{37} - \frac{7}{2}(\frac{30}{37}) = -\frac{5}{74}$$

and

$$x_3 = \frac{5}{2} - \frac{1}{2}(\frac{30}{37}) = \frac{155}{74}.$$

Thus, the optimal solution is

$$\bar{\mathbf{x}} = [-\frac{5}{74}, -\frac{5}{37}, \frac{155}{74}, \frac{30}{37}]^T.$$

Note that the main difference between this method and the Lagrange multipliers method is that this method is easy to solve for several constraint equations simultaneously if they are linear. The gradient of the objective function is called the reduced gradient, and the method is therefore called the reduced-gradient method.

Note that pivoting in a linear programming tableau can also be used to obtain the above result as follows:

x_1	x_2	x_3	x_4	constants
①	2	3	5	10
1	2	5	6	15

x_1	x_2	x_3	x_4	constants
1	2	3	5	10
0	0	②	1	5

x_1	x_2	x_3	x_4	constants
1	2	0	$\frac{7}{2}$	$\frac{5}{2}$
0	0	1	$\frac{1}{2}$	$\frac{5}{2}$

which gives

$$x_1 = \frac{5}{2} - 2x_2 - \frac{7}{2}x_4$$

$$x_3 = \frac{5}{2} - \frac{1}{2}x_4,$$

the same as above.   ●

Nonlinear Constraints

In the previous example, we solved the problem having linear constraints
by the gradient method. Now we will deal with a problem having nonlinear
constraints and consider the possibility of approximating such a problem
by a problem with linear constraints. To do this we expand each nonlinear
constraint function in a Taylor series and then truncate terms beyond the
linear one.

Example 7.50 *Minimize the function*

$$z = x_1 + 2x_2^2 + x_3 + x_4^2,$$

subject to the nonlinear constraints

$$x_1 + x_2^2 + x_3 + 2x_4 = 2$$
$$2x_1 + 2x_2 + x_3 + x_4^2 = 4.$$

Solution. *Given the objective function*

$$f(\mathbf{x}) = x_1 + 2x_2^2 + x_3 + x_4^2$$

and the nonlinear constraints

$$g_1(\mathbf{x}) = x_1 + x_2^2 + x_3 + 2x_4 - 2$$
$$g_2(\mathbf{x}) = 2x_1 + 2x_2 + x_3 + x_4^2 - 4,$$

*we know that the Taylor series approximation for a function of several
variables about given point* $\mathbf{x}^*$ *can be written as*

$$f(\mathbf{x}) = f(\mathbf{x}^*) + \nabla f(\mathbf{x}^*)^T \Delta \mathbf{x} + \frac{1}{2}\Delta \mathbf{x}^T \nabla^2 f(\mathbf{x}^*)\Delta \mathbf{x} + \cdots.$$

If we truncate terms after the second term, we obtain the linear approximation

$$f(\mathbf{x}) \approx f(\mathbf{x}^*) + \nabla f(\mathbf{x}^*)^T (\Delta \mathbf{x} \mathbf{x}^*),$$

which can also be written as

$$f(\mathbf{x}) \approx f(\mathbf{x}^*) + \nabla f(\mathbf{x}^*)^T (\mathbf{x} - \mathbf{x}^*).$$

With the help of this formula we can replace the inequality constraints by equality constraints that approximate the true constraints in the vicinity of the point $\mathbf{x}^*$ at which the linearization is performed:

$$g_1(\mathbf{x}) \approx [x_1^* + x_2^{*2} + x_3^* + 2x_4^* - 2] + [1, 2x_2^*, 1, 2] \begin{bmatrix} x_1 - x_1^* \\ x_2 - x_2^* \\ x_3 - x_3^* \\ x_4 - x_4^* \end{bmatrix}$$

or, it can be written as

$$g_1(\mathbf{x}) \approx x_1 + (2x_2^*)x_2 + x_3 + 2x_4 - [2x_2^{*2} + 2].$$

Doing the similar approximation for the second constraint function, we obtain

$$g_2(\mathbf{x}) \approx [2x_1^* + 2x_2^* + x_3^* + x_4^{*2} - 4] + [2, 2, 1, 2x_4^*] \begin{bmatrix} x_1 - x_1^* \\ x_2 - x_2^* \\ x_3 - x_3^* \\ x_4 - x_4^* \end{bmatrix}$$

or, it can be written as

$$g_2(\mathbf{x}) \approx 2x_1 + 2x_2 + x_3 + (2x_4^*)x_4 - [x_4^{*2} + 4].$$

This process is called the generalized reduced-gradient method and is used to solve a sequence of subproblems, each of which uses a linear approximation of the constraints. The first step of the method is to start with the initial given point and then, at each iteration of the method, the constraint linearization is recalculated at the point gotten from the previous iteration. After each iteration the approximated solution comes closer to the optimal point and the linearized constraints of the subproblems become better approximations to the original nonlinear constraints in the neighborhood of the optimal point. At the optimal point, the linearized problem has the same solution as the original nonlinear problem.

To apply the generalized reduced-gradient method, first we have to pick the starting point $\mathbf{x}^0 = [2, -1, 1, -1]^T$ *at which the first linearization can be performed. In the second step we use the approximation formulas already given to linearize the constraint functions at the starting point* $\mathbf{x}^0$ *and form the first approximate problem as follows:*

$$f^{(0)}(\mathbf{x}) = x_1 + 2x_2^2 + x_3 + x_4^2,$$

and the linear constraints are

$$g_1(\mathbf{x}) = x_1 - 2x_2 + x_3 + 2x_4 - 4$$
$$g_2(\mathbf{x}) = 2x_1 + 2x_2 + x_3 - 2x_4 - 5.$$

Now we solve the equality constraints of the approximate problem to express two of the variables in terms of the others. By selecting x_1 *and* x_3 *to be basic variables, we solve the linear system to write them in terms of the other variables* x_2 *and* x_4 *(nonbasic variables):*

$$x_1 = 1 - 4x_2 + 4x_4$$
$$x_3 = 3 + 6x_2 - 6x_4.$$

Putting the expressions for x_1 *and* x_3 *in the objective function, we get*

$$f^{(0)}(\mathbf{x}) = (1-4x_2+4x_4)+2x_2^2+(3+6x_2-6x_4)+x_4^2 = 4+2x_2+2x_2^2-2x_4+x_4^2.$$

Solving this unconstrained minimization by putting the first partial derivatives with respect to the nonbasic variables x_2 *and* x_4 *equal to zero, we obtain*

$$\frac{\partial f^{(0)}}{\partial x_2} = 2 + 4x_2 = 0 \quad gives \quad x_2 = -\frac{1}{2}$$

$$\frac{\partial f^{(0)}}{\partial x_4} = -2 + 2x_4 = 0 \quad gives \quad x_4 = 1.$$

Using these values in the above x_1 equation and x_3 equation, we get

$$x_1 = 1 - 4\left(-\frac{1}{2}\right) + 4(1) = 1 + 2 + 4 = 7$$

$$x_3 = 3 + 6\left(-\frac{1}{2}\right) - 6(1) = 3 - 3 - 6 = -6.$$

Thus, we get the new point, $\mathbf{x}^1 = [7, -\frac{1}{2}, -6, 1]^T$.

Similarly, using this new point $\mathbf{x}^1$, *we obtain the second approximate problem as follows:*

$$f^{(1)}(\mathbf{x}) = x_1 + 2x_2^2 + x_3 + x_4^2,$$

and the linear constraints are

$$g_1(\mathbf{x}) = x_1 - x_2 + x_3 + 2x_4 - 2.5$$
$$g_2(\mathbf{x}) = 2x_1 + 2x_2 + x_3 + 2x_4 - 5.$$

Again by selecting x_1 and x_3 to be basic variables, we solve the linear system and write them in terms of other nonbasic variables x_2 and x_4 as follows:

$$x_1 = 2.5 - 3x_2$$
$$x_3 = 4x_2 - 2x_4.$$

Putting the expressions for x_1 and x_3 in the objective function, we get

$$f^{(1)}(\mathbf{x}) = (2.5 - 3x_2) + 2x_2^2 + (4x_2 - 2x_4) + x_4^2 = 2.5 + x_2 + 2x_2^2 - 2x_4 + x_4^2.$$

Solving this unconstrained minimization, we get

$$\frac{\partial f^{(1)}}{\partial x_2} = 1 + 4x_2 = 0 \quad gives \quad x_2 = -\frac{1}{4}$$

$$\frac{\partial f^{(1)}}{\partial x_4} = -2 + 2x_4 = 0 \quad gives \quad x_4 = 1.$$

Using the values of x_2 and x_4, we get the values of the other variables as

$$x_1 = 2.5 - 3\left(-\frac{1}{4}\right) = 2.5 + 0.75 = 3.25$$

$$x_3 = 4\left(-\frac{1}{4}\right) - 2(1) = -1 - 2 = -3.$$

Thus, we get the other new point, $\mathbf{x}^2 = [3.5, -\frac{1}{4}, -3, 1]^T$.

Continuing to convert nonlinear constraint functions into linear functions at the new point, use the resulting system of linear equations to express two of the variables in terms of the other, substituting into the objective function to get a new reduced problem, solving the reduced problem for $\mathbf{x}^3$*, and so forth.*

We can also solve this problem by converting the given constraints for two of the variables in terms of the other two. So solving for x_1 *and* x_3 *in terms of* x_2 *and* x_4*, we obtain*

$$x_1 = 2 - x_2^2 - x_3 - 2x_4$$

and

$$x_3 = 4 - 2x_1 - 2x_2 - x_4^2.$$

Putting the x_1 *equation in the* x_3 *equation, we get*

$$x_3 = 2x_2 - 2x_2^2 - 4x_4 + x_4^2.$$

Now putting this value of x_3 *in the* x_1 *equation, we obtain*

$$x_1 = 2 + x_2^2 - 2x_2 + 2x_4 - x_4^2.$$

Using these new values of x_1 *and* x_3*, the given objective function becomes*

$$f(\mathbf{x}) = (2 + x_2^2 - 2x_2 + 2x_4 - x_4^2) + 2x_2^2 + (2x_2 - 2x_2^2 - 4x_4 + x_4^2) + x_4^2$$

or

$$f(\mathbf{x}) = 2 + x_2^2 - 2x_4 + x_4^2.$$

Setting the first partial derivatives with respect to x_2 *and* x_4 *equal to zero gives*

$$\frac{\partial f}{\partial x_2} = 2x_2 = 0$$

and

$$\frac{\partial f}{\partial x_4} = -2 + 2x_4 = 0.$$

By solving these two equations, we get

$$x_2 = 0 \quad and \quad x_4 = 1.$$

Using the values of x_2 and x_3, we get

$$x_1 = 2 + (0)^2 - 2(0) + 2(1) - (1)^2 = 2 + 2 - 1 = 3$$

and

$$x_3 = 2(0) - 2(0)^2 - 4(1) + (1)^2 = -4 + 1 = -3.$$

Thus, the optimal solution is

$$\bar{\mathbf{x}} = [3, 0, -3, 1]^T,$$

and the minimum value of the function is

$$f(\mathbf{x}) = (3) + 2(0)^2 + (-3) + (1)^2 = 1.$$

●

Procedure 7.6 (Generalized Reduced-Gradient Method)

1. *Start with the initial point $\mathbf{x}^{(0)}$.*

2. *Convert the nonlinear constraint functions into linear constraint functions using*

$$f(\mathbf{x}) \approx f^{(k)}(\mathbf{x}^{(k)}) + \nabla f^{(k)}(\mathbf{x}^{(k)})^T (\mathbf{x} - \mathbf{x}^{(k)}).$$

3. *Solve the linear constraint equations for the basic variables in terms of the other (nonbasic) variables.*

4. *Solve the unconstrained reduced problem for nonbasic variables.*

5. *Find the basic variables using the nonbasic variables from the linear constraints equations.*

6. *Repeat all the previous steps unless you get the desired accuracy.*

7.10   Separable Programming

In separable programming NLP problems are solved by approximating the nonlinear functions with piecewise linear functions and then solving the optimization problem through the use of a modified simplex algorithm of linear programming and, in special cases, the ordinary simplex algorithm.

Definition 7.17 (Separable Programming)

In using separable programming, a basic condition is that all given functions in the problem be separable, i.e.,

$$f(x_1, x_2, \ldots, x_n) = f_1(x_1) + f_2(x_2) + \cdots + f_n(x_n).$$

For example, the function

$$f(x_1, x_2) = x_1^3 - 4x_1^2 + 2x_1 + 2x_2^2 - 3x_2$$

is separable because

$$f(x_1, x_2) = f_1(x_1) + f_2(x_2),$$

where
$$f_1(x_1) = x_1^3 - 4x_1^2 + 2x_1 \quad and \quad f_2(x_2) = 2x_2^2 - 3x_2.$$

But the function

$$f(x_1, x_2) = x_1^2 - \cos(x_1 + x_2) + x_1 e^{x_2} + 2x_2^2$$

is not separable.

Sometimes the given nonlinear functions are not separable, but they can be made separable by approximate substitution. For example, the given nonlinear programming problem

$$maximize \quad z = x_1 e^{x_2}$$

is not separable, but it can be made separable by letting

$$w = x_1 e^{x_2},$$

then

$$\ln w = \ln x_1 + x_2.$$

Thus,

$$maximize \quad z = w,$$

subject to

$$\ln w = \ln x_1 + x_2$$

is called a separable programming problem. Note that the substitution assumes that x_1 is a positive variable. •

There are different ways to deal with the separable programming problems, but we will solve the problems by the McMillan method.

McMillan states that any continuous nonlinear and separable function $f(x_1, x_2, \ldots, x_n)$ can be approximated by a piecewise linear function and solved using linear programming solving techniques provided that the following condition is applied:

$$f(\mathbf{x}) \approx \bar{f} = \sum_{k=0}^{d} \lambda_{k1} f_{k1} + \sum_{k=0}^{d} \lambda_{k2} f_{k2} + \cdots + \sum_{k=0}^{d} \lambda_{kn} f_{kn},$$

where

$$\mathbf{x}_n = \sum_{k=0}^{d} \lambda_{kn} x_{kn},$$

and d is any suitable integer representing the number of segments into which the domain of x is divided, given that

1. $\displaystyle\sum_{k=0}^{d} \lambda_{kj} = 1, \quad j = 1, 2, \ldots, n,$

2. $\lambda_{kj} \geq 0, \quad k = 0, 1, \ldots d, \quad j = 1, 2, \ldots, n,$ and

3. no more than two of the λs that are associated with any one variable j are greater than zero, and if two are greater than zero, they must be adjacent.

Example 7.51 *Consider the NLP problem*

$$minimize \quad z = (x_1 - 2)^2 + 4(x_2 - 6)^2,$$

subject to

$$g(\mathbf{x}) \;=\; 6x_1 + 3(x_2 + 1)^2 \le 12$$
$$x_1 \;\ge\; 0, \quad x_2 \ge 0.$$

Solution. *Both the objective function and the constraint are separable functions because*

$$f(\mathbf{x}) = f_1(x_1) + f_2(x_2),$$

where

$$f_1(x_1) = (x_1 - 2)^2 \quad and \quad f_2(x_2) = 4(x_2 - 6)^2,$$

and also

$$g(\mathbf{x}) = g_1(x_1) + g_2(x_2),$$

where

$$g_1(x_1) = 6x_1 \quad and \quad g_2(x_2) = 4(x_2 - 6)^2.$$

Notice that both x_1 and x_2 enter the problem nonlinearly in the objective function and the constraint. Thus we must write both x_1 and x_2 in terms of the λs. But if the variables are linear throughout the entire problem, it is not necessary to write in terms of the λs; they can be used as the variables themselves.

First, we determine the domain d of interest for the variables x_1 and x_2. From the given constraints, the possible values for x_1 and x_2 are

$$0 \le x_1 \le 2 \quad and \quad 0 \le x_2 \le 1,$$

respectively. Dividing the domain of interest for x_1 and x_2 arbitrarily into two segments each and obtaining the grid points, the piecewise linear function to be used to approximate f is

$$\bar{f} = \sum_{k=0}^{2} \lambda_{k1} f_{k1} + \sum_{k=0}^{2} \lambda_{k2} f_{k2},$$

and the approximation function for g is

$$\bar{g} = \sum_{k=0}^{2} \lambda_{k1} g_{k1} + \sum_{k=0}^{2} \lambda_{k2} g_{k2}.$$

Evaluating both approximate functions gives

k	x_{k1}	f_{k1}	g_{k1}	x_{k2}	f_{k2}	g_{k2}
0	0	4	0	0	144	3
1	1	1	6	$\frac{1}{2}$	121	$\frac{27}{3}$
2	2	0	12	1	100	12

Now we solve the NLP problem

$$minimize \quad z = 4\lambda_{01} + \lambda_{11} + 144\lambda_{02} + 121\lambda_{12} + 100\lambda_{22},$$

subject to

$$
\begin{aligned}
6\lambda_{11} + 12\lambda_{21} + 3\lambda_{02} \ + \ \frac{27}{4}\lambda_{12} + 12\lambda_{22} &\leq 12 \\
\lambda_{01} + \lambda_{11} + \lambda_{21} &= 1 \\
\lambda_{02} + \lambda_{12} + \lambda_{22} &= 1 \\
\lambda_{ij} &\geq 0, \ i,j = 0,1,2.
\end{aligned}
$$

Note that this approximating problem to our original nonlinear problem is linear. Thus, we can solve this problem using the simplex algorithm of linear programming if we modified it to ensure that in any basic solution no more than two of the λs that are associated with either of the x_j variables are greater than zero and if two (rather than one) are greater than zero, then they must be adjacent.

λ_{01}	λ_{11}	λ_{21}	λ_{02}	λ_{12}	λ_{22}	$s1$	$a2$	$a3$	rhs
1	1	1	1	1	1	0	0	0	$w = 2$
0	6	12	3	$\frac{27}{4}$	12	1	0	0	$s1 = 12$
①	1	1	0	0	0	0	1	0	$a2 = 1$
0	0	0	1	1	1	0	0	1	$a3 = 1$

λ_{01}	λ_{11}	λ_{21}	λ_{02}	λ_{12}	λ_{22}	$s1$	$a3$	rhs
0	0	0	1	1	1	0	0	$w = 1$
0	6	12	3	$\frac{27}{4}$	12	1	0	$s1 = 12$
1	1	1	0	0	0	0	0	$\lambda_{01} = 1$
0	0	0	①	1	1	0	1	$a3 = 1$

λ_{01}	λ_{11}	λ_{21}	λ_{02}	λ_{12}	λ_{22}	$s1$	rhs
0	0	0	0	0	0	0	$w = 0$
0	6	12	0	$\frac{15}{4}$	9	1	$s1 = 9$
1	1	1	0	0	0	0	$\lambda_{01} = 1$
0	0	0	1	1	1	0	$\lambda_{02} = 1$

λ_{01}	λ_{11}	λ_{21}	λ_{02}	λ_{12}	λ_{22}	$s1$	rhs
-4	-1	0	-144	-121	-100	0	
0	6	12	0	$\frac{15}{4}$	9	1	$s1 = 9$
1	1	1	0	0	0	0	$\lambda_{01} = 1$
0	0	0	1	1	1	0	$\lambda_{02} = 1$

λ_{01}	λ_{11}	λ_{21}	λ_{02}	λ_{12}	λ_{22}	$s1$	rhs
0	3	4	0	23	44	0	$z = 148$
0	6	12	0	$\frac{15}{4}$	9	1	$s1 = 9$
1	1	1	0	0	0	0	$\lambda_{01} = 1$
0	0	0	1	1	①	0	$\lambda_{02} = 1$

λ_{01}	λ_{11}	λ_{21}	λ_{02}	λ_{12}	λ_{22}	$s1$	rhs
0	3	4	-44	-21	0	0	$z = 104$
0	6	⑫	-9	$-\frac{21}{4}$	0	1	$s1 = 0$
1	1	1	0	0	0	0	$\lambda_{01} = 1$
0	0	0	1	1	1	0	$\lambda_{22} = 1$

λ_{01}	λ_{11}	λ_{21}	λ_{02}	λ_{12}	λ_{22}	$s1$	rhs
0	1	0	-41	$-\frac{77}{4}$	0	$-\frac{1}{3}$	$z = 104$
0	$\boxed{\frac{1}{2}}$	1	$-\frac{3}{4}$	$-\frac{7}{16}$	0	$\frac{1}{2}$	$\lambda_{21} = 0$
1	$\frac{1}{2}$	0	$\frac{3}{4}$	$\frac{7}{16}$	0	$\frac{1}{12}$	$\lambda_{01} = 1$
0	0	0	1	1	1	0	$\lambda_{22} = 1$

λ_{01}	λ_{11}	λ_{21}	λ_{02}	λ_{12}	λ_{22}	$s1$	rhs
0	0	-2	$-\frac{79}{2}$	$-\frac{147}{8}$	0	$-\frac{1}{2}$	$z = 104$
0	1	2	$-\frac{3}{2}$	$-\frac{7}{8}$	0	$\frac{1}{6}$	$\lambda_{11} = 0$
1	0	-1	$\frac{6}{4}$	$\frac{14}{16}$	0	0	$\lambda_{01} = 1$
0	0	0	1	1	1	0	$\lambda_{22} = 1$

Since

$$\lambda_{11} = 0, \quad \lambda_{01} = 1, \quad \lambda_{22} = 1,$$

we have

$$
\begin{aligned}
x_1 &= \lambda_{01}x_{01} + \lambda_{11}x_{11} + \lambda_{21}x_{21} = 0 \\
x_2 &= \lambda_{02}x_{02} + \lambda_{12}x_{12} + \lambda_{22}x_{22} = 1.
\end{aligned}
$$

Hence,

$$f(\mathbf{x}) = (x_1 - 2)^2 + 4(x_2 - 6)^2 = 4 + 100 = 104.$$

7.11 Quadratic Programming

Quadratic programming is a technique that has a quadratic objective function. For example, for the two variables, the objective function must contain only terms in x_1^2, x_2^2, x_1, x_2, $x_1 x_2$, and a constant term. The constraints can be linear inequalities or equalities.

An *NLP* problem whose constraints are linear and whose objective function is the sum of the terms of the form $x_1^{k_1} x_2^{k_2} \ldots x_n^{k_n}$ (with each term having a degree of $2, 1$, or 0) is a *quadratic programming* problem.

There are several algorithms that can be used to solve quadratic programming problems. For solving such programming problems we describe Wolfe's method. The basic approach of this method is that all the variables must be nonnegative.

Example 7.52 *Solve the quadratic programming problem*

$$maximize \quad z = 6x_1 + 3x_2 - 4x_1x_2 - 2x_1^2 - 3x_2^2,$$

subject to

$$
\begin{aligned}
x_1 + x_2 &\leq 1 \\
2x_1 + 3x_2 &\leq 4 \\
x_1, x_2 &\geq 0.
\end{aligned}
$$

Solution. *The KT conditions may be written as:*

$$
\begin{aligned}
6 - 4x_1 - 4x_2 - \lambda_1 - 2\lambda_2 &\leq 0 \\
3 - 4x_1 - 6x_2 - \lambda_1 - 3\lambda_2 &\leq 0 \\
x_1 + x_2 &\leq 1 \\
2x_1 + 3x_2 &\leq 4 \\
x_1, x_2 &\geq 0.
\end{aligned}
$$

The objective function may be shown to be concave, so any point satisfying the KT conditions will solve this quadratic programming problem. Applying the excess variable e_1 for the first constraint (called the x_1 constraint), the excess variable e_2 for the second constraint (called the x_2 constraint), and the slack variable s_1 for the third constraint, and the slack variable s_2 for *the last constraint, we have*

$$
\begin{aligned}
6 - 4x_1 - 4x_2 - \lambda_1 - 2\lambda_2 - e_1 &= 0 \\
3 - 4x_1 - 6x_2 - \lambda_1 - 3\lambda_2 - e_2 &= 0 \\
x_1 + x_2 + s_1 &= 1 \\
2x_1 + 3x_2 + s_2 &= 4.
\end{aligned}
$$

All variables are nonnegative:

$$
\begin{aligned}
e_1 x_1 &= 0 \\
e_2 x_2 &= 0 \\
\lambda_1 s_1 &= 0 \\
\lambda_2 s_2 &= 0.
\end{aligned}
$$

Observe that with the exception of the last four equations, the KT conditions are all linear or nonnegative constraints. The last four equations are the complementary slackness conditions for this quadratic programming problem.

For general quadratic programming problems, the complementary slackness conditions may be verbally expressed by

"e_i from x_i constraints and x_i cannot both be slack, or the excess variable for the ith constraint and λ_i cannot both be positive."

To find a point satisfying the KT conditions (except for the complementary slackness conditions), Wolfe's method simply applies a modified version of Phase I of the Two-Phase simplex method. We first add an artificial variable to each constraint in the KT conditions that does not have an obvious basic variable, and then we attempt to minimize the sum of the artificial variables.

To ensure that the final solution (with all the artificial variables equal to zero) satisfies the above slackness conditions, Wolfe's method is modified by the simplex choice of the entering variable as follows:

1. *Never perform a pivotal that would make the e_i from the above jth constraint and x_i both basic variables.*

2. *Never perform a pivot that would make the slack (or excess) variable for the ith constraint and λ_i both basic variables.*

To apply Wolfe's method to the given problem, we have to the solve the LP problem

$$\text{minimize} \ \ w = a_1 + a_2,$$

subject to

$$4x_1 + 4x_2 + \lambda_1 + 2\lambda_2 - e_1 + a_1 \ = \ 6$$
$$4x_1 + 6x_2 + \lambda_1 + 3\lambda_2 - e_2 + a_2 \ = \ 3$$
$$x_1 + x_2 + s_1 \ = \ 1$$
$$2x_1 + 3x_2 + s_2 \ = \ 4$$
$$e_1x_1 = e_2x_2 = \lambda_1 s_1 \ = \ \lambda_2 s_2 = 0$$
$$x_1, x_2 \geq 0, \ \lambda_i \ \geq \ 0, \ i = 1, 2.$$

x_1	x_2	λ_1	λ_2	e_1	e_2	s_1	s_2	a_1	a_2	rhs
8	10	2	5	-1	-1	0	0	0	0	$w=9$
4	4	1	2	-1	0	0	0	1	0	$a_1=6$
4	(6)	1	3	0	-1	0	0	0	1	$a_2=3$
1	1	0	0	0	0	1	0	0	0	$s_1=1$
2	3	0	0	0	0	0	1	0	0	$s_2=4$

x_1	x_2	λ_1	λ_2	e_1	e_2	s_1	s_2	a_1	a_2	rhs
$\frac{4}{3}$	0	$\frac{1}{3}$	0	-1	$\frac{2}{3}$	0	0	0		$w=4$
$\frac{4}{3}$	0	$\frac{1}{3}$	0	-1	$\frac{2}{3}$	0	0	1		$a_1=4$
$(\frac{2}{3})$	1	$\frac{1}{6}$	$\frac{1}{2}$	0	$-\frac{1}{6}$	0	0	0		$x_2=\frac{1}{2}$
$\frac{1}{3}$	0	$-\frac{1}{6}$	$-\frac{1}{2}$	0	$\frac{1}{6}$	1	0	0		$s_1=\frac{1}{2}$
0	0	$-\frac{1}{2}$	$-\frac{3}{2}$	0	$\frac{1}{2}$	0	1	0		$s_2=\frac{5}{2}$

x_1	x_2	λ_1	λ_2	e_1	e_2	s_1	s_2	a_1	a_2	rhs
0	-2	0	-1	-1	1	0	0	0		$w=3$
0	-2	0	-1	-1	1	0	0	1		$a_1=3$
1	$\frac{3}{2}$	$\frac{1}{4}$	$\frac{3}{4}$	0	$-\frac{1}{4}$	0	0	0		$x_1=\frac{3}{4}$
0	$-\frac{1}{2}$	$-\frac{1}{4}$	$-\frac{3}{4}$	0	$\left(\frac{1}{4}\right)$	1	0	0		$s_1=\frac{1}{4}$
0	0	$-\frac{1}{2}$	$-\frac{3}{2}$	0	$\frac{1}{2}$	0	1	0		$s_2=\frac{5}{2}$

x_1	x_2	λ_1	λ_2	e_1	e_2	s_1	s_2	a_1	a_2	rhs
0	0	1	2	-1	0	-4	0	0		$w=2$
0	0	1	(2)	-1	0	-4	0	1		$a_1=2$
1	1	0	0	0	0	1	0	0		$x_1=1$
0	-2	-1	-3	0	1	4	0	0		$e_2=1$
0	1	0	0	0	0	-2	1	0		$s_2=2$

x_1	x_2	λ_1	λ_2	e_1	e_2	s_1	s_2	rhs
0	0	0	0	0	0	0	0	$w=2$
0	0	$\left(\frac{1}{2}\right)$	1	$-\frac{1}{2}$	0	-2	0	$\lambda_2=1$
1	1	0	0	0	0	1	0	$x_1=1$
0	-2	$\frac{1}{2}$	0	$-\frac{3}{2}$	1	-2	0	$e_2=4$
0	1	0	0	0	0	-2	1	$s_2=2$

x_1	x_2	λ_1	λ_2	e_1	e_2	s_1	s_2	rhs
0	0	0	0	0	0	0	0	$w=0$
0	0	1	2	-1	0	-4	0	$\lambda_1=2$
1	1	0	0	0	0	1	0	$x_1=1$
0	-2	0	-1	-1	1	0	0	$e_2=3$
0	1	0	0	0	0	-2	1	$s_2=2$

We note from the last tableau that $w = 0$, so we have found a solution that satisfies the KT conditions and is optimal for the quadratic programming problem. Thus, the optimal solution to the quadratic programming problem is

$$x_1 = 1 \quad and \quad x_2 = 0.$$

We also note that

$$\lambda_1 = 2, \quad \lambda_2 = 0, \quad e_1 = 0, \quad e_2 = 3, \quad s_1 = 0, \quad s_2 = 2,$$

which satisfies

$$e_1 x_1 = 0, \quad e_2 x_2 = 0, \quad \lambda_1 s_1 = 0, \quad \lambda_2 s_2 = 0.$$

Note that Wolfe's method is guaranteed to obtain the optimal solution to a quadratic programming problem if all the leading principals of minors of the objective function's Hessian are positive. Otherwise, the method may not converge in a finite number of pivots. •

7.12   Summary

Nonlinear programming is a very vast subject and in this chapter we gave a brief introduction to the idea of nonlinear programming problems. We started with a review of differential calculus. Classical optimization theory uses differential calculus to determine points of extrema (maxima and minima) for unconstrained and constrained problems. The methods may not be suitable for efficient numerical problems, but they provide the theory that is the basis for most nonlinear programming methods. The solution methods for the nonlinear programming problem were discussed, including direct and indirect methods. For the one-dimensional optimization problem solution we used three indirect numerical methods. First, we used one of the direct search methods called golden-section search, which helped us identify the interval of uncertainty that is known to include the optimum solution point. This method locates the optimum by iteratively decreasing the interval of uncertainty to any given accuracy. The other two one-dimensional methods we discussed are the quadratic interpolation method and Newton's method. Both are fast convergence methods compared with the golden-section search method. In the case of direct methods, we discussed gradient methods where the maximum (minimum) of a problem is found following the fastest rate of increase (decrease) of the objective function.

We also discussed necessary and sufficient conditions for determining extremum, the Lagrange method for problems with equality constraints, and the Karush–Kuhn–Tucker (KT) conditions for problems with inequality constraints. The KT conditions provide the most unifying theory for all nonlinear programming problems. In indirect methods, the original problem is replaced by an auxiliary one from which the optimum is determined. For such cases we used quadratic programming (Wolfe's method) and separable programming (McMillan method).

This chapter contained many examples which we solved numerically and graphically and using MATLAB.

7.13 Problems

1. Find the following limits as x approaches 0:

   (a) $\dfrac{\sin x^2}{x \tan x}$.

   (b) $\dfrac{1 + \cos^2 x + 3x^2 - 2\cos x}{x^2}$.

   (c) $\dfrac{\sqrt{x+2} - \sqrt{2}}{x}$.

   (d) $\dfrac{\sin x - 1 + \cos x}{x}$.

2. Find the constants a and b so that the following function is continuous on the entire real line:

(a)

$$f(x) = \begin{cases} 3x + 1, & \text{if } x \leq -1 \\ ax + b, & \text{if } -1 < x < 4 \\ x + 4, & \text{if } x \geq 4. \end{cases}$$

(b)

$$f(x) = \begin{cases} x + 1, & \text{if } 1 < x < 3 \\ x^2 + ax + b, & \text{if } |x - 2| \geq 1. \end{cases}$$

(c)

$$f(x) = \begin{cases} ax - b, & \text{if } x \leq -2 \\ \dfrac{x^2 - 4}{x + 2}, & \text{if } -2 < x < 1 \\ x^2 + b, & \text{if } x \geq 1. \end{cases}$$

(d)

$$f(x) = \begin{cases} \dfrac{2\sin 2x}{x}, & \text{if } x < 0 \\ a - 3x, & \text{if } x > 0 \\ a = b, & \text{if } x = 0. \end{cases}$$

3. Find the third derivatives of the following functions:

   (a) $f(x) = x^{3/4} + x^2 + 5x + 3.$

   (b) $f(x) = \left(x - \dfrac{1}{x}\right).$

   (c) $f(x) = x^3 + x\cos x + x \ln x + 3.$

   (d) $f(x) = x^2 + \tan x + e^{3x}.$

4. Find the local extrema using the second derivative test, and find the point of inflection of the following functions:

(a) $f(x) = x^3 - 2x^2 + x + 3.$

(b) $f(x) = x^4 - 2x^2.$

(c) $f(x) = x^3 \left(1 - \dfrac{12}{x^2} + \dfrac{1}{x^3} \right).$

(d) $f(x) = 2 - 15x + 9x^2 - x^3.$

5. Find the second partial derivatives of the following functions:

(a) $z = f(x, y) = e^{-xy} \cos y.$

(b) $z = f(x, y) = \dfrac{1}{2} (e^x - e^{-y}) \sin xy.$

(c) $z = f(x, y, z) = 4x^2y - 6xyz^2 + 3yz^2.$

(d) $z = f(x, y, z) = e^{-2x} \sin yz.$

6. Find the directional derivative of the following functions at the indicated point in the direction of the indicated vector:

(a) $z = f(x, y, z) = x^2 + y^2 + z^2 + xz + yz,\ P(1, 1, 1),\ \mathbf{u} = (2, -1, 3).$

(b) $z = f(x, y, z) = \cos xy + e^{2z},\ P(0, -1, 2),\ \mathbf{u} = (0, 2, 3).$

(c) $z = f(x, y, z) = x^3 + y^3 + z^3 + xyz,\ P(2, 1, 2),\ \mathbf{u} = (0, -1, 1).$

(d) $z = f(x, y, z) = \sin x + e^{yz},\ P(0, 3, 3),\ \mathbf{u} = (2, -1, 2).$

7. Find the gradient of the following functions at the indicated points:

(a) $z = f(x, y, z) = x^3 - yz^2 + x^2z + y^2z^3,\ (1, 1, 1).$

(b) $z = f(x, y, z) = \cos xy + xye^{2z},\ (0, 1, 2).$

(c) $z = f(x, y, z) = \ln(x^3 + y^3 + z^3), (1, 2, 3)$.

(d) $z = f(x, y, z) = \tan x e^{yz}, (0, 1, 1)$.

8. Find the Hessian matrix for the following functions:

   (a) $f(x, y) = (x - 3)^2 + 2y^2 + 5$.

   (b) $f(x, y) = x^4 + 3y^4 - 2x^2 y + xy^2 - x - y$.

   (c) $f(x, y, z) = x^2 + 2y^2 + 4z^2 - 2x - 3yz$.

   (d) $f(x, y, z) = x^3 + 3y^2 + 2z^2 - 2xy + yz$.

9. Find the Hessian matrix for the following functions at the indicated points:

   (a) $f(x, y) = x^4 + y^4 + 5xy, P(1, -1)$.

   (b) $f(x, y) = 3x^5 + y^5 + 3x^3 y^3 - x^2 y + 2x - 3y, P(2, 1)$.

   (c) $f(x, y, z) = x^3 + 4y^3 + 2z^3 + 12x + 5yz, P(1, 2, 2)$.

   (d) $f(x, y, z) = 2x^4 + 6y^4 + 2z^4 - 2x^2 y + 3y^2 z, P(-2, 3, -2)$.

10. Find the linear and quadratic approximations of the following functions at the given point (a, b) using Taylor's series formulas:

    (a) $f(x, y) = \ln(x^2 + y^2), (1, 1)$.

    (b) $f(x, y) = x^2 + xy + y^2, (1, 1)$.

    (c) $f(x, y) = x^2 + y^2 + e^{xy}, (1, 1)$.

    (d) $f(x, y) = \cos xy, (0, 0)$.

11. Find the quadratic forms of the associated matrices:

$$\text{(a)} \quad A = \begin{pmatrix} 2 & 3 \\ 3 & 4 \end{pmatrix}. \qquad \text{(b)} \quad A = \begin{pmatrix} 4 & -1 \\ -1 & 3 \end{pmatrix}.$$

$$\text{(c)} \quad A = \begin{pmatrix} 2 & 4 & -2 \\ 4 & 6 & 0 \\ -2 & 0 & 5 \end{pmatrix}. \qquad \text{(d)} \quad A = \begin{pmatrix} 4 & 3 & -1 \\ 3 & 5 & 1 \\ -1 & 1 & 3 \end{pmatrix}.$$

12. Find the matrices associated with each of the following quadratic forms:

(a) $q(x, y) = 5x^2 + 7y^2 + 12xy.$

(b) $q(x, y) = 3x^2 + 2y^2 - 4xy.$

(c) $q(x, y, z) = 7x^2 + 6y^2 + 2z^2 - 3xy - 3xz - 3yz.$

(d) $q(x, y, z) = x^2 + 2y^2 + 3z^2 - 2xy + 2xz + 2yz.$

13. Classify each of the following quadratic forms as positive-definite, negative-definite, indefinite, or none of these:

(a) $q(x, y) = 3x^2 + 4y^2 - 2xy.$

(b) $q(x, y) = 4x^2 + 2y^2 + 4xy.$

(c) $q(x, y, z) = 3x^2 + 4y^2 + 5z^2 - 2xy - 2xz - 2yz.$

(d) $q(x, y, z) = 4x^2 + 4y^2 + 4z^2 - 4xy + xz + 2yz.$

14. Use the bisection method to find solutions accurate to within 10^{-4} on the indicated interval of the following functions:

(a) $f(x) = x^5 - 2x^2 + x + 1, [-1, 1].$

(b) $f(x) = x^5 - 4x^2 + x + 1, [0, 1].$

(c) $f(x) = x^6 - 7x^4 - 2x^2 + 1, [2, 3]$.

(d) $f(x) = e^x - 3x^2 + x + 1, [3, 4]$.

15. Use the bisection method to find solutions accurate to within 10^{-4} on the indicated intervals of the following functions:

(a) $f(x) = x^3 - 8x^2 + 4, [7, 8]$.

(b) $f(x) = x^3 + 2x^2 + x - 3, [0, 1]$.

(c) $f(x) = x^4 + 2x^3 - 7x^2 - x + 2, [1, 2]$.

(d) $f(x) = \ln x + 2x^5 + x - 3, [0.5, 1.5]$.

16. Use Newton's method to find a solution accurate to within 10^{-4} of Problem 14 using the suitable initial approximation.

17. Use Newton's method to find a solution accurate to within 10^{-4} using the given initial approximations of the following functions:

(a) $f(x) = x^3 + 2x^2 - 5, x_0 = 1.5$.

(b) $f(x) = x^3 - 5x^2 + 3x - 2, x_0 = 4.5$.

(c) $f(x) = x^4 - 3x^3 - 4x^2 + 3x + 5, x_0 = 3.5$.

(d) $f(x) = e^x - x^2/2 + x + 1, x_0 = -0.5$.

18. Use the fixed-point iteration to find a solution accurate to within 10^{-4} using the given intervals and initial approximations of the following functions:

(a) $f(x) = 2x^3 - 5x + 2, [1, 2], x_0 = 1.5$.

(b) $f(x) = 3x^3 - 7x^2 - 2x + 1, [0, 1], x_0 = 0.5$.

(c) $f(x) = x^4 - 3x^3 + 5x^2 - 6x + 1$, $[1, 2]$, $x_0 = 1.5$.

(d) $f(x) = e^x - 4x + 1$, $[0, 1]$, $x_0 = 0.5$.

19. Solve the following system by Newton's method using the indicated initial approximation (x_0, y_0) and stop when successive iterates differ by less than 10^{-4}:

(a) $(x_0, y_0) = (1, 1)$

$$
\begin{aligned}
2x^3 + y &= 4, \\
x^2 y &= 2.
\end{aligned}
$$

(b) $(x_0, y_0) = (-1, -1)$

$$
\begin{aligned}
x + y^2 &= -2, \\
2x^2 + 2y &= -4.
\end{aligned}
$$

(c) $(x_0, y_0) = (-1, 1)$

$$
\begin{aligned}
x^2 + y^2 &= 4, \\
x^2 + y &= 3.
\end{aligned}
$$

(d) $(x_0, y_0) = (0, 3)$

$$
\begin{aligned}
x + e^y &= 40, \\
\sin x - y &= -2.
\end{aligned}
$$

20. Solve the following system by fixed-point iterations using the indicated initial approximation (x_0, y_0) and stop when successive iterates differ by less than 10^{-4}:

(a) $(x_0, y_0) = (3, -1.5)$

$$
\begin{aligned}
2x^3 + y &= 4, \\
xy &= 1.
\end{aligned}
$$

(b) $(x_0, y_0) = (1, 1)$

$$
\begin{aligned}
x^2 - 2y^2 &= 12, \\
xy^2 &= 2.
\end{aligned}
$$

(c) $(x_0, y_0) = (0, 0.5)$

$$\begin{aligned} x^2 + y &= 4, \\ x + y^2 &= 3. \end{aligned}$$

(d) $(x_0, y_0) = (0, 1)$

$$\begin{aligned} 3x + e^y &= 4, \\ x^2 + y &= 2. \end{aligned}$$

21. Find the maximum value of the following functions using accuracy $\epsilon = 0.005$ by the golden-section search:

    (a) $f(x) = 3\cos x - 2x^3$, $[0, 2]$.

    (b) $f(x) = \sin x - \dfrac{x^2}{5}$, $[0, 2]$.

    (c) $f(x) = \dfrac{1}{5}x^3 - 2x^2 - 1.75x + 1.5$, $[0, 1]$.

    (d) $f(x) = x^6 - 5x^5 + 3x^4 + 2x^2 + 4$, $[2, 4]$.

22. Find the extrema of the following functions using accuracy $\epsilon = 0.005$ by the quadratic interpolation method for optimization:

    (a) Minimize $f(x) = x^3 + \dfrac{3}{2}x^2 - 6x + 5$, $x_0 = 0.0$, $x_1 = 0.5$, $x_2 = 1.5$.

    (b) Maximize $f(x) = x^3 - 3x^2 - 9x + 2$, $x_0 = -2.0$, $x_1 = 0.0$, $x_2 = 1.0$.

    (c) Maximize $f(x) = x^4 - 8x^3 + 18x^2 - 16x + 5$, $x_0 = 3.9$, $x_1 = 4.1$, $x_2 = 4.5$.

    (d) Minimize $f(x) = x^5 - 2x^4 - x^3$, $x_0 = 0.2$, $x_1 = 0.5$, $x_2 = 1.0$.

23. Find the extrema of the following functions using accuracy $\epsilon = 0.005$ by the quadratic interpolation method for optimization:

    (a) Minimize $f(x) = x^3 + 3x^2 - 15x - 5$, $x_0 = 2.0$, $x_1 = 2.5$, $x_2 = 3.5$.

(b) Maximize $f(x) = \dfrac{1}{3}x^3 + 2x^2 - 12x + 1$, $x_0 = -4.0$, $x_1 = -5.5$, $x_2 = -7.0$.

(c) Maximize $f(x) = x^4 - 2x^2 - 12$, $x_0 = 1.0$, $x_1 = 1.5$, $x_2 = 2.5$.

(d) Minimize $f(x) = x^5 - 20x^3 + 1$, $x_0 = 1.0$, $x_1 = 1.5$, $x_2 = 2.5$.

24. Find the extrema of the following functions using accuracy $\epsilon = 0.005$ by Newton's method for optimization:

(a) Minimize $f(x) = x^3 + 3x^2 - 45x + 8$, $x_0 = 2.5$.

(b) Maximize $f(x) = x^3 + 3x^2 - 24x + 6$, $x_0 = -3.5$.

(c) Maximize $f(x) = x^4 - 4x^3 - 20x^2 + 4x + 8$, $x_0 = 2.5$.

(d) Minimize $f(x) = x^{2/3}(x - 6)^2 + 4$, $x_0 = 5.5$.

25. Find the extrema of the following functions using accuracy $\epsilon = 0.005$ by Newton's method for optimization:

(a) Minimize $f(x) = 5 - 6x^2 - 2x^3$, $x_0 = 0.5$.

(b) Minimize $f(x) = x^3 - 13x^2 - 10x + 1$, $x_0 = -1.0$.

(c) Minimize $f(x) = x^4 - 8x^2 + 1$, $x_0 = -0.5$.

(d) Maximize $f(x) = x(x - 4)^3$, $x_0 = 0.5$.

26. Find the extrema and saddle points of each of the following functions:

(a) $z = f(x, y) = 2x^3 - 3x^2y + 6x^2 - 6y^2$.

(b) $z = f(x, y) = xy + \ln x + y^2 - 10$.

(c) $z = f(x, y) = 3x^2 + x^2y + 4xy + y^2$.

(d) $z = f(x, y) = 1 + 4xy - 2x^2 - 2y^2.$

27. Find the extrema and saddle points of each of the following functions:

    (a) $z = f(x, y) = 4xy - x^4 - y^4.$

    (b) $z = f(x, y) = x^2y - 6y^2 - 3x^2.$

    (c) $z = f(x, y) = x^3 + y^3 - 6y^2 - 3x + 9.$

    (d) $z = f(x, y) = x^2 + y^2 - 3xy + 4x - 2y + 5.$

28. Use the method of steepest ascent to approximate (up to given iterations) the optimal solution to the following problems:

    (a)   Maximize $z = x^2 - 2y^2 + 2xy + 3y;$ $(1.0, 1.0), iter. = 2.$

    (b)   Maximize $z = 2x^2 + 2y^2 - 2xy - 2x;$ $(0.5, 0.5), iter. = 2.$

    (c)   Maximize $z = -(x - 3)^2 + (y + 1)^2 - 2x;$ $(1.5, 1.0), iter. = 2.$

    (d)   Maximize $z = (x - 2)^2 - 4y^2 - 3xy + y;$ $(2.0, 1.5), iter. = 2.$

29. Use the method of steepest descent to approximate (up to the given iterations) the optimal solution to the following problems:

    (a)   Minimize $z = 2x^2 + 2y^2 + 2xy - y;$ $(1.0, 1.0), iter. = 2.$

    (b)   Minimize $z = x^2 - 3y^2 - 4xy + 2x - y;$ $(2.5, 1.5), iter. = 2.$

    (c)   Minimize $z = (x + 2)^2 + (y - 1)^2 - y;$ $(0.0, 1.0), iter. = 2.$

    (d)   Minimize $z = (x - 1)^2 + 2y^2 + xy - x;$ $(0.5, 1.5), iter. = 2.$

30. Solve Problem 29 using Newton's method.

31. Use Lagrange multipliers to find the extrema of function f subject to the stated constraints:

(a) Maximize $z = f(x, y) = \sqrt{6 - x^2 - y^2}$
     Subject to $g(x, x) = x + y - 2 = 0$

(b) Maximize $z = f(x, x) = x^{2/3} y^{1/3}$
     Subject to $g(x, y) = x + y - 4 = 0$

(c) Minimize $w = f(x, y, z) = xy + 2xz + 2yz$
     Subject to $g(x, y, z) = xyz - 12 = 0$

(d) Minimize $w = f(x, y, z) = x^4 + y^4 + z^4$
     Subject to $g(x, y, z) = x + y + z - 1 = 0$

32. Use Lagrange multipliers to find the extrema of function f subject to the stated constraints:

(a) Minimize $z = f(x, y) = x^2 - y^2$
     Subject to $g(x, y) = x - 2y + 6 = 0$

(b) Maximize $z = f(x, y) = 4x^3 + y^2$
     Subject to $g(x, y) = 2x^2 + y^2 - 1 = 0$

(c) Minimize $w = f(x, y, z) = 2x + y + 2z$
     Subject to $g(x, y, z) = x^2 + y^2 + z^2 - 4 = 0$

(d) Minimize $z = f(x, y, z) = x^2 + y^2 + z^2$
     Subject to $g(x, y, z) = x + y + z - 6 = 0$

33. Use Lagrange multipliers to find the extrema of function f subject to the stated constraints:

(a) $w = f(x, y, z) = x^2 + y^2 + z^2$
     Subject to
$$g_1(x, y, z) = x + y + z - 12 = 0$$
$$g_2(x, y, z) = x^2 + y^2 - z = 0$$

(b) $w = f(x, y, z) = x + y + z$
Subject to
$$g_1(x, y, z) = x^2 - y^2 - z = 0$$
$$g_2(x, y, z) = x^2 + z^2 - 4 = 0$$

(c) $w = f(x, y, z) = x + 2y$
Subject to
$$g_1(x, y, z) = x + y + z - 1 = 0$$
$$g_2(x, y, z) = x^2 + z^2 - 4 = 0$$

(d) $w = f(x, y, z) = xy + yz$
Subject to
$$g_1(x_1, x_2, x_3) = xy - 1 = 0$$
$$g_2(x, y, z) = y^2 + z^2 - 1 = 0$$

34. Use Lagrange multipliers to find the extrema of function f subject to the stated constraints:

(a) Minimize $w = f(x, y, z) = x^2 + y^2 + z^2$
Subject to
$$g_1(x, y, z) = x + 2y + 3z - 6 = 0$$
$$g_2(x, y, z) = y + z = 0$$

(b) Maximize $w = f(x, y, z) = xyz$
Subject to
$$g_1(x, y, z) = x + y + z - 4 = 0$$
$$g_2(x, y, z) = x + y - z = 0$$

(c) Maximize $w = f(x, y, z) = x^2 + yz$
Subject to
$$g_1(x, y, z) = x + y + z - 5 = 0$$
$$g_2(x, y, z) = x + y - z - 2 = 0$$

(d) Minimize $w = f(x, y, z) = y^2 e^{(x^2 - z)}$
Subject to

$$g_1(x, y, z) = 9x^2 + 4y^2 + 36z^2 - 36 = 0$$
$$g_2(x_1, x_2, x_3) = xy + yz - 1 = 0$$

35. Use Lagrange multipliers to find the extrema of function f subject to the stated constraints:

   (a) $w = f(x, y, z) = x^2 + y^2 + z^2$
   Subject to
   $$g_1(x, y, z) = x - y - 1 = 0$$
   $$g_2(x, y, z) = y^2 - z^2 - 1 = 0$$

   (b) $w = f(x, y, z) = x^2 + y^2 + z^2$
   Subject to
   $$g_1(x, y, z) = x^2 + y^2 - z^2 = 0$$
   $$g_2(x, y, z) = x - 4z + 5 = 0$$

   (c) $w = f(x, y, z) = x^2 + y^2 + z^2$
   Subject to
   $$g_1(x, y, z) = x^2 + y^2 + 2z - 16 = 0$$
   $$g_2(x, y, z) = x + y - 4 = 0$$

   (d) $w = f(x, y, z) = 4x - 2y - 5z$
   Subject to
   $$g_1(x, y, z) = x + y - z - 1 = 0$$
   $$g_2(x_1, x_2, x_3) = x^2 + y^2 + 2z^2 - 2 = 0$$

36. Use the KT conditions to find a solution to the following nonlinear programming problems:

   (a) Minimize $z = f(x, y) = -x - 2y + y^2$
   Subject to
   $$x + y \le 1$$
   $$x \ge 0, \quad y \ge 0$$

   (b) Maximize $z = f(x, y) = x + 3y^2$
   Subject to
   $$x - y \le 2$$

$$x \geq 0, \quad y \geq 0$$

(c) Maximize $w = f(x, y, z) = x + y - z$
Subject to
$$x^2 + y^2 + z \leq 1$$
$$x \geq 0, \quad y \geq 0, \quad z \geq 0$$

(d) Minimize $w = f(x, y, z) = y^2 e^{(x^2 - z)}$
Subject to
$$g_1(x, y, z) = 9x^2 + 4y^2 + 36z^2 - 36 = 0$$
$$g_2(x, y, z) = xy + yz - 1 = 0$$
$$x \geq 0, \quad y \geq 0, \quad z \geq 0$$

37. Use the KT conditions to find a solution to the following nonlinear programming problems:

(a) Minimize $z = f(x, y) = x^2 + 4y^2 - 4xy - x + 12y$
Subject to
$$x + y \geq 4$$
$$x \geq 0, \quad y \geq 0$$

(b) Maximize $z = f(x, y) = x^2 - y^2$
Subject to
$$-(x - 2)^2 - y^2 \leq -4$$
$$x \geq 0, \quad y \geq 0$$

(c) Maximize $z = f(x, y) = -3x + 0.5y$
Subject to
$$x^2 + y^2 \leq 1$$
$$-x \leq 0$$
$$-y \leq 0$$

(d) Minimize $w = f(x, y, z) = (x - 10)^2 + (y - 10)^2 + (z - 10)^2$
Subject to
$$g_1(x, y, z) = x^2 + y^2 + z^2 - 12 = 0$$
$$g_2(x, y, z) = -x - y + 2z = 0$$

$$x \geq 0, \quad y \geq 0, \quad z \geq 0$$

38. Use the reduced-gradient method to find the extrema of function f subject to the stated constraints:

(a) Minimize $u = f(x, y, z, w) = x^2 + y + z^2 + w$
Subject to
$$g_1(x, y, z, w) = x + y - 2z + 2w - 2 = 0$$
$$g_2(x, y, z, w) = -x + y + 4z - 4w + 4 = 0$$

(b) Minimize $u = f(x, y, z, w) = 2x^2 + 3y + z + w^2$
Subject to
$$g_1(x, y, z, w) = 2x - y - 3z + w - 3 = 0$$
$$g_2(x, y, z, w) = x - y + 2z - 2w + 2 = 0$$

(c) Minimize $u = f(x, y, z, w) = 3x + y^2 + z^2 + 2w$
Subject to
$$g_1(x, y, z, w) = x - y - z + w + 1 = 0$$
$$g_2(x, y, z, w) = 3x - y + 2z - w - 2 = 0$$

(d) Minimize $u = f(x, y, z, w) = x^2 - y + z^2 - w$
Subject to
$$g_1(x, y, z, w) = x + 2y - 3z + 2w - 5 = 0$$
$$g_2(x, y, z, w) = 2x - y - 2z - 4w - 2 = 0$$

39. Convert the following problems into separable forms:

(a) Minimize $z = f(x, y) = 2x^2 + y^2$
Subject to
$$x + y \leq 3$$
$$x \geq 0, \quad y \geq 0$$

(b) Maximize $z = f(x, y) = 3x^3 - 2x^2 + 2x_1 + 5y^2 - 4y$
Subject to
$$2x + 3y \leq 1$$
$$x \geq 0, \quad y \geq 0$$

(c)   Maximize $w = f(x, y, z) = x^2 + y^2 - z^2$
Subject to

$$x^2 + 2y^2 + 3z \leq 3$$
$$x - y + z \leq 1$$
$$x \geq 0, \quad y \geq 0, \quad z \geq 0$$

(d)   Minimize $w = f(x, y, z) = x^2 + 2y^2 + z^2 - 2xz$
Subject to

$$5x^2 + 4y^2 + 6z^2 \leq 13$$
$$x + 3y - 2z \leq 2$$
$$x \geq 0, \quad y \geq 0, \quad z \geq 0$$

40. Solve each of the following quadratic programming problems using Wolfe's method:

(a)   Maximize $z = f(x, y) = 2x^2 + 2xy + 4y^2 - x + y$
Subject to

$$x + y \leq 2$$
$$x - y \leq 1$$
$$x \geq 0, \quad y \geq 0$$

(b)   Maximize $z = f(x, y) = -x - y + 0.5x^2 + y^2 - xy$
Subject to

$$x + y \leq 3$$
$$2x + 3y \geq 6$$
$$x \geq 0, \quad y \geq 0$$

Appendix A

Number Representations and Errors

A.1 Introduction

Here, we study in broad outline the floating-point representation used in computers for real numbers and the errors that result from the finite nature of this representation. We give a good general overview of how the computer represents and manipulates numbers. We see later that such considerations affect the choice of design of computer algorithms for solving higher-order problems. We introduce several definitions and concepts that may be unfamiliar. The reader should not spend time trying to master all these immediately but should rather try to acquire a rough idea of the sorts of difficulties that can arise from computer solutions of mathematical problems. We describe methods for representing numbers on computers and the errors introduced by these representations. In addition, we exam other sources of various types of computational errors.

917

A.2   Number Representations and the Base of Numbers

The number system we use daily is called the *decimal system*. The *base* of the decimal number system is **10**. The familiar decimal notation for numbers employs the digits $0, 1, 2, 3, 4, 5, 6, 7, 8, 9$. When we write down a whole number such as 478325, the individual digits represent coefficients of powers of 10 as follows:

$$478325 \;=\; 5 + 20 + 300 + 8000 + 70000 + 400000$$

$$=\; 5 \times 10^0 + 2 \times 10^1 + 3 \times 10^2 + 8 \times 10^3 + 7 \times 10^4 + 4 \times 10^5.$$

Thus, in general, a string of digits represents a number according to the formula

$$a_n a_{n-1} \cdots a_1 a_0 = a_0 \times 10^0 + a_1 \times 10^1 + \cdots + a_{n-1} \times 10^{n-1} + a_n \times 10^n. \quad \text{(A.1)}$$

This takes care of the positive whole numbers. A number between 0 and 1 is represented by a string of digits to the right of a decimal point. For example,

$$0.8543 = 8 \times 10^{-1} + 5 \times 10^{-2} + 4 \times 10^{-3} + 3 \times 10^{-4}.$$

Thus, in general, a string of digits represents a number according to the formula

$$0.b_1 b_2 b_3 \cdots = b_1 \times 10^{-1} + b_2 \times 10^{-2} + b_3 \times 10^{-3} + \cdots . \quad \text{(A.2)}$$

For a real number of the form

$$a_n a_{n-1} \cdots a_1 a_0 . b_1 b_2 \cdots = \sum_{k=0}^{n} a_k 10^k + \sum_{k=1}^{\infty} b_k 10^{-k}, \quad \text{(A.3)}$$

the *integer part* is the first summation in the expansion and the *fractional part* is the second. Computers, however, don't use the decimal system in computations and memory but use the *binary system*. The binary system is natural for computers because computer memory consists of a huge number of electronic and magnetic recording devices, of which each element has only "on" and "off" statues. In the binary system the base is 2, and the integer coefficients may take the values 0 or 1. The digits 0 and 1 are called *bits*, which is short for binary digits. For example, the number 1110.11 in the binary system represents the number

$$1 \times 2^4 + 1 \times 2^3 + 1 \times 2^2 + 0 \times 2^1 + 1 \times 2^{-1} + 1 \times 2^{-2}$$

in the decimal system.

There are other base systems used in computers, particularly, the *octal* and *hexadecimal* systems. The base for the octal system is **8** and for the hexadecimal it is **16**. These two systems are close relatives of the binary system and can be translated to and from binary easily. Expressions in octal or hexadecimal form are shorter than in binary form, so they are easier for humans to read and understand. Hexadecimal form also provides more efficient use of memory space for real numbers. If we use another base, say, β, then numbers represented in the β system look like this:

$$(a_n a_{n-1} \cdots a_1 a_0 . b_1 b_2 \cdots)_\beta = \sum_{k=0}^{n} a_k \beta^k + \sum_{k=1}^{\infty} b_k \beta^{-k}, \qquad (\text{A.4})$$

and the digits are $0, 1, 2, \ldots, \beta - 1$ in this representation. If $\beta > 10$, it is necessary to introduce symbols for $10, 11, \ldots, \beta - 1$. In this system based on 16, we use A, B, C, D, E, F for $10, 11, 12, 13, 14, 15$, respectively. Thus, for example,

$$(2BED)_{16} = D + E \times 16 + B \times 16^2 + 2 \times 16^3 = 11245.$$

The base of a number system is also called the *radix*. The base of a number is denoted by a subscript, for example, $(4.445)_{10}$ is 4.445 in base 10 (decimal), $(1011.11)_2$ is 1011.11 in base 2 (binary), and $(18C7.90)_{16}$ is 18C7.90 in base 16 (hexadecimal).

The conversion of an integer from one system to another is fairly simple and can probably best be presented in terms of an example. Let $k = 275$ in decimal form, i.e., $k = (2 \times 10^2) + (7 \times 10^1) + (5 \times 10^0)$. Now $(k/16^2) > 1$ but $(k/16^3) < 1$, so in hexadecimal form k can be written as $k = (\alpha_2 \times 16^2) + (\alpha_1 \times 16^1) + (\alpha_0 \times 16^0)$. Now, $275 = 1(16^2) + 19 = 1(16^2) + 1(16) + 3$, and so the decimal integer, 275, can be written in hexadecimal form as 113, i.e.,

$$275 = (275)_{10} = (113)_{16}.$$

The reverse process is even simple. For example,

$$(5C3)_{16} = 5(16^2) + 12(16) + 3 = 1280 + 192 + 3 = (1475)_{10}.$$

Conversion of a hexadecimal fraction to a decimal is similar. For example,

$$(0.2A8)_{16} = (2/16) + (A/16^2) + (8/16^3) = (2(16^2) + 10(16) + 8)/16^3$$
$$= 680/4096 = 0.166,$$

carrying only three digits in decimal form. Conversion of a decimal fraction to a hexadecimal (or a binary) proceeds as in the following example. Consider the number $r_1 = 1/10 = 0.1$ (decimal form). Then there exist constants $\{\alpha_k\}_{k=1}^{\infty}$ such that

$$r_1 = 0.1 = \alpha_1/16 + \alpha_2/16^2 + \alpha_3/16^3 + \cdots.$$

Now

$$16r_1 = 1.6 = \alpha_1 + \alpha_2/16 + \alpha_3/16^2 + \cdots.$$

Thus, $\alpha_1 = 1$ and

$$r_2 \equiv .6 = \alpha_2/16 + \alpha_3/16^2 + \alpha_4/16^3 + \cdots.$$

Again,

$$16r_2 = 9.6 = \alpha_2 + \alpha_3/16 + \alpha_4/16^2 + \cdots,$$

so $\alpha_2 = 9$, and

$$r_2 \equiv .6 = \alpha_3/16 + \alpha_4/16^2 + \cdots$$

From this stage on we see that the process will repeat itself, and so we have $(0.1)_{10}$ equals the infinitely repeating hexadecimal fraction, $(0.1999\cdots)_{16}$. Since $1 = (0001)_2$ and $9 = (1001)_2$ we also have the infinite binary expansion

$$r_1 = (0.1)_{10} = (0.1999\cdots)_{16} = (0.000110011001\cdots)_2.$$

Example A.1 *The conversion from one base to another base is:*

$$\begin{array}{llll}(17)_{10} & = (10001)_2 & = (21)_8 & = (11)_{16} \\ (13.25)_{10} & = (1101.01)_2 & = (15.2)_8 & = (D.4)_{16}.\end{array}$$

A.2.1 Normalized Floating-Point Representations

Unless numbers are specified to be integers, they are stored in the computers in what is known as normalized floating-point form. This form is similar to the scientific notation used as a compact form for writing very small or very large numbers. For example, the number 0.0000123 may be written in scientific notation as 0.123×10^{-4}.

In general, every nonzero real number x has a floating-point representation

$$x = \pm\mathbf{M} \times \mathbf{r}^{\mathbf{e}}, \quad \text{where} \quad \frac{1}{\mathbf{r}} \leq \mathbf{M} < 1, \quad \text{or} \quad -\frac{1}{\mathbf{r}} \geq \mathbf{M} > -1,$$

where

$$\mathbf{M} = \sum_{k=1}^{t} \mathbf{d}_k \mathbf{r}^{-k}.$$

Here, $\mathbf{M}$ is called the *mantissa*, $\mathbf{e}$ is an integer called the **exponent**, $\mathbf{r}$ the *base*, $\mathbf{d_k}$ is the value of the *kth* digit and $\mathbf{t}$ is the maximum number of digits allowed in the number. When $r = 10$, then the nonzero real number x has the normalized floating-point decimal representation

$$x = \pm\mathbf{M} \times 10^{\mathbf{e}},$$

where the normalized mantissa $\mathbf{M}$ satisfies $\frac{1}{10} \leq \mathbf{M} < 1$. Normalization consists of finding the exponent $\mathbf{e}$ for which $|x|/10^{\mathbf{e}}$ lies on the interval $[\frac{1}{10}, 1)$, then taking $\mathbf{M} = |x|/10^{\mathbf{e}}$. This corresponds to "floating" the decimal point to the left of the leading significant digit of x's decimal representation, then adjusting $\mathbf{e}$ as needed. For example,

$$
\begin{array}{llll}
-12.75 & \text{has representation} & -0.1275 \times 10^2 & (\mathbf{M} = 0.1275, \mathbf{e} = 2) \\
0.1 & \text{has representation} & +0.1 \times 10^0 & (\mathbf{M} = 0.1, \mathbf{e} = 0) \\
\frac{1}{15} = 0.0\overline{6} & \text{has representation} & (\frac{1}{15} \times 10) \times 10^{-1} & (\mathbf{M} = 0.\overline{6}, \mathbf{e} = -1).
\end{array}
$$

A *machine number* for a calculator is a real number that it stores exactly in normalized floating-point form. For the calculator storage, a nonzero x is a machine number, if and only if its normalized floating decimal point representation is of the form

$$x = \pm\mathbf{M} \times 10^{\mathbf{e}},$$

where

$$\begin{aligned}
\mathbf{M} &= 0.d_1 d_2 \cdots d_k (d_k = 0, 1, 2, \ldots, 9), \quad \text{with} \quad d_1 \neq 0 \\
\mathbf{e} &= -100, -99, \cdots, +99.
\end{aligned}$$

The condition $d_1 \neq 0$ ensures normalization (i.e., $\mathbf{M} \geq \frac{1}{10}$).

Computers use a normalized floating-point binary representation for real numbers. The computer stores a binary approximation to x as

$$x = \pm \mathbf{M} \times 2^{\mathbf{e}}.$$

Normalization in this case consists of finding the unique exponent e for which $|x|/2^e$ lies on the interval $(\frac{1}{2}, 1)$, and then taking $|x|/2^e$ as $\mathbf{M}$. For example,

$-12.75 = -\frac{51}{4}$ can be represented as $-(\frac{51}{4} \cdot \frac{1}{16}).2^4$ $(\mathbf{M} = \frac{51}{64}, \mathbf{e} = 4)$

$0.1 = \frac{1}{10}$ can be represented as $+(\frac{1}{10} \cdot 8).2^{-3}$ $(\mathbf{M} = \frac{4}{5}, \mathbf{e} = -3)$

$\frac{1}{15} = 0.06666\cdots$ can be representation $(\frac{1}{15} \cdot 8).2^{-3}$ $(\mathbf{M} = \frac{8}{15}, \mathbf{e} = -3)$.

Computers have both an integer mode and a floating-point mode for representing numbers. The *integer mode* is used for performing calculations that are known to be integer values and have limited usage for numerical analysis. *Floating-point numbers* are used for scientific and engineering applications. It must be understood that any computer implementation of equations $x = \pm \mathbf{M} \times 2^{\mathbf{e}}$ places restrictions on the number of digits used in the mantissa $\mathbf{M}$, and the range of the possible exponent $\mathbf{e}$ must be limited. Computers that use 32 bits to represent single-precision real numbers use 8 bits for the exponent and 24 bits for the mantissa. They can represent real numbers whose magnitude is in the range $2.938736E - 39$ to $1.701412E + 38$ (i.e., 2^{-128} to 2^{127}), with six decimal digits of numerical precision (for example, $(2^{-23} = 1.2 \times 10^{-7})$.

Computers that use 48 bits to represent single-precision real numbers might use 8 bits for the exponent and 40 bits for the mantissa. They can represent real numbers in the range $2.9387358771E - 39$ to $1.701418346E + 38$ (i.e., 2^{-128} to 2^{127}) with 11 decimal digits of precision (for example,

$2^{-39} = 1.8 \times 10^{-12}$). If the computer has 64 bit double-precision real numbers, it might use 11 bits for the exponent and 53 bits for the mantissa. They can represent real numbers in the range $5.56284646268003 \times 10^{-309}$ to $8.988465674311580 \times 10^{307}$ (i.e., 2^{-1024} to 2^{1023}) with about 16 decimal digits of precision (for example, $2^{-52} = 2.2 \times 10^{-16}$).

The most commonly used floating-point representations are the 1EEE binary floating-point standards. There are two such formats: single and double precision. 1EEE single precision uses a 32-bit word to represent the sign, exponent, and mantissa. Double precision uses 64 bits. These bits are distributed as shown below.

No. of Bits	Single Precision	Double Precision
Sign	1	1
Exponent	8	11
Mantissa	24	53

In all essential respects, MATLAB uses only one type of number—1EEE double precision floating-point. In other words, MATLAB uses pairs of these to represent double floating-point complex numbers, but that will not affect much of what we do here. Integers are stored in MATLAB as "floating integer" which means, in essence, that integers are stored in their floating-point representations.

MATLAB Variable	Meaning	Value
eps	Machine unit	$2^{-52} \approx 2.22044604925031e-016$
realmin	Smallest positive number	$2.22507385850723 - 308$
realmax	Largest positive number	$1.79769313486232e + 308$

From this we see that the above representation uses 11 bits for the binary exponent, which therefore ranges from about -2^{10} to 2^{10}. (The actual range is not exactly this because of special representations for small numbers and for $\pm\infty$.) The mantissa has 53 bits including the implicit bit. If $x = M \times 2^e$ is a normalized MATLAB floating-point number, then $M \in [1, 2)$

is represented by

$$M = 1 + \sum_{k=1}^{52} d_k 2^{-k}.$$

Since $2^{10} = 1024 \approx 10^3$, these 53 significant bits are equivalent to approximately 16 significant decimal digits accuracy in MATLAB representation. The fact that the mantissa has 52 bits after the binary point means that the next machine number greater than 1 is $1 + 2^{-52}$. This gap is called *machine epsilon.*

In MATLAB, neither underflow nor overflow cause a program to stop. Underflow is replaced by a zero, while overflow is replaced by $\pm\infty$. This allows subsequent instructions to be executed and may permit meaningful results. Frequently, however, it will result in meaningless answers such as $\pm\infty$ or NaN, which stands for Not-a-Number. NaN is the result of indeterminate arithmetic operations such as $0/0, \infty/\infty, 0.\infty, \infty - \infty$, etc.

There are two commonly used ways of translating a given real number x into a k- digits floating-point number, *rounding* and *chopping*, which we shall discuss in the following section.

A.2.2 Rounding and Chopping

When one gives the number of digits in a numerical value, one should not include zeros in the beginning of the number, as these zeros only help to denote where the decimal point should be. If one is counting the number of decimals, one should be off course and include leading zeros to the right of the decimal point. For example, the number 0.00123 is given with three digits but has five decimals. The number 11.44 is given with four digits but has two decimals. If the magnitude of the error in approximate number p does not exceed $1/2 \times 10^{-k}$, then p is said to have k *correct decimals.* The digits in p, which occupy positions where the unit is greater than or equal to 10^{-k}, are called, then, *significant digits* (any initial zeros are not counted). For example, 0.001234 ± 0.000004 has five correct decimals and three significant digits, while 0.0012342 ± 0.000006 has four correct decimals and two significant digits. The number of correct decimals gives one an idea of the magnitude of the absolute error, while the number of significant digits gives a rough idea of the magnitude of the relative error.

There are two ways of rounding off number s to a given number (k) of decimals. In *chopping*, one simply leaves off all the decimals to the right of the *kth*. That way of abridging a number is not recommended since the error has, systematically, the opposite sign of the number itself. Also, the magnitude of the error can be large as 10^{-k}. A surprising number of computers use chopping on the results of every arithmetical operation. This usually does not do much harm, because the number of digits used in the operations is generally far greater than the number of significant digits in the data. In *rounding* (sometimes called "correct rounding"), one chooses among the numbers that are closest to the given number. Thus, if the part of the number which stands to the right of the *kth* decimal is less than $1/2 \times 10^{-k}$ in magnitude, then one should leave the *kth* decimal unchanged. If it is greater than $1/2 \times 10^{-k}$, then one raises the *kth* decimal by 1. In the boundary case, when the number that stands to the right of the *kth* decimal is exactly $\frac{1}{2} \times 10^{-k}$, one should raise the *kth* decimal if it is odd or leave it unchanged if it is even. In this way, the error is positive or negative about equally often. Most computers that perform rounding always, in the boundary case mentioned above, raise the number by $1/2 \times 10^{-k}$ (or the corresponding operation in a base other than 10), because this is easier to realize technically. Whichever convention one chooses in the boundary case, the error in the rounding will always lie on the interval $[-1/2 \times 10^{-k}, 1/2 \times 10^{-k}]$. For example, shorting to three decimals:

0.2397	rounds to	0.240	(is chopped to	0.239)
−0.2397	rounds to	− 0.240	(is chopped to	− 0.239)
0.23750	rounds to	0.238	(is chopped to	0.237)
0.23650	rounds to	0.236	(is chopped to	0.236)
0.23652	rounds to	0.237	(is chopped to	0.236)

A.3   Error

An approximate number p is a number that differs slightly from an exact number α. We write

$$p \approx \alpha.$$

By error E of an approximate number p, we mean the difference between the exact number α and its computed approximation p. Thus, we define

$$E = \alpha - p. \tag{A.5}$$

If $\alpha > p$, the error E is positive, and if $\alpha < p$, the error E is negative. In many situations, the sign of the error may not be known and might even be irrelevant. Therefore, we define *absolute error* as

$$|E| = |\alpha - p|. \tag{A.6}$$

The *relative error* RE of an approximate number p is the ratio of the absolute error of the number to the absolute value of the corresponding exact number α. Thus,

$$RE = \frac{|\alpha - p|}{|\alpha|}, \qquad \alpha \neq 0. \tag{A.7}$$

If we approximate $\frac{1}{3}$ by 0.333, we have

$$E = \frac{1}{3} \times 10^{-3} \qquad \text{and} \qquad RE = 10^{-3}.$$

Note that the relative error is generally a better measure of the extent of error than the actual error. But one should also note that the relative error is undefined if the exact answer is equal to zero. Generally, we shall be interested in E (or sometimes $|E|$) rather than RE, but when the true value of a quantity is very small or very large, relative errors are more meaningful. For example, if the true value of a quantity is 10^{15}, an error of 10^6 is probably not serious, but this is more meaningfully expressed by saying that $RE = 10^{-9}$. In actual computation of the relative error, we shall often replace the unknown true value by the computed approximate value. Sometimes the quantity

$$\frac{|\alpha - p|}{|\alpha|} \times 100\% \tag{A.8}$$

is defined as *percentage error*. From the above example, we have

$$PE = 0.001 \times 100 = 0.1\%.$$

In investigating the effect of the total error in various methods, we shall often mathematically derive an error called the *error bound*, which is a limit on how large the error can be. We shall have reason to compute error bounds in many situations. This applies to both absolute and relative errors. Note that the error bound can be much larger than the actual error and that this is often the case in practice. Any mathematically derived error bound must account for the worst possible case that can occur and is often based upon certain simplifying assumptions about the problem which in many practical cases cannot be actually tested. For the error bound to be used in any practical way, the user must have a good understanding of how the error bound was derived in order to know how crude it is, i.e., how likely it is to overestimate the actual error. Of course, whenever possible, our goal is to eliminate or lessen the effects of errors, rather than trying to estimate them after they occur.

A.4 Sources of Errors

In analyzing the accuracy of numerical results, one should be aware of the possible sources of error in each stage of the computational process and of the extent to which these errors can affect the final answer. We will consider that there are three types of errors that occur in a computation. We discuss them step-by-step as follows.

A.4.1 Human Errors

These types of errors arise when the equations of the mathematical model are formed, due to sources such as the idealistic assumptions made to simplify the model, inaccurate measurements of data, miscopying of figures, the inaccurate representation of mathematical constants (for example, if the constant π occurs in an equation, we must replace π by 3.1416 or 3.141593), etc.

A.4.2 Truncation Errors

This type of error is caused when we are forced to use mathematical techniques that give approximate, rather than exact, answers. For example, suppose that we use Maclaurin's series expansion to represent $\sin x$, so that

$$\sin x = x - \frac{x^3}{3!} + \frac{x^5}{5!} - \frac{x^7}{7!} + \cdots.$$

If we want a number that approximates $\sin(\pi/2)$, we must terminate the expansion in order to obtain

$$\sin(\frac{\pi}{2}) = \frac{\pi}{2} - \frac{(\pi/2)^3}{3!} + \frac{(\pi/2)^5}{5!} - \frac{(\pi/2)^7}{7!} + E,$$

where E is the truncation error introduced in the calculation. Truncation errors in numerical analysis usually occur because many numerical methods are iterative in nature, with the approximations theoretically becoming more accurate as we take more iterations. As a practical matter, we must stop the iterations after a finite number of steps, thus introducing a truncation error. Taylor's series is the most important means used to derive numerical schemes and analysis truncation errors.

A.4.3 Round-off Errors

These errors are associated with the limited number of digits numbers in a computer. For example, by rounding off 1.32463672 to six decimal places to give 1.324637, any further calculation involving such a number will also contain an error. We round-off numbers according to the following rules:

1. If the first discarded digit is less than 5, leave the remaining digits of the number unchanged. For example,

$$48.47263 \approx 48.4726.$$

2. If the first discarded digit is exceeds 5, add 1 to the retained digit. For example,
$$48.4726 \approx 48.473.$$

3. If the first discarded digit is exactly 5 and there are nonzero among those discarded, add 1 to the last retained digit. For example,

$$3.0554 \approx 3.06.$$

4. If the first discarded digit is exactly 5 and all other discarded digits are zero, the last retained digit is left unchanged if it is *even*, otherwise, 1 is added to it. For example,

$$3.05500 \approx 3.06$$
$$3.04500 \approx 3.04.$$

With these rules, the error is never larger in magnitude than one-half unit of the place of the *nth* digit in the rounded number.

To understand the nature of round-off errors, it is necessary to learn the ways numbers are stored and additions and subtractions are performed in a computer. •

A.5 Effect of Round-off Errors in Arithmetic Operations

Here, we discuss the effect of rounding off errors in calculations in detail. Let a_r be the rounded off value a_e, which is the exact value of a number which is not necessarily known. Similarly, b_r, b_e, c_r, c_e, etc., are the corresponding values for other numbers. The number $E_A = a_r - a_e$ is called the error in the number a_r. Similarly, E_B is the error in b_r, etc. The error E_A will be positive or negative accordingly as a_r is greater or less than a_e. It is, however, usually impossible to determine the sign of E_A. Therefore, it is normal to consider only the value of $|E_A|$, called the absolute error of number a_r. To indicate that a number has been rounded off to two significant figures or four decimal places it is followed by $2S$ or $4dp$ as appropriate.

A.5.1 Round-off Errors in Addition and Subtraction

Let a_r and b_r be two approximate numbers and c_r be their sum, which have been rounded off, be represented by

$$c_r = a_r + b_r, \tag{A.9}$$

which is an approximation for

$$c_e = a_e + b_e. \tag{A.10}$$

Then by subtracting (A.10) from (A.9), we have

$$
\begin{aligned}
c_r - c_e &= (a_r - a_e) + (b_r - b_e) \\
E_C &= E_A + E_B.
\end{aligned}
\tag{A.11}
$$

Then

$$
|E_C| \leq |E_A| + |E_B|,
\tag{A.12}
$$

i.e., the absolute error of the sum of two numbers is less than or equal to the sum of the absolute error of the two numbers. Note that this can be extended to the sum of any number. One should follow a similar argument for the error involved in the *difference* of two rounded off numbers, i.e.,

$$
c_r = a_r - b_r, \qquad \text{with} \quad a_r > b_r,
\tag{A.13}
$$

and one should find that the same result is obtained, which is

$$
|E_C| \leq |E_A| + |E_B|.
\tag{A.14}
$$

This can also be extended to any number of terms. For example, consider the error in the numbers $1015 + 0.3572$, where both numbers have been rounded off. The first number $1.015(a_r)$ has been rounded off to 3dp so that the exact value must lie between 1.0145 and 1.0155, which implies that $-0.0005 \leq E_A \leq 0.0005$. This means that the absolute error is never greater than 0.0005 or $\frac{1}{2} \times 10^{-3}$, i.e., $|E_A| \leq \frac{1}{2} \times 10^{-3}$. Note that if a number is rounded off to n decimal places, then the absolute error is less than or equal to $\frac{1}{2} \times 10^{-n}$. Similarly, if the other given number $0.3572(b_r)$ has been rounded off to 4dp, then $E_B \leq \frac{1}{2} \times 10^{-4}$.

Since

$$
c_r = a_r + b_r,
$$

then

$$
c_r = 1.015 + 0.3572 = 1.3722
$$

but

$$
\begin{aligned}
|E_C| &\leq |E_A| + |E_B| \\[2mm]
&\leq \left(\tfrac{1}{2} \times 10^{-3}\right) + \left(\tfrac{1}{2} \times 10^{-4}\right) \\[2mm]
&\leq (0.5 + 0.05) \times 10^{-3} \\
&\leq 0.55 \times 10^{-3}.
\end{aligned}
$$

So the exact value of this sum must be in the range

$$1.3722 \pm 0.55 \times 10^{-3},$$

i.e., between 1.37165 and 1.37275, so this result may be correctly rounded off to 1.37, i.e., to only 2dp.

A.5.2 Round-off Errors in Multiplication

Let a_r and b_r be the rounded off values and c_r be the product of these numbers, i.e.,

$$c_r = a_r b_r,$$

the number which approximates to the exact number

$$c_e = a_e b_e.$$

Then

$$c_r - E_C = (a_r - E_A)(b_r - E_B)$$

$$= a_r b_r - E_A b_r - E_B a_r + E_A E_B,$$

since $c_r = a_r b_r$, so

$$E_C = E_A b_r + E_B a_r - E_A E_B$$

and

$$\frac{E_C}{c_r} = \frac{E_A b_r + E_B a_r - E_A E_B}{a_r b_r}$$

$$= \frac{E_A}{a_r} + \frac{E_b}{b_r} - \frac{E_A E_B}{a_r b_r}.$$

The last term has as its numerator the product of two very small numbers, both of which will also be small compared with a_r and b_r so we neglect the last term, then we have

$$\frac{E_C}{c_r} = \frac{E_A}{a_r} + \frac{E_b}{b_r}. \tag{A.15}$$

The number E_A/a_r is called the relative error in a_r. Then from (A.15), we have

$$\left| \frac{E_C}{c_r} \right| \leq \left| \frac{E_A}{a_r} \right| + \left| \frac{E_b}{b_r} \right|. \tag{A.16}$$

Hence, the relative error modulus of a product is less than or equal to the sum of the relative error moduli of the factors of the product. Having found the relative error modulus of a product from this result the absolute error is usually then obtained by multiplying the relative error modulus by a_r, i.e.,

$$RE = |E_A|/|a_r|.$$

This result can be extended to the product of more than two numbers and simply increases the number of terms on the right-hand side of the formula. For example, consider the error in 1.015×0.3573 where both numbers have been rounded off. Then

$$1.015 \times 0.3573 = 0.3626595 \approx 0.363.$$

So the relative error modulus is given by

$$\left|\frac{E_C}{0.363}\right| \leq \frac{\frac{1}{2} \times 10^{-3}}{1.015} + \frac{\frac{1}{2} \times 10^{-4}}{0.3573}$$

$$\leq (0.49 \times 10^{-3}) + (1.4 \times 10^{-4}).$$

Hence,

$$\left|\frac{E_C}{0.363}\right| \leq (0.49 \times 10^{-3}) + (0.14 \times 10^{-3})$$

$$\leq 0.63 \times 10^{-3}.$$

So, we have

$$|E_C| \leq 0.63 \times 0.363 \times 10^{-3}$$
$$\leq 0.23 \times 10^{-3}.$$

Hence, the exact value of this product lies in the range

$$0.3626595 \pm 0.00023,$$

i.e., between 0.3624295 and 0.3628895, so that this result may be correctly rounded off to 0.36, i.e., to 2dp.

A.5.3 Round-off Errors in Division

Let a_r and b_r be rounded off values and c_r be the division of these numbers, i.e.,

$$c_r = a_r/b_r,$$

the number which approximates to the exact number

$$c_e = a_e/b_e.$$

Then

$$
\begin{aligned}
c_r - E_C &= \frac{(a_r - E_A)}{(b_r - E_B)} \\[2ex]
&= \frac{a_r(1 - E_A/a_r)}{b_r(1 - E_B/b_r)} \\[2ex]
&= \frac{a_r}{b_r}\left(1 - \frac{E_A}{a_r}\right)\left(1 - \frac{E_B}{b_r}\right)^{-1}.
\end{aligned}
$$

The number

$$\left(1 - \frac{E_B}{b_r}\right)^{-1}$$

is expanded using the binomial series and neglecting those terms involving powers of the relative error E_B/b_r. Thus,

$$
\begin{aligned}
c_r - E_C &= \frac{a_r}{b_r}\left(1 - \frac{E_A}{a_r}\right)\left(1 + \frac{E_B}{b_r} + \cdots\right) \\[2ex]
&= \frac{a_r}{b_r}\left(1 - \frac{E_A}{a_r} + \frac{E_B}{b_r}\right) \qquad \left(\text{neglecting } \frac{E_A E_B}{a_r b_r}\right) \\[2ex]
&= \frac{a_r}{b_r} - \frac{E_A}{b_r} + \frac{E_B a_r}{b_r^2},
\end{aligned}
$$

which implies that

$$E_C = \frac{E_A}{b_r} - \frac{E_B a_r}{b_r^2}$$

and

$$\frac{E_C}{c_r} = \frac{E_A}{b_r} - \frac{E_B a_r}{b_r^2} \div \frac{a_r}{b_r}$$

$$= \frac{E_A}{a_r} - \frac{E_B}{b_r}.$$

Hence,

$$\left|\frac{E_C}{c_r}\right| \leq \left|\frac{E_A}{a_r}\right| + \left|\frac{E_B}{b_r}\right|,$$

which gives the same result as for the product of the two numbers. It follows that it is possible to extend this result to quotients with two or more factors in the numerator or denominator by simply increasing the number of terms on the right-hand side. For example, consider the error in $17.28 \div 2.136$, where both numbers have been rounded off. Then

$$17.28/2.136 = 8.0898876 \approx 8.09.$$

Therefore,

$$\left|\frac{E_C}{8.09}\right| \leq \frac{\frac{1}{2} \times 10^{-2}}{17.28} + \frac{\frac{1}{2} \times 10^{-3}}{2.136}$$

$$\leq (0.029 \times 10^{-2}) + (0.23 \times 10^{-3})$$
$$\leq (0.29 \times 10^{-3} + 0.23 \times 10^{-3})$$
$$\leq 0.52 \times 10^{-3},$$

so that

$$|E_C| \leq 4.2 \times 10^{-3}$$
$$\leq 0.42 \times 10^{-2}.$$

Hence, the exact value of this quotient lies in the range

$$8.08989 \pm 0.000432,$$

i.e., between 8.08569 and 8.09409, so that this result may be correctly rounded off to 8.09, i.e., to 2dp. The value of $|E_C|$ suggested this directly. This could be given to 3dp as 8.090, but with a large error of up to 5 units in the third decimal place.

Example A.2 *Consider the error in* $5.381 + (5.96 \times 17.89)$, *where all numbers have been rounded off. We first find the absolute error in* $|E_C|$. *So*

$$5.96 \times 17.89 = 106.6244 \approx 106.6,$$

then

$$\left| \frac{E_C}{106.6} \right| \leq \frac{\frac{1}{2} \times 10^{-2}}{5.96} + \frac{\frac{1}{2} \times 10^{-2}}{17.89}$$

$$\leq (0.084 \times 10^{-2}) + (0.028 \times 10^{-2})$$
$$\leq 0.112 \times 10^{-2},$$

which gives

$$|E_C| \leq 0.12.$$

The absolute error for 5.381 *is* $1/2 \times 10^{-3}$, *so that the maximum absolute error for the sum is*

$$0.12 + 0.0005 = 0.1205.$$

But by the calculator

$$5.381 + (5.96 \times 17.89) = 112.0054,$$

the exact value lies in the range

$$112.0054 \pm 0.1205,$$

i.e., between 111.8849 *and* 112.1259. *This means that the result may be correctly rounded off to 3S or 0dp as an error of 0.1205 as suggested or could be given as 112.0 with an error of up to 1 unit in the first decimal place.* ●

A.5.4 Round-off Errors in Powers and Roots

Let a_r and b_r be rounded off values and

$$b_r = (a_r)^p,$$

where the power is exact and may be rational. This approximates to the exact number

$$b_e = (a_e)^p.$$

Using $a_e = a_r - E_A$, we have

$$b_r - E_B \;=\; (a_r - E_A)^p$$

$$=\; (a_r)^p \left(1 - E_A/a_r\right)^p$$

$$=\; (a_r)^p \left(1 - \frac{pE_A}{a_r} + \cdots\right).$$

Using the binomial series and neglecting those terms involving powers of the relative error E_A/a_r gives

$$b_r - E_B = (a_r)^p - pE_A(a_r)^{p-1},$$

which implies that

$$E_B = pE_A(a_r)^{p-1},$$

and so

$$\frac{E_B}{b_r} \;=\; \frac{pE_A(a_r)^{p-1}}{(a_r)^p}$$

$$=\; p\frac{E_A}{a_r}.$$

Hence,

$$\left|\frac{E_B}{b_r}\right| = |p|\left|\frac{E_A}{a_r}\right|,$$

i.e., the relative error modulus of a power of a number is equal to the product of the modulus of the power and the relative error modulus of the number. For example, consider $\sqrt{8.675}$, where 8.675 has been rounded off. Here, $p = 1/2$ and by the calculator $\sqrt{8.675} = 2.9453$, retaining 4dp. Thus,

$$\left|\frac{E_B}{2.945}\right| \;\leq\; \frac{\frac{1}{2} \times 10^{-3}}{8.675}$$

$$\leq\; 0.029 \times 10^{-3},$$

so that

$$|E_B| \leq 0.85 \times 10^{-4}.$$

This means that $\sqrt{8.675}$ may be correctly rounded off to 2.945, i.e., to 3dp or may be given to 4dp with an error of up to 1 unit in the fourth decimal place.

●

A.6 **Summary**

In this chapter, we discussed the storage and arithmetic of numbers on a computer. Efficient storage of numbers in computer memory requires allocation of a fixed number of bits to each value. The fixed bit size translates to a limit on the number of decimal digits associated with each number, which limits the range of numbers that can be stored in computer memory.

The three number systems most commonly used in computing are binary (base 2), decimal (base 10), and hexadecimal (base 16). Techniques were developed for transforming back and forth between the number systems. Binary numbers are a natural choice for computers because they correspond directly to the underlying hardware, which features transistors that are switched on and off.

The absolute and relative errors were discussed as measures of difference between exact x and approximate $\hat{x}$. They were applied to the storage mechanisms of chopping and rounding to estimate the maximum error introduced when on storing a number. Rounding is somewhat more accurate than chopping (ignoring excess digits), but chopping is typically used because it is simpler to implement in hardware.

Round-off error is one of the principal sources of error in numerical computations. Mathematical operations on floating-point values introduce round-off errors because the results must be stored with a limited number of decimal digits. In numerical calculations involving many operations, round-off gradually corrupts the least significant digits of the results.

The other main source of error in numerical computations is called truncation error. Truncation error is the error that arises when approximations to exact mathematical expressions are used, such as the truncation of an infinite series to a finite number of terms. Truncation error is independent of round-off errors, although these two sources of error combine to affect the accuracy of a computed result. Truncation error considerations are important in many procedures and are discussed throughout the book. •

A.7   Problems

1. Convert the following binary numbers to decimal form:

$$(1010)_2, \ (100101)_2, \ (.1100011)_2.$$

2. Convert the following binary numbers to decimal form:

$$(101101)_2, \ (1010)_2, \ (100101)_2, \ (10000001)_2.$$

3. Write down the following ordinary numbers in terms of power of 10:

$$8383, 285.625, 413.14159265 \cdots .$$

4. Write down the following ordinary numbers in terms of power of 10:

$$769825, 654285.2625, 29873.3087045 \cdots .$$

5. Express the base of natural logarithms e as a normalized floating-point number, using both chopping and symmetric rounding for each of the following systems:
   (a) base 10, with 4 significant figures.
   (b) base 10, with 7 significant figures.
   (c) base 2, with 10 significant figures.

6. Write down the normalized binary floating-point representations of $\frac{1}{3}, \frac{1}{5}, \frac{1}{7}, \frac{1}{9}$, and $\frac{1}{10}$. Use enough bits in the mantissa to see the recurring patterns.

7. Find the first five binary digits of $(0.1)_{10}$. Obtain values for the absolute and relative errors in yours results.

8. Convert the following:
   (a) decimal numbers to binary numbers form.

   $$165, \ 3433, \ 111, \ 2345, \ 278.5, \ 347.45$$

   (b) decimal numbers to hexadecimal decimal numbers.

   $$1025, \ 278.5, \ 14.09375, \ 1445, \ 347.45$$

(c) hexadecimal numbers to both decimal and binary.

$$1F.C, \ FFF.118, \ 1A4.C, \ 1023, \ 11.1$$

9. If $a = 111010, b = 1011$, then evaluate $a + b, a - b, ab$, and a/b.

10. Find the following expressions in binary form:
(a) $101 + 11 + 110110 + 110101 - 1101 - 1010$.
(b) $111^2 - 110^2$.
(c) $(11011)(101101)$.
(d) $(101111001)/(10111)$.

11. What is the absolute error in approximating $\frac{1}{3}$ by 0.3333? What is the corresponding relative error?

12. Evaluate the absolute error in each of the following calculations and give the answer to a suitable degree of accuracy:
(a) $9.01 + 9.96$.
(b) $4.65 - 3.429$.
(c) 0.7425×0.7199.
(d) $0.7078 \div 0.87$.

13. Find the absolute and relative errors in approximating π by 3.1416. What are the corresponding errors in the approximation $100\pi \approx 314.16$?

14. Calculate the error, relative error, and number of significant digits in the following approximations, with $p \approx x$:
(a)
$$x = 25.234, \qquad p = 25.255.$$
(b)
$$x = e, \qquad p = 19/7.$$
(c)
$$x = \sqrt{2}, \qquad p = 1.414.$$

15. Write each of the following numbers in (decimal) floating-point form, starting with the word length m and the exponent e:
$$13.2, \quad -12.532, \quad 2/125.$$

16. Find absolute error in each of the following calculations (all numbers are rounded):

    (a)
    $$187.2 + 93.5.$$

    (b)
    $$0.281 \times 3.7148.$$

    (c)
    $$\sqrt{28.315}.$$

    (d)
    $$\sqrt{(6.2342 \times 0.82137)/27.268}.$$

Appendix B

Mathematical Preliminaries

This appendix presents some of the basic mathematical concepts that are used frequently in our discussion. We start with the concept of vector space, which is useful for the discussion of matrices and systems of linear equations. We also give a review of complex numbers and how they can be used in linear algebra. This appendix is also devoted to general inner product spaces and how the different notations and processes generalize.

B.1 The Vector Space

In dealing with systems of linear equations we notice that solutions to linear systems can be points in the plane if the equations have two variables, three-space if they are equations in three variables, points in four-space if they have four variables, and so on. The solutions make up subsets of large spaces. Here, we set out to investigate the spaces and their subsets and to develop mathematical structures on them. The spaces that we construct are called *vector spaces* and they arise in many areas of mathematics.

A vector space **V** is intuitively a set together with the operations of addition and multiplication by scalars. If we restrict the scalars to be the set of real numbers, then the vector space **V** is called a *vector space over the real numbers*. If the scalars are allowed to be complex numbers, then it is called *a vector space over the complex numbers*.

Many quantities in the physical sciences are vectors because they have both a magnitude and a direction associated with them. Examples are velocity, force, and angular momentum. We start a discussion of vectors in two dimensions because their properties are easy to visualize and the results are readily extended to three (or more) dimensions.

B.1.1   Vectors in Two Dimensions

This dimension vectors can be defined as ordered pairs of real numbers (a, b) that obey certain algebraic rules. The numbers a and b are called *components* of the vector. The vector (a, b) can be represented geometrically by a directed line segment (arrow) from the origin of a coordinate system to the point. As shown by Figure B.1, we use $\vec{PQ}$ to denote the vector with initial point P and terminal point Q and indicate the direction of the vector by placing an arrow head at Q. The *magnitude* of $\vec{PQ}$ is the length of the segment $\vec{PQ}$ and is denoted by $\|\vec{PQ}\|$. Vectors that have the same length and the same direction are equal.

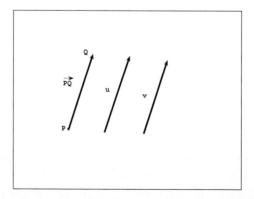

Figure B.1: Geometric representation of vectors.

Definition B.1 (Magnitude of a Vector)

The magnitude (norm or length) of a vector $\mathbf{u} =< u_1, u_2 >$ *is denoted by* $\|\mathbf{u}\|$ *and is defined as*

$$\|\mathbf{u}\| = \| < u_1, u_2 > \| = \sqrt{u_1^2 + u_2^2}.$$

For example, if $\mathbf{u} =< -4, 3 >$*, then*

$$\|\mathbf{u}\| = \| < -4, 3 > \| = \sqrt{(-4)^2 + 3^2} = \sqrt{25} = 5,$$

which is called the magnitude of the given vector.   ●

The norm of a vector can be obtained using MATLAB command window as follows:

```
>> u = [3 − 4];
>> v = norm(u);
```

Operations on Vectors

Let $\mathbf{u} =< u_1, u_2 >$ and $\mathbf{v} =< v_1, v_2 >$ be two vectors, then:

1. We can multiply a vector $\mathbf{u}$ by a scalar k, the result being

$$k\mathbf{u} = k < u_1, u_2 >=< ku_1, ku_2 > .$$

   Geometrically, the magnitude of $\mathbf{u}$ is changed by this operation. If $k > 0$, the length of $\mathbf{u}$ is scaled by a factor of k; if $k < 0$, then the direction of $\mathbf{u}$ is reversed, the magnitude of $\mathbf{u}$. If $k = 0$, we have the *zero vector*, i.e., $\mathbf{0} =< 0, 0 > .$

2. The addition of two vectors is defined as

$$\mathbf{u} + \mathbf{v} =< u_1, u_2 > + < v_1, v_2 >=< u_1 + v_1, u_2 + v_2 > .$$

   The $\mathbf{u} + \mathbf{v}$ is the vector connecting the tail of $\mathbf{u}$ to the head on $\mathbf{v}$.

3. The subtraction of two vectors is defined as

$$\mathbf{u} - \mathbf{v} =< u_1, u_2 > - < v_1, v_2 >=< u_1 - v_1, u_2 - v_2 > .$$

4. The vector addition is commutative and associative

$$\mathbf{u} + \mathbf{v} = \mathbf{v} + \mathbf{u}$$

$$(\mathbf{u} + \mathbf{v}) + \mathbf{w} = \mathbf{u} + (\mathbf{v} + \mathbf{w}).$$

5. The two vectors $\mathbf{i} =< 1, 0 >$ and $\mathbf{j} =< 0, 1 >$ have magnitude 1 and can be used to obtain another way of denoting vectors as

$$\mathbf{u} =< u_1, u_2 >= u_1\mathbf{i} + u_2\mathbf{j}.$$

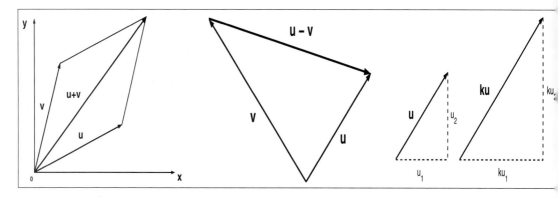

Figure B.2: Operations on vectors.

Definition B.2 (Unit Vector)

If $\mathbf{a} \neq \mathbf{0}$, then the unit vector $\mathbf{u}$ that has the same direction as $\mathbf{a}$ is defined as

$$\mathbf{u} = \frac{\mathbf{a}}{\|\mathbf{a}\|}.$$

For example, the unit vector $\mathbf{u}$ that has the same direction as $4\mathbf{i} - 3\mathbf{j}$ is

$$\mathbf{u} = \frac{4\mathbf{i} - 3\mathbf{j}}{\sqrt{4^2 + (-3)^2}} = \frac{4\mathbf{i} - 3\mathbf{j}}{5} = \frac{4}{5}\mathbf{i} - \frac{3}{5}\mathbf{j},$$

called the unit vector.   ●

The unit vector can be obtained using the MATLAB Command Window as follows:

```
>> a = [4 − 3];
>> u = a/norm(a);
```

Now we define two useful concepts that involve vectors **u** and **v**—the dot product, which is a scalar, and the cross product, which is a vector. First, we define the dot product of two vectors as follows.

Definition B.3 (Dot Product of Vectors)

The multiplication for two vectors $\mathbf{u} =< u_1, u_2 >$ *and* $\mathbf{v} =< v_1, v_2 >$ *is called the dot product (or scalar product) and is symbolized by* **u.v**. *It is defined as*

$$\mathbf{u.v} =< u_1, u_2 > . < v_1, v_2 >= u_1 v_1 + u_2 v_2.$$

For example, if $\mathbf{u} =< 3, -4 >$ *and* $\mathbf{v} =< 2, 3 >$ *are two vectors, then*

$$\mathbf{u.v} =< 3, -4 > . < 2, 3 >= (3)(2) + (-4)(3) = 6 - 12 = -6$$

is their dot product.　　　　　　　　　　　　　　　　　　　　　　　●

The dot product of two vectors can be obtained using the MATLAB Command Window as follows:

```
>> u = [3 − 4];
>> v = [2 3];
>> dot(u, v);
```

Theorem B.1 *If* **u, v**, *and* **w** *are vectors and k is a scalar, then these properties hold:*

1.　　**u.v = v.u.**

2.　　**u.(v + w) = u.v + u.w.**

3.　　$k(\mathbf{u.v}) = (k\mathbf{u}).\mathbf{v} = \mathbf{u}.(k\mathbf{v})$.

4.　　**0.u = u.0 = 0.**

5. $\mathbf{u}.\mathbf{u} = \|\mathbf{u}\|.$ ●

The dot product of two vectors can be also defined as follows.

Definition B.4 (Dot Product of Vectors)

If $\mathbf{u}$ and $\mathbf{v}$ are nonzero vectors, and θ is the angle between them, the dot product of $\mathbf{u}$ and $\mathbf{v}$ is defined as follows:

$$\mathbf{u}.\mathbf{v} = \|\mathbf{u}\|\|\mathbf{v}\|\cos\theta.$$

By the angle between vectors $\mathbf{u}$ and $\mathbf{v}$, we mean the smallest nonnegative angle between them, so $0 \le \theta \le \pi$.

For example, finding the angle between $\mathbf{u} =< 4, 3 >$ and $\mathbf{v} = (2, 5)$, we do the following:

$$< 4, 3 > . < 2, 5 >= \| < 4, 3 > \|\| < 2, 5 > \|\cos\theta$$

or

$$\cos\theta = \frac{< 4, 3 > . < 2, 5 >}{\| < 4, 3 > \|\| < 2, 5 > \|},$$

which gives

$$\theta = \cos^{-1}\left(\frac{23}{5\sqrt{29}}\right) \approx 0.5468 \ radian = 31.3287 \ degree,$$

which is called the angle between the given vectors.   ●

The angle between two vectors can be obtained using the MATLAB Command Window as follows:

```
>> u = [4 3];
>> v = [2 5];
>> T = a cos(dot(u, v)/(norm(u) * norm(v)));
>> (360 * T)/(2 * pi);
```

Theorem B.2 *Let $\mathbf{u}$ and $\mathbf{v}$ be two vectors and k is a scalar, then:*

1. $\mathbf{u}$ *and* $\mathbf{v}$ *are orthogonal if and only if* $\mathbf{u}.\mathbf{v} = 0.$

2. **u** *and* **v** *are parallel if and only if* $\mathbf{v} = k\mathbf{u}$.

For example, the vectors $\mathbf{u} =< 4, 3 >$ *and* $\mathbf{v} =< 3, -4 >$ *are orthogonal vectors because*

$$\mathbf{u}.\mathbf{v} =< 4, 3 > . < 3, -4 >= 12 - 12 = 0,$$

while the vectors $\mathbf{u} =< 1, -2 >$ *and* $\mathbf{v} =< 2, -4 >$ *are parallel vectors because*

$$\mathbf{v} =< 2, -4 >= \frac{1}{2} < 1, -2 >= \frac{1}{2}\mathbf{u}.$$

•

B.1.2 Vectors in Three Dimensions

Three-dimensional vectors can be treated as ordered triplets of three numbers and obey rules very similar to those obeyed by two-dimensional vectors. We represent three-dimensional vectors by arrows and the geometric interpretation of the addition and subtraction of these vectors follows a parallelogram rule just as it does in two dimensions. We define unit vectors $\mathbf{i}, \mathbf{j}$, and $\mathbf{k}$ along the x, y, and z axes of a cartesian coordinate system and express three-dimensional vectors as

$$\mathbf{u} = u_1\mathbf{i} + u_2\mathbf{j} + u_3\mathbf{k},$$

in terms of ordered triplets of the real numbers

$$\mathbf{i} =< 1, 0, 0 >, \quad \mathbf{j} =< 0, 1, 0 >, \quad \mathbf{k} =< 0, 0, 1 > .$$

Note that:
$$\begin{aligned} \mathbf{i}.\mathbf{i} &= \mathbf{j}.\mathbf{j} = \mathbf{k}.\mathbf{k} \\ \mathbf{i}.\mathbf{j} &= \mathbf{j}.\mathbf{i} = \mathbf{i}.\mathbf{k} = \mathbf{k}.\mathbf{i} = \mathbf{j}.\mathbf{k} = \mathbf{k}.\mathbf{j} = 0. \end{aligned}$$

Definition B.5 (Distance Between Points)

The distance between two points $P_1(x_1, y_1, z_1)$ *and* $P_2(x_2, y_2, z_2)$ *is denoted by* $d(P_1, P_2)$ *and is defined as*

$$d(P_1, P_2) = \sqrt{(x_2 - x_1)^2 + (y_2 - y_1)^2 + (z_2 - z_1)^2}.$$

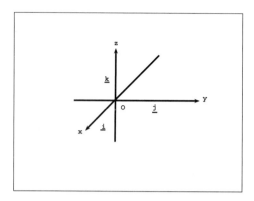

Figure B.3: Geometric representation of three-dimensional space.

If the points P_1 and P_2 are on the xy-plane so that $z_1 = z_2 = 0$, then the above distance formula reduces to the two-dimensional distance formula

$$d(P_1, P_2) = \sqrt{(x_2 - x_1)^2 + (y_2 - y_1)^2}.$$

Example B.1 *Find the distance between $P_1(-3, 2, 4)$ and $P_2(3, 5, 2)$.*

Solution. *Using the above distance formula, we have*

$$
\begin{aligned}
d(P_1, P_2) &= \sqrt{(3+3)^2 + (5-2)^2 + (2-4)^2} \\
&= \sqrt{36 + 9 + 4} \\
&= \sqrt{49} = 7,
\end{aligned}
$$

which is the required distance between the given points.   ●

The definitions stated for two-dimensional extend to three-dimensional, the only change is the inclusion of a third component for each vector. Magnitude of a vector, vector addition, and scalar multiples of vectors are defined as follows:

$$
\begin{aligned}
\|\mathbf{u}\| &= \sqrt{u_1^2 + u_2^2 + u_3^2} \\
\mathbf{u} + \mathbf{v} &= <u_1 + v_1, u_2 + v_2, u_3 + v_3> \\
k\mathbf{u} &= k <u_1, u_2, u_3> = <ku_1, ku_2, ku_3> \\
-\mathbf{u} &= - <u_1, u_2, u_3> = <-u_1, -u_2, -u_3> \\
\mathbf{0} &= <0, 0, 0>.
\end{aligned}
$$

Example B.2 *If* $\mathbf{u} =< 3, 4, -2 >$ *and* $\mathbf{v} =< -5, 7, 6 >$, *then find* $\mathbf{u} + \mathbf{v}$, $4\mathbf{u} - 3\mathbf{v}$, *and* $\|\mathbf{u}\|$.

Solution. *Using the given vectors, we have*

$$
\begin{aligned}
\mathbf{u} + \mathbf{v} &= \; < 3, 4, -2 > + < -5, 7, 6 >=< -2, 11, 4 > \\
4\mathbf{u} - 3\mathbf{v} &= \; 4 < 3, 4, -2 > -3 < -5, 7, 6 >=< 12, 16, -8 > - < -15, 21, 18 > \\
&= \; < 27, -5, -26 > \\
\|\mathbf{u}\| &= \; \sqrt{3^2 + 4^2 + (-2)^2} = \sqrt{9 + 16 + 4} = \sqrt{29},
\end{aligned}
$$

which is the required operations on the given vectors.　　　　　　　　　●

Also, if $\mathbf{u} =< u_1, u_2, u_3 >$ *and* $\mathbf{v} =< v_1, v_2, v_3 >$, *then their dot product is defined as*

$$
\mathbf{u}.\mathbf{v} = u_1 v_1 + u_2 v_2 + u_3 v_3,
$$

and the magnitude of the vectors $\mathbf{u}$ *and* $\mathbf{v}$ *is defined as*

$$
\|\mathbf{u}\| \; = \; \| < u_1, u_2, u_3 > \| = \sqrt{u_1^2 + u_2^2 + u_3^3}
$$

$$
\|\mathbf{v}\| \; = \; \| < v_1, v_2, v_3 > \| = \sqrt{v_1^2 + v_2^2 + v_3^3}.
$$

Example B.3 *If* $\mathbf{u} =< 5, -7, 8 >$ *and* $\mathbf{v} =< -3, 6, 5 >$, *then find the dot product of the vectors and the angle between the vectors.*

Solution. *Using the given vectors, we have*

$$
\begin{aligned}
\mathbf{u}.\mathbf{v} &= \; < 5, -7, 8 > . < -3, 6, 5 > \\
&= \; (5)(-3) + (-7)(6) + (8)(5) = -15 - 42 + 40 \\
&= \; -17,
\end{aligned}
$$

which is the required dot product of the given vectors.
The angle between the given vectors is defined as

$$
\begin{aligned}
\cos \theta \; &= \; \frac{\mathbf{u}.\mathbf{v}}{\|\mathbf{u}\|\|\mathbf{v}\|} \\
&= \; \frac{-17}{\sqrt{25 + 49 + 64}\sqrt{9 + 36 + 25}} \\
&= \; \frac{-17}{\sqrt{138}\sqrt{70}} = \frac{-17}{\sqrt{9660}}.
\end{aligned}
$$

Hence,

$$\theta = \cos^{-1}\left(\frac{-17}{\sqrt{9660}}\right) \approx 99.96^{\circ},$$

which is the required angle between the given vectors.   •

Definition B.6 (Direction Angles and Cosines)

The smallest nonnegative angles $\alpha, \beta,$ and γ between a nonzero vector $\mathbf{u}$ and the basis vectors $\mathbf{i}, \mathbf{j},$ and $\mathbf{k}$ are called the direction angles of $\mathbf{u}$. The cosines of these direction angles, $\cos \alpha, \cos \beta,$ and $\cos \gamma$, are called the direction cosines of the vector $\mathbf{u}$.
If $\mathbf{u} = u_1 \mathbf{i} + u_2 \mathbf{j} + u_3 \mathbf{k}$, then

$$\mathbf{u}.\mathbf{i} = \|\mathbf{u}\|\|\mathbf{i}\| \cos \alpha = \|\mathbf{u}\| \cos \alpha$$

and

$$\mathbf{u}.\mathbf{i} = < u_1, u_2, u_3 > . < 1, 0, 0 > = u_1,$$

and it follows that

$$\cos \alpha = \frac{u_1}{\|\mathbf{u}\|}.$$

By similar reasoning with the basis vectors $\mathbf{j}$ and $\mathbf{k}$, we have

$$\cos \beta = \frac{u_2}{\|\mathbf{u}\|} \quad and \quad \cos \gamma = \frac{u_3}{\|\mathbf{u}\|},$$

where $\alpha, \beta,$ and γ, are respectively, are the angles between $\mathbf{u}$ and $\mathbf{i}$, $\mathbf{u}$ and $\mathbf{j}$, and $\mathbf{u}$ and $\mathbf{k}$.
Consequently, any nonzero vector $\mathbf{u}$ in space has the normalized form

$$\frac{u}{\|\mathbf{u}\|} = \frac{u_1}{\|\mathbf{u}\|}\mathbf{i} + \frac{u_2}{\|\mathbf{u}\|}\mathbf{j} + \frac{u_3}{\|\mathbf{u}\|}\mathbf{k} = \cos \alpha + \cos \beta + \cos \gamma,$$

and because $\dfrac{u}{\|\mathbf{u}\|}$ is a unit vector, it follows that

$$\cos^2 \alpha + \cos^2 \beta + \cos^2 \gamma = 1.$$

Note that the vector $< \cos \alpha, \cos \beta, \cos \gamma >$ is a unit vector with the same direction as the original vector $\mathbf{u}$.   •

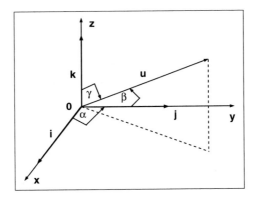

Figure B.4: Direction angles of a vector.

Example B.4 *Find the direction cosines and angles for the vector* $\mathbf{u} = 3\mathbf{i} + 6\mathbf{j} + 2\mathbf{k}$, *and show that* $\cos^2 \alpha + \cos^2 \beta + \cos^2 \gamma = 1$.

Solution. *Because*

$$\|\mathbf{u}\| = \sqrt{3^2 + 6^2 + 2^2} = \sqrt{9 + 36 + 4} = \sqrt{49} = 7,$$

we can write it as

$$\cos \alpha = \frac{u_1}{\|\mathbf{u}\|} = \frac{3}{7},$$

and it gives

$$\alpha = \cos^{-1} \frac{3}{7} \approx 64.62^o.$$

Similarly,

$$\cos \beta = \frac{u_2}{\|\mathbf{u}\|} = \frac{6}{7}, \quad \beta = \cos^{-1} \frac{6}{7} \approx 31.00^o$$

and

$$\cos \gamma = \frac{u_3}{\|\mathbf{u}\|} = \frac{2}{7}, \quad \beta = \cos^{-1} \frac{2}{7} \approx 73.40^o.$$

Furthermore, the sum of the squares of the direction cosines is

$$\cos^2 \alpha + \cos^2 \beta + \cos^2 \gamma = \frac{9}{49} + \frac{36}{49} + \frac{4}{49} = 1.$$

●

Definition B.7 (Component of Vector Along a Vector)

Let $\mathbf{u}$ *and* $\mathbf{v}$ *be nonzero vectors. Then the component of a vector (also called the scalar projection)* $\mathbf{u}$ *along vector* $\mathbf{v}$ *is denoted by* $comp_\mathbf{v}\mathbf{u}$ *and defined as*

$$comp_\mathbf{v}\mathbf{u} = \mathbf{u}.\frac{\mathbf{v}}{\|\mathbf{v}\|} = \frac{\mathbf{u}.\mathbf{v}}{\|\mathbf{v}\|} = \frac{\|\mathbf{u}\|\|\mathbf{v}\|\cos\theta}{\|\mathbf{v}\|} = \|\mathbf{u}\|\cos\theta.$$

Note that if $\mathbf{u} = u_1\mathbf{i} + u_2\mathbf{j} + u_3\mathbf{k}$, *then by the above definition*

$$comp_\mathbf{i}\mathbf{u} = \mathbf{u}.\mathbf{i} = u_1$$
$$comp_\mathbf{j}\mathbf{u} = \mathbf{u}.\mathbf{j} = u_2$$
$$comp_\mathbf{k}\mathbf{u} = \mathbf{u}.\mathbf{k} = u_3.$$

Thus, the components of $\mathbf{u}$ *along* $\mathbf{i}, \mathbf{j},$ *and* $\mathbf{k}$ *are the same as the components* $u_1, u_2,$ *and* u_3 *of the vector* $\mathbf{u}$.   ●

Example B.5 *If* $\mathbf{u} = 3\mathbf{i} + 2\mathbf{j} - 6\mathbf{k}$ *and* $\mathbf{v} = 2\mathbf{i} + 2\mathbf{j} + \mathbf{k}$, *then find* $comp_\mathbf{v}\mathbf{u}$ *and* $comp_\mathbf{u}\mathbf{v}$.

Solution. *Using the above definition, we have*

$$comp_\mathbf{v}\mathbf{u} = \mathbf{u}.\frac{\mathbf{v}}{\|\mathbf{v}\|} = (3\mathbf{i} + 2\mathbf{j} - 6\mathbf{k}).\frac{1}{3}(2\mathbf{i} + 2\mathbf{j} + \mathbf{k}),$$

since

$$\|\mathbf{v}\| = \sqrt{2^2 + 2^2 + 1^2} = \sqrt{9} = 3.$$

Thus,

$$comp_\mathbf{v}\mathbf{u} = \frac{(3)(2) + (2)(2) + (-6)(1)}{3} = \frac{4}{3}.$$

Similarly, we compute

$$comp_\mathbf{u}\mathbf{v} = \mathbf{v}.\frac{\mathbf{u}}{\|\mathbf{u}\|} = (2\mathbf{i} + 2\mathbf{j} + \mathbf{k}).\frac{1}{7}(3\mathbf{i} + 2\mathbf{j} - 6\mathbf{k})(2\mathbf{i} + 2\mathbf{j} + \mathbf{k}),$$

where

$$\|\mathbf{u}\| = \sqrt{3^2 + 2^2 + (-6)^2} = \sqrt{49} = 7.$$

Thus,

$$comp_{\mathbf{u}}\mathbf{v} = \frac{(2)(3) + (2)(2) + (1)(-6)}{7} = \frac{4}{7},$$

the required solution.   ●

To get the results of Example B.5, we use the MATLAB Command Window as follows:

```
>> u = [3 2 6];
>> v = [2 2 1];
>> compu = (dot(u, v))/norm(v);
>> compv = (dot(u, v))/norm(u);
```

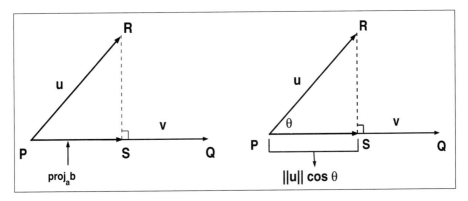

Figure B.5: Projections of a vector onto a vector.

Definition B.8 (Projection of a Vector onto a Vector)

If $\mathbf{u}$ *and* $\mathbf{v}$ *are nonzero vectors, then the projection of vector* $\mathbf{u}$ *onto vector* $\mathbf{v}$ *is denoted by* $proj_{\mathbf{v}}\mathbf{u}$ *and is defined as*

$$proj_{\mathbf{v}}\mathbf{u} = \left(\frac{\mathbf{u}.\mathbf{v}}{\|\mathbf{v}\|^2}\right)\mathbf{v}.$$

Note that the projection of $\mathbf{u}$ *onto* $\mathbf{v}$ *can be written as a scalar multiple of a unit vector in the direction of* $\mathbf{v}$, *i.e.,*

$$proj_{\mathbf{v}}\mathbf{u} = \left(\frac{\mathbf{u}.\mathbf{v}}{\|\mathbf{v}\|^2}\right)\mathbf{v} = \frac{(\mathbf{u}.\mathbf{v})}{\|\mathbf{v}\|}\frac{\mathbf{v}}{\|\mathbf{v}\|} = K\frac{\mathbf{v}}{\|\mathbf{v}\|},$$

where

$$K = \frac{\mathbf{u}.\mathbf{v}}{\|\mathbf{v}\|} = \|\mathbf{u}\| \cos \theta$$

is called the component of $\mathbf{u}$ *in the direction of* $\mathbf{v}$.   ●

Example B.6 *If* $\mathbf{u} = 4\mathbf{i} - 5\mathbf{j} + 3\mathbf{k}$ *and* $\mathbf{v} = 6\mathbf{i} - 3\mathbf{j} + 2\mathbf{k}$, *then find the projection of* $\mathbf{u}$ *onto* $\mathbf{v}$.

Solution. *Since*

$$\mathbf{u}.\mathbf{v} = (4)(6) + (-5)(-3) + (3)(2) = 24 + 15 + 6 = 45$$

and

$$\|\mathbf{v}\| = \sqrt{6^2 + (-3)^2 + 2^2} = \sqrt{36 + 9 + 4} = \sqrt{49} = 7,$$

using the above definition, we have

$$proj_{\mathbf{v}}\mathbf{u} = \left(\frac{\mathbf{u}.\mathbf{v}}{\|\mathbf{v}\|^2}\right)\mathbf{v} = \left(\frac{45}{49}\right)(6\mathbf{i} - 3\mathbf{j} + 2\mathbf{k})$$

or

$$proj_{\mathbf{v}}\mathbf{u} = \frac{270}{49}\mathbf{i} - \frac{135}{49}\mathbf{j} + \frac{90}{49}\mathbf{k},$$

which is the required projection of $\mathbf{u}$ *onto* $\mathbf{v}$.   ●

To get the results of Example B.6, we use the MATLAB Command Window as follows:

```
>> u = [4  − 5 3];
>> v = [6  − 3 2];
>> K = dot(u, v)/norm(v);
>> X = v/norm(v);
>> Proj = K * X;
```

Definition B.9 (Work Done)
The work done by a constant force $\mathbf{F} = \vec{PR}$ *as its point of application moves along the vector* $\mathbf{D} = \vec{PQ}$ *(displacement vector) and is defined as*

$$W = \mathbf{F}.\mathbf{D} = \|\mathbf{F}\|\|\mathbf{D}\| \cos \theta.$$

Thus, the work done by a constant force is the dot product of the vectors.●

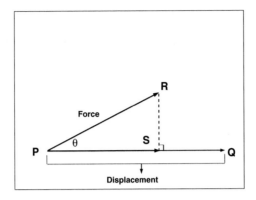

Figure B.6: Work done by a force.

Example B.7 *A force is given by a vector* $\mathbf{F} = 6\mathbf{i} + 4\mathbf{j} + 7\mathbf{k}$ *and moves a particle from the point* $P(2, -3, 4)$ *to the point* $Q(5, 4, -2)$*. Find the work done.*

Solution. *The vector* $\mathbf{D}$ *that corresponds to* $\vec{PQ}$ *is*

$$\mathbf{D} = <5 - 2, 4 + 3, -2 - 4> = <3, 7, -6>.$$

If $\vec{PR}$ *corresponds to* $\mathbf{F}$*, then the work done is*

$$\begin{aligned} W = \vec{PR}.\vec{PQ} = \mathbf{F}.\mathbf{D} &= <6, 4, 7> . <3, 7, -6> \\ &= 18 + 28 - 42 = 4. \end{aligned}$$

If, for example, the distance is in feet and the magnitude of the force is in pounds, then the work done is 4 ft-lb. If the distance is in meters and the force is in Newtons, then the work done is 4 joules. ●

Now we define the cross product of two vectors in three-dimensional space as follows.

Definition B.10 (Cross Product of Vectors)

The other way to multiply two vectors $\mathbf{u} =< (u_1, u_2, u_3 >$ *and* $\mathbf{v} =< v_1, v_2, v_3 >$ *is known as the cross product (or vector product) and is symbolized by* $\mathbf{u} \times \mathbf{v}$. *It is defined as*

$$\mathbf{u} \times \mathbf{v} = \begin{vmatrix} \mathbf{i} & \mathbf{j} & \mathbf{k} \\ u_1 & u_2 & u_3 \\ v_1 & v_2 & v_3 \end{vmatrix} = \begin{vmatrix} u_2 & u_3 \\ v_2 & v_3 \end{vmatrix} \mathbf{i} - \begin{vmatrix} u_1 & u_3 \\ v_1 & v_3 \end{vmatrix} \mathbf{j} + \begin{vmatrix} u_1 & u_2 \\ v_1 & v_2 \end{vmatrix} \mathbf{k}.$$

For example, if $\mathbf{u} =< 1, -1, 2 >$ *and* $\mathbf{v} =< 2, -1, -2 >$ *are two vectors, then their cross product is defined as*

$$\mathbf{u} \times \mathbf{v} = \begin{vmatrix} \mathbf{i} & \mathbf{j} & \mathbf{k} \\ 1 & -1 & 2 \\ 2 & -1 & -2 \end{vmatrix} = \begin{vmatrix} -1 & 2 \\ -1 & -2 \end{vmatrix} \mathbf{i} - \begin{vmatrix} 1 & 2 \\ 2 & -2 \end{vmatrix} \mathbf{j} + \begin{vmatrix} 1 & -1 \\ 2 & -1 \end{vmatrix} \mathbf{k}.$$

By evaluating these determinants, we get

$$\mathbf{u} \times \mathbf{v} = 4\mathbf{i} + 6\mathbf{j} + \mathbf{k},$$

the cross product of the vectors, which is also the vector.   ●

Theorem B.3 *Let* $\mathbf{u}$ *and* $\mathbf{v}$ *be vectors in three dimensions and* θ *be the angle between them, then:*

1. $\mathbf{u}.(\mathbf{u} \times \mathbf{v}) = \mathbf{v}.(\mathbf{u} \times \mathbf{v}) = 0.$

2. $\|\mathbf{u} \times \mathbf{v}\| = \|\mathbf{u}\|\|\mathbf{v}\| \sin \theta.$ ●

Note that:
$$\begin{array}{rll} \mathbf{i} \times \mathbf{j} &= \mathbf{k}, & \mathbf{j} \times \mathbf{k} = \mathbf{i}, \quad \mathbf{k} \times \mathbf{i} = \mathbf{j} \\ \mathbf{j} \times \mathbf{i} &= -\mathbf{k}, & \mathbf{k} \times \mathbf{j} = -\mathbf{i}, \quad \mathbf{i} \times \mathbf{k} = -\mathbf{j} \\ \mathbf{i} \times \mathbf{i} &= \mathbf{j} \times \mathbf{j} = \mathbf{k} \times \mathbf{k} = 0. \end{array}$$

Theorem B.4 *Two vectors* $\mathbf{u}$ *and* $\mathbf{v}$ *are parallel, if and only if*

$$\mathbf{u} \times \mathbf{v} = 0.$$

For example, the vectors $\mathbf{u} =< -6, -10, 4 >$ *and* $\mathbf{v} =< 3, 5, -2 >$ *are parallel because*

$$\mathbf{u} \times \mathbf{v} = \begin{vmatrix} \mathbf{i} & \mathbf{j} & \mathbf{k} \\ -6 & -10 & 4 \\ 3 & 5 & -2 \end{vmatrix} = \begin{vmatrix} -10 & 4 \\ 5 & -2 \end{vmatrix} \mathbf{i} - \begin{vmatrix} -6 & 4 \\ 3 & -2 \end{vmatrix} \mathbf{j} + \begin{vmatrix} -6 & -10 \\ 3 & 5 \end{vmatrix} \mathbf{k},$$

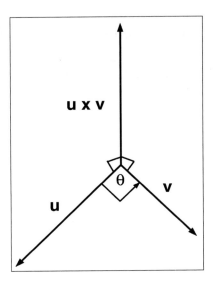

Figure B.7: Cross product of the vectors.

and it gives

$$\mathbf{u} \times \mathbf{v} = (20 - 20)\mathbf{i} - (12 - 12)\mathbf{j} + (-30 + 30)\mathbf{k} = 0.$$

•

Note that the length of the cross product $\mathbf{u} \times \mathbf{v}$ is equal to the *area of the parallelogram* determined by the vectors $\mathbf{u}$ and $\mathbf{v}$, i.e.,

$$\text{Area of the parallelogram} = A = \|\mathbf{u} \times \mathbf{v}\|.$$

Also, the area of the triangle is half of the area of the parallelogram, i.e.,

$$\text{Area of triangle} = A = \frac{1}{2}\|\mathbf{u} \times \mathbf{v}\|.$$

Example B.8 *Find the area of the parallelogram made by $\vec{PQ}$ and $\vec{PR}$, where $P(3, 1, 2), Q(2, -1, 1),$ and $R(4, 2, -1)$ are the points in the plane.*

Solution. *Since*

$$\vec{PQ} = (2 - 3)\mathbf{i} + (-1 - 1)\mathbf{j} + (1 - 2)\mathbf{k} = -\mathbf{i} - 2\mathbf{j} - \mathbf{k}$$

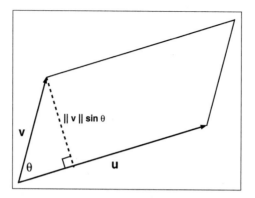

Figure B.8: Length of the cross product of the vectors.

and

$$\vec{PR} = (4-3)\mathbf{i} + (2-1)\mathbf{j} + (-1-2)\mathbf{k} = \mathbf{i} + \mathbf{j} - 3\mathbf{k},$$

their cross product is defined as follows:

$$\vec{PQ} \times \vec{PR} = \begin{vmatrix} \mathbf{i} & \mathbf{j} & \mathbf{k} \\ -1 & -2 & -1 \\ 1 & 1 & -3 \end{vmatrix} = \begin{vmatrix} -2 & -1 \\ 1 & -3 \end{vmatrix}\mathbf{i} - \begin{vmatrix} -1 & -1 \\ 1 & -3 \end{vmatrix}\mathbf{j} + \begin{vmatrix} -1 & -2 \\ 1 & 1 \end{vmatrix}\mathbf{k}$$

$$= (-6+1)\mathbf{i} - (-3+1)\mathbf{j} + (-1+2)\mathbf{k}$$

$$= -5\mathbf{i} + 2\mathbf{j} + \mathbf{k}.$$

Thus,

$$A = \|\vec{PQ} \times \vec{PR}\| = \sqrt{(-5)^2 + 2^2 + 1^2} = \sqrt{30} \ unit^2$$

is the required area of the parallelogram. ●

Example B.9 *Find a vector perpendicular to the plane that passes through the points* $P(2, 1, 4), Q(-3, 4, -2),$ *and* $R(2, -2, 1).$

Solution. *The* $\vec{PQ} \times \vec{PR}$ *is perpendicular to both* $\vec{PQ}$ *and* $\vec{PR}$ *and therefore perpendicular to the plane through* $P, Q,$ *and* R. *Since*

$$\vec{PQ} = (-3-2)\mathbf{i} + (4-1)\mathbf{j} + (-2-4)\mathbf{k} = -5\mathbf{i} + 3\mathbf{j} - 6\mathbf{k}$$

and

$$\vec{PR} = (2-2)\mathbf{i} + (-2-1)\mathbf{j} + (1-4)\mathbf{k} = 0\mathbf{i} - 3\mathbf{j} - 3\mathbf{k},$$

their cross product is defined as follows:

$$\vec{PQ} \times \vec{PR} = \begin{vmatrix} \mathbf{i} & \mathbf{j} & \mathbf{k} \\ -5 & 3 & -6 \\ 0 & -3 & -3 \end{vmatrix} = \begin{vmatrix} 3 & -6 \\ -3 & -3 \end{vmatrix} \mathbf{i} - \begin{vmatrix} -5 & -6 \\ 0 & -3 \end{vmatrix} \mathbf{j} + \begin{vmatrix} -5 & 3 \\ 0 & -3 \end{vmatrix} \mathbf{k}$$

$$= (-9 - 18)\mathbf{i} - (-15 - 0)\mathbf{j} + (15 - 0)\mathbf{k}$$

$$= -27\mathbf{i} - 15\mathbf{j} + 15\mathbf{k}.$$

Thus, the vector $(-27, -15, 15)$ *is perpendicular to the given plane. Any nonzero scalar multiple of this vector, such as* $(-9, 5, 5)$, *is also perpendicular to the plane.* •

Example B.10 *Find the area of the triangle with vertices* $P(2, 1, 1)$, $Q(3, 1, 2)$, *and* $R(1, -2, 1)$.

Solution. *Since the area of the triangle is half of the area of the parallelogram, we compute first the area of the parallelogram. The area of the parallelogram with adjacent sides PQ and PR is the length of the cross product* $\vec{PQ} \times \vec{PR}$, *therefore, we find the vectors* $\vec{PQ}$ *and* $\vec{PR}$ *as follows:*

$$\vec{PQ} = (3 - 2)\mathbf{i} + (1 - 1)\mathbf{j} + (2 - 1)\mathbf{k} = \mathbf{i} + 0\mathbf{j} + \mathbf{k},$$

and

$$\vec{PR} = (1 - 2)\mathbf{i} + (-2 - 1)\mathbf{j} + (1 - 1)\mathbf{k} = -\mathbf{i} - 3\mathbf{j} + 0\mathbf{k}.$$

Now we compute the cross product of these two vectors as follows:

$$\vec{PQ} \times \vec{PR} = \begin{vmatrix} \mathbf{i} & \mathbf{j} & \mathbf{k} \\ 1 & 0 & 1 \\ -1 & -3 & 0 \end{vmatrix} = \begin{vmatrix} 0 & 1 \\ -3 & 0 \end{vmatrix} \mathbf{i} - \begin{vmatrix} 1 & 1 \\ -1 & 0 \end{vmatrix} \mathbf{j} + \begin{vmatrix} 1 & 0 \\ -1 & -3 \end{vmatrix} \mathbf{k}$$

$$= (0 + 3)\mathbf{i} - (0 + 1)\mathbf{j} + (-3 - 0)\mathbf{k}$$

$$= 3\mathbf{i} - \mathbf{j} - 3\mathbf{k}.$$

Thus, the length of this cross product is

$$\|\vec{PQ} \times \vec{PR}\| = \sqrt{3^2 + (-1)^2 + (-3)^2} = \sqrt{19} \ unit^2,$$

which is the area of the parallelogram. The area A of the triangle PQR is half the area of this parallelogram, i.e.,

$$A = \frac{\sqrt{19}}{2} \ unit^2$$

is the required area of the triangle. •

Theorem B.5 *Let* **u**, **v**, *and* **w** *be three vectors and* k *is a scalar, then:*

1. $\mathbf{u} \times \mathbf{v} = -(\mathbf{v} \times \mathbf{u})$.

2. $(k\mathbf{u}) \times \mathbf{v} = k(\mathbf{u} \times \mathbf{v}) = \mathbf{u} \times (k\mathbf{v})$.

3. $\mathbf{u} \times (\mathbf{v} + \mathbf{w}) = (\mathbf{u} \times \mathbf{v}) + (\mathbf{u} \times \mathbf{w})$.

4. $(\mathbf{u} + \mathbf{v}) \times \mathbf{w} = (\mathbf{u} \times \mathbf{w}) + (\mathbf{v} \times \mathbf{w})$.

5. $(\mathbf{u} \times \mathbf{v}).\mathbf{w} = \mathbf{u}.(\mathbf{v} \times \mathbf{w})$.

6. $\mathbf{u} \times (\mathbf{v} \times \mathbf{w}) = (\mathbf{u}.\mathbf{w})\mathbf{v} - (\mathbf{u}.\mathbf{v})\mathbf{w}$.   ●

Note that the product $\mathbf{u}.(\mathbf{v} \times \mathbf{w})$ that occurs $5th$ in Theorem B.5 is called the *scalar triple product* of the vectors **u**, **v**, and **w**. We can write the scalar triple product of the vectors as a determinant:

$$\mathbf{u}.(\mathbf{v} \times \mathbf{w}) = \begin{vmatrix} u_1 & u_2 & u_3 \\ v_1 & v_2 & v_3 \\ w_1 & w_2 & w_3 \end{vmatrix}.$$

Example B.11 *Find the scalar triple product of the vectors* $\mathbf{u} = 2\mathbf{i} + \mathbf{j} + 3\mathbf{k}$, $\mathbf{v} = 3\mathbf{i} + 2\mathbf{j} + 4\mathbf{k}$, *and* $\mathbf{w} = 4\mathbf{i} + 3\mathbf{j} + 5\mathbf{k}$.

Solution. *We use the following determinant to compute the scalar triple product of the given vectors as follows:*

$$\begin{aligned} \mathbf{u}.(\mathbf{v} \times \mathbf{w}) &= \begin{vmatrix} 2 & 2 & 3 \\ 3 & 2 & 4 \\ 4 & 3 & 5 \end{vmatrix} = 2\begin{vmatrix} 2 & 4 \\ 3 & 5 \end{vmatrix} - 2\begin{vmatrix} 3 & 4 \\ 4 & 5 \end{vmatrix} + 3\begin{vmatrix} 3 & 2 \\ 4 & 3 \end{vmatrix} \\ &= 2(10 - 12) - 2(15 - 16) + 3(9 - 8) \\ &= -4 + 2 + 3 = 1, \end{aligned}$$

which is the required scalar triple product of the given vectors.   ●

To get the scalar triple product of the given vectors of Example B.11, we use the MATLAB Command Window as follows:

```
>> u = [2 1 3];
>> v = [3 2 4];
>> w = [4 3 5];
>> x = cross(v, w);
>> y = dot(u, x);
```

Note that the *volume of the parallelepiped* determined by the vectors $\mathbf{u}, \mathbf{v}$, and $\mathbf{w}$ is the magnitude of their scalar triple product:

$$\text{Volume of the parallelepiped} = V = |\mathbf{u}.(\mathbf{v} \times \mathbf{w})|.$$

Example B.12 *Find the volume of the parallelepiped having adjacent sides AB, AC, and AD, where*

$$A(0, 1, 0), \ B(2, -2, 3), \ C(1, 1, -1), \ and \ D(4, -1, -1).$$

Solution. *Since*

$$\mathbf{u} = \vec{AB} = (2 - 0)\mathbf{i} + (-2 - 1)\mathbf{j} + (3 - 0)\mathbf{k} = 2\mathbf{i} - 3\mathbf{j} + 3\mathbf{k}$$
$$\mathbf{v} = \vec{AC} = (1 - 0)\mathbf{i} + (1 - 1)\mathbf{j} + (-1 - 0)\mathbf{k} = \mathbf{i} + 0\mathbf{j} - \mathbf{k}$$
$$\mathbf{w} = \vec{AD} = (4 - 0)\mathbf{i} + (-1 - 1)\mathbf{j} + (-1 - 0)\mathbf{k} = 4\mathbf{i} - 2\mathbf{j} - \mathbf{k},$$

use the following determinant to compute the scalar triple product of the given vectors as follows:

$$\mathbf{u}.(\mathbf{v} \times \mathbf{w}) = \begin{vmatrix} 2 & -3 & 3 \\ 1 & 0 & -1 \\ 4 & -2 & -1 \end{vmatrix} = 2 \begin{vmatrix} 0 & -1 \\ -2 & -1 \end{vmatrix} + 3 \begin{vmatrix} 1 & -1 \\ 4 & -1 \end{vmatrix} + 3 \begin{vmatrix} 1 & 0 \\ 4 & -2 \end{vmatrix}$$
$$= 2(0 - 2) + 3(-1 + 4) + 3(-2 - 0)$$
$$= -4 + 9 - 6 = -1,$$

which is the scalar triple product of the given vectors. Thus,

$$V = |\mathbf{u}.(\mathbf{v} \times \mathbf{w})| = |-1| = 1$$

is the volume of the parallelepiped. ●

To get the volume of the parallelepiped of Example B.12, we use the MATLAB Command Window as follows:

```
>> u = [2 − 3 3];
>> v = [1 0 − 1];
>> w = [4 − 2 − 1];
>> x = cross(v, w);
>> y = dot(u, x);
>> v = abs(y);
```

Note that if the volume of the parallelepiped determined by the vectors **u**, **v**, and **w** is zero, then the vectors must lie in the same plane; i.e., they are *coplanar*.

Example B.13 *Use the scalar triple product to show that the vectors* **u** = 4**i** + 6**j** + 2**k**, **v** = 2**i** − 2**j**, *and* **w** = 14**i** + 6**j** + 4**k** *are coplanar.*

Solution. *Given*

$$
\begin{aligned}
\mathbf{u} &= 4\mathbf{i} + 6\mathbf{j} + 2\mathbf{k} \\
\mathbf{v} &= 2\mathbf{i} - 2\mathbf{j} \\
\mathbf{w} &= 14\mathbf{i} + 6\mathbf{j} + 4\mathbf{k},
\end{aligned}
$$

we use the following determinant to compute the scalar triple product of the given vectors as follows:

$$
\begin{aligned}
\mathbf{u}.(\mathbf{v} \times \mathbf{w}) &= \begin{vmatrix} 4 & 6 & 2 \\ 2 & -2 & 0 \\ 14 & 6 & 4 \end{vmatrix} = 4 \begin{vmatrix} -2 & 0 \\ 6 & 4 \end{vmatrix} - 6 \begin{vmatrix} 2 & 0 \\ 14 & 4 \end{vmatrix} + 2 \begin{vmatrix} 2 & -2 \\ 14 & 6 \end{vmatrix} \\
&= 4(-8 - 0) - 6(8 - 0) + 2(12 + 28) \\
&= -32 - 48 + 80 = 0.
\end{aligned}
$$

Since

$$
V = |\mathbf{u}.(\mathbf{v} \times \mathbf{w})| = 0,
$$

the volume of the parallelepiped determined by the given vectors **u**, **v**, *and* **w** *is zero. This means that* **u**, **v**, *and* **w** *are coplanar.* ●

Note that the product **u** × (**v** × **w**) that occurs 6*th* in Theorem B.5 is called the *triple vector product* of the vectors **u**, **v**, and **w**. We can write the triple vector product of the vectors in dot product form as

$$
\mathbf{u} \times (\mathbf{v} \times \mathbf{w}) = (\mathbf{u}.\mathbf{w})\mathbf{v} - (\mathbf{u}.\mathbf{v})\mathbf{w},
$$

and the result of the triple vector product of the vectors is a vector.

Example B.14 *Find the triple vector product of the vectors* $\mathbf{u} = 3\mathbf{i} - \mathbf{j}$, $\mathbf{v} = 2\mathbf{i} + \mathbf{j} + \mathbf{k}$, *and* $\mathbf{w} = \mathbf{i} - \mathbf{j} + \mathbf{k}$.

Solution. *To find the triple vector product of* $\mathbf{u} = <3, -1, 0>$, $\mathbf{v} = <2, 1, 1>$, *and* $\mathbf{w} = <1, -1, 1>$, *we compute the following dot products:*

$$\mathbf{u}.\mathbf{w} = <3, -1, 0> . <1, -1, 1> = (3)(1) + (-1)(-1) + (0)(1) = 3 + 1 + 0 = 4$$

and

$$\mathbf{u}.\mathbf{v} = <3, -1, 0> . <2, 1, 1> = (3)(2) + (-1)(1) + (0)(1) = 6 - 1 + 0 = 5.$$

Thus,

$$
\begin{aligned}
\mathbf{u} \times (\mathbf{v} \times \mathbf{w}) &= 4\mathbf{v} - 5\mathbf{w} \\
&= 4 <2, 1, 1> - 5 <1, -1, 1> \\
&= <8, 4, 4> - <5, -5, 5> \\
&= <8 - 5, 4 + 5, 4 - 5> \\
&= <3, 9, -1>,
\end{aligned}
$$

which is the required triple vector product of the given vectors.

We can also find the triple vector product of the given vectors directly by first taking the cross product of the vectors $\mathbf{v} \times \mathbf{w} = \mathbf{x}$ *and then taking one more time the cross product of the vectors* $\mathbf{u} \times \mathbf{x}$ *as follows:*

$$
\begin{aligned}
\mathbf{x} = \mathbf{v} \times \mathbf{w} &= \begin{vmatrix} \mathbf{i} & \mathbf{j} & \mathbf{k} \\ 2 & 1 & 1 \\ 1 & -1 & 1 \end{vmatrix} = \begin{vmatrix} 1 & 1 \\ -1 & 1 \end{vmatrix} \mathbf{i} - \begin{vmatrix} 2 & 1 \\ 1 & 1 \end{vmatrix} \mathbf{j} + \begin{vmatrix} 2 & 1 \\ 1 & -1 \end{vmatrix} \mathbf{k} \\
&= (1+1)\mathbf{i} - (2-1)\mathbf{j} + (-2-1)\mathbf{k} \\
&= 2\mathbf{i} - \mathbf{j} - 3\mathbf{k}
\end{aligned}
$$

and

$$\mathbf{u} \times (\mathbf{v} \times \mathbf{w}) \quad = \quad \mathbf{u} \times \mathbf{x} = \begin{vmatrix} \mathbf{i} & \mathbf{j} & \mathbf{k} \\ 3 & -1 & 0 \\ 2 & -1 & -3 \end{vmatrix}$$

$$= \begin{vmatrix} -1 & 0 \\ -1 & -3 \end{vmatrix} \mathbf{i} - \begin{vmatrix} 3 & 0 \\ 2 & -3 \end{vmatrix} \mathbf{j} + \begin{vmatrix} 3 & -1 \\ 2 & -1 \end{vmatrix} \mathbf{k}$$

$$= (3 - 0)\mathbf{i} - (-9 - 0)\mathbf{j} + (-3 + 2)\mathbf{k}$$

$$= 3\mathbf{i} + 9\mathbf{j} - \mathbf{k},$$

the triple vector product of the given vectors.   •

To get the triple vector product of the given vectors of Example B.14, we use the MATLAB Command Window as follows:

```
>> u = [3 − 1 0];
>> v = [2 1 1];
>> w = [1 − 1 1];
>> x = cross(v, w);
>> y = cross(u, x);
```

B.1.3   Lines and Planes in Space

Here, we discuss parametric equations of lines in space which is important because they generally provide the most convenient form for representing lines algebraically. Also, we will use vectors to derive equations of planes in space, which we will use to solve various geometric problems.

Lines in Space

Let us consider a line that passes through the point $P_1 = (x_1, y_1, z_1)$ and is parallel to the position vector $\mathbf{a} = (a_1, a_2, a_3)$. For any other point $P = (x, y, z)$ on the line, the vector $\vec{P_1 P}$ must be parallel to $\mathbf{a}$, i.e.,

$$\vec{P_1 P} = t\mathbf{a}$$

for some scalar t. Since

$$\vec{P_1P} = (x - x_1, y - y_1, z - z_1)$$

and

$$t\mathbf{a} = t(a_1, a_2, a_3) = (a_1 t, a_2 t, a_3 t),$$

we have

$$(x - x_1, y - y_1, z - z_1) = (a_1 t, a_2 t, a_3 t).$$

Two vectors are equal, if and only if all of their components are equal, so

$$x - x_1 = a_1 t, \quad y - y_1 = a_2 t, \quad z - z_1 = a_3 t,$$

which are called the *parametric equations* for the line, where t is the parameter.

Note that if all the components of the vector $\mathbf{a}$ are nonzero, then we can solve for the parameter t in each of the three equations as follows:

$$\frac{x - x_1}{a_1} = \frac{y - y_1}{a_2} = \frac{z - z_1}{a_3},$$

which are the *symmetric equations* of the line.

Example B.15 *Find the parametric and symmetric equations of the line passing through the points* $(1, 3, -2)$ *and* $(3, -2, 5)$.

Solution. *Begin by letting* $P_1 = ((1, 3, -2)$ *and* $P_2 = (3, -2, 5)$, *then a direction vector for the line passing through* P_1 *and* P_2 *is given by*

$$\mathbf{a} = \vec{P_1P_2} = (3 - 1, -2 - 3, 5 + 2) = (2, -5, 7),$$

which is parallel to the given line and taking either point will give us an equation for the line. So using direction number $a_1 = 2, a_2 = -5$, *and* $a_3 = 7$, *with the point* $P_1 = ((1, 3, -2)$, *we can obtain the parametric equations of the form*

$$x - 1 = 2t, \quad y - 3 = -5t, \quad z + 2 = 7t.$$

Similarly, the symmetric equations of the line are

$$\frac{x - 1}{2} = \frac{y - 3}{-5} = \frac{z + 2}{7}.$$

●

Neither the parametric equations nor the symmetric equation of a given line are unique. For instance, in Example B.15, by taking parameter $t = 1$ in the parametric equations we would obtain the point $(3, -2, 5)$. Using this point with the direction numbers $a_1 = 2$, $a_2 = -5$, and $a_3 = 7$ produces the parametric equations as follows:

$$x - 3 = 2t, \quad y + 2 = -5t, \quad z - 5 = 7t.$$

Definition B.11 *Let l_1 and l_2 be two lines in $\mathbf{R}^3$, with parallel vectors $\mathbf{a}$ and $\mathbf{b}$, respectively, and let θ be the angle between $\mathbf{a}$ and $\mathbf{b}$. Then:*

1. *Lines l_1 and l_2 are parallel.*

2. *If lines l_1 and l_2 intersect, then:*
   (i) the angle between l_1 and l_2 is θ.
   (ii) the lines l_1 and l_2 are orthogonal whenever $\mathbf{a}$ and $\mathbf{b}$ are orthogonal.

Example B.16 *Find the angle between lines l_1 and l_2, where*

$$l_1 : x = 1 + 3t, y = 2 - 3t, z = -1 + 2t, \quad l_2 : x = -3 + t, y = 2 - 2t, z = 2 + 3t.$$

Solution. *Given that lines l_1 and l_2 are parallel, respectively, to the vectors*

$$\mathbf{u} = <3, -3, 2> \quad and \quad \mathbf{v} = <1, -2, 3>,$$

if θ is the angle between $\mathbf{u}$ and $\mathbf{v}$, then

$$\cos \theta = \frac{<3, -3, 2> \, . \, <1, -2, 3>}{\| <3, -3, 2> \| \| <1, -2, 3> \|},$$

which gives

$$\theta = \cos^{-1} \left(\frac{15}{\sqrt{308}} \right) \approx 0.0149 \ radian,$$

the angle between the given lines.

    Note that the angle between lines l_1 and l_2 is defined for either intersecting or nonintersecting lines. ●

Note that nonparallel, nonintersecting lines are called skew lines. ●

Example B.17 *Show that lines l_1 and l_2 are skew lines:*

$$l_1 : x = 1 + 3t, y = 5 - 3t, z = -1 + 5t, \quad l_2 : x = 2 + 7t, y = 4 - 3t, z = 5 + t.$$

Solution. *Line l_1 is parallel to the vector $3\mathbf{i} - 3\mathbf{j} + 5\mathbf{k}$, and line l_2 is parallel to the vector $7\mathbf{i} - 3\mathbf{j} + \mathbf{k}$. These vectors are not parallel since neither is a scalar multiple of the other. Thus, the linear lines are not parallel.*

For l_1 and l_2 to intersect at some point (x_0, y_0, z_0) these coordinates would have to satisfy the equations of both lines, i.e., there exist values t_1 and t_2 for the parameters such that

$$x_0 = 1 + 3t, y_0 = 5 - 3t, z_0 = -1 + 5t$$

and

$$x_0 = 2 + 7t, y_0 = 4 - 3t, z_0 = 5 + t.$$

This leads to three conditions on t_1 and t_2:

$$
\begin{aligned}
1 + 3t_1 &= 2 + 7t_2 \\
5 - 3t_1 &= 4 - 3t_2 \\
-1 + 5t_1 &= 5 + t_2.
\end{aligned}
$$

Adding the first two equations of the above system, we get

$$6 = 6 + 4t_2, \quad \text{which gives} \quad t_2 = 0.$$

We can find t_1 by putting the value of t_2 in the first equation as

$$1 + 3t_1 = 2 + 7(0), \quad \text{which gives} \quad t_1 = 1/3.$$

With these values of t_1 and t_2, the third equation of the above system is not satisfied, so the lines do not intersect. Thus, the given linear lines are skew lines. $\bullet$

Planes in Space

As we have seen, an equation of a line in space can be obtained from a point on the line and a vector parallel to it. A plane in space is determined by specifying a vector $\mathbf{n} = <a, b, c>$ that is normal (perpendicular) to the plane (i.e., orthogonal to every vector lying in the plane), and a point $P_1 = (x_1, y_1, z_1)$ lying in the plane.

In order to find an equation of the plane, let $P = (x, y, z)$ represent any point in the plane. Then, since P and P_1 are both points in the plane, the vector

$$\vec{P_1P} = (x - x_1, y - y_1, z - z_1)$$

lies in the plane and so must be orthogonal to $\mathbf{n}$, i.e.,

$$
\begin{aligned}
\mathbf{n}.\vec{P_1P} &= 0 \\
(a, b, c).(x - x_1, y - y_1, z - z_1) &= 0 \\
a(x - x_1) + b(y - y_1) + c(z - z_1) &= 0
\end{aligned}
$$

The above third equation is called the equation of the plane in *standard form* or sometimes called the *point-normal form* of the equation of the plane.

Let us rewrite the equation as

$$ax - ax_1 + by - by_1 + cz - cz_1 = 0$$

or

$$ax + by + cy - ax_1 - by_1 - cz_1 = 0.$$

Since the last three terms are constant, combine them into one constant d and write

$$ax + by + cy + d = 0.$$

This is called the *general form* of the equation of the plane.

Given the general form of the equation of the plane, it is easy to find a normal vector to the plane. Simply use the coefficients of $x, y,$ and z and write $\mathbf{n} = <a, b, c>$.

Example B.18 *Find an equation of the plane through the point $(3, -4, 3)$ with normal vector $\mathbf{n} = < 3, -4, 5 >$.*

Solution. *Using the direction number for $\mathbf{n} = < 3, -4, 5 > = < a, b, c >$ and the point $(x_1, y_1, z_1) = (3, -4, 3)$, we can obtain*

$$
\begin{aligned}
a(x - x_1) + b(y - y_1) + c(z - z_1) &= 0 & \\
3(x - 3) - 4(y + 4) + 5(z - 3) &= 0 & \text{(standard form)} \\
3x - 4y + 5z - 40 &= 0, & \text{(general form)}
\end{aligned}
$$

the equation of the plane. Observe that the given point $(3, -4, 3)$ *satisfies this equation.* •

Example B.19 *Find the general equation of the plane containing the three points* $(2, -1, 3), (3, 1, 2),$ *and* $(4, 5, -3)$.

Solution. *To find the equation of the plane, we need a point in the plane and a vector that is normal to the plane. There are three choices for the point, but no normal vector is given. To find a normal vector, use the vector product of vectors* **a** *and* **b** *extending from the point* $P_1(2, -1, 3)$ *to the points* $P_2(3, 1, 2)$ *and* $P_3(4, 5, -3)$*. The component forms of* **a** *and* **b** *are as follows:*

$$\mathbf{a} = \vec{P_1P_2} = <3-2, 1+1, 2-3> = <1, 2, -1>$$
$$\mathbf{b} = \vec{P_2P_3} = <4-3, 5-1, -3-2> = <1, 4, -5>.$$

So, one vector orthogonal to both **a** *and* **b** *is the vector product*

$$\mathbf{n} = \mathbf{a} \times \mathbf{b} = \begin{vmatrix} \mathbf{i} & \mathbf{j} & \mathbf{k} \\ 1 & 2 & -1 \\ 1 & 4 & -5 \end{vmatrix} = \begin{vmatrix} 2 & -1 \\ 4 & -5 \end{vmatrix} \mathbf{i} - \begin{vmatrix} 1 & -1 \\ 1 & -5 \end{vmatrix} \mathbf{j} + \begin{vmatrix} 1 & 2 \\ 1 & 4 \end{vmatrix} \mathbf{k}.$$

Solving this, we get

$$\begin{aligned} \mathbf{n} &= (-10+4)\mathbf{i} - (-5+1)\mathbf{j} + (4-2)\mathbf{k} \\ &= -6\mathbf{i} + 4\mathbf{j} + 2\mathbf{k} \\ &= <-6, 4, 2> = <a, b, c>, \end{aligned}$$

the vector which is normal to the given plane. Using the direction number for **n** *and the point* $(x_1, y_1, z_1) = (2, -1, 3)$*, we can obtain an equation of the plane to be*

$$\begin{aligned} a(x - x_1) + b(y - y_1) + c(z - z_1) &= 0 \\ -6(x-2) + 4(y+1) + 2(z-3) &= 0 \quad \text{(standard form)} \\ -6x + 4y + 2z + 10 &= 0. \quad \text{(general form)} \end{aligned}$$

Note that each of the given points $(2, -1, 3), (3, 1, 2),$ *and* $(4, 5, -3)$ *satisfies this plane equation.* •

Note that:

1. Two planes are *parallel* if their normal vectors are parallel.

2. Two planes are *orthogonal* if their normal vectors are orthogonal.

3. The angle between planes is

$$\cos\theta = \frac{\mathbf{n}_1.\mathbf{n}_2}{\|\mathbf{n}_1\|\|\mathbf{n}_2\|},$$

   where $\mathbf{n}_1$ and $\mathbf{n}_2$ are normal vectors of the planes.

Example B.20 *Show that the planes $2x - 2y + 3z - 2 = 0$ and $8x - 8y + 12z - 5 = 0$ are parallel.*

Solution. *The normal vectors to the given planes, respectively, are*

$$\mathbf{n}_1 = <2, -2, 3> \quad and \quad \mathbf{n}_2 = <8, -8, 12>.$$

Since

$$\mathbf{n}_2 = <8, -8, 12> = 4<2, -2, 3> = 4\mathbf{n}_1,$$

the vectors $\mathbf{n}_1$ and $\mathbf{n}_2$ are parallel, and so are the planes. •

Theorem B.6 (Distance Between a Plane and a Point)

The distance between a plane and a point R (which is not in the plane) is defined as

$$D = \frac{|\vec{PR}.\mathbf{n}|}{\|\mathbf{n}\|},$$

where P is a point in the plane and $\mathbf{n}$ is normal to the plane. •

To find a point in the plane $ax + by + cz + d = 0$ $(a \neq 0)$, take $y = 0$ and $z = 0$, then we get

$$ax + d = 0.$$

It gives $x = -d/a$, so the point in the plane will be $(/a, 0, 0)$. •

Example B.21 *Find the distance between the point* $R = (3, 7, -3)$ *and the plane given by* $4x - 3y + 5z = 8$.

Solution. *The vector*

$$\mathbf{n} = < 4, -3, 5 >$$

is normal to the given plane. Now to find a point P *in the plane, let* $y = 0, z = 0$, *and we obtain the point* $P = (2, 0, 0)$. *The vector from* P *to* R *is given by*

$$\vec{PR} = < 3 - 2, 7 - 0, -3 - 0 > = < 1, 7, -3 >.$$

Using the above distance formula, we have

$$
\begin{aligned}
D = \frac{|\vec{PR}.\mathbf{n}|}{\|\mathbf{n}\|} &= \frac{| < 1, 7, -3 > . < 4, -3, 5 > |}{\sqrt{4^2 + (-3)^2 + 5^2}} \\
&= \frac{|4 - 21 - 15|}{\sqrt{16 + 9 + 25}} \\
&= \frac{32}{\sqrt{50}},
\end{aligned}
$$

which is the required distance between the point and the plane. ●

From Theorem B.6, we can determine the distance between the point $R = (x_0, y_0, z_0)$, and the plane given by $ax + by + cz + d = 0$ is

$$D = \frac{|a(x_0 - x_1) + b(y_0 - y_1) + c(z_0 - z_1)|}{\sqrt{a^2 + b^2 + c^2}}.$$

It can be written as

$$D = \frac{|ax_0 + by_0 + cz_0 + d|}{\sqrt{a^2 + b^2 + c^2}},$$

where $P = (x_1, y_1, z_1)$ is a point in the plane and $d = -(ax_1 + by_1 + cz_1)$.

Example B.22 *Find the distance between the point* $P(1, -2, -3)$ *and the plane* $6x - 2y + 3z = 2$.

Solution. *Given the equation of the plane*

$$6x - 2y + 3z - 2 = 0,$$

we obtain

$$a = 6, \ b = -2, \ c = 3, \ d = -2.$$

Using these values, we get

$$D = \frac{|(6)(1) + (-2)(-2) + (3)(-3) - 2|}{\sqrt{6^2 + (-2)^2 + 3^2}}$$

or

$$D = \frac{|-1|}{\sqrt{49}} = \frac{1}{7},$$

which is the distance from the given point to the given plane. ●

Example B.23 *Find the distance between the parallel planes $9x + 3y - 3z = 9$ and $3x + y - z = 2$.*

Solution. *First, we note that the planes are parallel because their normal vectors $< 9, 3, -3 >$ and $< 3, 1, -1 >$ are parallel, i.e.,*

$$< 9, 3, -3 > = 3 < 3, 1, -1 > .$$

To find the distance between the planes, we choose any point on one plane, say $(x_0, y_0, z_0) = (1, 0, 0)$ is a point in the first plane, then, from the second plane, we can find

$$a = 3, \ b = 1, \ c = -1, \ d = -2.$$

Using these values, the distance is

$$
\begin{aligned}
D &= \frac{|(3)(1) + (1)(0) + (-1)(0) - 2|}{\sqrt{3^2 + 1^2 + (-1)^2}} \\
&= \frac{3}{\sqrt{11}} \\
&\approx 0.9045,
\end{aligned}
$$

which is the distance between the given planes. ●

Example B.24 *Show that the following system of equations has no solution:*

$$\begin{aligned}
x_1 - x_2 + 4x_3 &= 1 \\
-2x_1 + 2x_2 - 8x_3 &= 3 \\
x_1 + x_2 + 3x_3 &= 2.
\end{aligned}$$

Solution. *Consider the general form of the equation of the plane*

$$ax + by + cy + d = 0,$$

where the vector (a, b, c) is normal to this plane. Interpret each of the given equations as defining a plane in $\mathbf{R}^3$. On comparison with the general form, it is seen that the following vectors are normal to these three planes:

$$(1, -1, 4), \quad (-2, 2, -8), \quad (1, 1, 3).$$

Note that

$$(-2, 2, -8) = -2(1, -1, 4),$$

which shows that the normals to the first two planes are parallel. Thus, these two planes are parallel and are distinct. Thus, three planes have no points in common and, therefore, the given system has no solution. ●

Theorem B.7 (Distance Between a Point and a Line in Space)

The distance between a point R and a line in a space is defined as

$$D = \frac{\|\vec{PR} \times \mathbf{u}\|}{\|\mathbf{u}\|},$$

where $\mathbf{u}$ is the direction vector for the line and P is a point on the line. ●

To find a point in the plane $ax + by + cz + d = 0$ ($a \neq 0$), take $y = 0$ and $z = 0$, then we get

$$ax + d = 0,$$

which gives $x = -d/a$, so the point in the plane will be $(/a, 0, 0)$. ●

Example B.25 *Find the distance between the point* $R = (4, -2, 5)$ *and the line given by*

$$x = -1 + 3t, \quad y = 2 - 5t, \quad and \quad z = 3 + 7t.$$

Solution. *Using the direction numbers* $3, -5, 7$, *we have the direction vector for the line, which is*

$$\mathbf{u} = <3, -5, 7>.$$

So to find a point P *on the line, let* $t = 0$, *and we get the point* $P = (-1, 2, 3)$. *Thus, the vector from* P *to* R *is given by*

$$\vec{PR} = <4 + 1, -2 - 2, 5 - 3> = <5, -4, 2>,$$

and we can form the vector product as

$$\vec{PR} \times \mathbf{u} = \begin{vmatrix} \mathbf{i} & \mathbf{j} & \mathbf{k} \\ 5 & -4 & 2 \\ 3 & -5 & 7 \end{vmatrix} = \begin{vmatrix} -4 & 2 \\ -5 & 7 \end{vmatrix} \mathbf{i} - \begin{vmatrix} 5 & 2 \\ 3 & 7 \end{vmatrix} \mathbf{j} + \begin{vmatrix} 5 & -4 \\ 3 & -5 \end{vmatrix} \mathbf{k}.$$

Solving this, we get

$$
\begin{aligned}
\vec{PR} \times \mathbf{u} &= (-28 + 10)\mathbf{i} - (35 - 6)\mathbf{j} + (-25 + 12)\mathbf{k} \\
&= -18\mathbf{i} - 29\mathbf{j} - 13\mathbf{k} \\
&= <-18, -29, -13>.
\end{aligned}
$$

Thus, the distance between the point R *and the given line is*

$$
\begin{aligned}
D = \frac{\|\vec{PR} \times \mathbf{u}\|}{\|\mathbf{u}\|} &= \frac{\| <-18, -29, -13> \|}{\| <3, -5, 7> \|} \\
&= \frac{\sqrt{(-18)^2 + (-29)^2 + (-13)^2}}{\sqrt{3^2 + (-5)^2 + 7^2}} \\
&= \frac{\sqrt{1334}}{\sqrt{83}} \\
&= \frac{36.5240}{9.1104} \\
&= 4.0090,
\end{aligned}
$$

which is the required distance between the given point and the line. ●

Example B.26 *Show that the lines*

$$l_1: \quad x = 1 + 2t, \qquad y = 3 - 2t, \qquad z = 5 + t$$
$$l_2: \quad x = 2 + 3s, \qquad y = 2 + s, \qquad z = -4 + 2s$$

are skew. Find the distance between them.

Solution. *Since the two lines l_1 and l_2 are skew, they can be viewed as lying on two parallel planes P_1 and P_2. The distance between l_1 and l_2 is the same as the distance between P_1 and P_2. The common normal vector to both planes must be orthogonal to both $\mathbf{u}_1 = <2, -2, 1>$ (the direction of l_1) and $\mathbf{u}_2 = <3, 1, 2>$ (the direction of l_2). So a normal vector is*

$$\mathbf{n} = \mathbf{u}_1 \times \mathbf{u}_2 = \begin{vmatrix} \mathbf{i} & \mathbf{j} & \mathbf{k} \\ 2 & -2 & 1 \\ 3 & 1 & 2 \end{vmatrix} = \begin{vmatrix} -2 & 1 \\ 1 & 2 \end{vmatrix} \mathbf{i} - \begin{vmatrix} 2 & 1 \\ 3 & 2 \end{vmatrix} \mathbf{j} + \begin{vmatrix} 2 & -2 \\ 3 & 1 \end{vmatrix} \mathbf{k}.$$

Solving this, we get

$$\begin{aligned} \mathbf{n} &= (-4 - 1)\mathbf{i} - (4 - 3)\mathbf{j} + (2 + 6)\mathbf{k} \\ &= -5\mathbf{i} - \mathbf{j} + 8\mathbf{k} \\ &= <-5, -1, 8>. \end{aligned}$$

If we put $s = 0$ in the equations of l_2, we get the point $(2, 2, -4)$ on P_2, and so the equation for P_2 is

$$-5(x - 2) - (y - 2) + 8(z + 4) = 0,$$

which can also be written as

$$-5x - y + 8z + 44 = 0.$$

If we set $t = 0$ in the equations of l_1, we get the point $(1, 3, 5)$ on P_1. So the distance between l_1 and l_2 is the same as the distance from $(1, 3, 5)$ to $-5x - y + 8z + 44 = 0$. Thus, the distance is

$$\begin{aligned} D &= \frac{|(-5)(1) + (-1)(3) + (8)(5) + 44|}{\sqrt{(-5)^2 + (-1)^2 + 8^2}} \\ &= \frac{76}{\sqrt{90}} \\ &\approx 8.0111, \end{aligned}$$

which is the required distance between the skew lines. ●

B.2   Complex Numbers

Although physical applications ultimately require real answers, complex numbers and complex vector spaces play an extremely useful, if not essential, role in the intervening analysis. Particularly in the description of periodic phenomena, complex numbers and complex exponentials help to simplify complicated trigonometric formulae.

Complex numbers arise naturally in the course of solving polynomial equations. For example, the solutions of the quadratic equation

$$ax^2 + bx + c = 0$$

are given by the quadratic formula

$$x = \frac{-b \pm \sqrt{b^2 - 4ac}}{2a},$$

which are complex numbers, if $b^2 - 4ac < 0$. To deal with the problem that the equation $x^2 = -1$ has no real solution, mathematicians of the eighteenth century invented the "imaginary" number

$$\mathbf{i} = \sqrt{-1},$$

which is assumed to have the property

$$\mathbf{i}^2 = (\sqrt{-1})^2 = -1,$$

but which otherwise has the algebraic properties of a real number.

A *complex number* z is of the form

$$z = a + \mathbf{i}b, \tag{B.1}$$

where a and b are real numbers; a is called the *real part* of z and is denoted by $\mathrm{Re}(z)$; and b is called the *imaginary part* of z and is denoted by $\mathrm{Im}(z)$.

We say that two complex numbers $z_1 = a_1 + \mathbf{i}b_1$ and $z_2 = a_2 + \mathbf{i}b_2$ are *equal*, if their real and imaginary parts are equal, i.e., if

$$a_1 = a_2 \quad \text{and} \quad b_1 = b_2.$$

Note that:

1. Every real number a is a complex number with its imaginary part zero; $a = a + \mathbf{i}0$.

2. The complex number $z = 0 + \mathbf{i}0$ corresponds to zero.

3. If $a = 0$ and $b \neq 0$, then $z = \mathbf{i}b$ is called the imaginary number, or a purely imaginary number.

B.2.1   Geometric Representation of Complex Numbers

A complex number $z = a + \mathbf{i}b$ may be regarded as an ordered pair (a, b) of real numbers. This ordered pair of real numbers corresponds to a point in the plane. Such a correspondence naturally suggests that we represent

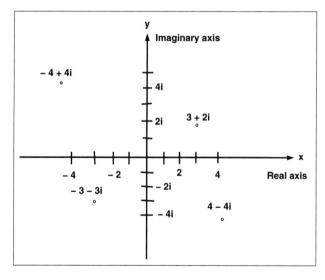

Figure B.9: Geometric representation of a complex number.

$a + \mathbf{i}b$ as a point in the complex plane (Figure B.9), where the *horizontal axis* (also called the *real axis*) is used to represent the real part of z and the *vertical axis* (also called the *imaginary axis*) is used to represent the imaginary part of the complex number z.

B.2.2 Operations on Complex Numbers

Complex numbers are added, subtracted, and multiplied in accordance with the standard rules of algebra but $\mathbf{i}^2 = -1$.

If $z_1 = a_1 + \mathbf{i}b_1$ and $z_2 = a_2 + \mathbf{i}b_2$ are two complex numbers, then their *sum* is

$$z_1 + z_2 = (a_1 + a_2) + \mathbf{i}(b_1 + b_2),$$

and their *difference* is

$$z_1 - z_2 = (a_1 - a_2) + \mathbf{i}(b_1 - b_2).$$

The *product* of z_1 and z_2 is

$$z_1 z_2 = (a_1 + \mathbf{i}b_1)(a_2 + \mathbf{i}b_2) = (a_1 a_2 - b_1 b_2) + \mathbf{i}(a_1 b_2 + a_2 b_1).$$

This multiplication formula is obtained by expanding the left side and using the fact that $\mathbf{i}^2 = -1$.

One can multiply a complex number by a real number α according to

$$\alpha z = \alpha a + \mathbf{i}\alpha b.$$

Finally, *division* is obtained in the following manner:

$$\frac{z_1}{z_2} = \frac{a_1 + \mathbf{i}b_1}{a_2 + \mathbf{i}b_2} = \frac{(a_1 + \mathbf{i}b_1)(a_2 - \mathbf{i}b_2)}{(a_2 + \mathbf{i}b_2)(a_2 - \mathbf{i}b_2)}$$

$$= \left(\frac{a_1 a_2 + b_1 b_2}{a_2^2 + b_2^2}\right) + \mathbf{i}\left(\frac{b_1 a_2 - a_1 b_2}{a_2^2 + b_2^2}\right).$$

An important quantity associated with complex number z is its *complex conjugate*, defined by

$$\overline{z} = a - \mathbf{i}b.$$

Note that

$$z\overline{z} = (a + \mathbf{i}b)(a - \mathbf{i}b) = a^2 + b^2$$

is an intrinsically positive quantity (unless $a = b = 0$).

The MATLAB built-in function **conj** can be used to find the complex conjugate as follows:

$$>> z_1 = conj(z);$$

We call $\sqrt{z\overline{z}}$ the *modulus, absolute value,* or the magnitude of z and write

$$|z| = |a + \mathbf{i}b| = \sqrt{z\overline{z}} = a^2 + b^2.$$

This also tells us that

$$\frac{1}{z} = \frac{\overline{z}}{|z|^2}.$$

Note that a complex number cannot be ordered in the sense that the inequality $z_1 < z_2$ has no meaning. Nevertheless, the absolute values of complex numbers, being real numbers, can be ordered. Thus, for example, $|z| < 1$ means that z is such that $\sqrt{a^2 + b^2} < 1$.

Note that:

1. $\overline{\overline{z}} = z$.

2. $\overline{z_1 + z_2} = \overline{z_1} + \overline{z_2}$.

3. $\overline{z_1 z_2} = \overline{z_1}\,\overline{z_2}$.

4. If $z_2 \neq 0$, then $\overline{\left(\dfrac{z_1}{z_2}\right)} = \dfrac{\overline{z_1}}{\overline{z_2}}$.

5. z is real, if and only if $\overline{z} = z$.

A *complex vector space* is defined in exactly the same manner as its real counterpart, the only difference being that we replace real scalars by *complex scalars.* The terms complex vector space and real vector space emphasize the set from which the scalars are chosen. The most basic example is the n-dimensional complex vector space $\mathbb{C}^n$ consisting of all column vectors $\mathbf{z} = (z_1, z_2, \ldots, z_n)^n$ that have n complex entries $z_1, z_2, \ldots, z_n$ in $\mathbb{C}^n$. Note that

$$\mathbf{z} \in \mathbb{R}^n \subset \mathbb{C}^n$$

is a real vector, if and only if $z = \overline{z}$.

B.2.3   Polar Forms of Complex Numbers

As we have seen, the complex number $z = a + \mathbf{i}b$ can be represented geometrically by the point (a, b). This point can also be expressed in terms of *polar coordinates* (r, θ), where $r \geq 0$, as shown in Figure B.10. We have

$$a = r \cos \theta \quad \text{and} \quad b = r \sin \theta,$$

so

$$z = a + \mathbf{i}b = r \cos \theta + \mathbf{i}r \sin \theta.$$

Thus, any complex number can be written in the polar form

$$z = r(\cos \theta + \mathbf{i} \sin \theta),$$

where

$$r = |z| = \sqrt{a^2 + b^2} \quad \text{and} \quad \tan \theta = \frac{b}{a}.$$

The angle θ is called an *argument* of z and is denoted $\mathbf{arg}z$. Observe

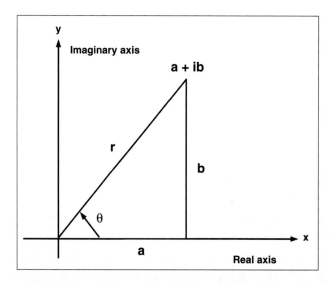

Figure B.10: Polar form of a complex number.

that $\mathbf{arg}z$ is not unique. Adding or subtracting any integer multiple of 2π

gives another argument of z. However, there is only one argument θ that satisfies

$$-\pi < \theta \leq \pi.$$

This is called the *principal argument* of z and is denoted $\mathbf{Arg}z$. Note that

$$z_1 z_2 = [r_1(\cos\theta_1 + \mathbf{i}\sin\theta_1)][r_2(\cos\theta_2 + \mathbf{i}\sin\theta_2)],$$

which can be written as (after using the trigonometric identities)

$$z_1 z_2 = r_1 r_2 [\cos(\theta_1 + \theta_2) + \mathbf{i}\sin(\theta_1 + \theta_2)],$$

which means that to multiply two complex numbers, we multiply their absolute values and add their arguments. Similarly, we can get

$$\frac{z_1}{z_2} = \frac{r_1}{r_2}\left[\cos(\theta_1 - \theta_2) + \mathbf{i}\sin(\theta_1 - \theta_2)\right], \quad z_2 \neq 0, \tag{B.2}$$

which means that to divide two complex numbers, we divide their absolute values and subtract their arguments.

As a special case of (B.2), we obtain a formula for the reciprocal of a complex number in polar form. Setting $z_1 = 1$ (and therefore $\theta_1 = 0$) and $z_2 = z$ (and therefore $\theta_2 = 0$), we obtain:
If $z = r(\cos\theta + \mathbf{i}\sin\theta)$ is nonzero, then

$$\frac{1}{z} = \frac{1}{r}\left(\cos\theta - \mathbf{i}\sin\theta\right).$$

In the following we give some well-known theorems concerning the polar form of a complex number.

Theorem B.8 (De Moivre's Theorem)

If $z = r(\cos\theta + \mathbf{i}\sin\theta)$ and n is a positive integer, then

$$z^n = r^n(\cos n\theta + \mathbf{i}\sin n\theta).$$

Theorem B.9 (Euler's Formula)

For any real number α,

$$e^{\mathbf{i}\alpha} = \cos\alpha + \mathbf{i}\sin\alpha.$$

Using Euler's formula, we see that the polar form of a complex number can be written more compactly as

$$z = r(\cos\theta + \mathbf{i}\sin\theta) = re^{\mathbf{i}\theta}$$

and $\bar{z} = re^{-\mathbf{i}\theta}$.   ●

Theorem B.10 (Multiplication Rule)

If $z_1 = r_1 e^{\mathbf{i}\theta_1}$ and $z_2 = r_1 e^{\mathbf{i}\theta_2}$ are complex numbers in polar form, then

$$z_1 z_2 = r_1 r_2 e^{\mathbf{i}(\theta_1 + \theta_2)}.$$

●

In the following we give other important theorems concerning complex numbers.

Theorem B.11 *If α_1 and α_2 are roots of the quadratic equation*

$$x^2 + ux + v = 0,$$

then $\alpha_1 + \alpha_2 = -u$ and $\alpha_1 \alpha_2 = v$.   ●

Theorem B.12 (Fundamental Theorem of Algebra)

Every polynomial function $f(x)$ of positive degree with complex coefficients has a complex root.   ●

Theorem B.13 *Every complex polynomial of degree $n \geq 1$ has the form*

$$f(x) = u(x - u_1)(x - u_2) \cdots (x - u_n),$$

where $u_1, u_2, \ldots, u_n$ are the roots of $f(x)$ (and need not all be distinct) and u is the coefficient of x^n.   ●

Theorem B.14 *Every polynomial $f(x)$ of positive degree with real coefficients can be factored as a product of linear and irreducible quadratic factors.* •

Theorem B.15 (Nth Roots of Unity)

If $n \geq 1$ is an integer, the nth roots of unity (i.e., the solution to $z^n = 1$) are given by

$$z = e^{2\pi k i}, \quad k = 0, 1, \ldots, n - 1.$$

•

B.2.4 Matrices with Complex Entries

If the entries of a matrix are complex numbers, we can perform the matrix operations of addition, subtraction, multiplication, and scalar multiplication in the same manner as for real matrices. The validity of these operations can be verified using the properties of complex arithmetic and just imitating the proofs for real matrices as discussed in Chapter 1. For example, consider the following matrices:

$$A = \begin{pmatrix} 1+i & 3i \\ 5+i & 6i \end{pmatrix}, \quad B = \begin{pmatrix} 2-2i & 4i \\ 1+3i & 2i \end{pmatrix}, \quad C = \begin{pmatrix} 1+i & 2+i \\ 5+i & 4+5i \\ 2-4i & 1+2i \end{pmatrix}.$$

Then

$$A + B = \begin{pmatrix} 1+i & 3i \\ 5+i & 6i \end{pmatrix} + \begin{pmatrix} 2-2i & 4i \\ 1+3i & 2i \end{pmatrix} = \begin{pmatrix} 3-i & 7i \\ 6+4i & 8i \end{pmatrix}$$

and

$$A - B = \begin{pmatrix} 1+i & 3i \\ 5+i & 6i \end{pmatrix} - \begin{pmatrix} 2-2i & 4i \\ 1+3i & 2i \end{pmatrix} = \begin{pmatrix} -1+3i & -i \\ 4-2i & 4i \end{pmatrix}.$$

Also,

$$CA = \begin{pmatrix} 1+i & 2+i \\ 5+i & 4+5i \\ 2-4i & 1+2i \end{pmatrix} \begin{pmatrix} 1+i & 3i \\ 5+i & 6i \end{pmatrix} = \begin{pmatrix} 9-9i & -9+15i \\ 19+35i & -30+39i \\ 9+9i & 12i \end{pmatrix}$$

and

$$3iA = 3i \begin{pmatrix} 1+i & 3i \\ 5+i & 6i \end{pmatrix} = \begin{pmatrix} -3+3i & -9i \\ -3+15i & -18 \end{pmatrix}.$$

There are special types of complex matrices, like Hermitian matrices, unitary matrices, and normal matrices which we discussed in Chapter 3.

B.2.5   Solving Systems with Complex Entries

The results and techniques dealing with the solutions of linear systems that we developed in Chapter 2 carry over directly to linear systems with complex coefficients. For example, the solution of the linear system

$$\begin{aligned} 3ix_1 &+& 4x_2 &=& 5+15i \\ 5-ix_1 &+& 3-4ix_2 &=& 24+5i \end{aligned}$$

can be obtained by using the Gauss–Jordan method as follows:

$$[A|b] = \begin{pmatrix} 3i & 4 & \vdots & 5+15i \\ 5-i & 3-4i & \vdots & 24+5i \end{pmatrix} \sim \begin{pmatrix} 3i & 4 & \vdots & 5+15i \\ 0 & \frac{13}{3}+\frac{8}{3}i & \vdots & \frac{2}{3}+\frac{55}{3}i \end{pmatrix}$$

$$\sim \begin{pmatrix} 3i & 4 & \vdots & 5+15i \\ 0 & 1 & \vdots & 2+3i \end{pmatrix} \sim \begin{pmatrix} 3i & 0 & \vdots & -3+3i \\ 0 & 1 & \vdots & 2+3i \end{pmatrix}$$

$$\sim \begin{pmatrix} 1 & 0 & \vdots & 1+i \\ 0 & 1 & \vdots & 2+3i \end{pmatrix}.$$

Thus, the solution to the given system is $x_1 = 1+i$ and $x_2 = 2+3i$.   ●

B.2.6   Determinants of Complex Numbers

The definition of a determinant and all its properties derived in Chapter 1 applies to matrices with complex entries. For example, the determinant of the matrix

$$A = \begin{pmatrix} 1+i & 3i \\ 5+i & 6i \end{pmatrix}$$

can be obtained as

$$|A| = \begin{vmatrix} 1+\mathrm{i} & 3\mathrm{i} \\ 5+\mathrm{i} & 6\mathrm{i} \end{vmatrix} = (6\mathrm{i} + 6\mathrm{i}^2) - (15\mathrm{i} + 3\mathrm{i}^2) = -9\mathrm{i} - 3.$$

●

B.2.7 Complex Eigenvalues and Eigenvectors

Let A be an $n \times n$ matrix. The complex number λ is an eigenvalue of A, if there exists a nonzero vector $\mathbf{x}$ in $\mathbb{C}^n$ such that

$$A\mathbf{x} = \lambda\mathbf{x}. \tag{B.3}$$

Every nonzero vector $\mathbf{x}$ satisfying (B.3) is called an eigenvector of A associated with the eigenvalue λ. The relation (B.3) can be rewritten as

$$(A - \lambda\mathbf{I})\mathbf{x} = \mathbf{0}. \tag{B.4}$$

This homogeneous system has a nonzero solution $\mathbf{x}$, if and only if

$$\det(A - \lambda\mathbf{I}) = 0$$

has a solution. As in Chapter 5, $\det(A - \lambda\mathbf{I})$ is called the characteristic polynomial of the matrix A, which is a complex polynomial of degree n in λ. The eigenvalues of the complex matrix A are the complex roots of the characteristic polynomial. For example, let

$$A = \begin{pmatrix} 0 & 1 \\ -1 & 0 \end{pmatrix},$$

then

$$\det(A - \lambda\mathbf{I}) = \lambda^2 + 1 = 0$$

gives the eigenvalues $\lambda_1 = \mathrm{i}$ and $\lambda_2 = -\mathrm{i}$ of A. One can easily find the eigenvectors

$$\mathbf{x}_1 = [1,\ \mathrm{i}]^T \quad \text{and} \quad \mathbf{x}_2 = [-1,\ \mathrm{i}]^T,$$

associated with eigenvalues i and $-\mathrm{i}$, respectively. ●

B.3 Inner Product Spaces

Now we study a more advanced topic in linear algebra called inner product spaces. Inner products lie at the heart of linear (and nonlinear) analysis, both in finite-dimensional vector spaces and infinite-dimensional function spaces. It is impossible to overemphasize its importance for both theoretical developments, practical application, and in the design of numerical solution techniques. Inner products are widely used from theoretical analysis to applied signal processing. Here, we discuss the basic properties of inner products and give some important theorems.

Definition B.12 (Inner Product)

An inner product on a vector space V is an operation that assigns to every pair of vectors $\mathbf{u}$ and $\mathbf{v}$ in V a real number $<\mathbf{u},\mathbf{v}>$ such that the following properties hold for all vectors $\mathbf{u},\mathbf{v}$, and $\mathbf{w}$ in V and all scalars α:

1. $<\mathbf{u},\mathbf{v}> = <\mathbf{v},\mathbf{u}>$.

2. $<\mathbf{u},\mathbf{v}+\mathbf{w}> = <\mathbf{u},\mathbf{v}> + <\mathbf{u},\mathbf{w}>$.

3. $<\alpha\mathbf{u},\mathbf{v}> = \alpha<\mathbf{u},\mathbf{v}>$.

4. $<\mathbf{u},\mathbf{u}> \geq 0$ and $<\mathbf{u},\mathbf{u}> = 0$, if and only if $\mathbf{u}=\mathbf{0}$.

A vector space with an inner product is called an inner product space. The most basic example of an inner product is the familiar dot product

$$<\mathbf{u},\mathbf{v}> = \mathbf{u}.\mathbf{v} = u_1 v_1 + u_2 v_2 + \cdots + u_n v_n = \sum_{j=1}^{n} u_j v_j,$$

between (column) vectors

$$\mathbf{u} = (u_1, u_2, \ldots, u_n)^T \quad and \quad \mathbf{v} = (v_1, v_2, \ldots, v_n)^T,$$

lying in the Euclidean space $\mathbb{R}^n$. ●

B.3.1 **Properties of Inner Products**

The following theorem summarizes some additional properties that follow from the definition of an inner product.

Theorem B.16 *Let* $\mathbf{u}, \mathbf{v}$, *and* $\mathbf{w}$ *be vectors in an inner product space* V *and let* α *be a scalar:*

1. $<\mathbf{u} + \mathbf{v}, \mathbf{w}> = <\mathbf{u}, \mathbf{w}> + <\mathbf{u}, \mathbf{v}>$.

2. $<\mathbf{u}, \alpha\mathbf{v}> = \alpha <\mathbf{u}, \mathbf{v}>$.

3. $<\mathbf{u}, \mathbf{0}> = <\mathbf{0}, \mathbf{v}> = \mathbf{0}$.

•

In an inner product space, we can define the *length of a vector*, the distance between vectors, and orthogonal vectors.

Definition B.13 (Length of a Vector)

Let $\mathbf{v}$ *be a vector in an inner product space* V. *Then the length (or norm) of* $\mathbf{v}$ *is defined as*

$$\|\mathbf{v}\| = \sqrt{<\mathbf{v}, \mathbf{v}>}.$$

•

Theorem B.17 (Inner Product Norm Theorem)

If V *is a real vector space with an inner product* $<\mathbf{u}, \mathbf{v}>$, *then the function* $\|\mathbf{u}\| = \sqrt{<\mathbf{u}, \mathbf{u}>}$ *is a norm on* V.   •

Definition B.14 (Distance Between Vectors)

Let $\mathbf{u}$ *and* $\mathbf{v}$ *be vectors in an inner product space* V. *Then the* **distance** *between* $\mathbf{u}$ *and* $\mathbf{v}$ *is defined as*

$$d(\mathbf{u}, \mathbf{v}) = \|\mathbf{u} - \mathbf{v}\|.$$

Note that:

$$d(\mathbf{u}, \mathbf{0}) = d(\mathbf{0}, \mathbf{u}) = \|\mathbf{u}\|.$$

A vector with norm 1 is called a unit vector. The set S *of all unit vectors is called a unit circle or unit sphere* $S = \{\mathbf{u} | \mathbf{u} \in V \ and \ \|\mathbf{u}\| = 1\}$.   •

The following theorem summarizes the most important properties of a distance function.

Theorem B.18 *Let d be a distance function defined on a normed linear space V. The following properties hold for all $\mathbf{u}, \mathbf{v}$, and $\mathbf{w}$ vectors in V:*

   1. $d(\mathbf{u}, \mathbf{v}) \geq 0$, and $d(\mathbf{u}, \mathbf{v}) = 0$, if and only if $\mathbf{u} = \mathbf{v}$.

   2. $d(\mathbf{u}, \mathbf{v}) = d(\mathbf{v}, \mathbf{u})$.

   3. $d(\mathbf{u}, \mathbf{w}) \leq d(\mathbf{u}, \mathbf{v}) + d(\mathbf{v}, \mathbf{w})$.   ●

Definition B.15 (Orthogonal Vectors)

Let $\mathbf{u}$ and $\mathbf{v}$ be vectors in an inner product space V. Then $\mathbf{u}$ and $\mathbf{v}$ are orthogonal if

$$< \mathbf{u}, \mathbf{v} > = 0.$$

      ●

In the following we give some well-known theorems concerning inner product spaces.

Theorem B.19 (Pythagoras' Theorem)

Let $\mathbf{u}$ and $\mathbf{v}$ be vectors in an inner product space V. Then $\mathbf{u}$ and $\mathbf{v}$ are orthogonal if and only if

$$\|\mathbf{u} + \mathbf{v}\|^2 = \|\mathbf{u}\|^2 + \|\mathbf{v}\|^2.$$

      ●

Theorem B.20 (Orthogonality Test for Linear Independence)

Nonzero orthogonal vectors in an inner product space are linearly independent.   ●

Theorem B.21 (Normalization Theorem)

For every nonzero vector $\mathbf{u}$ in an inner product space V, the vector $\mathbf{v} = \mathbf{u}/\|\mathbf{u}\|$ is a unit vector.   ●

Theorem B.22 (Cauchy–Schwarz Inequality)

Let $\mathbf{u}$ *and* $\mathbf{v}$ *be vectors in an inner product space* V. *Then*

$$|<\mathbf{u}, \mathbf{v}>| \leq \|\mathbf{u}\|\|\mathbf{v}\|,$$

with the inequality holding, if and only if $\mathbf{u}$ *and* $\mathbf{v}$ *are scalar multiples of each other.* •

Theorem B.23 (Triangle Inequality)

Let $\mathbf{u}$ *and* $\mathbf{v}$ *be vectors in an inner product space* V. *Then*

$$\|\mathbf{u} + \mathbf{v}\| \leq \|\mathbf{u}\| + \|\mathbf{v}\|.$$

•

Theorem B.24 (Parallelogram Law)

Let V *be an inner product space. For any vectors* $\mathbf{u}$ *and* $\mathbf{v}$ *of* V, *we have*

$$\|\mathbf{u} + \mathbf{v}\|^2 + \|\mathbf{u} - \mathbf{v}\|^2 = 2\|\mathbf{u}\|^2 + 2\|\mathbf{v}\|^2.$$

•

Theorem B.25 (Polarization Identity)

Let V *be an inner product space. For any vectors* $\mathbf{u}$ *and* $\mathbf{v}$ *of* V, *we have*

$$<\mathbf{u}, \mathbf{v}> = \frac{1}{4}\|\mathbf{u} + \mathbf{v}\|^2 - \frac{1}{4}\|\mathbf{u} - \mathbf{v}\|^2.$$

•

Theorem B.26 *Let* V *be an inner product space. For any vectors* $\mathbf{u}$ *and* $\mathbf{v}$ *of* V, *we have*

$$\|\mathbf{u} + \mathbf{v}\|^2 = \|\mathbf{u}\|^2 + \|\mathbf{v}\|^2 + 2<\mathbf{u}, \mathbf{v}>.$$

•

Theorem B.27 *Every inner product on $\mathbb{R}^n$ is given by*

$$< \mathbf{u}, \mathbf{v} > = \mathbf{u}^T A \mathbf{v}, \quad for \quad \mathbf{u}, \mathbf{v} \in \mathbb{R}^n,$$

where A is a symmetric, positive-definite matrix. ●

Theorem B.28 *All Gram matrices are positive semidefinite. The Gram matrix*

$$A = \begin{pmatrix} < \mathbf{u_1}, \mathbf{u_1} > & < \mathbf{u_1}, \mathbf{u_2} > & \cdots & < \mathbf{u_1}, \mathbf{u_n} > \\ < \mathbf{u_2}, \mathbf{u_1} > & < \mathbf{u_2}, \mathbf{u_2} > & \cdots & < \mathbf{u_2}, \mathbf{u_n} > \\ \vdots & \vdots & \ddots & \vdots \\ < \mathbf{u_n}, \mathbf{u_1} > & < \mathbf{u_n}, \mathbf{u_2} > & \cdots & < \mathbf{u_n}, \mathbf{u_n} > \end{pmatrix}$$

(where $\mathbf{u_1}, \mathbf{u_2}, \ldots, \mathbf{u_n}$ are vectors in the inner product space V) is positive-definite, if and only if $\mathbf{u_1}, \mathbf{u_2}, \ldots, \mathbf{u_n}$ are linearly independent. ●

B.3.2 Complex Inner Products

Certain applications of linear algebra require **complex-valued inner products**.

Definition B.16 (Complex Inner Product)

An inner product on a complex vector space V is a function that associates a complex number $< \mathbf{u}, \mathbf{v} >$ with each pair of vectors $\mathbf{u}$ and $\mathbf{v}$ in such a way that the following axioms are satisfied for all vectors $\mathbf{u}, \mathbf{v},$ and $\mathbf{w}$ in V and all scalars α:

1. $< \mathbf{u}, \mathbf{v} > = < \overline{\mathbf{v}, \mathbf{u}} > .$

2. $< \mathbf{u} + \mathbf{v}, \mathbf{w} > = < \mathbf{u}, \mathbf{w} > + < \mathbf{v}, \mathbf{w} > .$

3. $< \alpha\mathbf{u}, \mathbf{v} > = \alpha < \mathbf{u}, \mathbf{v} > .$

4. $< \mathbf{v}, \mathbf{v} > \geq 0$ and $< \mathbf{v}, \mathbf{v} > = 0,$ if and only if $\mathbf{v} = \mathbf{0}$.

The scalar $< \overline{\mathbf{v}, \mathbf{u}} >$ is the complex conjugate of $< \mathbf{v}, \mathbf{u} >$. Complex inner products are no longer symmetric since $< \mathbf{v}, \mathbf{u} >$ is not always equal to its complex conjugate. A complex vector space with an inner product is called a complex inner product space or unitary space. ●

The following additional properties follow immediately from the four inner product axioms:

1. $< \mathbf{0}, \mathbf{u} > \, = \, < \mathbf{v}, \mathbf{0} > \, = 0.$

2. $< \mathbf{u}, \mathbf{v} + \mathbf{w} > \, = \, < \mathbf{u}, \mathbf{v} > + < \mathbf{u}, \mathbf{w} > .$

3. $< \mathbf{u}, \alpha \mathbf{v} > \, = \, \overline{\alpha} < \mathbf{u}, \mathbf{v} > .$

An inner product can then be used to define the norm, orthogonality, and distance for a real vector space.

Let $\mathbf{u} = (u_1, u_2, \ldots, u_n)$ and $\mathbf{v} = (v_1, v_2, \ldots, v_n)$ be elements of $\mathbb{C}^n$. The most useful inner product for $\mathbb{C}^n$ is

$$< \mathbf{u}, \mathbf{v} > = u_1 \overline{v_1} + u_2 \overline{v_2} + \cdots + u_n \overline{v_n}.$$

It can be shown that this definition satisfies the inner product axioms for a complex vector space.

This inner product leads to the following definitions of norm, distance, and orthogonality for $\mathbb{C}^n$:

1. $\|\mathbf{u}\| = \sqrt{u_1 \overline{u_1} + u_2 \overline{u_2} + \cdots + u_n \overline{u_n}}.$

2. $d(\mathbf{u}, \mathbf{v}) = \|\mathbf{u} - \mathbf{v}\|.$

3. $\mathbf{u}$ is orthogonal to $\mathbf{v}$, if $< \mathbf{u}, \mathbf{v} > \, = 0.$ ●

B.4   Problems

1. Compute $\mathbf{u} + \mathbf{v}, \mathbf{u} - \mathbf{v}$ and their $\|.\|$ for each of the following:
   (a) $\mathbf{u} = <4, -5>,\quad \mathbf{v} = <3, 4>$.
   (b) $\mathbf{u} = <-3, -7>,\quad \mathbf{v} = <-4, 5>$.
   (c) $\mathbf{u} = <4, 5, 6>,\quad \mathbf{v} = <1, -3, 4>$.
   (d) $\mathbf{u} = <-7, 15, 26>,\quad \mathbf{v} = <11, -13, 24>$

2. Compute $\mathbf{u} + \mathbf{v}, \mathbf{u} - \mathbf{v}$ and their $\|.\|$ for each of the following:
   (a) $\mathbf{u} = 3\mathbf{i} + 2\mathbf{j} + 5\mathbf{k},\quad \mathbf{v} = 2\mathbf{i} - 5\mathbf{j} - 7\mathbf{k}$.
   (b) $\mathbf{u} = \mathbf{i} - 12\mathbf{j} - 13\mathbf{k},\quad \mathbf{v} = 8\mathbf{i} + 11\mathbf{j} + 16\mathbf{k}$.
   (c) $\mathbf{u} = 19\mathbf{i} - 22\mathbf{j} - 35\mathbf{k},\quad \mathbf{v} = -12\mathbf{i} + 32\mathbf{j} + 18\mathbf{k}$.
   (d) $\mathbf{u} = 34\mathbf{i} - 35\mathbf{k},\quad \mathbf{v} = -31\mathbf{i} + 25\mathbf{j} - 27\mathbf{k}$.

3. Find a unit vector that has the same direction as a vector $\mathbf{a}$:
   (a) $\mathbf{a} = <-7, 15, 26>$.
   (b) $\mathbf{a} = 2 <6, -11, 15>$.
   (c) $\mathbf{a} = -3 <11, -33, 46>$.
   (d) $\mathbf{a} = 5 <12, -23, -33>$.

4. Find a unit vector that has the same direction as a vector $\mathbf{a}$:
   (a) $\mathbf{a} = \mathbf{i} - 5\mathbf{j} + 4\mathbf{k}$.
   (b) $\mathbf{a} = 3\mathbf{i} + 7\mathbf{j} - 3\mathbf{k}$.
   (c) $\mathbf{a} = 25\mathbf{i} - 17\mathbf{j} + 22\mathbf{k}$.
   (d) $\mathbf{a} = 33\mathbf{i} + 45\mathbf{j} - 51\mathbf{k}$.

5. Find the dot product of each of the following vectors:
   (a) $\mathbf{u} = <-3, 4, 2>,\quad \mathbf{v} = <5, -3, 4>$.
   (b) $\mathbf{u} = <2, -1, 4>,\quad \mathbf{v} = <6, 9, 12>$.
   (c) $\mathbf{u} = <6, 7, 8>,\quad \mathbf{v} = <-8, -11, 15>$.
   (d) $\mathbf{u} = <-23, 24, 33>,\quad \mathbf{v} = <26, -45, 51>$.

6. Find the dot product of each of the following vectors:
   (a) $\mathbf{u} = \mathbf{i} - 3\mathbf{j} + 2\mathbf{k},\quad \mathbf{v} = -2\mathbf{i} + 3\mathbf{j} - 5\mathbf{k}$.
   (b) $\mathbf{u} = 5\mathbf{i} + 7\mathbf{j} - 8\mathbf{k},\quad \mathbf{v} = 6\mathbf{i} + 10\mathbf{j} + 14\mathbf{k}$.
   (c) $\mathbf{u} = -21\mathbf{i} + 13\mathbf{j} - 26\mathbf{k},\quad \mathbf{v} = 17\mathbf{i} - 33\mathbf{j} + 56\mathbf{k}$.
   (d) $\mathbf{u} = 55\mathbf{i} - 63\mathbf{j} + 72\mathbf{k},\quad \mathbf{v} = 33\mathbf{i} - 43\mathbf{j} - 75\mathbf{k}$.

7. Find the angle between each of the following vectors:
   (a) $\mathbf{u} =< 2, 3, 1 >,$ $\mathbf{v} =< 4, -2, 5 > .$
   (b) $\mathbf{u} =< -3, 0, 7 >,$ $\mathbf{v} =< 5, -8, 4 > .$
   (c) $\mathbf{u} =< 7, -9, 11 >,$ $\mathbf{v} =< 6, -13, 10 > .$
   (d) $\mathbf{u} =< 22, -29, 31 >,$ $\mathbf{v} =< 27, 41, 57 > .$

8. Find the angle between each of the following vectors:
   (a) $\mathbf{u} = 2\mathbf{i} + 4\mathbf{j} - 3\mathbf{k},$ $\mathbf{v} = 3\mathbf{i} + \mathbf{j} - 6\mathbf{k}.$
   (b) $\mathbf{u} = \mathbf{i} - 11\mathbf{j} - 12\mathbf{k},$ $\mathbf{v} = 5\mathbf{i} - 13\mathbf{j} - 16\mathbf{k}.$
   (c) $\mathbf{u} = -11\mathbf{i} + 23\mathbf{j} + 32\mathbf{k},$ $\mathbf{v} = 7\mathbf{i} - 43\mathbf{j} - 26\mathbf{k}.$
   (d) $\mathbf{u} = 25\mathbf{i} - 36\mathbf{j} + 47\mathbf{k},$ $\mathbf{v} = 31\mathbf{i} + 24\mathbf{j} + 15\mathbf{k}.$

9. Find the value of α such that the following vectors are orthogonal:
   (a) $\mathbf{u} =< 4, 5, -3 >,$ $\mathbf{v} =< \alpha, 4, 0 > .$
   (b) $\mathbf{u} =< 5, \alpha, -4 >,$ $\mathbf{v} =< 5, -3, 4 > .$
   (c) $\mathbf{u} =< 2\sin x, 2, -\cos x >,$ $\mathbf{v} =< -\sin x, \alpha, 2\cos x > .$
   (d) $\mathbf{u} =< \sin x, \cos x, -2 >,$ $\mathbf{v} =< \cos x, -\sin x, \alpha > .$

10. Show that the following vectors are orthogonal:
   (a) $\mathbf{u} =< 3, -2, 1 >,$ $\mathbf{v} =< 4, 5, -2 > .$
   (b) $\mathbf{u} =< 4, -1, -2 >,$ $\mathbf{v} =< 2, -2, 5 > .$
   (c) $\mathbf{u} =< \sin x, \cos x, -2 >,$ $\mathbf{v} =< \cos x, -\sin x, 0 > .$
   (d) $\mathbf{u} =< 2\sin x, 2, -\cos x >,$ $\mathbf{v} =< -\sin x, 2, 2\cos x > .$

11. Find the direction cosines and angles for the vector $\mathbf{u} = 1\mathbf{i} + 2\mathbf{j} + 3\mathbf{k}$, and show that $\cos^2 \alpha + \cos^2 \beta + \cos^2 \gamma = 1$.

12. Find the direction cosines and angles for the vector $\mathbf{u} = 9\mathbf{i} - 13\mathbf{j} + 22\mathbf{k}$, and show that $\cos^2 \alpha + \cos^2 \beta + \cos^2 \gamma = 1$.

13. Find $comp_\mathbf{v}\mathbf{u}$ and $comp_\mathbf{u}\mathbf{v}$ of each of the following vectors:
   (a) $\mathbf{u} =< 2, 1, 1 >,$ $\mathbf{v} =< 3, 2, 2 > .$
   (b) $\mathbf{u} =< 3, 2, 2 >,$ $\mathbf{v} =< 1, 2, 4 > .$
   (c) $\mathbf{u} =< 3, 5, -2 >,$ $\mathbf{v} =< 4, -1, 4 > .$
   (d) $\mathbf{u} =< 5, 7, 8 >,$ $\mathbf{v} =< 10, 11, 12 > .$

14. Find $comp_\mathbf{v}\mathbf{u}$ and $comp_\mathbf{u}\mathbf{v}$ of each of the following vectors:
   (a) $\mathbf{u} =< 2, 4, -3 >,$ $\mathbf{v} =< 2, 2, 7 > .$

(b) $\mathbf{u} =< 3, -3, -4 >,$ $\mathbf{v} =< -4, 2, 2 > .$
(c) $\mathbf{u} =< 7, 5, 5 >,$ $\mathbf{v} =< 8, -10, 14 > .$
(d) $\mathbf{u} =< 9, 7, 11 >,$ $\mathbf{v} =< 15, 17, 13 > .$

15. Find the projection of a vector $\mathbf{u}$ and $\mathbf{v}$ of each of the following vectors:
    (a) $\mathbf{u} =< 2, 3, 4 >,$ $\mathbf{v} =< 3, 5, 2 > .$
    (b) $\mathbf{u} =< 5, -4, 2 >,$ $\mathbf{v} =< 5, 3, 1 > .$
    (c) $\mathbf{u} =< 6, 4, -2 >,$ $\mathbf{v} =< 2, -1, 3 > .$
    (d) $\mathbf{u} =< 9, 7, -5 >,$ $\mathbf{v} =< 10, -11, 9 > .$

16. Find the projection of a vector $\mathbf{u}$ and $\mathbf{v}$ of each of the following vectors:
    (a) $\mathbf{u} =< 3, 4, 3 >,$ $\mathbf{v} =< 3, 2, 5 > .$
    (b) $\mathbf{u} =< 3, 3, 7 >,$ $\mathbf{v} =< -4, 6, 5 > .$
    (c) $\mathbf{u} =< 7, 8, 5 >,$ $\mathbf{v} =< 6, -10, 11 > .$
    (d) $\mathbf{u} =< 8, 7, 10 >,$ $\mathbf{v} =< 12, 13, 11 > .$

17. A force is given by a vector $\mathbf{F} = 12\mathbf{i} - 9\mathbf{j} + 11\mathbf{k}$ and move a particle from the point $P(9, 7, 5)$ to the point $Q(14, 22, 17)$. Find the work done.

18. A force is given by a vector $\mathbf{F} = 4\mathbf{i} + 5\mathbf{j} + 7\mathbf{k}$ and moves a particle from point $P(2, 1, 3)$ to point $Q(5, 4, 3)$. Find the work done.

19. Find the cross product of each of the following vectors:
    (a) $\mathbf{u} =< 2, -3, 4 >,$ $\mathbf{v} =< 3, -2, 6 > .$
    (b) $\mathbf{u} =< -3, 2, -2 >,$ $\mathbf{v} =< -1, 2, -4 > .$
    (c) $\mathbf{u} =< 3, 5, -2 >,$ $\mathbf{v} =< 2, -1, 4 > .$
    (d) $\mathbf{u} =< 12, -31, 21 >,$ $\mathbf{v} =< 14, 17, -19 > .$

20. Use the cross product to show that each of the following vectors are parallel:
    (a) $\mathbf{u} =< 2, -1, 4 >,$ $\mathbf{v} =< -6, 3, -12 > .$
    (b) $\mathbf{u} =< -3, -2, 1 >,$ $\mathbf{v} =< 6, 4, -2 > .$
    (c) $\mathbf{u} =< 3, 4, 2 >,$ $\mathbf{v} =< -6, -8, -4 > .$
    (d) $\mathbf{u} =< -6, -10, 4 >,$ $\mathbf{v} =< 3, 5, -2 > .$

21. Find the cross product of each of the following vectors:
    (a) $\mathbf{u} = 12\mathbf{i} - 8\mathbf{j} + 11\mathbf{k}$, $\mathbf{v} = -9\mathbf{i} + 17\mathbf{j} + 13\mathbf{k}$.
    (b) $\mathbf{u} = 13\mathbf{i} - 17\mathbf{j} + 3\mathbf{k}$, $\mathbf{v} = 22\mathbf{i} - 13\mathbf{j} + 12\mathbf{k}$.
    (c) $\mathbf{u} = 18\mathbf{i} - 19\mathbf{j} + 12\mathbf{k}$, $\mathbf{v} = 14\mathbf{i} - 9\mathbf{j} + 13\mathbf{k}$.
    (d) $\mathbf{u} = 15\mathbf{i} - 13\mathbf{j} + 4\mathbf{k}$, $\mathbf{v} = 8\mathbf{i} - 11\mathbf{j} + 15\mathbf{k}$.

22. Find the area of the parallelogram with vertices P, Q, R, and S:
    (a) $P(2, 3, 1)$, $Q(3, 4, 2)$, $R(5, -2, 1)$, $S(4, -6, 3)$.
    (b) $P(2, 1, 1)$, $Q(5, 2, 2)$, $R(4, 2, 3)$, $S(7, 5, 2)$.
    (c) $P(2, 1, 1)$, $Q(6, 1, 5)$, $R(5, -3, 4)$, $S(9, -5, 4)$.
    (d) $P(2, 1, 1)$, $Q(3, 5, 3)$, $R(2, 5, 9)$, $S(8, -7, 6)$.

23. Find the area of the parallelogram made by $\vec{PQ}$ and $\vec{PR}$, where P, Q, and R are the points in the plane:
    (a) $P(4, -5, 2)$, $Q(2, 4, 7)$, $R(-4, -2, 6)$.
    (b) $P(2, 1, 1)$, $Q(5, 2, -5)$, $R(4, -2, 5)$.
    (c) $P(2, 1, 1)$, $Q(4, 4, 5)$, $R(5, -3, 4)$.
    (d) $P(2, 1, 1)$, $Q(9, 5, 3)$, $R(4, 7, 9)$.

24. Find the area of the triangle with vertices P, Q, and R:
    (a) $P(2, 3, 1)$, $Q(3, 4, 2)$, $R(5, -2, 1)$.
    (b) $P(2, 1, 1)$, $Q(5, 2, 2)$, $R(4, 2, 3)$.
    (c) $P(2, 1, 1)$, $Q(6, 1, 5)$, $R(5, -3, 4)$.
    (d) $(2, 1, 1)$, $Q(3, 5, 3)$, $R(2, 5, 9)$.

25. Find the area of the triangle with vertices P, Q, and R:
    (a) $P(4, -5, 2)$, $Q(2, 4, 7)$, $R(-4, -2, 6)$.
    (b) $P(2, 1, 1)$, $Q(5, 2, -5)$, $R(4, -2, 5)$.
    (c) $P(2, 1, 1)$, $Q(4, 4, 5)$, $R(5, -3, 4)$.
    (d) $P(2, 1, 1)$, $Q(9, 5, 3)$, $R(4, 7, 9)$.

26. Find the scalar triple product of each of the following vectors:
    (a) $\mathbf{u} = <2, 0, 1>$, $\mathbf{v} = <3, -4, 2>$, $\mathbf{w} = <3, -2, 0>$.
    (b) $\mathbf{u} = <5, 3, 0>$, $\mathbf{v} = <1, -2, 5>$, $\mathbf{w} = <3, -2, 7>$.
    (c) $\mathbf{u} = <5, -3, 2>$, $\mathbf{v} = <2, 1, 6>$, $\mathbf{w} = <4, 0, -5>$.
    (d) $\mathbf{u} = <5, -4, 9>$, $\mathbf{v} = <7, -4, 4>$, $\mathbf{w} = <6, 1, 5>$.

27. Find the scalar triple product of each of the following vectors:
    (a) $\mathbf{u} = 3\mathbf{i} - 5\mathbf{j} + 2\mathbf{k}$, $\mathbf{v} = 6\mathbf{i} + 3\mathbf{j} + 4\mathbf{k}$, $\mathbf{w} = 3\mathbf{i} - 8\mathbf{j} + \mathbf{k}$.
    (b) $\mathbf{u} = 4\mathbf{i} - 3\mathbf{j} + 6\mathbf{k}$, $\mathbf{v} = -4\mathbf{i} + 3\mathbf{j} + 5\mathbf{k}$, $\mathbf{w} = -3\mathbf{i} + 9\mathbf{j} + 2\mathbf{k}$.
    (c) $\mathbf{u} = 17\mathbf{i} - 25\mathbf{j} + 10\mathbf{k}$, $\mathbf{v} = 5\mathbf{i} + 9\mathbf{j} + 13\mathbf{k}$, $\mathbf{w} = 4\mathbf{i} + 5\mathbf{j} + 8\mathbf{k}$.
    (d) $\mathbf{u} = 25\mathbf{i} + 24\mathbf{j} + 15\mathbf{k}$, $\mathbf{v} = 13\mathbf{i} - 11\mathbf{j} + 17\mathbf{k}$, $\mathbf{w} = -9\mathbf{i} + 18\mathbf{j} + 27\mathbf{k}$.

28. Find the volume of the parallelepiped determined by each of the following vectors:
    (a) $\mathbf{u} = <1, -1, 2>$, $\mathbf{v} = <3, 2, 1>$, $\mathbf{w} = <2, -2, 1>$.
    (b) $\mathbf{u} = <2, 3, 4>$, $\mathbf{v} = <3, 2, 5>$, $\mathbf{w} = <3, -2, 3>$.
    (c) $\mathbf{u} = <5, -3, 2>$, $\mathbf{v} = <2, 1, 6>$, $\mathbf{w} = <4, 3, 7>$.
    (d) $\mathbf{u} = <3, 6, 8>$, $\mathbf{v} = <5, 7, 9>$, $\mathbf{w} = <4, -2, 5>$.

29. Find the volume of the parallelepiped determined by each of the following vectors:
    (a) $\mathbf{u} = \mathbf{i} - 4\mathbf{j} + 2\mathbf{k}$, $\mathbf{v} = 2\mathbf{i} + 3\mathbf{j} + 4\mathbf{k}$, $\mathbf{w} = 3\mathbf{i} - 5\mathbf{j} + \mathbf{k}$.
    (b) $\mathbf{u} = 3\mathbf{i} - 2\mathbf{j} + 5\mathbf{k}$, $\mathbf{v} = -3\mathbf{i} + 2\mathbf{j} + 5\mathbf{k}$, $\mathbf{w} = -2\mathbf{i} + 4\mathbf{j} + 3\mathbf{k}$.
    (c) $\mathbf{u} = 7\mathbf{i} - 5\mathbf{j} + 10\mathbf{k}$, $\mathbf{v} = \mathbf{i} + 9\mathbf{j} + 3\mathbf{k}$, $\mathbf{w} = 2\mathbf{i} + 6\mathbf{j} + 7\mathbf{k}$.
    (d) $\mathbf{u} = 2\mathbf{i} + 4\mathbf{j} + 5\mathbf{k}$, $\mathbf{v} = 5\mathbf{i} - 11\mathbf{j} + 6\mathbf{k}$, $\mathbf{w} = -\mathbf{i} + 2\mathbf{j} + \mathbf{k}$.

30. Find the volume of the parallelepiped with adjacent sides PQ, PR, and PS:
    (a) $P(2, -1, 2)$, $Q(4, 2, 1)$, $R(3, -2, 1)$, $S(5, -2, 1)$.
    (b) $P(3, -2, 4)$, $Q(3, 2, 5)$, $R(2, 1, 5)$, $S(4, -3, 3)$.
    (c) $P(2, 4, 2)$, $Q(4, 2, 3)$, $R(3, 4, 6)$, $S(3, 2, 5)$.
    (d) $P(10, 3, 3)$, $Q(4, 2, 5)$, $R(7, 11, 9)$, $S(13, 12, 15)$.

31. Find the triple vector product by using each of the following vectors:
    (a) $\mathbf{u} = <3, 2, 1>$, $\mathbf{v} = <3, -2, 1>$, $\mathbf{w} = <-2, -2, 1>$.
    (b) $\mathbf{u} = <5, 3, -4>$, $\mathbf{v} = <3, -2, 4>$, $\mathbf{w} = <3, -2, 2>$.
    (c) $\mathbf{u} = <5, 6, 7>$, $\mathbf{v} = <6, 8, 9>$, $\mathbf{w} = <14, 13, 17>$.
    (d) $\mathbf{u} = <17, 21, 18>$, $\mathbf{v} = <15, 7, 12>$, $\mathbf{w} = <14, -12, 15>$.

32. Find the triple vector product by using each of the following vectors:
    (a) $\mathbf{u} = 2\mathbf{i} - 2\mathbf{j} + 2\mathbf{k}$, $\mathbf{v} = 3\mathbf{i} + 5\mathbf{j} + 4\mathbf{k}$, $\mathbf{w} = 4\mathbf{i} - 3\mathbf{j} + 2\mathbf{k}$.
    (b) $\mathbf{u} = \mathbf{i} - 3\mathbf{j} + 2\mathbf{k}$, $\mathbf{v} = 4\mathbf{i} + 6\mathbf{j} + 5\mathbf{k}$, $\mathbf{w} = 2\mathbf{i} + 4\mathbf{j} + 3\mathbf{k}$.
    (c) $\mathbf{u} = 3\mathbf{i} + 4\mathbf{j} + 7\mathbf{k}$, $\mathbf{v} = 4\mathbf{i} + 9\mathbf{j} + 11\mathbf{k}$, $\mathbf{w} = 5\mathbf{i} + 7\mathbf{j} + 12\mathbf{k}$.
    (d) $\mathbf{u} = 8\mathbf{i} + 9\mathbf{j} + 15\mathbf{k}$, $\mathbf{v} = 10\mathbf{i} - 14\mathbf{j} + 16\mathbf{k}$, $\mathbf{w} = -16\mathbf{i} - 22\mathbf{j} + 15\mathbf{k}$.

33. Find the parametric equations for the line through point P parallel to vector $\mathbf{u}$:
    (a) $P(3, 2, -4)$, $\quad \mathbf{u} = 2\mathbf{i} + 2\mathbf{j} + 2\mathbf{k}$.
    (b) $P(2, 0, 3)$, $\quad \mathbf{u} = -2\mathbf{i} - 3\mathbf{j} + \mathbf{k}$.
    (c) $P(1, 2, 3)$, $\quad \mathbf{u} = 2\mathbf{i} + 4\mathbf{j} + 6\mathbf{k}$.
    (d) $P(2, 2, -3)$, $\quad \mathbf{u} = 4\mathbf{i} + 5\mathbf{j} - 6\mathbf{k}$.

34. Find the parametric equations for the line through points P and Q :
    (a) $P(4, -3, 5)$, $\quad Q(3, 5, 2)$.
    (b) $P(-2, 2, -3)$, $\quad Q(5, 8, 9)$.
    (c) $P(3, 2, 4)$, $\quad Q(-7, 2, 4)$.
    (d) $P(6, -5, 3)$, $\quad Q(3, -3, -4)$.

35. Find the angles between the lines l_1 and l_2 :
    (a) $l_1 : x = 1 + 2t, y = 3 - 4t, z = -2 + t$;
    $l_2 : x = -5 - t, y = 2 - 3t, z = 4 + 3t$.
    (b) $l_1 : x = 2 + 5t, y = 1 - 7t, z = 7 + 3t$;
    $l_2 : x = 5 - 6t, y = 9 - 2t, z = 8 + 11t$.
    (c) $l_1 : x = 6 + 2t, y = 8 - 4t, z = 2 + 4t$;
    $l_2 : x = -4 + 6t, y = -3 + 2t, z = 4 - 3t$.
    (d) $l_1 : x = -3 - 4t, y = 2 - 7t, z = -3 + 5t$;
    $l_2 : x = 2 + 3t, y = 4 - 5t, z = 2 - 3t$.

36. Determine whether the two lines l_1 and l_2 intersect, and if so, find the point of intersection:
    (a) $l_1 : x = 1 + 3t, y = 2 - 5t, z = 4 - t$;
    $l_2 : x = 1 - 6v, y = 2 + 3v, z = 1 + v$.
    (b) $l_1 : x = -3 + 3t, y = 2 - 2t, z = 4 - 4t$;
    $l_2 : x = 1 - v, y = 2 + 2v, z = 3 + 3v$.
    (c) $l_1 : x = 4 + 3t, y = 12 - 15t, z = 14 - 13t$;
    $l_2 : x = 11 - 16v, y = 12 + 13v, z = 10 + 10v$.
    (d) $l_1 : x = 9 + 5t, y = 10 - 11t, z = 9 - 21t$;
    $l_2 : x = 16 - 22v, y = 13 + 23v, z = 11 - 15v$.

37. Find an equation of the plane through the point P with normal vector $\mathbf{u}$:
    (a) $P(4, -3, 5)$, $\quad \mathbf{u} = 2\mathbf{i} + 3\mathbf{j} + 4\mathbf{k}$.

(b) $P(3, 5, 6)$, $\mathbf{u} = \mathbf{i} - \mathbf{j} + 2\mathbf{k}$.
(c) $P(5, -1, 2)$, $\mathbf{u} = \mathbf{i} + 5\mathbf{j} + 4\mathbf{k}$.
(d) $P(9, 11, 13)$, $\mathbf{u} = 6\mathbf{i} - 9\mathbf{j} + 8\mathbf{k}$.

38. Find an equation of the plane determined by the points P, Q, and R:
(a) $P(3, 3, 1)$, $Q(2, 4, 2)$, $R(5, 3, 4)$.
(b) $P(2, 5, 6)$, $Q(5, 2, 5)$, $R(3, 2, 6)$.
(c) $P(5, -4, 3)$, $Q(6, -3, 7)$, $R(5, -3, 4)$.
(d) $P(12, 11, 11)$, $Q(8, 6, 11)$, $R(12, 15, 19)$.

39. Find the distance from point P to the plane:
(a) $P(1, -4, -3)$, $2x - 3y + 6z + 1 = 0$.
(b) $P(3, 5, 7)$, $4x - y + 5z - 9 = 0$.
(c) $P(3, 0, 0)$, $2x + 4y - 4z - 7 = 0$.
(d) $P(9, 10, 11)$, $12x - 23y + 11z - 64 = 0$.

40. Show that the two planes are parallel and find the distance between the planes:
(a) $8x - 4y + 12z - 6 = 0$, $-6x + 3y - 9z - 4 = 0$.
(b) $x + 2y - 2z - 3 = 0$, $2x + 4y - 4z - 7 = 0$.
(c) $2x - 2y + 2z - 4 = 0$, $x - y + z - 1 = 0$.
(d) $-4x + 2y + 2z - 1 = 0$, $6x - 3y - 3z - 4 = 0$.

41. Perform the indicated operation on each of the following:
(a) $(3 + 4i) + (7 - 2i) + (9 - 5i)$.
(b) $4(6 + 2i) - 7(4 - 3i) + 11(6 - 8i)$.
(c) $(1 - 2i)(1 + 2i) + (3 - 5i)(5 - 3i) + (7 - 3i)(8 - 2i)$.
(d) $(-4 - 12i)(-13 + 5i) + (21 + 15i)(-11 - 23i) + (13 + i)(-4 + 7i)$.

42. Perform the indicated operation on each of the following:
(a) $(-2 + 7i) - (9 + 12i) - (11 - 15i)$.
(b) $3(-7 - 5i) + 9(-3 + 5i) - 8(6 + 17i)$.
(c) $(3 + 2i)(5 - 3i) - (13 + 10i)(13 - 10i) + (-12 + 5i)(7 - 11i)$.
(d) $(17 + 21i)(31 - 26i) - (15 - 22i)(10 + 22i) - (25 - i)(9 - 15i)$.

43. Convert each of the following complex numbers to its polar form:
(a) $4 + 3i$.
(b) $5\sqrt{3} + 7i$.

    (c) $-3 + \sqrt{5}i$.

    (d) $-7\sqrt{3} - 5i$.

44. Convert each of the following complex numbers to its polar form:

    (a) $11 - 8i$.

    (b) $-15\sqrt{7} + 22i$.

    (c) $-45 + \sqrt{19}i$.

    (d) $24\sqrt{18} - 25i$.

45. Compute the conjugate of each of the following:

    (a) $5 - 3i$.

    (b) $-7 + 9i$.

    (c) $e^{-2\pi i}$.

    (d) $11e^{5\pi i/4}$.

46. Use the polar forms of the complex numbers $z_1 = -1 - \sqrt{3}i$ and $z_2 = \sqrt{3} + i$ to compute $z_1 z_2$ and z_1/z_2.

47. Find $A + B, A - B$, and CA using the following matrices:

$$A = \begin{pmatrix} 2+i & i \\ 3-i & 2i \end{pmatrix}, \quad B = \begin{pmatrix} 1-i & -2i \\ 1+2i & 3i \end{pmatrix}, \quad C = \begin{pmatrix} i & 1-i \\ 4-i & 3+2i \\ 2-3i & 1-2i \end{pmatrix}.$$

48. Find $2A + 5B, 3A - 7B$, and $4CA$ using the following matrices:

$$A = \begin{pmatrix} 1+i & 3i & 1+i \\ 4-i & 2i & 1-3i \\ i & 2i & i \end{pmatrix}, \quad B = \begin{pmatrix} -3i & 4i & 1-5i \\ -5+2i & 8i & 2+7i \\ -i & 1+4i & -7i \end{pmatrix},$$

$$C = \begin{pmatrix} 1+i & 1-i & 4i \\ 29-12i & 1-44i & 13i \\ 52-63i & 31+21i & -42i \end{pmatrix}.$$

49. Solve each of the following systems:

    (a)

$$\begin{aligned} 2ix_1 - 3x_2 &= 1 - i \\ 4ix_1 + (5+i)x_2 &= 2i. \end{aligned}$$

(b)

$$(1-i)x_1 + (2+i)x_2 = 3i$$
$$(3+i)x_1 + 4ix_2 = 1-5i.$$

(c)

$$-6ix_1 + 9x_2 = 4-i$$
$$(1-4i)x_1 - 6ix_2 = -7i.$$

(d)

$$5ix_1 - 3ix_2 = 1-i$$
$$7ix_1 + 8ix_2 = 3i.$$

50. Solve each of the following systems:
    (a)

$$ix_1 - 7x_2 + (1-2i)x_3 = 4-3i$$
$$2ix_1 - 5ix_2 + 4ix_3 = 1-2i$$
$$ix_1 + 11ix_2 - 9ix_3 = 5i.$$

(b)

$$ix_1 + 5x_2 + 3i)x_3 = 6i$$
$$2ix_1 - 9ix_2 - 7ix_3 = 7i$$
$$3ix_1 - 8ix_2 + 10ix_3 = 8i.$$

(c)

$$13ix_1 + 12x_2 + (1-2i)x_3 = 4-3i$$
$$21ix_1 - 11ix_2 - 12ix_3 = 4-17i$$
$$33ix_1 - 41ix_2 - 29ix_3 = 27i.$$

(d)

$$
\begin{aligned}
1 - 3\mathrm{i}x_1 - 13\mathrm{i}x_2 + (1 + 21\mathrm{i})x_3 &= 6\mathrm{i} \\
-5\mathrm{i}x_1 + 14\mathrm{i}x_2 - 24\mathrm{i}x_3 &= 1 + 11\mathrm{i} \\
7\mathrm{i}x_1 + 15\mathrm{i}x_2 + 23\mathrm{i}x_3 &= 18\mathrm{i}.
\end{aligned}
$$

51. Find the determinant of each of the following matrices:

(a)

$$
A = \begin{pmatrix} 2 - 3\mathrm{i} & 2\mathrm{i} \\ 6 + 7\mathrm{i} & 1 - 3\mathrm{i} \end{pmatrix}.
$$

(b)

$$
A = \begin{pmatrix} 12 + 7\mathrm{i} & 2\mathrm{i} \\ 11 - 5\mathrm{i} & 3 + 4\mathrm{i} \end{pmatrix}.
$$

(c)

$$
A = \begin{pmatrix} 12 + 23\mathrm{i} & 1 - 5\mathrm{i} \\ 16 - 27\mathrm{i} & 21 - 13\mathrm{i} \end{pmatrix}.
$$

(d)

$$
A = \begin{pmatrix} 10 - 27\mathrm{i} & 8\mathrm{i} \\ 13 - 15\mathrm{i} & 11 - 24\mathrm{i} \end{pmatrix}.
$$

52. Find the determinant of each of the following matrices:

(a)

$$
A = \begin{pmatrix} 1 - \mathrm{i} & 2\mathrm{i} & 4\mathrm{i} \\ 4\mathrm{i} & 3\mathrm{i} & -2\mathrm{i} \\ \mathrm{i} & 5\mathrm{i} & 4\mathrm{i} \end{pmatrix}.
$$

(b)

$$
A = \begin{pmatrix} 3\mathrm{i} & 1 - 5\mathrm{i} & 7\mathrm{i} \\ \mathrm{i} & 2\mathrm{i} & 5\mathrm{i} \\ 6\mathrm{i} & -2\mathrm{i} & 3\mathrm{i} \end{pmatrix}.
$$

(c)

$$
A = \begin{pmatrix} 1 - 2\mathrm{i} & 7\mathrm{i} & 3 + 5\mathrm{i} \\ 2 + 7\mathrm{i} & 8\mathrm{i} & 1 + 7\mathrm{i} \\ 3 - 5\mathrm{i} & 9\mathrm{i} & 5 - 8\mathrm{i} \end{pmatrix}.
$$

(d)

$$A = \begin{pmatrix} 13i & 11i & 15i \\ 23i & -25i & 5-15i \\ 33i & 33i & 18i \end{pmatrix}.$$

53. Find the determinant of each of the following matrices:

(a)

$$A = \begin{pmatrix} i & i & -1+3i \\ 3-5i & 2+7i & 2-2i \\ i & 3i & 5i \end{pmatrix}.$$

(b)

$$A = \begin{pmatrix} 2-3i & 11-15i & 12-7i \\ 5-8i & 6-13i & 3-15i \\ 8-25i & 10-2i & 3i \end{pmatrix}.$$

(c)

$$A = \begin{pmatrix} 1+2i & 1-7i & 3-6i \\ 2-7i & 4-8i & 11-5i \\ 13+5i & 3-9i & 5+17i \end{pmatrix}.$$

(d)

$$A = \begin{pmatrix} 32-13i & 8+11i & 4-15i \\ 42+23i & 9-25i & 5-27i \\ 52-33i & 10-31i & 13-18i \end{pmatrix}.$$

54. Find the real and imaginary part of each of the following matrices:

(a)

$$A = \begin{pmatrix} 3i & 2 \\ 2-5i & 1-i \end{pmatrix}.$$

(b)

$$A = \begin{pmatrix} 2-5i & i \\ 1-3i & 2-3i \end{pmatrix}.$$

(c)

$$A = \begin{pmatrix} 13-21i & 8-25i \\ 15+27i & 10+13i \end{pmatrix}.$$

(d)

$$A = \begin{pmatrix} 10 - 33i & 3 - 9i \\ 41 - 17i & 10 \end{pmatrix}.$$

55. Find the real and imaginary part of each of the following matrices:

(a)

$$A = \begin{pmatrix} 4 - 2i & 3i & 2 + 3i \\ 3 - 5i & 2 + 7i & 3 - 7i \\ 2i & 3i & 1 - 5i \end{pmatrix}.$$

(b)

$$A = \begin{pmatrix} 2 + 5i & 11 - 15i & 12 - 7i \\ 5 - 7i & 6 - 13i & 3 - 15i \\ 3 - 2i & 9 - 12i & 4 - 2i \end{pmatrix}.$$

(c)

$$A = \begin{pmatrix} 3i & 3 - 4i & 2 - 5i \\ 2 - 7i & 4 - 8i & 11 - 5i \\ 11 + 2i & 6 & 3 + 7i \end{pmatrix}.$$

(d)

$$A = \begin{pmatrix} i & 4 + 14i & 24 - 35i \\ 11 + 13i & 9 - 25i & 25 - 17i \\ 22 + 23i & 12 - 31i & 23 - 38i \end{pmatrix}.$$

56. Find the eigenvalues and the corresponding eigenvectors of each of the following matrices:

(a)

$$A = \begin{pmatrix} 1 & 2 & 0 \\ 4 & 5 & 6 \\ 7 & 0 & 9 \end{pmatrix}.$$

(b)

$$A = \begin{pmatrix} 0 & 2 & 0 \\ 4 & 5 & -6 \\ 7 & 0 & 1 \end{pmatrix}.$$

(c)

$$A = \begin{pmatrix} 1 & 0 & 1 \\ 3 & 0 & 0 \\ 1 & 1 & 1 \end{pmatrix}.$$

(d)

$$A = \begin{pmatrix} 2 & 0 & -5 \\ 3 & 1 & 0 \\ 1 & 5 & 1 \end{pmatrix}.$$

57. Find the eigenvalues and the corresponding eigenvectors of each of the following matrices:

(a)

$$A = \begin{pmatrix} 7 & 2 & 1 \\ 3 & 3 & 0 \\ 1 & 2 & 1 \end{pmatrix}.$$

(b)

$$A = \begin{pmatrix} -2 & 12 & 1 \\ -23 & 13 & 1 \\ 6 & -2 & 21 \end{pmatrix}.$$

(c)

$$A = \begin{pmatrix} 11 & -22 & 11 \\ 12 & -24 & 42 \\ 13 & -26 & -63 \end{pmatrix}.$$

(d)

$$A = \begin{pmatrix} -29 & 18 & -11 \\ 6 & 7 & 12 \\ 10 & 16 & -6 \end{pmatrix}.$$

58. Find $\overline{A \pm B}, \overline{AB}$, and $\overline{A^T}$ by using each of the following matrices:

(a)

$$A = \begin{pmatrix} 2+i & i \\ 3-i & 2i \end{pmatrix}, \quad B = \begin{pmatrix} 1-i & -2i \\ 1+2i & 3i \end{pmatrix}.$$

(b)

$$A = \begin{pmatrix} 9+2i & 1-4i \\ 3+4i & 3-2i \end{pmatrix}, \quad B = \begin{pmatrix} 3-5i & 7-8i \\ 9+11i & 12-15i \end{pmatrix}.$$

(c)

$$A = \begin{pmatrix} 1-3i & 7i & 1-5i \\ 3-i & 4i & 4-3i \\ 2i & 2i & 4i \end{pmatrix}, \quad B = \begin{pmatrix} 1-3i & 3-4i & 6-5i \\ -5+2i & 8i & 3-7i \\ 1-3i & 5+4i & 2-6i \end{pmatrix}.$$

(d)

$$A = \begin{pmatrix} 1+5i & 7-3i & 1+i \\ 4-7i & 8-2i & 2-2i \\ 3i & 9-i & 3-3i \end{pmatrix}, \quad B = \begin{pmatrix} 9-10i & 11+12i & 14-15i \\ 10+11i & 11-12i & 13+14i \\ 11-12i & 13+14i & 15-16i \end{pmatrix}.$$

Appendix C

Introduction to MATLAB

C.1 Introduction

In this appendix, we discuss programming with the software package MATLAB. The name MATLAB is an abbreviation for "Matrix Laboratory." MATLAB is an extremely powerful package for numerical computing and programming. In MATLAB, we can give direct commands, as on a hand calculator, and we can write programs. MATLAB is widely used in universities and colleges in introductory and advanced courses in mathematics, science, and especially in engineering. In industry, software is used in research, development, and design. The standard MATLAB program has tools (functions) that can be used to solve common problem. Until recently, most users of MATLAB have been people who had previous knowledge of programming languages such as FORTRAN or C and switched to MATLAB as the software became popular.

MATLAB software exists as a primary application program and a large library of program modules called the standard toolbox. Most of the numerical methods described in this textbook are implemented in one form or another in the toolbox. The MATLAB toolbox contains an extensive library for solving many practical numerical problems, such as root finding, interpolation, numerical integration and differentiation, solving systems of linear and nonlinear equations, and solving ordinary differential equations.

The MATLAB package also consists of an extensive library of numerical routines, easily accessed two- and three-dimensional graphics, and a high-level programming format. The ability to quickly implement and modify programs makes MATLAB an appropriate format for exploring and executing the algorithms in this textbook.

The MATLAB is a mathematical software package based on matrices. It is a highly optimized and extremely reliable system for numerical linear algebra. Many numerical tasks can be concisely expressed in the language of linear algebra.

MATLAB is a huge program, therefore, it is impossible to cover all of it in this appendix. Here, we focus primarily on the foundations of MATLAB. It is believed that once these foundations are well understood, the student will be able to learn advanced topics easily by using the information in the *Help* menu.

The MATLAB program, like most other software, is continually being developed and new versions are released frequently. This appendix covers version 7.4, release 14. It should be emphasized, however, that this appendix covers the basics of MATLAB which do not change much from version to version. This appendix covers the use of MATLAB on computers that use the Windows operating system and almost everything is the same when MATLAB is used on other machines.

C.2   Some Basic MATLAB Operations

It is assumed that the software is installed on the computer and that the user can start the program. Once the program starts, the window that opens contains three smaller windows which are the *Command Window* (main window, enter variables, runs programs), the *Current Directory Window* (logs commands entered in the Command Window), and the *Command History Window* (shows the files in the current directory). Besides these, there are other windows, including the *Figure Window* (contains output from graphics commands), the *Editor Window* (creates and debugs script and function files), the *Help Window* (gives help information), and the *Workspace Window* (gives information about the variables that are used). The Command Window in MATLAB is the main window and can be used for executing commands, opening other windows, running programs written by users, and managing the software.

(1) Throughout this discussion we use $>>$ to indicate a MATLAB command statement. The command prompt $>>$ may vary from system to system. The command prompt $>>$ is given by the system and you only need to enter the MATLAB command.

(2) It is possible to include comments in the MATLAB workspace. Typing % before a statement indicates a comment statement. Comment statements are not executable. For example:

$>>$ % Find root of nonlinear equation $f(x) = 0$

(3) To get help on a topic, say, determinant, enter

$>>$ help determinant

(4) A semicolon placed at the end of an expression suppresses the computer output. For example:

$$>> a = 25;$$

Without ; a was displayed.

(5) If a command is too long to fit on one line, it can be continued to the next line by typing three periods $\cdots$ (called an *ellipsis*).   •

C.2.1   MATLAB Numbers and Numeric Formats

All numerical variables are stored in MATLAB in double-precision floating-point form. While it is possible to force some variables to be other types, this is not done easily and is unnecessary.

The default output to the screen is to have 4 digits to the right of the decimal point. To control the formatting of output to the screen, use the command format. The default formatting is obtained by using the following command:

```
>> format short
>> pi
ans =
    3.1416
```

To obtain the full accuracy available in a number, we can use the following (to have 14 digits to the right of the decimal point):

```
>> format long
>> pi
ans =
    3.14159265358979
```

The other format commands, called format short *e* and format long *e*, use 'scientific notation' for the output (4 decimal digits and 15 decimal digits):

```
>> format short e
>> pi
ans =
    3.1416e + 000
>> format long e
>> pi
ans =
    3.141592653589793e + 000
```

The other format commands, called *format short g* and *format long g*, use 'scientific notation' for the output (the best of 5-digit fixed or floating-point and the best of 15-digit fixed or floating-point):

```
>> format short g
>> pi
ans =
    3.1416
>> format long g
>> pi
ans =
    3.14159265358979
```

We can also use the other format command for the output, called *format bank* (to have 2 decimal digits):

```
>> format bank
>> pi
ans =
    3.14
```

There are two other format commands which can be used for the output, called *format compact* (which eliminates empty lines to allow more lines to be displayed on the screen) and *format loose* (which adds empty lines (opposite of compact)).

As part of its syntax and semantics, MATLAB provides for exceptional values. Positive infinity is represented by *Inf*, negative infinity by − *Inf*, and not a real number by *NAN*. These exceptional values are carried through the computations in a logically consistent way. ●

C.2.2 Arithmetic Operations

Arithmetic in MATLAB follows all the rules and uses standard computer
symbols for its arithmetic operation signs:

Symbol	Effect
$+$	*Addition*
$-$	*Subtraction*
$*$	*Multiplication*
$\backslash$	*Division*
$\wedge$	*Power*
$\prime$	*Conjugate transpose*
pi, e	*Constants*

In the present context, we shall consider these operations as scalar arithmetic operations, which is to say that they operate on 2 numbers in the
conventional manner:

```
>> (4 − 2 + 3 * pi)/2
ans =
    5.7124
>> a = 2; b = sin(a);
>> 2 * b^ 2
ans =
    1.6537
```

MATLAB's arithmetic operations are actually much more powerful than
this. We shall see just a little of this extra power later.

There are some arithmetic operations that require great care. The order
in which multiplication and division operations are specified is especially
important. For example:

```
>> a = 2; b = 3; c = 4;
>> a/b * c
```

Here, the absence of any parentheses results in MATLAB executing the
two operations from left-to-right so that:

First, a is divided by b, and then: The result is multiplied by c.

The result is therefore:

$$\boxed{ans = 2.6667}$$

This arithmetic is equivalent to $\frac{a}{b}c$ or as a MATLAB command:

$$\boxed{>> (a/b) * c}$$

Similarly, $a/b/c$ yields the same result as $\frac{a/b}{c}$ or $\frac{a}{b/c}$, which could be achieved with the MATLAB command:

$$\boxed{>> a/(b * c)}$$

Use parentheses to be sure that MATLAB does what you want.

MATLAB executes the calculations according to the order of precedence which is the same as used in most calculations:

Precedence	**Mathematical Operation**
First	Parentheses. For nested parentheses, the innermost are executed first.
Second	Exponentiation.
Third	Multiplication, division (equal precedence).
Fourth	Addition and subtraction.

Note that in an expression that has several operations, higher precedence operations are executed before lower precedence operations. If two or more operations have the same precedence, the expression is executed from left-to-right.

MATLAB can also be used as a calculator in the Command Window by typing a mathematical expression. MATLAB calculates the expression and responds by displaying $ans =$ and the numerical result of the expression in the next line. For example:

```
>> 25 + 10/5
ans =
    27
>> (25 + 10)/5
ans =
    7
>> 35^ (1/2)+12*4
ans =
    53.9161
>> 115^ (1/3)+(112*40)/12 - (0.87+3.25)/6
ans =
    377.5096
```

C.2.3   MATLAB Mathematical Functions

All of the standard mathematical functions—often called elementary functions—that we learned in our calculus courses are available in MATLAB using their usual mathematical names. The important functions for our purposes are:

Symbol	Effect
$abs(x)$	*Absolute value*
$sqrt x$	*Square root*
$\sin(x)$	*Sine function*
$\cos(x)$	*Cosine function*
$\tan(x)$	*Tangent function*
$\log(x)$	*Natural logarithmic function*
$\exp(x)$	*Exponential function*
$atan(x)$	*Inverse tangent function*
$acos(x)$	*Inverse cosine function*
$asin(x)$	*Inverse sine function*
$\cos h(x)$	*Hyperbolic cosine function*
$\sin h(x)$	*Hyperbolic sine function*

Note that the various trigonometric functions expect their argument to be in radian (or pure number) form but not in degree form. For example:

$$>> cos(pi/3)$$

gives the output:

$$ans = 0.5$$

C.2.4 Scalar Variables

A variable is a name made of a letter or a combination of several letters that is assigned a numerical value. Once a variable is assigned a numerical value, it can be used in mathematical expressions, in functions, and in any MATLAB statements and commands. Note that the variables appear to be scalars. In fact, all MATLAB variables are arrays. An important aspect of MATLAB is that it works very efficiently with arrays and the main tasks are best done with arrays.

A variable is actually the name of a memory location. When a new variable is defined, MATLAB allocates an appropriate memory space where the variable's assignment is stored. When the variable is used the stored data is used. If the variable is assigned a new value the content of the memory location is replaced:

$$>> x = 0.5$$
$$>> z = sin(x) + cos(x)\char94 2$$
$$z =$$
$$\qquad 1.2496$$

The following commands can be used to eliminate variables or to obtain information about variables that have been created:

Command	Outcome
clear	Remove all variables from the memory.
who	Display a list of variables currently in the memory.
whos	Display a list of variables currently in the memory and their size togather with information about their bytes and class. •

C.2.5 Vectors

In MATLAB the word *vector* can really be interpreted simply as a 'list of numbers.' Strictly, it could be a list of objects other than numbers but 'list of numbers' will fits our need's for now.

There are two basic kinds of MATLAB vectors: row and column vectors. As the names suggest, a row vector stores its numbers in a long 'horizontal list' such as

$$1, 2, 3, 1.23, -10.3, 1.2,$$

which is a row vector with 6 components. A column vector stores its numbers in a vertical list such as:

$$
\begin{array}{c}
1 \\
2 \\
3 \\
1.23 \\
-10.3 \\
2.1,
\end{array}
$$

which is a column vector with 6 components. In mathematical notation these arrays are usually enclosed in brackets [ ].

There are various convenient forms of these vectors for allocating values to them and accessing the values that are stored in them. The most basic method of accessing or assigning individual components of a vector is based on using an index, or subscript, which indicates the position of the particular component in the list. MATLAB notation for this subscript is to enclose it in parentheses ( ). For assigning a complete vector in a single statement, we can use the square brackets [ ] notation. For example:

$$
\begin{aligned}
&>> x = [1, 2, 3.4, 1.23, -10.3, 2.1] \\
&x = \\
&\quad 1.0000\ 2.0000\ 3.4000\ 1.2300\ -10.3000\ 2.1000 \\
&>> x(3) = x(1) + 3 * x(6) \\
&x = \\
&\quad 1.0000\ 2.0000\ 7.3000\ 1.2300\ -10.3000\ 2.1000
\end{aligned}
$$

Remember that when in entering values for a row vector, space could be used in place of commas. For the corresponding column vector, simply replace the commas with semicolons. To switch between column and row format for a MATLAB vector we use the transpose operator denoted by $\prime$. For example:

$$
\begin{aligned}
&>> x = x' \\
&x = \\
&\quad 1.0000 \\
&\quad 2.0000 \\
&\quad 7.3000 \\
&\quad 1.2300 \\
&\quad -10.3000 \\
&\quad 2.1000
\end{aligned}
$$

MATLAB has several convenient ways of allocating values to a vector where these values fit a simple pattern.

The *colon :* has a very special and powerful role in MATLAB. Basically, it allows an easy way to specify a vector of equally spaced numbers. There are two basic forms of the MATLAB colon notation.

The first one is that two arguments are separated by a colon as in:

$$ >> x = -2 : 4 $$

which generates a row vector with the first component –2, the last one 4, and others spaced at unit intervals.

The second form is that the three arguments separated by two colons has the effect of specifying the *starting value : spacing : final value*. For example:

```
>> x = -2 : 0.5 : 1
```

which generates

```
x =
    -2.0  -1.5  -1.0  -0.5  0.0  0.5  1.0
```

Also, one can use MATLAB colon notation as follows:

```
>> y = x(2 : 6)
```

which generates

```
y =
    -1.5  -1.0  -0.5  0.0  0.5
```

MATLAB has two other commands for conveniently specifying vectors. The first one is called the *linspace* function, which is used to specify a vector with a given number of equally spaced elements between specified start and finish points. For example:

```
>> x = linspace(0, 1, 10)
x =
    0.000 0.111 0.222 0.333 0.444 0.556 0.667 0.778 0.889 1.000
```

Using 10 points results in just 9 steps.

The other command is called the *logspace* function, which is similar to the linspace function, except that it creates elements that are logarithmically equally spaced. The statement:

```
>> lognspace(start value, endvalue, numpoints)
```

will create numpoints elements between $10^{start\ value}$ and $10^{end\ value}$. For example:

```
>> x = lognspace(1, 4, 4)
x =
    10 100 1000 10000
```

We can use MATLAB's vectors to generate tables of function values. For example:

```
>> x = linspace(0, 1, 11);
>> y = cos(x);
>> [x', y']
ans =
    0.0000   1.0000
    0.1000   0.9950
    0.2000   0.9801
    0.3000   0.9553
    0.4000   0.9211
    0.5000   0.8776
    0.6000   0.8253
    0.7000   0.7648
    0.8000   0.6967
    0.9000   0.6216
    1.0000   0.5403
```

Note the use of the transpose to convert the row vectors to columns, and the separation of these two columns by a comma.

Note also that the standard MATLAB functions are defined to operate on vectors of inputs in an element-by-element manner. The following example illustrates the use of the colon (:) notation and arithmetic within the argument of a function as:

```
>> y = sqrt(4 + 2 * (0 : 0.1 : 1)')
ans =
      2.0000
      2.0494
      2.0976
      2.1448
      2.1909
      2.2361
      2.2804
      2.3238
      2.3664
      2.4083
      2.4495
```

C.2.6 Matrices

A matrix is a two-dimensional array of numerical values that obeys the rules of linear algebra as discussed in Chapter 3.

To *enter* a matrix, list all the entries of the matrix with the first row, separating the entries by blank space or commas, separating two rows by a semicolon, and enclosing the list in square brackets. For example, to enter a 3×4 matrix A, we do the following:

```
>> A = [1 2 3 4; 3 2 1 4; 4 1 2 3]
```

and it will appears as follows:

```
A =
      1   2   3   4
      3   2   1   4
      4   1   2   3
```

There are also other options available when directly defining an array. To define a column vector, we can use the transpose operation. For example:

$$>> [1\ 2\ 5]'$$

result in the column vector:

$$ans =$$
$$1$$
$$2$$
$$5$$

The components (entries) of matrices can be manipulated in several ways. For example:

$$>> A = [1\ 2\ 3; 4\ 5\ 6; 7\ 8\ 9];$$
$$>> A(2,3)$$
$$ans =$$
$$6$$

Select a *submatrix* of A as follows:

$$>> A([1\ 3],[1\ 3])$$
$$ans =$$
$$1\quad 3$$
$$7\quad 9$$

or

$$>> A(1:2, 2:3)$$
$$ans =$$
$$2\quad 3$$
$$5\quad 6$$

An individual element or group of elements can be *deleted* from vectors and matrices by assigning these elements to the null (zero) matrix, [ ]. For example:

```
>> x = [1 2 3 4 5];
>> x(3) = [ ]
x =
    [1 2 4 5]
>> A = [1 2 3; 4 5 6; 7 8 9];
>> A(:, 1) = [ ]
ans =
    2  3
    5  6
    8  9
```

To *interchange* the two rows of a given matrix A, we type the following:

```
>> B = A([new order of rows separating the entries by commas], :)
```

For example, if the matrix A has three rows and we want to change rows 1 and 3, we type:

```
>> B = A([3, 2, 1], :)
```

For example:

```
>> A = [1 2 3; 4 5 6; 7 8 9]
>> B = A([3, 2, 1], :)
B =
    7  8  9
    4  5  6
    1  2  3
```

Note that the method can be used to change the order of any number of rows.

Similarly, one can interchange the columns easily by typing:

```
>> B = A(:, [new order of columns separating the entries by commas])
```

For example, if the matrix A has three columns and we want to change column 1 and 3, we type:

```
>> B = A(:, [3, 2, 1])
B =
      3   2   1
      6   5   4
      9   8   7
```

Note that the method can be used to change the order of any number of columns.

In order to *replace* the *kth* row of a matrix A, set $A(k, :)$ equal to the new entries of the row separated by a space and enclosed in square brackets, i.e., type:

```
>> A(k, :) = [New entries of kth row]
```

For example, to change the second row of a 3×3 matrix A to $[2, 2, 2]$, type the command:

```
>> A(2, :) = [2 2 2]
```

For example:

```
>> A = [1 2 3; 4 5 6; 7 8 9]
>> A(2, :) = [2 2 2]
A =
      1   2   3
      2   2   2
      7   8   9
```

Similarly, one can replace the *kth* column of a matrix A equal to the new entries of the column in square brackets separated by semicolons, i.e., type:

```
>> A(:, k) = [New entries of kth column]
```

For example, to change the second column of a 3×3 matrix A to $[2, 2, 2]'$, type the command:

$$>> A(:,2) = [2\ 2\ 2]$$

For example:

```
>> A = [1 2 3; 4 5 6; 7 8 9]
>> A(:,2) = [2; 2; 2]
A =
     1  2  3
     4  2  6
     7  2  9
```

C.2.7 Creating Special Matrices

There are several built-in functions for *creating* vectors and matrices.

- Create a zero matrix with m rows and n columns using *zeros* function as follows:

$$>> A = zeros(m,n)$$

or, one can create an $n \times n$ zero matrix as follows:

$$>> A = zeros(n)$$

For example:

```
>> A = zeros(3)
A =
     0  0  0
     0  0  0
     0  0  0
```

- Create an $n \times n$ ones matrix using the *ones* function as follows:

$$>> A = ones(n, n)$$

For example, the 3×3 ones matrix:

```
>> A = ones(3, 3)
A =
     1  1  1
     1  1  1
     1  1  1
```

Of course, the matrix need not be square:

```
>> A = ones(2, 4)
A =
     1  1  1  1
     1  1  1  1
```

Indeed, ones and zeros can be used to create row and column vectors:

```
>> u = ones(1, 4)
u =
     1  1  1  1
```

and

```
>> v = ones(1, 4)
v =
     1
     1
     1
     1
```

- Create an $n \times n$ identity matrix using the *eye* function as follows:

```
>> I = eye(n)
```

For example:

```
>> I = eye(3)
I =
     1  0  0
     0  1  0
     0  0  1
```

- Create an $n \times n$ *diagonal* matrix using the *diag* function, which either creates a matrix with specified values on the diagonal or it extracts the diagonal entries. Using the diag function, the argument must be a vector:

```
>> v = [4 5 6];
>> A = diag(v)
A =
     4  0  0
     0  5  0
     0  0  6
```

or it can be specified directly in the input argument as in:

```
>> A = diag([4 5 6])
```

To extract the diagonal entries of an existing matrix, the same *diag* function is used, but with the input being a matrix instead of a vector:

```
>> u = diag(A)
u =
      4
      5
      6
```

- Create the *length* function and *size* function which are used to determine the number of elements in vectors and matrices. These functions are useful when one is dealing with matrices of unknown or variable size, especially when writing loops. To define the length function, type:

```
>> u = 1 : 5
u =
      1  2  3  4  5
```

Then

```
>> n = length(u)
n = 5
```

Now to define the size command, which returns two values and has the syntax:

$$[nr, nc] = size(A),$$

where nr is the number of rows and nc is the number of columns in matrix A. For example:

```
>> A = eye(3, 4);
>> [nr, nc] = size(A)
nr = 3
nc = 4
>> B = ones(size(A))
B =
     1  1  1  1
     1  1  1  1
     1  1  1  1
```

- Creating a square root of a matrix A using the *sqrt* function means to obtain a matrix B with entries of the square root of the entries of matrix A. Type:

```
>> B = sqrt(A)
```

For example:

```
>> A = [1 4 5; 2 3 4; 4 7 8];
>> B = sqrt(A)
B =
     1.0000  2.0000  2.2361
     1.4142  1.7321  2.0000
     2.0000  2.6458  2.8284
```

- Create an upper triangular matrix for a given matrix A using the *triu* function as follows:

```
>> U = triu(A)
```

For example:

```
>> A = [1 2 3; 4 5 6; 7 8 9];
>> U = triu(A)
A =
    1  2  3
    0  5  6
    0  0  9
```

Also, one can create an upper triangular matrix from a given matrix A with a zero diagonal as:

```
>> W = triu(A, 1)
```

For example:

```
>> A = [1 2 3; 4 5 6; 7 8 9];
>> W = triu(A, 1)
W =
    0  2  3
    0  0  6
    0  0  0
```

- Create a lower triangular matrix A for a given matrix using the *tril* function as:

```
>> L = tril(A)
```

For example:

```
>> A = [1 2 3; 4 5 6; 7 8 9];
>> L = tril(A)
A =
    1  0  0
    4  5  0
    7  8  9
```

Also, one can create a lower triangular matrix from a given matrix A with a zero diagonal as follows:

```
>> V = tril(A, 1)
```

For example:

```
>> A = [1 2 3; 4 5 6; 7 8 9];
>> V = tril(A, 1)
V =
     0  0  0
     4  0  0
     7  8  0
```

- Create an $n \times n$ random matrix using the *rand* function as follows:

```
>> R = rand(n)
```

For example:

```
>> R = rand(3)
R =
     0.6038  0.0153  0.9318
     0.2722  0.7468  0.4660
     0.1988  0.4451  0.4186
```

- Create a reshape matrix of matrix A using the *reshape* function as follows:

```
>> B = reshape(A, newrows, newcols)
```

For example:

```
>> A = [1 2 3; 4 5 6; 7 8 9; 10 11 12]
>> B = reshape(A, 2, 6)
B =
    1   7   2   8   3   9
    4  10   4  11   6  12
```

and

```
>> c = reshape(A, 1, 12)
c =
    1   4   7  10   2   5   8  11   3   6   9  12
```

- Create an $n \times n$ Hilbert matrix using the *hilb* function as follows:

```
>> H = hilb(n)
```

For example:

```
>> H = hilb(3)
H =
    1.0000   0.5000   0.3333
    0.5000   0.3333   0.2500
    0.3333   0.2500   0.2000
```

- Create a Toeplitz matrix with a given column vector C as the first column and a given row vector R as the first row using the *toeplitz* function as follows:

```
>> U = toeplitz(C, R)
```

C.2.8 **Matrix Operations**

The basic arithmetic operations of addition, subtraction, and multiplication may be applied directly to matrix variables, provided that the particular operation is legal under the rules of linear algebra. When two matrices have the same size, we add and subtract them in the standard way matrices are added and subtracted. For example:

```
>> A = [3 2 − 3; 4 5 6; 7 6 7];
>> B = [1 2 3; 4 − 2 1; 7 5 − 4];
>> C = A + B
C =
      4    4   0
      8    3   7
     14   11   3
```

and the difference of A and B gives:

```
>> D = A − B
D =
      2   0   −6
      0   7    5
      0   1   11
```

Matrix multiplication has the standard meaning as well. Given any two compatible matrix variables A and B, MATLAB expression $A * B$ evaluates the product of A and B as defined by the rules of linear algebra. For example:

```
>> A = [2 3; −1 4];
>> B = [5 − 2 1; 3 8 − 6];
>> C = A * B
C =
     19   20   −16
      7   34   −25
```

Also,

```
>> A = [1 2; 3 4];
>> B = A';
>> C = 3 * (A * B)^ 3
C =
       13080   29568
       29568   66840
```

Similarly, if the two vectors are the same size, they can be added or subtracted from one other. They can be multiplied, or divided by a scalar, or a scalar can be added to each of their components.

Mathematically the operation of division by a vector does not make sense. To achieve the corresponding component-wise operation, we use the ./ operator. Similarly, for multiplication and powers we use .* and .∧, respectively. For example:

```
>> a = [1 2 3];
>> b = [2 − 1 4];
>> c = a. * b
c =
       2  −2  12
```

Also,

```
>> c = a./b
c =
       0.5  −2.0  0.75
```

and

```
>> c = a.^ 3
c =
       1  8  27
```

Similarly,

```
>> c = 2.^ a
c =
     2   4   8
```

and

```
>> c = b.^ b
c =
     2   1   64
```

Note that these operations apply to matrices as well as vectors. For example:

```
>> A = [1 2 3; 4 5 6; 7 8 9];
>> B = [9 8 7; 6 5 4; 3 2 1];
>> C = A.*B
C =
     9   16   21
    24   25   24
    21   16    9
```

Note that $A.*B$ is not the same as $A*B$.

```
>> C = A.^ 2
C =
     1    4    9
    16   25   36
    49   64   81
```

and

```
>> C = A.^ (1/2)
C =
    1.0000   1.4142   1.7321
    2.0000   2.2361   2.4495
    2.6458   2.8284   3.0000
```

Note that there are no such special operators for addition and subtraction.

•

C.2.9 **Strings and Printing**

Strings are matrices with character elements. In more advanced applications such as symbolic computation, string manipulation is a very important topic. For our purposes, however, we shall need only very limited skills in handling strings initially. One most important use might be to include your name.

Strings can be defined in MATLAB by simply enclosing the appropriate string of characters in single quotes such as:

$$>> first =' Rizwan';$$
$$>> last =' Butt';$$
$$>> name = [first, last]$$
$$name =$$
$$\quad Rizwan\ Butt$$

Since the transpose operator and the string delimiter are the same character (the single quote), creating a single column vector with a direct assignment requires enclosing the string literal in parentheses:

$$>> Last\ Name = ('Butt')'$$
$$Name =$$
$$\quad B$$
$$\quad u$$
$$\quad t$$
$$\quad t$$

String matrices can also be created as follows:

$$>> Name = ['Rizwan';' Butt']$$
$$Name =$$
$$\quad Rizwan$$
$$\quad Butt$$

There are two functions for text output, called *disp* and *fprintf*. The disp function is suitable for a simple printing task. The fprintf function provides fine control over the displayed information as well as the capability of directing the output to a file.

The *disp* function takes only one argument, which may be either a string matrix or a numerical matrix. For example:

```
>> disp('Hello')
   Hello
```

and

```
>> x = 0 : pi/5 : 2 * pi;
>> y = sin(x);
>> disp([x' y'])
   0.0000    1.0000
   0.6283    0.8090
   1.2566    0.3090
   1.8850   -0.3090
   2.5133   -0.8090
   3.1416   -1.0000
   3.7699   -0.8090
   4.3982   -0.3090
   5.0265    0.3090
   5.6549    0.8090
   6.2832    1.0000
```

More complicated strings can be printed using the fprintf function. This is essentially a C programming command that can be used to obtain a wide range of printing specifications. For example:

```
>> fprintf('My Name is \n Rizwan Butt \n')
My Name is
   Rizwan Butt
```

where the $\backslash n$ is the newline command.

The sprintf function allows specification of the number of digits in the display, as in:

> $\gg root2 = fprintf('The\ square\ root\ of\ 2\ is\ \%9.7f', (sqrt(2)))$
> $root2 =$
>   The square root of 2 is 1.4142136

or use of the exponential format:

> $\gg root2 = fprintf('The\ square\ root\ of\ 2is\ \%11.5e', (sqrt(2)))$
> $root2 =$
>   The square root of 2 is $1.41421e + 000$

C.2.10 Solving Linear Systems

MATLAB started as a linear algebra extension of Fortran. Since its early days, MATLAB has been extended beyond its initial purpose, but linear algebra methods are still one of its strongest features. To solve the linear system

$$A\mathbf{x} = \mathbf{b}$$

we can just set

> $\gg \mathbf{x} = A \backslash \mathbf{b}$

with A as a nonsingular matrix. For example:

> $\gg A = [1\ 1\ 1; 2\ 3\ 1; 1\ -1\ -2];$
> $\gg \mathbf{b} = [2; 3; -6];$
> $\gg \mathbf{x} = A \backslash \mathbf{b}$
> $\mathbf{x} =$
> -1
> 1
> 2

There are a small number of functions that should be mentioned.

- Reduce a given matrix A to reduced row echelon form by using the *rref* function as:

```
>> rref(A)
```

For example:

```
>> A = [1 1 1; 2 3 1; 1 -1 -2];
>> rref(A)
ans =
     1  0  0
     0  1  0
     0  0  1
```

- Find the determinant of a matrix A by using the *det* function as:

```
>> det(A)
```

For example:

```
>> A = [1 2 -1; 3 0 1; 4 2 1];
>> det(A)
ans = -6
```

- Find the rank of a matrix A by using the *rank* function as:

```
>> rank(A)
```

For example:

```
>> A = [1 4 5; 2 3 4; 4 7 8];
>> rank(A)
ans = 3
```

- Find the inverse of a nonsingular matrix A by using the *inv* function as:

$$>> inv(A)$$

For example:

```
>> A = [1 1 1; 1 2 4; 1 3 9];
>> inv(A)
ans =
      3.0000   -3.0000    1.0000
     -2.5000    4.0000   -1.5000
      0.5000   -1.0000    0.5000
```

- To find the augmenting matrix $[A\ \mathbf{b}]$, which is the combination of the coefficient matrix A and the right-hand side vector b of the linear system $A\mathbf{x} = \mathbf{b}$ and saving the answer in the matrix C, type:

$$>> C = [A\ \mathbf{b}];$$

For example:

```
>> A = [1 1 1; 1 2 4; 1 3 9];
>> b = [2; 3; 4];
>> C = [A b]
C =
     1  1  1  2
     1  2  3  3
     1  4  9  4
```

- The LU decomposition of a matrix A can be computed by using the *lu* function as:

```
>> [L, U] = lu(A)
```

For example:

```
>> A = [1 4 5; 2 3 4; 4 7 8];
>> [L, U] = lu(A)
L =
      0.2500    1.0000    0.0000
      0.5000   -0.2222    1.0000
      1.0000    0.0000    0.0000
```

and

```
U =
      4.0000    7.0000    8.0000
      0.0000    2.2500    3.0000
      0.0000    0.0000    0.6667
```

- Using indirect LU decomposition, one can compute as:

```
>> [L, U, P] = lu(A)
```

For example:

```
>> A = [1 4 5; 2 3 4; 4 7 8];
>> [L, U, P] = lu(A)
L =
         1       0   0
      0.25       1   0
       0.5    -0.2   1
```

and

$$
U =
\begin{array}{ccc}
4 & 7 & 8 \\
0 & 2.25 & 3 \\
0 & 0 & 0.67
\end{array}
$$

and

$$
P =
\begin{array}{ccc}
0 & 0 & 1 \\
1 & 0 & 0 \\
0 & 1 & 0
\end{array}
$$

- One can compute the various norms of the vectors and matrices by using the *norm* function. The expression $norm(A, 2)$ or $norm(A)$ gives the Euclidean norm or l_2-norm of A while $norm(A,\text{Inf})$ gives the maximum or l_∞-norm. Here, A can be a vector or a matrix. The l_1-norm of a vector or matrix can be obtained by $norm(A,1)$. For example, the different norms of the vector can be obtained as:

$$
\begin{aligned}
&>> a = [6, 7, 8] \\
&V1 >> norm(a) \\
&V1 = 12.8841 \\
&V2 >> norm(a, 1) \\
&V2 = 22 \\
&V3 >> norm(a, Inf) \\
&V3 = 8
\end{aligned}
$$

Similarly, to find the different norms of matrix A type:

```
>> A = [1 1 1; 1 2 4; 1 3 9]
M1 >> norm(A)
M1 = 10.6496
M2 >> norm(A, 1)
M2 = 14
M3 >> norm(A, Inf)
M3 = 13
```

- The condition number of a matrix A can be obtained by using the *cond* function as cond(A). This is equivalent to $norm(A, Inf)*norm(inv(A), I$
  For example:

```
>> A = [1 1 1; 1 2 4; 1 3 9]
>> B = inv(A) = [3 -3 1; -2.5 4 -1; 0.5 -1 0.5]
N1 >> norm(A, Inf)
N1 = 13
N2 >> norm(B, Inf)
N2 = 8
```

Thus, the condition number of a matrix A is computed by cond(A) as:

```
>> cond(A) = N1 * N2
cond(A) = 104
```

- The root of polynomial $p(x)$ can be obtained by using the *roots* function roots(p). For example, if $p(x) = 3x^2 + 5x - 6$ is a polynomial, enter:

```
>> p = [3 5 -6];
>> r = roots(p)
r =
    -2.4748
     0.8081
```

- Use the *polyvar* function to evaluate a polynomial $p_n(x)$ at a particular point x. For example, to find the polynomial function $p_3(x) = x^3 - 2x + 12$ at given point $x = 1.5$, type:

```
>> ceof = [1 0 − 2 12];
>> sol = polyvar(coef, 1.5)
sol = 12.3750
```

- Create eigenvalues and eigenvectors of a given matrix A by using the *eig* function as follows:

```
>> [U, D] = eig(A)
```

Here, U is a matrix with columns as eigenvectors and D is a diagonal matrix with eigenvalues on the diagonal. For example:

```
>> A = [1 1 2; −1 2 1; 0 1 3];
>> [U, D] = eig(A)
U =
        0.4082   −0.5774   0.7071
        0.8165    0.5774   0.0000
       −0.4082   −0.5774   0.7071
```

and

```
D =
    1   0   0
    0   2   0
    0   0   3
```

which shows that $1, 2$, and 3 are eigenvalues of the given matrix.

C.2.11   Graphing in MATLAB

Plots are a very useful tool for presenting information. This is true in any field, but especially in science and engineering where MATLAB is mostly used. MATLAB has many commands that can be used for creating different types of plots. MATLAB can produce two- and three-dimensional plots of curves and surfaces. The *plot* command is used to generate graphs of two-dimensional functions. MATLAB's plot function has the ability to plot many types of 'linear' two-dimensional graphs from data which is stored in vectors or matrices. For producing two-dimensional plots we have to do the following:

- Divide the interval into subintervals of equal width. To do this, type:

$$>> x = a : d : b;$$

  where a is the lower limit, d is the width of each subinterval, and b is the upper limit of the interval.

- Enter the expression for y in term of x as:

$$>> y = f(x);$$

- Create the plot by typing:

$$>> plot(x, y)$$

For example, to graph the function $y = e^x + 10$, type:

$$>> x = -2 : 0.1 : 2;$$
$$>> y = exp(x) + 10;$$
$$>> plot(x, y)$$

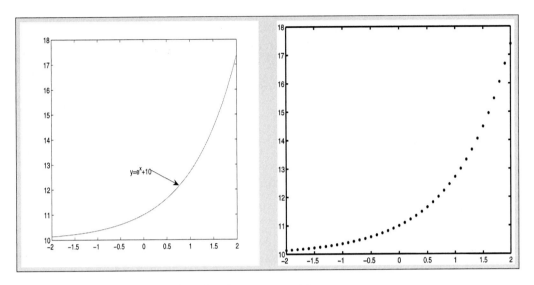

Figure C.1: Graph of $y = e^x + 10$.

By default, the plot function connects the data with a solid line. The markers used for points in a plot may be any of the following:

Symbol	Effect
•	*Point*
○	*Circle*
×	*Cross*
⋆	*Star*

For example, to put a marker for points in the above function plot using the following commands, we get:

```
>> x = −2 : 0.1 : 2;
>> y = exp(x) + 10;
>> plot(x, y,' o')
```

To plot several graphs using the *hold on*, *hold off* commands, one graph is plotted first with the plot command. Then the hold on command is typed. It keeps the Figure Window with the first plot open, including its axis properties and formatting if any was done. Additional graphs can be added with plot commands that are typed next. Each plot command creates a graph that is equal to that figure. To stop this process, the hold off command can be used. For example:

```
>> x = [-3 : 0.01 : 5];
>> y = x.^2 + x - cos(x);
>> plot(x, y, '-')
>> hold on
>> dy = 2*x + sin(x);
>> plot(x, dy, '--')
>> ddy = 2 + cos(x);
>> plot(x, ddy, '.')
>> dddy = -sin(x);
>> plot(x, dddy, '.-')
>> hold off
```

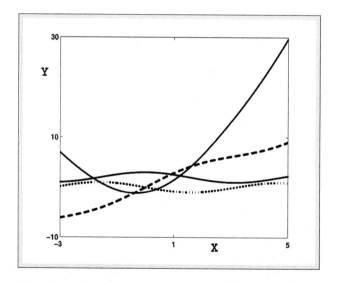

Figure C.2: Graph of function and its first three derivatives.

Also, we can used the *fplot command*, which plots a function with the form $y = f(x)$ between specified limits. For example, to plot the function

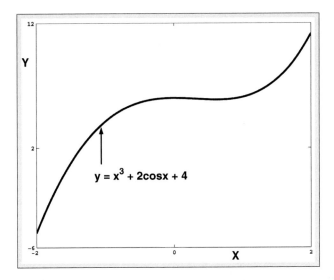

Figure C.3: A plot of the function $y = x^3 + 2 \cos x + 4$.

$f(x) = x^3 + 2 \cos x + 4$ in the domain $-2 \leq x \leq 2$ in the Command Window type:

$$>> fplot('x.\hat{}\ 3 + 2 * cos(x) + 4', [-2, 2])$$

Three-dimensional surface plots are obtained by specifying a rectangular subset of the domain of a function with the *meshgrid* command and then using the *mesh* or *surf* commands to obtain a graph. For a three-dimensional graph, we do the following:

For the function of two variables $Z = f(X, Y)$ and three-dimensional plots, use the following procedure:

- Define the scaling vector for X. For example, to divide the interval $[-2, 2]$ for x into subintervals of width 0.1, enter:

$$>> x = -2 : 0.1 : 2;$$

- Define the scaling vectors for Y. In order to use the same scaling for

y, enter:

```
>> y = x;
```

One may, however, use a different scaling for y.

- Create a meshgrid for the x and y axis:

```
>> [X, Y] = meshgrid(x, y);
```

- Compute the function $Z = f(X, Y)$ at the points defined in the first two steps. For example, if $f(X, Y) = -3X + Y$, enter:

```
>> Z = -3 * X + Y;
```

- To plot the graph of $Z = f(X, Y)$ in three dimensions, type:

```
>> mesh(X, Y, Z)
```

For example, to create a surface plot of $z = \sqrt{x^2 + y^2 + 1}$ on the domain $-5 \leq x \leq 5$, $-5 \leq y \leq 5$, we type the following:

```
>> x = linspace(-5, 5, 20);
>> y = linspace(-5, 5, 20);
>> [X, Y] = meshgrid(x, y);
>> R = sqrt(X.^2+Y.^2+1) + eps;
>> Z = sin(R)./R;
>> surf(X, Y, Z)
```

Adding *eps* (a MATLAB command that returns the smallest floating-point number on your system) avoids the indeterminate $0/0$ at the origin.

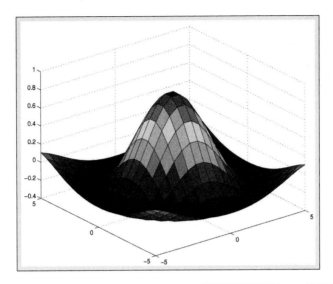

Figure C.4: Surface plot of $z = \sin(\sqrt{x^2 + y^2 + 1})/\sqrt{x^2 + y^2 + 1}$.

Subplots

Often, it is in our interest to place more than one plot in a single figure window. This is possible with the graphic command called the *subplot function*, which is always called with three arguments as in:

$$ >> subplot(nrows, ncols, thisplot) $$

where nrows and ncols define a visual matrix of plots to be arranged in a single figure window and thisplot indicates the number of subplots that is being currently drawn. This plot is an integer that counts across rows and then columns. For a given arrangement of subplots in a figure window, the nrows and ncols arguments do not change. Just before each plot in the matrix is drawn, the subplot function is issued with the appropriate value of thisplot. The following figure shows four subplots created with the following statements:

```
>> x = linspace(0, 2 * pi);
>> subplot(2, 2, 1);
>> plot(x, cos(x));        axis([0 2 * pi  -1.5 1.5]);    title('cos(x)');
>> subplot(2, 2, 2);
>> plot(x, cos(2 * x));    axis([0 2 * pi  -1.5 1.5]);    title('cos(2x)');
>> subplot(2, 2, 3);
>> plot(x, cos(3 * x));    axis([0 2 * pi  -1.5 1.5]);    title('cos(3x)');
>> subplot(2, 2, 4);
>> plot(x, cos(4 * x));    axis([0 2 * pi  -1.5 1.5]);    title('cos(4x)');
```

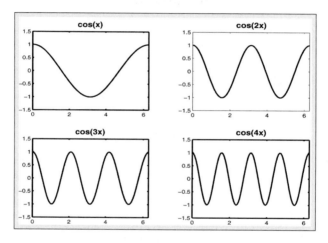

Figure C.5: Four subplots in a figure window.

Similarly, one can use the subplots function for creating surface plots by using the following command:

```
>> x = linspace(-5, 5, 20); y = linspace(-5, 5, 20);
>> [X, Y] = meshgrid(x, y); Z = 2 + (X.^ 2+Y.^ 2);
>> subplot(2, 2, 1);    mesh(x, y, Z);     title('meshplot');
>> subplot(2, 2, 2);    surf(x, y, Z);     title('surfplot');
>> subplot(2, 2, 3);    surfc(x, y, Z);    title('surfcplot');
>> subplot(2, 2, 4);    surfl(x, y, Z);    title('surflplot');
```

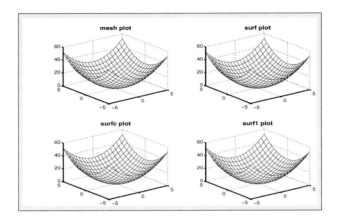

Figure C.6: Four types of surface plots.

C.3 Programming in MATLAB

Here, we discuss the structure and syntax of MATLAB programs. There are many similarities between MATLAB and other high-level languages. The syntax is similar to Fortran, with the same ideas borrowed from *C*. MATLAB has loop and conditional execution constructs. Several important features of MATLAB differentiate it from other high-level languages. MATLAB programs are tightly integrated into an interactive environment. MATLAB programs are interpreted, not compiled. All MATLAB variables are sophisticated data structures that manifest themselves to the user as matrices. MATLAB automatically manages dynamic memory allocation for matrices, which affords convenience and flexibility in the development of algorithms. MATLAB provides highly optimized, built-in routines for multiplication, adding, and subtracting matrices, along with solving linear systems and computing eigenvalues.

C.3.1 Statements for Control Flow

The commands *for*, *while*, and *if* define decision-making structures called control flow statements for the execution of parts of a script based on various conditions. Each of the three structures is ended by an *end* command. The statements that we use to control the flow are called relations.

The repetitions can be handled in MATLAB by using a *for loop* or a *while loop*. The syntax is similar to the syntax of such loops in any programming languages. In the following, we discuss such loops.

C.3.2 **For Loop**

This loop enables us to have an operation repeat a specified number of times. This may be required in summing terms of a series, or specifying the elements of a nonuniformly spaced vector such as the first terms of a sequence defined recursively.

The syntax includes a counter variable, initial value of the counter, the final value of the counter, and the action to be performed, written in the following format:

$$>> for\ counter\ name = initial\ value : final\ value, action; end$$

For example, in order to create the 1×4 row vector x with entries according to formula $x(i) = i$; type:

```
>> for i = 1 : 4
       x(i) = i
end
```

The action in this loop will be performed once for each value of counter name i beginning with the initial value 1 and increasing by each time until the actions are executed for the last time with the final value $i = 4$.

C.3.3 **While Loop**

This loop allows the number of times the loop operation is performed to be determined by the results. It is often used in iterative processes such as approximations to the solution of an equation.

The syntax for a while loop is as follows:

$$>> while \ (condition) \ action; increment \ action; end$$

The loop is executed until the condition (the statement in parentheses) is evaluated. Note that the counter variable must be initialized before using the above command and the increment action gives the increment in the counter variable. For example:

$$
\begin{aligned}
&>> x = 1; \\
&while \ x > 0.01 \\
&\qquad x = x/2; \\
&end \\
&>> disp(x)
\end{aligned}
$$

which generates:

$$
\begin{aligned}
x = \ & \\
& 0.5000 \\
& 0.2500 \\
& 0.1250 \\
& 0.0625 \\
& 0.0313 \\
& 0.0156 \\
& 0.0078
\end{aligned}
$$

C.3.4 Nested for Loops

In order to have a nest of for loops or while loops, each type of loop must have a separate counter. The syntax for two nested for loops is:

$$
\begin{aligned}
&>> for \ counter1 = initial \ value1 : final \ value1, \\
&>> for \ counter2 = initial \ value2 : final \ value2, \\
&\qquad\qquad action; \\
&\qquad end \\
&end
\end{aligned}
$$

For example, in order to create a 5×4 matrix A by the formula $A(i,j) =$

$i + j$, type:

```
>> for i = 1 : 5,
        for j = 1 : 4,
            A(i, j) = i + j;
end,
end
```

which generates a matrix of the form:

```
ans =
    2  3  4  5
    3  4  5  6
    4  5  6  7
    5  6  7  8
    6  7  8  9
```

C.3.5   Structure

Finally, we introduce the basic structure of MATLAB's logical branching commands. Frequently, in programs, we wish for the computer to take different actions depending on the value of some variables. Strictly speaking these are logical variables, or, more commonly, logical expressions similar to those we saw when defining while loops.

Two types of decision statements are possible in MATLAB, one-way decision statements and two-way decision statements.

The syntax for the one-way decision statement is:

```
>> if (condition), action, end
```

in which the statements in the action block are executed only if the condition is satisfied (true). If the condition is not satisfied (false) then the action block is skipped. For example:

```
>> if x > 0, disp('x is positive number'); end
```

For a two-way decision statement we define its syntax as:

$$>> if \ (condition), action, \ else \ action, end$$

in which the first set of instructions in the action block is executed if the condition is satisfied while the second set, the action block, is executed if the condition is not satisfied. For example, if x and y are two numbers and we want to display the value of the number, we type:

$$>> if \ (x > y), disp(x), else \ disp(y), end$$

MATLAB also contains a number of logical and relational operators.

The logical operations are represented by the following:

Symbol	Effect
$\&$	*and*
$|$	*or*
$\sim$	*not*

However, these operators not only apply to side variables but such operators will also work on vectors and matrices when the operation is valid.

The relational operators in MATLAB are:

Symbol	Effect
$==$	is equal to
$<=$	is less than or equal
$>=$	is greater than or equal
$\sim=$	is not equal to
$<$	is less than
$>$	is greater than

The relational operators are used to compare values or elements of arrays. If the relationship is true, the result is a logical variable whose value is one. Otherwise, the value is zero if the relationship is false.

C.4   Defining Functions

A simple function in mathematics, $f(x)$, associates a unique number to each value of x. The function can be expressed in the form $y = f(x)$, where $f(x)$ is usually a mathematical expression in terms of x. Many functions are programmed inside MATLAB as built-in functions and can be used in mathematical expressions simply by typing their names with an argument; examples are $\tan(x), sqrt(x)$, and $exp(x)$. A user-defined function is a MATLAB program that is created by the user, saved as a function file, and then can be used like a built-in function.

MATLAB allows us to define their functions by constructing an m-file in the m-file editor. If the m-file is to be a function m-file, the first word of the file is function, and we must also specify names for its input and output. The last two of these are purely local variable names.

The first line of the function has the form:

> function y = function name(input arguments)

For example, to define the function

$$f(x) = e^x - \frac{2x}{(1 + x^3)},$$

type:

> $function y = fn1(x)$
> $y = exp(x) - 2*x./(1 + x.\hat{\ } 3);$

Once this function is saved as an m-file named fn1.m, we can use the MATLAB Command Window to compute function at any given point. For example:

```
>> x = (0 : 0.2 : 2)';
>> fx = fn1(x);
>> [x', fx']
```

generates the following table:

```
ans =
    0.0000    1.0000
    0.2000    0.8246
    0.4000    0.7399
    0.6000    0.8353
    0.8000    1.1673
    1.0000    1.7183
    1.2000    2.4404
    1.4000    3.3073
    1.6000    4.3251
    1.8000    5.5227
    2.0000    6.9446
```

MATLAB provides the option of using *inline functions*. An inline function is defined with computer code (not as a separate file like a function file) and is then used in the code. Inline functions are created with the inline command according to the following format:

```
name = inline('expression')
```

For example, the function $f(x) = \dfrac{x^2}{\sqrt{x^2 + 1}}$ can be defined in the MATLAB Command Window as follows:

```
>> y = inline('x^ 2 /sqrt(x^ 2+1)')
>> y =
    Inline function :
    y(x) = x^ 2/sqrt(x^ 2+1)
```

The function can be calculated for different values of x. For example,

```
>> y(2) =
ans =
      1.7889
```

If x is expected to be an array and the function is calculated for each element, then the function must be modified for element-by-element calculations:

```
>> y = inline('x.^ 2./sqrt(x.^ 2+1)')
>> x = ([1 2 3])
ans =
      0.7071  1.7889  2.8460
```

If the inline function has two or more independent variables it can be written in the following format:

```
>> y = inline('x.^ 2+y.^ 2+4. * x. * y')
>> z(1, 2)
ans =
      13
```

In MATLAB we can use the *feval* (function evaluate) command to evaluate the values of a function for a given value (or values) of the function's argument.

```
variable = feval('function name', argument value)
```

For example:

```
>> y = feval('sqrt', 169)
y =
      13
```

Note that *feval* command can be used in the user-defined function.   ●

C.5 MATLAB Built-in Functions

Listed below are some of the MATLAB built-in functions grouped by subject area:

Built-in Function   Definition

Built-in Function	Definition
abs	absolute value
cos	cosine function
sin	sine function
tan	tangent function
cosh	cosine hyperbolic function
sinh	sine hyperbolic function
tanh	tangent hyperboic function
acos	inverse cosine function
asin	inverse sine function
atan	inverse tangent function
erf	error function $erf(x) = (2/\sqrt{\pi}) \int_0^x e^{-t^2} \, dt$
exp	exponential function
expm	matrix exponential
log	natural logarithm
log10	common logarithm
sqrt	calculate square root
sqrm	calculate square root of a matrix
sort	arrange elements in asending order
std	calculate standard deviation
mean	calculate mean value
median	calculate median value
sum	calculate sum of elements
angle	calculate phase angle
fix	round toward zero
floor	round toward $-\infty$
ceil	round toward ∞
sign	signum function
round	round to nearest integer
dot	dot product of two vectors

Built-in Function   Definition

cross	cross product of two vectors
dist	distance between two points
frac	return the rational approximation
max	return maximum value
min	return minimum value
factorial	factorial function
rref	reduce echelon form
zeros	generates a matrix of all zeros
ones	generates a matrix of all ones
eye	generates an identity matrix
hilb	generates a Hilbert matrix
reshape	rearranges a matrix
det	calculate a determinant of a matrix
eig	calculate a eigenvalues of a matrix
rank	calculate a rank of a matrix
norm(v)	calculate a the Euclidean norm of a vector v
norm(v,inf)	calculate a maximum norm of vector v
cond(A,2)	calculate a condition number of matrix using Euclidean norm
cond(A,inf)	calculate a condition number of matrix using maximum norm
toeplitz	creates a Toeplitz matrix
inv	find the inverse of a matrix
pinv	find the pseudoinverse of a matrix
diag	create a diagonal matrix
length	number of elements in a vector
size	size of an array
qr	create QR-decomposition of a matrix
svd	calculate singular value decomposition of a matrix
polyval	calculate the value of a polynomial
roots	calculate the roots of a polynomial
conv	multiplies two polynomials
deconv	divide two polynomials
polyder	calculate derivative of a polynomial

Built-in Function	Definition
polyint	calculate integral of a polynomial
polyfit	calculate coefficients of a polynomial
fzero	solve an equation with one variable
quad	integrate a function
linspace	create equally spaced vector
logspace	create logarithically spaced elements
axis	sets limts to axes
plot	create a plot
pie	create a pie plot
polar	create a polar plot
hist	create a histogram
bar	create a vertical bar plot
barh	create a horizontal bar plot
fplot	plot a function
bar3	create a vertical 3-D bar plot
contour	create a 2-D contour plot
contour3	create a 3-D contour plot
cylinder	create a cylinder
mesh	create a mesh plot
meshc	create a mesh and a contour plot
surf	create a surface plot
surfc	create a surface and a contour plot
surf1	create a surface plot with lighting
sphere	create a sphere
subplot	create multiple plot on one page
title	add a title to a plot
xlabel	add label to x-axis
ylabel	add label to y-axis
grid	add grid to a plot

C.6 Symbolic Computation

In this appendix we discuss symbolic computation which is an important and complementary aspect of computing. As we have noted, MATLAB uses floating-point arithmetic for its calculations. But one can also do exact arithmetic with symbolic expressions. Here, we will give many examples to get the exact arithmetic. The starting point for symbolic operations is symbolic objects. Symbolic objects are made of variables and numbers that, when used in mathematical expressions, tell MATLAB to execute the expression symbolically. Typically, the user first defines the symbolic variables that are needed and then uses them to create symbolic expressions that are subsequently used in symbolic operations. If needed, symbolic expressions can be used in numerical operations.

Many applications in mathematics, science, and engineering require symbolic operations, which are mathematical operations with expressions that contain symbolic variables. Symbolic variables are variables that don't have specific numerical values when the operation is executed. The result of such operations is also mathematical expression in terms of the symbolic variables. Symbolic operations can be performed by MATLAB when the Symbolic Math Toolbox is installed. The Symbolic Math Toolbox is included in the student version of the software and can be added to the standard program. The Symbolic Math Toolbox is a collection of MATLAB functions that are used for execution of symbolic operations. The commands and functions for the symbolic operations have the same style and syntax as those for the numerical operations.

Symbolic computations are performed by computer programs such as Derive®, Maple®, and Mathematica®. MATLAB also supports symbolic computation through the Symbolic Math Toolbox, which uses the symbolic routines of Maple. To check if the Symbolic Math Toolbox is installed on a computer, one can type:

```
>> ver
```

In response, MATLAB displays information about the version that is used as well as a list of the toolboxes that are installed.

Using the MATLAB Symbolic Math Toolbox, we can carry out algebraic or symbolic calculations such as factoring polynomials or solving algebraic equations. For example, to add three numbers $\frac{3}{4}$, $\frac{1}{4}$, and $\frac{5}{4}$ symbolically, we do the following:

$$>> sym('3/4') + \text{sym}('1/4') + \text{sym}('5/4')$$
$$ans =$$
$$9/4$$

Symbolic computations can be performed without the approximations that are necessary for numerical calculations. For example, to evaluate $\sqrt{5}\sqrt{5}-5$ symbolically, we type:

$$>> sym(sqrt(5)) * sym(sqrt(5)) - 5$$
$$ans =$$
$$0$$

But when we do the same calculation numerically, we have:

$$>> sqrt(5) * sqrt(5) - 5$$
$$ans =$$
$$8.8818e - 016$$

In general, numerical results are obtained much more quickly with numerical computation than with numerical evaluation of a symbolic calculation. To perform symbolic computations, we must use *syms* to declare the variables we plan to use to be symbolic variables. For example, the quadratic formula can be defined in terms of a symbolic expression by the following kind of commands:

$$>> syms\ x\ a\ b\ c$$
$$>> sym(sqrt(a * x\hat{}\ 2 + b * x + c))$$
$$ans =$$
$$(a * x\hat{}\ 2 + b * x + c)\ \hat{}\ (1/2)$$

A symbolic object that is created can also be a symbolic expression written in terms of variables that have not been first created as symbolic objects. For example, the above quadratic formula can be created as a symbolic object by using the following *sym* command:

$$\boxed{\begin{aligned} &>> sym('sqrt(a*x\hat{\ }\,2+b*x+c)') \\ &ans = \\ &\qquad (a*x\hat{\ }\,2+b*x+c)\,\hat{\ }\,(1/2) \end{aligned}}$$

The *double (x)* command can be used to convert a symbolic expression (object) x, which is written in an exact numerical form. (The name double comes from the fact that the command returns a double-precision floating-point number representing the value of x.) For example:

$$\boxed{\begin{aligned} &>> x = sym(12)*sin(7*pi/4) \\ &x = \\ &\qquad -6*2\hat{\ }\,(1/2) \\ &x1 = double(x) \\ &x1 = \\ &\qquad -8.4853 \end{aligned}}$$

Symbolic expressions that already exist can be used to create new symbolic expressions, and this can be done by using the name of the existing expression in the new expression. For example:

$$\boxed{\begin{aligned} &>> syms\ x\ y\ z \\ &>> w = x\hat{\ }\,2+y\hat{\ }\,2 \\ &w = \\ &\qquad x\hat{\ }\,2+y\hat{\ }\,2 \\ &>> u = x\hat{\ }\,2*y\hat{\ }\,2 \\ &u = \\ &\qquad x\hat{\ }\,2*y\hat{\ }\,2 \\ &>> v = sqrt(w+u) \\ &v = \\ &\qquad (x\hat{\ }\,2+y\hat{\ }\,2+x\ \hat{\ }\,2*y\hat{\ }\,2)\hat{\ }\,(1/2) \end{aligned}}$$

C.6.1   Some Important Symbolic Commands

Symbolic expressions are either created by the user or by MATLAB as the result of symbolic operations. the expressions created by MATLAB might not be in the simplest form or in a form that the user prefers. The form of an existing symbolic expression can be changed by collecting terms with the same power, by expanding products, by factoring out common multipliers, by using mathematical and trigonometric identities, and by many other operations. Now we define several commands that can be used to change the form of an existing symbolic expression.

The Collect Command

This command collects the terms in the expression that have the variable with the same power. In the new expression, the terms will be ordered in decreasing order of power. The form of this command is:

$$>> collect(f)$$

or

$$>> collect(f, var_name)$$

For example, if $f = (2x^2 + y^2)(x + y^2 + 3)$, then use the following commands:

```
>> syms x y
>> f = (2 * x^2 + y^2) * (x + y^2 + 3)
>> collect(f)
ans =
    2 * x^3 + (2 * y^2 + 6) * x^2 + y^2 * x + y^2 * (y^2 + 3)
```

But if we take y as a symbolic variable, then we do the following:

```
>> syms x y
>> f = (2 * x^2 + y^2) * (x + y^2 + 3)
>> collect(f, y)
ans =
    y^4 + (2 * x^2 + x + 3) * y^2 + 2 * x^2 * (x + 3)
```

The Factor Command

This command changes an expression that is a polynomial to be a product of polynomials of lower degree. The form of this command is:

>> $factor(f)$

For example, if $f = x^3 - 3x^2 - 4x + 12$, then use the following commands:

>> *syms x*
>> $f = x\char`\^ 3 - 3 * x\char`\^ 2 - 4 * x + 12$
>> $factor(f)$
ans =
    $(x - 2) * (x - 3) * (x + 2)$

The Expand Command

This command multiplies the expressions. The form of this command is:

>> $expand(f)$

For example, if $f = (x^3 - 3x^2 - 4x + 12)(x - 3)^2$, then use the following commands:

>> *syms x*
>> $f = (x\char`\^ 3 - 3 * x\char`\^ 2 - 4 * x + 12) * (x - 3)\char`\^ 3$
>> $expand(f)$
ans =
    $x\char`\^ 6 - 12 * x\char`\^ 5 + 50 * x\char`\^ 4 - 60 * x\char`\^ 3 - 135 * x\char`\^ 2 + 432 * x - 324$

The Simplify Command

This command is used to generate a simpler form of the expression. The form of this command is:

>> $simplify(f)$

For example, if $f = (x^3 - 3x^2 - 4x + 12)/(x - 3)^2$, then use the following commands:

```
>> syms x
>> f = (x^ 3−3 ∗ x^ 2−4 ∗ x + 12)/(x − 3)^ 3
>> simplify(f)
ans =
    (x^ 2−4)/(x − 3)^ 2
```

The Simple Command

This command finds a form of the expression with the fewest number of characters. The form of this command is:

```
>> simple(f)
```

For example, if $f = (\cos x \cos y + \sin x \sin y)$, then use the **simplify** command, and we get:

```
>> syms x y
>> f = (cos(x) ∗ cos(y) + sin(x) ∗ sin(y))
>> simplify(f)
ans =
    cos(x) ∗ cos(y) + sin(x) ∗ sin(y)
```

But if we use the **simple** command, we get:

```
>> syms x y
>> f = (cos(x) ∗ cos(y) + sin(x) ∗ sin(y))
>> simple(f)
ans =
    cos(x − y)
```

The Pretty Command

This command displays a symbolic expression in a format in which expressions are generally typed. The form of this command is:

```
>> pretty(f)
```

For example, if $f = \sqrt{x^3 - 3x^2 - 4x + 12}$, then use the following commands:

```
>> syms x
>> f = sqrt(x^ 3−3 * x^ 2−4 * x + 12)
>> pretty(f)
ans =
    (x³ − 3x² − 4x + 12)^{1/2}
```

The findsym Command

To determine what symbolic variables are used in an expression, we use the **findsym** command. For example, the symbolic expressions $f1$ and $f2$ are defined by:

```
>> syms a b c x y z
>> f1 = a * x^ 2+b * x + c
>> f2 = x * y * z
>> findsym(f1)
ans =
    a, b, c
>> findsym(f2)
ans =
    x, y, z
```

The Subs Command

We can substitute a numerical value for a symbolic variable using the **subs** command. For example, to substitute the value $x = 2$ in the $f = x^3 y + 12xy + 12$, we use the following commands:

```
>> syms x y
>> f = x^ 3*y + 12 * x * y + 12
>> subs(f, 2)
ans =
    32 * y + 12
```

Note that if we do not specify a variable to substitute for, MATLAB chooses a default variable according to the following rule. For one-letter variables, MATLAB chooses the letter closest to x in the alphabet. If there are two letters equally close to x, MATLAB chooses the one that comes later in the alphabet. In the preceding function, subs(f,2) returns the same answer as subs(f,x,2). One can also use the findsym command to determine the default variable. For example:

```
>> syms u v
>> f = u * v
>> findsym(f, 1)
ans =
    v
```

C.6.2   Solving Equations Symbolically

We can find the solutions of certain equations symbolically by using the MATLAB command **solve**. For example, to solve the nonlinear equation $x^3 - 2x - 1 = 0$ we define the symbolic variable x and the expression $f = x^3 - 2x - 1$ with the following commands:

```
>> syms x
>> f = x^ 3−2 * x − 1
>> solve(f)
ans =
    [−1]
    [1/2 * 5 ^ (1/2) + 1/2]
    [1/2 − 1/2 * 5^ (1/2)]
```

Note that the equation to be solved is specified as a string; i.e., it is surrounded by single quotes. The answer consists of the exact(symbolic) solutions $-1, 1/2 \pm 1/2\sqrt{5}$. To get the numerical solutions, type **double**(ans):

```
>> double(ans)
ans =
    -1.0000
     1.6180
    -0.6180
```

or type **vpa**(ans):

```
>> vpa(ans)
ans =
    [-1.]
    [1.6180339887498949025257388711907]
    [-.6180339887498949025257388711907]
```

The command **solve** can also be used to solve polynomial equations of higher degrees, as well as many other types of equations. It can also solve equations involving more than one variable. For example, to solve the two equations $3x + 3y = 2$ and $x + 2y^2 = 1$, we do the following:

```
>> syms x y
>> [x, y] = solve('3 * x + 3 * y = 2',' x + 2 * y^ 2= 1')
x =
    [5/12 - 1/12 * 33^ (1/2)]
    [5/12 + 1/12 * 33^ (1/2)]

y =
    [1/4 + 1/12 * 33^ (1/2)]
    [1/4 - 1/12 * 33^ (1/2)]
```

Note that both solutions can be extracted with $x(1), y(1), x(2)$, and $y(2)$. For example, type:

```
>> x(1)
ans =
    [5/12 - 1/12 * 33^ (1/2)]
```

and

```
>> y(1)
ans =
    [1/4 + 1/12 * 33^ (1/2)]
```

If we want to solve $x + xy^2 + 3xy = 3$ for y in terms of x, then we have to specify the equation as well as the variable y as a string:

```
>> syms x y
>> solve('x + x * y^ 2+3 * x * y = 3',' y')
ans =
    [1/2/x * (-3 * x + (5 * x^2 + 12 * x)^ (1/2))]
    [1/2/x * (-3 * x - (5 * x^2 + 12 * x)^ (1/2))]
```

C.6.3 Calculus

The Symbolic Math Toolbox provides functions to do the basic operations of calculus. Here, we describe these functions.

Symbolic Differentiation

This can be performed by using the **diff** command as follows:

```
>> diff(f)
```

or

```
>> diff(f, var)
```

where the command diff(f, var) is used for differentiation of expressionss with several symbolic variables. For example, to find the first derivative of $f = x^3 + 3x^2 + 20x - 12$, we use the following commands:

```
>> syms x
>> f = x^ 3+3 * x^ 2+20 * x − 12
>> diff(f)
ans =
    3 * x^ 2+6 * x + 20
```

Note that if $f = x^3 + x \ln y + y e^{x^2}$ is taken, then MATLAB differentiates f with respect to x (default symbolic variable) as:

```
>> syms x y
>> f = x^ 3+x * log(y) + y * exp(x^ 2)
>> diff(f)
ans =
    3 * x^ 2+log(y) + 2 * y * x * exp(x^ 2)
```

If we want to differentiate $f = x^3 + x \ln y + y e^{x^2}$ with respect to y, then we use the MATLAB diff(f, y) command as:

```
>> syms x y
>> f = x^ 3+x * log(y) + y * exp(x^ 2)
>> diff(f, y)
ans =
    x/y + exp(x^ 2)
```

Find the numerical value of the symbolic expression by using the MATLAB **subs** command. For example, to find the derivative of $f = x^3 + 3x^2 + 20x − 12$ at $x = 2$, we do the following:

```
>> syms x
>> f = x^ 3+3 * x^ 2+20 * x − 12
>> df = diff(f)
>> subs(df, x, 2)
ans =
    44
```

We can also find the second and higher derivative of expressions by using the following command:

$$>> diff(f, n)$$

or

$$>> diff(f, var, n)$$

where n is a positive integer: $n = 2$ and $n = 3$ mean the second and third derivative, respectively. For example, to find the second derivative of $f = x^3 + x \ln y + y e^{x^2}$ with respect to y, we use the MATLAB diff(f, y, 2) command as:

```
>> syms x y
>> f = x^ 3 + x * log(y) + y * exp(x^ 2)
>> diff(f, y, 2)
ans =
    -x/y^ 2
```

Symbolic Integration

Integration can be performed symbolically by using the **int** command. This command can be used to determine indefinite integrals and definite integrals of expression f. For indefinite integration, we use:

$$>> int(f)$$

or

$$>> int(f, var)$$

If in using the **int**(f) command the expression contains one symbolic variable, then integration took place with respect to that variable. But if the expression contains more than one variable, then the integration is performed with respect to the default symbolic variable. For example, to find the indefinite integral (antiderivative) of $f = x^3 + x \ln y + y e^{x^2}$ with respect to y, we use the MATLAB **int(f, y)** command as:

```
>> syms x y
>> f = x^3+x*log(y) + y*exp(x^2)
>> int(f,y)
ans =
    x^3*y + x*y*log(y) - x*y + 1/2*y^2*exp(x^2)
```

Similarly, for the case of a definite integral, we use the following command:

```
>> int(f,a,b)
```

or

```
>> int(f,var,a,b)
```

where a and b are the limits of integration. Note that the limits a and b may be numbers or symbolic variables. For example, to determine the value of $\int_0^1 (x^2 + 3e^x + x\ln y)\,dx$, we use the following commands:

```
>> syms x y
>> f = x^2+3*exp(x) + x*log(y)
>> int(f,0,1)
ans =
    -8/3 + 3*exp(1) + 1/2*log(y)
```

We can also use symbolic integration to evaluate the integral when f has some parameters. For example, to evaluate the $\int_{-\infty}^{\infty} e^{-ax^2}\,dx$, we do the following:

```
>> syms a positve
>> syms x
>> f = exp(-a*x^2)
>> int(f,x,-inf,inf)
ans =
    1/(a)^2(1/2)*pi^2(1/2)
```

Note that if we don't assign a value to a, then MATLAB assumes that a

represents a complex number and therefore gives a complex answer. If a is any real number, then we do the following:

```
>> syms a real
>> syms x
>> f = exp(-a * x^ 2)
>> int(f, x, -inf, inf)
ans =
    PIECEWISE([1/a^ 2*pi^ 2(1/2), signum(a) = 1], [Inf, otherwise])
```

Symbolic Limits

The Symbolic Math Toolbox provides the **limit** command, which allows us to obtain the limits of functions directly. For example, to use the definition of the derivative of the function

$$f'(x) = \lim_{h \to 0} \frac{f(x+h) - f(x)}{h}, \quad \text{provided limits exist,}$$

and for finding the derivative of the function $f(x) = x^2$, we use the following commands:

```
>> syms h x
>> f = (x + h)^ 2 - x^ 2
>> limit(f/h, h, 0)
ans =
    2 * x
```

We can also find one-sided limits with the Symbolic Math Toolbox. To find the limit as x approaches a from the left, we use the commands:

```
>> syms a real
>> syms x
>> limit(f, x, a, 'left')
```

and to find the limit as x approaches a from the right, we use the commands:

```
>> syms a real
>> syms x
>> limit(f, x, a,' right')
```

For example, to find the limit of $\frac{|x-3|}{x-3}$ when x approaches 3, we need to calculate

$$\lim_{x \to 3^-} \frac{|x-3|}{x-3} \quad \text{and} \quad \lim_{x \to 3^+} \frac{|x-3|}{x-3}.$$

Now to calculate the left-side limit, we do as follows:

```
>> syms a real
>> syms x
>> a = 3
>> f = abs(x - 3)/(x - 3)
>> limit(f, x, a,' left')
ans =
    -1
```

and to calculate the right-side limit, we use the commands:

```
>> syms a real
>> syms x
>> a = 3
>> f = abs(x - 3)/(x - 3)
>> limit(f, x, a,' right')
ans =
    1
```

Since the limit from the left does not equal the limit from the right, the limit does not exist. It can be checked by using the following commands:

```
>> syms a real
>> syms x
>> a = 3
>> f = abs(x − 3)/x − 3
>> limit(f, x, a)
ans =
     NaN
```

Taylor Polynomial of a Function

The Symbolic Math Toolbox provides the **taylor** command, which allows us to obtain the analytical expression of the Taylor polynomial of a given function. In particular, having defined in the string the function f on which we want to operate **taylor**(f, x, n+1) the associated Taylor polynomial of degree n expanded about $x_0 = 0$. For example, to find the Taylor polynomial of degree three for $f(x) = e^x \sin x$ expanded about $x_0 = 0$, we use the following commands:

```
>> syms x
>> f = exp(x) * sin(x)
>> taylor(f, x, 4)
ans =
     x + x^ 2+1/3 * x^ 3
```

C.6.4 Symbolic Ordinary Differential Equations

Like differentiation and integration, an ordinary differential equation can be solved symbolically by using the **dsolve** command. This command can be used to solve a single equation or a system of differential equations. This command can also be used in getting a general solution or a particular solution of an ordinary differential equation. For first-order ordinary differential equations, we use:

```
>> dsolve('eq')
```

or

$$>> dsolve('eq',' var')$$

For example, in finding the general solution of the ordinary differential equation

$$\frac{dy}{dt} = t + \frac{3y}{t},$$

we use the following commands:

```
>> syms t y
>> f = t + 3 * y/t
>> dsolve('Dy = f')
ans =
    -t^2 + t^3 * C1
```

For finding a particular solution of a first-order ordinary differential equation, we use the following command:

$$>> dsolve('eq',' cond1')$$

For example, in finding the particular solution of the ordinary differential equation

$$\frac{dy}{dt} = t + \frac{3y}{t},$$

with the initial condition $y(1) = 4$, we do the following:

```
>> syms t y
>> f = x + 3 * y/t
>> dsolve('Dy = f',' y(1) = 4',' t')
ans =
    -t^2 + 5 * t^3
```

Similarly, the higher-order ordinary differential equation can be solved symbolically using the following command:

$$\boxed{>> dsolve('eq',' cond1',' cond2', \cdots,' var')}$$

For example, the second-order ordinary differential equation

$$\frac{d^2y}{dx^2} - 4\frac{dy}{dx} - 5y = 0, \quad y(1) = 0, \ y'(1) = 2$$

can be solved by using the following commands:

> $>> syms\ x\ y$
> $>> dsolve('D2y - 4*Dy - 5*y = 0',' y(1) = 0',' Dy(1) = 2',' x')$
> $ans =$
> $\quad 1/3/exp(1)\hat{}\ 5*exp(5*x) - 1/3*exp(1)*exp(-x)$

C.6.5 Linear Algebra

Consider the matrix

$$A = \begin{bmatrix} 3 & 2 \\ x & y \end{bmatrix} \quad \text{and} \quad \mathbf{b} = \begin{bmatrix} 1 \\ y \end{bmatrix}.$$

Since the matrix A is symbolic expressions, we can calculate the determinant and the inverse of A, and also solve the linear system using the vector b:

> $>> syms\ x\ y$
> $>> A = [3\ 2; x\ y])$
> $>> det(A)$
> $ans =$
> $\quad 3*y - 2*x$
> $>> inv(A)$
> $ans =$
> $\quad [-y/(-3*y + 2*x), 2/(-3*y + 2*x)]$
> $\quad [x/(-3*y + 2*x), -3/(-3*y + 2*x)]$
> $>> b = [1; x]$
> $>> A\backslash b$
> $ans =$
> $\quad [(2*x - y/(-3*y + 2*x)]$
> $\quad [-2*x/(-3*y + 2*x)]$

C.6.6 Eigenvalues and Eigenvectors

To find a characteristic equation of the matrix

$$A = \begin{bmatrix} 3 & -1 & 0 \\ -1 & 2 & -1 \\ 0 & -1 & 3 \end{bmatrix},$$

we use the following commands:

```
>> A = [3 − 1 0; −1 2 − 1; 0 − 1 3]
>> poly(sym(A))
ans =
    x^ 3−8 ∗ x^ 2+19 ∗ x − 12

>> factor(ans)
ans =
    (x − 1) ∗ (x − 3) ∗ (x − 4)
```

We can also get the eigenvalues and eigenvectors of a square matrix A symbolically by using the **eig**(sym(A)) command. The form of this command is as follows:

```
>> [X, D] = eig(sym(A))
```

For example, to find the eigenvalues and eigenvectors of the matrix A, we use the following commands:

```
>> A = [3  − 1 0; −1 2  − 1; 0  − 1 3]
>> [X, D] = eig(sym(A))
X =
    [1, −1, 1]
    [−1, 0, 2]
    [1, 1, 1]

D =
    [4, 0, 0]
    [0, 3, 0]
    [0, 0, 1]
```

where the eigenvector in the first column of vector X corresponds to the eigenvalue in the first column of D, and so on.

C.6.7 Plotting Symbolic Expressions

We can easily plot a symbolic expression by using the **ezplot** command. To plot a symbolic expression Z that contains one or two variables, the **ezplot** command is:

```
>> ezplot(Z)
```

or

```
>> ezplot(Z, [min, max])
```

or

```
>> ezplot(Z, [xmin, xmax, ymin, ymax])
```

For example, we can plot a graph of the symbolic expression $Z = (2x^2 + 2)/(x^2 − 6)$ using the following commands:

```
>> syms x
>> Z = (2 * x^ 2+2)/(x^ 2−64)
>> ezplot(Z)
```

and we obtain Figure C.7.

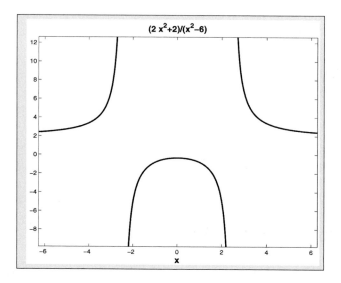

Figure C.7: Graph of $Z = (2x^2 + 2)/(x^2 - 6)$.

Note that **ezplot** can also be used to plot a function that is given in a parametric form. For example, when $x = \cos 2t$ and $y = \sin 4t$, we use the following commands:

```
>> syms t
>> x = cos(2 * t)
>> x = sin(4 * t)
>> ezplot(x, y)
```

and we obtain Figure C.8.

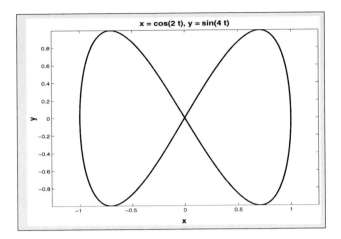

Figure C.8: Graph of function in a parametric form.

C.7   Symbolic Math Toolbox Functions

Listed below are some of the Symbolic Math Toolbox functions:

Symbolic Math Toolbox Function	Definition
diff	differentiate
int	integration
limit	limit of an expression
symsum	summation of series
taylor	Taylor's series expansion
det	determinant
diag	create or extract diagonals
eig	eigenvalues and eigenvectors
inv	inverse of a matrix
expm	exponential of a matrix
rref	reduced row echelon form
svd	singular value decomposition

Symbolic Math Toolbox Function	Definition
poly	characteristic polyonmial
rank	rank of a matrix
tril	lower triangle
triu	upper triangle
collect	collect common terms
expand	expand polynomials and elementary functions
factor	factor a expression
simplify	simplification
simple	search for shortest form
pretty	pretty print of symbolic expression
findsym	determine symbolic variables
subexpr	rewrite in terms of subexpresions
numden	numerator and denominator
compose	functional composition
solve	solution of algebraic equations
desolve	solution of differetial equations
finverse	functional inverse
sym	create symbolic object
syms	shortcut for creating multiple symbolic objects
real	real part of an imaginary number
latex	LaTeX representation of a symbolic expression
fortran	Fortran representation of a symbolic expression
imag	imaginary part of a complex number
conj	complex conjugate
resums	Riemann sums
taylortool	Taylor's seriecs calculator
funtool	Tfunction calculator
digits	set variable precision accuracy
vpa	variable precision arithmetic
double	convert symbolic matrix to double
char	convert sym object to string
poly2sym	function calculator
sym2poly	symbolic polynomial to coefficient vector

Symbolic Math Toolbox Function	Definition
fix	round toward zero
floor	round toward minus infinity
ceil	round toward plus infinity
int8	convert symbolic matrix to signed 8-bit integer
int16	convert symbolic matrix to signed 16-bit integer
int32	convert symbolic matrix to signed 32-bit integer
int64	convert symbolic matrix to signed 64-bit integer
uint8	convert symbolic matrix to unsigned 8-bit integer
uint16	convert symbolic matrix to unsigned 16-bit integer
dirac	dirac delta function
zeta	Riemann zeta function
cosint	cosine integral
sinint	sine integral
fourier	Fourier transform
ifourier	inverse fourier transform
laplace	Laplace transform
ilaplace	inverse laplace transform
ztrans	z-transform
iztrans	inverse z-transform
ezplot	function plotter
ezplot3	3-D curve plotter
ezpolar	polar coordinate plotter
ezcontour	contour plotter
ezcontourf	filled contour plotter
ezmesh	mesh plotter
ezmeshc	combined mesh and contour plotter
ezsurf	surface plotter
ezsurfc	combine surface and contour plotter

C.8   Index of MATLAB Programs

In this section we list all the MATLAB functions supplied with this book. These functions are contained in a CD included with this book. The CD-ROM includes a MATLAB program for each of the methods presented. Every program is illustrated with a sample problem or example that is closely correlated to the text. The programs can be easily modified for other problems by making minor changes. All the programs are designed to run on a minimally configured computer. Minimal hard disk space plus the MATLAB package are all that is needed. All the programs are given as ASCII files called m-files with the .m extension. They can be altered using any word processor that creates a standard ASCII file. The m-files can be run from within MATLAB by entering the name without the .m extension. For example, fixpt.m can be run using **fixpt** as the command. The files should be placed in MATLAB\work subdirectory of MATLAB.

MATLAB Function	Definition	Chapter 1
INVMAT	Inverse of a matrix	program 1.1
CofA	Minor and cofactor of a matrix	program 1.2
CofExp	Determinant by cofactor expansion	program 1.3
Adjoint	Adjoint of a matrix	program 1.4
CRule	Cramer's rule	program 1.5
WP	Gauss elimination method	program 1.6
PP	G.E. with partial pivoting	program 1.7
TP	G.E. with total pivoting	program 1.8
GaussJ	Gauss–Jordan method	program 1.9
lu-guass	LU decomposition method	program 1.10
Dolittle	Doolittle's method	program 1.11
Crout	Crout's method	program 1.12
Cholesky	Cholesky method	program 1.13
TridLU	Tridiagonal system	program 1.14
RES	Calculate residual vector	program 1.15

MATLAB Function	**Definition**	**Chapter 2**
JacobiM	Jacobi iterative method	program 2.1
GaussSM	Gauss–Seidel iterative method	program 2.2
SORM	SOR iterative method	program 2.3
CONJG	Conjugate gradient method	program 2.4

MATLAB Function	**Definition**	**Chapter 3**
trac	Trace of a matrix	program 3.1
EigTwo	Eigenvalues of a 2×2 matrix	program 3.2
Chim	Cayley–Hamilton theorem	program 3.3
sourian	Sourian frame theorem	program 3.4
BOCH	Bocher's theorem	program 3.5

MATLAB Function	**Definition**	**Chapter 4**
POWERM1	Power method	program 4.1
INVERSEPM1	Inverse power method	program 4.2
ShiftedIPM1	Shifted inverse power method	program 4.3
DEFLATION	Deflation method	program 4.4
JOBM	Jacobi method for eigenvalues	program 4.5
SturmS	Sturm sequence method	program 4.6
Given	Given's method	program 4.7
HHHM	Householder method	program 4.8
QRM	QR method's	program 4.9
hes	Upper Hessenberg form	program 4.10

MATLAB Function	Definition	Chapter 5
Lint	Lagrange method	program 5.1
DiviDiff	Divided differences of a function	program 5.2
NDiviD	Newton's divided differences formula	program 5.3
Aitken1	Aitken's method	program 5.4
ChebP	Chehbev polynomial	program 5.5
ChebYA	Chehbev polynomial approximation	program 5.6
linefit	Linear least squares fit	program 5.7
polyfit	Polynomial least squares fit	program 5.8
ex1fit	Nonlinear least squares fit	program 5.9
ex2fit	Nonlinear least squares fit	program 5.10
planefit	Least squares plane fit	program 5.11
overd	Overdetermined	program 5.12
underd	Underdetermined	program 5.13

MATLAB Function	Definition	Chapter 7
Hessian	Hessian Matrix	program 7.1
bisect	Bisection method	program 7.2
fixpt	Fixed-point method	program 7.3
newton	Newton's method	program 7.4
newton2	Newton's method for a nonlinear system	program 7.5
golden	Golden-section search method	program 7.6
Quadratic2	Quadratic interpolation method	program 7.7
newtonO	Newton's method for optimization	program 7.8

C.9 **Summary**

MATLAB has a wide range of capabilities. In this book, we used only a small portion of its features. We found that MATLAB's command structure is very close to the way one writes algebraic expressions and linear algebra operations. The names of many MATLAB commands closely parallel those of the operations and concepts of linear algebra. We gave descriptions of commands and features of MATLAB that related directly to this course. A more detailed discussion of MATLAB commands can be found in the MATLAB user guide that accompanies the software and in the following books:

Experiments in computational Matrix Algebra by David R. Hill (New York, Random House, 1988).

Linear Algebra LABS with MATLAB, second edition by David R. Hill and David E. Zitaret (Prentice-Hall, Inc., 1996).

For a very complete introduction to MATLAB graphics, one can use the following book:

Graphics and GUIs with MATLAB, 2nd ed., by P. Marchand (CRC Press, 1999).

There are many websites to help you learn MATLAB and you can locate many of those by using a web search engine. Alternatively, MATLAB software provides immediate on-screen descriptions using the *Help* command or one can contact Mathworks at: www.mathworks.com

C.10   Problems

1. Solve each of the following expressions in the Command window:

   (a)
   $$(15 + 17)^2 + \frac{(165)^{3/2}}{4} + \frac{(765)^2}{24}.$$

   (b)
   $$(2.55 + \ln(4))^3 + \frac{(\sqrt{(45)^3 + 23})^3}{17} + \frac{(101/21)^2}{15}.$$

   (c)
   $$(e^2 + 245)^3 + \frac{(165 + 2e^{3/2})^2}{134} + \frac{(\sqrt{876} + 234)^4}{342}.$$

   (d)
   $$(e^{4/5} + \ln(7))^2 + \frac{(\sqrt{(788)^5 + 120})^2}{111} + \frac{(e^4 + 333)^4}{254}.$$

2. Solve each of the following expressions in the Command Window:

   (a)
   $$(\sin(\frac{4\pi}{3}) + 0.7757)^2 + \cos^2(\frac{2\pi}{3}) + 2\cos(\frac{2\pi}{3})\sin(\frac{5\pi}{3}).$$

   (b)
   $$(\sin^2(\frac{7\pi}{4}) + \ln(2.5))^3 + e^3\cos(\frac{12\pi}{5}) + \cos(\frac{\pi}{3})\tan(\frac{\pi}{4}).$$

   (c)
   $$(\tan(\frac{\pi}{3}) + e^{0.5})^{1/2} + (\sin^3(\frac{\pi}{6}) + 3\cos(\frac{\pi}{6}))/4 + \sin^3(\frac{\pi}{6}).$$

   (d)
   $$(e^{3/2} + \ln(3.5))^2 + \frac{(\sqrt{(22)} + 12\cos(4\pi))}{24} + \frac{(e^2 + \ln(4.5)\sin(0.75))^2}{12}.$$

3. Solve each of the following expressions in the Command Window:

   (a)
   $$\cos^2(\frac{3\pi}{4})\sin(\frac{12\pi}{5}) + 2\tan(\frac{\pi}{4}).$$

(b)

$$\sec(\frac{5\pi}{4})\tan^2(\frac{5\pi}{5}) + 2\cos^2(\frac{\pi}{4})\sin(\frac{7\pi}{4}).$$

(c)

$$\cot^2(\frac{5\pi}{4})\sin^2(\frac{3\pi}{5}) + 3[\csc^2(\frac{\pi}{4})\sec^2(\frac{9\pi}{4})]^2.$$

(d)

$$[\tan(\frac{\pi}{6})\sin(\frac{\pi}{4})]^2 + \cos^3(\frac{\pi}{6})[\csc^2(\frac{\pi}{6})\sec^2(\frac{\pi}{6})]^3.$$

4. Define variable x and calculate each of the following in the Command Window:

(a)

$$f(x) = x^4 + 23x^3 + 19x^2 + 2x + 32; \quad x = 2.5.$$

(b)

$$f(x) = x^4 + \ln(x^2+2) + \frac{\sqrt{x^3 + 23x^2 + 1.2x}}{x} + \frac{((x^4 - 12)/13)^2}{5}; \quad x = 4.5.$$

(c)

$$f(x) = (e^{(x^2+1)} + 2x^3 + 5x)^4 + \frac{(15x^3 + 2e^{x/2})^4}{x} + \frac{(\sqrt{\sin(x)} + x)^2}{2}, \quad x12.5.$$

(d)

$$f(x) = 5(e^{(x+3)} + \ln(x^3-2))^3 + \frac{(\sqrt{(x+18)^6 + \sin(x+1)})^3}{33}, \quad x = 35.5.$$

5. Define variables x, y, z and solve each of the following in the Command Window:

(a)

$$w = x^2 y^3 + 3x^3 yz^4 + 9xy^3 z + 2xyz + 32y; \quad x = 0.5, \ y = 2.7, \ z = 13.5.$$

(b)

$$w = \ln(xy^3) + 2.5xyz + \sin(xy) + 2x^3 y^6 z^8; \quad x = 12.5, \ y = 22.5, \ z = 33.5.$$

(c)

$$w = \sqrt{\cos(xy+z)} + \ln(x^2+y^3+z^4) + \tan(xy); \quad x = 5.5, \ y = 6.5, \ z = 8.5.$$

(d)

$$w = \ln(\sqrt{x+y^3}) + \cos(x^3 y) + 15x^2 y z^4; \quad x = 11.0, \ y = 12.0, \ z = 13.0.$$

(d)

$$f(x) = 5(e^{(x+3)} + \ln(x^3-2))^3 + \frac{(\sqrt{(x+18)^6 + sin(x+1)})^3}{33}, \quad x = 35.5$$

6. Create a vector that has the following elements using the Command Window:

   (a)
   $$19, 4, 31, e^{25}, 63, \cos(\frac{\pi}{6}), \ln(3).$$

   (b)
   $$\pi, 44, 101, e^2, 116, \sin(\frac{11\pi}{4}), \ln(2).$$

   (c)
   $$35, 40, 321, e^7, 406, \cos^3(\frac{5\pi}{4}), 2\ln(7).$$

   (d)
   $$60, 4\ln(4), 17, 1+e^3, 83\sin(\frac{5\pi}{6}).$$

7. Plot the function $f(x) = \dfrac{x^3 - 4x + 1}{x^3 - x + 2}$ for the domain $-5 \le x \le 5$.

8. Plot the function $f(x) = 4x\cos x - 3x$ and its derivative, both on the same plot, for the domain $-2\pi \le x \le 2\pi$.

9. Plot the function $f(x) = 4x^4 - 3x^3 + 2x^2 - x + 1$, and its first and second derivatives, for the domain $-3 \le x \le 5$, all in the same plot.

10. Make two separate plots of the function

$$f(x) = x^4 - 3\sin x + \cos x + x,$$

one plot for $-2 \le x \le 2$ and the other for $-3 \le x \le 3$.

11. Use the fPlot command to plot the function $f(x) = 0.25x^4 - 0.15x^3 + 0.5x^2 - 1.5x + 3.5$, for the domain $-3 \le x \le 3$.

12. The position of a moving particle as a function of time is given by

$$
\begin{aligned}
x &= (1 - 2\sin t)\cos t \\
y &= (1 - 2\sin t)\sin t \\
z &= 3t^2.
\end{aligned}
$$

Plot the position of the particle for $3 \le t \le 10$.

13. The position of a moving particle as a function of time is given by

$$
\begin{aligned}
x &= 1 + \sin t \\
y &= 1 + \cos t \\
z &= 3t^4.
\end{aligned}
$$

Plot the position of the particle for $0 \le t \le 10$.

14. Make a 3-D surface plot and contour plot (both in the same figure) of the function $z = (x + 2)^2 + 3y^2 - xy$ in the domain $-5 \le x \le 5$ and $-5 \le x \le 5$.

15. Make a 3-D mesh plot and contour plot (both in the same figure) of the function $z = (x - 2)^2 + (y - 2)^2 + xy$ in the domain $-5 \le x \le 5$ and $-5 \le x \le 5$.

16. Define x as a symbolic variable and create the two symbolic expressions:
$P_1 = x^4 - 6x^3 + 12x^2 - 9x + 3$ and
$P_2 = (x + 2)^4 + 5x^3 + 17(x + 3)^2 + 12x - 20$.
Use symbolic operations to determine the simplest form of the following expressions:

(i) $P_1.P_2$.

(ii) $P_1 + P_2$.

(iii) $\dfrac{P_1}{P_2}$.

(iv) Use the subs command to evaluate the numerical value of the results for $x = 15$.

17. Define x as a symbolic variable and create the two symbolic expressions:

$P_1 = \ln(\sin(x+2) + x^2) - 12\sqrt{x^3 + 12} - e^{x+2} - 15x + 10$ and
$P_2 = (x-2)^3 + 11x^2 + 9x - 15$.

Use symbolic operations to determine the simplest form of the following expressions:

(i) $P_1.P_2$.

(ii) $P_1 + P_2$.

(iii) $\dfrac{P_1}{P_2}$.

(iv) Use the subs command to evaluate the numerical value of the results for $x = 9$.

18. Define x as a symbolic variable.

(i) Show that the roots of the polynomial

$$f(x) = x^5 - 12x^4 + 15x^3 + 200x^2 - 276x - 1008$$

are $-3, -2, 4, 6$, and 7 by using the *factor* command.

(ii) Derive the equation of the polynomial that has the roots $x = -5, x = 4, x = 2$, and $x = 1$.

19. Define x as a symbolic variable.

(i) Show that the roots of the polynomial

$$f(x) = x^5 - 30x^4 + 355x^3 - 2070x^2 + 5944x - 6720$$

are $4, 5, 6, 7$, and 8 by using the *factor* command.

(ii) Derive the equation of the polynomial that has the roots $x = 3, x = 2, x = 1$, and $x = 0$.

20. Find the fourth-degree Taylor polynomial for the function $f(x) = (x^3 + 1)^{-1}$, expanded about $x_0 = 0$.

21. Find the fourth-degree Taylor polynomial for the function $f(x) = x + 2\ln(x+2)$, expanded about $x_0 = 0$.

22. Find the fourth-degree Taylor polynomial for the function $f(x) = (x+1)e^x + \cos x$, expanded about $x_0 = 0$.

23. Find the general solution of the ordinary differential equation

$$y' = 2(y+1).$$

Then find its particular solution by taking the initial condition $y(0) = 1$ and plot the solution for $-2 \le x \le 2$.

24. Find the general solution of the second-order ordinary differential equation

$$y'' + xy' - 3y = x^2.$$

Then find its particular solution by taking the initial conditions $y(0) = 3$, $y'(0) = -6$ and plot the solution for $-4 \le x \le 4$.

25. Find the inverse and determinant of the matrix

$$A = \begin{pmatrix} 3 & -1 & 5 \\ p & 4 & 2q \\ q & 3p & 5 \end{pmatrix}.$$

Use $\mathbf{b} = [1, 2, 3]^T$ to solve the system $A\mathbf{x} = \mathbf{b}$.

26. Find the inverse and determinant of the matrix

$$A = \begin{pmatrix} 3 & -1 & 2 & -3 \\ 2p & 6 & -q & 3 \\ 3 & q & 3p & 5 \\ 4 & -5 & 7 & 8 \end{pmatrix}.$$

Use $\mathbf{b} = [3, -2, 4, 5]^T$ to solve the system $A\mathbf{x} = \mathbf{b}$.

27. Find the characteristic equation of the matrix

$$A = \begin{pmatrix} 4 & 2 & 1 \\ 2 & 8 & 0 \\ 1 & 0 & 8 \end{pmatrix}.$$

Then find its roots by using the *factor* command. Also, find the eigenvalues and the eigenvectors of A.

28. Find the characteristic equation of the matrix

$$A = \begin{pmatrix} 1 & 0 & 1 & 0 \\ 1 & 0 & 1 & 0 \\ 1 & 0 & 1 & 0 \\ 0 & 1 & 1 & 0 \end{pmatrix}.$$

Then find its roots by using the *factor* command. Also, find the eigenvalues and the eigenvectors of A.

29. Determine the solution of the nonlinear equation $x^3 + 2x^2 - 4x = 8$ using the *solve* command and plot the graph of the equation for $-4 \le x \le 4$.

30. Determine the solution of the nonlinear equation $\cos x + 3x^2 = 20$ using the *solve* command and plot the graph of the equation for $-2 \le x \le 2$.

Appendix D

Answers to Selected Exercises

D.0.1 Chapter 1

1. $C = \begin{pmatrix} 1 & -5 & -1 \\ -2 & 1 & -3 \\ -2 & -1 & -8 \end{pmatrix}$.

3. $|AB| = \begin{vmatrix} 1 & -5 & -1 \\ -2 & 1 & -3 \\ -2 & -1 & -8 \end{vmatrix} = 0$.

5. $x = -3, \quad y = 2$.

7. (a) B, (c) E.

9. (a) *Row E.F.* = $\begin{pmatrix} 2 & 3 & 4 \\ 0 & 1 & 2 \\ 0 & 0 & -3 \end{pmatrix}$, $\mathbf{x} = [4/3, 1/3, -2/3]^T$.

(c) *Row E.F.* = $\begin{pmatrix} 3 & 0 & 1 \\ 0 & -1 & 0 \\ 0 & 0 & 1 \end{pmatrix}$, $\mathbf{x} = [-1/3, -3, 2]^T$.

11. (a) $\mathbf{x} = [5/4, 5/2, -3/4]^T$.
(c) $\mathbf{x} = [-2, 0, 2, 0]^T$.

13. $\det(A) = a_{21}c_{21} + a_{22}c_{22} + a_{23}c_{23} = 0 + 3(-33) + 5(37) = 86.$
$\det(B) = a_{13}c_{13} + a_{23}c_{23} + a_{33}c_{33} = 4(-152) + 6(-107) + 12(-8) = -1346.$
$\det(A) = a_{12}c_{12} + a_{22}c_{22} + a_{32}c_{32} = -8(-52) + 1(-45) + 10(94) = 1311.$

15. $\det(A) = \begin{vmatrix} 3 & 1 & -1 \\ 0 & -2/3 & 14/3 \\ 0 & 0 & -36 \end{vmatrix} = (3)(-2/3)(-36) = 72.$

$\det(B) = \begin{vmatrix} 4 & 1 & 6 \\ 0 & 27/4 & 17/2 \\ 0 & 0 & 83/27 \end{vmatrix} = (4)(27/4)(83/27) = 83.$

$\det(C) = \begin{vmatrix} 17 & 46 & 7 \\ 0 & -87/17 & -4/17 \\ 0 & 0 & 10 \end{vmatrix} = (17)(-87/17)(10) = -870.$

17. $\det(A) = x^3 - 3x^2 - 12x + 31$, $det(A) = 0$, $x = [-3.3485, 4.0787, 2.2698]^T$.

19. $\det(A) = 17x^3 - 7x^2 - 882x - 2052$, $det(A) = 0$, $x = [8.3530, -5.1174, -2.8238]^T$

21. (a) $Adj(A) = \begin{pmatrix} 4 & -2 \\ 3 & 1 \end{pmatrix}$, $A^{-1} = \begin{pmatrix} 2/5 & -1/5 \\ 3/10 & 1/10 \end{pmatrix}.$

(c) $Adj(A) = \begin{pmatrix} -1 & -1 & 1 \\ -1 & 1 & 1 \\ 1 & -1 & -1 \end{pmatrix}$, $A^{-1} = \begin{pmatrix} 1/2 & 1/2 & -1/2 \\ 1/2 & -1/2 & 1/2 \\ -1/2 & 1/2 & 1/2 \end{pmatrix}.$

23.

$$Adj(A) = \begin{pmatrix} 12 & 16 & -27 \\ -5 & 9 & 13 \\ -10 & -11 & 19 \end{pmatrix},$$

$$(Adj(A))^{-1} = \begin{pmatrix} -314/707 & 1/101 & -451/707 \\ 5/101 & 6/101 & 3/101 \\ -145/707 & 4/101 & -188/707 \end{pmatrix},$$

$\det(Adj(A)) = -707.$

$$Adj(B) = \begin{pmatrix} 32 & -10 & 26 \\ 20 & 26 & -16 \\ -41 & 37 & 7 \end{pmatrix},$$

$$(Adj(B))^{-1} = \begin{pmatrix} 1/86 & 2/129 & -1/129 \\ 1/129 & 5/258 & 2/129 \\ 7/258 & -1/86 & 2/129 \end{pmatrix},$$

$\det(Adj(B)) = 66564.$

$$Adj(C) = \begin{pmatrix} 4 & 2 & -16 \\ -1 & -11 & 4 \\ -11 & 5 & 2 \end{pmatrix},$$

$$(Adj(C))^{-1} = \begin{pmatrix} -1/42 & -1/21 & -2/21 \\ -1/42 & -2/21 & 0 \\ -1/14 & -1/42 & -1/42 \end{pmatrix},$$

$\det(Adj(C)) = 1764.$

25.

$$Adj(A) = \begin{pmatrix} -2 & 11 & -3 \\ -8 & -10 & 15 \\ 7 & 2 & -3 \end{pmatrix}, \quad \det(A) = 27$$

$$A^{-1} = \begin{pmatrix} -2/27 & 11/27 & -1/9 \\ -8/27 & -10/27 & 5/9 \\ 7/27 & 2/27 & -1/9 \end{pmatrix}.$$

$$Adj(B) = \begin{pmatrix} 48 & -8 & 34 \\ 41 & 18 & -2 \\ -19 & 28 & 30 \end{pmatrix}, \quad \det(B) = 298$$

$$B^{-1} = \begin{pmatrix} 24/149 & -4/149 & 17/149 \\ 41/298 & 9/149 & -1/149 \\ -19/298 & 14/149 & 15/149 \end{pmatrix}.$$

$$Adj(C) = \begin{pmatrix} 74 & 108 & 44 & -80 \\ -190 & 112 & -128 & -20 \\ 4 & -96 & -208 & 176 \\ 90 & 66 & -232 & 68 \end{pmatrix}, \quad \det(C) = -1112$$

$$C^{-1} = \begin{pmatrix} -37/556 & 27/278 & -11/278 & 10/139 \\ 95/556 & -14/139 & 16/139 & 5/278 \\ -1/278 & 12/139 & 26/139 & -22/139 \\ -45/556 & -8/139 & 29/139 & -17/278 \end{pmatrix}.$$

27. (a) $\alpha = 1$.

(c) $\alpha = 11$.

29. (a) $A^{-1} = \begin{pmatrix} 0 & 1/7 & 1/7 \\ 1/5 & -3/35 & 4/35 \\ -2/5 & -4/35 & 17/35 \end{pmatrix},$

$\mathbf{x} = [1, 2, 3]^T.$

(c) $A^{-1} = \begin{pmatrix} 3/16 & 1/16 & 1/16 \\ -1/64 & 5/64 & -11/64 \\ -17/192 & 7/64 & 5/192 \end{pmatrix},$

$\mathbf{x} = [-1/2, 3/8, -5/24]^T.$

31. (a) $A^{-1} = \begin{pmatrix} -157/5 & 203/5 & 2/5 & 36/5 \\ 17 & -22 & 0 & -4 \\ 13 & -17 & 0 & -3 \\ 81/5 & -104/5 & -1/5 & -18/5 \end{pmatrix},$

$\mathbf{x} = [1277/5, -137, -106, -641/5]^T$.

(c) $A^{-1} = \begin{pmatrix} -34/2451 & 39/1634 & 13/258 & 139/4902 \\ -397/2451 & 143/1634 & 19/258 & -35/4902 \\ 898/2451 & -165/1634 & -55/258 & -211/4902 \\ -751/4902 & 501/3268 & -5/516 & 403/9804 \end{pmatrix}$,

$\mathbf{x} = [1557/751, 227/4902, -2273/4902, 2284/2485]^T$.

33. (a) $\det(A) = -50$, $\det(A1) = -58$, $\det(A2) = -44$, $\det(A3) = -66$,

$\mathbf{x} = [29/25, 22/25, 33/25]^T$.

(c) $\det(A) = 168$, $\det(A1) = 108$, $\det(A2) = -383$, $\det(A3) = -325$,

$\mathbf{x} = [9/14, -383/168, -325/168]^T$.

35. (a) $\det(A) = 2$, $\det(A1) = -7$, $\det(A2) = 4$, $\det(A3) = 3$,

$\mathbf{x} = [-7/2, 2, 3/2]^T$.

(c) $\det(A) = 166$, $\det(A1) = -105$, $\det(A2) = -21$, $\det(A3) = 101$,

$\mathbf{x} = [-105/166, -21/166, 101/166]^T$.

37. (a) $U = \begin{pmatrix} 3 & 4 & 5 \\ 0 & -2 & -4 \\ 0 & 0 & 3 \end{pmatrix}$, $\mathbf{x} = [7/6, -5/6, 1/6]^T$.

(c) $U = \begin{pmatrix} 6 & 7 & 8 \\ 0 & 53/6 & 26/3 \\ 0 & 0 & 45/53 \end{pmatrix}$, $\mathbf{x} = [-2/9, -5/9, 7/9]^T$.

39. (a) $U = \begin{pmatrix} 2 & 5 & -4 \\ 0 & -3 & 3 \\ 0 & 0 & -5/2 \end{pmatrix}$, $\mathbf{x} = [-4/3, 3, 7/3]^T$.

(c) $U = \begin{pmatrix} 1 & 2 & 0 \\ 0 & 2 & -2 \\ 0 & 0 & 2 \end{pmatrix}$, $\mathbf{x} = [-1, 2, 3]^T$.

41. (a) $\alpha \neq 3, \alpha \neq 5$.

(c) $\alpha \neq 0, \alpha \neq \pm 1$.

43.

$$A^{-1} = \begin{pmatrix} -1/7 & 3/7 & -1/7 \\ -4/7 & 3/14 & 3/7 \\ 6/7 & -4/7 & -1/7 \end{pmatrix}, \quad B^{-1} = \begin{pmatrix} 46/11 & 3/11 & -2 \\ -3/11 & 1/11 & 0 \\ -2 & 0 & 1 \end{pmatrix},$$

$$C^{-1} = \begin{pmatrix} 10/31 & -2/31 & -5/31 \\ 7/31 & 11/31 & -19/31 \\ -11/31 & -4/31 & 21/31 \end{pmatrix}.$$

45. $\text{rank}(A) = 3, \quad \text{rank}(B) = 3, \quad \text{rank}(C) = 2$.

47. (a) $U = \begin{pmatrix} 4 & -1 & 4 \\ 0 & 7/4 & 0 \\ 0 & 0 & -1 \end{pmatrix}$, $\mathbf{x} = [1, 3, 0]^T$.

(c) $U = \begin{pmatrix} 4 & 9 & 16 \\ 0 & -23/4 & -11 \\ 0 & 0 & 18/23 \end{pmatrix}$, $\mathbf{x} = [3/2, -5, 5/2]^T$.

49. $A = \begin{pmatrix} 1 & 1/2 & 1/3 & 1/4 \\ 1/2 & 1/3 & 1/4 & 1/5 \\ 1/3 & 1/4 & 1/5 & 1/6 \\ 1/4 & 1/5 & 1/6 & 1/7 \end{pmatrix}, \quad U = \begin{pmatrix} 1 & 1/2 & 1/3 & 1/4 \\ 0 & 1/12 & 4/45 & 1/12 \\ 0 & 0 & -1/180 & -1/120 \\ 0 & 0 & 0 & 1/2800 \end{pmatrix},$

$\mathbf{x} = [-64, 900, -2520, 1820]^T$.

51. (a) $(x) = [3/2, -7/2, 3/2]^T$.
(b) $(x) = [-14/9, 34/9, -13/9]^T$.
(c) $(x) = [-7/6, 23/6, -17/6]^T$.

53. $A^{-1} = \begin{pmatrix} 16 & -120 & 240 & -140 \\ -120 & 1200 & -2700 & 1680 \\ 240 & -2700 & 6480 & -4200 \\ -140 & 1680 & -4200 & 2800 \end{pmatrix}$, $\quad \mathbf{x} = [-64, 900, -2520, 1820]^T$.

55. (a)

$L = \begin{pmatrix} 1 & 0 & 0 & 0 \\ -1 & 1 & 0 & 0 \\ 2/3 & -11/15 & 1 & 0 \\ 7/3 & 1/3 & -1/3 & 1 \end{pmatrix}$, $\quad U = \begin{pmatrix} 3 & -2 & 1 & 1 \\ 0 & 5 & 5 & -2 \\ 0 & 0 & 6 & 28/15 \\ 0 & 0 & 0 & 133/45 \end{pmatrix}$,

$\mathbf{y} = [3, 5, 8/3, -52/9]^T$.
$\mathbf{x} = [99/133, -111/133, 20/19, -260/133]^T$.

(c)

$L = \begin{pmatrix} 1 & 0 & 0 & 0 \\ 5 & 1 & 0 & 0 \\ 1 & -3/8 & 1 & 0 \\ 1/2 & 5/8 & -9 & 1 \end{pmatrix}$, $\quad U = \begin{pmatrix} 2 & 2 & 3 & -2 \\ 0 & -8 & -2 & 21 \\ 0 & 0 & 1/4 & 127/8 \\ 0 & 0 & 0 & 551/4 \end{pmatrix}$,

$\mathbf{y} = [10, -36, -25/2, -86]^T$.
$\mathbf{x} = [7967/551, 3003/551, -5706/551, -344/551]^T$.

(e)

$L = \begin{pmatrix} 1 & 0 & 0 & 0 \\ 12 & 1 & 0 & 0 \\ 22 & -53/5 & 1 & 0 \\ 8 & -32/5 & 628/1119 & 1 \end{pmatrix}$,

$U = \begin{pmatrix} 1 & -1 & 10 & 8 \\ 0 & -5 & -109 & -74 \\ 0 & 0 & -6812/5 & -4807/5 \\ 0 & 0 & 0 & 2541/232 \end{pmatrix}$,

$\mathbf{y} = [-2, 31, 1893/5, 727/105]^T$.
$\mathbf{x} = [758/1851, 391/1723, -501/692, 342/541]^T$.

57. (a) $U = \begin{pmatrix} 2 & 3 & -1 \\ 0 & 1/2 & 3/2 \\ 0 & 0 & 1 \end{pmatrix}$, $\det(A) = (2)(1/2)(1) = 1$.

(c) $U = \begin{pmatrix} 2 & 4 & 1 \\ 0 & -3 & 1/2 \\ 070 & & 5/6 \end{pmatrix}$, $\det(A) = (2)(-3)(5/6) = -5$.

(e) $U = \begin{pmatrix} 1 & -1 & 0 \\ 0 & 1 & 1 \\ 0 & 0 & 1 \end{pmatrix}$, $\det(A) = (2)(-3)(5/6) = -5$.

59. (a)

$$A = LDV = \begin{pmatrix} 1 & 0 & 0 \\ 2/3 & 1 & 0 \\ 1/3 & 5 & 1 \end{pmatrix} \begin{pmatrix} 3 & 0 & 0 \\ 0 & 1/3 & 0 \\ 0 & 0 & 1 \end{pmatrix} \begin{pmatrix} 1 & 4 & 3 \\ 0 & 1 & 1 \\ 0 & 0 & 1 \end{pmatrix}.$$

$$B = LDV = \begin{pmatrix} 1 & 0 & 0 \\ 5/4 & 1 & 0 \\ 1 & 10/9 & 1 \end{pmatrix} \begin{pmatrix} 4 & 0 & 0 \\ 0 & 9/2 & 0 \\ 0 & 0 & 21/2 \end{pmatrix} \begin{pmatrix} 1 & -2 & 3 \\ 0 & 1 & -27/4 \\ 0 & 0 & 1 \end{pmatrix}.$$

(c)

$$A = LDV = \begin{pmatrix} 1 & 0 & 0 \\ 2 & 1 & 0 \\ 3 & 17/13 & 1 \end{pmatrix} \begin{pmatrix} 1 & -5 & 4 \\ 0 & 13 & -12 \\ 0 & 0 & 126/13 \end{pmatrix} \begin{pmatrix} 1 & -5 & 4 \\ 0 & 13 & -12 \\ 0 & 0 & 126/13 \end{pmatrix}.$$

$$B = LDV = \begin{pmatrix} 1 & 0 & 0 \\ 5/4 & 1 & 0 \\ 3/2 & 58/15 & 1 \end{pmatrix} \begin{pmatrix} 4 & 7 & -6 \\ 0 & -15/4 & 5/2 \\ 0 & 0 & 25/3 \end{pmatrix} \begin{pmatrix} 4 & 7 & -6 \\ 0 & -15/4 & 5/2 \\ 0 & 0 & 25/3 \end{pmatrix}$$

61. (a)

$$L = \begin{pmatrix} 2 & 0 & 0 \\ -3 & 5/2 & 0 \\ 1 & -1/2 & 3/5 \end{pmatrix}, \quad U = \begin{pmatrix} 1 & -1/2 & 1/2 \\ 0 & 1 & 1/5 \\ 0 & 0 & 1 \end{pmatrix},$$

$\mathbf{y} = [2, 22/5, 31/3]^T$.
$\mathbf{x} = [-2, 7/3, 31/3]^T$.

(c)

$$L = \begin{pmatrix} 2 & 0 & 0 \\ 1 & 1 & 0 \\ 3 & 0 & 1 \end{pmatrix}, \quad U = \begin{pmatrix} 1 & 1 & 1 \\ 0 & 1 & 0 \\ 0 & 0 & 1 \end{pmatrix},$$

$\mathbf{y} = [0, -4, 1]^T$.
$\mathbf{x} = [3, -4, 1]^T$.

(e)

$$L = \begin{pmatrix} 1 & 0 & 0 \\ 2 & 1 & 0 \\ 270 & -1 \end{pmatrix}, \quad U = \begin{pmatrix} 1 & -1 & 0 \\ 0 & 1 & 1 \\ 0 & 0 & 1 \end{pmatrix},$$

$\mathbf{y} = [2, 0, 1]^T$.
$\mathbf{x} = [1, -1, 1]^T$.

63. (a)

$$L = \begin{pmatrix} 1 & 0 & 0 \\ -1 & 2 & 0 \\ 1 & 0 & 3 \end{pmatrix}, \quad L^T = \begin{pmatrix} 1 & -1 & 1 \\ 0 & 2 & 0 \\ 0 & 0 & 3 \end{pmatrix},$$

$\mathbf{y} = [2, 2, 0]^T$.
$\mathbf{x} = [3, 1, 0]^T$.

(c)

$$L = \begin{pmatrix} 2 & 0 & 0 \\ 1 & 4 & 0 \\ 3/2 & -1/8 & 253/153 \end{pmatrix}, \quad L^T = \begin{pmatrix} 2 & 1 & 3/2 \\ 0 & 4 & -1/8 \\ 0 & 0 & 253/153 \end{pmatrix},$$

$\mathbf{y} = [1/2, 3/8, 1042/401]^T$.
$\mathbf{x} = [-1, 1/7, 11/7]^T$.

65. (a)

$$L = \begin{pmatrix} 1 & 0 & 0 \\ -1/3 & 1 & 0 \\ 0 & -3/8 & 1 \end{pmatrix}, \quad U = \begin{pmatrix} 3 & -1 & 0 \\ 0 & 8/3 & -1 \\ 0 & 0 & 21/8 \end{pmatrix},$$

$\mathbf{y} = [1, 4/3, 3/2]^T.$
$\mathbf{x} = [4/7, 5/7, 4/7]^T.$

(c)

$$L = \begin{pmatrix} 1 & 0 & 0 & 0 \\ -1/4 & 1 & 0 & 0 \\ 0 & -4/15 & 1 & 0 \\ 0 & 0 & -15/56 & 1 \end{pmatrix}, \quad U = \begin{pmatrix} 4 & -1 & 0 & 0 \\ 0 & 15/4 & -1 & 0 \\ 0 & 0 & 56/15 & -1 \\ 0 & 0 & 0 & 209/56 \end{pmatrix},$$

$\mathbf{y} = [1, 5/4, 4/3, 19/14]^T.$
$\mathbf{x} = [4/11, 5/11, 5/11, 4/11]^T.$

67. (a) $\|\mathbf{x}\|_1 = 12, \quad \|\mathbf{x}\|_2 = 7.0711, \quad \|\mathbf{x}\|_\infty = 6.$
(b) $\|\mathbf{x}\|_1 = 4.3818, \quad \|\mathbf{x}\|_2 = 3.1623, \quad \|\mathbf{x}\|_\infty = 3, \quad (k = 1).$

69. (a) $\|\mathbf{A}^3\|_1 = 2996, \quad \|\mathbf{A}^3\|_\infty = 23.$

(c) $\|\mathbf{B}C\|_1 = 427, \quad \|\mathbf{B}C\|_\infty = 531.$

71. (a) $K(A) = 3.969,$ well-conditioned.

(c) $K(A) = 42.2384,$ ill-conditioned.

73. $K(A) = 100$ (ill-conditioned), $\mathbf{r} = [-0.02, 0.02]^T,$
relative error $= 1.$

75. $K(A) \geq \dfrac{6.06}{0.24} = 25.25.$

77. $\delta\mathbf{x} = [-100, 101]^T, \quad K(A) = 404.01,$ ill-conditioned.

79. $\delta\mathbf{x} = [1.9976, 0.6675]^T, \quad K(A) = 80002,$ ill-conditioned.

81. $\delta\mathbf{x} = [-17, 20]^T, \quad K(A) = 8004,$ ill-conditioned.

83. (a) $y = 4 - 3x + x^2$.
    (c) $y = 3 + 2x$.
    (e) $y = 5 + x + 2x^2$.

85. (a) $l_1 = 3, l_2 = 1, l_3 = 2$.

87. $x_2 = 0$.

89. $x_1 = \dfrac{10}{7}, x_2 = \dfrac{12}{7}$ and $x_3 = \dfrac{10}{7}$.

91. (a) $4FeS_2 + 11O_2 \longrightarrow 2Fe_2O_3 + 8SO_2$.
    (c) $2C_4H_{10} + 13O_2 \longrightarrow 8CO_2 + 10H_2O$.

93. $x = \dfrac{55}{8}, \quad y = \dfrac{25}{4}, \quad z = \dfrac{5}{8}$.

95. 80,000 yen,   900 francs,   1200 marks.

97. $x = 100$ (bacteria of strain **I**).
    $y = 350$ (bacteria of strain **II**).
    $z = 350$ (bacteria of strain **III**).

D.0.2   Chapter 2

1. (a) $\mathbf{x}^{(13)} = [2, 4, 3]^T$,   **(c)** $\mathbf{x}^{(12)} = [-0.102, 0.539, 0.328]^T$.
    (e) $\mathbf{x}^{(14)} = [-0.673, 0.096, 0.711]^T$,   **(g)** $\mathbf{x}^{(12)} = [0.588, -0.147, 0.559]^T$.

3. (a) $\mathbf{x}^{(8)} = [2, 4, 3]^T$,   **(c)** $\mathbf{x}^{(6)} = [-0.102, 0.539, 0.328]^T$,
    (e) $\mathbf{x}^{(8)} = [-0.673, 0.096, 0.711]^T$ **(g)** $\mathbf{x}^{(7)} = [0.588, -0.147, 0.559]^T$.

5. (a) Divergent because $\rho(A) = 4 > 1$.

7. (a) $\mathbf{x}^{(8)} = [2, 4, 3]^T$,   **(c)** $\mathbf{x}^{(6)} = [-0.102, 0.539, 0.328]^T$,
    (e) $\mathbf{x}^{(8)} = [-0.673, 0.096, 0.711]^T$,   **(g)** $\mathbf{x}^{(8)} = [0.588, -0.147, 0.559]^T$.

9. Optimal choice $\omega = 1.172$, $\quad \mathbf{x}^{(9)} = [0.375, 1.750, 2.375]^T$.
The Gauss–Seidel method needed 15 iterations and the Jacobi method needed 26 iterations to get the same solution as the SOR method.

11. $\rho_J = 0.809017$, $\quad \rho_{GS} = 0.654508$, quad $\rho_{SOR} = 0.25962$.

13. **(a)** $\mathbf{x}^{(2)} = [0.51814, -0.72359, -1.94301]^T$,
$\mathbf{r}^{(2)} = [1.28497, -0.80311, 0.64249]^T$.
(c) $\mathbf{x}^{(2)} = [0.90654, 0.46729, -0.33645, -0.57009]^T$,
$\mathbf{r}^{(2)} = [-1.45794, -0.59813, -0.26168, -2.65421]^T$.

15. Simple Gaussian elimination $\mathbf{x}^{(1)} = [-10, 1.01]^T$,

Residual Corrector method $\mathbf{x}^{(2)} = [10, 1]^T$.

D.0.3   Chapter 3

1. **(a)** $p(\lambda) = \lambda^3 - 7\lambda^2 + 4\lambda + 12$, $\lambda_1 = 6, \lambda_2 = 2, \lambda_3 = -1$

$$\mathbf{x1} = \begin{pmatrix} 1 \\ 1 \\ 1 \end{pmatrix}, \quad \mathbf{x2} = \begin{pmatrix} 7 \\ -1 \\ -5 \end{pmatrix}, \quad \mathbf{x3} = \begin{pmatrix} 1 \\ -5/2 \\ 1 \end{pmatrix}.$$

(c) $p(\lambda) = \lambda^3 - 6\lambda^2 + 11\lambda - 6$, $\lambda_1 = 2, \lambda_2 = 1, \lambda_3 = 3$,

$$\mathbf{x1} = \begin{pmatrix} 1 \\ 1 \\ -1 \end{pmatrix}, \quad \mathbf{x2} = \begin{pmatrix} 0 \\ -1 \\ 1 \end{pmatrix}, \quad \mathbf{x3} = \begin{pmatrix} 1 \\ 1 \\ 0 \end{pmatrix}.$$

(e) $p(\lambda) = \lambda^3 - 2\lambda^2 - \lambda + 2$, $\lambda_1 = 2, \lambda_2 = 1, \lambda_3 = -1$,

$$\mathbf{x1} = \begin{pmatrix} 0 \\ 1 \\ 1 \end{pmatrix}, \quad \mathbf{x2} = \begin{pmatrix} 1 \\ 0 \\ 1 \end{pmatrix}, \quad \mathbf{x3} = \begin{pmatrix} 1 \\ -1 \\ 1 \end{pmatrix}.$$

3. **(a)** $k = 1$, **(c)** $k \neq 1/3$.

5. (a) Diagonalizable, real distinct eigenvalues $2, -2, -1$
(c) Not diagonalizable, real repeated eigenvalues $-1, 2, 2$
(e) Diagonalizable, real distinct eigenvalues $-11, 7, -1$

7. (a) $\begin{pmatrix} 1 & 1 & 2 \\ 1 & 2 & 3 \\ 2 & 4 & 1 \end{pmatrix}$, **(c)** $\begin{pmatrix} -5 & 1 & 0 \\ -2 & -8 & 1/2 \\ 1 & -10 & 1 \end{pmatrix}$, **(e)** $\begin{pmatrix} 0 & 1 & 1 \\ 1 & -1 & 0 \\ 1 & 1 & 1 \end{pmatrix}$

9. Similar matrices because have they have the same eigenvalues,
$\lambda = 1, 1, 1$.

11. (a) $\begin{pmatrix} 2 & 0 & 0 \\ 0 & 1 & 0 \\ 0 & 0 & -1 \end{pmatrix}$, **(c)** $\begin{pmatrix} 3 & 0 & 0 \\ 0 & -4 & 0 \\ 0 & 0 & -1 \end{pmatrix}$.

13. (a) $Q = \begin{pmatrix} 1 & 0 & 0 \\ 0 & 1/\sqrt{2} & 1/\sqrt{2} \\ 0 & 1/\sqrt{2} & -1/\sqrt{2} \end{pmatrix}$, $D = \begin{pmatrix} -1 & 0 & 0 \\ 0 & 1 & 0 \\ 0 & 0 & 1 \end{pmatrix}$.

(c) $Q = \begin{pmatrix} -2/3 & 1/3 & 2/3 \\ 2/3 & 2/3 & 1/3 \\ 1/3 & -2/3 & 2/3 \end{pmatrix}$, $D = \begin{pmatrix} 0 & 0 & 0 \\ 0 & 3 & 0 \\ 0 & 0 & 6 \end{pmatrix}$.

(e) $Q = \begin{pmatrix} 2/3 & -2/3 & 1/3 \\ 1/3 & 2/3 & 2/3 \\ 2/3 & 1/3 & -2/3 \end{pmatrix}$, $D = \begin{pmatrix} 0 & 0 & 0 \\ 0 & 9 & 0 \\ 0 & 0 & 9 \end{pmatrix}$

15. (a) $\lambda^2 - 7\lambda + 22 = 0$, $A^3 = \begin{pmatrix} -100 & 81 \\ -108 & -19 \end{pmatrix}$, $A^4 = \begin{pmatrix} -524 & 105 \\ -140 & -419 \end{pmatrix}$,

$A^{-1} = \begin{pmatrix} 5/22 & -3/22 \\ 2/11 & 1/11 \end{pmatrix}$, $A^{-2} = \begin{pmatrix} 13/484 & -21/484 \\ 7/121 & -2/121 \end{pmatrix}$.

(c) $\lambda^3 - \lambda^2 + 3 = 0$, $A^3 = \begin{pmatrix} -3 & 1 & 1 \\ -3 & -3 & 0 \\ -3 & -2 & -2 \end{pmatrix}$, $A^4 = \begin{pmatrix} -6 & -2 & 1 \\ 0 & -3 & -3 \\ 3 & -5 & -2 \end{pmatrix}$

$A^{-1} = \begin{pmatrix} 1/3 & 0 & -1/3 \\ 2/3 & 0 & 1/3 \\ 1/3 & 1 & -1/3 \end{pmatrix}$, $A^{-2} = \begin{pmatrix} 0 & -1/3 & 0 \\ 1/3 & 1/3 & -1/3 \\ 2/3 & -1/3 & 1/3 \end{pmatrix}$.

17. (a) $\lambda^3 - 2\lambda^2 - \lambda + 2 = 0,$ $A^{-1} = \begin{pmatrix} -5 & -1/2 & 5/2 \\ -4 & -1/2 & 3/2 \\ -12 & -1 & 6 \end{pmatrix}$

(c) $\lambda^3 + 2\lambda^2 - 11\lambda - 12 = 0,$ $A^{-1} = \begin{pmatrix} -3/4 & -1/4 & -1 \\ -1/2 & -1/2 & 1 \\ 1/12 & -1/12 & 1/3 \end{pmatrix}.$

19. (a) $\lambda^3 - 5\lambda^2 - 19\lambda + 65 = 0,$ $\det(A) = -65,$

$$adj(A) = \begin{pmatrix} -5 & -20 & 5 \\ -10 & 12 & -3 \\ 0 & 39 & -26 \end{pmatrix}, \quad A^{-1} = \begin{pmatrix} 1/3 & 4/13 & -1/13 \\ 2/13 & -12/65 & 3/65 \\ 0 & -3/5 & 2/5 \end{pmatrix}.$$

(c) $\lambda^4 + 6\lambda^3 - 3\lambda^2 - 1 = 0,$ $\det(A) = -1,$

$$adj(A) = \begin{pmatrix} 0 & -1 & 0 & -2 \\ -1 & 1 & 2 & -2 \\ 0 & -1 & -3 & 3 \\ 2 & -2 & -3 & 3 \end{pmatrix}, \quad A^{-1} = \begin{pmatrix} 0 & 1 & 0 & 2 \\ 1 & -1 & -2 & 2 \\ 0 & 1 & 3 & -3 \\ -2 & 2 & 3 & -3 \end{pmatrix}.$$

21.

$$f_1(t) = -\frac{2}{3}e^{4t} + \frac{5}{3}e^{-2t}$$

$$f_1(t) = -\frac{2}{3}e^{4t} - \frac{1}{3}e^{-2t}.$$

23. (a)

$$f_1(t) = -\frac{1}{2}e^{2t} + \frac{3}{2}e^{4t}$$

$$f_2(t) = e^{2t}$$

$$f_3(t) = -\frac{1}{2}e^2 + \frac{3}{2}e^{4t}.$$

(c)

$$
\begin{aligned}
f_1(t) &= -7e^{-3t} + 8e^{t} \\
f_2(t) &= -3e^{-3t} + 4e^{t} \\
f_3(t) &= 5e^{-3t} - 4e^{t}.
\end{aligned}
$$

25. (a) $a_n = \dfrac{1}{3}(2^{n+1} + 2^n),\ a_{20} = 1048576.$

(c) $a_n = (-1)^n,\ a_{15} = -1.$

D.0.4   Chapter 4

1. (a) $\lambda^{(4)} = 5.7320,\quad \mathbf{X} = [0.8255, 1.0000, 0.9225]^T.$

(c) $\lambda^{(4)} = 3.3103,\quad \mathbf{X} = [0.4583, 0.0729, 1.0000]^T.$

3. (a) $\lambda^{(4)} = 4.3731,\quad \mathbf{X} = [0.8897, -0.2962, 1.0000]^T.$

(c) $\lambda^{(4)} = 4.4143,\quad \mathbf{X} = [0.3333, 0.1381, 1.0000]^T.$

5. (a) The three real eigenvalues satisfies $0 \le \lambda \le 6.$

(c) The three real eigenvalues satisfies $-2 \le \lambda \le 4.$

7. $Q = \begin{pmatrix} 1 & 0 & 0 \\ 0 & 1 & 0 \\ 0 & 0 & 1 \end{pmatrix}, \qquad B = \begin{pmatrix} 2 & -3 & 6 \\ 0 & 3 & -4 \\ 0 & 2 & -3 \end{pmatrix}$

$C = \begin{pmatrix} 3 & -4 \\ 2 & -3 \end{pmatrix}.$

C has eigenvalues $1, -1$ with eigenvectors $[1, 1/2]^T$ and $[1, 1]^T$, respectively. The eigenvectors of B are $[0, 1, 1/2]^T$, $[-1, 1, 1]^T$, which are also the remaining eigenvectors of matrix A.

9. (a) $\lambda = 13.6138, -0.1635, -5.4502,$

$$\mathbf{X} = \begin{pmatrix} 0.7000 & -0.6414 & -0.3140 \\ 0.4357 & 0.7319 & -0.5240 \\ 0.5659 & 0.2300 & 0.79118 \end{pmatrix}.$$

(c) $\lambda = 11.5915, 6.6240, -7.2154,$

$$\mathbf{X} = \begin{pmatrix} 0.7933 & -0.1710 & -0.5843 \\ 0.4934 & 0.7429 & 0.4525 \\ 0.3567 & -0.6472 & 0.6737 \end{pmatrix}.$$

(e) $\lambda = -1.6073, 4.9988, 14.5592, 4.0494,$

$$\mathbf{X} = \begin{pmatrix} 0.6068 & 0.4539 & 0.4156 & -0.5031 \\ -0.4738 & 0.5814 & 0.5323 & 0.3928 \\ -0.4513 & -0.4775 & 0.5215 & -0.5444 \\ 0.4513 & -0.4775 & 0.5215 & 0.5444 \end{pmatrix}.$$

11. (a) The upper Hessenberg form of the matrix is

$$\begin{pmatrix} 11.0000 & 100.5000 & 45.0000 \\ 12.0000 & 55.5000 & 23.0000 \\ 0 & -14.7500 & -3.5000 \end{pmatrix}.$$

The 12th QR iteration is

$$\begin{pmatrix} 69.5837 & 99.9605 & 35.8569 \\ 0 & -9.7509 & -2.1207 \\ 0 & 0 & 3.1673 \end{pmatrix},$$

and the approximation of the eigenvalues of matrix A are

$$\lambda = 69.5837, -9.7509, 3.1673.$$

(c) The upper Hessenberg form of the matrix is

$$\begin{pmatrix} 14.0000 & 33.8000 & 1.9337 & 1.0000 \\ 5.0000 & 24.2000 & 5.1327 & -2.0000 \\ 0 & -92.2400 & -20.7564 & 11.4000 \\ 0 & 0 & -1.8283 & 7.5564 \end{pmatrix}.$$

The 42th QR iteration is

$$\begin{pmatrix} 26.3046 & -30.0959 & 76.9405 & 19.6782 \\ 0.0000 & -13.4367 & 53.4580 & 9.0694 \\ 0.0000 & -3.0615 & 5.7097 & -2.2534 \\ 0.0000 & -0.0000 & -0.0000 & 6.4224 \end{pmatrix},$$

and the approximation of eigenvalues of matrix A are

$$\lambda = 26.3046, \ -13.4367, \ 5.7097, \ 6.4224.$$

13. (a) $T = \begin{pmatrix} 3.0000 & 2.2361 & 0 \\ 2.2361 & 3.0000 & 0 \\ 0 & 0 & 3.0000 \end{pmatrix}$, $\lambda = 5.2361, 3.0000, 0.7639.$

(c) $T = \begin{pmatrix} 1 & -2 & 0 \\ -2 & 1 & -2 \\ 0 & -2 & 1 \end{pmatrix}$, $\lambda = 3.8284, -1.8284, 1.0000.$

15. (a) $T = \begin{pmatrix} 2.0000 & -5.0000 & 0.0000 \\ -5.0000 & 10.0800 & 0.4400 \\ 0.0000 & 0.4400 & -0.0800 \end{pmatrix}$,

$$\lambda = 12.4807, -0.4807, 0.0000.$$

(c) $T = \begin{pmatrix} 4.0000 & 4.5826 & 0.0000 & 0.0000 \\ 4.5826 & 5.0476 & -3.3873 & 0.0000 \\ 0.0000 & -3.3873 & 7.1666 & -1.1701 \\ 0.0000 & 0.0000 & -1.1701 & 5.7858 \end{pmatrix},$$

$$\lambda = 11.1117, 6.7240, 4.9673, -0.8030.$$

17. (a) $A^{(15)} = \begin{pmatrix} 5.4713 & -0.7373 & 2.1419 \\ -0.0097 & -3.4713 & 1.9705 \\ 0.0000 & 0.0000 & 1.0000 \end{pmatrix}$,

$\lambda = 5.4713, -3.4713, 1.0000.$

(c) $A^{(15)} = \begin{pmatrix} 3.8284 & -0.0001 & 0.0000 \\ -0.0001 & -1.8284 & -0.0002 \\ 0.0000 & -0.0002 & 1.0000 \end{pmatrix}$,

$\lambda = 3.8284, -1.8284, 1.0000.$

19. (a) $\lambda = 4.3028, 0.70, 0,$ Iterations $= 8.$

(c) $\lambda = 3, 2, 1,$ Iterations $= 24.$

21. (a) The upper Hessenberg form of the matrix is

$$\begin{pmatrix} 11.0000 & 100.5000 & 45.0000 \\ 12.0000 & 55.5000 & 23.0000 \\ 0 & -14.7500 & -3.5000 \end{pmatrix}.$$

The 12 QR iteration is

$$\begin{pmatrix} 69.5837 & 99.9605 & 35.8569 \\ 0.0000 & -9.7509 & -2.1207 \\ -0.0000 & -0.0000 & 3.1673 \end{pmatrix},$$

and the approximation of eigenvalues of matrix A are

$$\lambda = 69.5837, -9.7509, 3.1673.$$

(c) The upper Hessenberg form of the matrix is

$$\begin{pmatrix} 14.0000 & 33.8000 & 1.9337 & 1.0000 \\ 5.0000 & 24.2000 & 5.1327 & -2.0000 \\ 0 & -92.2400 & -20.7564 & 11.4000 \\ 0 & 0 & -1.8283 & 7.5564 \end{pmatrix}.$$

The 42 QR iteration is

$$
\begin{pmatrix}
26.3046 & -30.0959 & 76.9405 & 19.6782 \\
0.0000 & -13.4367 & 53.4580 & 9.0694 \\
0.0000 & -3.0615 & 5.7097 & -2.2534 \\
0.0000 & -0.0000 & -0.0000 & 6.4224
\end{pmatrix},
$$

and the approximation of the eigenvalues of matrix A are

$$\lambda = 26.3046,\ -13.4367,\ 5.7097,\ 6.4224.$$

23. (a) $\sigma_1 = \sqrt{5}, \quad \sigma_2 = 2, \quad \sigma_3 = 0.$

(c) $\sigma_1 = \sqrt{5}, \quad \sigma_2 = \sqrt{3}, \quad \sigma_3 = \sqrt{3}.$

25. Let A be an orthogonal matrix. Then $A^T = A^{-1}$. Moreover, the singular values of A are the eigenvalues of the matrix $A^T A = A^{-1}A = I$, since the only eigenvalue of an identity matrix is 1.

27. (a)

$$A = UDV^T = \begin{pmatrix} 0 & 1 \\ 1 & 0 \end{pmatrix} \begin{pmatrix} 3 & 0 \\ 0 & 2 \end{pmatrix} \begin{pmatrix} -1 & 0 \\ 0 & -1 \end{pmatrix}.$$

(c)

$$A = UDV^T = \begin{pmatrix} 0.5774 & 0 & 0.8165 \\ 0.5774 & 0.7071 & -0.4082 \\ -0.5774 & 0.7071 & 0.4082 \end{pmatrix} \begin{pmatrix} 1.7321 & 0 \\ 0 & 1.4142 \\ 0 & 0 \end{pmatrix} \begin{pmatrix} 0 & 1 \\ 1 & 0 \end{pmatrix}$$

29. (a) $\mathbf{x} = [x_1, x_2]^T = [0.25, -0.25]^T.$
(c) $\mathbf{x} = [x_1, x_2, x_3]^T = [-0.6154, 2.6923, 1.3846]^T.$

D.0.5 Chapter 5

1. $p_2(x) = (-4x^2 + 32x + 42)/210; \quad f(1.5) \approx p_2(1.5) = 0.3857;$
$Error = 0.0011.$

3. $A = -0.12, \quad B = 0.84, \quad C = 0.28.$

5. $\alpha = 1/2.$

7. $p_3(x) = -14.269x^3 + 8.726x^2 + 1.935x + 1.057; \quad f(0.3) \approx 2.0375;$
$Error = 0.0075;$ Error Bound $= 0.0124.$

9. Divided difference table is:

2.1 1.6669	0	0	0	0	0
3.2 6.3513	4.2585	0	0	0	0
4.3 23.1166	15.2412	4.9921	0	0	0
5.4 81.5715	53.1409	17.2271	3.7076	0	0
6.5 281.4813	181.7362	58.4524	12.4925	1.9966	0
7.6 955.0494	612.3346	195.7266	41.5982	6.6149	0.8397

11. (a) Divided difference table is:

2.0000	0	0	0	0
2.2361	0.2361	0	0	0
2.4495	0.2134	−0.0113	0	0
2.6458	0.1963	−0.0086	0.0009	0
2.8284	0.1826	−0.0069	0.0006	−0.0001

(b) $p_3(5.9) = 2.4290$ and $p_4(5.9) = 2.4290.$
(c) $E_3 = 0.0005742$ and $E_4 = 0.00000004.$

13. All three divided differences can be expanded as

$$\frac{(x_2 - x_1)f_0 - (x_2 - x_0)f_1 + (x_1 - x_0)f_2}{(x_2 - x_1)(x_2 - x_0)(x_1 - x_0)}.$$

15. $f[0, 1, 0] = 0.7183.$

17. $T_4(x) = 2xT_3(x) - T_2(x) = 8x^4 - 8x^2 + 1.$

19. $p_3(x) = 0.0491x^3 - 0.1434x^2 + 0.4991x + 0.6955;$
Error Bound $= 0.0313.$

21. $P_{0123}(1.5) = 2.0000 \approx f(1.5)$.

23. $P_{01234}(2.5) = 0.0673 \approx f(2.5)$.

25. **(a)** $a = 1, b = 3, \ y = x + 3, \ E(a, b) = 0.0$.
   (c) $a = 5.4, b = -11, \ y = 5.4x - 11, \ E(a, b) = 1.2000$.
   (e) $a = 6.24, b = 1.45, \ y = 6.24x + 1.45, \ E(a, b) = 4.0120$.

27. **(a)** $a = -1.782, b = 0.073, c = 1.818,$
      $y = -1.782x^2 + 0.073x + 1.818, \ E(a, b) = 0.146$.
   (c) $a = 28.125, b = 0.3, c = -3.125,$
      $y = 28.125x^2 + 0.3x - 3.125, \ E(a, b) = 0.2$.
   (e) $a = -0.6, b = -0.1, c = 2.5,$
      $y = -0.6x^2 - 0.1x + 2.5, \ E(a, b) = 1.6$.

29. **(a)** $a = 4.961, b = 0.037, \ y = 4.961e^{0.037x}, \ E(a, b) = 0.452$.
   (c) $a = 0.980, b = 0.530, \ y = 0.980e^{0.530x}, \ E(a, b) = 21.185$.
   (e) $a = 3.867, b = -0.508, \ y = 3.867e^{-0.508x}, \ E(a, b) = 0.071$.

31. **(a)** $a = 1.333, \ b = 0.167, \ c = 1.333, \ Z = 1.333x + 0.167y + 1.333$.
   (c) $a = 1.11, \ b = -0.44, \ c = 2.11, \ Z = 1.11x - 0.44y + 2.11$.

33. **(a)** $p(x) = 38.1975 + 16.3611 \cos x - 1.2156 \sin x$.
   (c) $p(x) = 8.6690 - 3.3430 \cos x + 1.7330 \sin x$.

35. **(a)** $\hat{\mathbf{x}} = [x_1, x_2]^T = [0.8460, -0.2051]^T$.
   (c) $\hat{\mathbf{x}} = [x_1, x_2, x_3]^T = [2.3041, 0.0986, 1.1107]^T$.

37. **(a)** $\hat{\mathbf{x}} = [x_1, x_2, x_3]^T = [-0.1827, -0.1068, 0.8733]^T$.
   (c) $\hat{\mathbf{x}} = [x_1, x_2, x_3, x_4]^T = [0.9316, -0.1110, -1.5383, 1.4928]^T$.

39. **(a)** $\hat{\mathbf{x}} = [x_1, x_2, x_3]^T = [1.4933, 0.5467, 0.4667]^T$.
   (c) $\hat{\mathbf{x}} = [x_1, x_2, x_3, x_4]^T = [-0.0799, 2.3216, 1.0979, 0.0356]^T$.

41. **(a)** $\hat{\mathbf{x}} = [x_1, x_2, x_3]^T = [0.9554, -0.8005, 2.1858]^T$.
   (c) $\hat{\mathbf{x}} = [x_1, x_2, x_3, x_4]^T = [0.3780, -0.0863, 0.8664, 0.1846]^T$.

43. (a)
$$A^+ = \begin{pmatrix} -1/75 & -2/15 & 43/75 \\ 3/25 & 1/5 & -4/25 \end{pmatrix}.$$

(c)
$$A^+ = \begin{pmatrix} -13/159 & -25/318 & 89/318 \\ 29/159 & 34/159 & -32/159 \end{pmatrix}.$$

45. (a) $A^+ = \begin{pmatrix} 11/17 & -5/17 \\ -1/17 & 2/17 \end{pmatrix}$, $\quad \hat{\mathbf{x}} = [19/17, 6/17]^T.$

(c) $A^+ = \begin{pmatrix} -38/29 & -9/29 & 17/29 \\ -5/29 & -5/29 & 3/29 \\ 41/29 & 12/29 & -13/29 \end{pmatrix}$, $\quad \hat{\mathbf{x}} = [-30/29, -7/29, 40/29]^T.$

47. $\hat{\mathbf{x}} = [x_1, x_2]^T = [-4, 9/5]^T.$

49. $\hat{\mathbf{x}} = [x_1, x_2]^T = [5/3, -2]^T.$

51. $\hat{\mathbf{x}} = [x_1, x_2]^T = [1, 0]^T.$

53. $\hat{\mathbf{x}} = [x_1, x_2]^T = [1, -2]^T.$

55. $\hat{\mathbf{x}} = [x_1, x_2, x_3]^T = [-0.3334, 0.3333, 0.6666]^T.$

57. (a) $\hat{\mathbf{x}} = [x_1, x_2]^T = [0.4000, 0.2000]^T.$
(c) $\hat{\mathbf{x}} = [x_1, x_2]^T = [0.4000, 0.2000]^T.$

59. (a) $\hat{\mathbf{x}} = [x_1, x_2]^T = [0.9250, 0.0250]^T.$
(c) $\hat{\mathbf{x}} = [x_1, x_2]^T = [0.8833, -0.0250]^T.$

D.0.6 Chapter 6

1. maximize $Z = 20x_1 + 15x_2$
Subject to the constraints

$$\begin{array}{rcrcl} & & x_2 & \leq & 8 \\ 2x_1 & - & x_2 & \leq & 0 \\ 2x_1 & + & x_2 & \leq & 12.5 \end{array}$$

$$x_1 \geq 0, \quad x_2 \geq 0.$$

3. minimize $Z = 20x_1 + 20x_2 + 31x_3 + 11x_4 + 12x_5$
subject to the constraints

$$
\begin{array}{rcll}
x_1 \quad + \quad x_3 + x_4 + 2x_5 & \geq & 21 & \text{(Vitamin A constraint)} \\
x_2 + 2x_3 + x_4 + \quad x_5 & \geq & 12 & \text{(Vitamin B constraint)}
\end{array}
$$

$$x_1 \geq 0, \; x_2 \geq 0, \; x_3 \geq 0, \; x_4 \geq 0, \; x_5 \geq 0.$$

5. maximize $Z = 6(9x_1 + 5x_2) + 9(7x_1 + 9x_2)$
subject to the constraints

$$
\begin{array}{rcl}
9x_1 + 5x_2 & \geq & 500 \\
7x_1 + 9x_2 & \geq & 300 \\
5x_1 + 3x_2 & \leq & 1500 \\
7x_1 + 9x_2 & \leq & 1900 \\
2x_1 + 4x_2 & \leq & 1000
\end{array}
$$

$$x_1 \geq 0, \quad x_2 \geq 0.$$

7. 550 Containers from Company A
300 Containers from Company B
Maximum Shipping Charges $= \$2110$

9. Company manufactures 400 of A
Company manufactures 300 of B
Maximum profit $= \$6400$

11. 0.4 pounds of Ingredient A
2.4 pounds of Ingredient B
Minimum Cost $= 24.8$ cents

13. **(a)** Maximum $= 24$ at $(6, 12)$,
(c) Maximum $= 6$ at $(1, 2)$

15. **(a)** Minimum $= -350$ at $(100, 50)$,
(c) Minimum $= -4$ at $(4, 0)$

17. **(a)** Maximum $= 14$ when $x_1 = 4$ and $x_2 = 1$,

   (c) Maximum $= 1600$ when $x_1 = 160$ and $x_2 = 0$

19. **(a)** No Maximum,

   (c) Maximum $= 22,500$ when $x_1 = 0, x_2 = 100$, and $x_3 = 50$

21. **(a)** Minimum $= 1$ when $x_1 = x_2 = 0, x_3 = 1$,

   (c) Minimum $= -24$ when $x_1 = 4$ and $x_2 = 6$

23. **(a)** minimize $V = 3y_1 + 5y_2$

   subject to the constraints

$$
\begin{aligned}
y_1 + y_2 &\geq -5 \\
-y_1 + 3y_2 &\geq 2
\end{aligned}
$$

$$y_1 \geq 0, \quad y_2 \geq 0.$$

   (c) maximize $V = 2y_1 + 5y_2$

   subject to the constraints

$$
\begin{aligned}
6y_1 + 3y_2 &\leq 6 \\
-3y_1 + 4y_2 &\leq 3 \\
y_1 + y_2 &\leq 0
\end{aligned}
$$

$$y_1 \geq 0, \quad y_2 \geq 0.$$

D.0.7   Chapter 7

1. **(a)** 1.

   (c) $\dfrac{\sqrt{2}}{4}$.

3. **(a)** $\dfrac{15}{16x^{9/4}}$.

   (c) $6 - 3\cos x + x \sin x - \dfrac{1}{x^2}$.

5. (a) $\dfrac{\partial^2 f}{\partial^2 x} = y^2 e^{-xy} \cos y.$

   (c) $\dfrac{\partial^2 f}{\partial^2 x} = (x^2 - 1)e^{-xy} \cos y + 2xe^{-xy} \sin y.$

7. (a) $[5, 1, 2]^T.$

   (c) $\left[\dfrac{3}{\ln 36}, \dfrac{12}{\ln 36}, \dfrac{27}{\ln 36} \right]^T.$

9. (a) $H(1, -1) = \begin{pmatrix} 12 & 5 \\ 5 & 12 \end{pmatrix},$ **(c)** $H(1, 2, 2) = \begin{pmatrix} 6 & 0 & 0 \\ 0 & 48 & 5 \\ 0 & 5 & 24 \end{pmatrix}.$

11. (a) $\mathbf{x}^T A \mathbf{x} = 2x^2 + 6xy + 4y^2,$ **(c)** $\mathbf{x}^T A \mathbf{x} = 2x^2 + 4y^2 + 5z^2 + 8xy - 4xz.$

13. (a) Positive-definite.
   (c) Positive-definite.

15. (a) $C_{18} = 7.9365.$
   (c) $C_{17} = 1.8187.$

17. (a) $x_4 = 1.2419.$
   (c) $x_5 = 3.7578.$

19. (a) $[1.2599, -0.0000],$ *iter.* $= 6.$
   (c) $[-1.1756, 1.6180],$ *iter.* $= 5.$

21. (a) $f(0.01) = 2.9998,$ *iter.* $= 10.$
   (c) $f(0.081) = 1.4856,$ *iter.* $= 9.$

23. (a) $f(2.25) = -42.5469,$ *iter.* $= 2.$
   (c) $f(1.25) = -12.6863,$ *iter.* $= 2.$

25. (a) $f(5.1200e - 006) = 5.0,$ *iter.* $= 3.$
   (c) $f(5.9729e - 012) = 1.0,$ *iter.* $= 2.$

27. (a) *Max* : $(1, 1)$ *and* $(-1, -1),$ *saddle* $(0, 0).$
   (c) *Min* : $(-1, 0)$ *Max* : $(1, 4)$ *saddle* $(1, 0), (-1, 4).$

29. (a) $(-0.0685, 0.2042).$
   (c) $(-2.0, 1.5).$

31. (a) $f(1,1) = 2.$
   (c) $f(2\sqrt[3]{3}, 2\sqrt[3]{3}, \sqrt[3]{3}) = 12(3)^{2/3}.$

33. (a) $Min: f(2,2,8) = 72, \quad max: f(-3,-3,18) = 342.$
   (c) $Min: f(1, \sqrt{2}, -\sqrt{2}) = 1 - 2\sqrt{2},$
      $max: f(1, -\sqrt{2}, \sqrt{2}) = 1 + 2\sqrt{2}.$

35. (a) $Min: f(0,-1,0) = 1, \quad min: f(2,1,0) = 5.$
   (c) $Min: f(2+\sqrt{3}, 2-\sqrt{3}, 1) = 15, \quad max: f(2,2,4) = 24.$

37. (a) $\mathbf{x}^* = [61/18, 11/18]^T, \quad z = 311/36.$
   (a) $\mathbf{x}^* = [1,0]^T, \quad z = -3.$

39. (a) $f_1(x) = 2x^2, f_2(y) = y^2$ and $g_1(x) = x, g_2(y) = y.$
   (c) $f_1(x) = x^2, f_2(y) = y, f_3(z) = -z^2$ and
      $g_1(x) = x^2, g_2(y) = 2y^2, g_3(z) = 3z$
      $g_4(x) = x, g_5(y) = y, g_6(z) = z.$

D.0.8 Appendix A

1. $10, \quad 37, \quad \dfrac{99}{128}.$

3. $8(10)^3 + 3(10)^2 + 8(10) + 3$
   $2(10)^2 + 8(10) + 5 + 6(10)^{-1} + 2(10)^{-2} + 5(10)^{-3}$
   $4(10)^2 + (10) + 3 + (10)^{-1} + 4(10)^{-2} + (10)^{-3} + \cdots.$

5. $(.00011)_2, \quad 0.00625, \quad 0.0025.$

7. $1000101, \quad 10111, \quad 1001111110, \quad 101.01000101\cdots.$

9. (a) Chopping: $e = 2.718 \times 10^0$, Rounding: $e = 2.718 \times 10^0.$
   (b) Chopping: $e = 2.718281 \times 10^0$, Rounding: $e = 2.718282 \times 10^0.$
   (c) Chopping: $e = 1.010110111 \times 2^1$,
      Rounding: $e = 1.010111000 \times 2^1.$

11. *Absolute error* $= 1/3 \times 10^{-4}$, *Relative error* $= 10^{-4}.$

13. (a) Absolute error $= 7.346 \times 10^{-6}$, Relative error $= 2.3383 \times 10^{-6}$.
   (b) Absolute error $= 7.346 \times 10^{-4}$, Relative error $= 2.3383 \times 10^{-4}$.

15. 0.132×10^2 $(m = 3, e = 2)$, -0.12532×10^2 $(m = 5, e = 2)$,
   0.16×10^{-1} $(m = 2, e = -1)$.

D.0.9 **Appendix B**

1. (a) $< 7, -1 >$, $< 1, -9 >$, 7.0711, 9.0554.
   (c) $< 5, 2, 10 >$, $< 3, 8, 2 >$, 11.3578, 8.7750.

3. (a) $< -444/1955, 1332/2737, 399/473 >$.
   (c) $< -70/367, 210/367, -4359/5465 >$.

5. (a) -19 **(c)** -5.

7. (a) 1.2882 radian.
   (c) 0.2363 radian.

9. (a) $\alpha = -5$.
   (c) $\alpha = 1$.

11. $\alpha \approx 74^0$, $\beta \approx 58^0$, $\gamma \approx 37^o$.

13. 27 joules.

15. (a) 2.4254, 4.0825.
   (c) -0.1741, -0.1622.

17. (a) $< 2.2895, 3.8158, 1.5263 >$.
   (c) $< 0.2857, -0.1429, 0.4286 >$.

19. (a) $< -10, 0, 5 >$.
   (c) $< 18, -16, -13 >$.

21. (a) $< -291 - 255132 >$.
   (c) $< -139, -66, 104 >$.

23. (a) 59.2621.
  (c) 113.9210.

25. (a) 38.1477.
  (c) 15.4110.

27. (a) -39.
  (c) -4771.

29. (a) 55.
  (c) 200.

31. (a) $< -15, 30, -15 >$.
  (c) $< -372, 303, 6 >$.

33. (a) $x = 3 + 2t, y = 2 + 2t, z = -4 + 2t4$.
  (c) $x = 1 + 2t, y = 2 + 4t, z = 3 + 6t$.

35. (a) $\theta \approx 0.86 \ radian$.
  (c) $\theta \approx 1.38 \ radian$.

37. (a) $d = 3/7$.
  (c) $d = 1/6$.

39. (a) $2x + 3y + 4z - 19 = 0$.
  (c) $x + 5y + 4z - 8 = 0$.

41. (a) $19 - 3\mathbf{i}$.
  (c) $55 - 72\mathbf{i}$.

43. (a) $z = 5(\cos(\frac{4}{3}) + \mathbf{i}\sin(\frac{4}{3}))$.
  (c) $z = \sqrt{14}(\cos(\frac{\sqrt{5}}{3}) + \mathbf{i}\sin(\frac{\sqrt{5}}{3}))$.

45. (a) $5 + 3\mathbf{i}$.
  (c) $e^{2\pi\mathbf{i}}$.

47.

$$\begin{pmatrix} 3 - 2\mathbf{i} & -\mathbf{i} \\ 4 + \mathbf{i} & 5\mathbf{i} \end{pmatrix}, \quad \begin{pmatrix} 1 & 3\mathbf{i} \\ 2 - 3\mathbf{i} & -\mathbf{i} \end{pmatrix}, \quad \begin{pmatrix} 3 - 2\mathbf{i} & 1 + 2\mathbf{i} \\ 18 - 3\mathbf{i} & -3 + 10\mathbf{i} \\ 2 - 15\mathbf{i} & 7 + 4\mathbf{i} \end{pmatrix}.$$

49. (a) $[-1/61 - 11/61\mathbf{i}, -13/61 - 21/61\mathbf{i}]^T$.
   (c) $[-42/41 + 137/123\mathbf{i}, -110/61 - 211/369\mathbf{i}]^T$.

51. (a) $7 - 21\mathbf{i}$.
   (c) $670 + 434\mathbf{i}$.

53. (a) $-34 - 5\mathbf{i}$.
   (c) $-317 + 1225\mathbf{i}$.

55. (a)

$$Re(A) = \begin{pmatrix} 4 & 0 & 2 \\ 3 & 2 & 3 \\ 0 & 0 & 1 \end{pmatrix}, \quad Im(A) = \begin{pmatrix} -2 & 3 & 3 \\ -5 & 7 & -7 \\ 2 & 3 & -5 \end{pmatrix}.$$

   (c)

$$Re(A) = \begin{pmatrix} 0 & 3 & 2 \\ 2 & 4 & 11 \\ 11 & 6 & 3 \end{pmatrix}, \quad Im(A) = \begin{pmatrix} 3 & -4 & -5 \\ -7 & -8 & -5 \\ 2 & 0 & 7 \end{pmatrix}.$$

57. (a)

$$\begin{aligned} \lambda_1 &= 8.3972 \\ \mathbf{x}_1 &= [0.8480, 0.4714, 0.2421]^T. \end{aligned}$$

$$\begin{aligned} \lambda_2 &= 1.3014 + 0.6707\mathbf{i} \\ \mathbf{x}_2 &= [0.2163 + 0.1536\mathbf{i}, -0.2378 - 0.3653\mathbf{i}, -0.8600]^T. \end{aligned}$$

$$\begin{aligned} \lambda_3 &= 1.3014 - 0.6707\mathbf{i} \\ \mathbf{x}_3 &= [0.2163 - 0.1536\mathbf{i}, -0.2378 + 0.3653\mathbf{i}, -0.8600]^T. \end{aligned}$$

   (c)

$$\begin{aligned} \lambda_1 &= 0.0000 \\ \mathbf{x}_1 &= [-0.8944, -0.4472, 0.0000]^T. \end{aligned}$$

$$\begin{aligned} \lambda_2 &= -38.0000 + 18.0000\mathbf{i} \\ \mathbf{x}_2 &= [0.4020 + 0.0783\mathbf{i}, 0.7732, -0.3726 + 0.3090\mathbf{i}]^T. \end{aligned}$$

$$\begin{aligned} \lambda_3 &= -38.0000 - 18.0000\mathbf{i} \\ \mathbf{x}_3 &= [0.4020 - 0.0783\mathbf{i}, 0.7732, -0.3726 - 0.3090\mathbf{i}]^T. \end{aligned}$$

D.0.10   Appendix C

1. **(a)** $2.9417e + 005.$
   (c) $3.0194e + 007.$

3. **(a)** $2.4755.$
   (c) $24.9045.$

5. **(a)** $8.0392e + 023.$
   (c) $6.2206e + 008.$

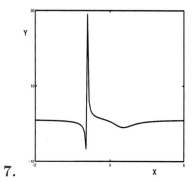

7.

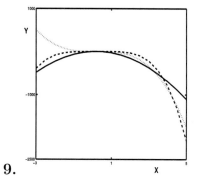

9.

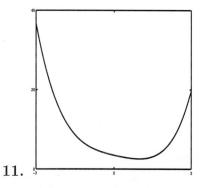

11.

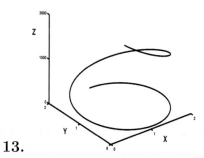

13.

17. **(i)** $(log(sin(x+2)+x\hat{\ }2) - 12*(x\hat{\ }3+12)\hat{\ }(1/2) - exp(x+2) - 15*x + 10)*((x-2)\hat{\ }3 + 11*x^2 + 9*x - 15).$

   (ii) $log(sin(x+2)+x\hat{\ }2) - 12*(x\hat{\ }3+12)\hat{\ }(1/2) - exp(x+2) - 6*x - 5 + (x-2)\hat{\ }3 + 11*x\hat{\ }2.$

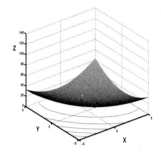

15.

(**iii**) $(log(sin(x+2)+x\hat{\ }2)-12*(x\hat{\ }3+12)\hat{\ }(1/2)-exp(x+2)-$
$15*x+10)/((x-2)\hat{\ }3+11*x\hat{\ }2+9*x-15).$
(**iv**) $-7.8418e+007,\quad -5.9021e+004,\quad -46.4011.$

19. (**i**) $(x-5)*(x-6)*(x-7)*(x-8)*(x-4)$
(**ii**) $x\hat{\ }4-6x\hat{\ }3+11x\hat{\ }2-6x$

21. $3*x-x\hat{\ }2+2/3*x\hat{\ }3-1/2*x\hat{\ }4.$

23. $y(x)=e^{2x+c}-1,\ y(x)=e^{2x}-1.$

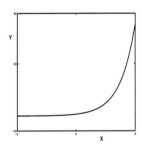

25. $60-18*q*p+5*p+15*p\hat{\ }2-2*q\hat{\ }2-20*q,$
$[-2*(-10+3*q*p)/(60-18*q*p+5*p+15*p\hat{\ }2-2*q\hat{\ }$
$2-20*q),5*(1+3*p)/(60-18*q*p+5*p+15*p\hat{\ }2-2*q\hat{\ }$
$2-20*q),-2*(q+10)/(60-18*q*p+5*p+15*p\hat{\ }2-2*q\hat{\ }$
$2-20*q)]$
$[-(5*p-2*q\hat{\ }2)/(60-18*q*p+5*p+15*p\hat{\ }2-2*q\hat{\ }$
$2-20*q),-5*(-3+q)/(60-18*q*p+5*p+15*p\hat{\ }2-2*q\hat{\ }$

$2 - 20 * q), (-6 * q + 5 * p)/(60 - 18 * q * p + 5 * p + 15 * p\hat{} \ 2 - 2 * q\hat{}$
$2 - 20 * q)]$
$[(3 * p\hat{} \ 2 - 4 * q)/(60 - 18 * q * p + 5 * p + 15 * p\hat{} \ 2 - 2 * q\hat{} \ 2 -$
$20 * q), -(9 * p + q)/(60 - 18 * q * p + 5 * p + 15 * p\hat{} \ 2 - 2 * q\hat{}$
$2 - 20 * q), (12 + p)/(60 - 18 * q * p + 5 * p + 15 * p\hat{} \ 2 - 2 * q\hat{} \ 2 - 20 * q)]$
$[-6 * (q * p - 5 * p + 5 + q)/(60 - 18 * q * p + 5 * p + 15 * p\hat{} \ 2 - 2 * q\hat{}$
$2 - 20 * q)]$
$[2 * (5 * p + 15 - 14 * q + q\hat{} \ 2)/(60 - 18 * q * p + 5 * p + 15 * p\hat{} \ 2 - 2 * q\hat{}$
$2 - 20 * q)]$
$[3 * (-5 * p + 12 + p\hat{} \ 2 - 2 * q)/(60 - 18 * q * p + 5 * p + 15 * p\hat{} \ 2 - 2 * q\hat{}$
$2 - 20 * q)].$

27. $x\hat{} \ 3 - 20 * x\hat{} \ 2 + 123 * x - 216$
$(x - 3) * (x - 8) * (x - 9)$
$X =$
$[0, -5, 1]$
$[1, 2, 2]$
$[-2, 1, 1]$
$D =$
$[8, 0, 0]$
$[0, 3, 0]$
$[0, 0, 9].$

29. $[2]$
$[-2]$
$[-2]$

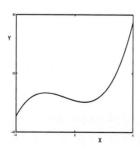

Bibliography

[1] Abramowitz M. and I. A. Stegum I. A.(eds): Handbook of Mathematical Functions, National Bureau of Standards, 1972.

[2] Achieser, N. I.: Theory of Approximation, Dover, New York, 1993.

[3] Ahlberg, J., E. Nilson, and J. Walsh: The Theory of Splines and Their Application, Academic Press, New York, 1967.

[4] Akai, T. J.: Applied Numerical Methods for Engineers, John Wiley & Sons, New York, 1993.

[5] Allgower, E. L., K. Glasshoff, and H. O. Peitgen (eds.): Numerical Solutions of Nonlinear Equations, LNM**878**, Springer–Verlag, 1981.

[6] Atkinson, K. E. and W. Han: An Introduction to Numerical Analysis, 3rd ed., John Wiley & Sons, New York, 2004.

[7] Axelsson, O.: Iterative Solution Methods, Cambridge University Press, New York, 1994.

[8] Ayyub, B. M. and R. H. McCuen: Numerical Methods for Engineers, Prentice–Hall, Upper Saddle River, NJ, 1996.

[9] Bellman, R.: Introduction to Matrix Analysis, 2nd ed., McGraw–Hill, New York, 1970.

[10] Bazaraa, M., H. Sherall, C. Shetty: Nonlinear Programming, Theory and Algorithms, 2nd ed., John Wiley & Sons, New York, 1993.

[11] Beale, E. M. L.: Numerical Methods in Nonlinear Programming, (Ed. J. Abadie), North–Holland, Amsterdam, 1967.

[12] Beale, E. M. L.: Mathematical Programming in Practice, Pitman, London, 1968.

[13] Bertsetkas, D.: Nonlinear Programming, Athena Publishing, Cambridge, MA, 1995.

[14] Beightler, C., D. Phillips, and D. Wilde: Foundations of Optimization, 2nd ed., Prentice–Hall, Upper Saddle River, New Jersey, 1979.

[15] Bradley, S. P., A. C. Hax and T. L. Magnanti: Applied Mathematical Programming, Addison–Wesley, Reading, MA, 1977.

[16] Bender, C. M. and S. A. Orszag: Advanced Mathematical Methods for Scientists and Engineers, McGraw–Hill, New York, 1978.

[17] Blum, E. K.: Numerical Analysis and Computation: Theory and Practice, Addison–Wesley, Reading, MA, 1972.

[18] Borse, G. H.: Numerical Methods with MATLAB, PWS, Boston, 1997.

[19] Bronson, R.: Matrix Methods—An Introduction, Academic Press, New York, 1969.

[20] Buchanan, J. L. and P. R. Turner: Numerical Methods and Analysis, McGraw–Hill, New York, 1992.

[21] Burden, R. L. and J. D. Faires: Numerical Analysis, 8th ed., Brooks/Cole Publishing Company, Boston, 2005.

[22] Butcher, J.: The Numerical Analysis of Ordinary Differential equations, John Wiley & Sons, New York, 1987.

[23] Carnahan, B., A. H. Luther, and J. O. Wilkes: Applied Numerical Methods, John Wiley & Sons, New York, 1969.

[24] Chapra, S. C. and R. P. Canale: Numerical Methods for Engineers, 3rd ed., McGraw–Hill, New York, 1998.

[25] Cheney, E. W.: Introduction to Approximation Theory, McGraw–Hill, New York, 1982.

[26] Chv'atal, V.: Linear Programming. W. H. Freeman, New York, 1983.

[27] Ciarlet, P. G.: Introduction to Numerical Linear Algebra and Optimization, Cambridge University Press, Cambridge, MA, 1989.

[28] Coleman, T. F. and C. Van Loan: Handbook for Matrix Computations, SAIM, Philadelphia, 1988.

[29] Conte, S. D. and C. de Boor: Elementary Numerical Analysis, 3rd ed., McGraw–Hill, New York, 1980.

[30] Daellenbach, H. G. and E. J. Bell: User's Guide to Linear Programming, Prentice–Hall, Englewood Cliffs, NJ, 1970.

[31] Dahlquist, G. and A. Bjorck: Numerical Methods, Prentice–Hall, Englewood Cliffs, NJ, 1974.

[32] Daniels, R. W.: An Introduction to Numerical Methods and Optimization Techniques, North–Holland, New York, 1978.

[33] Dantzig, G. B.: Minimization of a Linear Function of Variables Subject to Linear Inequalities. In Activity Analysis of Production and Allocation, Koopmans, T. C., ed., John Wiley & Sons, New York, Chapter XXI, pp. 339-347, 1951(a).

[34] Dantzig, G. B.: Application of the Simplex Method to a Transportation Problem. In Activity Analysis of Production and Allocation, Koopmans, T. C., ed., John Wiley & Sons, New York, Chapter XXIII, pp. 35993-373, 1951(b).

[35] Dantzig, G. B.: Linear Programming and Extensions, Princeton University Press, Princeton, NJ, 1963.

[36] Dantzig, G. B. and M. N. Thapa: Linear Programming 1: Introduction, Springer–Verlag, New York, 1997.

[37] Datta, B. N.: Numerical Linear Algebra and Application, Brook/Cole, Pacific Grove, CA, 1995.

[38] Davis, P. J.: Interpolation and Approximation, Dover, New York, 1975.

[39] Davis, P. J. and P. Rabinowitz: Methods of Numerical Integration, 2nd ed., Academic Press, 1984.

[40] Demmel, J. W.: Applied Numerical Linear Algebra, SIAM, Philadelphia, 1997.

[41] Driebeek, N. J.: Applied Linear Programming, Addison–Wesley, Reading, MA, 1969.

[42] Duff, I. S., A. M. Erisman and J. K. Reid: Direct Methods for Sparse Matrices, Oxford University Press, Oxford, England, 1986.

[43] Epperson, J. F.: An Introduction to Numerical Methods and Analysis, John Wiley & Sons, Chichester, 2001.

[44] Etchells, T. and J. Berry: Learning Numerical Analysis Through Derive, Chartwell–Bratt, Kent, 1997.

[45] Etter, D. M. and D. C. Kuncicky: Introduction to MATLAB, Prentice–Hall, Englewood Cliffs, NJ, 1999.

[46] Evans, G.: Practical Numerical Analysis, John Wiley & Sons, Chichester, England, 1995.

[47] Fang, S. C. and Puthenpura, S.: Linear Optimization and Extensions, AT&T Prentice–Hall, Englewood Cliffs, NJ, 1993.

[48] Fatunla, S. O.: Numerical Methods for Initial-Value Problems in Ordinary Differential Equations, Academic Press, New York, 1988.

[49] Ferziger, J. H.: Numerical Methods for Engineering Application, John Wiley & Sons, New York, 1981.

[50] Fiacco, A. V.: Introduction to Sensitivity and Stability Analysis in Numerical Programming, Academic Press, New York, 1983.

[51] Fletcher, R.: Practical Methods of Optimization, 2nd ed., John Wiley & Sons, New York, 1987.

[52] Forsythe, G. E. and C. B. Moler: Computer Solution of Linear Algebraic Systems, Prentice–Hall, Englewood Cliffs, NJ, 1967.

[53] Fox, L.: Numerical Solution of Two-Point Boundary-Value Problems in Ordinary Differential Equations, Dover, New York, 1990.

[54] Fox, L.: An Introduction to Numerical Linear Algebra, Oxford University Press, New York, 1965.

[55] Fröberg, C. E.: Introduction to Numerical Analysis, 2nd ed., Addison–Wesley, Reading, MA, 1969.

[56] Fröberg, C. E.: Numerical Mathematics: Theory and Computer Application, Benjamin/Cummnings, Menlo Park, CA, 1985.

[57] Gass, S. I.: An Illustrated Guide to Linear Programming, McGraw–Hill, New York, 1970.

[58] Gass, S. I.: Linear Programming, *4th* ed., McGraw–Hill, New York, 1975.

[59] Gerald, C. F. and P. O. Wheatley: Applied Numerical Analysis, 7th ed., Addison–Wesley, Reading, MA, 2004.

[60] Gilat A.: MATLAB—An Introduction with Applications, John Wiley & Sons, New York, 2005.

[61] Gill, P. E., W. Murray, and M. H. Wright: Numerical Linear Algebra and Optimization, Addison–Wesley, Reading, MA, 1991.

[62] Gill, P. E., W. Murray, and M. H. Wright: Practical Optimization, Academic Press, New York, 1981.

[63] Goldstine, H. H.: A History of Numerical Analysis From the 16th Through the 19th Century, Springer–Verlag, New York, 1977.

[64] Golub, G. H.: Studies in Numerical Analysis, MAA, Washington, DC, 1984.

[65] Golub, G. H. and J. M. Ortega: Scientific Computing and Differential Equations, Academic Press, New York, 1992.

[66] Golub, G. H. and C. F. van Loan: Matrix Computation, 3rd ed., Johns Hopkins University Press, Baltimore, MD, 1996.

[67] Goldstine, H. H.: A History of Numerical Analysis From the 16th Through the 19th Century, Springer–Verlag, New York, 1977.

[68] Greenbaum, A.: Iterative Methods for Solving Linear Systems, SIAM, Philadelphia, 1997.

[69] Greenspan, D. and V. Casulli: Numerical Analysis for Applied Mathematics, Science and Engineering, Addison–Wesley, New York, 1988.

[70] Greeville, T. N. E.: Theory and Application of Spline Functions, Academic Press, New York, 1969.

[71] Griffiths, D. V. and I. M. Smith: Numerical Methods for Engineers, CRC Press, Boca Raton, FL, 1991.

[72] Hadley, G.: Linear Algebra, Addison–Wesley, Reading, MA, 1961.

[73] Hadley, G.: Linear Programming, Addison–Wesley, Reading, MA, 1962.

[74] Hageman, L. A. and D. M. Young: Applied Iterative Methods, Academic Press, New York, 1981.

[75] Hager, W. W.: Applied Numerical Linear Algebra, Prentice–Hall, Englewood Cliffs, NJ, 1988.

[76] Hamming, R. W.: Introduction to Applied Numerical Analysis, McGraw–Hill, New York, 1971.

[77] Hamming, R. W.: Numerical Methods for Scientists and Engineers, 2nd ed., McGraw–Hill, New York, 1973.

[78] Harrington, S.: Computer Graphics: A Programming Approach, McGraw–Hill, New York, 1987.

[79] Hayhurst, G.: Mathematical Programming Applications, Macmillan, New York, 1987.

[80] Henrici, P. K.: Elements of Numerical Analysis, John Wiley & Sons, New York, 1964.

[81] Higham, N. J.: Accuracy and Stability of Numerical Algorithms, SIAM, Philadelphia, 1996.

[82] Higham, D. J. and N. J. Higham: MATLAB Guide, SIAM, Philadelphia, 2000.

[83] Hildebrand, F. B.: Introduction to Numerical Analysis, 2nd ed., McGraw–Hill, New York, 1974.

[84] Himmelblau, D. M.: Applied Nonlinear Programming, McGraw–Hill, New York, 1972.

[85] Hoffman, Joe. D.: Numerical Methods for Engineers and Scientists, McGraw–Hill, New York, 1993.

[86] Hohn, F. E.: Elementary Matrix Algebra, 3rd ed., Macmillan, New York, 1973.

[87] Horn, R. A. and C. R. Johnson: Matrix Analysis, Cambridge University Press, Cambridge, 1985.

[88] Hornbeck, R. W. Numerical Methods, Prentice–Hall, Englewood Cliffs, NJ, 1975.

[89] Householder, A. S.: The Theory of Matrices in Numerical Analysis, Dover Publications, New York, 1964.

[90] Householder, A. S.: The Numerical Treatment of a Single Non-linear Equation, McGraw–Hill, New York, 1970.

[91] Hultquist, P. E.: Numerical Methods for Engineers and Computer Scientists, Benjamin/Cummnings, Menlo Park, CA, 1988.

[92] Hunt, B. R., R. L. Lipsman, and J. M. Rosenberg: A Guide to MATLAB for Beginners and Experienced Users, Cambridge University Press, Cambridge, MA, 2001.

[93] Isaacson, E. and H. B. Keller: Analysis of Numerical Methods, John Wiley & Sons, New York, 1966.

[94] Jacques, I. and C. Judd: Numerical Analysis, Chapman and Hall, New York, 1987.

[95] Jahn, J.: Introduction to the Theory of Nonlinear Optimization, 2nd ed., Springer–Verlag, Berlin, 1996.

[96] Jennings, A.: A Matrix Computation for Engineers and Scientists, John Wiley & Sons, London, 1977.

[97] Jeter, M. W.: Mathematical Programming: An Introduction to Optimization, Marcel Dekker, New York, 1986.

[98] Johnson, L. W. and R. D. Riess: Numerical Analysis, 2nd ed., Addison–Wesley, Reading, MA, 1982.

[99] Johnston, R. L.: Numerical Methods—A Software Approach, John Wiley & Sons, New York, 1982.

[100] Kahanger, D., C. Moler, and S. Nash: Numerical Methods and Software, Prentice–Hall, Englewood Cliffs, NJ, 1989.

[101] Kharab, A. and R. B. Guenther: An Introduction to Numerical Methods—A MATLAB Approach, Chapman & Hall/CRC, New York, 2000.

[102] Kelley, C. T.: iterative Methods for Linear and Nonlinear Equations, SIAM, Philadelphia, 1995.

[103] Kincaid, D. and W. Cheney: Numerical Analysis—Mathematics of Scientific Computing, 3rd ed., Brooks/Cole Publishing Company, Boston, 2002.

[104] King, J.T.: Introduction to Numerical Computation, McGraw–Hill, New York, 1984.

[105] Knuth, D. E.: Seminumerical Algorithms, 2nd ed., Vol. 2 of The Art of Computer Programming, Addison–Wesley, Reading, MA, 1981.

[106] Kolman, B. and R. E. Beck: Elementary Linear Programming with Applications, Academic Press, New York, 1980.

[107] Lambert, J. D.: Numerical Methods for Ordinary Differential Systems, John Wiley & Sons, Chichester, 1991.

[108] Lancaster, P.: "Explicit Solution of Linear Matrix Equations," SIAM Review 12, 544-66, 1970.

[109] Lancaster, P. and M. Tismenetsky: The Theory of Matrices, 2nd ed., Academic Press, New York, 1985.

[110] Lastman, G. J. and N. K. Sinha: Microcomputer-Based Numerical Methods for Science and Engineering, Saunders, New York, 1989.

[111] Lawson, C. L. and R. J. Hanson: Solving Least Squares Problems, SIAM, Philadelphia, 1995.

[112] Leader, J. J.: Numerical Analysis and Scientific Computation, Addison–Wesley, Reading, MA, 2004.

[113] Linear, P.: Theoretical Numerical Analysis, John Wiley & Sons, New York, 1979.

[114] Linz, P. and R. L. C. Wang: Exploring Numerical Methods—An Introduction to Scientific Computing Using MATLAB, Jones and Bartlett Publishers, Boston, 2002.

[115] Luenberger, D. G.: Linear and Nonlinear Programming, 2nd ed., Addison–Wesley, Reading, MA, 1984.

[116] Mangasarian, O.: Nonlinear Programming, McGraw–Hill, New York, 1969.

[117] Marcus, M.: Matrices and Matlab, Prentice–Hall, Upper Saddle River, NJ, 1993.

[118] Marcus, M. and H. Minc: A Survey of Matrix Theory and Matrix Inequalities, Allyn and Bacon, Boston, 1964.

[119] Maron, M. J. and R. J. Lopez: Numerical Analysis: A Practical Approach, 3rd ed., Wadsworth, Belmont, CA, 1991.

[120] Mathews, J. H.: Numerical Methods for Mathematics, Science and Engineering, 2nd ed., Prentice–Hall, Englewood Cliffs, NJ, 1987.

[121] McCormick, G. P.: Nonlinear Programming Theory, Algorithms, and Applications, John Wiley & Sons, New York, 1983.

[122] Meyer, C. D.: Matrix Analysis and Applied Linear Algebra, SIAM, Philadelphia, 2000.

[123] Mirsky, L.: An Introduction to Linear Algebra, Oxford University Press, Oxford, 1963.

[124] Modi, J. J.: Parallel Algorithms and Matrix Computation, Oxford University Press, Oxford, 1988.

[125] Moore, R. E.: Mathematical Elements of Scientific Computing, Holt, Reinhart & Winston, New York, 1975.

[126] Mori, M. and R. Piessens (eds.): Numerical Quadrature, North Holland, New York, 1987.

[127] Morris, J. L.: Computational Methods in Elementary Theory and Application of Numerical Analysis, John Wiley & Sons, New York, 1983.

[128] Murty, K. G.: Linear Programming, John Wiley & Sons, New York, 1983.

[129] Nakamura, S.: Applied Numerical Methods in C, Prentice–Hall, Englewood Cliffs, NJ, 1993.

[130] Nakos, G. and D. Joyner: Linear Algebra With Applications, Brooks/Cole Publishing Company, Boston, 1998.

[131] Nash, S. G. and A. Sofer: Linear and Nonlinear Programming, McGraw–Hill, New York, 1998.

[132] Nazareth, J. L.: Computer Solution of Linear Programs, Oxford University Press, New York, 1987.

[133] Neumaier, A.: Introduction to Numerical Analysis, Cambridge University Press, Cambridge, MA, 2001.

[134] Nicholson, W. K.: Linear Algebra With Applications, 4th ed., McGraw–Hill Ryerson, New York, 2002.

[135] Noble, B. and J. W. Daniel: Applied Linear Algebra, 2nd ed., Prentice–Hall, Englewood Cliffs, NJ, 1977.

[136] Olver, P. J. and C. Shakiban: Applied Linear Algebra, Pearson Prentice–Hall, Upper Saddle River, NJ, 2005.

[137] Ortega, J. M.: Numerical Analysis—A Second Course, Academic Press, New York, 1972.

[138] Ostrowski, A. M.: Solution of Equations and Systems of Equations, Academic Press, New York, 1960.

[139] Parlett, B. N.: The Symmetric Eigenvalue Problem, Prentice–Hall, Englewood Cliffs, NJ, 1980.

[140] Peressini, A. L., F. E. Sullivan, and J. J. Uhl Jr.: The Mathematics of Nonlinear Programming, Springer–Verlag, New York, 1988.

[141] Pike, R. W.: Optimization for Engineering Systems, Van Nostrand Reinhold, New York, 1986.

[142] Polak, E.: Computational Methods in Optimization, Academic Press, New York, 1971.

[143] Polak, E.: Optimization Algorithm and Consistent Approximations, Computational Methods in Optimization, Springer–Verlag, New York, 1997.

[144] Powell, M. J. D.: Approximation Theory and Methods, Cambridge University Press, Cambridge, MA, 1981.

[145] Pshenichnyj, B. N.: The Linearization Method for Constrained Optimization, Springer–Verlag, Berlin, 1994.

[146] Quarteroni, Alfio: Scientific Computing with MATLAB, Springer–Verlag, Berlin Heidelberg, 2003.

[147] Ralston, A. and P. Rabinowitz: A First Course in Numerical Analysis, 2nd ed., McGraw–Hill, New York, 1978.

[148] Rao, S. S.: Engineering Optimization: Theory and Practice, 3rd ed., John Wiley & Sons, New York, 1996.

[149] Ravindran, A.: Linear Programming, in Handbook of Industrial Engineering, (G. Salvendy ed.), Chapter 14 (pp 14.2.1-14.2-11), John Wiley & Sons, New York, 1982.

[150] Rardin, R. L.: Optimization in Operational Research, Prentice–Hall, Upper Saddle River, NJ, 1998.

[151] Recktenwald, G.: Numerical Methods with MATLAB—Implementation and Application, Prentice–Hall, Englewood Cliffs, NJ, 2000.

[152] Rice, J. R.: Numerical Methods, Software and Analysis, McGraw–Hill, New York, 1983.

[153] Rivlin, T. J.: An Introduction to the Approximation of Functions, Dover Publications, New York, 1969.

[154] Rorrer, C. and H. Anton: Applications of Linear Algebra, John Wiley & Sons, New York, 1977.

[155] Saad, Y.: Numerical Methods for Large Eigenvalue Problems: Theory and Algorithms, John Wiley & Sons, New York, 1992.

[156] Saad, Y.: Iterative Methods for Sparse Linear Systems, PWS Publishing Co., Boston, 1996.

[157] Scales, L. E.: Introduction to Nonlinear Optimization, Macmillan, London, 1985.

[158] Scarborough, J. B.: Numerical Mathematical Analysis, 6th ed., The Johns Hopkins University Press, Baltimore, MA, 1966.

[159] Schatzman, M.: Numerical Analysis—A Mathematical Introduction, Oxford University Press, New York, 2002.

[160] Scheid, F.: Numerical Analysis, McGraw–Hill, New York, 1988.

[161] Schilling, R. J. and S. L. Harris: Applied Numerical Methods for Engineers using MATLAB and C, Brooks/Cole Publishing Company, Boston, 2000.

[162] Shampine, L. F. and R. C. Allen: Numerical Computing—An Introduction, Saunders, Philadelphia, 1973.

[163] Shapiro, J. F.: Mathematical Programmin Structures and Algorithms, John Wiley & Sons, New York, 1979.

[164] Steward, B. W.:Introduction to Matrix Computations, Academic Press, New York, 1973.

[165] Stewart, G. W.: Afternotes on Numerical Analysis, SIAM, Philadelphia, 1996.

[166] Store, J. and R. Bulirsch: Introduction to Numerical Analysis, Springer–Verlag, New York, 1980.

[167] Strang, G.: Linear Algebra and Its Applications 3rd ed., Brooks/Cole Publishing Company, Boston, 1988.

[168] Stroud, A. H. and D. Secrest: Gaussian Quadrature Formulas, Prentice–Hall, Englewood Cliffs, NJ, 1966.

[169] Suli, E. and D. Mayers: An Introduction to Numerical Analysis, Cambridge University Press, Cambridge, MA, 2003.

[170] The Mathworks, Inc.: Using MATLAB, The Mathworks, Inc., Natick, MA, 1996.

[171] The Mathworks, Inc.: Using MATLAB Graphics, The Mathworks, Inc., Natick, MA, 1996.

[172] The Mathworks, Inc.: MATLAB Language Reference Manual. The Mathworks, Inc., Natick, MA, 1996.

[173] Trefethen, L. N. and D. Bau III: Numerical Linear Algebra, SIAM, Philadelphia, 1997.

[174] Traub, J. F.: Iterative Methods for the Solution of Equations, Prentice–Hall, Englewood Cliffs, NJ, 1964.

[175] Turnbull, H. W. and A. C. Aitken: An Introduction to the Theory of Canonical Matrices, Dover Publications, New York, 1961.

[176] Turner, P. R.: Guide to Scientific Computing, 2nd ed., Macmillan Press, Basingstoke, 2000.

[177] Ueberhuber, C. W.: Numerical Computation 1: Methods, Software, and Analysis, Springer–Verlag, New York, 1997.

[178] Usmani, R. A.: Numerical Analysis for Beginners, D and R Texts Publications, Manitoba, 1992.

[179] Vandergraft, J. S.: Introduction to Numerical Computations, Academic Press, New York, 1978.

[180] Verga, R. S.: Matrix Iterative Analysis, Prentice–Hall, Englewood Cliffs, NJ, 1962.

[181] Wagner, H. M.: Principles of Operational Research, 2nd ed., Prentice–Hall, Englewood Cliffs, NJ, 1975.

[182] Walsh, G. R.: Methods of Optimization, John Wiley & Sons, London, 1975.

[183] Wilkinson, J. H.: Rounding Errors in Algebraic Processes, Prentice–Hall, Englewood Cliffs, NJ, 1963.

[184] Wilkinson, J. H.: The Algebraic Eigenvalue Problem, Clarendon Press, Oxford, England, 1965.

[185] Wilkinson, J. H. and C. Reinsch: Linear Algebra, vol. II of Handbook for Automatic Computation, Springer–Verlag, New York, 1971.

[186] Williams, G.: Linear Algebra With Applications, 4th ed., Jones and Bartlett Publisher, UK, 2001.

[187] Wolfe, P.: Methods of Nonlinear Programming, in Nonlinear Programming (Ed. J. Abadie), North–Holland, Amsterdam, 1967.

[188] Wood, A.: Introduction to Numerical Analysis, Addison–Wesley, Reading, MA, 1999.

[189] Young, D. M.: Iterative Solution of Large Linear Systems, Academic Press, New York, 1971.

[190] Zangwill, W.: Nonlinear Programming, Prentice–Hall, Englewood Cliffs, NJ, 1969.

Index